Modeling of Atmospheric Chemistry

Mathematical modeling of atmospheric composition is a formidable scientific and computational challenge. This comprehensive presentation of the modeling methods used in atmospheric chemistry focuses on both theory and practice, from the fundamental principles behind models, through to their applications in interpreting observations. An encyclopedic coverage of methods used in atmospheric modeling, including their advantages and disadvantages, makes this a one-stop resource with a large scope. Particular emphasis is given to the mathematical formulation of chemical, radiative, and aerosol processes; advection and turbulent transport; emission and deposition processes; as well as major chapters on model evaluation and inverse modeling. The modeling of atmospheric chemistry is an intrinsically interdisciplinary endeavor, bringing together meteorology, radiative transfer, physical chemistry, and biogeochemistry. This book is therefore of value to a broad readership. Introductory chapters and a review of the relevant mathematics make the book instantly accessible to graduate students and researchers in the atmospheric sciences.

Guy P. Brasseur is a Senior Scientist and former Director at the Max Planck Institute for Meteorology in Hamburg, Germany, and a Distinguished Scholar at the National Center for Atmospheric Research in Boulder, USA. He received his doctor's degree at the University of Brussels and has conducted research in Belgium, the USA, and Germany. He was Professor at the Universities of Brussels and Hamburg. His scientific interests include questions related to atmospheric chemistry and air pollution, biogeochemical cycles, climate change, and upper atmosphere chemistry and dynamics. He has chaired several international research programs, and is associated with national academies in Hamburg, Germany, Brussels, Belgium, and Oslo, Norway.

Daniel J. Jacob is the Vasco McCoy Family Professor of Atmospheric Chemistry and Environmental Engineering at Harvard University. He received his PhD from Caltech in 1985 and joined the Harvard faculty in 1987. His research covers a wide range of topics in atmospheric composition, with focus on model development and applications to interpretation of observations. Among his professional honors are the NASA Distinguished Public Service Medal (2003), the AGU Macelwane Medal (1994), and the Packard Fellowship for Science and Engineering (1989). Jacob has published over 350 research papers and trained over 80 PhD students and postdocs in atmospheric chemistry modeling over the course of his career.

Modeling of Atmospheric Chemistry

GUY P. BRASSEUR

Max Planck Institute for Meteorology and
National Center for Atmospheric Research

DANIEL J. JACOB

Harvard University

CAMBRIDGE
UNIVERSITY PRESS

CAMBRIDGE
UNIVERSITY PRESS

University Printing House, Cambridge CB2 8BS, United Kingdom

One Liberty Plaza, 20th Floor, New York, NY 10006, USA

477 Williamstown Road, Port Melbourne, VIC 3207, Australia

4843/24, 2nd Floor, Ansari Road, Daryaganj, Delhi – 110002, India

79 Anson Road, #06–04/06, Singapore 079906

Cambridge University Press is part of the University of Cambridge.

It furthers the University's mission by disseminating knowledge in the pursuit of
education, learning, and research at the highest international levels of excellence.

www.cambridge.org
Information on this title: www.cambridge.org/9781107146969

First published 2017

Printed in the United States of America by Sheridan Books, Inc.

A catalogue record for this publication is available from the British Library.

Library of Congress Cataloging-in-Publication Data
Names: Brasseur, Guy. | Jacob, Daniel J., 1958–
Title: Modeling of atmospheric chemistry / Guy P. Brasseur, Max Planck Institute for Meteorology,
Hamburg, Daniel J. Jacob, Harvard University.
Description: Cambridge : Cambridge University Press, 2017. | Includes bibliographical references
and index.
Identifiers: LCCN 2016040128 | ISBN 9781107146969 (Hardback : alk. paper)
Subjects: LCSH: Atmospheric chemistry–Mathematical models. | Atmospheric
diffusion–Mathematical models.
Classification: LCC QC879 .B6974 2017 | DDC 551.51/1–dc23 LC record available at
https://lccn.loc.gov/2016040128

ISBN 978-1-107-14696-9 Hardback

Contents

Preface

Modern science dealing with complex dynamical systems increasingly makes use of mathematical models to formalize the description of interactive processes and predict responses to perturbations. Models have become fundamental tools in many disciplines of natural sciences, engineering, and social sciences. They describe the essential aspects of a system using mathematical concepts and languages and they can in this manner provide powerful approximations of reality. They are used to analyze observations, understand relationships, test hypotheses, and project future evolution. Disagreements between models and observations often lead to important advances in theoretical understanding. Models also play a critical role in the development of policy options and in decision-making.

In atmospheric science, mathematical models have long been central tools for weather prediction and climate research. They are now also used extensively to describe the chemistry of the atmosphere. The corresponding model equations describe the factors controlling atmospheric concentrations of chemical species as a function of emissions, transport, chemistry, and deposition. Chemical species are often coupled through intricate mechanisms, and the corresponding differential equations are then also coupled. Simulation of aerosol particles needs to account in addition for microphysical processes governing particle size and composition, as well as interactions with the hydrological cycle through cloud formation. The difficulty of modeling atmospheric composition is compounded by the need to resolve a continuum of temporal and spatial scales stretching over many orders of magnitude from microseconds to many years, from local to global, and involving coupling of transport and chemistry on all scales.

Mathematical modeling of atmospheric chemistry is thus a formidable scientific and computational challenge. It integrates elements of meteorology, radiative transfer, physical chemistry, and biogeochemistry. Solving the large systems of coupled nonlinear partial differential equations that characterize the atmospheric evolution of chemical species requires advanced numerical algorithms and pushes the limits of supercomputing resources.

The purpose of this book is to provide insight into the methods used in models of atmospheric chemistry. The book is designed for graduate students and professionals in atmospheric chemistry, but also more broadly for researchers interested in atmospheric models, numerical methods, and optimization theory.

The book is divided into three parts. The first part presents background material. Chapter 1 introduces the reader to the concept of model and provides a historical perspective on the development of atmospheric and climate models, leading to the development of atmospheric chemistry models. It reviews the different types of atmospheric chemistry models and highlights their role as components of observing systems.

Fundamentals of atmospheric dynamics and chemistry are presented in Chapters 2 and 3. Chapter 2 describes the vertical structure of the atmosphere, defines key parameters that characterize the dry and the wet atmosphere, and introduces the concept of static stability and geostrophic balance. It goes on to describe the general circulation of the atmosphere. Chapter 3 provides a summary survey of the chemical processes relevant to the atmosphere as well as the microphysical processes controlling the evolution of aerosol particles. Chapter 4 presents the fundamental mathematical equations on which atmospheric models are based and gives an introduction to the numerical methods used to solve these equations.

The second part of this book focuses on the formulation of model processes and reviews the numerical algorithms used to solve the model equations. Chapter 5 covers the formulation of radiative transfer, chemical kinetics, and aerosol microphysics. Chapter 6 reviews numerical methods to solve the stiff systems of nonlinear ordinary differential equations that describe atmospheric chemistry mechanisms. Chapter 7 presents numerical algorithms used to solve the advection equation describing transport by resolved winds. The formulation of small-scale (parameterized) transport processes including turbulent mixing, organized convection, plumes, and boundary layer dynamics is addressed in Chapter 8. Chapter 9 reviews formulations of emissions to the atmosphere, deposition to the surface, and two-way coupling between the atmosphere and surface reservoirs.

The third part of this book deals with the role of models as components of the atmospheric observing system. Chapter 10 focuses on model evaluation and presents different metrics for this purpose. It illustrates the importance of models for the interpretation of observational data. Chapter 11 covers fundamental concepts of inverse modeling and data assimilation. It shows how chemical transport models can be integrated with atmospheric observations through optimization theory to provide best estimates of the chemical state of the system and of the driving variables.

At the end of the volume, the reader will find several appendices with numerical values of physical constants and other quantities, unit conversions, and a list of important chemical reactions with corresponding rate constants. Some basic mathematical definitions and relations are also provided.

Over the years, both of us have benefited from numerous discussions with our colleagues, students, and postdoctoral fellows. Several of them have contributed to this book by reviewing chapters, making suggestions, and providing scientific material. We are deeply indebted to them. We would like to thank in particular Helen Amos, Alexander Archibald, Jerome Barre, Mary Barth, Cathy Clerbaux, Jim Crawford, Louisa Emmons, Rolando Garcia, Paul Ginoux, Claire Granier, Alex Guenther, Colette Heald, Jan Kazil, Patrick Kim, Douglas Kinnison, Monika Kopacz, Jean-François Lamarque, Peter Lauritzen, Sasha Madronich, Daniel Marsh, Iain Murray, Vincent-Henri Peuch, Philip Rasch, Brian Ridley, Anne Smith, Piotr Smolarkiewicz, Alex Turner, Xuexi Tie, Stacy Walters, Kevin Wecht, Christine Wiedinmyer, Lin Zhang, and Peter Zoogman. We would like also to acknowledge Sebastian Eastham, Emilie Ehretsmann, Natasha Goss, Lu Hu, Rajesh Kumar, Eloise

Marais, Jost Müsse, Elke Lord, Barbara Petruzzi, Jianxiong Sheng, and Natalia Sudarchikova for their technical assistance during the preparation of the manuscript. A substantial fraction of this volume was written by one of us (G. P. B) at the National Center for Atmospheric Research, which is sponsored by the US National Science Foundation.

Symbols

The symbols used in the different chapters of this book are listed below with their corresponding units in the MKSA system. When no units are given, the quantity is either dimensionless or has no intrinsic dimensions. Appendix B gives further information on units, prefixes, and conversion factors. In some cases, when no confusion exists, the same symbols are used to characterize different variables. Scalars are represented as italics (alphabet letters) or as regular font (Greek and other symbols). Vectors and matrices are represented by lowercase and uppercase bold fonts, respectively.

A

a	Earth's radius [m]
A	Surface area density of atmospheric particles [$m^2\ m^{-3}$]
$\mathbf{A}$	Averaging kernel matrix

B

B	Blackbody radiative emission flux [$W\ m^{-2}$]
B_λ	Spectral density of blackbody emission flux (Planck function) [$W\ m^{-2}\ nm^{-1}$]

C

c	One-dimensional constant flow velocity [$m\ s^{-1}$]
c	Speed of light in vacuum [$m\ s^{-1}$]
$c*$	Phase velocity of a wave [$m\ s^{-1}$]
c_g*	Group velocity of a wave [$m\ s^{-1}$]
c_p	Specific heat at constant pressure [$J\ K^{-1}\ kg^{-1}$]
c_v	Specific heat at constant volume [$J\ K^{-1}\ kg^{-1}$]
C_c	Slip correction factor
C_D	Drag coefficient
C_i	Mole fraction or molar mixing ratio of species i
$CRMSE$	Centered root-mean-square-error

D

d	Displacement height [m]
D	Divergence of the flow [s^{-1}]
Da	Damköhler number
D_d	Detrainment rate associated with downdrafts in convective systems [$kg\ m^{-3}\ s^{-1}$]
D_i	Molecular diffusion coefficient for species i [$m^2\ s^{-1}$]
D_p	Particle diameter [m]

D_u Detrainment rate associated with updrafts in convective systems [kg m^{-3} s^{-1}]
$DOFS$ Degrees of freedom for signal

E
e Water vapor partial pressure [Pa]
e_s Saturation water vapor pressure [Pa]
e   Eigenvector
E Emission flux [kg m^{-2} s^{-1}]
E   Eliassen–Palm Flux [components E_φ and E_z in kg s^{-2}]
E   Matrix of eigenvectors arranged by columns
$E(k)$ Spectral distribution of turbulent energy for a given wavenumber k [m^3 s^{-2}]
E_a Activation energy [J mol^{-1}]
E_d Entrainment rate associated with downdraft in convective systems [kg m^{-3} s^{-1}]
E_u Entrainment rate associated with updraft in convective systems [kg m^{-3} s^{-1}]

F
f Coriolis factor [s^{-1}]
f_A Fractional area of a model grid cell experiencing precipitation
f_A Fractional area of land suitable for saltation
$f_{i,I}$ Fraction of soluble compound i partitioned in ice water
$f_{i,L}$ Fraction of soluble compound i partitioned in liquid water
F Mass flux [kg m^{-2} s^{-1}]
F Radiative flux [W m^{-2}]
$\mathcal{F}$ Air mass factor
F   Force vector with its three components F_x, F_y, and F_z [N]
F   Forward model
$F_{D,i}$ Deposition flux of species i [kg m^{-2} s^{-1}]
F_λ Spectral density of the radiative flux [W m^{-2} nm^{-1}]

G
g Vector of gravitational acceleration [m s^{-2}]
g Amplitude of gravitational acceleration [m s^{-2}]
g Amplification function in numerical methods
g Asymmetry factor
g Gain factor
G Green function
G Gravity wave drag [m s^{-2}]
G   Gain matrix
G_m Grade of model m

H
h Mixing depth [m]
H Atmospheric scale height [m]
$\mathcal{H}$ Effective (constant) scale height [m]
H_i Dimensionless Henry's law constant for species i

I

i	Unit vector in the zonal (x) direction
I	Light intensity [W m^{-2}]
I	Identity matrix
I_{AB}	Segregation ratio for chemical compounds A and B
I_i	Condensation growth rate of species i [m^3 s^{-1}]

J

j	Unit vector in the meridional (y) direction
j	Radiative source term [Wm^{-2} sr^{-1} nm^{-1} m^{-1}]
J	Radiative source function [Wm^{-2} sr^{-1} nm^{-1}]
J	Photodissociation (photolysis) frequency [s^{-1}]
J	Cost function
J	Jacobian matrix
$J_{i,j}$	Coagulation rate between particles i and j [m^{-3} s^{-1}]
J_0	Nucleation rate [m^{-3} s^{-1}]

K

k	Unit vector in the vertical (z) direction
k	Wavenumber [m^{-1}]
k	Boltzmann's constant (1.38×10^{-23} J K^{-1})
k	von Karman's constant (0.35)
k	Chemical rate constant [first order: s^{-1}; second order: cm^3 s^{-1}; third order: cm^6 s^{-2}]
k_{ext}	Mass extinction cross-section [m^2 kg^{-1}]
$k_{G,i}$	Conductance for vertical transfer of species i in the gas phase [m s^{-1}]
$k_{W,i}$	Conductance for vertical transfer of species i in the water phase [m s^{-1}]
K	Eddy diffusion coefficient [m^2 s^{-1}]
K	Equilibrium constant
K	Henry's law constant [M atm^{-1}]
K^*	Effective Henry's law constant [M atm^{-1}]
K	Eddy diffusion tensor
K	Jacobian matrix (Chapter 11)
K_a	Acid dissociation constant
K_m	Eddy viscosity coefficient [m^2 s^{-1}]
Kn	Knudsen number
K_i	Air–sea exchange velocity for species i [m s^{-1}]
K_θ	Eddy diffusivity of heat [m^2 s^{-1}]

L

l	Mixing length [m]
ℓ_i	Loss rate constant or loss coefficient of species i [s^{-1}]
L	Characteristic length [m]
L	Liquid water content [kg water/kg air]
L	Monin–Obukhov length [m]

L	Lagrange function
L_i	Loss rate of species i [m^{-3} s^{-1}]
L_{vap}	Latent heat of vaporization of liquid water [J kg^{-1}]
L_λ	Spectral density of the radiance at wavelength λ [W m^{-2} sr^{-1} nm^{-1}]

M

m	Mean molecular mass of air (4.81×10^{-26} kg)
m	Refraction index
m	Wavenumber
M_a	Molar mass of air (28.97×10^{-3} kg mol^{-1})
M_d	Mean vertical downdraft convective flux of air [kg m^{-2} s^{-1}]
M_e	Mean subsidence flux compensating for convective fluxes [kg m^{-2} s^{-1}]
M_i	Molar mass of species i [kg mol^{-1}]
M_k	Moment of order k for a given aerosol distribution
M_u	Mean vertical updraft convective flux of air [kg m^{-2} s^{-1}]
M_w	Molar mass of water (18.01×10^{-3} kg mol^{-1})
MAD	Mean absolute deviation
MAE	Mean absolute error
MFB	Mean fractional bias
MFE	Mean fractional error
$MNAE$	Mean normalized absolute error
MNB	Mean normalized bias

N

$\mathbf{n}$	Unit outward vector normal to a surface
n_a	Number density for air [m^{-3}]
n_i	Number density for species i [m^{-3}]
n_N	Particle number size distribution function [m^{-4}]
n_S	Particle surface distribution function [m^2 m^{-4}]
n_V	Particle volume distribution function [m^3 m^{-4}]
$\mathcal{N}_A$	Avogadro number (6.022×10^{23} molecules per mole)
NMB	Normalized mean bias

P

p	Pressure [Pa]
p_d	Pressure of dry air [Pa]
p_i	Production rate of species i [kg m^{-3} s^{-1}]
p_s	Surface pressure [Pa]
P	Phase function for scattered radiation
P	Ertel potential vorticity [m^2 s^{-2} K kg^{-1}]
P	Probability density function
$\mathcal{P}$	Steric factor
Pe	Péclet number
P_i	Production rate of species i [m^{-3} s^{-1}]
P_l^m	Associated Legendre polynomial
Pr	Prandtl number

Q

q	Specific humidity [kg water vapor/kg of air]
q	Diabatic heating expressed in K day^{-1}
q	Actinic flux [photons m^{-2} s^{-1}]
q_k	Water concentration in hydrometeor of type k
q_λ	Photon flux density [photons m^{-2} s^{-1} nm^{-1}]
Q	Diabatic heating rate [J kg^{-1} s^{-1} or W m^{-3}]
Q_{abs}	Absorption efficiency
Q_{ext}	Extinction efficiency
Q_s	Saltation flux [kg m^{-1} s^{-1}]
Q_{scat}	Scattering efficiency

R

r	Geometric distance from the center of the Earth
$\mathbf{r}$	Position vector
r	Particle radius [m]
r_w	Mass mixing ratio of water vapor [kg kg^{-1}]
r	Pearson correlation coefficient
R	Gas constant for air [J K^{-1} kg^{-1}]
$\mathcal{R}$	Universal gas constant (8.3143 J K^{-1} mol^{-1})
R^2	Coefficient of determination
R_A	Aerodynamic resistance [s m^{-1}]
$R_{B,i}$	Boundary resistance for species i [s m^{-1}]
$R_{C,i}$	Surface resistance for species i [s m^{-1}]
R_d	Gas constant for dry air (287 J K^{-1} kg^{-1})
Re	Reynolds number
RH	Relative humidity [percent]
Ri	Richardson number
R_i	Total resistance to dry deposition of species i [s m^{-1}]
$RMSE$	Root mean square error
R_w	Gas constant for water vapor (461.5 J K^{-1} kg^{-1})

S

s_i	Source rate of species i (in mass) [kg m^{-3} s^{-1}]
S	Solar energy flux [W m^{-2}] or solar constant (approx. 1368 W m^{-2})
$\mathbf{S}$	Error covariance matrix
$\mathbf{S'}$	Error correlation matrix
$\mathbf{S_a}$	Aggregation error covariance matrix
$\mathbf{S_A}$	Prior error covariance matrix
$\mathbf{S_I}$	Instrument error covariance matrix
$\mathbf{S_M}$	Forward model error covariance matrix
$\mathbf{S_O}$	Observational error covariance matrix
$\mathbf{S_R}$	Representation error covariance matrix
$\hat{\mathbf{S}}$	Posterior error covariance matrix
Sc_i	Schmidt number for species i

T

t	Time [s]
t	Student's variable for the t-test
T	Transmission of radiation
T	Absolute temperature [K]
T_E	Effective temperature of the Earth [K]
TKE	Turbulent kinetic energy [$m^2\ s^{-1}$]
T_s	Effective temperature of the Sun
T_v	Virtual temperature [K]

U

u	Zonal component of wind velocity [$m\ s^{-1}$]
u	Path length [$kg\ m^{-2}$]
u_*	Friction velocity [$m\ s^{-1}$]
u^*	Residual zonal wind velocity [$m\ s^{-1}$]
u^A	Anti-diffusion velocity [$m\ s^{-1}$]
u_g	Zonal component of the geostrophic wind [$m\ s^{-1}$]
u_{10}	Wind velocity 10 m above the surface [$m\ s^{-1}$]

V

v	Meridional component of wind velocity [$m\ s^{-1}$]
v^*	Residual meridional wind velocity [$m\ s^{-1}$]
$\mathbf{v}$	Wind velocity vector in Earth's rotating frame [$m\ s^{-1}$]
v_g	Meridional component of the geostrophic wind [$m\ s^{-1}$]
v_i	Mean thermal velocity [$m\ s^{-1}$]
V	Molar volume [$m^3\ mol^{-1}$]
V	Aerosol volume density [$m^3\ m^{-3}$]
$\mathbf{V}$	Wind velocity in inertial frame [$m\ s^{-1}$]
V_T	Translational Earth's rotation velocity [$m\ s^{-1}$]

W

w	Vertical component of wind velocity [$m\ s^{-1}$]
w^*	Residual vertical wind velocity [$m\ s^{-1}$]
w^*	Convective velocity scale [$m\ s^{-1}$]
$w_{D,i}$	Surface deposition velocity of species i [$m\ s^{-1}$]
w_s	Terminal settling velocity [$m\ s^{-1}$]

X

x	Geometric distance in the zonal direction [m]
$\mathbf{x}$	State vector (often refers to the true value)
$\hat{\mathbf{x}}$	Optimal estimate of state vector
$\mathbf{x_A}$	Prior estimate of state vector

Y

y	Geometric distance in the meridional direction [m]
$\mathbf{y}$	Observation vector

Z

z Geometric altitude [m]

$z_{0,m}$ Aerodynamic roughness length [m]

Z Log pressure altitude [m]

Z Potential vorticity [s^{-1} m^{-1}]

Z_{AB} Collision frequency for molecules A and B [s^{-1}]

α

α Albedo

α Aerosol particle size parameter

α Mass accommodation coefficient

α Courant number

α_T Thermal diffusion factor

β

β Fourier number

β_{ext} Aerosol extinction coefficient [m^{-1}]

β_{abs} Aerosol absorption coefficient [m^{-1}]

β_{scat} Aerosol scattering coefficient [m^{-1}]

$\beta_{i,j}$ Coagulation coefficient for particles i and j [m^3 s^{-1}]

γ

γ Reactive uptake coefficient for heterogeneous chemical process

γ Regularization factor

γ_c Coefficient for non-local turbulent transfer

Γ Actual atmospheric lapse rate [K m^{-1}]

Γ Mean age of air [s]

Γ_d Dry adiabatic lapse rate [K m^{-1}]

Γ_w Wet adiabatic lapse rate [K m^{-1}]

$\mathbf{\Gamma}_{\varpi}$ Aggregation matrix

δ

δ Dirac function

ΔH Enthalpy of dissolution [J mol^{-1}]

ε

ε_A Quantum efficiency (or yield) for the photolysis of molecule A

$\boldsymbol{\varepsilon_O}$ Observational error vector

$\boldsymbol{\varepsilon_a}$ Aggregation error vector

$\boldsymbol{\varepsilon_A}$ Prior estimate error vector

$\boldsymbol{\varepsilon_I}$ Instrument error vector

$\boldsymbol{\varepsilon_M}$ Forward model error vector

$\boldsymbol{\varepsilon_R}$ Representation error vector

ζ

ζ Relative vorticity of the flow [s^{-1}]

η
η   Step mountain coordinate (eta coordinate)

θ
θ   Zenithal direction [radians]
θ   Potential temperature [K]
θ_v Virtual potential temperature [K]

λ
λ   Longitude [radians]
λ   Wavelength [m]
λ   Mean free path of air molecules [m]
λ   Lyapunov exponent [s^{-1}]
λ_i Eigenvalue associated with eigenvector $\mathbf{e}_i$
Λ   Leaf area index (LAI) [m^2 m^{-2}]

μ
μ   Cosine of zenithal direction (θ)
μ   Molecular dynamic viscosity coefficient [Pa s or kg m^{-1} s^{-1}]
μ_i Mass mixing ratio of species i [kg kg^{-1}]
μ_w Mass mixing ratio of water vapor [kg kg^{-1}]

ν
ν   Kinematic viscosity [m^2 s^{-1}]
ν   Asselin-filter parameter
ν   Frequency [Hz]
ν_{ion} Ion-neutral collision frequency [s^{-1}]

π
π   3.14159

ρ
ρ_a Mass density of air [kg m^{-3}]
ρ_d Mass density of dry air [kg m^{-3}]
ρ_i Mass density of species i [kg m^{-3}]
ρ_p Mass density of particles or drops [kg m^{-3}]
ρ_w Mass density of water vapor [kg m^{-3}]

σ
σ   Stefan-Boltzmann constant (5.67×10^{-8} W m^{-2} K^{-4})
σ   Standard deviation
σ   Normalized pressure coordinate (sigma coordinate)
$\tilde{\sigma}$ Pseudo density in isentropic coordinates
σ_A Absorption cross-section for molecule A [m^2]

τ

τ	Optical depth
τ	Lifetime [s]
$\boldsymbol{\tau}$	Stress tensor
$\tau_{i,j}$	Element of the stress tensor

φ

φ	Latitude [radians]
φ	Azimuthal direction
ϕ	Radial basis function
Φ	Geopotential [$m^2\ s^{-2}$]
$\boldsymbol{\Phi}_{\infty}$	Solar flux at the top of the atmosphere [W m^{-2}]
Φ_k	Basis function in the spectral element method
Φ_λ	Spectral density of solar flux [W m^{-2} nm^{-1}]

χ

χ	Solar zenith angle
χ	Velocity potential

ψ

Ψ	Generic mathematical function or variable
Ψ	Streamfunction of the flow
Ψ	Montgomery function (isentropic coordinate system) [J kg^{-1} or $m^2\ s^{-2}$]

ω

ω	"Vertical" velocity in the pressure coordinate system [Pa s^{-1}]
ω	Single scattering albedo
Ω	Angular Earth rotation period (7.292×10^{-5} rad s^{-1})
Ω	Column concentration [molecules m^{-2}]
Ω_s	Slant column concentration [molecules m^{-2}]

The Concept of Model

1.1 Introduction

This book describes the foundations of mathematical models for atmospheric chemistry. Atmospheric chemistry is the science that focuses on understanding the factors controlling the chemical composition of the Earth's atmosphere. Atmospheric chemistry investigates not only chemical processes but also the dynamical processes that drive atmospheric transport, the radiative processes that drive photochemistry and climate forcing, the evolution of aerosol particles and their interactions with clouds, and the exchange with surface reservoirs, including biogeochemical cycling. It is a highly interdisciplinary science.

Atmospheric chemistry is a young and rapidly growing science, motivated by the societal need to understand and predict human perturbations to atmospheric composition. These perturbations have increased greatly over the past century due to population growth, industrialization, and energy demand. They are responsible for a range of environmental problems including degradation of air quality, damage to ecosystems, depletion of stratospheric ozone, and climate change. Quantifying the link between human activities and their atmospheric effects is essential to the development of sound environmental policy.

The three pillars of atmospheric chemistry research are laboratory studies, atmospheric measurements, and models. Laboratory studies uncover and quantify the fundamental chemical processes expected to proceed in the atmosphere. Atmospheric measurements probe the actual system in all of its complexity. Models simulate atmospheric composition using mathematical expressions of the driving physical and chemical processes as informed by the laboratory studies. They can be tested with atmospheric measurements to evaluate and improve current knowledge, and they can be used to make future projections for various scenarios. Models represent a quantitative statement of our current knowledge of atmospheric composition. As such, they are fundamental tools for environmental policy.

Atmospheric chemistry modeling has seen rapid improvement over the past decades, driven by computing resources, improved observations, and demand from policymakers. Thirty years ago, models were so simplified in their treatments of chemistry and transport that they represented little more than conceptual exercises. Today, state-of-science *chemical transport models* provide realistic descriptions of the 3-D transport and chemical evolution of the atmosphere. Although uncertainties remain large, these models are used extensively to interpret atmospheric observations and to make projections for the future. The state of the science is advancing rapidly, and atmospheric chemists 30 years from now may well scoff at the crude nature of present-day models. Nevertheless, we are now at a point where models can provide a credible, process-based mathematical representation of the atmosphere to serve the needs of science and policy. It is with this perspective of a mature yet evolving state of science that this book endeavors to describe the concepts and algorithms that provide the foundations of atmospheric chemistry models.

This chapter is intended to introduce the reader to the notion and utility of models, and to provide a broad historical perspective on the development of atmospheric chemistry models. It starts with general definitions and properties of mathematical models. It then covers the genesis and evolution of meteorological models, climate models, and finally atmospheric chemistry models, leading to the current state of science. It describes conceptually different types of atmospheric chemistry models and the value of these models as part of atmospheric observing systems. It finishes with a brief overview of the computational hardware that has played a crucial role in the progress of atmospheric modeling.

1.2 What is a Model?

A model is a simplified representation of a complex system that enables inference of the behavior of that system. The *Webster New Collegiate Dictionary* defines a model as a description or analogy used to help visualize something that cannot be directly observed, or as a system of postulates, data, and inferences presented as a mathematical description of an entity or state of affairs. The *Larousse Dictionary* defines a model as a formalized structure used to account for an ensemble of phenomena between which certain relations exist. Models are abstractions of reality, and are often associated with the concept of metaphor (Lakoff and Johnson, 1980). Humans constantly create models of the world around them. They observe, analyze, isolate key information, identify variables, establish the relationships between them, and anticipate how these variables will evolve in various scenarios.

One can distinguish between cognitive, mathematical, statistical, and laboratory models (Müller and von Storch, 2004). Cognitive models convey ideas and test simple hypotheses without pretending to simulate reality. For example, the Daisy-world model proposed by Lovelock (1989) illustrates the stability of climate through the insolation–vegetation–albedo feedback. This model calculates the changes in the geographical extent of imaginary white and black daisies covering a hypothetical planet in response to changes in the incoming solar energy. It shows that the biosphere can act as a planetary thermostat. Such apparently fanciful models can powerfully illustrate concepts. More formal mathematical models attempt to represent the complex intricacies of real-world systems, and describe the behavior of observed quantities on the basis of known physical, chemical, and biological laws expressed through mathematical equations. They can be tested by comparison to observations and provide predictions of events yet to be experienced. Examples are meteorological models used to perform daily weather forecasts. Statistical models describe the behavior of variables in terms of their observed statistical relationships with other variables, and use these relationships to interpolate or extrapolate

(a)   (b)

Figure 1.1   Prussian naturalist and explorer Alexander von Humboldt (a) and French mathematician and astronomer Pierre-Simon, Marquis de Laplace (b).

behavior. They are empirical in nature, as opposed to the physically based mathematical models. Laboratory models are physical replicas of a system, at a reduced or enlarged geometric scale, used to perform controlled experiments. They mimic the response of the real system to an applied perturbation, and results can be extrapolated to the actual system through appropriate scaling laws.

In his 1846 book *Kosmos*, German scientist Alexander von Humboldt (1769–1859, see Figure 1.1) states that the structure of the universe can be reduced to a problem of mechanics, and reinforces the view presented in 1825 by Pierre-Simon Laplace (1749–1827, see Figure 1.1). In the introduction of his *Essai Philosophique sur les Probabilités* (Philosophical Essay on Probabilities), Laplace explains that the present state of the Universe should be viewed as the consequence of its past state and the cause of the state that will follow. Once the state of a system is known and the dynamical laws affecting this system are established, all past and future states of the system can be rigorously determined. This concept, which applies to many aspects of the natural sciences, is extremely powerful because it gives humanity the tools to monitor, understand, and predict the evolution of the Universe.

Although von Humboldt does not refer explicitly to the concept of model, he attempts to describe the functioning of the world by isolating different causes, combining them in known ways, and asking whether they reinforce or neutralize each other. He states that, "by suppressing details that distract, and by considering only large masses, one rationalizes what cannot be understood through our senses." This effectively defines models as idealizations of complex systems designed to achieve understanding. Models isolate the system from its environment, simplify the relationships between variables, and make assumptions to neglect certain internal variables and external influences (Walliser, 2002). They are not fully objective tools because they emphasize the essential or focal aspects of a system as conceived by their authors. They are not universal because they include assumptions and simplifications that may be acceptable for some specific applications but not others. Indeed, the success of a model is largely the product of the skills and imagination of the authors.

During the twentieth century, models started to become central tools for addressing scientific questions and predicting the evolution of phenomena such as economic cycles, population growth, and climate change. They are extensively used today in many disciplines and for many practical applications of societal benefit, weather forecasting being a classic example. As computing power increases and knowledge grows, models are becoming increasingly elaborate and can unify different elements of a complex system to describe their interactions. In the case of Earth science, this is symbolized by the vision of a "virtual Earth" model to describe the evolution of the planet, accounting for the interactions between the atmosphere, ocean, land, biosphere, cryosphere, lithosphere, and coupling this natural system to human influences. Humans in this "virtual Earth" would not be regarded as external factors but as actors through whom environmental feedbacks operate.

1.3 Mathematical Models

Mathematical models strip the complexity of a system by identifying the essential driving variables and describing the evolution of these variables with equations based on physical laws or empirical knowledge. They provide a quantitative statement of our knowledge of the system that can be compared to observations. Models of natural systems are often expressed as mathematical applications of the known laws that govern these systems. As stated by Gershenfeld (1999), mathematical models can be rather general or more specific, they can be guided by *first principles* (physical laws) or by empirical information, they can be analytic or numerical, deterministic or stochastic, continuous or discrete, quantitative or qualitative. Choosing the best model for a particular problem is part of a modeler's skill.

Digital computers in the 1950s ushered in the modern era for mathematical models by enabling rapid numerical computation. Computing power has since been doubling steadily every two years ("Moore's law") and the scope and complexity of models has grown in concert. This has required in turn a strong effort to continuously improve the physical underpinnings and input information for the models. Otherwise we have "garbage in, garbage out." Sophisticated models enabled by high-performance computing can extract information from a system that is too complex to be fully understood or quantifiable by human examination. By combining a large amount of information, these models point to system behavior that may not have been anticipated from simple considerations. From this point of view, models generate knowledge. In several fields of science and technology, computer simulations have become a leading knowledge producer. In fact, this approach, which does not belong either to the theoretical nor to the observational domains, is regarded as a new form of scientific practice, a "third way" in scientific methodology complementing theoretical reasoning and experimental methods (Kaufmann and Smarr, 1993).

For a model to be useful it must show some success at reproducing past observations and predicting future observations. By definition, a model will always have some error that reflects the assumptions and approximations involved in its development. The question is not whether a model has error, but whether the error is small enough for the model to be useful. As the saying goes, "all models are wrong, but some are useful." A crucial task is to quantify the error statistics of the model, which can be done through error propagation analyses and/or comparison with observations. The choice of observational data sets and statistics to compare to the model is an important part of the modeler's skill, as is the interpretation of the resulting comparisons. Discrepancies with observations may be deemed acceptable, and used to compile model error statistics, but they may also point to important flaws in the founding assumptions or implementation of the model. The modeler must be able to recognize the latter as it holds the key to advancing knowledge. Some dose of humility is needed because the observations cannot sample all the possible realizations of a complex system. As a result, the error statistics of the model can never be characterized fully.

Many mathematical models are based on differential equations that describe the evolution in space and time of the variables of interest. These are often conservation equations, generalizing Newton's second law that the acceleration of an object is proportional to the force applied to that object. Atmospheric chemistry models are based on the *continuity equation* that describes mass conservation for chemical species. Consider an ensemble of chemical species ($i = 1, \ldots n$) with mole fractions (commonly called *mixing ratios*) assembled in a vector $\mathbf{C} = (C_1, \ldots C_n)^T$. The continuity equation for species i in a fixed (*Eulerian*) frame of reference is given by

$$\frac{\partial C_i}{\partial t} = -\mathbf{v} \cdot \nabla C_i + P_i(\mathbf{C}) - L_i(\mathbf{C}) \quad (i = 1, \ldots n) \tag{1.1}$$

Here, $\mathbf{v}$ is the 3-D wind vector, and P_i and L_i are total production and loss rates for species i that may include contributions from chemical reactions (coupling to other species), emissions, and deposition. The local change in mixing ratio with time ($\partial C_i / \partial t$) is expressed as the sum of transport in minus transport out (flux divergence term $\mathbf{v} \cdot \nabla C_i$) and net local production ($P_i - L_i$). Similar conservation equations are found in other branches of science. For example, replacing C_i with momentum yields the Navier–Stokes equation that forms the basis for models of fluid dynamics.

A system is said to be *deterministic* if it is uniquely and entirely predictable once initial conditions are specified. It is *stochastic* if randomness is present so that only probabilities can be predicted. Systems obeying the laws of classical mechanics are generally deterministic. The two-body problem (e.g., a satellite orbiting a planet or a planet orbiting the Sun), described by Newton's laws and universal gravitation, is a simple example of a deterministic system. An analytic solution of the associated differential equations can be derived with no random element. All trajectories derived with different initial conditions converge toward the same subspace called an *attractor*. By contrast, when trajectories starting from slightly different initial conditions diverge from each other at a sufficiently fast rate, the system is said to be *chaotic*. Meteorological models are a classic example. They are deterministic but exhibit chaotic behavior due to nonlinearity of the Navier–Stokes equation. This chaotic behavior is called *turbulence*. Chaotic systems evolve in a manner that is exceedingly dependent on the precise choice of initial conditions. Since initial conditions in a complex system such as the weather can never be exactly defined, the model results are effectively stochastic and multiple simulations (*ensembles*) need to be conducted to obtain model output statistics.

1.4 Meteorological Models

The basic ideas that led to the development of meteorological forecast models were formulated about a century ago. American meteorologist Cleveland Abbe (1838–1916) first proposed a mathematical approach in a 1901 paper entitled "The physical basis of long-range weather forecasting." A few years later, in 1904, in a paper entitled "Das Problem von der Wettervorhersage betrachtet vom Standpunkte der Mechanik und der Physik" (The problem of weather prediction from the standpoint of mechanics and physics), Norwegian meteorologist Vilhelm Bjerknes (1862–1951) argued that weather forecasting should be based on the well-established laws of physics and should therefore be regarded as a deterministic problem (see Figure 1.2). He wrote:

> If it is true, as every scientist believes, that subsequent atmospheric states develop from the preceding ones according to physical law, then it is apparent that the necessary and sufficient conditions for the rational solution of forecasting problems are the following:
>
> 1. A sufficiently accurate knowledge of the state of the atmosphere at the initial time;
> 2. A sufficiently accurate knowledge of the laws according to which one state of the atmosphere develops from another.

(a)   (b)

Figure 1.2   Norwegian meteorologist Vilhelm Bjerknes (a), and American meteorologist Cleveland Abbe (b). Source: Wikimedia Commons.

(a)   (b)   (c)

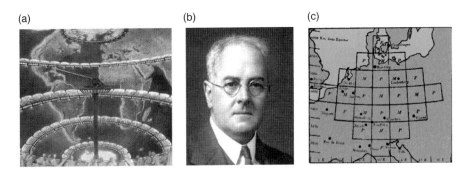

Figure 1.3   British meteorologist Lewis Fry Richardson (b), the map grid he used to make his numerical weather forecast (c), and an artist's view of a theater hall (a) imagined by Richardson to become a "forecast factory." Panel (a) reproduced with permission from "Le guide des cités" by François Schuiten and Benoît Peeters, © Copyright Casterman.

Bjerknes reiterated his concept in a 1914 paper entitled "Die Meteorologie als exakte Wissenschaft" (Meteorology as an exact science). He used the medical terms "diagnostics" and "prognostics" to describe the two steps shown. He suggested that the evolution of seven meteorological variables (pressure, temperature, the three wind components, air density, and water vapor content) could be predicted from the seven equations expressing the conservation of air mass and water vapor mass (continuity equations), the conservation of energy (thermodynamic equation, which relates the temperature of air to heating and cooling processes), as well as Newton's law of motion (three components of the Navier–Stokes equation), and the ideal gas law (which relates pressure to air density and temperature). Bjerknes realized that these equations could not be solved analytically, and instead introduced graphical methods to be used for operational weather forecasts.

During World War I, Lewis Fry Richardson (1881–1951; see Figure 1.3), who was attached to the French Army as an ambulance driver, attempted during his free time to create a numerical weather forecast model using Bjerknes' principles. He used a numerical algorithm to integrate by hand a simplified form of the meteorological equations, but the results were not satisfying. The failure of his method was later attributed to insufficient knowledge of the initial weather conditions, and to instabilities in the numerical algorithm resulting from an excessively long time step of six hours. Richardson noted that the number of arithmetic operations needed to solve the meteorological equations numerically was so high that it would be impossible for a single operator to advance the computation faster than the weather advances. He proposed then to divide the geographic area for which prediction was to be performed into several spatial domains, and to assemble for each of these domains a team of people who would perform computations in parallel with the other teams, and, when needed, communicate their information between teams. His fantasy led him to propose the construction of a "forecast factory" in a large theater hall (Figure 1.3), where a large number of teams would perform coordinated computations. This construction was a precursor vision of modern massively parallel supercomputers. The methodology used by Richardson to solve numerically the

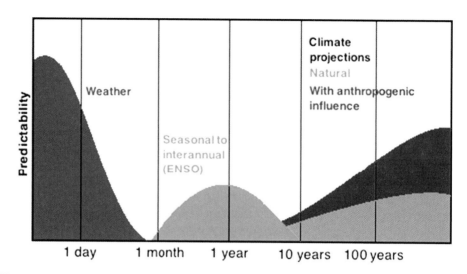

Figure 1.4 Qualitative representation of the predictability of weather, seasonal to interannual variability (El Nino – Southern Oscillation) and climate (natural variations and anthropogenic influences). Adapted from US Dept. of Energy, 2008.

meteorological equations was published in 1922 in the landmark book *Weather Prediction by Numerical Process*.

The first computer model of the atmosphere was developed in the early 1950s by John von Neumann (1903–1957) and Jule Charney (1917–1981), using the Electronic Numerical Integrator and Computer (ENIAC). The computation took place at about the same pace as the real evolution of the weather, and so results were not useful for weather forecasting. However, the model showed success in reproducing the large-scale features of atmospheric flow. Another major success of early models was the first simulation of cyclogenesis (cyclone formation) in 1956 by Norman Phillips at the Massachusetts Institute of Technology (MIT). Today, with powerful computers, meteorological models provide weather predictions with a high degree of success over a few days and some success up to ten days. Beyond this limit, chaos takes over and the accuracy of the prediction decreases drastically (Figure 1.4). As shown by Edward Lorenz (1917–2008), lack of forecasting predictability beyond two weeks is an unavoidable consequence of imperfect knowledge of the initial state and exponential growth of model instabilities with time (Lorenz, 1963, 1982). Increasing computer power will not relax this limitation. Lorenz's finding clouded the optimistic view of forecasting presented earlier by Bjerknes. Predictions on longer timescales are still of great value but must be viewed as stochastic, simulating (with a proper ensemble) the statistics of weather rather than any prediction of specific realization at a given time. The statistics of weather define the *climate*, and such long-range statistical weather prediction is called *climate modeling*,

Meteorological models include a so-called dynamical core that solves Bjerknes' seven equations at a spatial and temporal resolution often determined by available computing power. Smaller-scale turbulent features are represented through somewhat empirical parameterizations. Progress in meteorological models over the past decades has resulted from better characterization of the initial state, improvements in the formulation of physical processes, more effective numerical algorithms, and higher resolution enabled by increases in computer power. Today, atmospheric models may be used as *assimilation* tools, to help integrate observational data into a coherent theoretical framework; as *diagnostic* tools, to assist in the interpretation of observations and in the identification of important atmospheric processes; and as *prognostic* tools, to project the future evolution of the atmosphere on timescales of weather or climate.

Data assimilation plays a central role in weather forecasting because it helps to better define the initial state for the forecasts. Observations alone cannot define that state because they are not continuous and are affected by measurement errors. The meteorological model provides a continuous description of the initial state, but with model errors. Data assimilation blends the information from the model state with the information from the observations, weighted by their respective errors, to achieve an improved definition of the state. Early approaches simply nudged the model toward the observations by adding a non-physical term to the meteorological equations, relaxing the difference between model and observations. Optimal estimation algorithms based on Bayes' theorem were developed in the 1960s and provide a sounder foundation for data assimilation. They define a most likely state through minimization of an error-weighted least-squares cost function including information from the model state and from observations. Current operational forecast models use advanced methods to assimilate observations of a range of meteorological variables collected from diverse platforms and at different times. Four-dimensional *variational* data assimilation (4DVAR) methods ingest all observations within a time window to numerically optimize the 3-D state at the initial time of that window.

1.5 Climate Models

The climate represents the long-term statistics of weather, involving not only the atmosphere but also the surface compartments of the Earth system (atmosphere, oceans, land, cryosphere). It is a particularly complex system to investigate and to model. The evolution of key variables in the different compartments can be described by partial differential equations that represent fundamental physical laws. Solution of the equations involves spatial scales from millimeters (below which turbulence dissipates) to global, and temporal scales from milliseconds to centuries or longer. The finer scales need to be parameterized in order to focus on the evolution of the larger scales. Because of the previously described chaos in the solution to the equations of motion, climate model simulations are effectively stochastic. *Ensembles* of climate simulations conducted over the same time horizon but with slightly modified initial conditions provide statistics of model results that attempt to reproduce observed climate statistics.

The first climate models can be traced back to the French mathematician Joseph Fourier (1768–1830, see Figure 1.5), who investigated the processes that have maintained the mean Earth's temperature at a relatively constant value during its

(a)   (b)   (c)

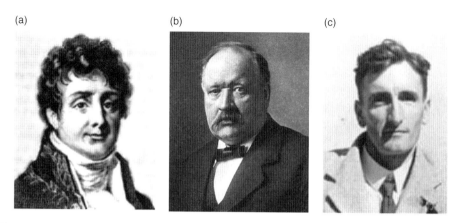

Figure 1.5 French mathematician and physicist Jean Baptiste Joseph Fourier (a), Swedish chemist Svante August Arrhenius (b), and British scientist Guy Stewart Callendar (c). Source of panel (c): G. S. Callendar Archive, University of East Anglia.

history. In 1896, the Swedish scientist Svante Arrhenius (1859–1927; see Figure 1.5) made the first estimate of the changes in surface temperature to be expected from an increase in the atmospheric concentration of CO_2. He did so by using measurements of infrared radiation emitted by the full Moon at different viewing angles to deduce the sensitivity of absorption to the CO_2 amount along the optical path, and then using the result in an energy balance equation for the Earth.

In 1938, Guy S. Callendar (1898–1964; see Figure 1.5) used a simple radiative balance model to conclude that a doubling in atmospheric CO_2 would warm the Earth surface by 2 °C on average, with considerably more warming at the poles. In the following decades, more detailed calculations were performed by 1-D (vertical) radiative–convective models allowing for vertical transport of heat as well as absorption and emission of radiation. Increasing computing power in the 1950s and 1960s paved the way for 3-D atmospheric climate models, called *general circulation models* (GCMs) for their focus on describing the general circulation of the atmosphere. Early GCMs were developed by Norman Phillips at MIT, Joseph Smagorinsky and Syukuro Manabe at the Geophysical Fluid Dynamics Laboratory (GFDL) in Princeton, Yale Mintz and Akio Arakawa at the University of California at Los Angeles (UCLA), and Warren Washington and Akira Kasahara at the National Center for Atmospheric Research (NCAR).

Climate models today have become extremely complex and account for coupling between the atmosphere, the ocean, the land, and the cryosphere. The Intergovernmental Panel on Climate Change (IPCC) uses these models to inform decision-makers about the climate implications of different scenarios of future economic development. Several state-of-science climate models worldwide contribute to the IPCC assessments, and yield a range of climate responses to a given perturbation. Attempts to identify a "best" model tend to be futile because each model has its strengths and weaknesses, and ability to reproduce present-day climate is not necessarily a gauge of how well the model can predict future climate. The IPCC uses instead the range of climate responses from the different models for a given scenario as a "wisdom of crowds" statistical ensemble to assess confidence in predictions of climate change.

1.6 Atmospheric Chemistry Models

Interest in developing chemical models for the atmosphere can be traced to the early twentieth century with the first observational inference by Fabry and Buisson (1913) of an ozone layer at high altitude. Subsequent ground-based measurements of the near-horizon solar spectrum in the 1920s established that this ozone layer was present a few tens of kilometers above the surface. Its origin was first explained in 1929 by British geophysicist Sydney Chapman (1888–1970; see Figure 1.6) as a natural consequence of the exposure of molecular oxygen (O_2) to ultraviolet (UV) radiation, producing oxygen atoms (O) that go on to combine with O_2 to produce ozone (O_3). Chapman's model produced an ozone maximum a few tens of kilometers above the surface, consistent with observations. It introduced several important new

(a)   (b)   (c)   (d)

(e)   (f)   (g)

Figure 1.6 From the top left to the bottom right ((a)–(g)): Sydney Chapman (Courtesy of the University Corporation for Atmospheric Research), Sir David Bates (Courtesy of Queen's University Belfast), Baron Marcel Nicolet, Paul Crutzen (Courtesy of Tyler Prize for Environmental Achievement), Mario Molina, (Tyler Prize for Environmental Achievement), Frank Sherwood (Sherry) Rowland (Tyler Prize for Environmental Achievement), and Susan Solomon.

concepts, including the interaction of radiation with chemistry (*photochemistry*) and the chemical cycling of short-lived species (oxygen atom and ozone), the usefulness of dynamical steady-state assumptions applied to short-lived species, and the negative feedback of ozone on itself through absorption of UV radiation.

By the 1940s and 1950s, attention had turned to the ionized upper atmosphere due to interest in the propagation of radio waves and the origin of the aurora. Models were developed to simulate the chemical composition of this region, and some were 1-D (vertical) to address conceptual issues of coupling between chemistry and transport. In 1950, British and Belgian scientists Sir David Bates (1916–1994) and Baron Marcel Nicolet (1912–1996) (Figure 1.6), who were studying radiative emissions (airglow) in the upper atmosphere, deduced from their photochemical model that hydrogen species produced by the photolysis of water vapor could destroy large amounts of ozone in the mesosphere (50–80 km). Such catalysis by hydrogen oxide radicals was found to also represent a significant sink for ozone in the stratosphere, adding to the ozone loss in the Chapman mechanism. The late 1960s and early 1970s saw the discoveries of additional catalytic cycles for ozone loss involving nitrogen oxide radicals ($NO_x \equiv NO + NO_2$) and chlorine radicals ($ClO_x \equiv Cl + ClO$) originating from biogenic nitrous oxide (N_2O) and industrial chlorofluorocarbons (CFCs), respectively. The NO_x-catalyzed cycle was found to be the dominant ozone-loss process in the natural stratosphere and this finally enabled a successful quantitative simulation of stratospheric ozone. The discovery of a CFC-driven ozone-loss cycle triggered environmental concern over depletion of the ozone layer. This work led to the awarding of the 1995 Nobel Prize in Chemistry to Dutch scientist Paul Crutzen, Mexican scientist Mario Molina, and American scientist Sherwood Rowland (Figure 1.6).

By the 1970s it was thought that our understanding of stratospheric ozone was mature, and global models coupling chemistry and transport began to be developed. These models were mostly two-dimensional (latitude–altitude), assuming uniformity in the longitudinal direction. Early three-dimensional models were also developed by Derek Cunnold at MIT and Michael Schlesinger and Yale Mintz at UCLA. A shock to the research community came in 1985 with the observational discovery of the Antarctic ozone hole, which had not been predicted by any of the models. This prompted intense research in the late 1980s and early 1990s to understand its origin. American scientist Susan Solomon (Figure 1.6) discovered that formation of polar stratospheric clouds (PSCs) under the very cold conditions of the wintertime Antarctic stratosphere enabled surface reactions regenerating chlorine radicals from their reservoirs, thus driving very fast ozone loss. The Antarctic ozone hole was a spectacular lesson in the failure of apparently well-established models when exposed to previously untested environments. Since then there have been no fundamental challenges to our understanding of stratospheric ozone, but continual improvement of models has led to a better understanding of ozone trends.

Rising interest in climate change in the 1980s and 1990s led the global atmospheric chemistry community to turn its attention to the troposphere, where most of the greenhouse gases and aerosol particles reside. In 1971, Hiram (Chip) Levy of the Harvard–Smithsonian Center for Astrophysics (Figure 1.7) used a radiative transfer model to show that sufficient UV-B radiation penetrates into the troposphere to

(a)   (b)   (c)

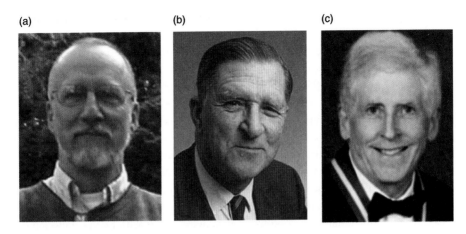

Figure 1.7   (a) Hiram (Chip) Levy, (b) Arie Haagen-Smit, and (c) John Seinfeld.

produce the hydroxyl radical OH, a strong radical oxidant that drives the removal of methane, carbon monoxide (CO), and many other important atmospheric gases. This upended the view of the global troposphere as chemically inert with respect to oxidation. As recently as 1970, a review of atmospheric chemistry in *Science* magazine had stated that "The chemistry of the troposphere is mainly that of a large number of atmospheric constituents and of their reactions with molecular oxygen . . . Methane and CO are chemically quite inert in the troposphere" (Cadle and Allen, 1970). Levy showed not only that fast oxidation by the OH radical takes place in the troposphere, but also that it drives intricate radical-propagated reaction chains. These chains provide the foundation for much of the current understanding of tropospheric oxidant chemistry.

Early global 3-D models of tropospheric chemistry were developed in the 1980s by Hiram Levy (by then at GFDL), Michael Prather (Harvard), and Peter Zimmermann (Max-Planck Institute for Chemistry in Mainz). Simulating the troposphere presented modelers with a new range of challenges. Transport is far more complex in the troposphere than in the stratosphere, and is closely coupled to the hydrological cycle through wet convection, scavenging, and clouds. Natural and anthropogenic emissions release a wide range of reactive chemicals that interact with transport on all scales and lead to a variety of chemical regimes. The surface also provides a sink through wet and dry deposition. The environmental issues in the troposphere are diverse and require versatility in models to simulate greenhouse gases, aerosols, oxidants, various pollutants, and deposition. Present-day global models of tropospheric chemistry typically include over 100 coupled species and a horizontal resolution of the order of tens to hundreds km. A number of issues remain today at the frontier of model capabilities, including aerosol microphysics, hydrocarbon oxidation mechanisms, formation of organic aerosols, coupling with the hydrological cycle, and boundary layer turbulent processes.

As the global atmospheric chemistry community gradually worked its way down from the upper atmosphere to the troposphere, a completely independent community with roots in engineering was working on the development of urban and regional air pollution models. Attention to air pollution modeling began in the 1950s. Prior to that, the sources of pollution were considered obvious (smokestacks and chimneys, industry, sewage, etc.) and their impacts immediate. Emergence of the Los Angeles smog in the 1940s shook this concept. The smog was characterized by decreased visibility and harmful effects on health and vegetation, but neither the causes nor the actual agents could be readily identified. The breakthrough came in the 1950s when Caltech chemist Arie Haagen-Smit (1900–1977, see Figure 1.7) showed that NO_x and volatile organic compounds (VOCs) emitted by vehicles could react in the atmosphere in the presence of solar radiation to produce ozone, a strong oxidant and toxic agent in surface air. This ozone production in surface air involved a totally different mechanism than in the stratosphere. Ozone was promptly demonstrated to be the principal toxic agent in Los Angeles smog. This introduced a new concept in air pollution; the pollution was worst not at the point of emission, but after atmospheric reaction some distance downwind. Additional toxicity and suppression of visibility was attributed to fine aerosol particles, also produced photochemically during transport in the atmosphere downwind from pollution sources. Similar mechanisms were found subsequently to be responsible for smog in other major cities of the world.

The discovery of photochemically generated ozone and aerosol pollutants in urban air spurred the development of air pollution models to describe the coupling of transport and chemistry. Initial efforts in the 1950s and 1960s focused on tracking the chemical evolution in transported air parcels (simple *Lagrangian* models) and describing the diffusion of chemically reactive plumes (*Gaussian plume* models). Three-dimensional air pollution models of the urban environment began to be developed in the 1970s. John Seinfeld of Caltech (Figure 1.7) was a pioneer with his development of airshed models for the Los Angeles Basin and of the underlying algorithms to simulate ozone and aerosols. By the 1970s, it also became apparent that long-range transport of ozone and aerosols caused significant pollution on the regional scale, and this together with concern over acid rain led in the 1980s and 1990s to development of 3-D regional models extending over domains of the order of 1000 km.

A major development over the past decade has been the convergence of the global atmospheric chemistry and air pollution modeling communities. This convergence has been spurred by issues of common interest: intercontinental transport of air pollution, climate forcing by aerosols and tropospheric ozone, and application of satellite observations to understanding of air pollution. Addressing these issues requires global models with fine resolution over the regions of interest. A new scientific front has emerged in bridging the scales of atmospheric chemistry models from urban to global.

Atmospheric chemistry modeling today is a vibrant field, with many challenges facing the research community when it comes to addressing issues of pressing environmental concern. We have discussed some of those challenges involving the representations of processes and the bridging across scales. There are a number of others. One is the development of whole-atmosphere models (from the surface to outer space) to study the response of climate to solar forcing and the response of the upper atmosphere to climate change. Another is the coupling of atmospheric chemistry to biogeochemical processes in surface reservoirs, which is emerging as a critical issue for modeling the nitrogen cycle and the fate of persistent pollutants such as mercury. Yet another challenge is the development of powerful chemical data assimilation tools to successfully manage the massive flow of atmospheric composition data from satellites. These tools are necessary for exploiting the data to test and improve current understanding of atmospheric processes, constrain surface fluxes through inverse modeling, and increase the capability of forecasts for both weather and air quality. Finally, a grand challenge is to integrate atmospheric chemistry into Earth System Models (ESMs) that attempt to fully couple the physics, chemistry, and biology in the different reservoirs of the Earth in order to diagnose interactions and feedbacks. Inclusion of atmospheric chemistry into ESMs has been lagging, largely because of the computational costs associated with the numerical integration of large chemical and aerosol mechanisms. Developing efficient and reliable algorithms is an important task for the future.

1.7 Types of Atmospheric Chemistry Models

The general objective of atmospheric chemistry models is to simulate the evolution of n interacting chemicals in the atmosphere. This is done by solving a coupled system of continuity equations, which in a fixed frame of reference can be written in the general form of equation (1.1). The solution of (1.1) depends on meteorological variables through the 3-D wind vector $\mathbf{v}$, generally including parameterizations to account for fine-scale turbulent contributions to the flux divergence term $\mathbf{v} \cdot \nabla C_i$. The local production and loss terms P_i and L_i may also depend on meteorological variables.

Many atmospheric chemistry models do not generate their own meteorological environment and instead use 3-D time-dependent data (including winds, humidity, temperature, etc.) generated by an external meteorological model. These are called "offline" models. The meteorological input data must define a mass-conserving airflow with consistent values for the different variables affecting transport, P_i, and L_i. By contrast, "online" atmospheric chemistry models are integrated into the parent meteorological model so that the chemical continuity equations are solved together with the meteorological equations for conservation of air mass, momentum, heat, and water. Online models have the advantage that they fully couple chemical transport with dynamics and with the hydrological cycle. They avoid the need for high-resolution meteorological archives, and they are not subject to time-averaging errors associated with the use of offline meteorological fields. They are not necessarily much more computer-intensive, since the cost of simulating many coupled chemical variables is often larger than the cost of the meteorological simulation. But they are far more complex to operate and interpret than offline models. The term *chemical transport model* (CTM) usually refers to offline 3-D models in the jargon of the atmospheric chemistry community. Here we will use the CTM terminology to refer to atmospheric chemistry models in general, since the methods are usually common to all models.

The meteorological model used to drive an atmospheric chemistry model can either be "free-running" or include assimilation of meteorological data. Data assimilation allows a meteorological model to simulate a specific observed meteorological year. A free-running model without data assimilation generates an ensemble of possible meteorological years, but not an actual observed year. Use of assimilated meteorological data is necessary to compare an atmospheric chemistry model to observations for a particular year. With a free-running meteorological model only climatological statistics can be compared. However, one advantage of using a free-running meteorological model is that winds and other meteorological variables are physically consistent. Data assimilation applies a non-physical correction to the model meteorology that can cause unrealistic behavior of non-assimilated variables called "data shock." For example, stratospheric models using assimilated meteorological data tend to suffer from excessive vertical transport because the assimilation of horizontal wind observations generates spurious vertical flow to enforce mass conservation. Advanced data assimilation schemes attempt to minimize these data shocks.

Another distinction can be made between *Eulerian* and *Lagrangian* models (Figure 1.8). A Eulerian model solves the continuity equations in a geographically

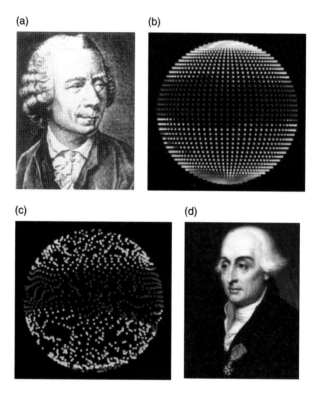

(a)

(b)

(c)

(d)

Figure 1.8   Global atmospheric distribution of trace species concentrations computed using Eulerian and Lagragian model representations. Concentrations are represented by colors (high values in blue, low values in orange). The Eulerian framework uses a fixed computational grid (b) while the Lagrangian framework uses an ensemble of points moving with the airflow (c). The figure also shows portraits of the Swiss mathematician and physicist Leonhard Paul Euler (1707–1783, a) and of the French mathematician Joseph Louis, Comte de Lagrange (1736–1813, d).

fixed frame of reference, while a Lagrangian model uses a frame of reference that moves with the atmospheric flow. The continuity equation as written in (1.1), including partial differentiation with respect to time and space, describes the evolution of concentrations in a fixed frame of reference and represents the Eulerian approach. Finite-difference approximations of the partial derivatives produce solutions on a fixed grid of points representing the model domain. By contrast, the Lagrangian approach solves the continuity equations for points moving with the flow; for these points we can rewrite (1.1) as

$$\frac{dC_i}{dt} = P_i(\mathbf{C}) - L_i(\mathbf{C}) \tag{1.2}$$

where $d/dt = \partial/\partial t + \mathbf{v} \cdot \nabla$ is the total derivative. From a mathematical standpoint, the Lagrangian approach reduces the continuity equation to a 1-D ordinary differential equation applied to points (0-D packets of mass) moving with the flow. The trajectory of each point still needs to be computed. The Eulerian approach is generally preferred in 3-D models because it guarantees a well-defined concentration field over the whole domain. In addition, Eulerian models deal better with nonlinear chemistry and mass conservation. On the other hand, Lagrangian models often have lower numerical transport errors and are better suited for tracking the transport of pollution plumes, as a large number of points can be released at the location of the pollution source. They are also generally the better choice for describing the source influence function contributing to observations made at a particular location (*receptor-oriented modeling*). In that case, a large number of points can be released at the location of the observations and transported backward in time in the Lagrangian framework.

We gave in Section 1.6 a brief history of the growing complexity of atmospheric chemistry models leading to the current generation of global 3-D models. At the other end of the complication spectrum, 0-D models remain attractive as simple tools for improving our understanding of processes. These models solve the continuity equation as $dC_i/dt = P_i(\mathbf{C}) - L_i(\mathbf{C})$, without consideration of spatial dimensions. They are called *box models* and are often appropriate to compute chemical steady-state concentrations of short-lived species, for which the effect of transport is negligible, or to compute the global budgets of long-lived chemical species for which uniform mixing within the domain can be assumed. Other simple models frequently used in atmospheric research include *Gaussian plume models* to simulate the fate of chemicals emitted from a point source and mixing with the surrounding background and 1-D models to simulate vertical mixing of chemicals in the atmospheric boundary layer (lowest few kilometers) assuming horizontal uniformity. Two-dimensional models (latitude–altitude) are also still used for stratospheric applications where longitudinal concentration gradients are generally small.

1.8 Models as Components of Observing Systems

The usefulness of a model is often evaluated by its ability to reproduce observations, but this can be misleading for two reasons. First, the observations themselves have errors, and models can in fact help to identify bad observations by demonstrating their inconsistency with independent knowledge. Second, the observations sample only a small domain of the space simulated by the model and may not be particularly relevant for testing the model predictions of interest. It is important to evaluate the model in the context of the problem to which it is applied. It is also important to establish whether the goal of the evaluation is to diagnose errors in the model physics or in the data used as input to the model.

A broad class of modeling applications involves the use of observations to quantify the variables driving the system (*state variables*) when these variables cannot be directly observed. Here, the observations probe the manifestations of the system (*observation variables*) while the model physics provide a prediction of the observation variables as a function of the state variables. *Inverting* the model then yields a prediction of the state variables for given values of the observational variables. One can think of this as fitting the model to observations in order to infer values for the state variables. Because of errors in the model and in the observations, the best that can be achieved in this manner is an *optimal estimate* for the state variables. Such an analysis is called *inverse modeling* and requires careful consideration of errors in the model, errors in the observations, and compatibility of results with prior knowledge. Model and observations are inseparable partners in inverse modeling. Having a very precise model is useless if the observations are not precise; having very precise observations is useless if the model is not precise.

This partnership between models and observations leads to the concept of *observing system* as the concerted combination of models and observations to address targeted monitoring or scientific goals. This concept has gained momentum with the dramatic growth in atmospheric observations, in particular from satellites generating massive amounts of data that are complicated to interpret. A model provides a continuous field of concentrations that can serve as a common platform for examining the complementarity and consistency of observations taken from different instruments at different locations, on different schedules, and for different species. Formal integration of the model and observational data can take the form of *chemical data assimilation* to produce optimized fields of concentrations, or inverse modeling to optimize state variables that are not directly observed.

Following on this concept of an observing system integrating observations and models, one may use models to compare the value of different observational data sets for addressing a particular problem, and to propose new observations that would be of particular value for that problem. *Observing system simulation experiments* (*OSSEs*) are now commonly conducted to quantify the benefit of a new source of observations (as from a proposed satellite) for addressing a quantifiable monitoring or scientific objective. Observing system simulation experiments use a CTM to generate a "true" synthetic atmosphere to be sampled by the ensemble of existing and proposed observing instruments. Pseudo-observations are made of that atmosphere along the instrument sampling paths and sampling schedules, with random error added following the instrument specifications. A second independent CTM is then used to invert these pseudo-observations and assess their value toward meeting the objective. A well-designed OSSE can tell us whether a proposed instrument will add significant information to the existing observing system.

1.9 High-Performance Computing

Modeling of atmospheric chemistry is a grand computational challenge. It requires solution of a large system of coupled 4-D partial differential equations (1.1). Relevant temporal and spatial scales range over many orders of magnitude. Atmospheric chemistry modeling is a prime application of *high-performance computing*, which describes mathematical or logical operations performed on *supercomputers* that are at the frontline of processing capacity and many orders of magnitude more powerful than desktop computers. The early supercomputers developed in the 1960s and 1970s by Seymour Cray spurred the development of weather and climate models. An important breakthrough in the 1980s was the development of *vector processors* able to run mathematical operations on multiple data elements simultaneously. The costs of these specialized vector platforms were still relatively high, so that in the 1990s the computer industry turned to developing high-performance machines based on mass-produced, less expensive components. This was accomplished through architectures that include a large number of *scalar micro-processing elements* operating in *massively parallel architectures*. Box 1.1 defines some of the relevant computing terminology.

Box 1.1 **High-Performance Computing Terminology**

Basic arithmetic operations are performed by *processing elements* or *processors*. Each processing element may have its own local *memory*, or can share a memory with other processors. The speed at which data transfer between memory and processor takes place is called the *memory bandwidth*. Computers with slow shared-memory bandwidth rely on small, fast-access local memories that hold data temporarily and are called *cache memory*. A *computational node* is a collection of processors with their shared memories. If data from a processor on one node can access directly a memory area in another node, the system is said to have *shared memory*. If messages have to be exchanged across the network to share data between nodes, the computer is said to have *distributed memory*. A *cluster* is a collection of nodes linked by a local high-speed network. When applications are performed in parallel, individual nodes are responsible only for a fraction of the calculations. Central to fully exploiting massively parallel architectures is the ability to divide and synchronize the computational burden effectively among the individual processors and nodes. The efficiency of the multi-node computation depends on the optimal use of the different processors, the memory bandwidth, and the bandwidth of the connection between nodes.

The effective use of supercomputers requires advanced programming. Fortran remains the language of choice because Fortran compilers generate faster code than other languages. Programming for parallel architectures may use the *Message Passing Interface* (*MPI*) protocol for loosely connected clusters with distributed memory and/or the *Open Multi-Processing* (*openMP*) protocol for shared-memory nodes. Massively parallel architectures require the use of distributed memory. *Grid computing* refers to a network of loosely coupled, heterogeneous, and geographically dispersed computers, offering a flexible and inexpensive resource to access a large number of processors or large amounts of data.

(Adapted in part from Washington and Parkinson, 2005)

The speed of a computer is measured by the number of floating point operations performed per second (called *Flops*). The peak performance of the Cray-1 installed in the 1970s at the Los Alamos National Laboratory (New Mexico) was 250 mega-flops (10^6 flops), while the performance of the Cray-2 installed in 1985 at the Lawrence Livermore National Laboratory (California) was 3.9 gigaflops (10^9 flops). The Earth Simulator introduced in Yokohama (Japan) in 2002, the largest computer in the world until 2004, provided 36 teraflops (10^{12} flops). This machine included 5120 vector processors distributed among 640 nodes. It was surpassed in 2004 by the IBM Blue Gene platform at the Lawrence Livermore National Laboratory with a performance that reached nearly 500 teraflops at the end of 2007. The performance of leading-edge supercomputers exceeded tens of petaflops (10^{15} flops) in 2015, and is predicted to be close to exaflops (10^{18} flops) by 2018. Enabling models to scale efficiently on such powerful platforms is a major engineering challenge.

References

Cadle R. D. and Allen E. R. (1970) Atmospheric photochemistry, *Science*, **167**, 243–249.

Fabry C. and Buisson M. (1913) L'absorption de l'ultraviolet par l'ozone et la limite du spectre solaire, *J. Phys. Rad.*, **53**, 196–206.

Gershenfeld N. (1999) *The Nature of Mathematical Modeling*, Cambridge University Press, Cambridge.

Kaufmann W. J. and Smarr L. L. (1993) *Supercomputing and the Transformation of Science*, Scientific American Library, New York.

Lakoff G. and Johnson M. (1980) *Metaphors We Live By*, Chicago University Press, Chicago, IL.

Lorenz E. (1963) Deterministic nonperiodic flow, *J. Atmos. Sci.*, **20**, 131–141.

Lorenz E. (1982) Atmospheric predictability experiments with a large numerical model, *Tellus*, **34**, 505–513.

Lovelock J. E. (1989) Geophysiology, the science of Gaia, *Rev. Geophys.*, **27**, 2, 215–222, doi: 10.1029/RG027i002p00215.

Müller P. and von Storch H. (2004) *Computer Modelling in Atmospheric and Oceanic Sciences: Building Knowledge*, Springer-Verlag, Berlin.

Walliser B. (2002) Les modèles économiques. In *Enquête sur le concept de modèle* (Pascal Nouvel, ed.), Presses Universitaires de France, Paris.

Washington W. M. and Parkinson C. L. (2005) *An Introduction to Three-Dimensional Climate Modeling*, University Science Book, Sausalito, CA.

Atmospheric Structure and Dynamics

2.1 Introduction

The atmosphere surrounding the Earth is a thin layer of gases retained by gravity (Figure 2.1). Table 2.1 lists the most abundant atmospheric gases. Concentrations are expressed as mole fractions, commonly called *mixing ratios*. The principal constituents are molecular nitrogen (N_2), molecular oxygen (O_2), and argon (Ar). Their mixing ratios are controlled by interactions with geochemical reservoirs below the Earth's surface on very long timescales. Water vapor is present at highly variable mixing ratios (10^{-6}–10^{-2} mol mol^{-1}), determined by evaporation from the Earth's surface and precipitation. In addition to these major constituents, the atmosphere contains a very large number of *trace gases* with mixing ratios lower than 10^{-3} mol mol^{-1}, including carbon dioxide (CO_2), methane (CH_4), ozone (O_3), and many others. It also contains solid and liquid *aerosol particles*, typically 0.01–10 μm in size and present at concentrations of 10^1–10^4 particles cm^{-3}. These trace gases and aerosol particles do not contribute significantly to atmospheric mass, but are of central interest for environmental issues and for atmospheric reactivity.

The mean atmospheric pressure at the Earth's surface is 984 hPa, which combined with the Earth's radius of 6378 km yields a total mass for the atmosphere of 5.14×10^{18} kg. As we will see, atmospheric pressure decreases quasi-exponentially with height: 50% of total atmospheric mass is found below 5.6 km altitude and 90% below 16 km. Atmospheric pressures are sufficiently low for the ideal gas law to be obeyed within 1% under all conditions. The global mean surface air temperature is 288 K, and the corresponding air density is 1.2 kg m^{-3} or 2.5×10^{19} molecules cm^{-3}; air density also decreases quasi-exponentially with height.

Figure 2.1 The Earth's atmosphere seen from space, with the Sun just below the horizon. Air molecules scatter solar radiation far more efficiently in the blue than in the red. The red sunset color represents solar radiation transmitted through the lower atmosphere. The blue color represents solar radiation scattered by the upper atmosphere. Cloud structures are visible in the lowest layers.

Table 2.1 Mixing ratios of gases in dry air[a]	
Gas	Mixing ratio (mol mol^{-1})
Nitrogen (N_2)	0.78
Oxygen (O_2)	0.21
Argon (Ar)	0.0093
Carbon dioxide (CO_2)	400×10^{-6}
Neon (Ne)	18×10^{-6}
Ozone (O_3)	$0.01 \text{--} 10 \times 10^{-6}$
Helium (He)	5.2×10^{-6}
Methane (CH_4)	1.8×10^{-6}
Krypton (Kr)	1.1×10^{-6}
Hydrogen (H_2)	500×10^{-9}
Nitrous oxide (N_2O)	330×10^{-9}

[a]excluding water vapor

We present in this chapter a general overview of the structure and dynamics of the atmosphere to serve as a foundation for atmospheric chemistry models. More detailed considerations on atmospheric physics and dynamics can be found in meteorological textbooks such as Gill (1982), Pedlosky (1987), Andrews *et al.* (1987), Zdunkowski and Bott (2003), Green (2004), Vallis (2006), Martin (2006), Mak (2011), Holton and Hakim (2013).

2.2 Global Energy Budget

The main source of energy for the Earth system is solar radiation. The Sun emits radiation as a blackbody of effective temperature $T_S = 5800$ K. The corresponding blackbody energy flux is $F = \sigma T_S^4$ where $\sigma = 5.67 \times 10^{-8}$ W m^{-2} K^{-4} is the Stefan-Boltzmann constant. This radiation extends over all wavelengths but peaks in the visible at 0.5 μm. The solar energy flux intercepted by the Earth's disk (surface perpendicular to the incoming radiation) is 1365 W m^{-2}. This quantity is called the *solar constant* and is denoted S. Thus, the mean solar radiation flux received by the terrestrial sphere is $S/4 = 341$ W m^{-2}. A fraction $\alpha = 30\%$ of this energy is reflected back to space by clouds and the Earth's surface; this is called the *planetary albedo*. The remaining energy is absorbed by the Earth–atmosphere system. This energy input is compensated by blackbody emission of radiation by the Earth at an *effective temperature* T_E. At steady state, the balance between solar heating and terrestrial cooling is given by

$$\frac{(1 - \alpha)S}{4} = \sigma T_E^4 \qquad (2.1)$$

The mean effective temperature deduced from this equation is $T_E = 255$ K. It is the temperature of the Earth that would be deduced by an observer in space from measurement of the emitted terrestrial radiation. The corresponding wavelengths of

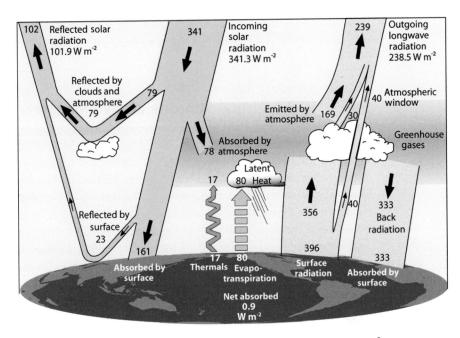

Figure 2.2 Global annual mean energy budget of the Earth for the 2000–2004 period. Units are W m^{-2}. From Trenberth *et al.* (2009). Copyright © American Meteorological Society, used with permission.

terrestrial emission are in the infrared (IR), peaking at 10 μm. The effective temperature is 33 K lower than the observed mean surface temperature, because most of the terrestrial radiation emitted to space originates from the atmosphere aloft where clouds and greenhouse gases such as water vapor and CO_2 absorb IR radiation emitted from below and re-emit it at a colder temperature. This is the essence of the greenhouse effect.

Figure 2.2 presents a more detailed description of the energy exchanges in the atmosphere. Of the energy emitted by the Earth's surface (396 W m^{-2}), only 40 W m^{-2} is directly radiated to space, while the difference (356 W m^{-2}) is absorbed by atmospheric constituents. Thus, the global heat budget of the atmosphere must include the energy inputs resulting from (1) the absorption of infrared radiation by clouds and greenhouse gases (356 W m^{-2}), (2) the *latent heat* released in the atmosphere by condensation of water (80 W m^{-2}), (3) the *sensible heat* from vertical transport of air heated by the surface (17 W m^{-2}), and (4) the absorption of solar radiation by clouds, aerosols, and atmospheric gases (78 W m^{-2}). Of this total atmospheric heat input (532 W m^{-2}), 199 W m^{-2} is radiated to space by greenhouse gases and clouds, while 333 W m^{-2} is radiated to the surface and absorbed. This greenhouse heating of the surface (333 W m^{-2}) is larger than the heating from direct solar radiation (161 W m^{-2}). At the top of the atmosphere, the incoming solar energy of 341 W m^{-2} is balanced by the reflected solar radiation of 102 W m^{-2} (corresponding to a planetary albedo of 0.30 with 23 W m^{-2} reflected by the surface and 79 W m^{-2} by clouds, aerosols and atmospheric gases) and by the IR terrestrial emission of 239 W m^{-2}. Note that the system as described here for the 2000–2004 period is slightly out of balance because of anthropogenic greenhouse gases: A net energy per unit area of 0.9 W m^{-2} is absorbed by the surface, producing a gradual warming.

2.3 Vertical Structure of the Atmosphere

Figure 2.3 shows the mean vertical profile of atmospheric temperature. Atmospheric scientists partition the atmosphere vertically on the basis of this thermal structure. The lowest layer, called the *troposphere*, is characterized by a gradual decrease of temperature with height due to solar heating of the surface. It typically extends to 16–18 km in the tropics and to 8–12 km at higher latitudes. It accounts for 85% of total atmospheric mass. Heating of the surface allows buoyant motions, called *convection*, to transport heat and chemicals to high altitude. During this rise the water cools and condenses, leading to the formation of clouds. The process of condensation releases heat, providing additional buoyancy to the rising air parcels that can result in thunderstorms extending to the top of the troposphere. The mean decrease of temperature with altitude (called the *lapse rate*) in the troposphere is

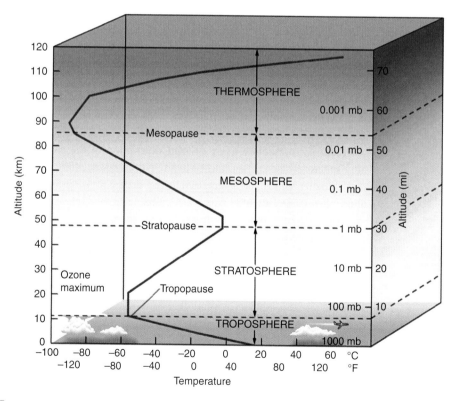

Figure 2.3 Mean vertical profile of air temperature and definition of atmospheric layers. From Aguado and Burt (2013), Copyright © Pearson Education.

6.5 K km^{-1}, reflecting the combined influences of radiation, convection, and the latent heat release from water condensation.

The top of the troposphere, defined by a temperature minimum (190–230 K), is called the *tropopause*. The layer above, called the *stratosphere*, is characterized by increasing temperatures with height to reach a maximum of about 270 K at the *stratopause* located at 50 km altitude. This warming is due to the absorption of solar UV radiation by ozone. A situation in which the temperature increases with altitude is called an *inversion*. Because heavier air is overlain by lighter air, vertical motions are strongly suppressed. The stratosphere is therefore very stable against vertical motions. Exchange of air with the troposphere is restricted, and vertical transport within the stratosphere is very slow. The residence time of tropospheric air against transport to the stratosphere is 5–10 years, and the residence time of air in the stratosphere ranges from a year to a decade. In summer, the *zonal* (longitudinal) mean temperature distribution in the stratosphere is determined primarily by radiative processes (solar heating by ozone absorption and terrestrial cooling by CO_2, water vapor, and ozone emission to space). In winter, radiation is weaker and the radiative equilibrium is perturbed by the propagation of planetary waves. This generates a large-scale *meridional* (latitudinal) circulation, called the *Brewer–Dobson circulation*, transporting air from low to higher latitudes.

The *mesosphere* extends from 50 km to the *mesopause* located at approximately 90–100 km altitude, where the mean temperature is about 160 K (120 K at the summer pole, which is the lowest temperature in the atmosphere). In this layer, where little ozone is available to absorb solar radiation, but where radiative cooling by CO_2 is still effective, the temperature decreases again with height. Turbulence is frequent and often results from the dissipation of vertically propagating gravity waves (see Section 2.11), when the amplitude of these waves becomes so large that the atmosphere becomes thermally unstable.

The *thermosphere* above 100 km is characterized by a dramatic increase in temperature with height resulting primarily from the absorption of strong UV radiation by molecular oxygen O_2, molecular nitrogen N_2, and atomic oxygen O. Collisions become rare so that a stable population of ions can be sustained, producing a plasma (ionized gas). The temperature above 200 km reaches asymptotic values of typically 500 to 2000 K, depending on the level of solar activity (Figure 2.4). This asymptotic behavior reflects the small heat content and the high heat conductivity of this low air density region. The corresponding altitude is called the *thermopause* and varies from 250 to 500 km altitude. Atmospheric pressure is sufficiently low above 100 km that vertical transport of atmospheric species occurs primarily by molecular diffusion. This process tends to separate with height the different chemical species according to their respective mass. As a result, the relative abundance of light species like atomic oxygen, helium, and hydrogen increases with height relative to species like molecular nitrogen and oxygen. Molecular nitrogen dominates up to 180 km, while the prevailing constituent between 180 and about 700 km is atomic oxygen. Helium is the most abundant constituent between 700 km and 1700 km, and atomic hydrogen at higher altitudes. Above the thermopause, atoms follow ballistic trajectories because of the rarity of collisions. In this region of the atmosphere, light atoms (hydrogen) can overcome the forces of gravity and

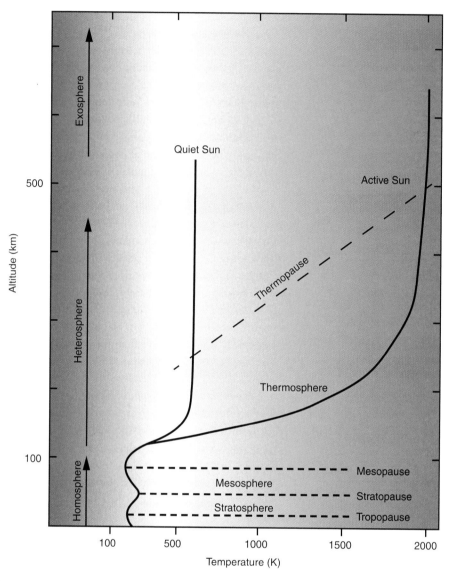

Figure 2.4 Vertical distribution of the mean temperature for two levels of solar activity with emphasis on the upper atmosphere layers. Adapted from Banks and Kockarts (1973).

escape to space if their velocity is larger than a threshold value (escape velocity). At that point the atmosphere effectively merges with outer space.

Air motions below 100 km are dominated by gravity and pressure forces, following the laws of *hydrodynamics*. Above 100 km, where ionization produces a plasma, the flow is affected by electromagnetic forces, and more complex equations from *magneto-hydrodynamics* must be applied. *Aeronomy* is the branch of science that describes the behavior of upper atmospheric phenomena (with emphasis on ionization and dissociation processes), while *meteorology* refers to the study of the lower levels of the atmosphere (with emphasis on dynamical and physical processes). The aeronomy literature has its own classification of atmospheric layers (see e.g., Prölss, 2004). For example, it refers to the troposphere as the *lower atmosphere*, to the stratosphere and mesosphere as the *middle atmosphere*, and to the thermosphere as the *upper atmosphere*. It defines the *homosphere* below 100 km as the region where vertical mixing is sufficiently intense to maintain constant the relative abundance of inert gases, and the heterosphere above 100 km as the region where gravitational settling becomes sufficiently important for the relative concentration of heavy gases to decrease more rapidly than that of lighter ones. The atmospheric region above 1700 km is often called the *geocorona*. It produces an intense glow resulting from the fluorescence of hydrogen excited by the solar Lyman-α radiation at 122 nm. In another nomenclature, one distinguishes between the barosphere, where air molecules are bound to the Earth by gravitational forces, and the exosphere in which the air density is so small that collisions can be neglected. The lower boundary of the exosphere, called the *exobase*, is located at 400–1000 km. Aeronomers refer to the *ionosphere* as the atmospheric region where ionization of molecules and atoms by extreme UV radiation (less than 100 nm) and energetic particle precipitation is a dominant process. Different ionospheric layers are distinguished: the *D-region* below 90 km altitude, the *E-region* between 90 and 170 km, the *F-region* between 170 and 1000 km, and the *plasmasphere* above 1000 km altitude. The region in which the magnetic field of the Earth controls the motions of charged particles is called the *magnetosphere*. Its shape is determined by the extent of the Earth's internal magnetic field, the solar wind plasma, and the interplanetary magnetic field.

2.4 Temperature, Pressure, and Density: The Equation of State

The state of the atmosphere is described by pressure p [Pa], temperature T [K], and chemical composition. Atmospheric pressure is sufficiently low that the ideal gas law is obeyed within 1% under all conditions. The equation of state can therefore be expressed as

$$pV = \mathcal{R}T \tag{2.2}$$

where V [m^3 mol^{-1}] represents the molar volume of air, and $\mathcal{R} = 8.3143$ J K^{-1} mol^{-1} is the universal gas constant. When expressed as a function of the number density $n_a = \mathcal{N}_A/V$ [molecules m^{-3}], where $\mathcal{N}_A = 6.022 \times 10^{23}$ molecules mol^{-1} is Avogadro's number, this expression becomes

$$p = n_a kT \tag{2.3}$$

where $k = \mathcal{R}/\mathcal{N}_A = 1.38066 \times 10^{-23}$ J K^{-1} is Boltzmann's constant.

The equation of state can also be expressed as a function of the *mass density* of air $\rho_a = n_a M_a/\mathcal{N}_A$ [kg m^{-3}]

$$p = \rho_a RT \tag{2.4}$$

where $R = \mathcal{R}/M_a$ is the specific gas constant for air and M_a [kg mol^{-1}] is the molar mass of air.

The molar mass of air is the weighted average of the mass of its components

$$M_a = \sum_i C_i M_i \tag{2.5}$$

where C_i and M_i are respectively the *mole fraction* (commonly called molar or volume *mixing ratio*) and the molar mass of constituent i. Since dry air can be closely approximated as a mixture of nitrogen N_2 (with $C_{N2} = 0.78$), oxygen O_2 (with $C_{O2} = 0.21$) and argon Ar (with $C_{Ar} = 0.01$), the molar mass for dry air is $M_d = 28.97 \times 10^{-3}$ kg mol^{-1}. Water vapor, which can account for up to a few percent of air in the lower troposphere, will make air slightly lighter.

2.5 Atmospheric Humidity

Because of the high variability of water vapor in air, meteorologists like to use separate equations of state for dry air and water vapor. This is legitimate following Dalton's law, which states that the total pressure of a mixture of gases is the sum of the partial pressures of its individual components. The equation of state for *dry* air is given by

$$p_d = \rho_d R_d T \tag{2.6}$$

where the specific gas constant for dry air is $R_d = \mathcal{R}/M_d = 8.314/(28.97 \times 10^{-3}) = 287 \, \text{J K}^{-1} \, \text{kg}^{-1}$. A similar equation can also be applied to water vapor (or any chemical constituent). The partial pressure of water vapor is commonly noted e and the equation of state is expressed by

$$e = \rho_w R_w T \tag{2.7}$$

where ρ_w is the water vapor mass density [kg m^{-3}] and $R_w = \mathcal{R}/M_w = 8.314/(18.01 \times 10^{-3}) = 461.6 \, \text{J K}^{-1} \, \text{kg}^{-1}$ is the specific gas constant for water vapor with a molar mass M_w of 18.01×10^{-3} kg mol^{-1}. Note that the total air pressure is $p = p_d + e$. The *volume mixing ratio* C_w and *mass mixing ratio* μ_w of water vapor are expressed by

$$C_w = \frac{n_w}{n_a} = \frac{e}{p} \tag{2.8}$$

and

$$\mu_w = \frac{\rho_w}{\rho_a} = \frac{M_w}{M_a} \frac{e}{p} = 0.622 \, C_w \tag{2.9}$$

where n_w and n_a are the number densities [molecules m^{-3}] of water and moist (total) air, respectively, $M_w = 18.01 \times 10^{-3}$ kg mol^{-1} is the molar mass of water, and M_a is the molar mass of moist air: $M_a = (1 - C_w)M_d + C_w M_w$. Meteorologists conventionally call μ_w the *specific humidity* (and write it q). They instead define the *water vapor mass mixing ratio* r_w as the ratio between the water vapor density ρ_w and the *dry* air density ρ_d, where $\rho_d = \rho_a - \rho_w$:

$$r_w = \frac{\rho_w}{\rho_d} = \frac{\rho_w}{\rho_a - \rho_w} = \frac{\mu_w}{1 - \mu_w} \tag{2.10}$$

The equation of state (2.6) for dry air can be applied to moist air if the temperature T is replaced by *the virtual temperature* T_v, the temperature at which dry air has the same pressure and density as moist air. Thus one writes

$$p = \rho_a R_d T_v \tag{2.11}$$

From the above equations, it follows that

$$\rho_a = \frac{p}{R_d T_v} = \frac{p - e}{R_d T} + \frac{e}{R_w T}$$

or

$$T_v = \frac{T}{1 - \frac{e}{p}\left(1 - \frac{R_d}{R_w}\right)} \tag{2.12}$$

with $R_d/R_w = M_w/M_d = 18/28.97 = 0.621$. A good approximation to this expression is provided by

$$T_v \approx (1 + 0.61\, r_w)T \tag{2.13}$$

Phase transitions of atmospheric water play a crucial role in meteorology. The *relative humidity RH* [percent] is expressed by

$$RH = 100\frac{e}{e_s} \tag{2.14}$$

where e_s is the *saturation* pressure at which water vapor is in equilibrium with the condensed phase (liquid or ice). For a saturated atmosphere ($e = e_s$), condensation and evaporation are in balance. One shows easily that the water mass mixing ratio r_w^{sat} corresponding to saturation is

$$r_w^{sat} = 0.622\frac{e_s}{p - e_s} \simeq 0.622\frac{e_s}{p} \tag{2.15}$$

Its value is inversely proportional to the total pressure and is a function of temperature because the saturation pressure e_s varies with temperature (see later). An atmosphere with $e < e_s$ is called *subsaturated* while one with $e > e_s$ is called *supersaturated*. A supersaturated atmosphere leads to cloud formation, contingent on the presence of suitable aerosol particles to provide pre-existing surfaces for condensation and overcome the energy barrier from surface tension. These particles are called *cloud condensation nuclei* (*CCN*) for liquid-water clouds and *ice nuclei* (*IN*) for ice clouds. Water-soluble particles greater than 0.1 μm in size are adequate CCN, and are sufficiently plentiful that liquid cloud formation takes place at supersaturations of a fraction of a percent. Ice nuclei are solid particles such as dust that provide templates for ice formation and are present at much lower concentrations than CCN. Because of the paucity of IN, clouds may remain liquid or mixed ice–liquid at temperatures as low as –40 °C; one then refers to the metastable liquid phase as *supercooled.*

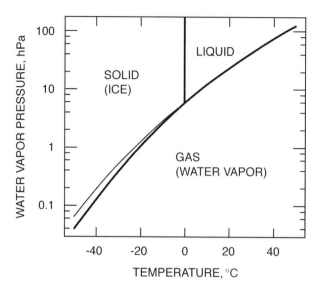

Figure 2.5 Phase diagram for water describing the stable phases present at equilibrium as a function of water vapor pressure and temperature. The thin line represents the metastable equilibrium between gas and liquid below 0 °C. Reproduced from Jacob (1999).

Phase equilibrium for water is defined by the *phase diagram* in Figure 2.5. Lines on this diagram represent equilibria between two phases. Equilibrium between vapor and condensed phases is expressed by the *Clausius–Clapeyron equation*

$$\frac{de_s}{dT} = e_s \frac{L}{R_w T^2} \tag{2.16}$$

where L represents the latent heat of vaporization or sublimation [J kg^{-1}]. Integration between a reference temperature T_0 and temperature T yields

$$e_s(T) = e_s(T_0) \exp\left(\frac{L}{R_w}\left[\frac{1}{T_0} - \frac{1}{T}\right]\right) \tag{2.17}$$

where L can be approximated as constant. For vaporization of liquid water, L has a value of 2.50×10^6 J kg^{-1} at 0 °C and varies with temperature T_C (in degrees Celsius) as (Rogers and Yau, 1989):

$$L(T) = \left[2500.79 - 2.36418\,T_C + 0.00158927\,T_C^2 - 0.0000614342\,T_C^3\right] \times 10^3 \tag{2.18}$$

For sublimation at 0 °C, L is 2.83×10^6 J kg^{-1}. Latent heat is released to the atmosphere (warming) when clouds condense from the gas phase; conversely, latent heat is absorbed from the atmosphere (cooling) when clouds evaporate.

The thin line in the phase diagram of Figure 2.5 represents the metastable phase equilibrium between water vapor and liquid water at temperatures below 0 °C. This equilibrium is relevant to the atmosphere because of supercooling of liquid clouds. When ice crystals do form in such clouds, the water vapor at equilibrium with the ice is lower than that at equilibrium with the supercooled liquid; thus, the liquid cloud droplets evaporate, transferring their water to the ice crystals. This transfer of water can also take place by collision between the supercooled liquid cloud droplets and the ice crystals (*riming*). In either case, the resulting rapid growth of the ice crystals promotes precipitation. The heat release associated with the conversion from liquid to ice also adds to the buoyancy of air parcels, fostering further rise and additional condensation and precipitation. Such precipitation formation in mixed-phase clouds is known as the *Bergeron process*.

2.6 Atmospheric Stability

2.6.1 The Hydrostatic Approximation

The vertical variation of atmospheric pressure can be deduced from *hydrostatic equilibrium*,

$$\frac{dp}{dz} = -\rho_a(z)g \tag{2.19}$$

which expresses that the downward gravitational force acting on a fluid parcel is balanced by an upward force exerted by the vertical pressure gradient that characterizes the fluid. Here $g = 9.81$ m s^{-2} is the acceleration of gravity, ρ_a [kg m^{-3}] the mass density of air, p [Pa] the atmospheric pressure, and z [m] is the altitude above the surface, often referred to as the *geometric altitude*. Equation (2.19) assumes that the fluid parcel is in vertical equilibrium between the gravitational and pressure-gradient forces, or more broadly that any vertical acceleration of the air parcel due to buoyancy is small compared to the acceleration of gravity. This is called the *hydrostatic approximation*. It is a good approximation for global models, but not for small-scale models attempting to resolve strong convective motions.

Because the atmosphere is thin relative to the Earth's radius (6378 km), g can be treated as constant with altitude. From the ideal gas law, we can rewrite equation (2.19) as

$$\frac{dp}{p} = -\frac{dz}{H(z)} \tag{2.20}$$

where

$$H(z) = \frac{RT(z)}{g} = \frac{kT(z)}{mg} \tag{2.21}$$

is a characteristic length scale for the decrease in pressure with altitude z and is called the *atmospheric scale height*. Its value is 8 ± 1 km in the troposphere and stratosphere. Here $k = 1.38 \times 10^{-23}$ J K^{-1} is Boltzmann's constant, $m = M_a/\mathcal{N}_A$ is the mean molecular mass of air ($28.97 \times 10^{-3}/6.022 \times 10^{23}$ kg $= 4.81 \times 10^{-26}$ kg), and $R = 287$ J K^{-1} kg^{-1} is the specific gas constant for air. By integrating (2.20), one finds the vertical dependence of atmospheric pressure

$$p(z) = p(0) \exp \int_0^z \frac{-dz'}{H(z')} \qquad (2.22)$$

where $p(0)$ is the surface pressure. Approximating H as constant yields the simple expression

$$p(z) \approx p(0) \exp \left(\frac{-z}{H} \right) \qquad (2.23)$$

which states that the air pressure decreases exponentially with altitude. Equation (2.23) is called the *barometric law*.

The hydrodynamic equations of the atmosphere are often expressed by using the pressure p rather than the altitude z as the vertical coordinate. It is then convenient to define the *log-pressure altitude Z*

$$Z = -\mathcal{H} \ln \left(\frac{p}{p_0} \right) \qquad (2.24)$$

as the vertical coordinate. Here $\mathcal{H}$ is a constant "effective" scale height (specified to be 7 km) and p_0 is a reference pressure (specified to be 1000 hPa). Thus Z depends solely on p. It is also convenient to introduce the *geopotential* Φ as the work required for raising a unit mass of air from sea level to geometric altitude z:

$$\Phi = \int_0^z g \, dz \qquad (2.25)$$

where g includes dependences on altitude and latitude, the latter due to the non-sphericity of the Earth (Section 2.7). g at Earth's surface varies from 9.76 to 9.83 m s^{-2}. The hydrostatic law relates Φ to $d\Phi = (RT/\mathcal{H}) \, dZ$. The *geopotential height* is defined as Φ/g_o, where $g_o = 9.81$ m s^{-2} is a constant called the *standard gravity*. Movement along a surface of uniform geopotential height involves no change in potential energy, i.e., no conversion between potential and kinetic energy. Meteorological weather conditions aloft are often represented as contour maps of geopotential heights at a given pressure. As we will see in Section 2.7, air motions tend to follow contour lines of geopotential heights, so this type of map is very useful.

2.6.2 Adiabatic Lapse Rate and Stability

Meteorologists use the concept of *air parcel* as a body of air sufficiently small to be defined by a single state (p, T), yet sufficiently large to preserve its identity during transport over some distance of interest. Applying the laws of thermodynamics to such an idealized air parcel gives valuable insight into atmospheric motions. The temperature of an air parcel changes as its pressure changes. A less variable measure of the heat content of an air parcel is the *potential temperature* θ [K]:

$$\theta = T \left[\frac{p_0}{p} \right]^{\kappa} \qquad (2.26)$$

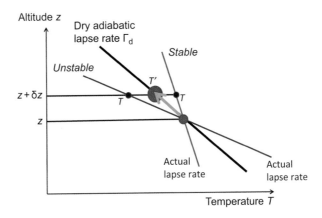

Figure 2.6   Schematic representation of the vertical temperature profile corresponding to a *dry adiabatic* lapse rate (9.8 K km^{-1} in black) and two hypothetical actual lapse rates (stable conditions in blue and unstable conditions in red). If an air parcel moves upwards under adiabatic conditions (green arrow), it will experience a buoyancy force that opposes the displacement if the temperature of the parcel T' is lower than the temperature T of the local environment ($T' < T$). In this case, the buoyancy force will restore the air parcel to its original position (stable conditions). If $T' > T$, the buoyancy force reinforces the displacement and drives the parcel further away from its original position (unstable conditions).

where p_0 = 1000 hPa is the reference pressure, $\kappa = R/c_p = 0.286$, and $c_p = 1005$ J K^{-1} kg^{-1} is the specific heat of dry air at constant pressure. The potential temperature is the temperature that an air parcel (p, T) would reach if it were brought adiabatically (without external input or loss of energy) to the reference pressure p_0. Under adiabatic conditions we have $d\theta/dz = 0$. One can show from a simple thermodynamic cycle analysis that under these adiabatic conditions the temperature must decrease linearly with altitude

$$\Gamma_d = -\frac{dT}{dz} = \frac{g}{c_p} = 9.8 \text{ K km}^{-1} \tag{2.27}$$

where Γ_d is called the *dry adiabatic lapse rate*. "Dry" refers to an air parcel sub-saturated with respect to water vapor; in the case of a saturated air parcel, latent heat release/loss during cloud condensation/evaporation complicates the analysis. The case of a saturated air parcel is discussed next.

An atmosphere left to evolve without exchanging energy with its surroundings will eventually achieve an adiabatic lapse rate due to the motion of air parcels up and down. Input or output of energy will force the actual lapse rate $\Gamma = -dT/dz$ to differ from the adiabatic lapse rate. In the stratosphere, for example, absorption of solar radiation by ozone causes the temperature to increase with altitude, a situation called a temperature *inversion*. The value of Γ relative to Γ_d diagnoses the *stability* of an air parcel relative to vertical motions (Figure 2.6).

Consider an air parcel located initially at an altitude z in an atmosphere with temperature $T(z)$ and lapse rate Γ. Let us apply an elemental push upward to this air parcel so that its altitude increases by δz. This motion takes place adiabatically so that the new temperature of the air parcel is $T'(z + \delta z) = T(z) - \Gamma_d\,\delta z$. The temperature of the surrounding air at that altitude is $T(z + \delta z) = T(z) - \Gamma\delta z$. If $\Gamma > \Gamma_d$, then the air parcel at $z + \delta z$ is warmer and hence lighter than the surrounding air; it is therefore accelerated upward by buoyancy, amplifying the initial upward displacement. A similar reasoning can be made if the air parcel is initially pushed downward; at altitude $z - \delta z$ it will be colder than its surroundings and buoyancy will accelerate the downward motion. Thus an atmosphere with $\Gamma > \Gamma_d$ is said to be *unstable* with respect to vertical motions. Rapid convective vertical mixing takes place in such an atmosphere. If on the contrary the lapse rate is smaller than the adiabatic lapse rate such that $\Gamma < \Gamma_d$, an air parcel displaced adiabatically toward higher altitude will become colder and denser than the surrounding air. As a result, it will return to its original level; the atmosphere is said to be *stable*. If $\Gamma = \Gamma_d$, the air parcel will continue its upward or downward motion with no acceleration, and the atmosphere is said to be *neutral*.

If air is saturated with water vapor, the stability conditions are modified: Ascending motion results in water condensation, which releases heat within the air parcel even under the adiabatic assumption. Similarly, in such an atmosphere, downward motion results in water evaporation and hence internal cooling. There results a decrease in stability. The lapse rate for saturated air parcels moving adiabatically up or down, called the *wet* (or *moist* or *saturated*) adiabatic lapse rate Γ_w, can be derived from the energy balance equation and the Clausius–Clapeyron equation

$$\Gamma_w = -\frac{dT}{dz} = \frac{g}{c_p}\frac{\left(1 + \frac{Lr_w}{RT}\right)}{\left(1 + \frac{L^2 r_w}{c_p R_w T^2}\right)} \tag{2.28}$$

Γ_w is smaller than the dry adiabatic lapse rate Γ_d. Its value depends on the water vapor condensation rate, which is determined by the water vapor mass mixing ratio r_w under the saturated conditions of the air parcel. Since r_w under these conditions is a strong function of temperature (Clausius–Clapeyron equation), it follows that Γ_w depends strongly on temperature. It typically ranges from 2 to 8 K km^{-1}. Under saturated conditions, stability requires that $\Gamma < \Gamma_w$. Buoyant motions in clouds occur when $\Gamma > \Gamma_w$ and are referred to as *wet convection*. The atmosphere is said to be *conditionally unstable* if $\Gamma_w < \Gamma < \Gamma_d$. Such an atmosphere is stable unless sufficient water vapor is supplied to it to make it saturated, in which case it becomes unstable (Figure 2.7).

Unstable conditions in the atmosphere can be triggered by solar heating of the ground. The heat is communicated by conduction from the ground to the overlying atmosphere, leading to $\Gamma > \Gamma_d$. Under such conditions, rapid vertical motions maintain an effective adiabatic lapse rate for the atmosphere, so that $\Gamma > \Gamma_d$ is not practically observed. In fact, observation of $\Gamma = \Gamma_d$ is generally a reliable diagnostic of an unstable atmosphere; the unstable lapse rate continually re-adjusts to Γ_d through the motion of air parcels up and down. Conversely, cooling of the ground at night produces a stable atmosphere ($\Gamma < \Gamma_d$). Particularly stable conditions are encountered when the temperature increases with height and produces a temperature inversion. In the troposphere, such an inversion often occurs during compressional heating associated with large-scale descent of air (a process called *subsidence*) or when the ground is particularly cold. In the stratosphere, the temperature increases

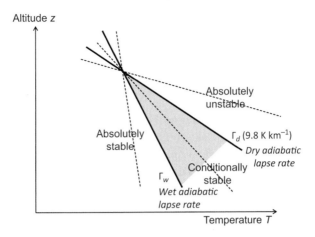

Altitude z

Figure 2.7   Effect of atmospheric humidity on the static stability of the atmosphere. The solid lines represent the dry (Γ_d) and wet (Γ_w) adiabatic lapse rates. As the actual lapse rate increases, the atmosphere evolves from a state of absolute stability to a state of conditional stability (shaded region for which saturated parcels are unstable but unsaturated parcels are not) and to a state of absolute instability.

with height due to the absorption of solar UV radiation by ozone. Thus the whole stratosphere is characterized by an inversion.

Buoyant convection as described above is the principal driver for vertical transport of trace constituents in the troposphere. Because it is driven by local temperature gradients, it occurs at scales too small to be resolved by regional or global atmospheric models. It therefore needs to be parameterized by using the model-scale information on temperature gradients and water vapor to estimate the resulting model-scale vertical motions. Such *convective parameterizations* rely on approximation of the actual physics and often include empirical or adjustable coefficients to better reproduce observations. They are crucial to the representation of vertical motions and cloud formation in atmospheric models. See Chapter 8 for further discussion.

2.7 Geostrophic Balance

We now turn to the forces driving horizontal motions in the atmosphere. Horizontal pressure gradients resulting from differential heating produce motions directed from high- to low-pressure areas. A complication is that the Earth is a rotating sphere, where different points have different translational velocities in a fixed frame of reference. The useful frame of reference for us is one that rotates with the sphere, since we measure all air motions with respect to this frame of reference. From the perspective of this rotating frame of reference, any motion taking place in the fixed frame of reference (such as driven by a pressure-gradient force) will be deflected due to the rotation. The deflection accelerates the air parcel away from its original direction and thus behaves as a fictitious force, called the *Coriolis* force. The Coriolis force operates in three dimensions but is negligible in the vertical relative to the acceleration of gravity. It is of critical importance for large-scale motions in the horizontal direction.

To understand the Coriolis effect, consider that the translational Earth's rotation velocity V_T (directed eastward)

$$V_T = a\,\Omega \cos \varphi \tag{2.29}$$

decreases with increasing latitude φ. With an Earth's radius $a = 6378$ km and an angular rotation velocity $\Omega = 7.292 \times 10^{-5}$ rad s^{-1} (or $2\,\pi$ rad d^{-1}), the velocity V_T is 1672 km h^{-1} at the Equator and 836 km h^{-1} at 60° latitude. If we consider an air parcel that is displaced poleward in the northern hemisphere, starting from latitude φ_1, the conservation of angular momentum in the absence of external forces requires that the product $\rho_a\,V(\varphi)\,a \cos \varphi$ remain constant at its initial value $\rho_a\,V_T(\varphi_1)\,a \cos \varphi_1$ during the displacement of the parcel. Here, V is the absolute eastward velocity of the air parcel in the fixed frame of reference, and ρ_a is the air density. Since $V_T(\varphi)$ decreases with latitude φ, this condition can only be fulfilled if, for an observer located at the Earth's surface, the air parcel acquires a gradually increasing eastward velocity. For the same reason, an air parcel moving toward the Equator in the northern hemisphere will be displaced westward (see Figure 2.8).

The same Coriolis effect also applies to motions in the longitudinal direction. In that case it can be understood in terms of the centrifugal force exerted on air parcels in the rotating frame of reference of the Earth. An air parcel at rest at a given latitude is subject to a centrifugal acceleration V_T^2/a that would make it drift toward the Equator were it not for the oblate geometry of the Earth (Figure 2.9). The resultant force of gravity (oriented toward the center of the Earth) and reaction (oriented normal to the surface) exactly cancels the centrifugal force, as shown in Figure 2.9. This should not be surprising considering that the oblate geometry is actually a consequence of the centrifugal force applied to the solid Earth. Consider now an eastward motion applied to the air parcel so that $V > V_T$. This motion increases the centrifugal force and deflects the air parcel equatorward. Conversely, a westward motion with $V < V_T$ weakens the centrifugal force and deflects the air parcel poleward. In both cases the deflection is to the right in the northern hemisphere and to the left in the southern hemisphere.

In summary, for an observer on the rotating Earth, air parcels moving horizontally are subject to a Coriolis force that is perpendicular to the direction of motion and proportional to the parcel's velocity; this force deflects air parcels to the right in the northern hemisphere and to the left in the southern hemisphere. It can be shown that the corresponding Coriolis acceleration is

$$\left[\frac{d\mathbf{v}}{dt}\right]_{Coriolis} = -2[\mathbf{\Omega} \times \mathbf{v}] \tag{2.30}$$

where $\mathbf{v}$ represents the velocity vector in the rotating frame of reference and $\mathbf{\Omega}$ is the Earth angular velocity vector directed from the south to the north pole. When expressed in Cartesian coordinates and considering the zonal and meridional wind components (u, v), the Coriolis acceleration becomes

$$\left[\frac{du}{dt}\right]_{Coriolis} = f\,v \tag{2.31}$$

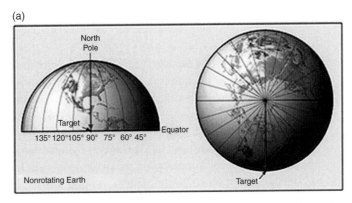

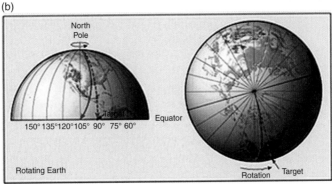

Figure 2.8   Trajectory of an object (such as an air parcel) directed from the north pole toward the Equator at 90° W. (a) Case of a non-rotating planet. (b) Deflection of the trajectory toward the right due to the rotation of the Earth. The arrival point at the Equator is displaced to the west of the original target point. Reproduced with permission from Lutgens *et al.* (2013). Copyright © Pearson Education, Inc.

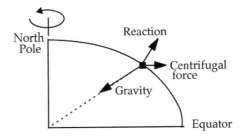

Figure 2.9   Equilibrium of forces for an air parcel at rest in the frame of reference of the rotating Earth. The centrifugal force is directed away from the axis of rotation, gravity is directed toward the center of the Earth, and the reaction force is perpendicular to the surface. The centrifugal force would cause the air parcel to drift toward the Equator if the Earth were a perfect sphere. The oblate geometry of the Earth (greatly exaggerated for the purpose of this figure) results in equilibrium in the triangle of forces. Reproduced from Jacob (1999).

$$\left[\frac{dv}{dt}\right]_{Coriolis} = -f\,u \tag{2.32}$$

where $f = 2\,\Omega\,\sin\varphi$ is the *Coriolis parameter*. It is positive in the northern hemisphere and negative in the southern hemisphere. Its amplitude increases with latitude. Thus, the Coriolis acceleration, which is zero at the Equator, increases with latitude and with the velocity of the flow. Its effect is substantial for large-scale motions (~1000 km, the *synoptic scale*).

For an observer attached to the rotating Earth, the large-scale motions in the extratropical atmosphere can be represented by a balance between the Coriolis and the pressure-gradient forces, called the *geostrophic approximation*:

$$2[\mathbf{\Omega} \times \mathbf{v}] = -\frac{1}{\rho_a}\nabla p \tag{2.33}$$

or in a Cartesian projection (x and y being the geometric distances in the zonal and meridional directions, respectively)

$$f\,v = \frac{1}{\rho_a}\frac{\partial p}{\partial x} \tag{2.34}$$

$$f\,u = -\frac{1}{\rho_a}\frac{\partial p}{\partial y} \tag{2.35}$$

From (2.33) we see that the geostrophic motions on a horizontal surface are parallel to the isobars (lines of constant pressure). In the northern hemisphere ($f > 0$), air parcels rotate clockwise around high pressure (anti-cyclonic) cells, and counter-clockwise around low pressure (cyclonic) cells (see Figure 2.10). The situation is reversed in the southern hemisphere ($f < 0$).

When formulated using pressure rather than geometric altitude as the vertical coordinate, the geostrophic balance takes the form:

$$f\,v = \frac{\partial \Phi}{\partial x} \tag{2.36}$$

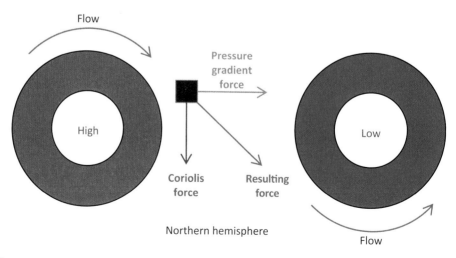

Figure 2.10   Flow of air in the northern hemisphere between anti-cyclonic (high) and cyclonic (low) regions. The motion originally directed from the high- to low-pressure cells is deflected to the right by the Coriolis force.

$$f u = -\frac{\partial \Phi}{\partial y} \tag{2.37}$$

where Φ is the geopotential. Thus, on isobaric surfaces, the geostrophic motions follow the contours of the geopotential fields. Replacing $d\Phi = (RT/\mathcal{H})dZ$ yields the *thermal wind equations*

$$f \frac{\partial v}{\partial Z} = \frac{R}{\mathcal{H}} \frac{\partial T}{\partial x} \tag{2.38}$$

$$f \frac{\partial u}{\partial Z} = -\frac{R}{\mathcal{H}} \frac{\partial T}{\partial y} \tag{2.39}$$

These show that the vertical shear in the horizontal (constant pressure level) wind field is proportional to the horizontal temperature gradient. In both hemispheres, the zonal wind component u increases with height when temperature decreases with latitude and decreases with height when temperature increases with latitude (Figure 2.11). The strong decrease of temperature with latitude in the troposphere produces intense subtropical *jet streams*, seen in Figure 2.11 as westerly wind maxima centered at about 40° latitude and 10 km altitude.

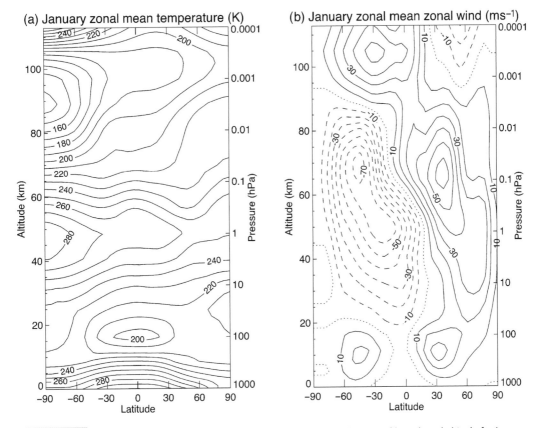

Figure 2.11 Zonal mean temperature (a) and zonal wind velocity (b) as a function of latitude and altitude for January, from the COSPAR International Reference Atmosphere (CIRA). Reprinted with permission from Shepherd (2003), Copyright © American Chemical Society.

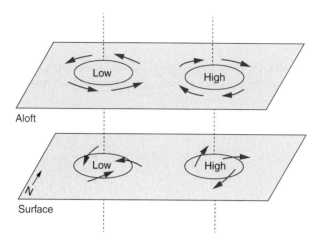

Winds around low- and high-pressure cells in the northern hemisphere. Geostrophic balance dominates aloft and the flow is directed along isobaric lines. Near the surface, friction deflects the flow toward low pressure. Reproduced from Ahrens (2000), Copyright © Cengage Learning EMEA.

Near the surface, the geostrophic flow is modified by *friction* resulting from the loss of momentum as the flow encounters obstacles (vegetation, ocean waves, buildings, etc.). The friction force is directed in the direction opposite to the flow (slowdown of the wind), effectively weakening the Coriolis force. This deflects the flow toward areas of low pressure (or low geopotential areas on isobaric surfaces), as shown in Figure 2.12.

2.8 Barotropic and Baroclinic Atmospheres

A *barotropic* atmosphere is one in which changes in air density are driven solely by changes in pressure. It is a good approximation in the tropics, where horizontal temperature gradients are small. In a barotropic atmosphere, isobaric (uniform pressure) surfaces coincide with isopycnic (uniform air density) surfaces. From the ideal gas law, they must also coincide with isothermal (uniform temperature) and isentropic (uniform potential temperature) surfaces. Since there is no temperature gradient on isobaric surfaces, the geostrophic wind is independent of height (see (2.38) and (2.39)). Under adiabatic conditions ($d\theta/dt = 0$), air parcels remain on isentropic surfaces, and since no pressure gradient exists along these surfaces to drive atmospheric motions, no potential energy is available for conversion into kinetic energy.

Outside the tropics, where meridional temperature gradients are large (Figure 2.11), the temperature varies along the isobars, and the atmosphere is said to be *baroclinic*. Isobars and isentropes do not coincide. In this case, pressure gradients can drive adiabatic displacement along isentropic surfaces. Conversion of potential energy into kinetic energy becomes possible. Temperature gradients along

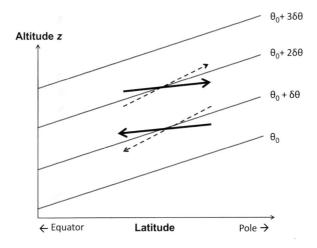

Altitude z

$\theta_0 + 3\delta\theta$

$\theta_0 + 2\delta\theta$

$\theta_0 + \delta\theta$

θ_0

← Equator   **Latitude**   Pole →

Figure 2.13 Baroclinic instability illustrated by parcel trajectories in the latitude–altitude plane with respect to isentropic surfaces. The isentropes slope upward and poleward. For quasi-horizontal parcel trajectories with slopes shallower than those of the background isentropes (solid arrows), unstable conditions arise even though the background atmosphere is stable ($\partial\theta/\partial z > 0$). By contrast, trajectories with slopes steeper than the isentropes (dashed arrows) are suppressed by stability.

isobars cause vertical shear in the geostrophic wind (see thermal wind equation), leading to a strong jet stream in the upper troposphere as discussed in Section 2.7. The axis of the jet stream is located in the $30°$–$60°$ latitudinal band characterized by a pronounced meridional temperature gradient separating cold and dense air of polar origin from warmer, less dense tropical air. In the presence of strong velocity shears, the jet stream may be unstable with respect to small perturbations, and disturbances may amplify, producing the so-called *baroclinic instability*.

Figure 2.13 illustrates baroclinic instability. The meridional gradient in temperature causes the isentropic surfaces (*isentropes*) to slope upward with increasing latitude. A poleward motion at constant altitude or with an upward slope shallower than the isentropes produces an unstable atmosphere even though the isentropes imply a vertically stable atmosphere ($\partial\theta/\partial z > 0$). Despite the stable conditions, potential energy from the flow can be converted into kinetic energy. Baroclinic instabilities drive the development of mid-latitude cyclones and associated frontal systems.

2.9 General Circulation of the Troposphere

Solar heating of the Earth must be balanced on a global basis by emission of terrestrial radiation to space. Both of these terms are a strong function of latitude (Figure 2.14). On an annual mean basis, the tropics receive much higher solar radiation than higher latitudes. Terrestrial emission to space peaks at about $20°$ latitude and drops at higher latitudes. On balance, the tropics have a surplus of radiative energy and the high latitudes have a deficit. Energy balance requires that

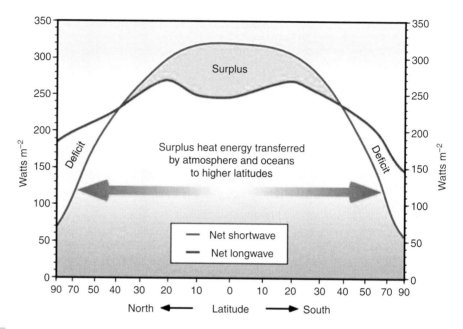

Annual mean balance between average net shortwave solar and longwave terrestrial radiation as a function of latitude. Reproduced with permission from Pidwirny (2006).

heat be transported from the tropics to the poles by atmospheric and oceanic motions. The *general circulation* of the atmosphere refers to the global wind systems that carry out this transport of energy.

In the absence of planetary rotation, the latitudinal gradient in surface heating would drive hemispheric circulation cells with upward motion in the tropics, high-altitude poleward flow, subsidence over the poles, and return equatorward flow near the surface (Figure 2.15, (a)). Planetary rotation complicates this simple picture due to the Coriolis force acting on air parcels as they travel meridionally. The poleward flow is deflected to the east and the equatorward flow is deflected to the west. The high-altitude poleward flow originating from the Equator becomes fully zonal at a latitude of about 30° and at that point no further meridional transport takes place. Thus the meridional circulation cells only extend from the Equator to 30° (Figure 2.15, (b)). These are called the *Hadley cells* after George Hadley (1685–1768), who first recognized the effect of planetary rotation on the general circulation of the atmosphere. Convergence between the southern and northern cells near the Equator defines the *intertropical convergence zone* (*ITCZ*) as a band of persistent precipitation. The ITCZ moves seasonally with *solar declination*, which is the angle between the Sun's rays and the equatorial plane of the Earth. Solar declination varies from +23° on June 21 to –23° on December 22. This defines the wet seasons of the tropics. Subsidence at 30° produces hot and dry conditions over land; the major deserts of the world are in that latitudinal band. Deflection to the west of the return equatorward flow near the surface produces the persistent tropical easterlies known as the *trade winds*.

(a)

(b)

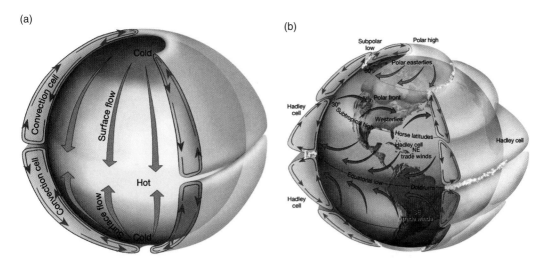

Figure 2.15   General circulation of the atmosphere. (a): Expected circulation in the absence of planetary rotation. (b): Circulation of the rotating troposphere. The Hadley cells feature rising air in the tropics and sinking air in the subtropics. The higher-latitude meridional cells are mainly conceptual as most of the meridional transport in the extratropics is driven by waves traveling longitudinally. From NASA, courtesy Barbara Summey, NASA Goddard VisAnalysis Laboratory; and from Lutgens and Tarbuck (2000).

Poleward of 30°, the movement of air masses is considerably modified by the Earth's rotation as the Coriolis force becomes stronger. The Coriolis force imposes a strong circumpolar flow, so that air travels zonally around the Earth on a timescale of weeks. The mid-latitude troposphere is strongly baroclinic, resulting in dynamical instabilities that spawn mid-latitude cyclones (Section 2.8). In the extratropics, most of the meridional heat exchange takes place by wave systems manifested by traveling weather disturbances (cyclones, anticyclones, and associated fronts between warm and cold air masses). The polar regions are characterized by cold and dry air with small weather disturbances and rare precipitation. Meridional mixing of air within a hemisphere takes place on a timescale of about three months, while mixing of air between the two hemispheres across the ITCZ takes place on a timescale of one year. The ITCZ is a major *dynamical barrier* for atmospheric mixing because of the weak thermal contrast across the Equator. Many long-lived gases such as CO_2 are well-mixed within each hemisphere, but feature an *interhemispheric gradient* maintained by the ITCZ.

The general circulation of the atmosphere is further influenced by the geographic distribution of continents and oceans. In the tropics, differences in surface heating between warm continents and cooler oceans drive zonal asymmetries in the circulation. Deep tropical convection takes place over the continents and the western equatorial Pacific (the *warm pool*, where sea surface temperatures are the highest in the world). Subsidence prevails over most of the tropical oceans, particularly where ocean currents maintain relatively cold surface temperatures (East Pacific, South Atlantic). Seasonal variations in land heating and cooling produce *monsoon* circulations, as illustrated in Figure 2.16 for South Asia. During winter the cold continental surface air flows toward the ocean, producing dry conditions over land.

(a)   (b)

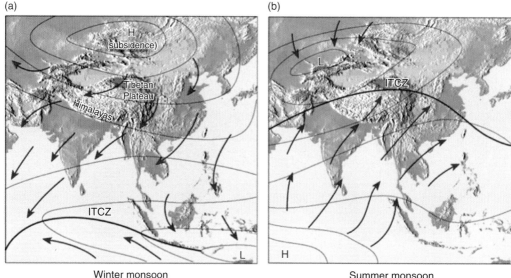

Winter monsoon   Summer monsoon

Figure 2.16   Surface circulation in Southeast Asia during northern winter (a) and summer (b), featuring the seasonal monsoons. The seasonal shift in the ITCZ is a consequence of the monsoon. Reproduced with permission from Lutgens *et al.* (2013), Copyright © Pearson Education, Inc.

During summer the moist ocean air flows over the heated land, resulting in heavy convective precipitation. Yet another effect of land on the general circulation is friction and topography. Thus the extensive land masses at northern mid-latitudes promote weather disturbances and meridional flow, facilitating the transport of heat to the Arctic.

Figure 2.17 illustrates the mean climatological distributions of surface pressure and winds in January and July. The seasonal shift of the ITCZ is apparent. Easterly trade winds prevail on both sides of the ITCZ. The subtropics are characterized by semi-permanent anti-cyclonic conditions that reflect the downwelling branches of the Hadley cells. Mid-latitude westerlies develop on the poleward side of these subtropical anti-cyclones and are far more steady in the southern hemisphere than in the north due to lack of ocean–land contrast. Meridional pressure gradients (shown by the isobars) are generally stronger in winter than summer, due to the greater meridional heating gradients, and this results in stronger winds.

Figure 2.18 shows global climatological distributions of precipitation in January and July. The band of intense precipitation near the Equator corresponds to the ITCZ. Seasonal shift in the ITCZ drives the wet and dry seasons in the tropics; in January the northern tropics are dry while the southern tropics are wet, and this is reversed in July. Subtropics are dry while mid-latitudes generally experience moderate precipitation in all seasons. Prominent *storm tracks* off the east coasts of Asia and North America play an important role in transport from northern mid-latitudes to the Arctic.

Different modes of interannual climatic variability are superimposed on this mean climatological description of the atmospheric circulation. The dominant mode in the tropics is the *El Niño–Southern Oscillation* (ENSO), a pattern of reversing ocean

(a)

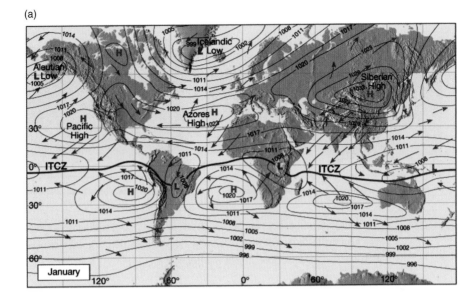

(b)

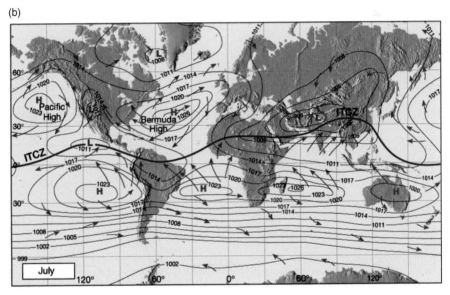

Figure 2.17 Climatological mean surface pressures (hPa) and winds in January (a) and July (b). The location of the intertropical convergence zone (ITCZ) is indicated. Major centers of high (H) and low (L) pressure are also shown. Reproduced from Lutgens and Tarbuck (2000).

temperatures between the eastern and western tropical Pacific that takes place every 3–8 years (Figure 2.19). During the normal *cold phase* of ENSO (also called *La Niña*), sea surface temperatures are cold in the eastern Pacific and warm in the western Pacific. There results strong subsidence and dry conditions in the east, and deep convection and wet conditions in the west. During the *warm phase* (also called *El Niño*), warm waters move from the western to the central and eastern Pacific, modifying considerably the tropical circulation with droughts over Oceania,

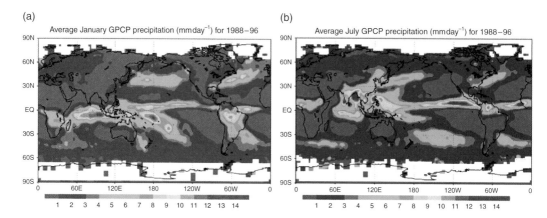

Figure 2.18 Precipitation rates [mm day^{-1}] in January (a) and in July (b), averaged between 1988 and 1996, based on data from the Global Precipitation Climatology Project (GPCP). From Xie and Arkin (1997), copyright © American Meteorological Society, used with permission.

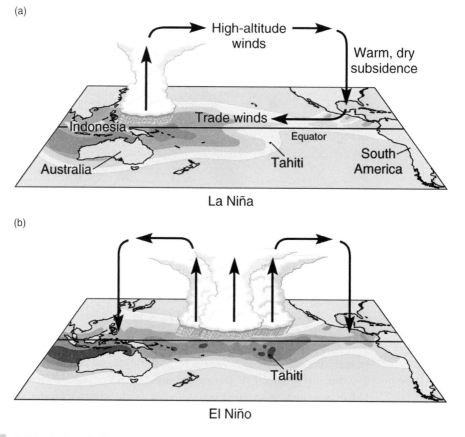

Figure 2.19 El Niño–Southern Oscillation (ENSO) mode of climatic variability, featuring La Niña conditions (*cold phase*, (a)) and El Niño conditions (*warm phase*, (b)). Reproduced with permission from Cunningham and Cunningham (2010).

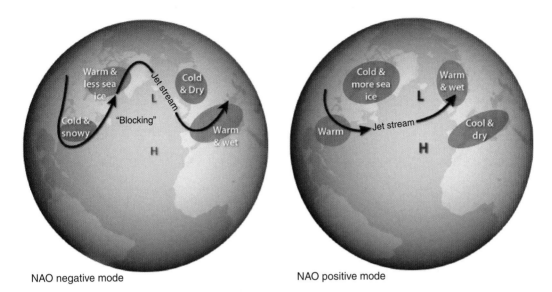

NAO negative mode   NAO positive mode

Figure 2.20   Schematic representation of the main dynamical patterns over the North Atlantic during negative and positive phases of the North Atlantic Oscillation. From the National Oceanic and Atmospheric Administration NOAA (www.climate.gov).

precipitation over eastern South America, and weakened trade winds. Beyond the Pacific, ENSO affects the climate of other regions of the world through complex teleconnections.

At higher latitudes, the major mode of interannual climate variability is the *Arctic Oscillation (AO)*, characterized by changing meridional pressure gradients between northern mid-latitudes and the Arctic. The *North Atlantic Oscillation (NAO)* is a regional manifestation of the AO (Figure 2.20) and its phase is measured by the pressure difference between the Azores high and the Icelandic low (*positive phase* when the pressure difference is large, *negative phase* when it is small). In the positive phase of the AO/NAO, high pressure at northern mid-latitudes pushes the jet stream northward, maintains strong surface westerlies, and restricts exchange of air with the Arctic. This leads to relatively warm and wet conditions in northern Europe and Alaska, and dry conditions in the eastern USA and Mediterranean region. In the negative phase of the AO/NAO there is more meandering of the jet stream and cold Arctic air can penetrate deep into northern mid-latitudes.

2.10 Planetary Boundary Layer

The *planetary boundary layer* (PBL) is the layer of the atmosphere that interacts with the surface on a timescale of a day or less (Figure 2.21). It typically extends up to 1–3 km above the surface. The air above the PBL is called the *free troposphere*. The free troposphere has a general slow sinking motion, balancing the few locations where deep convection or frontal lifting injects PBL air to high altitudes. The

(a)

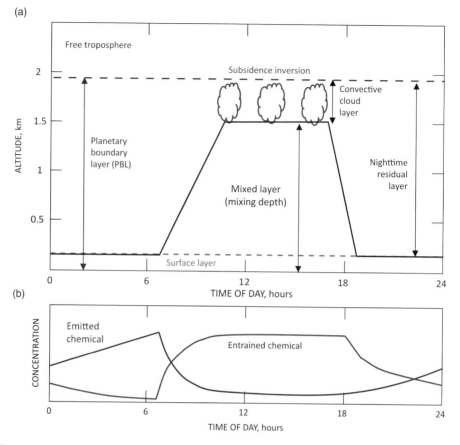

(b)

Diurnal evolution of the planetary boundary layer (PBL) over land (a) and implications for chemical concentrations in surface air (b).

compressional heating from this sinking air produces a semi-permanent *subsidence inversion* (Section 2.6.2) that caps the PBL and sharply restricts mixing between the PBL and the free troposphere.

PBL dynamics plays an important role in determining the fate of chemicals emitted at the surface and the resulting concentrations in surface air. Vertical mixing driven by solar heating of the surface can drive large diurnal cycles of concentrations within the PBL. Venting of the PBL to the free troposphere is critical for global dispersal of chemicals.

Vertical mixing within the PBL is driven by turbulent eddies. These eddies are generated at the surface by the action of the wind on rough surface elements (*mechanical turbulence*) and by buoyancy (*buoyant turbulence*). Over land, sensible heating of the surface during the day generates buoyant plumes that may rise up to the base of the subsidence inversion. Conversely, nighttime cooling of the land surface produces stable conditions that dampen the mechanical turbulence. Over the oceans, the large heat capacity of the ocean minimizes this diurnal cycle of heating and cooling and the PBL remains neutral throughout the day.

Figure 2.21 shows the diurnal evolution of the PBL structure over land. At night, mechanical turbulence usually maintains a shallow, well-mixed layer typically

10–100 m deep called the *surface layer*. Above that altitude, the atmosphere is stable because of surface cooling; this is the *residual layer*. After sunrise, surface heating erodes the stable residual layer from below, producing an unstable *mixed layer* that grows over the morning hours to eventually reach the full depth of the PBL. Clouds may develop in the upper part; these are the familiar *fair-weather cumuli* and the corresponding layer is called the *convective cloud layer* (*CCL*). The CCL tends to have moderate stability due to the latent heat release from cloud condensation, resulting in some separation from the mixed layer. The depth of the mixed layer (excluding any CCL) is called the *mixing depth*. Suppression of surface heating at sunset causes rapid collapse of the mixed layer and the nighttime conditions return.

The diurnal variation of PBL structure has important implications for the diurnal evolution of chemical concentrations in surface air, as shown in Figure 2.21. An inert chemical continuously emitted at the surface will accumulate in surface air over the course of the night, leading to high concentrations. During morning the concentration will decrease as growth of the mixed layer causes dilution. By contrast, a chemical originating in the free troposphere and removed by deposition to the surface will be depleted in surface air over the course of the night, and replenished during morning by entrainment from aloft as the mixed layer grows.

Over the ocean there is no diurnal cycle of surface heating and cooling, and neutral conditions prevail where vertical mixing is driven by mechanical turbulence. The mixed layer is called the *marine boundary layer* (*MBL*) and typically extends to about 1 km altitude with no diurnal variation. It is often capped by a shallow cloud layer, either cumulus clouds or stratus, capped in turn by the subsidence inversion.

Entrainment of air from the free troposphere into the PBL and *ventilation* of PBL air to the free troposphere are important processes for atmospheric chemistry, connecting the surface to the global atmosphere. Ventilation generally takes place by weather events, such as frontal systems or deep convective updrafts that force boundary layer air to the free troposphere. Entrainment, by contrast, generally takes place as a slow, steady process involving the large-scale sinking of the atmosphere to compensate for the convective updrafts. Typical downward entrainment velocities at the top of the PBL are of the order of $0.1–1$ cm s^{-1}, and this replaces the PBL air on a timescale of days to a week.

2.11 Middle Atmosphere Dynamics

Vertical motions in the stratosphere are strongly suppressed by the temperature inversion resulting from absorption of solar UV radiation by ozone. A first approximation of the thermal structure of the stratosphere can be made by assuming *radiative equilibrium* conditions, where the heating rate from UV absorption by ozone and O_2 is balanced by the cooling rate from emission of IR terrestrial radiation by CO_2, water vapor, and ozone. The resulting temperatures increase with latitude from the winter to the summer pole (Figure 2.11). Based on the thermal wind equation, the zonal wind is easterly in the summer hemisphere and westerly in the winter hemisphere. The polar stratosphere in winter features strong zonal winds that form a *polar vortex*, isolating it from lower latitudes.

Departure from radiative equilibrium conditions is induced by the dissipation of upward propagating waves generated at the Earth's surface. Wave breaking occurs when the amplitude of the wave becomes sufficiently large to render the disturbance unstable. This dissipation process tends to mix the medium through which the wave is propagating. Further, the momentum deposited by these waves as they break produces a torque that tends to decelerate the zonal wind and generate a meridional circulation. The resulting mean meridional temperature distribution arises from a balance between the net radiative heating/cooling described previously and the adiabatic heating/cooling associated by the compression/expansion of air produced by the wave-generated vertical motions.

Different types of waves are observed in the middle atmosphere. *Rossby waves* are planetary-scale disturbances in the zonal atmospheric flow that owe their existence to the latitudinal variation of the Coriolis effect. A familiar example is the meandering jet stream. These waves are generated by baroclinic instability and the forcing action of zonally asymmetric heating and topography. Upward wave propagation to the middle atmosphere is possible only when the wind is westerly (during winter) and for the longest waves with wavenumber 1–3 ("wavenumber" is the number of complete wave cycles along the longitude around the entire Earth). Shorter Rossby waves are confined to the troposphere, where they contribute to the formation of weather systems. Rossby wave breaking in the stratosphere takes place in a relatively large "surf zone" characterized by intense quasi-horizontal mixing of chemical species. The mean circulation produced by dissipation of the waves in the winter hemisphere is directed from the Equator to the pole (Figure 2.22) and is called the Brewer–Dobson circulation since it was inferred from observations of water vapor (by Alan Brewer) and of ozone (by Gordon Dobson) in the lower stratosphere. Occasional large amplification of Rossby waves in the northern hemisphere disrupts the stratospheric circulation and causes *sudden warming* events in the Arctic stratosphere that disrupt the polar vortex.

The change of sign in the Coriolis parameter at the Equator leads to a specific class of planetary-scale waves. *Kelvin waves* propagate eastward in the equatorial zone, which acts as a waveguide, and are trapped in the vicinity of the Equator. Their vertical wavelength is typically 10 km and their wavenumber 1 to 3. *Mixed Rossby–gravity* waves are also trapped waves, but with vertical wavelengths of 4–8 km in the vertical and wavenumber 3–5. They propagate westward.

Gravity waves are generated by local disturbances in the flow over mountain ranges or in relation to weather (frontal systems, convective storms). As these waves propagate upwards into progressively more rarified air, their amplitude increases until nonlinear effects cause the waves to break and to transfer momentum to the mean flow, predominantly in the mesosphere. This drives the mesosphere away from radiative equilibrium and generates a meridional flow directed from the summer to the winter hemisphere. The associated upwelling in the summer high latitudes leads to strong adiabatic cooling and explains the presence of a very cold summer mesopause (Figure 2.11).

Atmospheric tides are global-scale waves produced by the release of latent heat in the troposphere and the absorption of solar radiation by ozone and water vapor. They

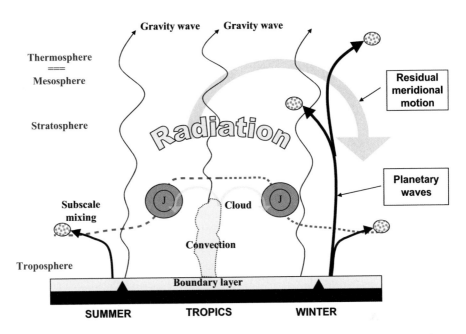

Figure 2.22 Schematic representation of the upward propagation of planetary waves in the winter middle atmosphere (thick black lines) and of gravity waves (thin black lines). The meridional circulation resulting from the dissipation of these waves in the stratosphere and mesosphere is shown by the large arrow. The circulation is directed from the tropics to the pole in the winter stratosphere and from the summer to the winter pole in the mesosphere. Radiative equilibrium prevails in the summer stratosphere. The position of the jets near the tropopause (dotted line) is shown. Personal communication from Richard Rood, University of Michigan.

propagate upwards and break primarily in the lower thermosphere. Migrating tides are Sun-synchronous and so propagate westward with the apparent motion of the Sun. Since solar forcing is nearly a square wave that is rich in harmonics, waves with periods shorter than 24 hours (e.g., semi-diurnal wave) also are observed. Non-migrating tides produced by the release of latent heat in the troposphere do not follow the motion of the Sun. They may be stationary, or propagate westward or eastward.

Oscillations in the tropical zonal winds, including the *quasi-biennial oscillation* (QBO with a period of 22 to 34 months) in the lower stratosphere and the *semi-annual oscillation* (SAO) near the stratopause, are the result of interactions between dissipating waves and the mean flow. The QBO is the major cause of interannual variance of the zonal wind in the equatorial stratosphere. The amplitude of the easterly phase is about twice as strong as that of the westerly phase. The momentum source that produces the oscillation in the zonal wind is provided by the dissipation of Kelvin and mixed Rossby–gravity waves. The *Arctic Oscillation* (AO), an oscillation in temperature and pressure between the Arctic and mid-latitudes discussed in Section 2.9, extends to the stratosphere, where it affects the strength of the polar vortex with associated effects on stratospheric ozone.

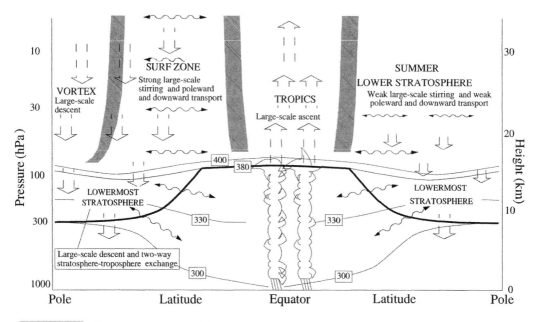

Figure 2.23 Schematic representation of the most important dynamical barriers in the stratosphere. Thin contour lines represent uniform potential temperature (isentropes). The thick solid line is the mean altitude of the tropopause. Reproduced from the World Meteorological Organization (WMO, 1999).

Observations of long-lived tracers in the stratosphere have highlighted the existence of *dynamical barriers* (Figure 2.23) hindering the exchange of air between different atmospheric regions. The mid-latitude stratosphere is fairly isolated from tropical influences through a barrier against meridional transport situated at 20°–30° latitude. The resulting upward motion confined to the tropical stratosphere is called the *tropical pipe*. The *polar vortex* is another dynamical barrier that separates the polar from the mid-latitude stratosphere. The photochemically produced Antarctic ozone hole is sustained as long as the polar vortex remains present, but disappears when the vortex breaks down and strong mixing of air masses takes place.

References

Aguado E. and Burt J. E. (2013) *Understanding Weather and Climate*, 6th edition, Pearson Education, Harlow.

Ahrens C. (2000) *Essentials of Meteorology*, Cengage Learning EMEA, Stamford, CT.

Andrews D. G., Holton J. R., and Leovy C. B. (1987) *Middle Atmosphere Dynamics*, Academic Press, New York.

Banks P. M. and Kockarts G. (1973) *Aeronomy*, Academic Press, New York.

Cunningham W. P. and Cunningham M. A. (2010) *Principles of Environmental Sciences, Inquiry and Applications*, McGraw Hill, New York.

Gill A. E. (1982) *Atmosphere–Ocean Dynamics*, Academic Press, New York.

Green J. (2004) *Atmospheric Dynamics*, Cambridge University Press, Cambridge.

Holton J. R. and Hakim G. J. (2013) *An Introduction to Dynamic Meteorology*, 5th edition, Elsevier Academic Press, Amsterdam.

Jacob D. J. (1999) *Introduction to Atmospheric Chemistry*, Princeton University Press, Princeton, NJ.

Lutgens F. K. and Tarbuck E. J. (2000), *The Atmosphere*, 8th edition, Prentice Hall, Englewood Cliffs, NJ.

Lutgens F. K., Tarbuck E. J., and Tasa D. (2013) *The Atmosphere: An Introduction to Meteorology*, 12th edition, Pearson Education, Harlow.

Mak M. (2011) *Atmospheric Dynamics*, Cambridge University Press, Cambridge.

Martin J. E. (2006) *Mid-Latitude Atmospheric Dynamics: A First Course*, Wiley, Chichester.

Pedlosky J. (1987) *Geophysical Fluid Dynamics*, Springer Verlag, Berlin.

Pidwirny M. (2006) *Fundamentals of Physical Geography*, 2nd edition, Physicalgeography.net.

Prölss G. W. (2004) *Physics of the Earth's Space Environment: An Introduction*, Springer Verlag, Berlin.

Rogers R. R. and Yau M. K. (1989) *A Short Course in Cloud Physics*, Pergamon Press, Oxford.

Shepherd T. G. (2003) Large-scale atmospheric dynamics for atmospheric chemists, *Chem Rev.*, **103** (12), 4509–4532, doi: 10.1021/cr020511z.

Trenberth K. E., Fasullo J. T., and Kiehl J. (2009) Earth's global energy budget, *Bull. Amer. Meteor. Soc.*, **90**, 311–323, doi: http://dx.doi.org/10.1175/2008BAMS2634.1.

Vallis G. K. (2006) *Atmospheric and Oceanic Fluid Dynamics: Fundamental and Large Scale Circulation*, Cambridge University Press, Cambridge.

World Meteorological Organization (WMO) (1999) *Scientific Assessment of Ozone Depletion: 1998, Global Ozone Research and Monitoring Project – Report No. 44*, World Meteorological Organization, Geneva.

Xie P. and Arkin P. A. (1997) Global precipitation: A 17-year monthly analysis based on gauge observations, satellite estimates and numerical model outputs, *Bull. Amer. Meteor. Soc.*, **78**, 2539–2558.

Zdunkowski W. and Bott A. (2003) *Dynamics of the Atmosphere: A Course in Theoretical Meteorology*, Cambridge University Press, Cambridge.

3   Chemical Processes in the Atmosphere

3.1 Introduction

Atmospheric chemistry models simulate the concentrations of chemical species as determined by emissions, transport, chemical production and loss, and deposition. Chemical production and loss are computed for an ensemble of reactions described by kinetic equations and often involve coupling between species. The ensemble of reactions is the *chemical mechanism* of the model. The purpose of this chapter is to give a primer of important atmospheric chemical processes as a basis for understanding the construction of atmospheric chemistry models. Model equations will be introduced in Chapter 4 and a more in-depth formulation of the kinetic equations will be presented in Chapter 5. Numerical methods for solving complex chemical mechanisms (*chemical solvers*) are presented in Chapter 6. Emission processes, including global emission budgets for major species, are described in Chapter 9. A sample chemical mechanism is given in Appendix D.

The major components of the atmosphere are molecular nitrogen (N_2), molecular oxygen (O_2), argon (Ar), and water vapor (H_2O). Argon has no chemical reactivity, but N_2, O_2, and H_2O react in the atmosphere to drive chemical processes. Many other species present in trace amounts also contribute to drive chemical processes as described in this chapter. A species directly emitted to the atmosphere is called *primary*, while a species chemically produced within the atmosphere is called *secondary*.

Fast chemistry generally involves *radical-assisted reaction chains*. Radicals are chemical species with unfilled electron orbitals. An orbital can contain two electrons, and having filled orbitals lowers the internal energy of an atom or molecule. An atom or molecule with an odd number of electrons has high reactivity due to its unfilled orbital. Reaction of a radical (odd number of electrons) with a non-radical (even number of electrons) necessarily produces a radical (since the sum of electrons in the product species is odd), thus propagating a chain reaction. Radicals originate in the atmosphere from cleavage of non-radicals, usually by solar radiation (*photolysis*). Energetic input from solar radiation is thus critical to driving the chemistry of the atmosphere. The ensemble of chemical reactions enabled by solar radiation is called *photochemistry*, and chemical mechanisms in models are often called *photochemical mechanisms*.

Our intent here is to give a compact summary of major processes relevant to atmospheric chemistry. Detailed presentations of chemical and aerosol processes in the atmosphere can be found, for example, in the books by Warneck (1999), Brasseur *et al.* (1999, 2003), Finlayson-Pitts and Pitts (2000), and Seinfeld and Pandis (2006).

3.2 Oxygen Species and Stratospheric Ozone

Molecular oxygen is photolyzed in the atmosphere by solar UV radiation:

$$O_2 + hv\,(\lambda < 242\,nm) \rightarrow O + O \tag{3.1}$$

where the reaction threshold of 242 nm corresponds to the minimum energy required to dissociate the molecule. The oxygen atoms combine with O_2 by a three-body reaction to produce ozone (O_3):

$$O + O_2 + M \rightarrow O_3 + M \tag{3.2}$$

The *third body* M is an inert molecule such as N_2 or O_2 that collides with the excited O_3* product of the collision of O and O_2 and takes up its internal energy, thus allowing stabilization of O_3* to ground-state O_3. The internal energy of the excited M* eventually dissipates as heat. A third body is needed for any reaction where two reactants combine to form a single product. It is standard practice to include M in the expression of a three-body reaction as it may play a limiting role in the kinetics. See Chapter 5 for discussion of the kinetics of three-body reactions.

Reaction (3.1) is the main source of ozone in the stratosphere. Solar radiation of wavelength shorter than 242 nm is efficiently absorbed by both O_2 and O_3 as it propagates down through the atmosphere, so the rate of (3.1) becomes negligible below 20 km altitude. Production of ozone in the troposphere takes place by a different mechanism, described in Section 3.6.

Ozone is loosely bound and is photolyzed rapidly in daytime:

$$O_3 + hv\,(\lambda < 1180\,nm) \rightarrow O + O_2 \tag{3.3}$$

The principal bands for absorption of solar radiation by ozone are the Hartley (200–290 nm), Huggins (310–400 nm) and Chappuis (400–850 nm) bands. Radiation of wavelengths longer than 320 nm produces the O atom in its electronically ground state (^3P). Radiation of shorter wavelengths produces the O atom in its electronically excited ^1D state (Figure 3.1):

$$O_3 + hv\,(\lambda < 320\,nm) \rightarrow O\left(^1D\right) + O_2 \tag{3.4}$$

Radiation at wavelengths shorter than 234 nm can produce an even more excited state of the O atom (^1S). Standard chemical notation omits mention of the spectroscopic state when the species is in its ground state (here $O(^3P) \equiv O$) and retains it when the species is in an excited state ($O(^1D)$, $O(^1S)$). Deactivation of excited

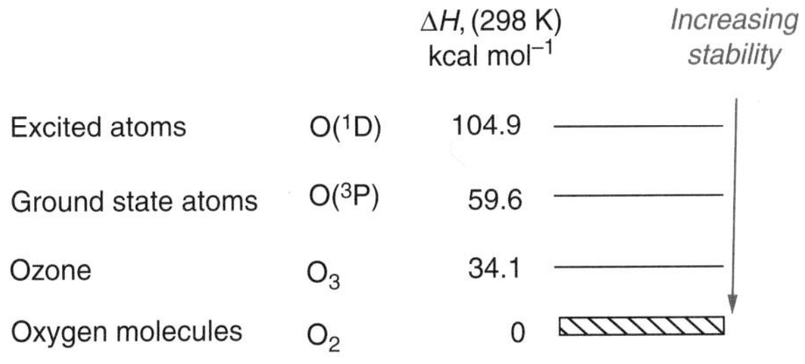

		ΔH, (298 K) kcal mol^{-1}	*Increasing stability*
Excited atoms	$O(^1D)$	104.9	
Ground state atoms	$O(^3P)$	59.6	
Ozone	O_3	34.1	
Oxygen molecules	O_2	0	

Figure 3.1 Enthalpy of formation [kcal mol^{-1}] of gas-phase oxygen species.

species occurs by collision with other molecules, a very rapid process in the lower atmosphere. The $O(^1D)$ atom is thus deactivated to the ground state:

$$O\left(^1D\right) + M \rightarrow O + M \tag{3.5}$$

A small fraction of $O(^1D)$ can also react with non-radical species to produce radicals and initiate radical-assisted reaction chains. This will be discussed in Sections 3.3 and 3.4.

The ground-state oxygen atom produced by (3.3) recombines with O_2 by (3.2) to regenerate ozone. It can also react with ozone to form two O_2 molecules:

$$O + O_3 \rightarrow O_2 + O_2 \tag{3.6}$$

Reactions (3.1), (3.2), (3.3), and (3.6) comprise the *Chapman mechanism* for stratospheric ozone, originally proposed by Sydney Chapman in 1930. Reactions (3.2) and (3.3) interconvert O and ozone. The lifetime of O against loss by (3.2) is less than a second in the stratosphere and troposphere, so that O and ozone are in photochemical equilibrium during daytime with $[O]/[O_3] \ll 1$. It follows that reaction (3.3) is not a true sink for ozone because O atoms will immediately return ozone by (3.2). Nor is (3.2) a true source of ozone if the reactant O atoms originated from ozone photolysis by (3.3). Ozone concentration is thus actually controlled by production in (3.1) and loss through (3.6). Accounting is aided by defining an "odd oxygen" family $(O_x \equiv O + O_3)$ produced by (3.1), lost by (3.6), and unaffected by (3.2) and (3.3). Since $[O]/[O_3] \ll 1$, the budget of ozone is actually that of O_x. The general concept of *chemical families* is important for atmospheric chemistry modeling and is described further in Box 3.1.

Box 3.1	Chemical Families

The concept of "chemical family" is central to atmospheric chemistry. It enables convenient accounting of the budgets of species cycling rapidly with each other. It is nothing more than an accounting device; it does not imply any similarity in the chemical properties of different members of the family. Consider an ensemble of species $\{A_1, \ldots A_n\}$ cycling with each other by chemical reactions. If this cycling is sufficiently fast, then a chemical equilibrium is established defining concentration ratios $[A_i]/[A_j]$. Consider now a chemical family A_x representing the ensemble of these species: $A_x \equiv A_1 + \cdots + A_n$ such that $[A_x] = \sum_{i=1}^{n} [A_i]$. Writing $[A_i] = [A_x]([A_i]/[A_x])$, we see that the budget of A_i can be defined from the budget of the family A_x and the chemical partitioning $[A_i]/[A_x]$ within the family. In the case where A_i is the dominant member of the family such that $[A_i]/[A_x] \approx 1$, the budget of A_i is solely defined by that of A_x. The chemical family is a useful accounting tool if the lifetime of A_x is longer than that of any individual family member, so that A_x is a more conserved quantity in the atmosphere. It is most useful when the family members are in equilibrium so that chemical partitioning within the family can be easily derived.

In the case of the Chapman mechanism described in Section 3.2, there is rapid cycling between O and ozone so that it is useful to group them into a chemical family. That chemical family is commonly called odd oxygen: $O_x \equiv O_3 + O$. The terminology "odd" simply refers to ozone and O having an odd number of O atoms. Since $[O]/[O_3] \ll 1$, the ozone budget is well approximated by the O_x budget. The budget of the O atom is defined by that of O_x together with the $[O]/[O_x] \approx [O]/[O_3]$ ratio from chemical equilibrium.

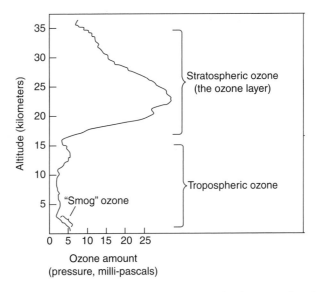

Figure 3.2   Typical vertical profile of atmospheric ozone measured by ozonesonde. About 90% of total atmospheric ozone is located in the stratosphere. The origin of the "smog" ozone near the surface is discussed in Section 3.6. Source: www.esrl.noaa.gov.

Following on the above, we can qualitatively explain the distribution of ozone in the stratosphere in terms of the odd oxygen budget. The source of odd oxygen from O_2 photolysis (3.1) peaks at about 40 km altitude, reflecting opposite trends in O_2 number density, which decreases with altitude, and the UV photon flux, which increases with altitude. The maximum ozone number density occurs at a somewhat lower altitude (Figure 3.2) because the sink of odd oxygen from reaction (3.6) increases with altitude as the O atom concentration increases (the O loss rate from (3.2) has a quadratic pressure dependence). The lifetime of odd oxygen in the upper stratosphere is less than a day, sufficiently short that the ozone concentration is determined by the local chemical steady state between production and loss of odd oxygen. Below 30 km, the lifetime of odd oxygen is sufficiently long that the distribution of ozone is affected by transport on a global scale. Coupling of chemistry and transport results in a minimum ozone column in the tropical stratosphere (Figure 3.3), as the Brewer–Dobson circulation carries low-ozone air from the troposphere upward (see Figure 2.23). Box 3.2 gives historical milestones in the development of our knowledge of stratospheric ozone.

3.3 Hydrogen Oxide Radicals

The importance of hydrogen oxide radicals for atmospheric chemistry was first recognized in studies of the middle atmosphere in the 1950s. The hydroxy radical

Monthly average GOME total ozone
Jun 2002

KNMI/ESA

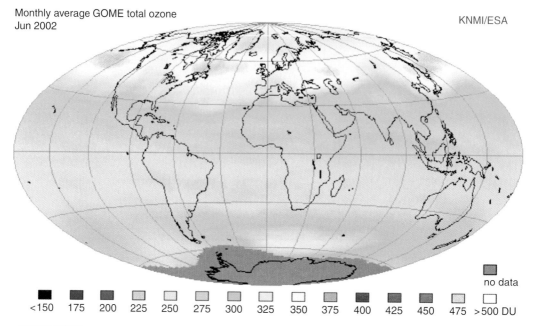

no data

| <150 | 175 | 200 | 225 | 250 | 275 | 300 | 325 | 350 | 375 | 400 | 425 | 450 | 475 | >500 DU |

Figure 3.3 Total ozone columns measured by the Global Ozone Monitoring Experiment (GOME) satellite instrument in June 2002. The column concentration is expressed in Dobson units (DU), where 1 DU is defined as 0.01 mm of pure ozone at standard conditions of temperature and pressure: 1 DU $= 2.69 \times 10^{16}$ molecules cm^{-2}.

Box 3.2   **Historical Milestones in our Understanding of Stratospheric Ozone**

The First Steps

- 1839: Christian Friedrich Schönbein (Basel, Switzerland) identifies a particular odor following electric discharges in air. He calls this "property" *ozone* from the Greek word $o\zeta\varepsilon\iota\nu$ (to smell), and recognizes that it represents a gas.
- 1845: Jean-Charles de Marignac and Auguste de la Rive (Geneva, Switzerland) suggest that this gas is produced by a transformation of oxygen.
- 1863: Jean-Louis Soret (Geneva, Switzerland) determines experimentally that ozone is made of three oxygen atoms.

The First Observations

- 1853: Schönbein detects ozone in the atmosphere.
- 1858: André Houzeau measures atmospheric ozone in Rouen, France.
- 1877–1907: Albert Levy conducts systematic observations of ozone at Parc Montsouris in the outskirts of Paris.

Laboratory Investigations and Spectroscopic Observations

- Starting in the 1870s, the spectroscopic properties of ozone are investigated and are used to observe ozone in the atmosphere. Important contributions are due to Alfred Cornu (1878), J. Chappuis (1880), Walter N. Hartley (1880/1881), William Huggins (1890), A. Fowler and R. J. Strutt (1917), Charles Fabry and Henri Buisson (1913–1920), Gordon M. B. Dobson (1920s), and F. W. P. Götz (1924).

Box 3.2
Figure 1 Christian F. Schoenbein, Gordon M. B. Dobson (from the Royal Meteorological Society), Harold Johnston (courtesy of Denis Galloway, UC Berkeley), Joseph Farman in his office, late 1980s (courtesy of the British Library).

Box 3.2 (*cont.*)

- 1920: Fabry and Buisson make the first quantitative observation of the ozone column abundance.
- 1920s: Dobson develops a spectrophotometer that remains one of the most accurate instruments to measure the ozone column.
- 1929: Götz uses Dobson's instrument in Spitzbergen to infer the vertical profile of ozone.

Photochemical Theories

- 1930: Sydney Chapman presents the first photochemical theory of ozone formation and destruction.
- 1950s: Arie Haagen-Smit discovers that urban ozone is formed from the by-products of fuel combustion.
- 1950: David Bates and Marcel Nicolet show that hydrogen radicals produced by photolysis of water vapor destroy ozone efficiently in the mesosphere.
- 1964: John Hampson demonstrates the importance of hydrogen radicals for the chemistry of ozone in the stratosphere.
- 1970: Paul Crutzen shows that the largest ozone destruction mechanism in the stratosphere is due to a catalytic cycle involving nitrogen oxides, thus reconciling theory and observations of stratospheric ozone abundances.
- 1971: Harold Johnston suggests that the nitrogen oxides released by a planned fleet of high-altitude supersonic aircraft could destroy considerable amounts of ozone in the stratosphere.
- 1974: Richard Stolarski and Ralph Cicerone report that a similar cycle with chlorine atoms could also efficiently destroy ozone. Steven Wofsy shows that bromine atoms could also catalytically destroy ozone.
- 1974: Mario Molina and Sherwood Rowland establish that the major source of stratospheric chlorine is provided by industrially manufactured chlorofluorocarbons.

Polar Ozone

- 1985: Joseph Farman and colleagues from the British Antarctic Survey observe very low ozone columns at the Antarctic station of Halley Bay during the austral spring, confirming earlier observations by S. Chubachi at the polar station of Syowa.
- 1986: Susan Solomon (NOAA) shows that chlorine activation on the surface of particles in polar stratospheric clouds explains the presence of a springtime ozone hole in the Antarctic.
- 1987: Luisa and Mario Molina show that formation and photolysis of the ClO dimer can account for most of the springtime ozone loss in Antarctica.
- 1995: Crutzen, Molina, and Rowland are awarded the Nobel Prize in Chemistry for their discoveries on the chemistry of ozone.

OH is produced by reaction of water vapor with the electronically excited oxygen atom $O(^1D)$ originating from photolysis of ozone:

$$O(^1D) + H_2O \rightarrow OH + OH \tag{3.7}$$

Hydroxyl reacts with ozone to produce the hydroperoxy radical HO_2, which goes on to react with ozone and return OH, leading to a catalytic cycle for ozone loss:

$$O_3 + OH \rightarrow HO_2 + O_2 \tag{3.8}$$

$$O_3 + HO_2 \rightarrow OH + 2O_2 \tag{3.9}$$

Hydrogen oxide radicals can also react with the oxygen atom O, producing the hydrogen atom and leading to additional catalytic cycles for odd oxygen (and hence ozone) loss. We define the *hydrogen oxides radical family* as $HO_x \equiv H + OH + HO_2$ and refer to the associated catalytic cycles destroying odd oxygen as *HO_x-catalyzed ozone loss.*

Hydroxyl is of most interest in atmospheric chemistry for its role as a strong oxidant. This role came to the fore in the 1970s with the realization that sufficient solar radiation in the 300–320 nm wavelength region penetrates the troposphere to produce $O(^1D)$. The resulting OH is the main agent for oxidizing reduced gases emitted from the surface. The most important reduced gases on a global scale are carbon monoxide (CO) and methane (CH_4), which have large emission fluxes. Oxidation of CO proceeds by:

$$CO + OH \xrightarrow{O_2} CO_2 + HO_2 \tag{3.10}$$

where O_2 above the reaction sign indicates a species that participates in the overall reaction but does not limit the kinetics. In this case, H produced by conversion of CO to CO_2 reacts rapidly with O_2 to produce HO_2. Oxidation of methane proceeds by:

$$CH_4 + OH \xrightarrow{O_2} CH_3O_2 + H_2O \tag{3.11}$$

where CH_3O_2 is the methylperoxy radical and should be viewed as an additional component of HO_x since it goes on to cycle with the other components. This chemistry will be discussed in Section 3.5. The atmospheric lifetimes of CO and methane against oxidation by OH are two months and ten years, respectively.

Conversion between HO_x radicals takes place sufficiently rapidly that photochemical equilibrium can be assumed in the daytime (except in the upper mesosphere and thermosphere, where collisions are infrequent). Production of HO_x is mostly by reaction (3.7). Loss of HO_x can take place by various pathways, the dominant one in the troposphere being the formation of hydrogen peroxide (H_2O_2):

$$HO_2 + HO_2 + M \rightarrow H_2O_2 + M \tag{3.12}$$

This loss is terminal if H_2O_2 is removed by deposition or is converted to water, as by reaction with OH:

$$H_2O_2 + OH \rightarrow HO_2 + H_2O \tag{3.13}$$

However, it is temporary if H_2O_2 is instead recycled to HO_x radicals by photolysis:

$$H_2O_2 + h\nu \rightarrow OH + OH \tag{3.14}$$

Thus H_2O_2 and other peroxides should be viewed as *reservoirs* for HO_x. It can be convenient to define a *hydrogen oxides family* $HO_y \equiv HO_x +$ peroxides to account for the exchange between HO_x and its reservoirs.

3.4 Nitrogen Oxide Radicals

Nitrogen oxide radicals ($NO_x \equiv NO + NO_2$) are of central importance for atmospheric chemistry. They are emitted to the troposphere by combustion, lightning, and microbial processes in soils. The largest source is combustion. A typical combustor mixes fuel and air at very high flame temperatures (about 2000 K). At these temperatures, O_2 from the air thermally dissociates to O atoms, which react with N_2 from the air to drive a catalytic cycle for formation of nitric oxide (NO) known as the *Zel'dovich mechanism*:

$$N_2 + O \rightarrow NO + N \tag{3.15}$$

$$N + O_2 \rightarrow NO + O \tag{3.16}$$

The NO_x generated in this manner is called *thermal* NO_x. The Zel'dovich mechanism is not efficient at low flame temperatures such as from open fires, but production of NO still takes place in those cases by oxidation of nitrogen present in the fuel. The NO_x generated in that manner is called *fuel* NO_x.

In the stratosphere, the main source of NO_x is the oxidation by $O(^1D)$ of nitrous oxide (N_2O), a long-lived gas emitted by microbial activity in soils:

$$O\left(^1D\right) + N_2O \rightarrow NO + NO \tag{3.17}$$

A dominant sink for NO in both the troposphere and stratosphere is reaction with ozone to form nitrogen dioxide (NO_2):

$$NO + O_3 \rightarrow NO_2 + O_2 \tag{3.18}$$

In the daytime, NO_2 is photolyzed on a timescale of a minute to return NO:

$$NO_2 + hv\,(\lambda < 400\ \text{nm}) \xrightarrow{O_2} NO + O_3 \tag{3.19}$$

The reaction cycle (3.18) + (3.19) has no effect on atmospheric composition; it is called a *null cycle*. However, it forces photochemical equilibrium between NO and NO_2 in the daytime.

Alternate reaction cycles for NO and NO_2 lead to production or loss of ozone. In the stratosphere, NO_2 can be converted back to NO by reaction with atomic oxygen:

$$NO_2 + O \rightarrow NO + O_2 \tag{3.20}$$

The reaction cycle (3.18) + (3.20) is a major catalytic sink for stratospheric ozone. The rate of ozone destruction is set by the rate of reaction (3.20), which competes with (3.19). Reaction (3.20) is called the *rate-limiting step* for ozone loss. It is unimportant in the troposphere, where oxygen atom concentrations are low. In the troposphere, however, ozone concentrations are sufficiently low that peroxy radicals can compete for reaction with NO. The reaction

$$HO_2 + NO \rightarrow OH + NO_2 \tag{3.21}$$

followed by (3.19) provides an important source of tropospheric ozone. The rate of ozone production is determined by the rate of (3.21) as the rate-limiting step. Similar mechanisms in which NO reacts with methyl peroxy (CH_3O_2) and other organic peroxy radicals (RO_2) lead to additional ozone production. This is discussed in Section 3.5.

Loss of NO_x takes place on a timescale of one day. It involves primarily the conversion of NO_2 to nitric acid (HNO_3). In the daytime, this conversion is by oxidation by OH:

$$NO_2 + OH + M \rightarrow HNO_3 + M \tag{3.22}$$

At night it takes place by oxidation by ozone, forming dinitrogen pentoxide (N_2O_5) that hydrolyzes to HNO_3 in aqueous aerosol particles:

$$NO_2 + O_3 \rightarrow NO_3 + O_2 \tag{3.23}$$
$$NO_2 + NO_3 + M \rightarrow N_2O_5 + M \tag{3.24}$$
$$N_2O_5 + H_2O \xrightarrow{aerosol} 2HNO_3 \tag{3.25}$$

In the daytime, the nitrate radical (NO_3) has a lifetime of less than a minute against photolysis back to NO_2. Thus loss of NO_x by (3.23)–(3.25) can operate only at night.

In the troposphere, the dominant sink of HNO_3 is deposition, including scavenging by precipitation (wet deposition) and direct reaction at the surface (dry deposition). In the stratosphere, however, deposition does not take place and HNO_3 is instead recycled to NO_x on a timescale of weeks through photolysis and reaction with OH:

$$HNO_3 + h\nu \rightarrow NO_2 + OH \tag{3.26}$$
$$HNO_3 + OH \rightarrow NO_3 + H_2O \tag{3.27}$$
$$NO_3 + h\nu \rightarrow NO_2 + O \tag{3.28}$$

Thus HNO_3 serves as a reservoir for NO_x and the concentration of NO_x is determined by photochemical equilibrium with HNO_3. Similarly to HO_x and HO_y, it is useful to define a chemical family NO_y as the sum of NO_x and its reservoirs. Loss of NO_y from the stratosphere is mainly by transport to the troposphere followed by HNO_3 deposition.

The short lifetime of NO_x in the troposphere, combined with the rapid removal of HNO_3 by deposition, results in strong concentration gradients between combustion source regions and the remote oceans (Figure 3.4). However, a low background NO_x concentration is sustained in the remote troposphere and plays a critical role for ozone and OH generation following (3.21). A major source of this background NO_x is the long-range transport and decomposition of the peroxyacetylnitrate reservoir (PAN, formula $CH_3C(O)O_2NO_2$). PAN is produced by reaction of NO_2 with peroxyacetyl radicals $CH_3C(O)O_2$ originating from the oxidation of various organic species (Section 3.5):

$$CH_3C(O)O_2 + NO_2 + M \rightleftharpoons CH_3C(O)O_2NO_2 + M \tag{3.29}$$

The main sink of PAN is thermal decomposition to the original reactants. The lifetime of PAN is one hour at 295 K but months at 250 K. PAN produced in NO_x source regions and lifted to high altitudes can be transported on global scales, eventually decomposing to deliver NO_x to the remote troposphere. Other organic

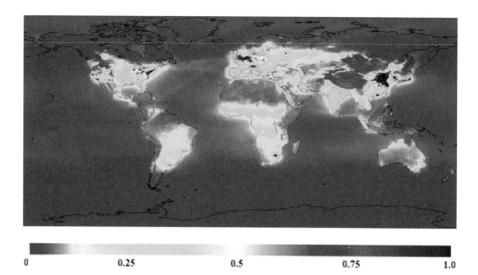

0   0.25   0.5   0.75   1.0

Figure 3.4   Global annual mean distribution of the tropospheric NO_2 column [10^{16} molecules cm^{-2}] observed from 2005 to 2008 by the Ozone Monitoring Instrument (OMI) on the National Aeronautics and Space Administration (NASA) Aura satellite. Source: Bas Mijling, Folkert Boersma, and Ronald van der A (KNMI).

nitrates can be similarly produced from the oxidation of organic species in the presence of NO_x, and this is discussed in Section 3.5. PAN is the most important because of its high yield and its wide range of atmospheric lifetimes enabling both long-range transport in cold air masses and quick release of NO_x when these air masses warm up (as from subsidence).

3.5 Volatile Organic Compounds and Carbon Monoxide

Atmospheric chemists refer to the ensemble of organic species present in the gas phase as *volatile organic compounds* (VOCs). VOCs are emitted by biogenic, combustion, and industrial processes, mainly as hydrocarbons (C_xH_y). Atmospheric oxidation of hydrocarbons produces a cascade of oxygenated VOC species eventually leading to CO and CO_2. The longest-lived VOC is methane, with a lifetime of ten years against oxidation by OH. Other VOCs have considerably shorter lifetimes. Isoprene, the dominant VOC emitted by vegetation, has a lifetime of only about one hour during summer daytime. Thus the VOCs are largely confined to the troposphere, and the short-lived non-methane VOCs influence mostly their region of emission. VOC chemistry involves a succession of steps as carbon is oxidized from its most reduced state –4 (hydrocarbons) to its most oxidized state +4 (CO_2). This chemistry is responsible for much of the complexity in chemical mechanisms of the troposphere. Only general rules will be presented here. Box 3.3 gives nomenclature for major VOCs.

   The main sink for most VOCs is oxidation by OH. Additional oxidants including ozone, NO_3, and halogen atoms can be important for some species. Photolysis is an

Box 3.3	Nomenclature of Major Atmospheric VOCs (Common Names in Parentheses)

Alkanes (C_nH_{2n+2})

CH_4	methane
C_2H_6 or $CH_3\text{-}CH_3$	ethane
C_3H_8 or $CH_3\text{-}CH_2\text{-}CH_3$	propane
C_4H_{10} (2 isomers)	butane
C_5H_{12} (3 isomers)	pentane

Alkenes (C_nH_{2n})

C_2H_4 or $CH_2{=}CH_2$	ethene (ethylene)
C_3H_6 or $CH_2{=}CHCH_3$	propene

Alkynes (C_nH_{2n-2})

C_2H_2 or $CH{\equiv}CH$	ethyne (acetylene)

Aromatics (benzene ring)

C_6H_6	benzene
$C_6H_5CH_3$	methylbenzene (toluene)
$C_6H_4(CH_3)_2$ (3 isomers)	dimethylbenzene (xylene)

Dienes (two C=C bonds)

C_5H_8 or $CH_2{=}C(CH_3)CH{=}CH_2$	2-methyl-1,3-butadiene (isoprene)

Terpenes (multiple isoprene units)

$C_{10}H_{16}$ (many isomers)	monoterpenes (α-pinene, β-pinene...)
$C_{15}H_{24}$ (many isomers)	sesquiterpenes (β-caryophyllene, α-humulene...)

Alcohols (hydroxy function –OH)

CH_3OH	methanol
CH_3CH_2OH	ethanol

Aldehydes (terminal carbonyl function –CHO)

CH_2O	methanal (formaldehyde)
CH_3CHO	ethanal (acetaldehyde)

Ketones (internal carbonyl function –C(O)–)

CH_3COCH_3	propanone (acetone)
$CH_3COCH_2CH_3$	butanone (methylethylketone, MEK)

Dicarbonyls (two carbonyl functions)

$CHOCHO$	glyoxal
$CH_3C(O)CHO$	methylglyoxal

Carboxylic acids (carboxylic function –C(O)OH)

$HCOOH$	methanoic acid (formic acid)
CH_3COOH	ethanoic acid (acetic acid)

Box 3.3 (*cont.*)

Organic peroxides (peroxide function –OO–)

CH_3OOH methylhydroperoxide

Organic nitrates (nitrate function –ONO$_2$)

CH_3ONO_2 methylnitrate

Peroxyacyl nitrates (peroxyacyl function –C(O)OONO$_2$)

$CH_3C(O)OONO_2$ nitroethaneperoxoate (peroxyacetylnitrate, PAN)

---

additional sink for carbonyl and peroxide species. VOCs with relatively low vapor pressures (called *semivolatile*) can partition into the aerosol and cloud phases with subsequent removal by deposition. They can also directly deposit to surfaces.

Oxidation of VOCs by OH can take place by abstraction of an H atom, as in the case of methane with (3.11), or by addition at an unsaturated bond as in ethylene:

$$CH_2 = CH_2 + OH \xrightarrow{O_2} CH_2(OH)CH_2OO \qquad (3.30)$$

In both cases the oxidation produces an organic radical, R, that subsequently adds oxygen to produce an organic peroxy radical RO_2. The RO_2 radicals react with NO in a manner analogous to HO_2 in (3.21):

$$RO_2 + NO \rightarrow RO + NO_2 \qquad (3.31)$$

This produces ozone from subsequent photolysis of NO_2 as discussed in Section 3.4. When NO_x concentrations are low, RO_2 radicals can react instead with HO_2 to form organic peroxides:

$$RO_2 + HO_2 \rightarrow ROOH + O_2 \qquad (3.32)$$

This does not produce ozone and instead provides a sink for HO_x radicals, analogous to the formation of H_2O_2 from the self-reaction of HO_2 radicals as given by (3.12). Additional minor sinks for RO_2 radicals include permutation reactions with other RO_2 radicals and isomerization. An atmosphere where RO_2 radicals react dominantly with NO is said to be in the *high-NO$_x$ regime*; oxidation of VOCs in that regime is a source of ozone. An atmosphere where RO_2 radicals do not react dominantly with NO is said to be in the *low-NO$_x$ regime*; VOC oxidation in that regime tends to scavenge HO_x radicals.

The oxy radicals RO produced by (3.31) can be oxidized by O_2, decompose, or isomerize. The organic peroxides $ROOH$ produced by (3.32) can be oxidized by OH or photolyze. These reactions generally produce carbonyl species including aldehydes and ketones (Box 3.3). The carbonyls further react with OH or photolyze, leading to production of multifunctional compounds and to breakage of chains producing simpler compounds. Successive oxidation steps ultimately lead to CO and on to CO_2, where carbon is in its highest oxidation state. Figure 3.5 gives a general schematic of the oxidation cascade.

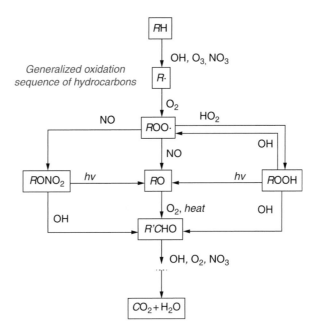

Figure 3.5 Generic oxidation scheme for a hydrocarbon RH.

We described in Section 3.4 the formation of PAN by reaction (3.29) as a reservoir for NO_x in the troposphere. More generally, RO_2 radicals can react with NO_2 to form peroxynitrates,

$$RO_2 + NO_2 + M \rightleftharpoons RO_2NO_2 + M \qquad (3.33)$$

but most of these peroxynitrates have lifetimes of less than a minute against thermal decomposition and therefore are not effective NO_x reservoirs. Peroxyacylnitrates $RC(O)OONO_2$ are an exception and PAN is the most abundant of these peroxyacylnitrates. The peroxyacetyl radical $CH_3C(O)OO$ that serves as precursor of PAN originates mainly from oxidation of acetaldehyde and from photolysis of acetone and methylglyoxal:

$$CH_3CHO + OH \xrightarrow{O_2} CH_3C(O)OO + H_2O \qquad (3.34)$$

$$CH_3C(O)CH_3 + hv \xrightarrow{O_2} CH_3C(O)OO + CH_3 \qquad (3.35)$$

$$CH_3C(O)CHO + hv \xrightarrow{O_2} CH_3C(O)OO + CHO \qquad (3.36)$$

Another class of organic nitrates is produced as a minor branch in the oxidation of RO_2 by NO:

$$RO_2 + NO + M \rightarrow RONO_2 + M \qquad (3.37)$$

These tend to be much more stable than the peroxynitrates. They may undergo further oxidation by OH, photolysis, fractionation into aerosol, or deposition. The organic nitrate yield by (3.37) in competition with (3.31) generally increases with the size of R. Organic nitrate formation can be a significant sink for NO_x in regions with large biogenic VOC emissions.

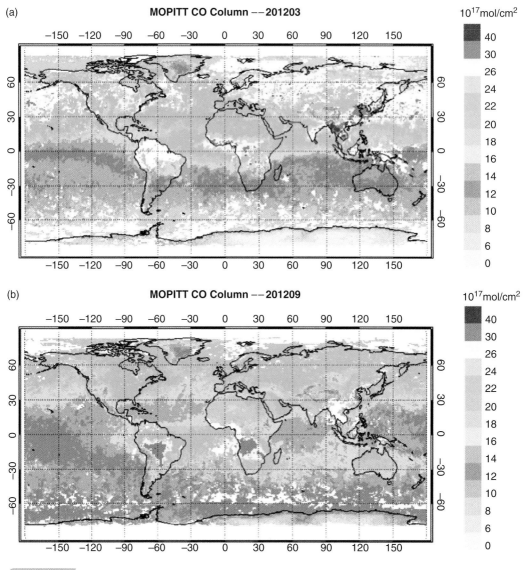

Figure 3.6 Global distribution of the tropospheric carbon monoxide (CO) column in March (a) and September (b) 2012, as observed by the MOPITT instrument aboard the National Aeronautics and Space Administration (NASA) Terra satellite. Courtesy of David Edwards, National Center for Atmospheric Research (NCAR).

Carbon monoxide is a general intermediate in the oxidation of VOCs to CO_2. It is also directly emitted by incomplete combustion, with particularly large emissions from open fires where the combustion process is uncontrolled and inefficient. Carbon monoxide has a mean lifetime of two months against oxidation by OH, its main sink. Figure 3.6 shows the global distribution of CO observed by satellite. Concentrations are highest over tropical regions during the burning season, and are also relatively high over northern mid-latitude continents. The lifetime of CO is sufficiently long to allow transport on intercontinental scales (Chapter 2), making CO a useful tracer for long-range transport of combustion plumes. Figure 3.6 shows that the northern hemispheric background of CO is elevated relative to the south because of combustion influence, and has a seasonality driven by the sink from photochemical oxidation by OH. The southern hemispheric background is mostly contributed by the global source from the oxidation of methane.

3.6 Tropospheric Ozone

Ozone is produced in the troposphere by photochemical oxidation of VOCs and CO in the presence of NO_x, In the simplest case of CO, this involves a sequence of reactions (3.10), (3.21), and (3.19):

$$CO + OH \xrightarrow{O_2} CO_2 + HO_2$$
$$HO_2 + NO \rightarrow OH + NO_2$$
$$NO_2 + hv \xrightarrow{O_2} NO + O_3$$
$$Net: CO + 2O_2 \rightarrow CO_2 + O_3$$

In the case of VOCs, the reaction sequence is similar but with RO_2 radicals reacting with NO following (3.31). The HO_x and NO_x radicals serve as catalysts for the oxidation of VOCs and CO by O_2, and ozone is produced in the process.

The above mechanism provides the dominant source of ozone in the troposphere (Table 3.1). Transport from the stratosphere is an additional minor source. The dominant sink of tropospheric ozone is photochemical loss, including photolysis in the presence of water vapor and reactions with HO_x radicals:

$$O_3 + hv\,(\lambda < 320\,nm) \rightarrow O\left(^1D\right) + O_2$$
$$O\left(^1D\right) + H_2O \rightarrow OH + OH$$
$$O_3 + OH \rightarrow HO_2 + O_2$$
$$O_3 + HO_2 \rightarrow OH + 2O_2$$

There is also a minor sink from deposition to the surface. The lifetime of ozone ranges from a few days in the boundary layer to months in the dry upper troposphere.

Table 3.1 Global present-day budget of tropospheric ozone	
	Best estimate, Tg O_3 a^{-1}
Sources	
Tropospheric chemical production	4500
Transport from stratosphere	500
Sinks	
Tropospheric chemical loss	4000
Deposition	1000

Estimates based on Wu *et al.* (2007)

This difference in lifetime results in a general pattern of net ozone production in the upper troposphere, balanced by net loss in the lower troposphere, driving a gradient of increasing ozone concentrations with altitude.

The rate of ozone production depends on the supply of NO_x, VOCs, and CO in a manner controlled by the cycling of HO_x radicals and competition with HO_x sinks. In most of the troposphere, the dominant HO_x sink is the conversion of peroxy radicals to peroxides following (3.12) and (3.32). In that regime OH radicals mainly react with VOCs or CO, and whether ozone production takes place depends on competition for the peroxy radicals between reaction with NO (producing ozone) and production of peroxides. Thus the ozone production rate increases linearly with the NO_x concentration but does not depend on the concentrations of VOCs and CO. This is called the *NO_x-limited regime.*

A different regime for ozone production applies when UV radiation is low (as in winter) or when NO_x concentrations are very high. In that case the dominant HO_x sink becomes the formation of nitric acid by reaction (3.22). There is no longer competition for peroxy radicals, because the low UV radiation suppresses peroxide formation by (3.12), which has a quadratic dependence on peroxy radical concentrations. Instead, the competition is for OH between reactions with VOCs and CO

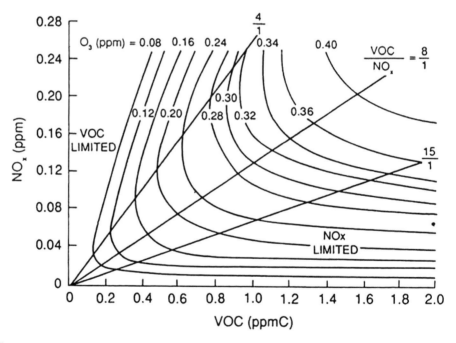

Figure 3.7 Ozone isopleths diagram showing the dependence of ozone concentration on NO_x and VOC concentrations for a simple box model calculation. NO_x- and VOC-limited regimes are indicated. This representation is often referred to as the *Empirical Kinetic Modeling Approach* or *EKMA* diagram. The NO_x-limited regions are typical of locations downwind of urban and suburban areas, whereas the VOC-limited regions are typical of highly polluted urban areas. Source: National Research Council (1991).

(producing ozone) and reaction with NO_2 to produce nitric acid. Thus the ozone production rate increases linearly with the VOCs and CO concentrations but inversely with the NO_x concentration. This is called the *VOC-limited* or *NO_x-saturated regime.*

Figure 3.7 shows a simple box model calculation of ozone isopleths (lines of constant mixing ratios) calculated as a function of NO_x and VOC concentrations using a standard chemical mechanism. The NO_x- and VOC-limited regimes identified on the diagram illustrate the different dependences of ozone production on NO_x and VOCs. The nonlinear dependences and transitions between regimes are readily apparent.

Figure 3.8 shows the global distribution of tropospheric ozone columns observed from satellite. The July data feature elevated ozone at northern mid-latitudes, reflecting high NO_x emissions and strong UV radiation. The lifetime of ozone is sufficiently long to allow transport on intercontinental scales. The October data feature elevated ozone downwind of South America and Africa, reflecting NO_x emissions from biomass burning during that time of year (end of dry season in the southern tropics). Ozone concentrations peak downwind of the source continents due to sustained production in the continental plume.

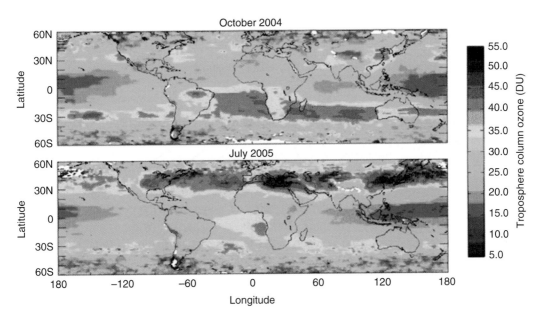

Figure 3.8   Tropospheric ozone column (Dobson units) derived from satellite observations by subtracting the stratospheric column measured by the Microwave Limb Sounder (MLS) from the total column measured by the Ozone Monitoring Instrument (OMI). Monthly mean values are shown for October 2004 (top) and July 2005 (bottom). Plumes are streaming from Africa and South America (ozone produced by precursors released from biomass burning in the tropics during the dry season) in October and from North America, Europe and China (summertime ozone formation from anthropogenic precursors) in July. From The National Aeronautics and Space Administration (NASA).

3.7 Halogen Radicals

Natural sources of atmospheric halogens include the marine biosphere, sea salt, and volcanoes. The marine biosphere emits a wide range of organohalogen gases, the simplest being methyl chloride, bromide, and iodide (CH_3Cl, CH_3Br, CH_3I). Since the 1950s, industrial sources have released to the atmosphere a number of long-lived organohalogens including chlorofluorocarbons and bromine-containing halons (anthropogenic sources of iodine are thought to be negligibly small). These long-lived compounds are transported to the stratosphere, where strong radiation triggers their photolysis to release halogen atoms, which destroy ozone through the catalytic cycle:

$$X + O_3 \rightarrow XO + O_2 \tag{3.38}$$

$$XO + O \rightarrow X + O_2 \tag{3.39}$$

where $X \equiv F$, Cl, Br, or I. One commonly defines the radical family $XO_x \equiv X + XO$ as the catalyst for ozone loss. Termination of the catalytic loss cycle requires conversion of the radicals to non-radical reservoirs including X_2, $XONO_2$, HOX, and HX:

$$XO + XO \rightarrow X_2 + O_2 \tag{3.40}$$

$$XO + NO_2 + M \rightarrow XONO_2 + M \tag{3.41}$$

$$XO + HO_2 \rightarrow HOX + O_2 \tag{3.42}$$

$$X + CH_4 \rightarrow HX + CH_3 \tag{3.43}$$

There are also significant cross-halogen reactions, such as between XO and another halogen oxide YO:

$$XO + YO \rightarrow XY + O_2 \tag{3.44}$$

These non-radical reservoirs can be recycled to the radicals by photolysis (for X_2, XY, $XONO_2$, HOX), thermolysis ($XONO_2$), hydrolysis ($XONO_2$), or reaction with OH (HX). In the troposphere they can also be removed by deposition, representing a terminal sink. One generally refers to the chemical family X_y (total inorganic X) as the sum of XO_x and the inorganic non-radical reservoirs (for example, $Br_y \equiv Br + BrO +$ inorganic non-radical reservoirs).

Halogen radicals are of interest as sinks of stratospheric and tropospheric ozone, and as oxidants for various species. Their concentrations are determined by the abundance of X_y and by the partitioning of X_y between radical and non-radical forms, i.e., the XO_x/X_y ratio. A major factor in the efficacy of halogen radical chemistry is the stability of the non-radical reservoir HX against oxidation by OH. This stability greatly decreases in the order HF > HCl > HBr > HI. Thus F_y in the atmosphere is almost entirely present as HF and there is no significant fluorine radical chemistry. At the other end, iodine in the atmosphere is present principally in radical form. Iodine and bromine are particularly efficient at destroying ozone but have much weaker sources than chlorine.

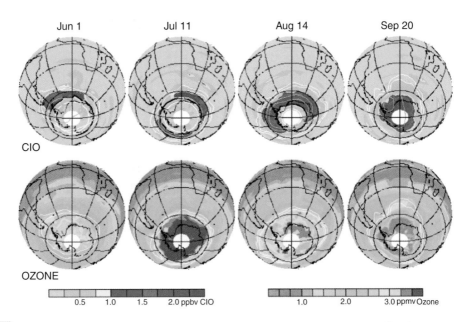

Jun 1   Jul 11   Aug 14   Sep 20

ClO

OZONE

| | 0.5 1.0 1.5 2.0 ppbv ClO | | 1.0 2.0 3.0 ppmv Ozone |

Figure 3.9 ClO and ozone concentrations in the Antarctic lower stratosphere in winter–spring 1992. Data from the Microwave Limb Sounder (MLS) on the National Aeronautics and Space Administration (NASA) Upper Atmosphere Research satellite (UARS). Source: Waters *et al.* (1993).

Chlorine radical concentrations are particularly high in the Antarctic lower stratosphere in spring, where they drive near-total ozone depletion (Figure 3.9). This involves unique chemistry taking place on *polar stratospheric cloud* (*PSC*) particles that form under the very cold conditions of the Antarctic stratosphere in winter and early spring. The PSCs consist of liquid supercooled ternary solutions (H_2SO_4–HNO_3–H_2O or STS), solid nitric acid trihydrate (HNO_3–$3H_2O$ or NAT), and ice crystals. Their formation takes place below a temperature threshold of about 189 K for ice particles, 192 K for STS, and 196 K for NAT. Polar stratospheric cloud surfaces enable fast conversion of non-radical chlorine reservoirs to chlorine radicals by

$$ClONO_2 + HCl \xrightarrow{\text{PSC}} Cl_2 + HNO_3 \qquad (3.45)$$

$$Cl_2 + h\nu \rightarrow 2Cl \qquad (3.46)$$

$$Cl + O_3 \rightarrow ClO + O_2 \qquad (3.47)$$

This converts most of Cl_y in Antarctic spring to ClO, as shown in Figure 3.9. High ClO concentrations are found in a ring around Antarctica in winter, filling in over the pole in spring, because a minimum of radiation is needed to photolyze Cl_2 and other weakly bound chlorine non-radical reservoirs.

High concentrations of ClO prime the Antarctic springtime stratosphere for rapid ozone destruction. However, the O atom concentration remains very low because of weak solar radiation, so that the ClO_x-catalyzed radical mechanism involving (3.38) + (3.39) is very slow. A different mechanism operates involving formation of the ClO dimer followed by photolysis:

$$ClO + ClO + M \rightarrow Cl_2O_2 + M \qquad (3.48)$$

$$Cl_2O_2 + h\nu \rightarrow ClOO + Cl \qquad (3.49)$$

$$ClOO + M \rightarrow Cl + O_2 + M \qquad (3.50)$$

The Cl atoms react again with ozone by (3.47), yielding a catalytic cycle for ozone destruction. The rate-limiting step for this cycle is (3.48), which is quadratic in ClO concentrations. Near-total ozone depletion can thus take place in a matter of weeks. A similar mechanism takes place in the Arctic stratosphere in winter–spring but is much less pronounced because Arctic temperatures are on average 10 K warmer (limiting PSC formation) and the polar vortex is considerably more perturbed by planetary-scale waves as discussed in Section 2.11.

3.8 Sulfur Species

Sulfate is a major component of atmospheric aerosol and an important contributor (as sulfuric acid) to acid deposition. It is produced in the atmosphere by oxidation of sulfur gases emitted from anthropogenic sources (mainly combustion and metallurgy), the marine biosphere, and volcanoes. Anthropogenic and volcanic emissions are mainly as SO_2. Biogenic emission is mostly as dimethylsulfide ($(CH_3)_2S$, commonly called DMS. There is also a small anthropogenic and marine source of carbonyl sulfide (COS), which is of interest because COS has a long enough atmospheric lifetime to be transported to the stratosphere, where it provides a background source of sulfate aerosol. Figure 3.10 shows the global distribution of SO_2 columns measured by satellite. The anthropogenic source is particularly large over China, reflecting coal combustion with limited emission controls.

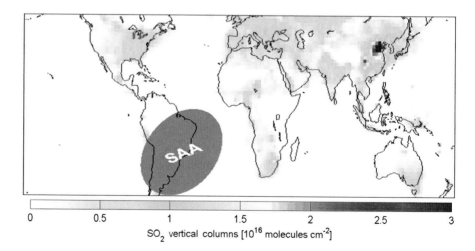

Figure 3.10 Annual mean SO_2 vertical columns from the Scanning Imaging Absorption spectrometer for Atmospheric chartography (SCIAMACHY) satellite instrument for 2006. The South Atlantic Anomaly (SAA) is subject to excessive measurement noise. From Lee *et al.* (2009).

The oxidation state of sulfur ranges from –2 in DMS to +6 in sulfate. The DMS is oxidized in the marine boundary layer by OH, NO_3, and halogen radicals to produce a cascade of sulfur species eventually leading to SO_2 as a major product. The COS is converted to SO_2 following oxidation by OH, and also in the stratosphere following oxidation by $O(^1D)$ and photolysis. And SO_2 is oxidized by OH to produce sulfuric acid (H_2SO_4) in the gas phase:

$$SO_2 + OH + M \rightarrow HSO_3 + M \tag{3.51}$$

$$HSO_3 + O_2 \rightarrow SO_3 + HO_2 \tag{3.52}$$

$$SO_3 + H_2O + M \rightarrow H_2SO_4 + M \tag{3.53}$$

Sulfuric acid has an extremely low vapor pressure over H_2SO_4–H_2O solutions and therefore condenses immediately, either on existing particles or by forming new particles.

The lifetime of SO_2 against gas-phase oxidation by OH is of the order of a week. In the lower troposphere, more rapid oxidation of SO_2 can take place in the aqueous phase in clouds. This involves dissolution of SO_2 into cloud water, followed by acid–base dissociation of sulfurous acid ($SO_2 \cdot H_2O$) to bisulfite (HSO_3^-) and sulfite (SO_3^{2-}):

$$SO_2(g) \rightleftharpoons SO_2 \cdot H_2O\,(aq) \tag{3.54}$$

$$SO_2 \cdot H_2O\,(aq) \rightleftharpoons HSO_3^- + H^+ (pK_1 = 1.9) \tag{3.55}$$

$$HSO_3^- \rightleftharpoons SO_3^{2-} + H^+ (pK_2 = 7.2) \tag{3.56}$$

The ions can then be oxidized rapidly in the aqueous-phase. Major oxidants are H_2O_2 and ozone, both dissolved from the gas phase:

$$HSO_3^- + H_2O_2(aq) + H^+ \rightarrow SO_4^{2-} + H_2O + 2H^+ \tag{3.57}$$

$$SO_3^{2-} + O_3(aq) \rightarrow SO_4^{2-} + O_2 \tag{3.58}$$

Oxidation by H_2O_2 is acid-catalyzed and therefore remains fast even as the cloud droplets become acidified. Oxidation of SO_3^{2-} by $O_3(aq)$ is extremely fast but is limited by the supply of SO_3^{2-} and shuts down as cloud droplets are acidified below pH 5. Other aqueous-phase SO_2 oxidants can be important in winter or in highly polluted conditions when H_2O_2 concentrations are low. Competition between gas-phase and aqueous-phase oxidation of SO_2 has important implications for aerosol formation because gas-phase H_2SO_4 is a major precursor for nucleation of new aerosol particles.

3.9 Aerosol Particles

The atmosphere contains suspended condensed particles ranging in size from ~0.001 μm (molecular cluster) to ~100 μm (small raindrop). Atmospheric chemists commonly refer to the ensemble of particles of a certain type as an *aerosol* (for example, *sulfate aerosol*) and to an ensemble of particles of different types as *aerosols*.

The ensemble of particles in the atmosphere is often called the *atmospheric aerosol.* This terminology has force of usage but departs from the dictionary definition of an aerosol as a suspension of dispersed particles in a gas (by that definition, the atmosphere itself would be an aerosol). Referring to *aerosol particles* removes the ambiguity. The air quality community refers to aerosols as *particulate matter* and uses the acronym *PM* to denote the aerosol mass concentration per unit volume of air. For example, $PM_{2.5}$ denotes the mass concentration [$\mu g\ m^{-3}$] of particles less than 2.5 μm in diameter. Aerosols are removed efficiently by precipitation and thus have atmospheric lifetimes of the order of a week, leading to large regional gradients.

3.9.1 Size Distribution

An aerosol particle is characterized by its shape, size, phase(s), and chemical composition. Liquid particles are spherical but solid particles can be of any shape. There is a continuous distribution of particle sizes. For the purpose of characterizing this distribution the particles are conventionally assumed to be spherical. Such an assumption is obviously incorrect for solid particles but can be viewed as an operational approximation where the solid particles behave as equivalent spheres for the purpose of sizing measurements or microphysical dynamics. An aerosol composed of particles of a single size is called *monodisperse*, while an aerosol composed of particles of multiple sizes is called *polydisperse*. Aerosols produced in the laboratory under carefully controlled conditions can be close to monodisperse. Aerosols in the atmosphere are polydisperse.

The aerosol size distribution can be characterized by the *number size distribution function* $n_N(r)$ [particles $cm^{-3}\ \mu m^{-1}$]

$$n_N(r) = \frac{dN}{dr} \tag{3.59}$$

such that $n_N(r)dr$ represents the number of particles per cm^3 of air in the radius size range $[r, r+dr]$, and N is the total particle number concentration. Other related measures of the aerosol size distribution are the *surface area size distribution function*

$$n_S(r) = 4\pi r^2 n_N(r) \tag{3.60}$$

and the *volume size distribution function*

$$n_V(r) = \frac{4}{3}\pi r^3 n_N(r) \tag{3.61}$$

Plots of the aerosol size distribution generally use a log scale to account for the variation of particle sizes over typically five orders of magnitude, from 10^{-3} to 10^2 μm:

$$n_N(\log r) = \frac{dN}{d(\log r)} = \ln(10)r n_N(r) \tag{3.62}$$

such that $n_N(\log r)d(\log r)$ represents the number concentration of particles in the size range $[\log r, \log r + d(\log r))]$. Similar expressions apply for $n_S(\log r)$ and $n_V(\log r)$.

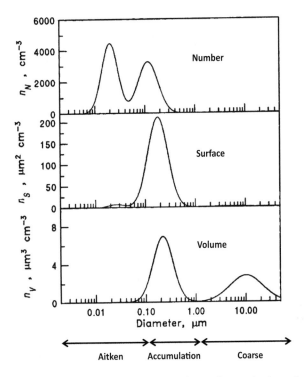

Idealized size distributions for an aerosol population by number, surface, and volume. From Seinfeld and Pandis (2006).

The integrals of these distributions yield the total aerosol number concentration N, surface concentration S, and volume concentration V:

$$N = \int_{-\infty}^{\infty} n_N(\log r)\, d\log r \qquad (3.63)$$

$$S = \int_{-\infty}^{\infty} n_S(\log r)\, d\log r = 4\pi \int_{-\infty}^{\infty} r^2 n_N(\log r)\, d\log r \qquad (3.64)$$

$$V = \int_{-\infty}^{\infty} n_V(\log r)\, d\log r = \frac{4\pi}{3} \int_{-\infty}^{\infty} r^3 n_N(\log r)\, d\log r \qquad (3.65)$$

Figure 3.11 shows the number, surface, and volume size distributions for a generic aerosol. There are three distinct modes, called the *Aitken mode* (diameter < 0.1 μm), the *accumulation mode* (0.1–1 μm) and the *coarse mode* (>1 μm). The Aitken mode is made up of freshly nucleated particles, which grow rapidly by gas condensation and coagulation to the accumulation mode. Further growth of accumulation mode particles above 1 μm is slow. The coarse mode is mostly composed of primary particles emitted mechanically from the Earth's surface, such as soil dust, sea salt, and pollen.

We see from Figure 3.11 that the number, surface, and volume size distributions for a given aerosol are very different, reflecting the r^2 and r^3 weighting of the surface

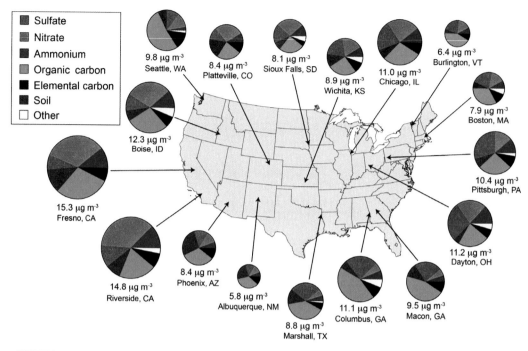

Legend:
- Sulfate
- Nitrate
- Ammonium
- Organic carbon
- Elemental carbon
- Soil
- Other

9.8 µg m⁻³ Seattle, WA
8.4 µg m⁻³ Platteville, CO
8.1 µg m⁻³ Sioux Falls, SD
8.9 µg m⁻³ Wichita, KS
11.0 µg m⁻³ Chicago, IL
6.4 µg m⁻³ Burlington, VT
7.9 µg m⁻³ Boston, MA
12.3 µg m⁻³ Boise, ID
10.4 µg m⁻³ Pittsburgh, PA
15.3 µg m⁻³ Fresno, CA
11.2 µg m⁻³ Dayton, OH
14.8 µg m⁻³ Riverside, CA
8.4 µg m⁻³ Phoenix, AZ
5.8 µg m⁻³ Albuquerque, NM
11.1 µg m⁻³ Columbus, GA
9.5 µg m⁻³ Macon, GA
8.8 µg m⁻³ Marshall, TX

Figure 3.12 Annual mean chemical composition of PM$_{2.5}$ at selected sites in the USA in 2013. Values are from the Chemical Speciation Network of the US Environmental Protection Agency. Figure produced by Eloïse Marais, Harvard University, used with permission.

and volume size distributions over five orders of magnitude in r. Thus the Aitken particles dominate the number size distribution but are unimportant for volume. The coarse particles contribute an insignificant number but make a major contribution to volume. Accumulation particles are important for all measures of the size distribution and especially for the surface size distribution, which is most relevant for aerosol optical properties.

3.9.2 Chemical Composition

The aerosol chemical composition is commonly classified following dominant *aerosol types* as (1) sulfate, (2) nitrate, (3) organic carbon (OC), (4) black carbon (BC), (5) soil dust, and (6) sea salt. There are other minor constituents, such as trace metals and pollen. Sulfate and nitrate are often associated with ammonium and one refers to *sulfate–nitrate–ammonium (SNA) aerosol* as the coupled thermodynamic system. Organic carbon and black carbon are sometimes grouped as *carbonaceous aerosol*. They are directly emitted by incomplete combustion, and OC also has an important secondary source from condensation of semivolatile organic compounds (SVOCs; Section 3.5). Figure 3.12 shows the mass concentrations of different aerosol types at surface sites in the United States. SNA and OC dominate. BC makes little contribution to mass but is of great interest because of its light-absorbing properties and effects on health. Figure 3.13 shows illustrative results from a global

Figure 3.13   Portrait of global aerosols produced by a simulation of the NASA Goddard Earth Observing System Model (GEOS) version-5 at a spatial resolution of 10 km. The figure highlights the relative abundance in different regions of the world of dust (red) lifted from arid soils, sea salt (blue) embedded in fronts and cyclones, smoke (green) from tropical wildfires, and sulfate particles (white) from fossil fuel and volcanic emissions. It also shows the influence of transport of aerosols by the atmospheric circulation. From W. Putman, National Aeronautics and Space Administration (NASA) Goddard Space Flight Center (GSFC): www.nasa.gov.

model simulation of different aerosol types, highlighting the dominance of different types in different regions.

We previously saw how sulfuric and nitric acids are produced in the atmosphere by oxidation of SO_2 and NO_x, respectively. Ammonia has natural biogenic sources and also a large anthropogenic source from agriculture. Sulfuric acid has very low vapor pressure over H_2SO_4–H_2O solutions and condenses immediately under all atmospheric conditions, including in the stratosphere, to produce concentrated sulfuric acid particles. Ammonia partitions into this aerosol following acid–base titration to produce ammonium bisulfate (NH_4HSO_4) and ammonium sulfate (($NH_4)_2SO_4$) aerosol depending on the ammonia to sulfuric acid ratio:

$$H_2SO_4(a) + NH_3(g) \rightarrow NH_4HSO_4(a) \tag{3.66}$$

$$NH_4HSO_4(a) + NH_3(g) \rightarrow (NH_4)_2SO_4(a) \tag{3.67}$$

Letovicite (($NH_4)_3H(SO_4)_2$) can also be produced. Here (g) denotes the gas phase and (a) the aerosol phase, which may be solid or aqueous depending on relative humidity. In aqueous aerosol the ammonium-sulfate salts dissociate to NH_4^+, SO_4^{2-}, and H^+. The thermodynamic driving force for reactions (3.66) and (3.67) is so large that complete titration of sulfuric acid is achieved if ammonia is in excess. In that case, the excess ammonia may go on to combine with nitric acid and form nitrate aerosol:

$$NH_3(g) + HNO_3(g) \rightleftharpoons NH_4NO_3(a) \tag{3.68}$$

with an equilibrium constant that increases with decreasing temperature and with increasing relative humidity. Again, the ammonium nitrate aerosol may be solid or aqueous (NH_4^+, NO_3^-), depending on relative humidity.

Organic carbon (OC) aerosol includes a very large number of species. Its composition is highly variable and not well understood. The traditional approach in models is to distinguish between *primary organic aerosol* (POA) directly emitted by combustion and *secondary organic aerosol* (SOA) formed in the atmosphere. SOA is formed from the atmospheric oxidation of VOCs, resulting in products with chemical functionalities (such as hydroxy, peroxy, carbonyl, carboxylic, and nitrate groups) that enable their uptake by pre-existing organic or aqueous aerosol. This uptake may be reversible or irreversible, in the latter case through subsequent oxidation or oligomerization in the aerosol.

3.9.3 Mixing State, Hygroscopicity, and Activation

Individual particles have compositions that reflect their origin and atmospheric history. Particles may originate as a specific aerosol type, such as sulfate or BC, but subsequent mixing with other aerosol types takes place as they age in the atmosphere. The degree of mixing is important for characterizing aerosol properties. It is convenient in models to consider two limiting cases: *external* and *internal mixing.* An external mixture is one in which different aerosol types do not mix, so that individual particles are of a single type; this is usually appropriate close to the source. An internal mixture is one in which all particles have the same composition; this is usually appropriate in a remote air mass. Aerosols in the atmosphere gradually evolve from an external to an internal mixture through particle coagulation, gas uptake, and cloud processing. Internal mixing usually assumes that the different chemical constituents are well-mixed within individual particles, but that does not hold if the particles are not fully liquid. For example, a common configuration for BC-sulfate internal mixing is for BC to form a solid core embedded within the aqueous sulfate solution. This *core–shell model* has important implications for calculating the optical properties of BC.

The *hygroscopicity* of an aerosol refers to its thermodynamic capacity for taking up water at a given relative humidity (RH). By Raoult's law, an aerosol particle behaving as an ideal aqueous solution has a water molar content $x_{H2O} = $ RH/100, where RH is expressed as a percentage. Such dissolution of the aerosol requires from precipitation equilibrium that x_{H2O} be sufficiently high (i.e., that RH be sufficiently high). Sulfuric acid is aqueous at all RH because its condensation takes place as a H_2SO_4–H_2O binary mixture. Other aerosols are dry at thermodynamic equilibrium in a low-RH atmosphere and become aqueous when the RH exceeds the level required by precipitation equilibrium. This RH level is called the *deliquescence relative humidity* (DRH); it represents a sharp particle transition from non-aqueous (usually solid) aerosol to aqueous. It is 40% for NH_4HSO_4, 62% for NH_4NO_3, 75% for NaCl, and 80% for $(NH_4)_2SO_4$. Additional water condenses as the RH continues to increase. Starting from high RH, a decrease in RH will similarly result in a sharp conversion of the aerosol from aqueous to non-aqueous. However, if the non-aqueous state involves crystallization (as in the case of SNA and sea salt particles) it may be retarded by the energy barrier for crystal formation. In that case, the *crystallization relative humidity* (CRH) is lower than the DRH, and for CRH < RH < DRH the aqueous phase is *metastable.*

Another important property of aerosol particles is their ability to serve as *cloud condensation nuclei* (*CCN*) for activation of cloud droplets under supersaturated conditions (RH > 100%). This involves overcoming the surface tension for growth of the gas–droplet interface. Particles larger than 0.1 μm and at least partly wettable are effective CCN at typical atmospheric supersaturations (100% < RH < 101%). Models often distinguish between *hydrophobic* particles as non-wettable and *hydrophilic* particles as wettable. For example, freshly emitted BC is typically hydrophobic and thus an ineffective CCN, but becomes hydrophilic in the atmosphere on a timescale of a day as it ages and mixes with other aerosol types.

3.9.4 Optical Properties

Aerosol particles interfere with the propagation of radiation by scattering and absorption. The resulting attenuation of radiation along a path through the atmosphere is described by Beer's law:

$$I = I_o \exp\left[-\tau\right] \tag{3.69}$$

where I_o is the incident radiation, I is the transmitted radiation through the path, and τ is the *optical path*. One can express this optical path as $\tau = \beta_{ext} L$ where L [m] is the physical path length and β_{ext} [m^{-1}] is an *extinction coefficient* characteristic of the aerosol. The extinction coefficient is the sum of a *scattering coefficient* β_{scat} [m^{-1}] and an *absorption coefficient* β_{abs} [m^{-1}]. For a monodisperse population of spherical particles of radius r with number density N [m^{-3}], the extinction coefficient is a function of the dimensionless *extinction efficiency* Q_{ext} (defined as the probability that a photon incident on the particle will be absorbed or scattered), as given by

$$\beta_{ext} = Q_{ext} N \pi r^2 \tag{3.70}$$

The extinction efficiency of a particle can be viewed as the sum of a scattering efficiency Q_{scat} and an absorption efficiency Q_{abs}. The *single-scattering albedo* ω is defined as the ratio

$$\omega = \frac{Q_{scat}}{Q_{ext}} = \frac{Q_{scat}}{Q_{scat} + Q_{abs}} \tag{3.71}$$

and measures the relative contributions of scattering and absorption to extinction. Absorption depends on the chemical properties of the molecule, while scattering depends on the *particle size parameter*

$$\alpha = \frac{2\pi r}{\lambda} \tag{3.72}$$

Scattering is most efficient for particles of radius equal to the radiation wavelength ($\alpha = 2\pi$). Since most of the aerosol particle area is typically in the 0.1–1 μm size range, we see that aerosols are efficient scatterers of solar radiation. See Figure 5.8 in Section 5.2.4 for dependences of Q_{scat} and Q_{abs} on particle size and refractive index.

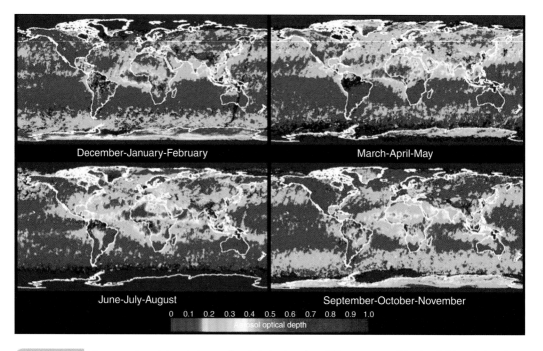

December-January-February

March-April-May

June-July-August

September-October-November

0   0.1   0.2   0.3   0.4   0.5   0.6   0.7   0.8   0.9   1.0

Aerosol optical depth

Figure 3.14   Average column-integrated aerosol optical depth at $\lambda = 558$ nm measured by the Multi-angle Imaging Spectro-Radiometer (MISR) satellite instrument from December 2001 to November 2002. Source: National Aeronautics and Space Administration/GSFC/LaRC/JPL, MISR Team.

The *aerosol optical depth* in the atmospheric column is defined as the optical path for radiation propagating vertically from the top of the atmosphere to the surface. It is given by

$$\tau(\lambda) = \int_0^\infty \beta_{ext}(\lambda, z)\, dz \qquad (3.73)$$

Here again, the total optical depth can be expressed as the sum of two terms that account for absorption and scattering: $\tau = \tau_{scat} + \tau_{abs}$. Figure 3.14 shows the global distribution of the total aerosol optical depth in the visible light as measured from space in different seasons. Elevated values are due to desert dust, biomass burning, anthropogenic pollution, and sea salt.

The detailed calculation of the radiative effects of aerosols is complicated. One distinguishes between three regimes: (1) the *Rayleigh scattering* regime in which the particles are much smaller than the wavelength of the incident radiation ($\alpha \ll 1$), (2) the *Mie scattering* regime in which the size of the particles is of the same order of magnitude as the wavelength ($\alpha \approx 1$), and (3) the *geometric scattering* regime, in which the particles are much larger than the wavelength ($\alpha \gg 1$). For small particles (Rayleigh regime), one can show that the scattering efficiency varies as λ^{-4}, while the absorption efficiency varies as λ^{-1}. Thus, light at shorter (bluer) wavelengths is scattered more effectively than light at longer (redder) wavelengths. White light passing through a layer of small aerosol particles becomes redder as it propagates toward an observer.

References

Brasseur G. P., Orlando J. J., and Tyndall G. S. (eds.) (1999) *Atmospheric Chemistry and Global Change*, Oxford University Press, Oxford.

Brasseur G. P., Prinn R. G. and Pszenny, A. A. P. (eds.) (2003) *Atmospheric Chemistry in a Changing World*, Springer, New York.

Finlayson-Pitts B. and Pitts J. N. (2000) *Chemistry of the Upper and Lower Atmosphere: Theory, Experiments and Applications*, Academic Press, New York.

Lee C., Martin R. V., van Donkelaar A., *et al.* (2009) Retrieval of vertical columns of sulphur dioxide from SCIAMACHY and OMI: Air mass factor algorithm development, validation, and error analysis, *J. Geophys. Res.* **114**, D22303, doi: 10.1029/2009JD012123

National Research Council (1991) *Rethinking the Ozone Problem in Urban and Regional Air Pollution*, National Academy Press, Washington, DC.

Seinfeld J. H. and Pandis S. N. (2006) *Atmospheric Chemistry and Physics: From Air Pollution to Climate Change*, Wiley, New York.

Warneck P. (1999) *Chemistry of the Natural Atmosphere*, Academic Press, New York.

Waters J. W., Froidevaux L., Read W. G., *et al.* (1993) Stratospheric ClO and ozone from the Microwave Limb Sounder on the Upper Atmosphere Research Satellite. *Nature*, **362**, 597–602.

Wu S., Mickley, L. J., Jacob D. J., *et al.* (2007) Why are there large differences between models in global budgets of tropospheric ozone?, *J. Geophys. Res.*, **112**, D05302, doi:10.1029/2006JD007801

4 Model Equations and Numerical Approaches

4.1 Introduction

Atmospheric chemistry focuses on understanding the factors that control the concentrations of chemical species in the atmosphere. These factors include processes of emissions, transport, chemical production and loss, and deposition. Here we present the general mathematical foundations for atmospheric chemistry and the corresponding model frameworks.

We begin in Section 4.2 by introducing the *continuity equation*, which is the fundamental mass conservation equation for atmospheric chemistry. The continuity equation expresses how the concentration of a chemical species changes with time in response to a sum of individual forcing terms describing emissions, transport, chemistry, and deposition. The continuity equation for aerosols also includes terms to describe microphysical growth of particles; this is presented in Section 4.3. The continuity equation is a differential equation in space and time, and its integration solves for the evolution of chemical concentrations as controlled by the ensemble of driving processes. Analysis of timescales over which the individual processes operate can be very useful to identify dominant terms; this is presented in Section 4.4. Computing transport terms requires solving the conservation equations for atmospheric dynamics that serve as foundations for meteorological models. These equations are presented in Section 4.5.

The continuity equations for atmospheric chemistry cannot be solved exactly (except for highly idealized problems) because of the complexity of the flow and because of chemical coupling between species. Numerical methods are needed that provide the foundations for atmospheric chemistry models. We present in this chapter the general frameworks for these methods as implemented in models including different coordinate systems, dimensionality, and grid geometries (Sections 4.6–4.7), as well as different approaches including Eulerian, Lagrangian, and statistical (Sections 4.8–4.13). Standard modeling strategies of operator splitting, numerical filtering, and remapping are presented in Sections 4.14–4.16. This chapter sets the stage for the following chapters where specific numerical algorithms will be presented to compute chemistry (Chapters 5 and 6), transport (Chapters 7 and 8), and emission and deposition (Chapter 9).

4.2 Continuity Equation for Chemical Species

4.2.1 Eulerian and Lagrangian Formulations

The continuity equation expresses mass conservation within an elemental volume of fluid. For a chemical species i with concentration measured by its mass density ρ_i [kg of species i per m^3 of air], the continuity equation is expressed in an Eulerian framework (Chapter 1) as

$$\frac{\partial \rho_i}{\partial t} + \nabla \cdot (\rho_i \mathbf{v}) = s_i \tag{4.1}$$

where $\mathbf{v} = (u, v, w)^T$ is the wind velocity vector, $\nabla \cdot (\rho_i \mathbf{v})$ is the flux divergence (flux out of the volume minus flux in), which represents the *transport term*, and s_i is the net local source of the species, which represents the *local term*. Equation (4.1) is the *Eulerian flux form* of the continuity equation. The transport term includes contributions from *advection*, which describes the flow by large-scale winds resolved on the scale of the model, and *turbulence*, which is not resolved on the model scale and must be represented stochastically. The turbulent component of the transport term is separated further into *turbulent mixing*, which is effectively random on the model scale, and *convection*, which has organized vertical structure on the model scale. The local term s_i includes contributions from chemistry, emissions, and wet and dry deposition. Surface exchange by emissions and dry deposition represents a flux boundary condition for the continuity equation but in a gridded model environment it is treated as a local term for the lowest model level.

Equation (4.1) can be applied to air itself as a conservation equation for the air density ρ_a. In that case there is no local term since changes in air density are driven solely by transport. Thus

$$\frac{\partial \rho_a}{\partial t} + \nabla \cdot (\rho_a \mathbf{v}) = 0 \tag{4.2}$$

Replacing the mass mixing ratio $\mu_i = \rho_i / \rho_a$ into (4.1) and expanding the derivative of the product, we obtain

$$\mu_i \frac{\partial \rho_a}{\partial t} + \rho_a \frac{\partial \mu_i}{\partial t} + \mu_i \nabla \cdot (\rho_a \mathbf{v}) + \rho_a \mathbf{v} \cdot \nabla \mu_i = s_i \tag{4.3}$$

Replacing (4.2) into (4.3) then yields

$$\frac{\partial \mu_i}{\partial t} + \mathbf{v} \cdot \nabla \mu_i = \frac{s_i}{\rho_a} \tag{4.4}$$

This is the *Eulerian advective form* of the continuity equation, expressing the concentration in terms of mixing ratio rather than density. The local term s_i/ρ_a is now in mixing ratio units. The velocity vector $\mathbf{v}$ is outside of the gradient operator because compression of air ($\nabla \cdot \mathbf{v} \neq 0$) does not change the mixing ratio.

Introducing now the total derivative for an air parcel moving with the flow:

$$\frac{d\mu_i}{dt} = \frac{\partial\mu_i}{\partial t} + \frac{\partial\mu_i}{\partial x}\frac{dx}{dt} + \frac{\partial\mu_i}{\partial y}\frac{dy}{dt} + \frac{\partial\mu_i}{\partial z}\frac{dz}{dt} = \frac{\partial\mu_i}{\partial t} + u\frac{\partial\mu_i}{\partial x} + v\frac{\partial\mu_i}{\partial y} + w\frac{\partial\mu_i}{\partial z} = \frac{\partial\mu_i}{\partial t} + \mathbf{v}\cdot\nabla\mu_i$$

(4.5)

we obtain

$$\frac{d\mu_i}{dt} = \frac{s_i}{\rho_a}$$

(4.6)

which is the *Lagrangian form* of the continuity equation, based on a frame of reference moving with the flow. The total derivative (4.5) is sometimes called the *Lagrangian derivative.* For an air parcel moving with the flow, transport does not change the mixing ratio and the only change is from the local term s_i/ρ_a. The Lagrangian form needs to be expressed as mixing ratio so that it is not affected by compression of air as the air parcel moves.

Atmospheric chemists often express the continuity equation in terms of the *number density* n_i [molecules cm^{-3}] and the *volume (or molar) mixing ratio* $C_i = n_i/n_a$ (where n_a is the air number density). These are related to the mass density by $\rho_i = n_i(M_i/\mathcal{N}_A)$ where M_i is the molecular mass of species i and $\mathcal{N}_A$ is Avogadro's number, and to the mass mixing ratio by $\mu_i = C_i(M_i/M_a)$ where M_a is the molecular mass of air. The Eulerian flux form of the continuity equation is then

$$\frac{\partial n_i}{\partial t} + \nabla\cdot(n_i\mathbf{v}) = s_i$$

(4.7)

where s_i is now in units of [molecules cm^{-3} s^{-1}]. The Eulerian advective form is

$$\frac{\partial C_i}{\partial t} + \mathbf{v}\cdot\nabla C_i = \frac{s_i}{n_a}$$

(4.8)

and the Lagrangian form is

$$\frac{dC_i}{dt} = \frac{s_i}{n_a}$$

(4.9)

The transport and local terms involve a number of different processes operating in the model environment. The continuity equation is thus usefully represented for model purposes as a sum of terms describing the different processes for which the model provides independent formulations. For example, the Eulerian form may be decomposed as

$$\frac{\partial\rho_i}{\partial t} = \left[\frac{\partial\rho_i}{\partial t}\right]_{adv} + \left[\frac{\partial\rho_i}{\partial t}\right]_{mix} + \left[\frac{\partial\rho_i}{\partial t}\right]_{conv} + \left[\frac{\partial\rho_i}{\partial t}\right]_{scav} + \left[\frac{\partial\rho_i}{\partial t}\right]_{chem} + \left[\frac{\partial\rho_i}{\partial t}\right]_{em} + \left[\frac{\partial\rho_i}{\partial t}\right]_{dep}$$

(4.10)

where the terms on the right-hand side represent successively the contributions of advection, turbulent mixing, convection, wet scavenging by precipitation, chemistry, emissions, and dry deposition. We describe the formulations for each of these terms in the following subsections. The Lagrangian form using the total derivative may be

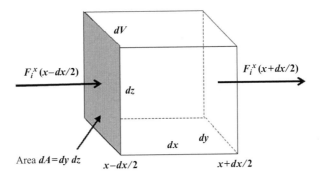

Figure 4.1 Flux F^x_i of species i in the x-direction through an elemental volume dV.

similarly decomposed but without the transport terms; a separate algorithm is needed to describe the Lagrangian transport of air parcels and this is also described below.

4.2.2 Advection

Advection describes transport by the wind resolved on the model scale. The wind velocity vector **v** is then a spatial and temporal average over the model grid and time step. The corresponding mass flux is $\mathbf{F}_i = \left(F^x_i, F^y_i, F^z_i\right)^T = \rho_i \mathbf{v} = \rho_i(u, v, w)^T$. Consider an elemental volume $dV = dx\, dy\, dz$ centered at (x, y, z), and a wind velocity component u in the x-direction. The corresponding mass flux for species i is $F^x_i = \rho_i u$ [kg m^{-2} s^{-1}]. The flow rate into the volume (kg s^{-1}) is $F^x_i(x - dx/2)dy\, dz$ and the flow rate out of the volume is $F^x_i(x + dx/2)dy\, dz$ (Figure 4.1). The change per unit time in the concentration ρ_i within the volume is then given by

$$\left[\frac{\partial \rho_i}{\partial t}\right]_{adv} = \frac{\left[F^x_i(x - dx/2) - F^x_i(x + dx/2)\right] dy\, dz}{dx\, dy\, dz} = -\frac{\partial F^x_i}{\partial x} = -\frac{\partial(\rho_i u)}{\partial x} \quad (4.11)$$

By adding similar contributions for the y and z directions (with wind components v and w, respectively), we obtain

$$\left[\frac{\partial \rho_i}{\partial t}\right]_{adv} = -\frac{\partial F^x_i}{\partial x} - \frac{\partial F^y_i}{\partial y} - \frac{\partial F^z_i}{\partial z} = -\frac{\partial(\rho_i u)}{\partial x} - \frac{\partial(\rho_i v)}{\partial y} - \frac{\partial(\rho_i w)}{\partial z} \quad (4.12)$$

or

$$\left[\frac{\partial \rho_i}{\partial t}\right]_{adv} = -\nabla \cdot \mathbf{F}_i = -\nabla \cdot (\rho_i \mathbf{v}) \quad (4.13)$$

which is the Eulerian flux form used in (4.1). It applies only to the model-resolved winds for which we have actual information on **v**. Smaller-scale motions are described by turbulence parameterizations presented next. Numerical methods for computing advection are presented in Chapter 7.

4.2.3 Turbulent Mixing

Fluctuating wind patterns that are not resolved on the grid scale of the model are called *turbulence*. By definition, the turbulent component of the wind has a mean value of zero when averaged over the model grid and time step. It can still cause significant chemical transport in the presence of a chemical gradient. Consider as analogy a commuter train operating at the morning rush hour between the city and the suburbs. The train is full as it travels from the suburbs to the city, and empty as it travels from the city back to the suburbs. The net motion of the train is zero, yet there is a net flow of commuters from the suburbs to the city. In the same way, a back-and-forth wind operating in a uniform chemical gradient will cause a net down-gradient flux from the region of high concentration to the region of low concentration. The flux is proportional to the gradient. This proportionality, called *Fick's law of diffusion*, is the foundation of transport by molecular diffusion. Therefore, a simple parameterization of small-scale turbulent transport (where "small-scale" is relative to the model grid) is to treat it as a diffusive process. Diffusive transport results in mixing and we refer to the parameterization as *turbulent mixing*.

The corresponding equation for the change in species concentration as a result of turbulent mixing is

$$\left[\frac{\partial \rho_i}{\partial t}\right]_{mix} = \mathbf{\nabla} \cdot \left(\mathbf{K} \, \rho_a \mathbf{\nabla}\left(\frac{\rho_i}{\rho_a}\right)\right) \tag{4.14}$$

where $\mathbf{K}$ is an empirical tensor describing the 3-D *turbulent diffusion*. The elements of that tensor are the *turbulent diffusion coefficients*. $\mathbf{K}$ is generally taken to be a diagonal matrix with diagonal elements K_x, K_y, K_z describing turbulent diffusion in the horizontal (x, y) and vertical (z) directions. The term $-\mathbf{K} \, \rho_a \mathbf{\nabla}(\rho_i/\rho_a)$ is the turbulent diffusion flux by analogy with Fick's law, and the right-hand side of (4.14) expresses the flux divergence as derived previously for the advection term. The concentration gradient $\mathbf{\nabla}(\rho_i/\rho_a)$ must be expressed in terms of mixing ratio to account for compressibility of air; a concentration gradient driven solely by air density changes does not drive a turbulent flux.

Turbulent diffusion is not a mechanistic description of turbulence; it is simply a parameterization that has some physical basis and describes relatively well the effect of small-scale turbulence on a chemical concentration field. It allows for consistent treatment of turbulent transport for all chemical species, because the turbulent diffusion coefficients are generally taken to be the same for all species; this is known as the *similarity assumption* for turbulence. Models often apply a turbulent mixing parameterization in the planetary boundary layer below ~2 km altitude, where the turbulence tends to be small in scale. Turbulent mixing is particularly important in the vertical direction, where mean winds are weak but strong turbulent motions can be generated by surface roughness or by buoyancy.

Numerical methods for computing turbulent diffusion as formulated by (4.14) are presented in Chapter 8. The formulation introduces a second-order derivative in the continuity equations that dampens local concentration gradients. In fact, introducing a diffusion term tends to stabilize the numerical solution of the continuity equation by reducing the magnitude of local gradients. This stabilization effect is discussed in Chapter 7.

Turbulent diffusion should not be confused with *molecular diffusion*, which is an actual physical process describing the random motion of molecules. Vertical turbulent diffusion coefficients in the planetary boundary layer are typically in the range 10^4–10^6 cm^2 s^{-1}, whereas the molecular diffusion coefficient is of the order of 10^{-1} cm^2 s^{-1}. Under these conditions, molecular diffusion is completely negligible relative to turbulent mixing. The molecular diffusion coefficient varies as the inverse of atmospheric pressure (lower pressure means longer mean free paths for molecules), so that at sufficiently high altitudes molecular diffusion becomes relevant for atmospheric motions. In the thermosphere above 90 km altitude, molecular diffusion is a major process for vertical transport of chemical species and a corresponding term must be added to the continuity equation (Chapman and Cowling, 1970; Banks and Kockarts, 1973).

4.2.4 Convection

Convection refers to buoyant vertical motion of sufficiently large scale to have grid-resolved vertical structure while still remaining subgrid on the horizontal scale. It is generally associated with cloud formation (*wet convection*), since water vapor condensation in a rising air parcel provides a local source of heat that accelerates the rise by buoyancy. Thunderstorms are a dramatic example and represent *deep wet convection*. Fair-weather cumuli are a more placid example and represent *shallow wet convection*. *Dry convection* refers to vertical buoyant motion not involving cloud formation, and is in general smaller in vertical extent than wet convection.

A wet convective system involves a *cloud updraft,* typically ~1 km in horizontal scale, that drives rapid upward transport from the base to the top of the convectively unstable column. This transport typically takes place on timescales of less than one hour, with updraft velocities of the order of 1–10 m s^{-1}. Models with sub-kilometer horizontal resolution can describe convective updrafts as advection; these are called *cloud-resolving models* or *large-eddy simulation* (*LES*) *models.* They are limited computationally to small domains, typically of the order of 100 km. Larger-domain models cannot resolve convective updrafts and these must therefore be parameterized. As part of the parameterization, the updraft must be balanced on the horizontal grid scale by *large-scale subsidence* so that there is no net vertical air motion on the model grid. This subsidence is typically modeled by grid-scale sinking of the convective outflow. Additional processes in convective transport parameterizations include *entrainment* into the updraft, *detrainment* from the updraft, and subgrid-scale *downdrafts*. Figure 4.2 shows a general schematic of wet convective transport identifying the individual processes. Different types of convective transport parameterizations are described in Section 8.7.

4.2.5 Wet Scavenging

Aerosols and water-soluble gases are efficiently scavenged by precipitation. One generally distinguishes in meteorological models between *convective precipitation* initiated by subgrid convection and *large-scale* or *stratiform precipitation* resolved by grid-scale motions. Scavenging takes place below the cloud through uptake by

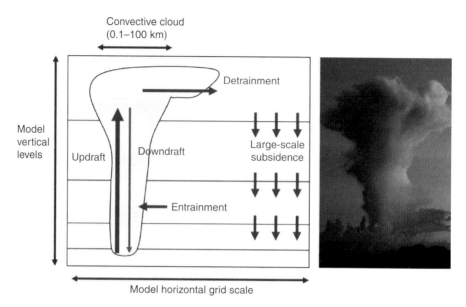

Figure 4.2   Wet convective transport processes.

precipitation (*below-cloud scavenging* or *washout*) and within the cloud through uptake by cloud droplets and ice crystals followed by precipitation (*in-cloud scavenging* or *rainout*). These processes are represented with varying degrees of complexity in models. They are generally first-order loss processes, with rates proportional to the concentration of the species being scavenged. However, in some cases the scavenging may be complicated by nonlinear chemistry; for example, scavenging of SO_2 by cloud droplets is contingent on the supply of oxidants (Section 3.8). In coupled aerosol–climate models, the aerosols may affect the precipitation resulting in nonlinear scavenging. Evaporation of precipitation below the cloud releases the water-soluble species to the atmosphere, in which case the effect of scavenging is downward transport in the atmosphere rather than deposition. Scavenging efficiencies are in general very different for liquid and solid precipitation, and the *retention efficiency* upon cloud freezing (*riming*) may be very uncertain. Figure 4.3 illustrates the physical processes involved. Details about the formulation of wet scavenging in chemical transport models are presented in Section 8.8.

4.2.6 Chemistry

Concentrations of chemical species change as a result of chemical production or loss in a manner determined by the rate laws for the elementary reactions. For example, if a molecule XY is photolyzed by solar radiation into fragments X and Y, the loss rate for XY and the corresponding production rate for X and Y is proportional to the concentration of XY, and the proportionality factor is the *photolysis frequency* J_{XY} [expressed in s^{-1}]. J_{XY} depends on the *actinic photon flux* (flux integrated over

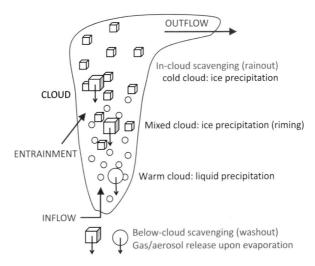

OUTFLOW

In-cloud scavenging (rainout)
cold cloud: ice precipitation

CLOUD

Mixed cloud: ice precipitation (riming)

ENTRAINMENT

Warm cloud: liquid precipitation

INFLOW

Below-cloud scavenging (washout)
Gas/aerosol release upon evaporation

Figure 4.3   Wet scavenging processes.

4π solid angle, i.e., photons coming from all directions), on the *absorption cross-section* measuring the probability that a photon intercepted by molecule XY will be absorbed, and on the *quantum yield* measuring the probability that photon absorption will result in photolysis. As another example, if two chemical species A and B react to form product species C and D, the rate of destruction of A and B (equal to the rate of formation of C and D) is proportional to the collision frequency and to the probability that collision will result in reaction. The rate of reaction is $k_{A+B}[A][B]$ where k_{A+B} is the reaction *rate constant* (also called *rate coefficient*) and [A] and [B] are in units of density so that their product is proportional to the collision frequency.

In the general case, the chemical tendency equation for species i takes the form

$$\left[\frac{\partial \rho_i}{\partial t}\right]_{chem} = \sum_{j\neq i}\left[J_j\,\rho_j + \sum_{k\neq i} k_{j+k}\,\rho_j\rho_k\right] - \left(J_i + \sum_j k_{i+j}\,\rho_j\right)\rho_i \qquad (4.15)$$

where the indices j and k refer to other species that produce or react with i. Equation (4.15) can be written more concisely as

$$\left[\frac{\partial \rho_i}{\partial t}\right]_{chem} = p_i - \ell_i\rho_i \qquad (4.16)$$

where p_i [kg m^{-3} s^{-1}] and ℓ_i [s^{-1}] represent the overall production rate and loss rate constant of species i, summing over all individual processes. If p_i and ℓ_i are independent of the density ρ_i, equation (4.16) is linear and has a simple exponential solution. However, p_i and ℓ_i may depend on ρ_i due to coupling with other species in the chemical mechanism. One then needs to solve equation (4.16) as part of a system of coupled equations, one for each species in the mechanism. Numerical methods for this purpose are presented in Chapter 6.

4.2.7 Surface Exchange

Exchange with the surface by emission and deposition represents a vertical flux boundary condition to the continuity equation, with flux F given by

$$F_i = E_i - D_i \tag{4.17}$$

where E_i is the emission flux of species i generally provided as external input to the model, and D_i is the deposition flux which is generally first-order dependent on the atmospheric concentration:

$$D_i = w_{D,i}(z)\, \rho_i(z) \tag{4.18}$$

Here, $w_{D,i}$ [m s^{-1}] is the *deposition velocity* of species i determining the deposition flux, D_i is often called the *dry deposition flux,* and $w_{D,i}$ the *dry deposition velocity,* to distinguish them from wet deposition which operates by precipitation through the atmospheric column (Section 4.2.5), $w_{D,i}$ is called a "velocity" because of its dimensions. If calculated at the actual surface ($z = 0$), $w_{D,i}$ depends solely on the chemical properties of i for uptake or reaction at the surface. In practice, however, the model does not resolve the concentration at the actual surface. The deposition flux must be calculated from knowledge of the concentration $\rho_i(z_1)$ at the lowest model level z_1, and in that case the calculation of $w_{D,i}(z_1)$ must account for turbulent transfer between z_1 and the surface. Model representation of emission and deposition processes is described in Chapter 9.

The surface flux boundary condition to the continuity equation is practically implemented in chemical transport models as a tendency term in the lowest model level. In an Eulerian model with a lowest model layer of thickness Δz, this is expressed as

$$\left[\frac{\partial \rho_i}{\partial t}\right]_{em} = \frac{E_i}{\Delta z} \tag{4.19}$$

and

$$\left[\frac{\partial \rho_i}{\partial t}\right]_{dep} = -\frac{w_{D,i}(z_1)\, \rho_i(z_1)}{\Delta z} \tag{4.20}$$

In a Lagrangian model this is expressed as source and sink terms for particles brought in sufficiently close contact with the surface. Emissions injected aloft such as from smokestacks, buoyant fires, aircraft, lightning, or volcanoes are handled by applying the tendency terms to the corresponding model levels. In the case of gravitational settling of very large particles, the deposition velocity represents an actual downward settling velocity that needs to be applied to all model levels. Again, this is readily done as a tendency term following the above formulation.

4.2.8 Green Function for Lagrangian Transport

Lagrangian models track the transport of individual air parcels within which local source and sink terms operate to describe chemistry, emissions, and deposition. Wind information to describe the transport of the individual parcels is provided as input, typically from a gridded meteorological data set. Subgrid turbulence must be described by an additional *stochastic* (probabilistic) motion applied to the air parcels.

Consider a Lagrangian model from which we wish to obtain a 3-D field of mixing ratios $\mu_i(\mathbf{r}, t)$ for a specified set of points $\mathbf{r} = (x, y, z)^T$ at time t. This can be derived as the superimposition of mixing ratios produced by an ensemble of pulses applied at points $\mathbf{r}'$ and times $t' < t$, with local source/sink terms applied over the $[t', t]$ trajectory. We define the *Green function* $G(\mathbf{r}, t; \mathbf{r}', t')$ as the normalized time-evolving spatial distribution of the species mixing ratio at points $\mathbf{r}$ and time t resulting from the injection of a pulse at location $\mathbf{r}'$ and time t'. From an Eulerian perspective, the Green function is the solution of the continuity equation (4.4) in which the source rate s_i/ρ_a is replaced by a Dirac function in space and time:

$$\frac{\partial G(\mathbf{r}, t; \mathbf{r}', t')}{\partial t} + \mathbf{v} \cdot \nabla G(\mathbf{r}, t; \mathbf{r}', t') = \delta^3(\mathbf{r} - \mathbf{r}') \, \delta(t - t') \tag{4.21}$$

and $\mathbf{v}$ includes turbulent components that can be described using the turbulent diffusion and convective parameterizations of Sections 4.2.3 and 4.2.4. The Green function as defined here represents the *transition probability density* that a parcel initially located at point $\mathbf{r}'$ at time t' will move to point $\mathbf{r}$ at time t. It is also sometimes called an *influence function*. It has the unit of inverse volume.

For an inert chemical (no local sources or sinks), the Green function provides a solution to the evolution of the mixing ratio over time $[0, t]$, starting from initial conditions defined over the model domain V:

$$\mu(\mathbf{r}, t) = \int_V G(\mathbf{r}, t; \mathbf{r}', 0) \, \mu(\mathbf{r}', 0) \, d^3\mathbf{r}' \tag{4.22}$$

The formulation can be readily extended to include a local source $p_i(\mathbf{r}', t')$ and linear loss rate constant ℓ_i [s^{-1}]. We then have

$$\begin{aligned} \mu(\mathbf{r}, t) = {} & \int_V G(\mathbf{r}, t; \mathbf{r}', 0) \, \mu(\mathbf{r}', 0) \exp[-\ell_i t] d^3\mathbf{r}' \\ & + \int_0^t \int_V G(\mathbf{r}, t; \mathbf{r}', t') \, p_i(\mathbf{r}', t') \exp[-\ell_i(t - t')] d^3\mathbf{r}' dt' \end{aligned} \tag{4.23}$$

where p_i and ℓ_i are applied over each trajectory arriving at point $\mathbf{r}$ at time t. Nonlinear chemistry cannot be accommodated in this framework because individual trajectories do not interact.

The Green function provides a general statement of source–receptor relationships and is used in a wide range of atmospheric chemistry applications. It is of specific use in *receptor-oriented problems* where we seek the contributions from a 2-D or 3-D source field to concentrations at a particular time and location. This forms the basis for simple inversion techniques relating observed atmospheric concentrations to surface fluxes (Enting, 2000). Approaches using a Green function are also used to characterize the transport history of air parcels. In the case of a single source at location $\mathbf{r}'$ affecting the entire modeling domain, we can express the mixing ratio anywhere in the domain with a Green function in unit of inverse time:

$$\mu(\mathbf{r}, \mathbf{r}', t) = \int_0^t G(\mathbf{r}, t; \mathbf{r}', t')\, \mu(\mathbf{r}', t - t')\, dt' \qquad (4.24)$$

The first moment of this Green function,

$$\Gamma(\mathbf{r}, \mathbf{r}') = \int_0^\infty t\, G(\mathbf{r}, t; \mathbf{r}', t')\, dt \qquad (4.25)$$

defines the *mean age of air* for transport from $\mathbf{r}'$ to $\mathbf{r}$. It has been used in particular to characterize transport times in the stratosphere for air originating at the tropical tropopause (Hall and Plumb, 1994).

4.2.9 Initial and Boundary Conditions

The Eulerian form of the continuity equation is a 4-D partial differential equation (PDE) in time and space, and solution requires specification of initial and spatial boundary conditions. The Lagrangian form expressed for a moving air parcel is a 1-D ordinary differential equation (ODE) in time, but spatial boundary conditions are still needed to describe the flow of air parcels.

Initial conditions describe the chemical concentrations over the 3-D domain at the beginning of the simulation and account therefore for the former evolution of these variables. They can be provided by a previous simulation using the same model, by a simulation from another model, by assimilated observations (Chapter 11), or by some mean climatological state. Initial conditions are often not well characterized and may not be consistent for the different species described in the model. The model simulation then carries that legacy of inconsistency during the initial phase of the simulation. Good practice is to initialize the model by conducting a simulation for some time period prior to the period of interest, sufficiently long that the inconsistency of initial conditions is dissipated. This is called *spinning up* the model. The length of the spin-up period depends on the characteristic timescales over which the species of interest respond. These timescales are discussed in Section 4.4.

Vertical boundary conditions are defined by imposed fluxes or concentrations at the boundaries of the domain. The lower boundary is usually the Earth's surface, so that emission and dry deposition fluxes are appropriate boundary conditions. The dry deposition flux is usually computed on the basis of the concentration at the lowest model level with a specified deposition velocity, as given by (4.18). We show in Chapter 9 that this represents in fact a zero-concentration boundary condition at the surface, and can be made into a two-way exchange when the surface concentration is not zero. Further discussion of surface boundary conditions is presented in Chapter 9. A pure advection problem in which the model domain extends to the top of the atmosphere would require no other boundary condition since the continuity equation is first-order in space. Representation of vertical transport as turbulent diffusion (4.14) makes the continuity equation second-order in space and requires an additional boundary condition. Also and in practice, models operate over a limited vertical domain and there is some inflow through the top into the model domain. Therefore an upper flux boundary condition at the model top is needed. The model top may be sufficiently high that a zero-flux boundary condition is acceptable, or concentrations may be imposed at the model top based on information from observations or another model.

Lateral boundary conditions depend on whether the model is global or of limited domain. In a global model, the lateral boundary conditions are periodic as concentrations at longitude λ and latitude φ must be reproduced at $(\lambda + 2\pi, \varphi + 2\pi)$. In a limited-domain model, lateral boundary conditions must be provided as fluxes or concentrations at the edges of the domain. Lateral boundary conditions are actually needed only for the upwind boundary of the domain, but since the wind varies in direction one needs in practice to prescribe boundary conditions at all edges of the domain. The superfluous boundary conditions at the downwind edges may lead to model noise if allowed to propagate into the model domain. It is standard to specify lateral boundary conditions as concentrations just outside the model domain, to be entrained into the model domain by the wind at the model boundaries. This ensures that only the upwind boundary conditions propagate into the model at any given time.

4.3 Continuity Equation for Aerosols

The continuity equation can be applied to different size classes and chemical components of aerosols in the same way as for gases. The transport terms are the same, since the particles are sufficiently small that they are advected by the wind in the same way as gases, except that one should add a sink term from gravitational settling in the case of very large particles (>10 μm in the troposphere, > 1 μm in the stratosphere). The chemical term in the continuity equation is of the same form as for gases if we only simulate the total mass concentrations of aerosol chemical components (without regard to their size distributions).

If we are interested in describing the evolution of the aerosol size distribution, however, we need to introduce additional terms in the continuity equation to account for aerosol nucleation, condensational growth, coagulation, and cloud interactions. This ensemble of processes is called *aerosol microphysics*. For accounting purposes it is more convenient to use volume rather than radius as the independent variable to characterize the aerosol size distribution. We thus define a *volume distribution function* $n_N(V)$ such that $n_N(V)\, dV$ represents the number concentration of particles in the volume range $[V, V + dV]$. The relationship between $n_N(V)$ and $n_V(r)$ defined in Chapter 3 can be derived by equaling the number of particles in the volume range $[V, V + dV]$ with that in the equivalent size range $[r, r + dr]$:

$$n_N(V)\, dV = n_N(r)\, dr \quad \Rightarrow \quad n_N(V) = \frac{dr}{dV} n_N(r) = \frac{1}{4\pi r^2} n_N(r) \qquad (4.26)$$

The local evolution of the aerosol size distribution can be written as

$$\frac{\partial n_N(V)}{\partial t} = \left[\frac{\partial n_N(V)}{\partial t}\right]_{nucleation} + \left[\frac{\partial n_N(V)}{\partial t}\right]_{condensation/evaporation} + \left[\frac{\partial n_N(V)}{\partial t}\right]_{coagulation}$$

(4.27)

where *nucleation* describes the formation of new particles from the gas phase, *condensation/evaporation* describes the exchange of mass between the gas phase and the particles, and *coagulation* describes the formation of larger particles from the collision of smaller particles by Brownian diffusion. Representation of aerosol–cloud interactions may require additional terms.

The nucleation term is dependent on the supersaturation of species in the gas phase and on the stability of successively larger molecular clusters formed from these gas-phase molecules. Nucleation acts as a source of new particles at the low end of the size distribution ($\sim 10^{-3}$ μm). The condensation/evaporation term can be calculated from knowledge of the *condensation growth rate* $I(V) = dV/dt$ of particles of volume V. Consider a volume element $[V, V + dV]$. The particles growing into that volume element over time dt are those that were in the volume element $[V, V - I(V) dt]$; their number concentration is $n_N(V) I(V) dt$. Similarly, the number concentration of particles growing out of the volume element is $n(V + dV) I(V + dV) dt$. Thus the change in the volume distribution function is

$$\left[\frac{\partial n_N(V)}{\partial t}\right]_{condensation/evaporation} = -\frac{\partial(I(V)n_N(V))}{\partial V}$$

(4.28)

The coagulation term is defined by the frequency of collisions between particles (collision usually results in coagulation). We characterize the collision frequency by a *coagulation coefficient* $\beta(V, V')$ [m^3 particle^{-1}s^{-1}] representing the rate constant at which particles of volume V collide with particles of volume V'. In this manner, the rate of production of particles of volume $(V + V')$ by collision of particles of volume V and V' is given by $\beta(V, V') n_N(V) n_N(V') (dV)^2$. To determine the change with time in the number concentration of particles in the volume element $[V, V + dV]$ due to coagulation processes we consider collisions with particles over the entire range of the size distribution and account for both production and loss of particles out of the volume element:

$$\left[\frac{\partial n_N(V)}{\partial t}\right]_{coagulation} = \frac{1}{2}\int_0^V \beta(V', V - V') n_N(V') n_N(V - V') dV'$$
$$- n_N(V)\int_0^\infty \beta(V, V') n_N(V') dV'$$

(4.29)

where the ½ coefficient on the first term of the right-hand side is to avoid double-counting. Physical formulations for $I(V)$, $\beta(V, V')$, and aerosol nucleation are presented by Seinfeld and Pandis (2006).

Standard numerical algorithms for aerosol microphysics approximate the size distribution by a series of square functions called sections or "bins." These are called *sectional models*. Accurate representation of aerosol microphysics generally requires more than 30 bins for each chemical component of the aerosol, so including multiple components can become computationally cumbersome. An alternative is to make some assumption about the shape of the size distribution and track the evolution of the low-order moments of that distribution. See Section 5.6 for additional information on the methods used in aerosol models.

4.4 Atmospheric Lifetime and Characteristic Timescales

4.4.1 Atmospheric Lifetime

The *atmospheric lifetime* of a chemical species is defined as the average time that the species remains in the atmosphere before it is removed by one of its sinks. The term *residence time* is equivalently used, most often when the sink involves deposition to the surface. If a species i has a local mass concentration ρ_i [kg m^{-3}] and loss rate L_i [kg m^{-3} s^{-1}], the local atmospheric lifetime is given by

$$\tau_i = \frac{\rho_i}{L_i} \tag{4.30}$$

We are often interested in the mean atmospheric lifetime $\overline{\tau_i}$ averaged over some atmospheric domain V and time period T. This is obtained by summation of the concentrations and loss rates:

$$\overline{\tau_i} = \frac{\int\limits_T \int\limits_V \rho_i \, dv \, dt}{\int\limits_T \int\limits_V L_i \, dv \, dt} \tag{4.31}$$

For example, the *global mean atmospheric lifetime* is defined by summing over the whole atmosphere and a full annual cycle.

The atmospheric lifetime of a species is a very useful thing to know because it characterizes the spatial and temporal variability of its concentration. A species with a short lifetime will have strong concentration gradients defined by fluctuations of its sources and sinks, while a species with a long lifetime will be well mixed. A species with a short lifetime will respond rapidly to changes in sources or sinks, while a species with a long lifetime will respond more slowly.

The loss rate of a species is usually proportional to its concentration (first-order loss). We then write $L_i = l_i \rho_i$, where l_i [s^{-1}] is a loss rate coefficient, and derive $\tau_i = 1/l_i$. If there is no compensating production, the evolution of the concentration is expressed by

$$\frac{d\rho_i}{dt} + l_i \, \rho_i = 0 \tag{4.32}$$

with solution

$$\rho_i(t) = \rho_i(0) \exp\left[-l_i \, t\right] = \rho_i(0) \exp\left[-t/\tau_i\right] \tag{4.33}$$

Thus we see that τ_i defines a characteristic e-folding timescale for decay: $\rho_i(\tau) = \rho_i(0)/e$ = 0.37 $\rho_i(0)$. τ_i is sometimes called the *e-folding lifetime* to distinguish it from the *half-life* $t_{1/2}$ used in the radiochemistry literature to describe 50% loss of an initial amount of radioactive material. For a first-order loss, τ and $t_{1/2}$ are related by $\tau = t_{1/2}/\ln 2 = 1.44\ t_{1/2}$.

Different processes can contribute to the removal of a species from the atmosphere, including chemical loss, transport, wet scavenging, and surface uptake, as described by the different terms in the continuity equation (Section 4.2). If several processes contribute to the sink and all are first-order, the overall loss coefficient l_i is the sum of the loss coefficients for the individual processes. Thus we have

$$l_i = \sum_q l_{i,q} \tag{4.34}$$

where $l_{i,q}$ is the loss coefficient associated with process q. One can similarly define a lifetime $\tau_{i,q} = 1/l_{i,q}$ associated with process q. The overall lifetime τ resulting from all loss processes is

$$\frac{1}{\tau_i} = \sum_q \frac{1}{\tau_{i,q}} \tag{4.35}$$

Thus the loss coefficients for individual processes add in series while the corresponding atmospheric lifetimes add in parallel.

We now examine the formulation of lifetimes for individual processes. Consider first the chemical loss of species i by reaction with species j with a rate constant k. The corresponding chemical lifetime is

$$\tau_{chem} = \frac{1}{k\,\rho_j} \tag{4.36}$$

For a more general case where species i reacts with several species j (rate constants k_j) and also photolyzes with a photolysis frequency J, the chemical lifetime is given by

$$\tau_{chem} = \frac{1}{J + \sum_j k_j\,\rho_j} \tag{4.37}$$

Chemical lifetimes in the atmosphere range from less than a second for highly reactive radicals to practically infinite for noble gases.

Consider now the timescales for transport of species i. For one-dimensional advection along direction x with a velocity u, the continuity equation is

$$\frac{\partial \rho_i}{\partial t} + \frac{\partial (u\,\rho_i)}{\partial x} = 0 \tag{4.38}$$

Let us view advection as representing a sink for species i, such as ventilation from a source region. In that case we can define a loss coefficient associated with advection as per (4.30): $l_{adv} = \partial(u\,\rho_i)/\partial x\ /\ \rho_i$. We approximate the divergence term as $\partial(u\,\rho_i)/\partial x \approx u\,\rho_i/D$, where D represents a characteristic distance over which the transport flux varies. The lifetime associated with advection is then given by

$$\tau_{adv} = \frac{D}{u} \tag{4.39}$$

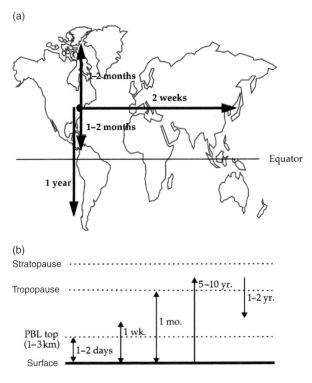

(a)

(b)

Stratopause

Tropopause

5–10 yr.

1–2 yr.

1 mo.

PBL top (1–3 km)

1 wk.

1–2 days

Surface

Figure 4.4   Typical time constants associated with global atmospheric transport. From Jacob (1999).

For a source region of dimension D, (4.39) defines the lifetime against ventilation out of the source region by the mean wind. In a more general case, advection acts as both a source and a sink for the concentration at a given location. τ_{adv} in (4.39) can be viewed more generally and usefully as a *transport timescale* for species i to be transported over a distance D.

For small-scale turbulent mixing parameterized with a turbulent diffusion coefficient K, the continuity equation is

$$\frac{\partial \rho_i}{\partial t} - \frac{\partial}{\partial x}\left[K\frac{\partial \rho_i}{\partial x}\right] = 0 \tag{4.40}$$

Again one can make an order-of-magnitude approximation of the diffusion term as $K\rho_i/D^2$, and the corresponding mixing timescale is

$$\tau_{mix} = \frac{D^2}{K} \tag{4.41}$$

Transport timescales can be derived from knowledge of the atmospheric circulation and also from observations of chemical tracers. Typical values for the troposphere are given in Figure 4.4. Horizontal transport is fastest in the longitudinal direction because of strong winds; thus air can circumnavigate the world in a given latitudinal band on a timescale of a month. Meridional mixing within a hemisphere takes longer, on a timescale of about three months, and exchange across hemispheres requires the order of a year because of the lack of a strong thermal forcing gradient across the Equator. Vertical transport in the troposphere is driven largely by buoyant mixing with characteristic timescales of a day for mixing of the planetary boundary layer (PBL, surface to 1–3 km) and a month for the full troposphere. Tropospheric air has a residence time of 5–10 years against transport to the stratosphere, while stratospheric air has on average a residence time of 1–2 years against transport to the troposphere.

The lifetime of species i against dry deposition (surface uptake) for a well-mixed atmospheric layer extending from the surface to altitude h is expressed as a function of the deposition velocity $w_{D,i}$ by

$$\tau_{dep} = \frac{h}{w_{D,i}} \tag{4.42}$$

For a gas removed efficiently at the surface, $w_{D,i}$ is of the order of 1 cm s^{-1} so that the lifetime against dry deposition in the PBL is of the order of a day.

The lifetime of water-soluble gases and aerosols against wet scavenging can be roughly estimated from the frequency of precipitation. It is typically of the order of a week in the lower troposphere, longer in the upper troposphere. The estimate is complicated by the coupling between atmospheric transport and precipitation and by differences in scavenging efficiencies for different forms of precipitation.

4.4.2 Relaxation Timescales in Response to a Perturbation

An important application of the concept of atmospheric lifetime is to determine the time required for the atmosphere to relax in response to a perturbation. Consider a species i initially at steady state to which an instantaneous perturbation $\Delta\rho_i(0)$ is applied at time $t = 0$. If the loss is linear with coefficient l_i, then the perturbation decays as $\Delta\rho_i(t) = \Delta\rho_i(0)\exp[-l_i t]$ and the corresponding timescale is defined by the species lifetime $\tau_i = 1/l_i$. The situation is more complicated if the loss is nonlinear due to chemical coupling between species. To analyze the behavior of a nonlinear coupled system, let us consider the general kinetics equation

$$\frac{d\boldsymbol{\rho}}{dt} = \mathbf{A}(\boldsymbol{\rho}) \tag{4.43}$$

in which $\boldsymbol{\rho}$ $(\rho_1, \rho_2, \dots \rho_N)^T$ represents the concentration vector for N chemical species and $\mathbf{A}(\boldsymbol{\rho})$ is an $N \times N$ matrix operator containing the kinetic expressions for the chemical production and destruction rates of each species. If a small perturbation $\Delta\boldsymbol{\rho}$ is applied to the system, the response of the perturbed state is determined by the linearized equation (Prather, 2007)

$$\frac{d(\boldsymbol{\rho} + \Delta\boldsymbol{\rho})}{dt} = \mathbf{A}(\boldsymbol{\rho} + \Delta\boldsymbol{\rho}) = \mathbf{A}(\boldsymbol{\rho}) + \mathbf{J}\,\Delta\boldsymbol{\rho} + \mathbf{O}\big(\Delta\boldsymbol{\rho}^2\big) \tag{4.44}$$

where the $(N \times N)$ *Jacobian matrix* $\mathbf{J}$ of operator $\mathbf{A}$ is the first term in the Taylor expansion of $\mathbf{A}(\boldsymbol{\rho} + \Delta\boldsymbol{\rho})$. We neglect the higher-order terms $\mathbf{O}(\Delta\boldsymbol{\rho}^2)$ in what follows. If the perturbation is an eigenvector of $\mathbf{J}$ with eigenvalue λ (Box 4.1), then

$$\frac{d\Delta\boldsymbol{\rho}}{dt} = \mathbf{J}(\boldsymbol{\rho})\Delta\boldsymbol{\rho} = \lambda\Delta\boldsymbol{\rho} \tag{4.45}$$

Box 4.1	Eigenvalues and Eigenvectors

The vector $\mathbf{a} = (a_1, \ldots a_N)^T$ is an *eigenvector* of the square matrix $\mathbf{A}$ ($N \times N$) if it satisfies the equation

$$\mathbf{Aa} = \lambda \mathbf{a}$$

for some scalar λ called an *eigenvalue* of $\mathbf{A}$. A non-degenerate matrix (determinant $|\mathbf{A}| \neq 0$) has N linearly independent eigenvectors forming a base and N eigenvalues (one for each eigenvector).

The above equation can be rearranged as

$$(\mathbf{A} - \lambda\mathbf{I})\mathbf{a} = 0$$

where $\mathbf{I}$ is the identity matrix. The equation must have an infinite number of solutions since any scalar multiplier of $\mathbf{a}$ is a solution. It follows that the matrix $\mathbf{A} - \lambda\mathbf{I}$ must be degenerate, so that its determinant must be zero:

$$|\mathbf{A} - \lambda\mathbf{I}| = 0$$

When expanded, this equation takes the form of a polynomial equation of Nth degree in λ. The N roots of this equation are the eigenvalues of the system. Replacing in $\mathbf{A}\,\mathbf{a} = \lambda\,\mathbf{a}$ then yields the corresponding eigenvector for each eigenvalue.

As an illustrative example, consider the matrix $\mathbf{A} = \begin{pmatrix} 2 & 1 \\ 1 & 2 \end{pmatrix}$

The eigenvalues are the roots of $\begin{vmatrix} 2-\lambda & 1 \\ 1 & 2-\lambda \end{vmatrix} = 0$ or

$$\lambda^2 - 4\lambda + 3 = 0$$

The roots of this quadratic equation are $\lambda = 1$ and $\lambda = 3$.

The eigenvectors $a = \begin{pmatrix} a_1 \\ a_2 \end{pmatrix}$ are obtained by solving the linear equation

$$\begin{pmatrix} 2 & 1 \\ 1 & 2 \end{pmatrix} \begin{pmatrix} a_1 \\ a_2 \end{pmatrix} = \lambda \begin{pmatrix} a_1 \\ a_2 \end{pmatrix}$$

for the two values of λ. We find $\begin{pmatrix} 1 \\ -1 \end{pmatrix}$ for $\lambda = 1$ and $\begin{pmatrix} 1 \\ 1 \end{pmatrix}$ for $\lambda = 3$.

and the solution is

$$\Delta\boldsymbol{\rho}(t) = \Delta\boldsymbol{\rho}(0) \exp(\lambda t) \tag{4.46}$$

If the N species are linearly independent, then $\mathbf{J}$ has full rank with N linearly independent eigenvectors $\mathbf{a}_j$ each with an eigenvalue λ_j ($j = 1, N$). In the general case, an initial perturbation $\Delta\boldsymbol{\rho}(0)$ can be decomposed on the basis of eigenvectors with coefficients α_j:

$$\Delta\boldsymbol{\rho}(0) = \sum_{j=1}^{N} \alpha_j \mathbf{a}_j \tag{4.47}$$

so that the time evolution of the perturbation is given by

$$\Delta\boldsymbol{\rho}(t) = \sum_{j=1}^{N} \alpha_j \mathbf{a}_j e^{\lambda_j t} \tag{4.48}$$

We see that the eigenvalues define the characteristic timescales for the responses to the perturbation, and the eigenvectors define the modes over which the timescales apply. If all eigenvalues have negative real components, then $\Delta\boldsymbol{\rho}$ will relax back to steady state over a suite of timescales $[-1/\lambda_1, \ldots -1/\lambda_N]$ that define the *characteristic timescales* of the system. This can be used in particular to identify the longest timescale for response. If any of the eigenvalues has a positive real component then an initial perturbation to that mode will grow exponentially with time and the system is unstable (it is *explosive*). If any of the eigenvalues has a non-zero imaginary component then a perturbation to that mode will induce an oscillation that may grow or decay depending on the sign of the real component. Eigenvalues for realistic mechanisms used in atmospheric chemistry models are all real and negative, so that the mechanisms are stable against perturbations. This should not be surprising as chemical systems generally follow Le Chatelier's principle ("any perturbation to equilibrium prompts an opposing response to restore equilibrium"). Oscillatory or unstable chemical behavior may occasionally occur under unusual conditions but these tend to be rapidly dissipated by model mixing.

4.5 Conservation Equations for Atmospheric Dynamics

Solution of the chemical continuity equation (4.1) requires information on winds and turbulence to compute the transport terms. This information must be provided by a meteorological model. We discussed in Chapter 1 how chemical transport models can operate either "online," integrated within the meteorological model, or "offline," using archived output from the meteorological model. Here we describe the conservation equations for atmospheric dynamics that form the basis of meteorological models.

The dynamical properties of the atmosphere are determined by the fundamental principles of mass, momentum, and energy conservation. Mass cannot be created nor destroyed, momentum can be changed only through the application of a force, and internal energy can be altered only through the existence of a heat source or sink, or by performance of work. The properties are described at any spatial location and time by six dependent variables: the pressure p [Pa], the density ρ_a [kg m^{-3}], the absolute temperature T [K], and the wind vector $\mathbf{v}$ [m s^{-1}] with its three components (u, v, and w).

Pressure, density, and temperature are related by the ideal gas law:

$$p = \rho_a R T \tag{4.49}$$

where R is the *gas constant* (287 J kg^{-1} K^{-1}). Solving the dynamical system requires five additional equations. These describe the evolution of mass, momentum (in the three spatial directions), and energy for the compressible fluid.

Since the Earth is rotating around its north–south axis, the most appropriate reference frame to describe air motions for an observer located on the Earth is not an inertial frame attached to the center of the planet, but rather a rotating coordinate system attached to the surface of the Earth. The geometric coordinates used in this case are denoted (x, y, z). Variables x and y represent geometric distances along parallels and meridians, respectively. Variable z is the geometric altitude. When spherical coordinates are used, the equations are expressed as a function of longitude λ and latitude φ rather than longitudinal and meridional distances x and y. We have

$$\frac{\partial}{\partial x} = \frac{1}{r \cos \varphi} \frac{\partial}{\partial \lambda} \tag{4.50}$$

$$\frac{\partial}{\partial y} = \frac{1}{r} \frac{\partial}{\partial \varphi} \tag{4.51}$$

$$\frac{\partial}{\partial z} = \frac{\partial}{\partial r} \tag{4.52}$$

where r represents the geometric distance from the center of the Earth. The three wind components, and specifically the curvilinear velocities along a latitude circle (u) and along a meridian (v), and the linear velocity along the vertical (w) are expressed as

$$u = r \cos \varphi \frac{d\lambda}{dt} \tag{4.53}$$

$$v = r \frac{d\varphi}{dt} \tag{4.54}$$

$$w = \frac{dr}{dt} \tag{4.55}$$

The total derivative of a quantity Ψ is expressed as

$$\frac{d\Psi}{dt} = \frac{\partial \Psi}{\partial t} + \mathbf{v} \cdot \nabla \Psi \tag{4.56}$$

In local geometric coordinates (x, y, z), the total derivative is written as

$$\frac{d\Psi}{dt} = \frac{\partial \Psi}{\partial t} + u \frac{\partial \Psi}{\partial x} + v \frac{\partial \Psi}{\partial y} + w \frac{\partial \Psi}{\partial z} \tag{4.57}$$

In spherical coordinates (λ, φ, z), an equivalent formulation of the total derivative is

$$\frac{d\Psi}{dt} = \frac{\partial \Psi}{\partial t} + \frac{u}{r \cos \varphi} \frac{\partial \Psi}{\partial \lambda} + \frac{v}{r} \frac{\partial \Psi}{\partial \varphi} + w \frac{\partial \Psi}{\partial z} \tag{4.58}$$

In many applications, the thickness of the atmosphere z is assumed to be small compared to the Earth's radius a. In this *shallow atmosphere approximation*, one assumes $z \ll a$ and $r = a$. For models extending to the upper atmosphere, this approximation may not be valid.

4.5.1 Mass

Mass conservation for air is expressed by the continuity equation. In flux form, it is written as

$$\frac{\partial \rho_a}{\partial t} + \nabla \cdot (\rho_a \mathbf{v}) = 0 \tag{4.59}$$

This equation can be expanded as

$$\frac{\partial \rho_a}{\partial t} + \mathbf{v} \cdot \nabla \rho_a + \rho_a \nabla \cdot \mathbf{v} = 0 \tag{4.60}$$

with the sum of the first two terms being equal to the total derivative of ρ_a. Thus we write

$$\frac{d\rho_a}{dt} + \rho_a \nabla \cdot \mathbf{v} = 0 \tag{4.61}$$

which shows that the change in the density of a fluid parcel is proportional to the velocity divergence. Under the shallow atmosphere approximation, (4.59) and (4.61) are expressed in spherical coordinates as

$$\frac{\partial \rho_a}{\partial t} + \frac{1}{a \cos \varphi} \left(\frac{\partial (\rho_a u)}{\partial \lambda} + \frac{\partial (\rho_a v \cos \varphi)}{\partial \varphi} \right) + \frac{\partial (\rho_a w)}{\partial z} = 0 \tag{4.62}$$

and

$$\frac{d\rho_a}{dt} + \frac{\rho_a}{a \cos \varphi} \left(\frac{\partial u}{\partial \lambda} + \frac{\partial (v \cos \varphi)}{\partial \varphi} \right) + \rho_a \frac{\partial w}{\partial z} = 0 \tag{4.63}$$

If the fluid is incompressible ($d\rho_a/dt = 0$), the continuity equation (4.61) becomes simply $\nabla \cdot \mathbf{v} = 0$. The velocity field is said to be non-divergent and the fluid density is conserved along the flow. Air is compressible and consequently this condition is not verified in the atmosphere.

4.5.2 Momentum

Newton's second law applied to a continuum medium (i.e., with infinitely divisible fluid parcels) states that the acceleration of a small fluid element results from (1) body forces f that affect the whole fluid element (gravity, and electromagnetic forces if the fluid is ionized) and (2) stress forces that act on the surface of the fluid element and represent interactions with the rest of the fluid. These stress forces are expressed as the gradient of a second-order stress tensor $\boldsymbol{\sigma}$. The balance of momentum in an inertial frame of reference (attached to the center of the Earth) is therefore provided by the differential equation (McWilliams, 2006; Neufeld and Hernandez-Garcia, 2010):

$$\rho_a \frac{d\mathbf{V}}{dt} = f + \nabla \cdot \boldsymbol{\sigma} \tag{4.64}$$

where ρ_a is the mass density of the fluid and $\mathbf{V}$ is the velocity in the inertial frame of reference. The only body force considered here is gravity ($f = \rho_a \, \mathbf{g_a}$). Further, one distinguishes between an isotropic term (diagonal elements of the stress tensor representing the pressure of the fluid: $\sigma_{ii} = -p$) and an anisotropic component expressed by a tensor $\boldsymbol{\tau}$ that accounts for interactions (viscosity) between the different fluid layers that move relative to each other. The resulting equation, known as the *Cauchy momentum equation*, is

$$\rho_a \frac{d\mathbf{V}}{dt} = f - \nabla p + \nabla \cdot \boldsymbol{\tau} \tag{4.65}$$

To solve this vector equation, an expression must relate the stress terms to the velocity of the flow. In the hypothesis of a Newtonian flow, the stress $\boldsymbol{\tau}$ is assumed to be proportional to the gradient of the velocity perpendicular to the shear. The proportionality coefficient μ [Pa s or kg m^{-1} s^{-1}] is known as the *dynamic viscosity* and is a scalar when the fluid is isotropic (otherwise, it is a tensor). If the fluid is incompressible with a constant scalar viscosity μ, the elements $\tau_{i,j}$ of the stress tensor are expressed as

$$\tau_{i,j} = \mu\left(\frac{\partial v_i}{\partial x_j} + \frac{\partial v_j}{\partial x_i}\right) \tag{4.66}$$

if x_j is the jth spatial coordinate and v_i is the fluid's velocity in the direction of axis i. The stress tensor is therefore related to the velocity vector by (Neufeld and Hernandez-Garcia, 2010):

$$\boldsymbol{\tau} = \mu\left(\boldsymbol{\nabla}\cdot\mathbf{V} + \boldsymbol{\nabla}\cdot\mathbf{V}^T\right) \tag{4.67}$$

The assumption of incompressibility ($\boldsymbol{\nabla}\cdot\mathbf{V} = 0$) is appropriate for compressible fluids such as air if the velocity of the flow corresponds to Mach numbers smaller than about 0.3. Under this assumption, one obtains the *Navier–Stokes equation* written here in an inertial frame of reference:

$$\rho_a\left(\frac{\partial\mathbf{V}}{\partial t} + (\mathbf{V}\cdot\boldsymbol{\nabla})\mathbf{V}\right) = \boldsymbol{f} - \boldsymbol{\nabla}p + \mu\boldsymbol{\nabla}^2\mathbf{V} \tag{4.68}$$

In the more general case of the compressible flows encountered in the atmosphere, the friction term takes a more complicated form, and the Navier-Stokes equation becomes

$$\rho_a\left(\frac{\partial\mathbf{V}}{\partial t} + (\mathbf{V}\boldsymbol{\cdot}\boldsymbol{\nabla}\mathbf{V})\right) = \boldsymbol{f} - \boldsymbol{\nabla}p + \mu\left[\boldsymbol{\nabla}^2\mathbf{V} + \frac{1}{3}\boldsymbol{\nabla}(\boldsymbol{\nabla}\boldsymbol{\cdot}\mathbf{V})\right]$$

with the viscosity μ assumed to have a constant value.

Here, we have expressed the material (total) derivative as the sum of the local acceleration $\partial\mathbf{V}/\partial t$ and the convective acceleration $(\mathbf{V}\cdot\boldsymbol{\nabla})\mathbf{V}$. This equation provides the three components of the velocity field $\mathbf{V}(\mathbf{r}, t)$ at a given point in space $\mathbf{r}$ and time t, and is used in fluid dynamics to address many different questions at various scales. The convective acceleration produced by spatial gradients in the velocity introduces a nonlinear component in the equation that can be the source of chaotic behavior known as turbulence (Box 4.2). The viscosity term operates as a linear diffusion term for momentum. The solution of this equation requires that appropriate conditions be prescribed at the boundary of the domain. The traditional condition adopted in many fluid dynamics problems is zero flow across solid boundaries and zero fluid velocity at the boundaries ("no-slip" condition).

For the rotating Earth, the velocity $\mathbf{V}$ of an air parcel at a location $\mathbf{r}$ and expressed in an inertial frame is equal to its velocity $\mathbf{v}$ relative to the Earth (the velocity that is measured by an observer located on the Earth) plus the velocity owing to the rotation of the Earth:

$$\mathbf{V} = \mathbf{v} + [\boldsymbol{\Omega} \times \mathbf{r}] \tag{4.69}$$

where $\boldsymbol{\Omega}$ is the Earth's angular velocity vector (directed from the south to the north pole and with an amplitude of 7.292×10^{-5} s^{-1}). We deduce that the absolute acceleration is

$$\frac{d\mathbf{V}}{dt} = \frac{d\mathbf{v}}{dt} + 2[\boldsymbol{\Omega} \times \mathbf{v}] + [\boldsymbol{\Omega} \times [\boldsymbol{\Omega} \times \mathbf{r}]] \tag{4.70}$$

Box 4.2   **The Navier–Stokes Equation, the Reynolds Number, and Turbulent Flows**

The nature of fluid flow can be assessed by comparing the importance of the inertial (or convective) term $(\mathbf{v}\cdot\nabla)\mathbf{v}$ with the viscosity (diffusive) term $\nu\nabla^2\mathbf{v}$ in the Navier–Stokes equation (4.68). Here $\nu = \mu/\rho_a$ represents the kinematic viscosity [m^2 s^{-1}]. To address this question, we define reduced quantities

$$\mathbf{v}' = \mathbf{v}/U \quad t' = tU/L \quad x_i' = x_i/L \quad p' = p/\rho U^2 \quad \mathbf{f}' = \mathbf{f}L/U^2$$

where L and U are characteristic length and velocity scales, and apply this transformation to the Navier–Stokes equation. Omitting the prime signs for clarity, we obtain the non-dimensional equation

$$\frac{\partial \mathbf{v}}{\partial t} + (\mathbf{v}\cdot\nabla)\mathbf{v} = -\frac{1}{\rho_a}\nabla p + \frac{1}{\rho_a}\mathbf{f} + \frac{1}{Re}\nabla^2\mathbf{v}$$

where

$$Re = \frac{(\mathbf{v}\cdot\nabla)\mathbf{v}}{\nu\nabla^2\mathbf{v}} \approx \frac{U^2/L}{\nu U/L^2} = \frac{UL}{\nu}$$

is the dimensionless *Reynolds number*, a measure of the ratio between the inertial force $U(U/L)$ and viscous forces ($\nu U/L^2$). Parameter $\nu = \mu/\rho$ [m^2 s^{-1}] denotes the kinematic viscosity with a value of 1.5×10^{-5} m^2 s^{-1} at 20 °C. Reynolds (1883) showed that as the value of Re increases, the motions become progressively more complex with a gradual transition from laminar to turbulent conditions. For low Reynolds numbers (less than 10), the steady-state flow results from a balance between pressure gradient and viscous forces. For larger Reynolds numbers, the nonlinear inertial term becomes dominant and chaotic eddies, vortices, and other instabilities are produced. Under this turbulent regime, viscosity is too small to dissipate the large-scale motions and the kinetic energy of the flow "cascades" progressively to smaller scales that are eventually dissipated by viscosity when their size becomes sufficiently small.

Atmospheric flows are characterized by very large Reynolds numbers. For a typical length scale of 1 km and a fluid velocity of 10 m s^{-1}, the Reynolds number is 7×10^8. The atmosphere can therefore be treated as a frictionless medium (inviscid fluid), prone to multi-scale turbulent motions. This is illustrated in Box 4.2 Figure 1 (b) with large-scale turbulence in the Jovian atmosphere. The solution of the momentum equation in this turbulent regime is very sensitive to small perturbations and to inaccuracies in the initial and boundary conditions, so that the predictability of the flow (i.e., weather) is limited to a few days. When describing small-scale motions such as those encountered at the interface between two fluid elements, the Reynolds number is considerably smaller; viscous dissipation leads to mixing between adjacent fluid elements, and the full Navier–Stokes equation needs to be considered. An example of turbulent mixing of initially separated chemical tracers is shown in Box 4.2 Figure 1 (a).

Deterministic representation of turbulent flow requires a direct numerical simulation (DNS) method in which the whole range of spatial and temporal scales is resolved with a ~1 mm-resolution grid and very small time steps. An alternative approach commonly adopted for PBL

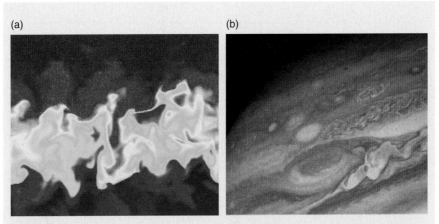

(a)   (b)

**Box 4.2
Figure 1**

(a): Concentration of two initially separated tracers mixing and reacting in an isotropic turbulent flow of a wind tunnel. The distribution is derived by the $512 \times 512 \times 1024$ grid point direct numerical simulation (DNS) of de Bruyn Kops *et al*. (2001). The image shows the wide range of length scales involved in the turbulent mixing process. See www.efluids.com/efluids/pages/gallery.htm. (b): Large-scale turbulent flow on Jupiter with irregular motions and the presence of vortices. From the National Aeronautics and Space Administration (NASA).

turbulence is the LES method in which large-scale eddy motions are resolved explicitly while the effects of small-scale eddies are parameterized. If we separate the small-scale, rapidly fluctuating components (eddy term denoted by prime) of the dependent variables (velocity, pressure, and density) from their large-scale, slowly varying components (mean term denoted by overbar) and ignore the Coriolis term, the three components of the Reynolds averaged Navier–Stokes (RANS) equation are written under the assumption of incompressibility as

$$\frac{\partial \bar{v}_i}{\partial t} + \sum_{j=1,2,3} \bar{v}_j \frac{\partial \bar{v}_i}{\partial x_j} = -\frac{1}{\rho_a}\frac{\partial \bar{p}}{\partial x_i} + \bar{F}_i + \nu \frac{\partial^2 \bar{v}_i}{\partial x_i^2} - \sum_{j=1,2,3} \frac{\partial \overline{v_j' v_i'}}{\partial x_j}$$

where $\bar{F}_i$ represents the components of the acceleration resulting from gravity and other body forces. The symmetric tensor $\overline{v_j' v_i'}$, called the Reynolds stress, characterizes the action of turbulent motions on the large-scale flow. To solve the equation, a "closure relation" that relates this eddy term to the mean flow must be specified. See Chapter 8 for more details.

Here $2[\mathbf{\Omega} \times \mathbf{v}]$ is the Coriolis acceleration, and $[\mathbf{\Omega} \times [\mathbf{\Omega} \times \mathbf{r}]] = -\Omega^2\mathbf{R}$ is the centripetal acceleration where $\mathbf{R}$ is the position vector perpendicular from the Earth's axis of rotation (Figure 4.5). When expressed relative to the rotating Earth, the Navier–Stokes equation (also called the *equation of motion*) becomes

$$\frac{\partial \mathbf{v}}{\partial t} + (\mathbf{v}\cdot\mathbf{\nabla})\mathbf{v} = -2[\mathbf{\Omega} \times \mathbf{v}] - \frac{1}{\rho_a}\mathbf{\nabla}p + \mathbf{g} + \mathbf{F}_{diss} \tag{4.71}$$

where $\mathbf{g} = \mathbf{g_a} + \Omega^2\mathbf{R}$ is the apparent gravitational acceleration (gravitational acceleration corrected by the centripetal force) and $\mathbf{F}_{diss}$ denotes the dissipation term resulting

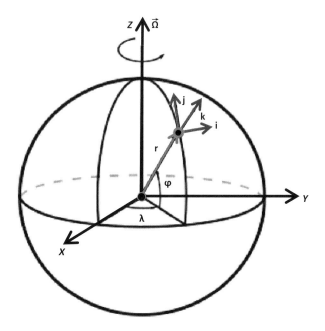

Coordinate systems: Inertial frame attached to the center of the Earth and rotating frame attached to the Earth's surface. A point in the atmosphere is determined by its longitude λ, its latitude φ, and the distance r from the center of the Earth. The altitude z is given by $r - a$, if a is the Earth's radius. The Earth rotation rate is Ω. From Brasseur and Solomon (2005).

from viscosity. In most atmospheric applications, the effect of molecular viscosity can be ignored and the corresponding equation, in which term $\mathbf{F}_{diss}$ is neglected, is referred to as the *Euler equation*. As we proceed, we will keep in the momentum equation a term $\mathbf{F}$ (vector with three components F_λ, F_φ, and F_z) to account for any possible momentum dissipation.

When expressed in spherical coordinates (λ, φ, z), the three components of the equation of motion become

$$\frac{du}{dt} = \left(2\Omega + \frac{u}{r\cos\varphi}\right)(v\sin\varphi - w\cos\varphi) - \left(\frac{1}{\rho_a\, r\cos\varphi}\right)\frac{\partial p}{\partial\lambda} + F_\lambda \qquad (4.72)$$

$$\frac{dv}{dt} = -\frac{uw}{r} - \left(2\Omega + \frac{u}{\cos\varphi}\right)u\sin\varphi - \left(\frac{1}{\rho_a\, r}\right)\frac{\partial p}{\partial\varphi} + F_\varphi \qquad (4.73)$$

$$\frac{dw}{dt} = \frac{u^2 + v^2}{r} + 2\Omega\; u\cos\varphi - g - \frac{1}{\rho_a}\frac{\partial p}{\partial z} + F_z \qquad (4.74)$$

It is important to note in the last equation (vertical projection of the momentum equation) that two terms (pressure gradient and gravity) are dominant, and this equation is often replaced by the *hydrostatic approximation* (see Section 2.6.1):

$$\frac{1}{\rho_a}\frac{\partial p}{\partial z} + g = 0 \qquad (4.75)$$

When neglecting the smaller terms and adopting the shallow atmosphere assumption (with $r = a$), we obtain

$$\frac{du}{dt} = \left[f + \frac{u \tan \varphi}{a} \right] v - \left(\frac{1}{\rho_a \, a \cos \varphi} \right) \frac{\partial p}{\partial \lambda} + F_\lambda \tag{4.76}$$

$$\frac{dv}{dt} = \left[f + \frac{u \tan \varphi}{a} \right] u - \left(\frac{1}{\rho_a a} \right) \frac{\partial p}{\partial \varphi} + F_\varphi \tag{4.77}$$

where $f = 2 \, \Omega \sin \varphi$ is the *Coriolis factor*, and a is the Earth's radius. The so-called metric terms in (4.72) to (4.74), uw/r, $[u^2 + v^2]/r$, $uv (\tan \varphi)/r$, $u^2 (\tan \varphi)/r$, arise from the spherical coordinate system. A scale analysis shows that these terms can be neglected when applied to mid-latitude systems. In this case, the equation for large-scale quasi-horizontal flows is expressed in the vector form

$$\frac{d\mathbf{v}_h}{dt} = -\frac{1}{\rho_a} \nabla_h p - f [\mathbf{k} \times \mathbf{v}_h] + \mathbf{F} \tag{4.78}$$

where $\mathbf{v}_h$ (u,v) represents the horizontal wind vector (i.e., wind along a surface of equal geometric height),

$$\mathbf{v}_h = u\mathbf{i} + v\mathbf{j} \tag{4.79}$$

$\mathbf{F}(F_\lambda, F_\varphi)$ is again the friction force, and ∇_h the horizontal gradient operator:

$$\nabla_h = \mathbf{i} \left(\frac{1}{a \cos \varphi} \right) \frac{\partial}{\partial \lambda} + \mathbf{j} \left(\frac{1}{a} \right) \frac{\partial}{\partial \varphi} \tag{4.80}$$

Here, $\mathbf{i}$, $\mathbf{j}$, and $\mathbf{k}$ are the unity vectors in the zonal (x), meridional (y), and vertical (z) directions.

By applying the $\mathbf{k} \cdot [\nabla_h \times]$ and $\nabla \cdot$ vector operators on the horizontal equations of motion, one obtains the *vorticity and divergence equations*, which are often used in atmospheric general circulation and weather prediction models as a replacement for the momentum equations. These are expressed by

$$\frac{\partial \zeta}{\partial t} = -\mathbf{v}_h \cdot \nabla_h (\zeta + f) - w \frac{\partial \zeta}{\partial z} - (\zeta + f) \nabla_h \cdot \mathbf{v}_h + \mathbf{k} \cdot \left[\nabla_h w \times \frac{\partial \mathbf{v}_h}{\partial z} \right]$$
$$+ \mathbf{k} \cdot \left[\nabla_h p \times \nabla_h \frac{1}{\rho_a} \right] + \mathbf{k} \cdot [\nabla_h \times \mathbf{F}] \tag{4.81}$$

and

$$\frac{\partial D}{\partial t} = -\mathbf{v}_h \bullet \{(\mathbf{v}_h \bullet \nabla_h) \mathbf{v}_h\} - \nabla_h \bullet [f \mathbf{k} \times \mathbf{v}_h] - \nabla_h w \bullet \frac{\partial \mathbf{v}_h}{\partial z} - w \frac{\partial D}{\partial z} - \nabla_h \bullet \left(\frac{1}{\rho_a} \nabla_h p \right) + \nabla_h \bullet \mathbf{F} \tag{4.82}$$

where the vertical projection of the curl of the horizontal velocity

$$\zeta = \mathbf{k} \bullet [\nabla_h \times \mathbf{v}_h] = \frac{\partial v}{\partial x} - \frac{\partial u}{\partial y} \tag{4.83}$$

is called the *relative vorticity* and

$$D = \mathbf{\nabla}_h \bullet \mathbf{v}_h = \frac{\partial u}{\partial x} + \frac{\partial v}{\partial y} \tag{4.84}$$

is the horizontal *divergence*. If we define the absolute vorticity as

$$\eta = \zeta + f \tag{4.85}$$

and note that the Coriolis parameter f is a function of latitude only, the rate of change of the absolute vorticity at a given point of the atmosphere is given by

$$\frac{\partial \eta}{\partial t} = -\mathbf{v}_h \cdot \mathbf{\nabla}_h \eta - w \frac{\partial \eta}{\partial z} - \eta \mathbf{\nabla}_h \cdot \mathbf{v}_h - \mathbf{k} \cdot \left[\mathbf{\nabla}_h w \times \frac{\partial \mathbf{v}_h}{\partial z} \right] + \mathbf{k} \cdot \left[\mathbf{\nabla}_h p \times \mathbf{\nabla}_h \frac{1}{\rho_a} \right] + \mathbf{k} \cdot [\mathbf{\nabla}_h \times \mathbf{F}]$$

$$\tag{4.86}$$

The first two terms on the right-hand side of the equation represent the horizontal and vertical advection of the absolute vorticity, respectively. The third term, called the *vortex stretching* term, describes the stretching or compression of vortex tubes in the vertical direction that results from a non-zero horizontal convergence or divergence of the flow. The fourth term, known as the *tilting term*, describes how the horizontal variations in the vertical velocity tend to tilt the horizontal components of the absolute vorticity vector toward the vertical direction. The fifth term, called the *solenoid term*, describes the effect of barocliniticity on the vertical component of the vorticity, and the last term represents the effects of friction.

Simpler forms of the vorticity equation ignore the smallest terms in the equation on the basis of a scale analysis or as a result of specific assumptions. For example, if one assumes that the fluid is incompressible (divergence of the velocity is equal to zero) and barotropic (no horizontal variation of the temperature, no vertical variation of wind velocities, so that surfaces of equal pressure coincide with surfaces of equal density; see Section 2.8), and if one neglects friction forces, one obtains the following equation

$$\frac{d\eta}{dt} = \frac{\partial \eta}{\partial t} + \mathbf{v}_h \cdot \mathbf{\nabla}_h \eta = -\eta \mathbf{\nabla}_h \cdot \mathbf{v}_h \tag{4.87}$$

which states that the rate of change of absolute vorticity following an air parcel is proportional to the convergence of the horizontal velocity. The first computer weather forecast by Charney *et al.* (1950) was based on the barotropic vorticity equation. Such models suffer from some severe restrictions (such as their inability to correctly generate cyclones), so that later models have been expressed by conserving the baroclinic terms in the governing equations.

Large-scale flows in the extratropics can also be represented by the *quasi-geostrophic* vorticity equation. This equation is established by decomposing the horizontal wind velocity $\mathbf{v}_h$ into a geostrophic component $\mathbf{v}_g$ (see Section 2.7 for its definition) and a remaining ageostrophic component $\mathbf{v}_{ag}$ of smaller amplitude. The actual wind in the momentum equation is approximated by its geostrophic component, except in the case of the divergence term. This exception (which explains why the approximation is known as the *quasi*-geostrophic approximation) is justified by the fact that the divergence of the geostrophic wind velocity is equal to zero and that, without accounting for the non-zero divergence of the ageostrophic wind component, no vertical wind could be generated. The resulting equation for the quasi-geostrophic relative vorticity, in which friction is ignored, is

$$\frac{\partial \zeta_g}{\partial t} + \mathbf{v}_g \cdot \boldsymbol{\nabla}_h \zeta_g + \beta \, \mathbf{v}_g = -f_0 \boldsymbol{\nabla}_h \cdot \mathbf{v}_{ag} \tag{4.88}$$

or, if the fluid is assumed to be incompressible,

$$\frac{\partial \zeta_g}{\partial t} + \mathbf{v}_g \cdot \boldsymbol{\nabla}_h \zeta_g + \beta \, \mathbf{v}_g = f_0 \frac{\partial w}{\partial z} \tag{4.89}$$

These two equations are based on the mid-latitude beta-plane approximation in which the Coriolis parameter $f = f_0 + \beta y$ is assumed to vary linearly with the geometric distance y measured along a meridian, and $f_0 = 2\,\Omega \sin \varphi_0$ denotes the value of parameter f at a point of reference at latitude φ_0 where $y = 0$.

Following Helmholtz's theorem, the horizontal wind vector $\mathbf{v}_h$ can be deduced from the vorticity and divergence by noting that it can be separated into two components

$$\mathbf{v}_h = [\mathbf{k} \times \boldsymbol{\nabla}_h \Psi] + \boldsymbol{\nabla}_h \chi \tag{4.90}$$

where Ψ is the *streamfunction* (which represents the nondivergent part of the flow whose value is constant along a streamline that follows the flow) and χ is the *velocity potential* (a scalar function whose gradient equals the velocity of an irrotational flow). It is straightforward to show that these scalar terms are related to the vorticity ζ and divergence D by

$$\zeta = \boldsymbol{\nabla}_h^2 \, \Psi \tag{4.91}$$

$$D = \boldsymbol{\nabla}_h^2 \, \chi \tag{4.92}$$

where the "horizontal" Laplacian operator is defined in spherical coordinates as

$$\boldsymbol{\nabla}_h^2 = \left(\frac{1}{a^2 \cos \varphi} \right) \frac{\partial}{\partial \varphi} \left(\cos \varphi \, \frac{\partial}{\partial \varphi} \right) + \left(\frac{1}{a^2 \cos^2 \varphi} \right) \frac{\partial^2}{\partial \lambda^2} \tag{4.93}$$

By combining relations (4.91) and (4.92) with the vorticity and divergence equations (4.81) and (4.82), one derives easily the two components of the horizontal wind velocity from the calculated streamfunction and velocity potential. The vertical wind component is obtained from the continuity equation. The *potential vorticity*, a conserved quantity in the absence of dissipative processes, is often used to diagnose atmospheric transport (see Box 4.3).

In the upper atmosphere, above approximately 100 km altitude, molecular viscosity (as measured by the viscosity coefficient μ) must be taken into account. Collisions between neutral particles and ions, whose motions are sensitive to electromagnetic fields, also affect the winds, especially at high latitudes. The effect of this *ion drag* is often assumed to be proportional to the difference between the neutral and ion winds. Thus, if $\mathbf{v}_{ion}$ represents the ion velocity (bulk motions and gyromotions generated by the electromagnetic fields) and ν_{ion} the ion-neutral collision frequency [s^{-1}], the momentum equation (4.71) becomes

$$\frac{d\mathbf{v}}{dt} = -2[\boldsymbol{\Omega} \times \mathbf{v}] - \frac{1}{\rho_a} \boldsymbol{\nabla} p + \mathbf{g} + \frac{\mu}{\rho_a} \frac{\partial^2 \mathbf{v}}{\partial z^2} + \nu_{ion}(\mathbf{v} - \mathbf{v}_{ion}) \tag{4.94}$$

The *relative vorticity* ζ defined in (4.83) is a measure of the spin of a fluid parcel relative to coordinates attached to the Earth. The *absolute vorticity* $\zeta + f$ is the sum of the spin of the fluid and the planetary vorticity (represented here by the Coriolis parameter f). It is not a conserved quantity since, even in the absence of dissipation, its tendency is proportional to the divergence of the motion field.

In the case of an incompressible fluid and in the absence of dissipative forces, the integration of the continuity equation over a height h separating two free surfaces of a fluid (often referred to as the *depth* of the fluid), shows that the *potential vorticity*

$$Z = \frac{(\zeta + f)}{h} = \frac{\eta}{h}$$

is conserved following the motion of a fluid parcel. This property implies that the relative vorticity must adjust in response to changes in the planetary vorticity (as the parcel is displaced in latitude) and to changes in the depth of the fluid.

For a compressible fluid such as air, it is possible to derive a similar conservation principle for the *Ertel potential vorticity* [$m^2\ s^{-2}\ K\ kg^{-1}$]

$$P = \frac{1}{\rho_a}[2\mathbf{\Omega} + \nabla \times \mathbf{v}]\cdot\nabla\theta$$

where ρ_a is the air density, $\mathbf{\Omega}$ is the angular velocity vector of the Earth's rotation, $\mathbf{v}$ the three-dimensional velocity field, and θ the potential temperature. Neglecting some minor terms, P is expressed in spherical coordinates as

$$P = \frac{1}{\rho_a}\left[-\frac{\partial v}{\partial z}\frac{\partial\theta}{a\cos\varphi\partial\lambda} + \frac{\partial u}{\partial z}\frac{\partial\theta}{a\partial\varphi} + \left(2\Omega\sin\varphi + \frac{\partial v}{a\cos\varphi\partial\lambda} - \frac{\partial(u\cos\varphi)}{a\cos\varphi\partial\varphi}\right)\frac{\partial\theta}{\partial z}\right]$$

where a is the Earth's radius. In the absence of friction and heat sources or sinks, the Ertel potential vorticity P is a materially conservative property. As a result, P is often used to diagnose transport processes in the atmosphere since, over relatively short timescales during which diabatic and other dissipative processes can be neglected, this quantity is an excellent tracer of fluid motion. It has become accepted to define $1.0 \times 10^{-6}\ m^2\ s^{-1}\ K\ kg^{-1}$ as one *potential vorticity unit* (1 PVU).

where, as above, $\mathbf{v}$ is the neutral wind velocity, $\mathbf{\Omega}$ the angular velocity of the Earth rotation vector, p the total pressure, and z the altitude.

4.5.3 Energy

The equation of energy is an expression of the first law of thermodynamics. It states that the energy supplied to an air parcel produces an increase in its internal energy ($c_v T$) or induces work by expansion. Thus per unit time, the energy conservation is expressed as

$$c_v \frac{dT}{dt} + p \frac{d}{dt}\left(\frac{1}{\rho_a}\right) = Q \qquad (4.95)$$

where Q [J kg^{-1} s^{-1}] is the *diabatic* heating term and c_v [J K^{-1} kg^{-1}] the specific heat at constant volume. Diabatic heating/cooling may be driven by the absorption/emission of radiative energy by atmospheric gases, or by condensation/evaporation of water vapor. The value of c_v for air at 0 °C is 717.5 J kg^{-1} K^{-1}. The second term in this expression accounts for the *work* done upon a unit mass of air by compression or expansion of the volume. The reciprocal density $1/\rho_a$ is termed the *specific volume* [m^3 kg^{-1}]. When using the ideal gas approximation, this energy conservation can be expressed as

$$c_p \frac{dT}{dt} - \left(\frac{1}{\rho_a}\right)\frac{dp}{dt} = Q \qquad (4.96)$$

where c_p [J K^{-1} kg^{-1}] is the specific heat at constant pressure ($c_p = c_v + R$ where R is the gas constant). Its value at 0 °C is thus 717.5 + 287 = 1004.5 J K^{-1} kg^{-1}. The second term expresses compression heating or expansion cooling associated with *adiabatic* processes taking place in the compressible fluid.

Making use of the hydrostatic equation (4.75), the energy equation becomes

$$\frac{dT}{dt} + w\frac{RT}{c_pH} = \frac{Q}{c_p} \qquad (4.97)$$

where H is the scale height. In spherical coordinates,

$$\frac{\partial T}{\partial t} + \frac{u}{r\cos\varphi}\frac{\partial T}{\partial \lambda} + \frac{v}{r}\frac{\partial T}{\partial \varphi} + w\left[\frac{\partial T}{\partial z} + \frac{RT}{c_pH}\right] = \frac{Q}{c_p} \qquad (4.98)$$

In the absence of significant *diabatic* processes ($Q = 0$) and horizontal temperature gradients, the vertical temperature profile is obtained from

$$\frac{\partial T}{\partial z} + \frac{RT}{c_pH} = 0 \qquad (4.99)$$

It is often useful to introduce the *potential temperature* θ, defined in Chapter 2,

$$\theta = T\left[\frac{p_0}{p}\right]^{\frac{R}{c_p}} \qquad (4.100)$$

whose vertical gradient $\partial\theta/\partial z = 0$ under adiabatic conditions. θ is a better marker of diabatic processes and atmospheric heat transport than the absolute temperature T. In this case, the energy equation takes the form

$$\frac{d\theta}{dt} = \left[\frac{p_0}{p}\right]^{\frac{R}{c_p}}\frac{Q}{c_p} \qquad (4.101)$$

where the ratio $R/c_p = 0.285$. This equation shows that, in the absence of diabatic processes, the potential temperature of an air parcel is a conserved quantity along the motion of the fluid. In spherical coordinates (4.101) becomes

$$\frac{\partial\theta}{\partial t} + \frac{u}{r\cos\varphi}\frac{\partial\theta}{\partial \lambda} + \frac{v}{r}\frac{\partial\theta}{\partial \varphi} + w\frac{\partial\theta}{\partial z} = \left[\frac{p_0}{p}\right]^{\frac{R}{c_p}}\frac{Q}{c_p} \qquad (4.102)$$

When heating and cooling are produced by radiative absorption and emission, the value of the net heating rate Q is derived from the divergence of the radiative flux calculated at each point of the atmosphere after spectral integration. See Chapter 5 for more details.

In the upper atmosphere, additional processes affect the energy budget and must be accounted for in the energy conservation equation. In the thermosphere, heat transport by conduction becomes important. Diabatic heating that must be considered explicitly includes absorption of shortwave solar radiation, chemical heating resulting from collisional deactivation of energetically excited species, and chemical heating from exothermic reactions. Joule heating, which arises from the dissipation of electric currents in the ionosphere, represents another important effect, especially during geomagnetic storms. Infrared emissions of atomic oxygen and nitric oxide are important cooling processes.

4.5.4 Primitive and Non-Hydrostatic Equations

Meteorological models are based on the equations presented in the previous sections. If, in these equations, one separates the horizontal and vertical components of the velocity vector ($\mathbf{v} = \mathbf{v}_h + w\mathbf{k}$), and assumes that the atmosphere is frictionless, the non-hydrostatic equations needed to derive the dependent variables $\mathbf{v}_h$ (u,v), w, p, ρ and T are

$$\frac{1}{\rho_a}\frac{d\rho_a}{dt} = -\left(\boldsymbol{\nabla}_h\mathbf{v}_h + \frac{\partial w}{\partial z}\right) \quad \text{continuity equation} \qquad (4.103)$$

$$\frac{d\mathbf{v}_h}{dt} = -\frac{RT}{p}\boldsymbol{\nabla}_h p - f[\mathbf{k}\times\mathbf{v}_h] \quad \text{horizontal projection of the momentum equation}$$

$$(4.104)$$

$$\frac{dw}{dt} = -\frac{RT}{p}\frac{\partial p}{\partial z} - g \quad \text{vertical projection of the momentum equation} \qquad (4.105)$$

$$c_p\frac{dT}{dt} - \frac{RT}{p}\frac{dp}{dt} = Q \quad \text{thermodynamic equation} \qquad (4.106)$$

$$p = \rho_a RT \quad \text{equation of state} \qquad (4.107)$$

This system of equations describes air motions in a dry atmosphere over a wide range of scales, including the propagation of planetary waves, inertia-gravity waves, and even acoustic (sound) waves. An additional equation expressing the mass conservation of water vapor is added to fully describe a moist atmosphere:

$$\frac{\partial(\rho_a\mu_w)}{\partial t} + \boldsymbol{\nabla}_h\cdot(\rho_a\mu_w\mathbf{v}_h) + \frac{\partial(\rho_a\mu_w w)}{\partial z} = \rho_a(E - C)$$

where μ_w is the mass mixing ratio of water vapor (usually referred to as specific humidity and denoted q in the meteorological literature). The source and sink terms E and C account for the evaporation and condensation processes and represent the influence of physical processes on the dynamics of the atmosphere.

With the exception of the equation of state, which is a *diagnostic* relation that relates the pressure, density, and temperature, all other equations are *prognostic* equations that can be integrated forward in time and predict the evolution of atmospheric variables if the initial meteorological situation is known.

When the vertical acceleration dw/dt can be neglected in comparison with other terms, the differential equation for the vertical velocity (4.105) is replaced by the diagnostic *hydrostatic balance approximation* (4.75). The resulting equations are referred to as the *primitive equations* because they are very close to the original system established in 1904 by Vilhelm Bjerknes, and on which numerical weather predictions were first attempted.

One of the difficulties encountered in the early integration of the dynamical equations was the generation of large-amplitude, fast-propagating waves (acoustic and gravity waves) of no meteorological significance that resulted from a lack of momentum balance in the adopted initial conditions. This issue explains why Lewis Fry Richardson failed in his attempt to provide a successful numerical weather forecast in the late 1910s: high-frequency acoustic waves generated large time derivatives in the numerical method that masked the time derivatives associated with the actual weather signal (see Section 1.4). Thirty years later, in 1950, Jule Charney, Agnar Fjörtoft, and John von Neumann solved the hydrostatic equations of motions by imposing an additional *quasi-geostrophic approximation*. This configuration leads to the elimination of acoustic and gravity waves, but retains larger patterns such as the Rossby (planetary) waves and mesoscale weather systems. Quasi-geostrophic models, known also as filtered-equation models, do not require that initial conditions be perfectly balanced.

The geostrophic approximation adopted in the early models is now regarded as inaccurate and unnecessarily restrictive; in addition, it is invalid in the tropics and therefore must be reserved for examining simple scientific questions in the extra-tropics. Many global models of the atmosphere are based on the primitive equations using the hydrostatic approximation, which is convenient for treating motions at horizontal scales larger than about 10 km. A disadvantage of the hydrostatic models is that they cannot be applied to simulate small-scale processes such as convection, storms, or mountain waves, which are characterized by large vertical accelerations. High-resolution weather prediction and even global atmospheric models are increasingly based on *non-hydrostatic* equations in which the full vertical momentum equation (4.105) is retained.

Non-hydrostatic models are considerably more complex and computationally more demanding than the hydrostatic models. In addition, they generate fast-propagating acoustic waves, whose propagation depends on the compressibility of the fluid. These waves must be eliminated in order to avoid the use of prohibitively small time steps to solve the equations. This can be achieved by adopting an incompressibility assumption ($d\rho_a/dt = 0$), which is appropriate for shallow atmospheric circulations (vertical depth of the motion considerably smaller than the typical horizontal scale of the circulation). In this case, the continuity equation reduces to the prognostic equation

$$\nabla_h \cdot \mathbf{v}_h + \partial w / \partial z = 0$$

(non-divergent wind). Equations for shallow layers are often considered as prototypes for the primitive equations, and used to test numerical methods.

Another approach to filter acoustic waves, not limited to shallow flows, is to assume that the air density varies with height (denoted $\rho_0(z)$), but does not change locally with time ($\partial\rho_a/\partial t = 0$). In this case, the continuity equation (4.103) for fully compressible (elastic) air is replaced by the diagnostic *anelastic* continuity equation

$$\nabla_h\cdot(\rho_0\mathbf{v}_h + \partial(\rho_0 w)/\partial z) = 0$$

This particular approximation has been widely adopted for non-hydrostatic atmospheric models.

Finally, in the *Boussinesq approximation*, which also eliminates acoustic waves, the variables representing density and pressure in the equations of motions are separated into a reference state in hydrostatic balance, and small perturbations from this basic state. In the resulting equations, these perturbations are neglected, except in the buoyancy (gravity) term of the momentum equation.

4.6 Vertical Coordinates

So far we have expressed the equations for chemical continuity and atmospheric dynamics using the geometric altitude z [m] as vertical coordinate. This is the obvious choice for vertical coordinate but generally not the best. Other vertical coordinates can be used that are single-valued, monotonic functions of z. This is the case, for example, of the atmospheric pressure p (under the hydrostatic approximation). The following rule (Kasahara, 1974) allows us to transform the equations to arbitrary coordinates $\eta(\lambda, \varphi, z, t)$ defined as a function of longitude λ, latitude φ, altitude z, and time t. For any dependent variable $\Psi(\lambda, \varphi, z, t)$, we write (Lauritzen *et al.*, 2011):

$$\left[\frac{\partial\Psi}{\partial s}\right]_z = \left[\frac{\partial\Psi}{\partial s}\right]_\eta + \frac{\partial\Psi}{\partial\eta}\left[\frac{\partial\eta}{\partial s}\right]_z \tag{4.108}$$

where s can be λ, φ, or t. We deduce that

$$\nabla_z\Psi = \nabla_\eta\Psi + \frac{\partial\Psi}{\partial\eta}\nabla_z\eta \tag{4.109}$$

Similarly

$$\nabla_\eta\Psi = \nabla_z\Psi + \frac{\partial\Psi}{\partial z}\nabla_\eta z \tag{4.110}$$

The total derivative of Ψ is expressed as

$$\frac{d\Psi}{dt} = \left[\frac{\partial\Psi}{\partial t}\right]_\eta + (\mathbf{v}_\eta\cdot\nabla_\eta)\Psi + \frac{d\eta}{dt}\frac{\partial\Psi}{\partial\eta} \tag{4.111}$$

where $\mathbf{v}_\eta$ is the projection of the velocity on a η-surface, $\dot{\eta} = d\eta/dt$ is the "vertical" velocity in the η coordinate system, and ∇_η is the "horizontal" gradient calculated on a surface of constant η. The total derivative $d\Psi/dt$ is a property of the fluid and is therefore independent of the coordinate system.

Different vertical coordinates that can be considered besides the geometric altitude include (1) pressure coordinates, particularly attractive in the case of hydrostatic models, (2) terrain-following coordinates in which the application of lower boundary conditions is facilitated, (3) hybrid terrain-following variants that are terrain-following in the lowest levels of the atmosphere and pressure-following earlier, and (4) isentropic coordinates that approximate Lagrangian coordinates when diabatic processes are weak. We review these different coordinate systems next.

4.6.1 Pressure Coordinate System

Meteorologists often use atmospheric pressure to express the vertical variations of the dependent variables. This is the *pressure* or *isobaric coordinate system*. It is the most universally used coordinate in dynamical models since, under hydrostatic assumptions, the forms of some of the dynamical equations become particularly simple. Already at the beginning of the twentieth century, Vilhelm Bjerknes was establishing his synoptic charts on isobaric surfaces, and today meteorological variables (temperature, geopotential, etc.) are often represented as a function of atmospheric pressure. The relation between pressure and altitude is expressed by the hydrostatic approximation

$$dp = -\rho_a g \ dz \tag{4.112}$$

and the vertical velocity ω in the pressure coordinate system is defined as

$$\omega = \frac{dp}{dt} \tag{4.113}$$

For any arbitrary vertical coordinate η, the "horizontal" gradient $\nabla_\eta \Psi$ of a function Ψ is related to the horizontal gradient of this function $\nabla_z \Psi$ expressed in the geometric coordinate framework z by (see (4.110))

$$\nabla_z \Psi = \nabla_\eta \Psi - \frac{\partial \Psi}{\partial z} \nabla_\eta z \tag{4.114}$$

We deduce when $\eta = p$ and when applying the hydrostatic approximation that

$$\nabla_z p = -\frac{\partial p}{\partial z} \nabla_p z = \rho_a g \nabla_p z \tag{4.115}$$

since $\nabla_p p = 0$. The continuity equation expressed in pressure coordinates takes a particularly elegant expression. Separating in the continuity equation (4.59) the contributions provided by the horizontal ($\mathbf{v}_z$) and vertical (w) winds

$$\frac{\partial \rho_a}{\partial t} + \nabla_z \cdot (\rho_a \mathbf{v}_z) + \frac{\partial (\rho_a w)}{\partial z} = 0 \tag{4.116}$$

and using the hydrostatic equilibrium relation to replace ρ_a by $(-1/g \ \partial p/\partial z)$ we find the remarkable result

$$\nabla_p \cdot \mathbf{v}_p + \frac{\partial \omega}{\partial p} = 0 \tag{4.117}$$

where $\mathbf{v}_p$ (u, v) is the "horizontal" wind (components along isobaric surfaces) and $\omega = dp/dt$ is the previously defined vertical velocity in the pressure coordinate system. This form of the continuity equation contains no reference to the air density and is purely diagnostic (no time derivative). Its simplicity is one of the advantages of the pressure coordinate system. In spherical coordinates, the continuity equation is written as

$$\frac{1}{a\cos\varphi}\left[\frac{\partial u}{\partial\lambda}+\frac{\partial(v\cos\varphi)}{\partial\varphi}\right]+\frac{\partial\omega}{\partial p}=0 \qquad (4.118)$$

By integrating (4.117), we find the *isobaric tendency equation* that provides the vertical velocity ω at any pressure level p by integrating the isobaric divergence of the horizontal wind:

$$\omega=-\int_0^p \mathbf{\nabla}_p\cdot\mathbf{v}_p \, dp \qquad (4.119)$$

In pressure coordinates, the state of the atmosphere is often represented by the geopotential Φ on isobaric surfaces (rather than the pressure on equal altitude levels). We note that

$$d\Phi=gdz=-\frac{1}{\rho_a}dp=-RT\frac{dp}{p}=\frac{RT}{H_g}dz \qquad (4.120)$$

In pressure coordinates, the momentum equation is expressed as

$$\frac{d\mathbf{v}}{dt}=-2[\mathbf{\Omega}\times\mathbf{v}]-\mathbf{\nabla}_p\Phi+\mathbf{g}+\mathbf{F} \qquad (4.121)$$

or in spherical coordinates

$$\frac{du}{dt}=\left(f+\frac{u\tan\varphi}{a}\right)v+\left(\frac{1}{a\cos\varphi}\right)\frac{\partial\Phi}{\partial\lambda}+F_\lambda \qquad (4.122)$$

$$\frac{dv}{dt}=-\left(f+\frac{u\tan\varphi}{a}\right)u+\frac{\partial\Phi}{a\partial\varphi}+F_\varphi \qquad (4.123)$$

where u and v represent here the wind along *isobaric surfaces.*

The general form of the energy equation

$$\frac{d\theta}{dt}=\left[\frac{p_0}{p}\right]^{\frac{R}{c_p}}\frac{Q}{c_p} \qquad (4.124)$$

remains unchanged in pressure coordinates. In this and in the momentum equation, the total derivative of a function Ψ under the shallow atmosphere approximation is expressed as

$$\frac{d\Psi}{dt}=\left[\frac{\partial\Psi}{\partial t}\right]_p+\mathbf{v}_p\cdot\mathbf{\nabla}_p\Psi+\omega\frac{\partial\Psi}{\partial p} \qquad (4.125)$$

where $\mathbf{v}_p(u, v)$ represents the "horizontal" velocity vector on an isobaric surface. In spherical coordinates, this expression becomes

$$\frac{d\Psi}{dt} = \left[\frac{\partial\Psi}{\partial t}\right]_p + \frac{u}{a\cos\varphi}\frac{\partial\Psi}{\partial\lambda} + \frac{v}{a}\frac{\partial\Psi}{\partial\varphi} + \omega\frac{\partial\Psi}{\partial p} \qquad (4.126)$$

Here, the derivatives $\partial/\partial\lambda$ and $\partial/\partial\varphi$ are calculated along isobaric surfaces rather than along a constant altitude surface.

4.6.2 Log-Pressure Altitude Coordinate System

The dynamical equations can also be simplified in the *log-pressure coordinate system* by defining a log-pressure altitude (Chapter 2)

$$Z = -\mathcal{H}\ln\left(\frac{p}{p_0}\right) \qquad (4.127)$$

where $\mathcal{H}$ is a constant *effective scale height*. This vertical coordinate is often used for stratospheric models. We have

$$\frac{dZ}{\mathcal{H}} = \frac{dz}{H(z)} \qquad (4.128)$$

where $H(z) = kT/mg$ is the atmospheric scale height (*m*). The total time derivative of function Ψ is now expressed as

$$\frac{d\Psi}{dt} = \left[\frac{\partial\Psi}{\partial t}\right]_Z + \mathbf{v}_Z\cdot\nabla_Z\Psi + \frac{1}{\rho_0}\frac{\partial(\rho_0\Psi)}{\partial Z} \qquad (4.129)$$

or

$$\frac{d\Psi}{dt} = \left[\frac{\partial\Psi}{\partial t}\right]_Z + \frac{u}{a\cos\varphi}\frac{\partial\Psi}{\partial\lambda} + \frac{v}{a}\frac{\partial\Psi}{\partial\varphi} + \frac{1}{\rho_0}\frac{\partial(\rho_0\Psi)}{\partial Z} \qquad (4.130)$$

where $\mathbf{v}_z$ is the wind vector on a constant pressure–altitude surface and

$$\rho_0(Z) = \rho_0(0)\exp\left(-\frac{Z}{\mathcal{H}}\right) \qquad (4.131)$$

Components u and v stand here for the wind components along isobaric surfaces or equivalently for given pressure altitude (Z) surfaces. By choosing $\mathcal{H}$ to be equal to 7 km, the value of Z (which corresponds to a given pressure level) is *approximately* equal to the value of the geometric altitude z.

When the log-pressure representation is used, the hydrostatic equation takes the form

$$\frac{\partial\Phi}{\partial Z} = \frac{RT}{\mathcal{H}} \qquad (4.132)$$

and the continuity equation becomes

$$\frac{1}{a\cos\varphi}\left(\frac{\partial u}{\partial\lambda} + \frac{\partial(v\cos\varphi)}{\partial\varphi}\right) + \frac{1}{\rho_0}\frac{\partial(\rho_0 w)}{\partial Z} = 0 \qquad (4.133)$$

Newton's second law is expressed as in (4.122) and (4.123).

4.6.3 Terrain-Following Coordinate Systems

A difficulty in using geometric height or pressure as the vertical coordinate is the treatment of surface topography. To address this issue, one can introduce a normalized pressure coordinate (commonly called *sigma coordinate*)

$$\sigma = \frac{p - p_{top}}{p_s - p_{top}} \tag{4.134}$$

where p_s is the pressure at the Earth's surface (which varies with the topography) and p_{top} is the pressure at the highest level of the model. The sigma coordinate is illustrated in Figure 4.6. The value of σ varies from zero at the top of the model ($p = p_{top}$) to unity at the surface ($p = p_s$), where a simple boundary condition $d\sigma/dt = 0$ is applied. In the original definition of the sigma coordinate by Phillips (1957), the top of the model was assumed to be the top of the atmosphere ($p_{top} = 0$) with therefore $\sigma = p/p_s$. The advantage of the sigma coordinate system is that it conforms to the natural terrain and therefore eliminates the problem of intersection with the ground when the terrain is not flat. It is particularly well suited for the boundary layer.

In the sigma coordinate system, the total derivative of a function Ψ is expressed as

$$\frac{d\Psi}{dt} = \left[\frac{\partial\Psi}{\partial t}\right]_\sigma + \mathbf{v}_\sigma\cdot\nabla_\sigma\Psi + \frac{d\sigma}{dt}\frac{\partial\Psi}{\partial\sigma} \tag{4.135}$$

where $\mathbf{v}_\sigma$ is the "horizontal" velocity along a sigma surface and $d\sigma/dt$ is the vertical velocity. Alternatively, we can write

Figure 4.6   Sigma coordinate levels above a region with variable topography. Courtesy of Martin Schultz, Forschungszentrum Jülich.

$$\frac{d\Psi}{dt} = \left[\frac{\partial\Psi}{\partial t}\right]_\sigma + \frac{u}{a\cos\varphi}\frac{\partial\Psi}{\partial\lambda} + \frac{v}{a}\frac{\partial\Psi}{\partial\varphi} + \frac{d\sigma}{dt}\frac{\partial\Psi}{\partial\sigma} \qquad (4.136)$$

where the derivatives $\partial\Psi/\partial\lambda$ and $\partial\Psi/\partial\varphi$ are taken along a constant sigma surface. If $\pi = p_s - p_{top}$, the horizontal equations of motion become

$$\frac{dv}{dt} = -f[\mathbf{k}\times\mathbf{v}] - \nabla_\sigma\Phi + \sigma\,\nabla_\sigma\ln\pi\frac{\partial\Phi}{\partial\sigma} + \mathbf{F} \qquad (4.137)$$

the hydrostatic equation

$$\frac{\partial\Phi}{\partial\sigma} = -\frac{\pi\,R\,T}{p_{top} + \pi\,\sigma} = -\frac{\pi\,R\,\theta}{\left(p_{top} + \pi\,\sigma\right)(p_0/p)^{\frac{R}{c_p}}} \qquad (4.138)$$

the continuity equation

$$\frac{\partial\pi}{\partial t} + \nabla_\sigma\cdot(\pi\mathbf{v}_\sigma) + \frac{\partial}{\partial\sigma}\left(\pi\frac{d\sigma}{dt}\right) = 0 \qquad (4.139)$$

and the thermodynamic energy equation

$$\frac{\partial\theta}{\partial t} + \mathbf{v}_\sigma\cdot\nabla_\sigma\theta + \frac{d\sigma}{dt}\frac{\partial\theta}{\partial\sigma} = \left(\frac{p_0}{p}\right)^{\frac{R}{c_p}}\frac{Q}{c_p} \qquad (4.140)$$

Even though sigma surfaces do not intersect the ground, the use of the sigma coordinate system requires some precautions. Large errors can occur in calculating pressure gradients in regions of complex topography because the constant-sigma surfaces are steeply sloped. To address this problem, Mesinger (1984) introduced the step-mountain coordinate, commonly called *eta coordinate*, as

$$\eta = \frac{p - p_{top}}{p_s - p_{top}}\frac{p_{ref}(z_s) - p_{top}}{p_{ref}(0) - p_{top}} = \sigma\frac{p_{ref}(z_s) - p_{top}}{p_{ref}(0) - p_{top}} \qquad (4.141)$$

where $p_{ref}(z)$ is a reference pressure defined as a function of the geometric height z (e.g., pressure in the standard atmosphere with $p_{ref}(0) = 1013$ hPa), and z_s is the local terrain elevation. The scaling factor applied to the sigma coordinate ensures that the η surfaces are quasi horizontal.

The influence of topography on the mean flow decreases with altitude, so that the sigma-coordinate system is less desirable in the upper troposphere and above the tropopause. Models frequently use *hybrid σ–p coordinate systems* (Figure 4.7) that follow terrain in the lower troposphere and transition gradually to follow pressure in the stratosphere. The pressure p_k at model vertical level k ($k = 1, K$) is given by

$$p_k = A_k p_0 + B_k p_s \qquad (4.142)$$

where coefficients A_k and B_k have values that depend only on k. Parameter p_0 is chosen to be equal to the pressure at sea level (1013 hPa). The surface pressure p_s varies along the topography of the Earth's surface and may also vary with time. At the surface ($k = 1$), $A_1 = 0$ and $B_1 = 1$, while at the top of the model domain ($k = K$ and $p_K = p_{top}$), one imposes $A_K = p_{top}/p_0$ and $B_K = 0$. The value of B_k (equal to the value of σ at the surface) decreases with height, typically down to zero in the upper troposphere or in the stratosphere.

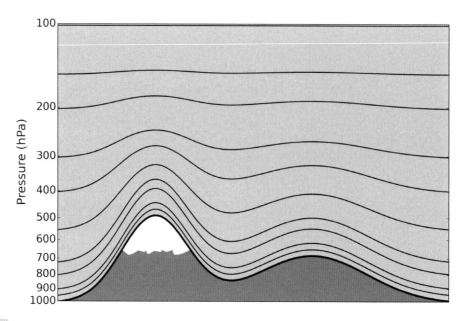

Figure 4.7 Representation of a hybrid sigma–pressure vertical coordinate system with coefficients $A_k = [0.0, 0.0, 0.0, 0.02, 0.1, 0.15, 0.18, 0.16, 0.14, 0.1, 0.05]$ and $B_k = [1.0, 0.95, 0.90, 0.80, 0.70, 0.45, 0.25, 0.12, 0.04, 0.01, 0.0, 0.0]$. Courtesy of Martin Schultz, Forschungszentrum Jülich.

4.6.4 Isentropic Coordinate System

In models that focus on the middle atmosphere, the vertical coordinate is sometimes chosen to be the potential temperature θ (Figure 4.8). The advantage is that, under adiabatic conditions, the flow follows isentropic surfaces and is therefore simple to analyze. In this *isentropic coordinate system*, the total derivative of a function Ψ is written as

$$\frac{d\Psi}{dt} = \left[\frac{\partial \Psi}{\partial t}\right]_\theta + \mathbf{v}_\theta \cdot \nabla_\theta \Psi + \frac{d\theta}{dt}\frac{\partial \Psi}{\partial \theta} \tag{4.143}$$

where $\mathbf{v}_\theta$ (u,v) is the "horizontal" wind vector on an isentropic surface, and in spherical coordinates

$$\frac{d\Psi}{dt} = \left[\frac{\partial \Psi}{\partial t}\right]_\theta + \frac{u}{a\cos\varphi}\frac{\partial \Psi}{\partial \lambda} + \frac{v}{a}\frac{\partial \Psi}{\partial \varphi} + \frac{d\theta}{dt}\frac{\partial \Psi}{\partial \theta} \tag{4.144}$$

In this case the "vertical velocity" $d\theta/dt$ is directly proportional to the net diabatic heating rate Q. The "horizontal" momentum equations are expressed as

$$\frac{du}{dt} = \left(f + \frac{u\tan\varphi}{a}\right)v - \left(\frac{1}{a\cos\varphi}\right)\frac{\partial \Psi}{\partial \lambda} + F_\lambda \tag{4.145}$$

$$\frac{dv}{dt} = -\left(f + \frac{u\tan\varphi}{a}\right)u - \frac{\partial \Psi}{a\partial \varphi} + F_\varphi \tag{4.146}$$

where $\Psi = c_p T + \Phi$ is now the *Montgomery streamfunction*.

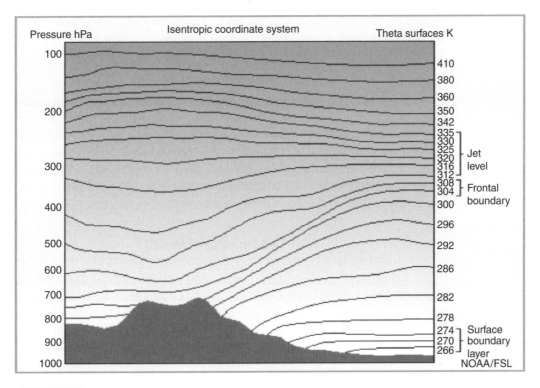

Figure 4.8   Isentropic coordinate system. From the National Oceanic and Atmospheric Administration (NOAA).

The hydrostatic relation takes the form

$$\frac{\partial \Psi}{\partial \varphi} = c_p \frac{T}{\theta} \tag{4.147}$$

and the continuity equation is written as

$$\frac{\partial \tilde{\sigma}}{\partial t} + \frac{1}{a \cos \varphi} \left(\frac{\partial (\tilde{\sigma}\, u)}{\partial \lambda} + \frac{\partial (\tilde{\sigma}\, v\, \cos \varphi)}{\partial \varphi} \right) + \frac{\partial (\tilde{\sigma}\, d\theta/dt)}{\partial \theta} = 0 \tag{4.148}$$

where

$$\tilde{\sigma} = -\frac{1}{g} \frac{\partial p}{\partial \theta} \tag{4.149}$$

is the *pseudo-density*.

Air motions under diabatic conditions are "horizontal" in the isentropic coordinate system. The surfaces of constant species mixing ratio (isopleths) tend to align themselves with the isentropic surfaces and the 3-D advection becomes essentially a 2-D problem. Isentropic coordinates are therefore convenient to analyze tracer motions over timescales of 1–2 weeks since air parcels remain close to their isentropes over such a period of time. Numerical noise is also reduced.

Although attractive for the stratosphere, the isentropic coordinate system is problematic in the lower troposphere because the flow is diabatic, θ is not a monotonous declining function of z, and the isentropes intersect the Earth's surface (see

Figure 4.8). A hybrid system with isentropic coordinate above the boundary layer and sigma coordinate near the Earth's surface is sometimes adopted.

4.7 Lower-Dimensional Models

The computational cost of a model can be decreased considerably by reducing its dimensionality. Although a 3-D representation of the atmosphere is most realistic and often necessary, there are many cases where simpler 2-D (zonal mean), 1-D(column), and 0-D (box) models can provide valuable insights (Figure 4.9). We discuss here the foundations of these lower-dimensional models and their applications.

4.7.1 Two-Dimensional Models

Global 2-D (latitude–altitude) models of the middle and upper atmosphere have been used extensively to simulate the meridional distribution of ozone and other chemical species. The motivation for their use is that zonal (longitudinal) gradients are generally weak, so that resolving the longitudinal dimension would add little information. In 2-D models the dependent variables Ψ such as concentrations, temperature, wind velocity, etc. are separated according to their zonally averaged value

$$\bar{\Psi}(\varphi, z, t) = \frac{1}{2\pi} \int_0^{2\pi} \Psi(\lambda, \varphi, z, t) \, d\lambda \qquad (4.150)$$

and the departure $\Psi'(\lambda, \varphi, z, t)$ from this mean value. Thus

$$\Psi(\lambda, \varphi, z, t) = \bar{\Psi}(\varphi, z, t) + \Psi'(\lambda, \varphi, z, t) \qquad (4.151)$$

By introducing this type of variable separation into the equations presented in Sections 4.5 and 4.6, we obtain the following continuity equation

$$\frac{1}{a \cos \varphi} \frac{\partial (\bar{v} \cos \varphi)}{\partial \varphi} + \frac{1}{\rho_0} \frac{\partial \rho_0 \bar{w}}{\partial Z} = 0 \qquad (4.152)$$

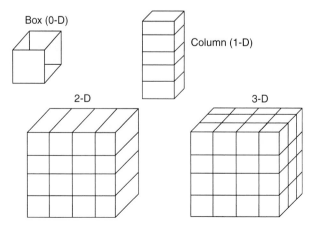

Box (0-D)

Column (1-D)

2-D

3-D

Figure 4.9   Conceptual representation of 0-D, 1-D, 2-D, and 3-D models. From Irina Sokolik, personal communication.

the zonal mean momentum equation

$$\frac{\partial \overline{u}}{\partial t} - \overline{v}\left\{ f - \left(\frac{1}{\cos\varphi}\right)\frac{\partial(\overline{u}\cos\varphi)}{a\partial\varphi} \right\} + \overline{w}\frac{\partial \overline{u}}{\partial Z} = -\frac{1}{a^2\cos\varphi}\frac{\partial\left(\overline{u'v'}\cos^2\varphi\right)}{a\partial\varphi} - \frac{1}{\rho_0}\frac{\partial\left(\rho_0\overline{u'v'}\right)}{\partial Z} \tag{4.153}$$

the zonal mean thermodynamic equation

$$\frac{\partial \overline{\theta}}{\partial t} + \overline{v}\frac{\partial \overline{\theta}}{a\partial\varphi} + \overline{w}\frac{\partial \overline{\theta}}{\partial Z} = -\frac{1}{\cos\varphi}\frac{\partial\left(\overline{v'\theta'}\cos\varphi\right)}{a\partial\varphi} - \frac{1}{\rho_0}\frac{\partial\left(\rho_0\overline{w'\theta'}\right)}{\partial Z} + \left[\frac{p_0}{p}\right]^{\frac{R}{c_p}}\frac{Q}{c_p} \tag{4.154}$$

and the zonal mean continuity equation for chemical species

$$\frac{\partial \overline{\mu}}{\partial t} + \overline{v}\frac{\partial \overline{\mu}}{a\partial\varphi} + \overline{w}\frac{\partial \overline{\mu}}{\partial Z} = -\frac{1}{\cos\varphi}\frac{\partial\left(\overline{v'\mu'}\cos\varphi\right)}{a\partial\varphi} - \frac{1}{\rho_0}\frac{\partial\left(\rho_0\overline{w'\mu'}\right)}{\partial Z} + \overline{S} \tag{4.155}$$

Here, $\rho_0(Z)$ is a standard vertical profile of the air density, Z the log-pressure altitude, a the Earth's radius, f the Coriolis factor, Q the net heating rate, and S the chemical source term. The correlation (eddy) terms such as $\overline{u'v'}$ or $\overline{v'\theta'}$ represent the effects of atmospheric waves on the zonal mean quantities. Closure relations expressing the eddy terms as a function of the mean quantities must be added to this system. In most cases, the eddy terms are parameterized as a function of empirical eddy diffusion coefficients. One can show that, for steady and conservative waves, the eddy flux divergence and the zonal mean advection terms cancel (*non-transport theorem*), so that errors in the specified eddy diffusion coefficients can have a large effect on the solution of the system, and specifically on the calculated distribution of chemical species.

To avoid the numerical problems associated with the quasi eddy-mean flow cancellation (small difference between two relatively large terms), it is useful to introduce the transformed Eulerian mean (TEM) velocities v^* and w^* (Boyd, 1976; Andrews and McIntyre, 1976):

$$\overline{v}^* = \overline{v} - \frac{1}{\rho_0}\frac{\partial}{\partial Z}\left\{ \frac{\rho_0\overline{v'\theta'}}{\partial\overline{\theta}/\partial Z} \right\} \tag{4.156}$$

$$\overline{w}^* = \overline{w} + \frac{1}{a}\frac{\partial}{\partial\varphi}\left\{ \frac{\overline{v'\theta'}\cos\varphi}{\partial\overline{\theta}/\partial Z} \right\} \tag{4.157}$$

It can be shown that, in the presence of conservative and steady waves, the TEM meridional velocity components are equal to zero. The quasi-compensation between large zonal mean and eddy terms that characterizes the classic Eulerian mean equations is therefore replaced by the concept of residual velocities that describes a *meridional residual circulation*. In the TEM framework, the continuity, momentum, and thermodynamic equations are expressed as

$$\frac{1}{a\cos\varphi}\frac{\partial(\overline{v}^*\cos\varphi)}{\partial\varphi} + \frac{1}{\rho_0}\frac{\partial(\rho_0\overline{w}^*)}{\partial Z} = 0 \tag{4.158}$$

$$\frac{\partial \overline{u}}{\partial t} - \overline{v}^* \left\{ f - \frac{1}{a \cos \varphi} \frac{\partial (\overline{u} \cos \varphi)}{\partial \varphi} \right\} + \overline{w}^* \frac{\partial \overline{u}}{\partial Z} = \overline{G}_u \qquad (4.159)$$

$$\frac{\partial \overline{\theta}}{\partial t} + \frac{\overline{v}^*}{a} \frac{\partial \overline{\theta}}{\partial \varphi} + \overline{w}^* \frac{\partial \overline{\theta}}{\partial Z} = \overline{q} + \overline{G}_\theta \qquad (4.160)$$

where

$$\overline{q} = \left[\frac{p_0}{p} \right]^{\frac{R}{c_p}} \frac{\overline{Q}}{c_p} \qquad (4.161)$$

represents the diabatic heating term expressed in K per unit time. With the TEM transformation, all eddy contributions are included in the forcing terms $\overline{G}_u$ and $\overline{G}_\theta$. Wave momentum forcing is assumed to be proportional to the divergence of the *Eliassen–Palm* (*EP*) *flux* **E**

$$\overline{G}_u = \frac{1}{\rho_0 a \cos \varphi} \boldsymbol{\nabla} \cdot \mathbf{E} \qquad (4.162)$$

For planetary waves under quasi-geostrophic scaling, the meridional and vertical components of this vector are proportional to the eddy momentum and heat fluxes, respectively. With a good approximation,

$$E_\varphi = -\rho_0 \, a \cos \varphi \; \overline{u'v'} \qquad (4.163)$$

$$E_z = \rho_0 \, a \cos \varphi \; f \, \frac{\overline{v'\theta'}}{\partial \overline{\theta}/\partial Z} \qquad (4.164)$$

The divergence is regarded as the wave stress that accelerates or decelerates the mean flow. It vanishes when the waves are steady and conservative. The contribution of eddies to the mean temperature tendency is given by

$$\overline{G}_\theta = -\frac{1}{\rho_0} \frac{\partial}{\partial Z} \left\{ \rho_0 \left(\overline{v'\theta'} \frac{\partial \overline{\theta}/a \partial \varphi}{\partial \overline{\theta}/\partial Z} + \overline{w'\theta'} \right) \right\} \qquad (4.165)$$

This term vanishes exactly for steady and conservative waves (Andrews and McIntyre, 1978), and is otherwise generally small and often ignored. In this case, to a good approximation, the vertical TEM velocity is directly proportional to the diabatic heating rate

$$\overline{w}^* = \frac{\overline{q}}{\partial \overline{\theta}/\partial Z} \qquad (4.166)$$

In the presence of gravity waves, the eddy term reduces to

$$\overline{G}_\theta = -\frac{1}{\rho_0} \frac{\partial}{\partial Z} \left(\rho_0 \overline{w'\theta'} \right) \qquad (4.167)$$

Finally, the zonally averaged continuity equation for chemical species is written in the TEM framework as

$$\frac{\partial \overline{\mu}}{\partial t} + \frac{\overline{v}^*}{a} \frac{\partial \overline{\mu}}{\partial \varphi} + \overline{w}^* \frac{\partial \overline{\mu}}{\partial Z} = \overline{S} + \overline{G}_\mu \qquad (4.168)$$

where

$$\overline{G}_\mu = -\frac{1}{a\cos\varphi}\frac{\partial\left(\overline{v'\mu'}^*\cos\varphi\right)}{\partial\varphi} - \frac{1}{\rho_0}\frac{\partial\left(\rho_0\overline{w'\mu'}^*\right)}{\partial Z} \qquad (4.169)$$

where the *net* eddy flux components $\overline{v'\mu'}^*$ and $\overline{w'\mu'}^*$ in the TEM framework are expressed as a function of the zonally average eddy mass and heat fluxes by (Garcia and Solomon, 1983; Andrews et al. 1987)

$$\overline{v'\mu'}^* = \overline{v'\mu'} - \frac{\overline{v'\theta'}}{\partial\overline{\theta}/\partial Z}\frac{\partial\overline{\mu}}{\partial Z}$$

$$\overline{w'\mu'}^* = \overline{w'\mu'} - \frac{\overline{v'\theta'}}{\partial\overline{\theta}/\partial Z}\frac{\partial\overline{\mu}}{a\partial\varphi}$$

Again, the term $\overline{G}_\mu$ vanishes if the waves are steady and conservative, and the tracer has no sources/sinks. Otherwise, it is generally small, and accounts for *chemical eddy transport*, i.e., the net meridional and vertical transport that occurs when reactive gases displaced by atmospheric waves encounter different photochemical environments. In this case, the eddy correlations $\overline{v'\mu'}^*$ and $\overline{w'\mu'}^*$ must be parameterized by closure relations, for example by diffusion terms

$$\overline{v'\mu'}^* = -\left\{K_{yy}\frac{\partial\overline{\mu}}{a\partial\varphi} + K_{yz}\frac{\partial\overline{\mu}}{\partial Z}\right\} \qquad (4.170)$$

$$\overline{w'\mu'}^* = -\left\{K_{zy}\frac{\partial\overline{\mu}}{a\partial\varphi} + K_{zz}\frac{\partial\overline{\mu}}{\partial Z}\right\} \qquad (4.171)$$

where the coefficients K_{ij} can be expressed as a function of the chemical lifetimes of the trace gases and of the time needed for an air parcel to move through a dynamical disturbance.

4.7.2 One-Dimensional Models

One-dimensional (1-D) models are valuable conceptual tools for vertical transport and chemistry in an atmospheric column. Uniformity is assumed in the horizontal plane; in other words, horizontal flux divergence is assumed to be negligible. Vertical transport is computed using an eddy (turbulent) diffusion parameterization. One-dimensional models may be designed for limited domains (such as the continental boundary layer over a source region) or for the global domain (vertical transport and chemistry in the atmospheric column). By averaging the continuity equation and denoting horizontal means as brackets $\langle\,\rangle$, we obtain the following equation for the density $\langle\rho_i\rangle$ of a chemical species i:

$$\frac{\partial\langle\rho_i\rangle}{\partial t} + \frac{\partial\langle\rho_i'w'\rangle}{\partial z} = \langle s_i\rangle \qquad (4.172)$$

where the net vertical flux $\langle\rho_i'w'\rangle$ is parameterized as

$$\langle\rho_i'w'\rangle = -K_z\,\rho_a\frac{\partial(\langle\rho_i\rangle/\rho_a)}{\partial z} \qquad (4.173)$$

Here, ρ_a is the mean air density and K_z is the vertical eddy diffusion coefficient. This coefficient accounts for the effects of motions at all scales. The vertical flux is assumed to be proportional to the vertical gradient in the mean mixing ratio. Values of K_z are specified as input. In 1-D models for the boundary layer, semi-empirical formulations of K_z are available as a function of the surface fluxes of momentum and heat that force a diurnal cycle of mixed layer growth and decay (see Seinfeld and Pandis, 2006). In global 1-D models, empirical values of K_z are derived from observations of chemical tracers (see, for example, Liu *et al.*, 1982).

In the thermosphere, vertical exchanges are dominated by molecular diffusion rather than eddy mixing. The 1-D continuity equation is then formulated in terms of the zonally and meridionally averaged number density $\langle n_i \rangle$ as

$$\frac{\partial \langle n_i \rangle}{\partial t} + \frac{\partial \langle \Phi_i \rangle}{\partial z} = \langle s_i \rangle \tag{4.174}$$

Here the vertical flux $\langle \Phi_i \rangle$ is expressed as

$$\langle \Phi_i \rangle = -D_i \left[\frac{\partial \langle n_i \rangle}{\partial z} + (1 + \alpha_T) \frac{\langle n_i \rangle}{\langle T \rangle} \frac{\partial \langle T \rangle}{\partial z} + \frac{\langle n_i \rangle}{\langle H_i \rangle} \right] \tag{4.175}$$

where D_i is the molecular diffusion coefficient, α_T is the thermal diffusion factor (–0.40 for helium, –0.25 for hydrogen, 0 for heavier species) and $H_i = kT/m_i g$ is the scale height of species i with molecular or atomic mass m_i. The other symbols have their usual meaning. An approximate expression for the molecular diffusion coefficient [here in cm^2 s^{-1}] is provided by Banks and Kockarts (1973):

$$D_i = 1.53 \cdot 10^{18} \left[\frac{1}{m_i} + \frac{1}{m} \right]^{\frac{1}{2}} \frac{T^{\frac{1}{2}}}{n_a} \tag{4.176}$$

where m_i and the mean molecular mass m are expressed in *atomic mass units (amu)*, the total air number density n_a is expressed in [molecules cm^{-3}], and T is in [K]. The value of the molecular diffusion coefficient increases rapidly with height as the air density decreases, which can lead to some numerical difficulties when solving the continuity equations.

4.7.3 Zero-Dimensional Models

Zero-dimensional models do not account for the effects of transport so that the continuity equations collapse to ordinary differential equations of the form $d\rho_i/dt = s_i$. Computing the chemical evolution of an ensemble of coupled species involves solving a system of coupled ODEs. Such models have three general classes of applications:

- Global box models for long-lived and therefore well-mixed species, such as CO_2 or methane, where long-term trends in total atmospheric mass can be computed from a balance between sources and sinks.
- Chemical evolution of an ensemble of short-lived species (such as a family of radicals) where the chemical source and sink terms are much larger than the transport terms. This is particularly useful to interpret field observations of radical chemistry. The concentrations of radicals are assumed to be in chemical steady

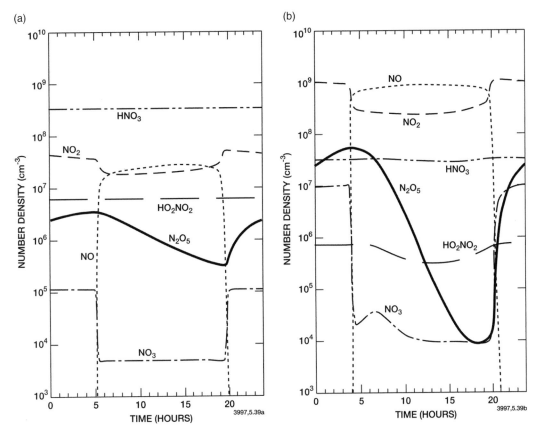

Figure 4.10 Diurnal evolution of nitrogen species calculated by a zero-dimensional box model for May 10 at 65° N at 20 km (a) and 40 km (b). From Brasseur and Solomon (2005).

state, while concentrations of longer-lived species are specified from observations or from climatology. Diurnal variation forced by photochemistry can be incorporated in the steady-state assumption by using periodic boundary conditions over the 24-hour diurnal cycle (Figure 4.10).

- Diagnostic studies to understand the nonlinear evolution of a chemical system. This is done, for example, to interpret the results of laboratory chamber experiments, or to understand the cascade of oxidation products resulting from the atmospheric oxidation of hydrocarbons.

4.8 Numerical Frameworks for Eulerian Models

Eulerian models derive the concentrations of chemical species at fixed locations on the Earth's sphere by numerical solution of the Eulerian chemical continuity equations (Section 4.2). Here, we present the general numerical frameworks used to solve these equations, including finite difference, finite volume, spectral, and finite element methods. In the finite difference method, each unknown function is described by its values at a set of discrete *grid points*, while in the finite volume method, the functions are represented by their value averaged over specified intervals called *grid cells* or *gridboxes*. In the spectral and finite element methods, the functions are expanded by a linear combination of orthogonal basis functions, and the coefficients appearing in these expansions become the unknowns. We use the generic symbol Ψ to represent non-negative scalar functions, which can be viewed as the concentration (density or mixing ratio) of chemical species.

4.8.1 Finite Difference (Grid Point) Methods

In the *finite difference method*, the partial derivatives in space $\partial\Psi/\partial x$ and time $\partial\Psi/\partial t$ of a function $\Psi(x, t)$ are approximated by finite difference analogs $\Delta\Psi/\Delta x$ and $\Delta\Psi/\Delta t$ in which the increments Δ are finite rather than infinitesimal. In this approach, a differential equation is replaced by a system of algebraic equations that can be solved by numerical methods. The solution of the system is obtained at a finite number (J) of grid points: $x_1, x_2, \ldots, x_{j-1}, x_j, x_{j+1}, \ldots x_J$ for discrete time levels $t_1, t_2, \ldots t_{n-1}$, $t_n, \ldots$ For simplicity, we first assume that the dependent variable Ψ is a function of variable x only (1-D problem) and we consider that the points are uniformly spaced, so that the distance Δx between points is constant. The finite difference forms are the same when the independent variable is time; in this case Δt represents the model *time step*.

The spatial resolution associated with a 1-D grid of constant spacing Δx can be estimated by noting that, if one decomposes function $\Psi(x)$ into several sinusoids of varying amplitudes, wavelengths, and phases (Fourier analysis), the smallest resolvable wavelength is $\lambda_{min} = 2\,\Delta x$. The resolution of the model is therefore twice the grid spacing. The maximum resolvable wavenumber is $k_{max} = 2\pi/\lambda_{min} = \pi/\Delta x$. Processes that occur at scales smaller than λ_{min} are referred to as *unresolved* or *subgrid-scale* processes; they are usually parameterized in terms of the resolved model variables (see Chapter 8).

To derive finite difference approximations to the derivatives of a function $\Psi(x)$ at point x_j, we expand the function as a Taylor series around that point:

$$\Psi\left(x_j + \Delta x\right) = \Psi\left(x_j\right) + \Delta x \left(\frac{\partial\Psi}{\partial x}\right)_{x_j} + \frac{\Delta x^2}{2!}\left(\frac{\partial^2\Psi}{\partial x^2}\right)_{x_j} + \frac{\Delta x^3}{3!}\left(\frac{\partial^3\Psi}{\partial x^3}\right)_{x_j} + \frac{\Delta x^4}{4!}\left(\frac{\partial^4\Psi}{\partial x^4}\right)_{x_j} + \cdots$$

$$(4.177)$$

Similarly, we can write

$$\Psi\left(x_j - \Delta x\right) = \Psi\left(x_j\right) - \Delta x \left(\frac{\partial\Psi}{\partial x}\right)_{x_j} + \frac{\Delta x^2}{2!}\left(\frac{\partial^2\Psi}{\partial x^2}\right)_{x_j} - \frac{\Delta x^3}{3!}\left(\frac{\partial^3\Psi}{\partial x^3}\right)_{x_j} + \frac{\Delta x^4}{4!}\left(\frac{\partial^4\Psi}{\partial x^4}\right)_{x_j} - \cdots$$

$$(4.178)$$

Function $\Psi(x)$ takes the values Ψ_{j-1}, Ψ_j, and Ψ_{j+1} at points $x_{j-1} = x_j - \Delta x$, x_j, and $x_{j+1} = x_j + \Delta x$, respectively. An approximate value of the first-order derivative $\partial\Psi/\partial x$ at point x_j is derived from (4.177):

$$\left(\frac{\partial\Psi}{\partial x}\right)_{x_j} = \frac{\Psi_{j+1} - \Psi_j}{\Delta x} + O(\Delta x)$$

$$(4.179)$$

The scheme is referred to as a *forward* difference scheme because the derivative is approximated by using information at points x_j and $x_j + \Delta x$. The error made by adopting this approximation, called the *truncation error*, is

$$O(\Delta x) = -\frac{\Delta x}{2!}\left(\frac{\partial^2 \Psi}{\partial x^2}\right)_{x_j} - \frac{\Delta x^2}{3!}\left(\frac{\partial^3 \Psi}{\partial x^3}\right)_{x_j} - \frac{\Delta x^3}{4!}\left(\frac{\partial^4 \Psi}{\partial x^4}\right)_{x_j} - \cdots \qquad (4.180)$$

From (4.178), one derives the *backward* scheme that uses information at points x_j and $x_j - \Delta x$

$$\left(\frac{\partial \Psi}{\partial x}\right)_{x_j} = \frac{\Psi_j - \Psi_{j-1}}{\Delta x} + O(\Delta x) \qquad (4.181)$$

with the truncation error

$$O(\Delta x) = \frac{\Delta x}{2!}\left(\frac{\partial^2 \Psi}{\partial x^2}\right)_{x_j} - \frac{\Delta x^2}{3!}\left(\frac{\partial^3 \Psi}{\partial x^3}\right)_{x_j} + \frac{\Delta x^3}{4!}\left(\frac{\partial^4 \Psi}{\partial x^4}\right)_{x_j} - \cdots \qquad (4.182)$$

These two algorithms are said to be *first-order* accurate because the largest term in the truncation error is proportional to Δx. By subtracting (4.178) from (4.177) we obtain

$$\Psi_{j+1} - \Psi_{j-1} = 2\Delta x \left(\frac{\partial \Psi}{\partial x}\right)_{x_j} + \frac{2\Delta x^3}{3!}\left(\frac{\partial^3 \Psi}{\partial x^3}\right)_{x_j} + \frac{2\Delta x^5}{5!}\left(\frac{\partial^5 \Psi}{\partial x^5}\right)_{x_j} + \cdots \qquad (4.183)$$

One deduces a *second-order* approximation (or *central* scheme) for the first derivative at point x_j

$$\left(\frac{\partial \Psi}{\partial x}\right)_{x_j} = \frac{\Psi_{j+1} - \Psi_{j-1}}{2\Delta x} + O(\Delta x^2) \qquad (4.184)$$

with the largest-magnitude term in the truncation error

$$O(\Delta x^2) = -\frac{\Delta x^2}{3!}\left(\frac{\partial^3 \Psi}{\partial x^3}\right)_{x_j} - \frac{\Delta x^4}{5!}\left(\frac{\partial^5 \Psi}{\partial x^5}\right)_{x_j} \qquad (4.185)$$

being proportional to Δx^2. Higher-order approximations to the derivative can similarly be obtained by addition and subtraction of the Taylor expansions to cancel error terms to higher order (Table 4.1). For example, if one multiplies the centered difference (4.183) obtained for a spacing Δx by a factor 4/3 and the similar expression derived for a grid spacing of $2\Delta x$ by a factor of 1/3, and if one substracts the two expressions, the second-order error disappears; one obtains the fourth-order accurate scheme

$$\left(\frac{\partial \Psi}{\partial x}\right)_{x_j} = \frac{4}{3}\left(\frac{\Psi_{j+1} - \Psi_{j-1}}{2\Delta x}\right) - \frac{1}{3}\left(\frac{\Psi_{j+2} - \Psi_{j-2}}{4\Delta x}\right) + O(\Delta x^4) \qquad (4.186)$$

Figure 4.11 shows a graphical representation of the first derivative for a continuous function $\Psi(x)$ at point x_j as the slope of the tangent to the function at that particular point (point B). It is approximated by the chords AB, BC, and AC in the case of the backward, forward, and central difference approximations, respectively. Higher order formulations involve more than two grid points around point B.

Table 4.1 Numerical approximations to the partial derivative $\partial \Psi / \partial x$

Approximation	Order	Expression
Forward	1	$\dfrac{\Psi_{j+1} - \Psi_j}{\Delta x}$
	2	$\dfrac{-3\Psi_j + 4\Psi_{j+1} - \Psi_{j+2}}{2\Delta x}$
	3	$\dfrac{-2\Psi_{j-1} - 3\Psi_j + 6\Psi_{j+1} - \Psi_{j+2}}{6\Delta x}$
	4	$\dfrac{-3\Psi_{j-1} - 10\Psi_j + 18\Psi_{j+1} - 6\Psi_{j+2} + \Psi_{j+3}}{12\Delta x}$
Backward	1	$\dfrac{\Psi_j - \Psi_{j-1}}{\Delta x}$
	2	$\dfrac{\Psi_{j-2} - 4\Psi_{j-1} + 3\Psi_j}{2\Delta x}$
	3	$\dfrac{\Psi_{j-2} - 6\,\Psi_{j-1} + 3\Psi_j + 2\Psi_{j+1}}{6\Delta x}$
	4	$\dfrac{-\Psi_{j-3} + 6\Psi_{j-2} - 18\Psi_{j-1} + 10\Psi_j + 3\Psi_{j+1}}{12\Delta x}$
Centered	2	$\dfrac{\Psi_{j+1} - \Psi_{j-1}}{2\Delta x}$
	4	$\dfrac{\Psi_{j-2} - 8\Psi_{j-1} + 8\Psi_{j+1} - \Psi_{j+2}}{12\Delta x}$

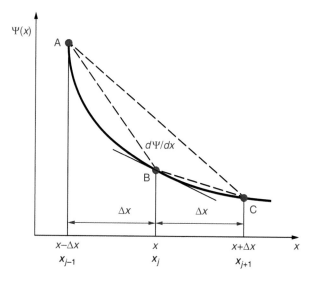

Figure 4.11 Finite difference approximation to derivative $d\Psi/dx$: forward, backward, and central approximations to slope of tangent at point B.

Table 4.2 Numerical approximations to the partial derivative $\partial^2 \Psi / \partial x^2$

Approximation	Order	Expression
Centered	2	$\dfrac{\Psi_{j+1} - 2\Psi_j + \Psi_{j-1}}{\Delta x^2}$
Centered	4	$\dfrac{-\Psi_{j-2} + 16\Psi_{j-1} - 30\Psi_j + 16\Psi_{j+1} - \Psi_{j+2}}{12\Delta x^2}$

By adding equations (4.178) and (4.177), we find the following expression

$$\Psi_{j+1} + \Psi_{j-1} = 2\Psi_j + \Delta x^2 \left(\frac{\partial^2 \Psi}{\partial x^2}\right)_{x_j} + \frac{\Delta x^4}{12} \left(\frac{\partial^4 \Psi}{\partial x^4}\right)_{x_j} + \cdots \tag{4.187}$$

from which we deduce a second-order accurate expression for the second derivative at point x_j

$$\left(\frac{\partial^2 \Psi}{\partial x^2}\right)_{x_j} = \frac{\Psi_{j+1} + \Psi_{j-1} - 2\Psi_j}{\Delta x^2} + O\left(\Delta x^2\right) \tag{4.188}$$

Approximations for the second derivative are summarized in Table 4.2.

Ordinary Differential Equations (ODEs)

To illustrate the use of the finite-difference method, we consider a system of initial-value ODEs that describes the evolution with time t of a vector valued function $\mathbf{\Psi}(t)$:

$$\frac{d\mathbf{\Psi}}{dt} = \mathbf{F}(\mathbf{\Psi}, t) \tag{4.189}$$

with an initial condition $\mathbf{\Psi}(t_0) = \mathbf{\Psi}^0$. The specified forcing function $\mathbf{F}$ is dependent on function $\mathbf{\Psi}$ and time t. In atmospheric problems, such equations describe the rate of changes in different physical quantities (mass, energy, momentum, concentration of chemical species, etc.) in response to known forcing factors.

From the first-order forward approximation of the derivative, and with $\mathbf{F}$ evaluated at time t_n, we approximate (4.189) by

$$\frac{\mathbf{\Psi}^{n+1} - \mathbf{\Psi}^n}{\Delta t} = \mathbf{F}(\mathbf{\Psi}^n, t_n) \tag{4.190}$$

where $\mathbf{\Psi}^n$ denotes an estimate of function $\mathbf{\Psi}$ at time t_n and where the time interval $\Delta t = t_{n+1} - t_n$ is the time step of the numerical method. From the knowledge of $\mathbf{\Psi}$ at time t_n, we obtain the solution for $\mathbf{\Psi}$ at time t_{n+1} as

$$\mathbf{\Psi}^{n+1} = \mathbf{\Psi}^n + \Delta t\, \mathbf{F}^n \tag{4.191}$$

where $\mathbf{F}^n$ stands for $\mathbf{F}(\mathbf{\Psi}^n, t_n)$. This is the explicit *Euler forward* method. It is called "explicit" because the right-hand side depends solely on the known value of $\mathbf{\Psi}$ at time t_n. Thus $\mathbf{\Psi}^{n+1}$ is readily calculated, and one can march in this manner forward from time step to time step.

An alternative approximation to (4.189) is to evaluate $\mathbf{F}$ at time t_{n+1}. This is the implicit *Euler backward* algorithm:

$$\mathbf{\Psi}^{n+1} = \mathbf{\Psi}^n + \Delta t\, \mathbf{F}^{n+1} \tag{4.192}$$

This algebraic equation is more difficult to solve since $\mathbf{F}$ is expressed as a function of the unknown solution at time t_{n+1}. In this latter case, the solution $\mathbf{\Psi}^{n+1}$ is generally determined by adopting a functional iteration process. In the Newton iterative procedure, for example, one solves a linearized version of the system

$$\mathbf{G}\left(\mathbf{\Psi}^{n+1}\right) = \mathbf{\Psi}^{n+1} - \mathbf{\Psi}^n - \Delta t\, \mathbf{F}^{n+1} = 0 \tag{4.193}$$

leading to the iterative relations

$$\mathbf{J}\Delta\mathbf{\Psi}^{n+1}_{(r)} = -\mathbf{G}\left(\mathbf{\Psi}^{n+1}_{(r)}\right)$$

$$\mathbf{\Psi}^{n+1}_{(r+1)} = \mathbf{\Psi}^{n+1}_{(r)} + \Delta\mathbf{\Psi}^{n+1}_{(r)}$$

with an initial guess, e.g., $\mathbf{\Psi}^{n+1}_{(0)} = \mathbf{\Psi}^n$. In these expressions,

$$\mathbf{J} = \frac{\partial \mathbf{G}}{\partial \mathbf{\Psi}} = \mathbf{I} - \Delta t\, \frac{\partial \mathbf{F}}{\partial \mathbf{\Psi}} \tag{4.194}$$

is the Jacobian matrix of $\mathbf{G}$, $\mathbf{I}$ is the identity matrix, and (r) is the iteration index. The iteration is interrupted when the absolute value of the correction $|\Delta\mathbf{\Psi}^{n+1}_{(r)}|$ becomes smaller than a user-prescribed tolerance.

A second-order algorithm is the implicit *trapezoidal* scheme

$$\mathbf{\Psi}^{n+1} = \mathbf{\Psi}^n + \frac{\Delta t}{2}\left[\mathbf{F}^{n+1} + \mathbf{F}^n\right] \tag{4.195}$$

The three algorithms presented above are single-step methods because the value of function $\mathbf{\Psi}$ at time t_{n+1} is calculated only as a function of $\mathbf{\Psi}$ at the previous time t_n. Higher-order *multi-step methods* provide the solution $\mathbf{\Psi}^{n+1}$ at time t_{n+1} by using information from s previous steps $t_n, t_{n-1}, t_{n-2}, \ldots, t_{n-s+1}$:

$$\mathbf{\Psi}^{n+1} = \sum_{j=0}^{s-1} a_j \mathbf{\Psi}^{n-j} + \Delta t \sum_{j=-1}^{s-1} b_j\, \mathbf{F}^{n-j} \tag{4.196}$$

The choice of coefficients $a_0, a_1, \ldots, a_{s-1}$ and $b_{-1}, b_0, b_1, \ldots, b_{s-1}$ defines the particular algorithm. If $b_{-1} = 0$, the method is explicit; otherwise, it is implicit. For $s = 1$ (single-step method) and $a_0 = 1$, the algorithm is a single-step Euler method (explicit if $b_0 = 1$ and $b_{-1} = 0$, and implicit if $b_0 = 0$ and $b_{-1} = 1$). For $b_{-1} = b_0 = 0.5$, one obtains the single-step trapezoidal method.

An example of a two-step algorithm ($s = 2$) is the *leapfrog* method

$$\mathbf{\Psi}^{n+1} = \mathbf{\Psi}^{n-1} + 2\Delta t\, \mathbf{F}^n \tag{4.197}$$

in which all parameters in expression (4.196) are chosen to be zero except $a_1 = 1$ and $b_0 = 2$. This explicit method is second-order accurate. A difficulty in the leapfrog algorithm is that the solutions obtained on the even time levels n, $n + 2$, $n + 4$, $\ldots$, and odd time levels $n + 1$, $n + 3$, $n + 5$, $\ldots$ are only weakly coupled (Figure 4.12).

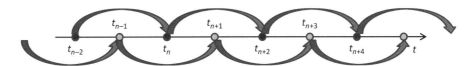

Figure 4.12 The leapfrog method. Numerical integration of the solution at t_{n-1}, t_{n+1}, t_{n+3}, ... time levels (in yellow) and t_{n-2}, t_n, t_{n+2}, ... time levels (in red).

This can lead to the appearance of a "computational mode" that gradually amplifies when treating nonlinear problems. The computational mode can be damped by applying a second-order time filter (see Section 4.15). Another corrective approach is to periodically apply over the course of the integration another algorithm, producing a permutation between the evolution at the odd and even time levels.

A classic multi-step algorithm is the explicit *Adams–Bashforth method*, in which the coefficients of expression (4.196) are defined as $a_0 = 1$, and a_1, a_2, ..., $a_{s-1} = 0$. A value of zero is imposed for coefficient b_{-1}, defining the explicit nature of the method. Coefficients b_0, b_1, ..., b_{s-1} are chosen so that the method has order s. For $s = 1$, the algorithm is the explicit Euler method. The two- and three-step Adams–Bashforth algorithms are expressed by

$$\mathbf{\Psi}^{n+1} = \mathbf{\Psi}^n + \frac{\Delta t}{2} \left[3 \, \mathbf{F}^n - \mathbf{F}^{n-1} \right] \tag{4.198}$$

and

$$\mathbf{\Psi}^{n+1} = \mathbf{\Psi}^n + \frac{\Delta t}{12} \left[23 \, \mathbf{F}^n - 16 \, \mathbf{F}^{n-1} + 5 \, \mathbf{F}^{n-2} \right] \tag{4.199}$$

respectively.

The *Adams–Moulton method* is an implicit multi-step algorithm. The values adopted for coefficients a_j are the same as in the Adams–Bashforth method (all values of a_i equal to zero, except $a_0 = 1$). In this implicit case, the restriction on b_{-1} is removed, and the values of coefficients b_j are chosen so that the method reaches order $s + 1$. The algorithm for $s = 1$ is the *trapezoidal rule*. For $s = 2$ and $s = 3$, it is expressed respectively as

$$\mathbf{\Psi}^{n+1} = \mathbf{\Psi}^n + \frac{\Delta t}{12} \left[5 \, \mathbf{F}^{n+1} + 8 \, \mathbf{F}^n - \mathbf{F}^{n-1} \right] \tag{4.200}$$

$$\mathbf{\Psi}^{n+1} = \mathbf{\Psi}^n + \frac{\Delta t}{24} \left[9 \, \mathbf{F}^{n+1} + 19 \, \mathbf{F}^n - 5 \, \mathbf{F}^{n-1} + \mathbf{F}^{n-2} \right] \tag{4.201}$$

Finally, the implicit *backward differentiation formula* (*BDF*) method, also called *Gear's method*, is defined by setting all coefficients b_0, b_1, ..., $b_{s-1} = 0$ and specifying a non-zero value only for b_{-1}. The resulting algorithm is generally written in the form

$$\mathbf{\Psi}^{n+1} = \sum_{j=0}^{s-1} a_j \mathbf{\Psi}^{n-j} + \Delta t \, b_{-1} \, \mathbf{F}^{n+1} \tag{4.202}$$

For example, the *two- and three-step BDFs* are

$$\mathbf{\Psi}^{n+1} = \frac{4}{3} \mathbf{\Psi}^n - \frac{1}{3} \mathbf{\Psi}^{n-1} + \frac{2}{3} \Delta t \, \mathbf{F}^{n+1} \tag{4.203}$$

$$\Psi^{n+1} = \frac{18}{11}\Psi^n - \frac{9}{11}\Psi^{n-1} + \frac{2}{11}\Psi^{n-2} + \frac{6}{11}\Delta t\, \mathbf{F}^{n+1} \qquad (4.204)$$

Higher-order expressions are provided in Section 6.4.3.

The choice of a particular algorithm to solve ODEs is determined first by the requirement that the method provide stable solutions for relatively large time steps. Accuracy is another important requirement. In the case of nonlinear initial value problems, no general numerical analysis is available for assessing the conditions under which a specific algorithm provides stable solutions. We consider therefore a simple *prototype* linear problem

$$\frac{d\Psi}{dt} = \beta\,\Psi \qquad (4.205)$$

for which different numerical schemes can be evaluated relative to a known analytic solution. The analysis of this linear problem provides guidance for the choice of integration algorithms to be adopted for more complex nonlinear problems. To keep our analysis simple, we assume that $\Psi(t)$ is a non-negative scalar function, and we examine the numerical stability of different algorithms discussed above by assuming that solution Ψ^{n+1} at time t_{n+1} is expressed as a function of the solution Ψ^n at time t_n by

$$\Psi^{n+1} = R(z)\,\Psi^n \qquad (4.206)$$

where function $R(z)$ is the so-called *stability function* and $z = \beta\Delta t$ is a complex variable $[z = (\lambda, \omega) = \lambda + i\,\omega]$. The real part of $R(z)$ determines the amplitude of the solution as time evolves. We consider here a problem whose solution decays with time, and assume therefore that $\lambda < 0$. Thus, in our discussion, only the left half of the complex plane with its axes $x = \lambda\Delta t$ and $y = \omega\Delta t$ is of particular interest. The imaginary part of $R(z)$ provides information on the phase of the solution, and is of relevance when treating fluid dynamical equations such as, for example, the momentum or the transport equation. When considering chemical kinetics equations, variable $z = (\lambda, 0)$ is real, and $R(z)$ is therefore a real function.

The stability function $R(z)$ associated with different algorithms is generally expressed as the ratio between two polynomials. This function approximates its analytic analog (exact solution of the prototype equation)

$$R(z) = e^z \qquad (4.207)$$

A numerical method is said to be *A-stable* if the amplitude of the stability function $|R(z)| \leq 1$ for $-\infty \leq \lambda \leq 0$. The area of stability covers therefore the entire left-half area of the complex plane for which $\lambda \leq 0$. An *A*-stable integration algorithm is stable for any value of the time step Δt. Any other type of stability introduces restrictions on z and therefore on the time step. The *A*-stability condition is met for the implicit (backward) Euler and the trapezoidal methods. The amplification functions of these two algorithms are

$$R(z) = \frac{1}{(1-z)} \text{ and } R(z) = \frac{1 + z/2}{1 - z/2}$$

respectively. It is not met in the case of the explicit (forward) Euler method, for which

$$R(z) = 1 + z$$

In this case, the stability region in the complex plane is limited to a circle of radius 1 centered around point ($\lambda = -1$, $\omega = 0$). In the case of a chemical kinetics equation (i.e., z is a real variable), the method is unstable unless time steps are smaller than $1/\lambda$. For a coupled system of several chemical species, the time step must be smaller than the lifetime of the shortest-lived species. This makes it inapplicable for *stiff* chemical kinetics systems of ODEs, as is typical of atmospheric chemistry problems, where the lifetimes of different species in the system vary over many orders of magnitude. The Adams–Bashforth and Adams–Moulton methods are not A-stable either. The one- and two-step BDF (Gear's) methods are A-stable, but the A-stability property is lost when the order of accuracy of the method becomes higher than 2 (this limit is called Dahlquist's second barrier). The values of z that spoil the A-stability conditions, however, are located only in a shallow area left of the imaginary axis in the complex domain (Durran, 2010). Gear's method is stable for accuracy orders up to 6 if z is real and negative. This algorithm is therefore particularly appropriate for treating stiff chemical kinetics systems. See Chapter 6 for further discussion.

A-stability may not be a sufficient condition for properly treating systems in which different components decay at very different rates. One introduces therefore the concept of L-stability: a method is said to be *L-stable* in the context of the prototype equation adopted here, if it is A-stable and, in addition, if $|R(z)| \rightarrow 0$ for large time steps ($\Delta t \rightarrow \infty$). L-stable algorithms are particularly useful for treating stiff systems. By adopting relatively large time steps, L-stable methods retain the slowly varying contributions of the solution (which are generally of most interest), while ignoring its rapidly decaying components (which are generally of little or no interest). The implicit Euler method is L-stable since $|R(z)|$ tends to zero for large time steps. On the contrary, the trapezoidal algorithm, which is A-stable, is not L-stable. Since no multi-step method is A-stable for orders higher than 2, *stricto sensu* only low-order BDF (Gear's) algorithms are L-stable.

A simple geometric interpretation of the numerical methods for ODEs presented above can be provided by integrating the scalar version of (4.189) over a time step $\Delta t = t_{n+1} - t_n$:

$$\Psi(t_{n+1}) = \Psi(t_n) + \int_{t_n}^{t_{n+1}} F(\Psi, t) \, dt \tag{4.208}$$

This equation does not suffer any approximation, so the accuracy of the solution is determined by the algorithm adopted to evaluate the integral of function F over a time step Δt. Different approaches are illustrated in Figure 4.13. In the simplest of them, the function F is approximated by a constant value over the time interval, equal to its value at time t_n or at time t_{n+1}. These approximations correspond to the first-order accurate explicit Euler forward and the implicit Euler backward methods, respectively. If function F is approximated by a linear function coincident with the value of F at times t_n and t_{n+1}, the accuracy of the method is improved and

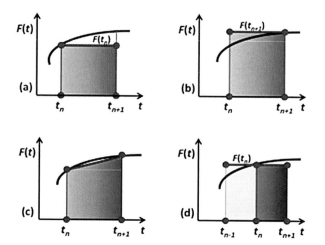

Figure 4.13 Geometric interpretation of numerical integration algorithms for ordinary differential equations: (a) Euler forward; (b) Euler backward; (c) trapezoidal and (d) leapfrog methods. Personal communication of Paul Ullrich, University of California Davis.

corresponds to the second-order trapezoidal rule. Finally, if the integration is performed over an interval $2\,\Delta t$ from time t_{n-1} to t_{n+1} and if function F is approximated over this interval by the value of F at time t_n, the method is also second-order accurate and corresponds to the leapfrog algorithm.

More accurate quadrature algorithms can be implemented to calculate the integral of function $F(\Psi, t)$. We can write, for example,

$$\Psi^{n+1} = \Psi^n + \Delta t \sum_{j=1}^{s} b_j F\big(\Psi(t_n + c_j \Delta t), t_n + c_j \Delta t\big) \qquad (4.209)$$

where $c_j < 1$ denotes nodes within the interval Δt, and b_j weights used for the quadrature. The values of $\Psi(t_n + c_j \Delta t)$ are not known, but they can be estimated numerically by a series of s preliminary calculations called *stages*. In the case of the s-stage explicit scheme, one computes sequentially the following quantities

$$\Phi_j = \Psi^n + \Delta t \sum_{k=1}^{j-1} a_{j,k} F(\Phi_k, t_n + c_k \Delta t) \qquad (4.210)$$

for $j = 2, 3, \ldots, s$ and with $\Phi_1 = \Psi^n$. Parameters s and $a_{j,k}$ are specific to a particular scheme. The solution at time t_{n+1} is then approximated by

$$\Psi^{n+1} = \Psi^n + \Delta t \sum_{j=1}^{s} b_j F\big(\Phi_j, t_n + c_j \Delta t\big) \qquad (4.211)$$

When considering implicit s-stage schemes, the upper bound $(j - 1)$ of the summation in expression (4.210) is replaced by a value equal or larger than j, for example by parameter s. Runge–Kutta and Rosenbrock methods are examples of multi-stage approaches and are discussed in Chapter 6.

Partial Differential Equations (PDEs)

We now illustrate the use of finite difference methods in the case of partial differential equations. We consider here the 1-D linear *advection equation*

$$\frac{\partial \Psi}{\partial t} + c\frac{\partial \Psi}{\partial x} = 0 \qquad (4.212)$$

where the scalar field $\Psi(x, t)$ represents, for example, the mixing ratio of a chemical species and c a constant velocity of the flow in the x direction. The initial condition is provided by a specified function of space x: $\Psi(x, 0) = \Psi_0(x)$. As will be shown in Section 7.3, the explicit *forward in time, centered in space (FTCS)* Euler solution

$$\frac{\Psi_i^{n+1} - \Psi_i^n}{\Delta t} + c\frac{\Psi_{i+1}^n - \Psi_{i-1}^n}{2\Delta x} = 0 \qquad (4.213)$$

is unstable for any time step. Here, Ψ_i^n represents the value of function $\Psi(x, t)$ at discrete time t_n and geometric location x_i. Other numerical algorithms that are at least conditionally stable (stable for a sufficiently small value of the time step) are presented in Chapter 7.

In the case of the combined 1-D *advection–diffusion equation* (with K representing a diffusion coefficient):

$$\frac{\partial \Psi}{\partial t} + c\frac{\partial \Psi}{\partial x} - K\frac{\partial^2 \Psi}{\partial x^2} = 0 \qquad (4.214)$$

the solution provided by the FTCS Euler forward scheme

$$\frac{\Psi_i^{n+1} - \Psi_i^n}{\Delta t} + c\frac{\Psi_{i+1}^n - \Psi_{i-1}^n}{2\Delta x} - K\frac{\Psi_{i+1}^n - 2\Psi_i^n + \Psi_{i-1}^n}{\Delta x^2} = 0 \qquad (4.215)$$

is stable for sufficiently small time steps through the addition of the diffusion term. The Lax–Wendroff numerical scheme makes use of this property when applied to the pure advection equation: A numerical diffusion term proportional to the chosen time step is added to the FTCS scheme (see Section 7.3).

If an *implicit* time stepping scheme is adopted for the advection–diffusion equation

$$\frac{\Psi_i^{n+1} - \Psi_i^n}{\Delta t} + c\frac{\Psi_{i+1}^{n+1} - \Psi_{i-1}^{n+1}}{2\Delta x} - K\frac{\Psi_{i+1}^{n+1} - 2\Psi_i^{n+1} + \Psi_{i-1}^{n+1}}{\Delta x^2} = 0 \qquad (4.216)$$

the solution is unconditionally stable (stable for all values of the velocity c and the diffusivity K). The system, however, is more difficult to solve. By rearranging the terms, (4.216) can be written as

$$\alpha_i \Psi_{i-1}^{n+1} + \beta_i \Psi_i^{n+1} + \gamma_i \Psi_{i+1}^{n+1} = \Psi_i^n \qquad (i = 1, 2, \ldots, I) \qquad (4.217)$$

where

$$\alpha_i = -\left(\frac{c}{2\Delta x} + \frac{K}{\Delta x^2}\right)\Delta t \qquad (4.218)$$

$$\beta_i = 1 + \frac{2K\Delta t}{\Delta x^2} \qquad (4.219)$$

Tridiagonal Matrix Algorithm

The Thomas algorithm, a simple form of Gaussian elimination, can be used to solve the tridiagonal system of equations that arises from the discretization of the 1-D diffusion equation

$$\alpha_i \Psi_{i-1} + \beta_i \Psi_i + \gamma_i \Psi_{i+1} = \delta_i \qquad (i = 1, 2, \ldots, n)$$

where $\alpha_1 = 0$ and $\gamma_n = 0$. If we first define the modified coefficients

$$\gamma_1' = \frac{\gamma_1}{\beta_1} \quad \delta_1' = \frac{\delta_1}{\beta_1}$$

$$\gamma_1' = \frac{\gamma_i}{\beta_i - \gamma_{i-1}'\alpha_i} \qquad (i = 2, \ldots, n-1)$$

$$\delta_1' = \frac{\delta_i - \delta_{i-1}'\alpha_i}{\beta_i - \gamma_{i-1}'\alpha_i} \qquad (i = 2, \ldots, n-1)$$

the solution is obtained by back substitution

$$\Psi_n = \delta_n'$$

$$\Psi_{i-1} = \delta_{i-1}' - \gamma_{i-1}'\Psi_i \qquad (i = n, n-1, \ldots, 2)$$

$$\gamma_i = \left(\frac{c}{2\Delta x} - \frac{K}{\Delta x^2} \right) \Delta t \qquad (4.220)$$

The system of I equations, whose matrix is tridiagonal, can be solved for each time t_{n+1} by a matrix decomposition and back-substitution method (Box 4.4). The approach described here can be generalized to more than one spatial dimension. Again, linear algebraic equations are derived and can be solved either directly in the explicit case or through more complex methods when an implicit algorithm is used. A more detailed description of the methods used to solve the transport equations is given in Chapter 7.

   The fluid dynamics equations, specifically the momentum equation, include non-linear terms of the form $u\, \partial u/\partial x$. The corresponding equation, expressed here along spatial dimension x,

$$\frac{\partial u}{\partial t} + u \frac{\partial u}{\partial x} = 0 \qquad (4.221)$$

produces interactions between atmospheric waves. As a result, new wave modes are generated, including waves shorter than can be explicitly represented by the model grid. The energy of these short waves is then folded back into longer waves, so that wave energy tends to spuriously accumulate in the spectral region near the cut-off wavelength. This process, called *aliasing*, can be a source of nonlinear numerical instability in dynamical models. Short waves can be eliminated by a filtering process (see Section 4.15).

4.8.2 Finite Volume (Grid Cell) Methods

The finite volume method provides another approach for solving PDEs. It is particularly well suited for applications in which mass and energy conservation is a critical consideration. In this method, the prognostic quantities are not defined on discrete nodes i (grid points), but are instead expressed as averages across specified finite control volumes. These control volumes correspond to the specified *grid cells* of the model. In the simple 1-D formulation, where the volume of cell i is replaced by the size of the spatial mesh $\Delta x_i = x_{i+\frac{1}{2}} - x_{i-\frac{1}{2}}$, the average $\langle \Psi_i(t) \rangle$ of a variable Ψ over cell i is given by

$$\langle \Psi_i(t) \rangle = \frac{1}{\Delta x_i} \int_{\Delta x_i} \Psi(x, t) \, dx \qquad (4.222)$$

Consider, for example, a 1-D version of the continuity equation (similar to (4.1) but for a single geometrical dimension x, and for no chemical source):

$$\frac{\partial \Psi}{\partial t} + \frac{\partial F}{\partial x} = 0 \qquad (4.223)$$

where Ψ represents the mass or number density of a chemical species and F its flux in the x direction. The value of Ψ at time $t_{n+1} = t_n + \Delta t$ is obtained from the value at time t_n by

$$\Psi(x, t_{n+1}) = \Psi(x, t_n) - \int_{\Delta t} \frac{\partial F}{\partial x} \, dt \qquad (4.224)$$

The equation for the average of Ψ over grid cell i (denoted $\langle \Psi_i \rangle$) is

$$\langle \Psi_i(t_{n+1}) \rangle = \langle \Psi_i(t_n) \rangle - \frac{1}{\Delta x_i} \int_{\Delta x_i} dx \int_{\Delta t} \frac{\partial F}{\partial x} \, dt \qquad (4.225)$$

or, after integration over Δx_i

$$\langle \Psi_i(t_{n+1}) \rangle = \langle \Psi_i(t_n) \rangle - \frac{1}{\Delta x_i} \int_{\Delta t} \left(F_{i+1/2} - F_{i-1/2} \right) dt \qquad (4.226)$$

The time derivative of this last expression is

$$\frac{d \langle \Psi_i \rangle}{dt} = -\frac{F_{i+1/2} - F_{i-1/2}}{\Delta x_i} \qquad (4.227)$$

where the value of flux F is estimated at both edges of grid cell i. Note that no approximation has been made to derive expression (4.227).

If (4.223) represents the continuity equation for density Ψ and the flux $F = \Psi u$ is expressed as the product of this density by the velocity u in the x direction, (4.227) can be rewritten as

$$\frac{d\langle\Psi_i\rangle}{dt} = \frac{\Psi_{i-1/2}\, u_{i-1/2} - \Psi_{i+1/2}\, u_{i+1/2}}{\Delta x_i} \tag{4.228}$$

If the velocities are prescribed at the edges of the grid cells, and the corresponding values of Ψ are constructed by interpolation of the mean values in the neighboring cells, then (4.228) can be expressed as

$$\frac{d\langle\Psi_i\rangle}{dt} = \frac{1}{2\Delta x_i}\left[u_{i-1/2}(\langle\Psi_{i-1}\rangle + \langle\Psi_i\rangle) - u_{i+1/2}(\langle\Psi_{i+1}\rangle + \langle\Psi_i\rangle)\right] \tag{4.229}$$

This first-order differential equation can easily be integrated numerically (see example in Section 4.8.1).

An important advantage of this method (compared to the finite difference method) is that it does not require a structured geometry, but can easily be applied to different types of grids (see Section 4.8.3). Consider the more general case of the hyperbolic problem

$$\frac{\partial\Psi}{\partial t} + \nabla\cdot\mathbf{F} = 0 \tag{4.230}$$

where the flux vector $\mathbf{F}$ has components in all spatial dimensions. The spatial domain under consideration can be subdivided into finite volumes V_i, and the equation can be integrated over the volume of a cell i

$$\int_{V_i}\frac{\partial\Psi}{\partial t}\,dV + \int_{V_i}\nabla\cdot\mathbf{F}\,dV = 0 \tag{4.231}$$

Noting that the first term is equal to $V_i\, d\langle\Psi_i\rangle/dt$ and applying the divergence (Stokes) theorem to the second term, we find

$$\frac{d\langle\Psi_i\rangle}{dt} + \frac{1}{V_i}\int_{S_i}(\mathbf{F}\cdot\mathbf{n})dS = 0 \tag{4.232}$$

Here, the integral is taken along the walls S_i of cell i and represents the material flowing across the boundaries of the cell; $\mathbf{n}$ is the unit vector perpendicular to wall element dS. Again, the values of $\mathbf{F}$ at the edge of the cell can be constructed by interpolation of average values in neighboring cells. As can be deduced from the equations, the finite volume method conserves the transported variables easily even for coarse grids: What is lost in one cell is gained by a neighboring cell. This approach is therefore well adapted to treat the advective transport of tracers on complex grids (see Chapter 7).

4.8.3 Model Grids

Different types of grids can be adopted to solve the finite difference or finite volume approximations to the model equations. The choice of grid must balance considerations regarding the nature of the problem to be solved, the need for accuracy and stability, and the computational resources at hand. Another important consideration in chemical transport models using offline meteorological fields is to ensure consistency with the grid of the parent meteorological model and thus to avoid interpolation errors.

Vertical Discretization

We first consider the vertical discretization of the continuity equation for scalar quantity Ψ (here again a mass or number density), and write, for example, for the vertical component

$$\left[\frac{\partial \Psi}{\partial t}\right]^{vert} = -\frac{\partial (\Psi\, w)}{\partial z} \tag{4.233}$$

where z is the vertical coordinate adopted in the model (not necessarily the geometric altitude). If quantity Ψ and the vertical wind component w are defined on the same discrete levels ($j-1, j, j+1$), (4.233) can be approximated as

$$\left[\frac{\partial \Psi}{\partial t}\right]^{vert}_j = -\frac{\Psi_{j+1} w_{j+1} - \Psi_{j-1}\, w_{j-1}}{2\Delta z} \tag{4.234}$$

If, on the other hand, the vertical layers are *staggered* (Box 4.5) as shown in Figure 4.14, and if quantity Ψ is defined at levels $j-1, j$, and $j+1$, while the

---

Box 4.5   **Grid Staggering**

When applying a numerical algorithm to solve a system of differential equations, all unknown variables are not necessarily defined at the same grid points. In a *staggered grid* model, the different dependent variables are provided on different grid points, usually offset by half the grid size (see, for example, Figure 4.14 in the case of vertical grid staggering). Staggered grids are often adopted because the accuracy of the calculated spatial derivatives is improved. Further, the short wavelength components of the solution are often more accurately represented (Durran, 2010). In an unstaggered grid, the derivative G_i at location x_i of a function Ψ is often calculated across a $2\Delta x$ interval

$$G_i = \left[\frac{d\Psi}{dx}\right]_i \approx \frac{(\Psi_{i+1} - \Psi_{i-1})}{2\Delta x}$$

while, in a staggered grid, it can be derived across a single Δx interval

$$G_i = \left[\frac{d\Psi}{dx}\right]_i \approx \frac{(\Psi_{i+1/2} - \Psi_{i-1/2})}{\Delta x}$$

In the latter case, the derivative G is calculated on one grid ($i-1, i, i+1$) while function Ψ is defined on the other grid ($i-1/2, i+1/2$). Staggered grids have the advantage of effectively halving the grid size, so that the truncation errors are reduced. A standard configuration is to define wind components on one grid and other dependent variables such as pressure, temperature, concentrations, etc. on the other grid. The utility of this can directly be seen in the finite-volume approximation (4.229) of the advection equation. Since the effective grid size is smaller than the nominal size Δx, the time step Δt adopted for integrating the equations must be correspondingly reduced to maintain the stability of the solution.

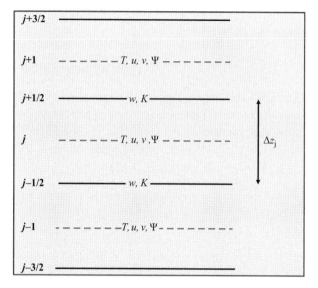

Figure 4.14 Example of a staggered vertical grid: temperature T, horizontal wind components (u, v), and function Ψ (e.g., trace species density ρ_i) are derived at the center of the cell (midpoint level) while the vertical wind component w and the eddy diffusion coefficient K are derived at the edge of the cell (interface level).

vertical wind component w is specified at levels, $j - \frac{1}{2}$ and $j + \frac{1}{2}$, the tendency resulting from the vertical component of the flux convergence is approximated by

$$\left[\frac{\partial\Psi}{\partial t}\right]_j^{vert} = -\frac{(\Psi w)_{j+1/2} - (\Psi w)_{j-1/2}}{\Delta z} \tag{4.235}$$

or, if function Ψ is linearly interpolated,

$$\left[\frac{\partial\Psi}{\partial t}\right]_j^{vert} = -\frac{(\Psi_j + \Psi_{j+1})w_{j+1/2} - (\Psi_j + \Psi_{j-1})w_{j-1/2}}{2\Delta z} \tag{4.236}$$

In general circulation models of the atmosphere, if the temperature and the vertical component of the wind are defined on intermediate levels between the geopotential and the horizontal wind components, the vertical staggering refers to the so-called Charney–Philips grid. If the geopotential, the temperature, and the horizontal wind components are provided on the same levels, while only the vertical wind component is defined on intermediate levels, the staggering is referred to as the Lorenz grid.

Horizontal Discretization

As the PDEs of the model are transformed into their finite difference or finite volume analogs, the variables at each level of the model have to be defined on a finite number of grid points or grid cells. Here again, different approaches can be used. The simplest of them is to define all variables on the same grid points and implement the discretization procedure accordingly. Such a grid is called *A-grid* by Arakawa and Lamb (1977). Despite its simplicity, it is little used because it often produces

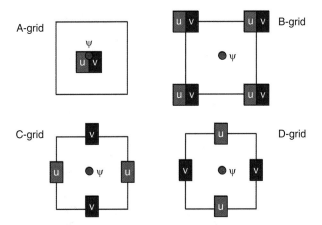

Figure 4.15 Unstaggered grid (A) and different staggered grids (B, C, and D) in two dimensions, using the Arakawa and Lamb (1977) classification. ψ is a scalar simulated by the model, for example chemical concentration; u and v are the zonal and meridional components of the wind velocity.

noise in the numerical solution of the equations. Alternate staggered configurations are shown in Figure 4.15. The *C-grid* is often adopted in atmospheric models because it is readily adapted to the finite-volume solution (4.229) of the continuity equation. A more detailed discussion of the different staggered grids is provided by Arakawa and Lamb (1977) and by Haltiner and Williams (1980).

Grid Geometries

The longitude–latitude grid is frequently adopted to represent the horizontal distribution of atmospheric variables (Figure 4.16). It has the advantage of being orthogonal so that the finite difference or finite volume approximations of the PDEs can be easily derived. A disadvantage is that the geometric spacing in longitude $\Delta x = a\, \Delta\lambda \cos \varphi$ decreases gradually from the Equator to the pole. (Here λ represents the longitude, φ the latitude, and a the Earth's radius). As discussed in Section 4.7.1, the time step Δt to be adopted in the model to ensure stability of the integration method is often bounded by the smallest geometric grid spacing. In the case of the advection equation, the time step applied in Eulerian schemes must be chosen in most cases to be smaller than the so-called *Courant number*, which is proportional to the grid size Δx (see Chapter 7 for definition and for examples). Thus, the existence of *singular points* at the poles introduces severe limitations on the adopted time step and leads to low computational performance.

Different approaches have been proposed to avoid the *pole problem*. The most straightforward is to apply either spatial or Fourier filters to eliminate the smaller-scale features (noise) appearing in the solutions near the poles. A second approach is to reduce the number of points in the longitudinal direction and hence to increase the longitudinal grid spacing in the vicinity of the poles. In Figure 4.17, a *reduced grid* is compared to a regular longitude–latitude grid. A third solution is to apply in the polar regions some backup algorithm that is not constrained by the Courant criteria (e.g., semi-Lagrangian methods, see Section 7.8).

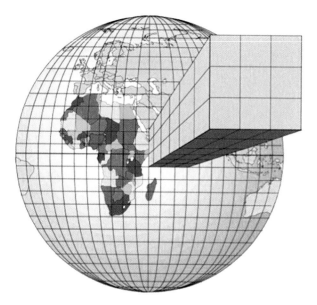

Figure 4.16 Latitude–longitude grid with representation of model levels and grid cells.

(a)   (b)

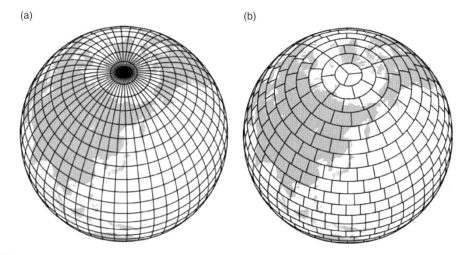

Figure 4.17 (a): Latitude–longitude grid. The convergence of the longitudinal lines toward the poles reduces gradually the geographical distances in the longitudinal direction, which in turn limits the time step required to solve the partial differential equations. (b): Reduced grid in which the number of longitudes per circle of latitude decreases as one approaches the poles. From Washington *et al.* (2009).

Other types of grids have been proposed to address the pole problem of the longitude–latitude grids. An increasingly frequent approach adopted in conjunction with finite volume methods (see Section 4.8.2) and referred to as "tiling" is to cover the surface of the sphere by some geometric shapes with no overlaps and no gaps. In the *Voronoi tessellation* approach, the surface of the Earth is partitioned into closed regions around pre-defined points, called *seeds* or *generators* of the grid. By definition, a Voronoi cell for a given seed includes all geometric points that are

(a)   (b)   (c)

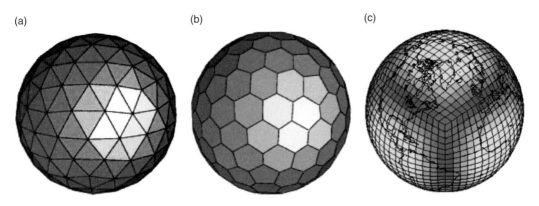

Figure 4.18 Examples of grids: icosahedral triangular (a), icosahedral hexagonal (b) and cubed sphere (c). Source: David Donofrio, Lawrence Berkeley National Laboratory (a and b).

Yang (N) system   **Yin (E) system**   **Yin–Yang** composition

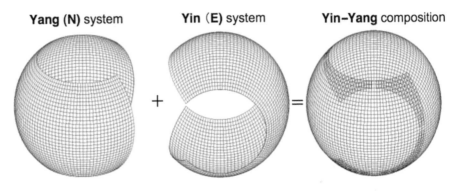

Figure 4.19 The Yin–Yang grid system. Reproduced From Qaddouri (2008).

closer to that particular point than to any other seed points in the model domain. The seed points can be specified so that the cells cover rather homogeneous areas of the surface (such as uniform ecosystems). They can also be chosen to have regular polygonal shapes (Figures 4.18 and 4.19). The *icosahedral grid*, for example, is characterized by a relatively uniform spatial resolution without any singularity. The grid elements can be triangular or hexagonal (with in this particular case 12 pentagonal cells). The *cubed sphere grid* also allows a relatively uniform resolution. In this approach, the existence of eight special "corner points" and plane boundaries requires special treatment. The Yin–Yang grid is the combination of two distinct longitude–latitude grid systems with mutually orthogonal axes and partial overlap (contact region). The advantage of homogeneous and highly isotropic model grids is that the spatial density of model results is relatively uniform, so that the data can be efficiently archived, analyzed, and remapped to other grid structures.

The choice of grid resolution in a model must be adapted to the scales of the processes that one wishes to resolve. Because of computational limitations, the use of a very high-resolution grid system covering the entire model domain may not be feasible; using a grid with fine resolution in areas that are presumed to be of interest, with coarser resolution elsewhere, may be an appropriate approach to address a given

(a)   (b)

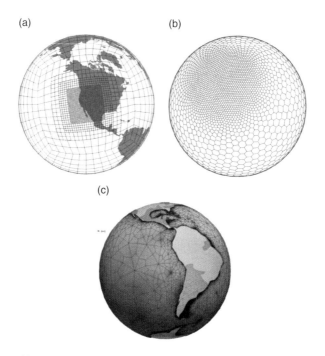

(c)

Figure 4.20   Model grids using (a) a cubed sphere with successive zooming capabilities for the Western USA; (b) a icosahedral system with a stretched grid and gradual zooming capability, and (c) an unstructured grid used here for tsunami simulations. From the National Center for Atmospheric Research (NCAR).

problem. *Mesh refinement* provides high resolution in selected areas of the model domain by embedding a higher resolution grid that resolves small-scale processes into a coarser grid that captures the large-scale features (Hubbard, 2002). A succession of nested grids (*zoom capability*) can be implemented to derive increasingly high-resolution features as one approaches a given point of the domain. The nesting can be *one-way*, with the coarser grid providing dynamic boundary conditions to the finer grid but no reverse effect of the finer grid on the coarser grid. It can be *two-way*, with full exchange of information between the two grids. Two-way nesting is far more difficult to implement because of numerical noise at the interface between grids. Figure 4.20 (a and b) shows two different approaches used in the development of multi-resolution models. *Unstructured grids* (Figure 4.20, c) in which the grid meshes (often triangles or tetrahedra) are distributed as irregular patterns, can also be adopted to enhance the spatial resolution in specified areas. They are often used in oceanic models. The location of the region of interest may also change with time, as in the case of the long-range transport of a pollution plume. A solution is to use a *dynamic adaptive grid*, in which the location of the nodes is constantly modified during the model integration to achieve high spatial resolution in areas where the gradients in the calculated fields are large (Odman *et al.*, 1997; Srivastava *et al.*, 2000). Adaptive grids used in atmospheric dynamical modeling (e.g., Dietachmayer and Droegemeier, 1992; Skamarock and Klemp, 1993) can also be adopted in chemical transport models to better represent the effects of multi-scale sources (Tomlin *et al.*, 1997; Box 4.6).

Box 4.6 — **Dynamically Adaptive Grids in Chemical Transport Models (Srivastava *et al.*, 2000; Odman *et al.*, 2002; Garcia-Menendez *et al.*, 2010)**

Large-scale models with limited spatial resolution may need to resolve small-scale processes in regions of high concentrations and strong concentration gradients. Static nested grids can provide high resolution over specific regions of interest (such as an urban center) but do not adjust to changing locations of these regions (as for long-range transport of a pollution plume). In the dynamic grid adaptation method, the grid resolution changes automatically at each time step to capture and follow small-scale features of interest such as plumes or frontal boundaries. The overall structure of the grid and the total number of grid points are fixed, but the locations of the grid points evolve through the simulation according to a user-defined weight function that determines in which areas grid nodes must be clustered. At a given grid point j, the weight function w_j can be expressed, for example, by

$$w_j = \sum_k \alpha_k \left[\nabla^2 n_k \right]_j$$

where k is an index for the chemical species, n_k is the concentration, and ∇^2 is the Laplacian operator which measures the curvature in the concentration field. Resolution requirements of different species can be accounted through the choice of the different coefficients α_k (Odman *et al.*, 2002). The new spatial position P_i^{new} of grid point i in a horizontal plane is calculated from

$$\mathbf{P}_i^{new} = \sum_{j=1}^{4} w_j \mathbf{P}_j \Big/ \sum_{j=1}^{4} w_j$$

where vector $\mathbf{P}_j$ ($j = 1$ to 4) represents the original position (before movement) of the centroids of the four grid cells that share grid point i (see Box 4.6 Figure 1, left panel). After the repositioning of the grid points, the concentration must be interpolated on the displaced grid nodes, and other parameters such as the emissions and meteorological variables must be redistributed on the adapted grid.

The two panels in Box 4.6 Figure 1 on the right show the spatial distribution of concentrations originating from a point source as simulated by a standard air quality model (CMAQ) at 1.33 km horizontal resolution (middle panel) and by the same model, but with a dynamically adapted grid (right panel). Sharper gradients with higher concentration peaks are obtained when the model grid is dynamically adapted.

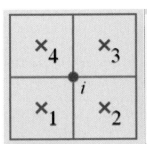 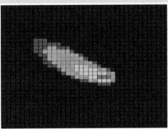 

Box 4.6 Figure 1 Development of a dynamically adaptive grid to represent the evolution of a plume. Sources: M. T. Odman, private communication; Steyn and Rao (2010).

Consistency Between Vertical and Horizontal Resolutions

To adequately resolve sloping features in atmospheric models including fronts and slantwise convective systems, and correctly represent the 3-D transport of chemical species associated with such dynamical systems, the grid increments in the vertical and horizontal directions should not be specified independently. Rather, the ratio between vertical and horizontal grid point spacings should be of the order of typically 0.005–0.02 m m^{-1} for an appropriate representation of mesoscale features (Warner, 2011). Thus, for a horizontal resolution of 100 km, typical of global models, the spacing between model layers should be of the order of 0.5–2 km. For a horizontal resolution of only 10 km, as often adopted in regional (limited-area) models, the vertical spacing should be reduced to 50–200 m. The lack of consistency between horizontal and vertical resolution in dynamical models can generate spurious waves during the simulation and thus undesired noisy fields.

4.9 Spectral Methods

An alternative approach to the finite difference and finite volume algorithms for global models is the *spectral method* developed by Orszag (1970) and Eliasen *et al.* (1970), and implemented for the first time by Bourke (1974). In this method, the horizontal distribution of the atmospheric variables is represented by a finite expansion of periodic functions (waves), called basis functions. The orthogonality of these functions (see Box 4.10) allows the derivation of coupled ODEs for the expansion coefficients, which vary with time and height. This method has been important for climate modeling but is rarely used in chemical transport models because chemical concentrations are strongly affected by local forcing processes that cannot be easily described by waves. In addition, spectral transport algorithms can produce negative concentrations.

To introduce the spectral method, we first consider a function $\Psi(x)$ defined on the spatial interval $[-\pi, +\pi]$. By applying a Fourier transform and assuming that the function repeats itself with a period 2π, we can write

$$\Psi(x) = a_0 + \sum_{k=1}^{\infty} \left(a_k \cos\left(kx\right) + b_k \sin\left(kx\right) \right) \tag{4.237}$$

where k represents the wavenumber. Expansion coefficients a_k and b_k are given by the following integrals over the $[-\pi, +\pi]$ interval

$$a_0 = \frac{1}{2\pi} \int_{-\pi}^{\pi} \Psi(x)\, dx \tag{4.238}$$

$$a_k = \frac{1}{\pi} \int_{-\pi}^{\pi} \Psi(x) \cos\left(kx\right) dx \tag{4.239}$$

$$b_k = \frac{1}{\pi} \int_{-\pi}^{\pi} \Psi(x) \sin\left(kx\right) dx \tag{4.240}$$

Box 4.7	An Example of Fourier Decomposition

We present here a simple illustration of the spectral decomposition method. A function $y(t)$ equal to 1 on interval [0, 0.5] and to -1 on interval [0.5, 1], as shown in Box 4.7 Figure 1, can be approximated by

$$y(t) = \frac{4}{\pi}\sin(2\pi t) + \frac{4}{3\pi}\sin(6\pi t) + \frac{4}{5\pi}\sin(10\pi t) + \cdots = \sum_{k=1}^{\infty} \frac{4}{k\pi}\sin(2k\pi t) \quad \text{for } k = 1, 3, 5, \ldots$$

The Fourier coefficients with an even numbered index disappear in this particular case.

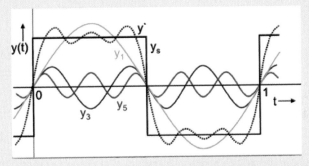

Box 4.7 Figure 1

Fourier decomposition of a square function y_s. The fundamental $y_1 = 3/\pi \sin 2\pi t$, and the two harmonics $y_3 = 4/(3\pi) \sin 6\pi t$ and $y_5 = 4/(5\pi)\sin(10\pi t)$ are shown together with the sum $y' = y_1 + y_2 + y_3$ (dotted line). Reproduced with permission From Cruse (2006).

Function $\Psi(x)$ may also be expressed as the sum of sine or cosine functions of different amplitudes d_k, wavenumbers k, and phases φ_k (Box 4.7):

$$\Psi(x) = \sum_{0}^{\infty} d_k \sin(kx + \varphi_k) \tag{4.241}$$

The component with the lowest frequency is called the *fundamental*, and any higher-frequency parts are referred to as *harmonic* components.

An equivalent, but often more convenient, form of the Fourier series is given by

$$\Psi(x) = \sum_{k=-\infty}^{\infty} c_k e^{ikx} \tag{4.242}$$

with

$$c_k = \frac{1}{2\pi} \int_{-\pi}^{\pi} \Psi(x)\, e^{-ikx}\, dx \tag{4.243}$$

This coefficient c_k is related to a_k and b_k in (4.237) by

$$c_k = \frac{1}{2}(a_k - ib_k) \quad \text{for } k > 0 \tag{4.244}$$

$$c_k = \frac{1}{2}\left(a_{|k|} + ib_{|k|}\right) \quad \text{for } k < 0 \tag{4.245}$$

with $c_0 = a_0$. In practical applications, the number of terms retained in the Fourier series is necessarily limited. A truncation is therefore applied when index k reaches a specified value K, and function $\Psi(x)$ is now approximated as

$$\Psi(x) \approx a_0 + \sum_{k=1}^{K}\left(a_k \cos(kx) + b_k \sin(kx)\right) \tag{4.246}$$

or

$$\Psi(x) \approx \sum_{k=-K}^{K} c_k e^{ikx} \tag{4.247}$$

Accuracy increases for higher values of K but so do computing costs.

The calculation of an approximation for the spatial derivative is straightforward:

$$\frac{d\Psi(x)}{dx} \approx \sum_{k=1}^{K}\left(\alpha_k \cos(kx) + \beta_k \sin(kx)\right) \tag{4.248}$$

or

$$\frac{d\Psi(x)}{dx} \approx \sum_{k=-K}^{K} \gamma_k e^{ikx} \tag{4.249}$$

with coefficients

$$\alpha_k = k\,b_k, \qquad \beta_k = -k\,a_k, \qquad \gamma_k = ik\,c_k \tag{4.250}$$

The calculation of the derivatives of a function is thus easy to perform once the Fourier expansion coefficients of the function have been derived. *Fast Fourier transform* (FFT) algorithms are routinely used in general circulation models to transfer information between the grid space and the spectral space (see Appendix E).

To solve PDEs such as the advection–diffusion equation (Box 4.8), the solution is approximated by an expansion such as (in one dimension)

Box 4.8   **Fourier Transform of the Advection–Diffusion Equation**

In the case of the 1-D advection–diffusion equation

$$\frac{\partial \Psi}{\partial t} + c\frac{\partial \Psi}{\partial x} - D\frac{\partial^2 \Psi}{\partial x^2} = 0$$

with the solution approximated by (4.251), we obtain the differential equations

$$\frac{da_k}{dt} + ik\,ca_k + k^2 Da_k = 0 \qquad (k = 0, K)$$

which are solved by an appropriate finite difference algorithm.

$$\Psi(x,t) \approx \sum_{k=0}^{K} a_k(t)\, e^{ikx} \tag{4.251}$$

in which the coefficients a_k are only a function of time t. When expression (4.251) is introduced in a PDE, one obtains for each value of k an ODE for a_k that is solved by a classic finite difference method. The algorithm requires that K differential equations be solved consecutively, which is usually computationally less expensive than solving the equation by a finite difference method at J grid points. Indeed, in most applications where function $\Psi(x, t)$ is relatively smooth, the value of K can be considerably smaller than the value of J. When function $\Psi(x)$ contains discontinuities or if too few terms are included in the Fourier series (too small value of K), the solution provided by the spectral decomposition method is characterized by large overshoots and undershoots (Box 4.9), and can produce undesired negative values. This is why the spectral approach does not provide satisfying results when applied to the transport of atmospheric species with strong gradients.

Application of the spectral method to the sphere can be done by expanding the functional forms $\Psi(\lambda, \mu, t)$ of the different variables as a function of longitude λ [0, 2π] and sine of latitude μ [-1, 1]) using normalized *spherical harmonics* $Y_n^m(\lambda, \mu)$ (see Figure 4.21):

$$\Psi(\lambda, \mu, t) = \sum_{m=-M}^{M} \sum_{n=|m|}^{N(m)} a_n^m(t)\, Y_n^m(\lambda, \mu) \tag{4.252}$$

where $a_n^m(t)$ are the spectral coefficients, which are the unknowns to be determined as a function of time t. The choice of parameters M and $N(m)$ define the truncation of

Box 4.9	Gibbs Phenomenon

The Gibbs phenomenon, named after J. Willard Gibbs, describes the peculiar manner in which the Fourier series of a piecewise continuously differentiable periodic function behaves at a jump discontinuity. The use of truncated series instead of infinite Fourier series leads to overshooting and undershooting of the true function. This effect can be illustrated by considering a step function whose value $\Psi(x)$ is -1 for $-\pi < x < 0$ and 1 for $0 < x < \pi$. This function can be expressed by the infinite series of periodic functions

$$\Psi(x) = \frac{4}{\pi}\left(\sin(x) + \frac{1}{3}\sin(3x) + \frac{1}{5}\sin(5x) + \cdots\right)$$

If the function $\Psi(x)$ is approximated by only the first term in the series, the maximum and minimum values of Ψ are $\pm 4/\pi$ (or ± 1.27), instead of 1 and -1, respectively. This misrepresentation of the field with overshoot and undershoot is called the Gibbs phenomenon. This effect decreases when more terms are included in the truncated series, and is less acute when the field contains no discontinuities.

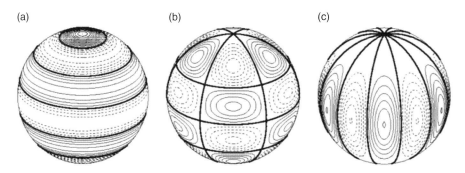

Representation of the characteristics of three spherical harmonics with total wavenumber $n = 6$. (a): zonal wavenumber $m = 0$, (b): $m = 3$ and (c): $m = 6$. From Williamson and Laprise (1998).

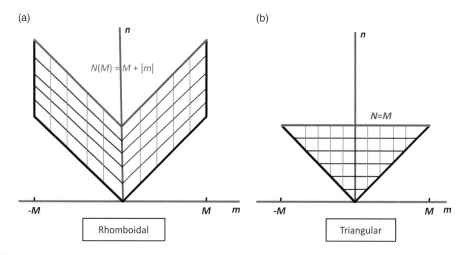

Rhomboidal (a) and triangular (b) truncations.

the expansion, and are discussed below. The spherical harmonics are the eigenfunctions of the Laplace operator on the sphere that verify the relation

$$\nabla^2 Y_n^m(\lambda, \mu) = -\frac{n(n+1)}{a^2} Y_n^m(\lambda, \mu) \tag{4.253}$$

They are expressed as a combination of sines and cosines (or equivalently by complex exponentials) to represent the periodic variations in the zonal direction, and by real associated Legendre functions of the first kind $P_n^m(\mu)$ (see Box 4.11) to account for the variations in the meridional direction. Thus,

$$Y_n^m(\lambda, \mu) = P_n^m(\mu) e^{im\lambda} \tag{4.254}$$

Here, index m represents the zonal wavenumber; its highest value M specifies the number of waves retained in the zonal direction. Index $n - |m|$ is called the meridional nodal number.

The type of truncation to be adopted for expression (4.252) is determined by the relation between the number of waves allowed in the zonal and the meridional directions. If N is chosen to be equal to M, the truncation is said to be *triangular*. If it is such that $N = M + |m|$, it is called *rhomboidal* (Figure 4.22). Triangular truncation is the universal choice for high-resolution models, while rhomboidal truncation is often adopted in the case of low-resolution atmospheric models.

We have seen that chemical concentrations are better described in the physical grid space than in the spectral space. *Pseudo-spectral* models allow certain processes to be treated on a physical grid while others are treated by the spectral method. In this approach, the governing equations are solved in the physical space by applying, for example, the finite difference method. The spatial derivatives, however, are calculated analytically after converting the physical quantities from the physical space to the spectral space. Rapid forward and inverse transforms of different variables between the physical and spectral space are thus required at each time step, and can be performed very efficiently by the FFT technique (see Appendix E). The aliasing problem arising from nonlinear terms in the transform process is avoided if the number of grid points in the physical grid is equal to $3M + 1$ in the zonal direction. The number of points in the meridional direction must be $(3M+1)/2$ for the triangular truncation and $5N/2$ for the rhomboidal truncation. A transformed grid that is often adopted is the *Gaussian grid* with grid points equally spaced along the longitudes, but not along the latitudes; their location in the meridional direction is defined by the roots μ_m ($m = 1, M$) of

$$P_M^0(\mu) = 0 \tag{4.255}$$

The Gaussian grid has no grid point at the pole. In a *reduced* Gaussian grid, the number of grid points in the zonal direction decreases toward the poles, which keeps the zonal distance between grid points approximately constant across the sphere.

One often wishes to characterize the spatial resolution of a spectral global model in terms of grid spacing L rather than by the highest wavenumber M. The definition of grid spacing that is equivalent to a given spectral resolution is not straightforward. Laprise (1992) suggests four possible approaches, whose corresponding grid spacing estimates differ by about a factor of 2. One simple approach is to calculate the average distance between grid points on the Gaussian grid (or equivalently the spacing L_1 [km] between longitudinal grid points at the Equator). With the triangular truncation, the equivalent grid spacing is

$$L_1 = \frac{2\pi a}{3M + 1} \approx \frac{13500 \text{ km}}{M} \tag{4.256}$$

where a (6378 km) represents the Earth's radius and M is again the largest wavenumber in the zonal direction. A second measure L_2 is half of the shortest resolved zonal wave at the Equator:

$$L_2 = \frac{\pi a}{M} \approx \frac{20000 \text{ km}}{M} \tag{4.257}$$

A third approach assumes that an equal area of the Earth's surface is assigned to every piece of information contained in the spherical series, with $(M+1)^2$ real coefficients in the case of a triangular truncation. This yields an area resolution $4\pi a^2/(M + 1)^2$, corresponding to a length L_3

$$L_3 = \frac{2\sqrt{\pi}a}{M + 1} \approx \frac{22600 \text{ km}}{M + 1} \tag{4.258}$$

Orthogonal Functions

Two functions Ψ_n and Ψ_m are said to be orthogonal if the integral of their product is zero. Thus, if δ_{nm} is the Kronecker delta (equal to zero if $m \neq n$ and one if $n = m$) and A is a normalization constant

$$\int_X \Psi_n(x)\Psi_m(x)\, dx = A\, \delta_{nm}$$

For example, when representing function $\Psi(x)$ by a series of elementary trigonometric functions as in (4.248),

$$\Psi(x) \approx a_0 + \sum_{k=1}^{K} (a_k \cos(kx) + b_k \sin(kx))$$

the orthogonal relationships are

$$\int_X \cos(nx)\sin(mx)\, dx = 0$$

$$\int_X \cos(nx)\cos(mx)\, dx = 0 \quad \text{if } m \neq n$$

$$\int_X \sin(nx)\sin(mx)\, dx = 0 \quad \text{if } m \neq n$$

The fourth definition is based on (4.253) that expresses the spherical harmonics $Y_n^m(\lambda, \mu)$ as the eigenfunctions of the Laplace operator on the sphere. By equating the eigenvalue of the highest resolved mode with the corresponding eigenvalue of Fourier modes in Cartesian geometry, Laprise (1992) deduces a fourth possible expression L_4 for the spatial resolution:

$$L_4 = \frac{\sqrt{2}\pi a}{M} \approx \frac{28000 \text{ km}}{M} \tag{4.259}$$

For example, the spatial resolution representative of a model with a T63 truncation is 1.8 degrees or 210 km along a latitude circle. It is 1.1 degrees or 125 km for a model with a T106 truncation.

Spectral methods have several advantages: The derivatives are accurately determined because they are calculated analytically with no related numerical diffusion. The method does not produce aliasing of the quadratic nonlinear terms and hence no nonlinear numerical instability, the use of staggered grids is avoided, and there is no pole problem. There are also disadvantages: the calculation of nonlinear terms is computationally expensive and the number of arithmetic operations increases faster with spatial resolution than in grid point models. Therefore, spectral methods are not well suited for implementation in massively parallel computing architectures. Spectral methods are also unsuitable for regional models where boundary conditions must

Box. 4.11   Associated Legendre Polynomials of the First Kind

Associated Legendre polynomials of degree n and order m are real functions of x and solution of the Legendre equation

$$\frac{d}{dx}\left[(1-x^2)\frac{dP_n^m(x)}{dx}\right] + \left[n(n+1) - \frac{m^2}{(1-x^2)}\right]P_n^m(x) = 0 \qquad \forall x \in [-1,+1]$$

They can be expressed analytically by

$$P_n^m(x) = \frac{(1-x^2)^{\frac{m}{2}}}{2^n n!}\frac{d^{n+m}}{dx^{n+m}}(x^2-1)^n$$

The first polynomials of degree $n(0$ to $5)$ for $m = 0$ are (Box 4.11 Figure 1)

$$P_0(x) = 1 \quad P_1(x) = x \quad P_2(x) = \frac{1}{2}(3x^2-1)$$

$$P_3(x) = \frac{1}{2}(5x^3-x) \quad P_4(x) = \frac{1}{8}(35x^4 - 30x^2 + 3)$$

$$P_5(x) = \frac{1}{8}(63x^5 - 70x^3 + 15x)$$

To satisfy the condition of orthonormality, the Legendre functions must obey

$$\frac{1}{2}\int\limits_{-1}^{+1}\left[P_n^m(x)\right]^2 dx = 1$$

and the normalized Legendre function is then expressed by

$$P_n^m(x) = \left[(2n+1)\frac{(n-m)!}{(n+m)!}\right]^{\frac{1}{2}}\frac{(1-x^2)^{\frac{m}{2}}}{2^n n!}\frac{d^{n+m}}{dx^{n+m}}(x^2-1)^n$$

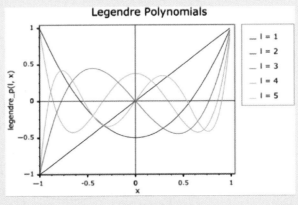

Legendre Polynomials

Box 4.11
Figure 1

First Legendre polynomials of degrees 1 to 5 displayed as a function of x. Copyright 2010 J. Maddock, P. A. Bristow, H. Holin, X. Zhang, B. Lalande, J. Råde, G. Sewani and T. van den Berg.

be specified at the edges of the modeling domain. Other disadvantages include non-conservation of mass and energy (Warner, 2011), spurious waves in the vicinity of large discontinuities (Gibbs phenomenon, see Box 4.9), and occurrence of negative values for the dependent variables.

4.10 Finite Element Method

The *finite element method* (Courant, 1943) is, to a certain extent, analogous to the spectral method. In this numerical technique, the solution of the partial differential equations is also expressed by a finite sum of spatially varying basis functions, but rather than being global as in the spectral method, they are expressed by low-order polynomials that are different from zero only in localized regions. Again, the PDEs are transformed into ordinary differential equations that are numerically integrated by standard techniques. To describe the method, we consider the PDE (in the 1-D space)

$$\mathcal{L}[\Psi] = F(x) \tag{4.260}$$

to be solved over the model domain $[a,b]$ with specified boundary conditions $\Psi(a)$ and $\Psi(b)$. Here, $\mathcal{L}$ represents a differential operator. For example, in the case of the linear 1-D advection equation with a velocity u, this operator can be expressed by $\partial/\partial t + u \, \partial/\partial x$, and $F(x) = 0$. In the *finite element method*, the solution $\Psi(x, t)$ of the PDE is approximated by a finite series of specified basis orthogonal functions $\Phi_k(x)$ that are non-zero only in a small part of the total domain called *finite element* (see below and Figure 4.23):

$$\Psi(x, t) = \sum_{k=1}^{K} a_k(t) \, \Phi_k(x) \tag{4.261}$$

The unknown coefficients a_k are a function of time t. As in the case of the spectral method, an advantage of this formulation is that the space derivative can be calculated exactly from the known basis functions.

The error resulting from the fact that the approximation in (4.261) retains only K terms is

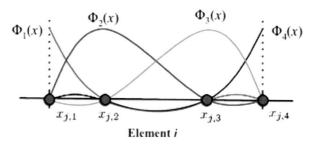

Figure 4.23 Example of four basis functions in a finite element j as a function of the spatial dimension x. Personal communication of Paul Ullrich, University of California Davis.

$$e_K = \mathcal{L} \left[\sum_{k=1}^{K} a_k(t)\, \Phi_k(x) \right] - F(x) \tag{4.262}$$

In the algorithm introduced by Galerkin, the residual e_K is required to be orthogonal to each basis function $\Phi_j(x)$, so that for all values of $j = 1, K$

$$\int_a^b e_K \Phi_j(x)\, dx = 0 \tag{4.263}$$

By substitution, we derive a system of K algebraic equations for the unknown coefficients $a_k(t)$

$$\int_a^b \Phi_j(x)\, \mathcal{L} \left[\sum_{k=1}^{K} a_k(t)\, \Phi_k(x) \right] dx - \int_a^b \Phi_j(x) F(x)\, dx = 0 \qquad (j = 1, \ldots K) \tag{4.264}$$

This system of K coupled ordinary equations can be solved to obtain the coefficients $a_k(t)$. For this purpose, the time derivatives included in operator $\mathcal{L}$ are usually approximated by finite differences. The basis functions inside each element are often expressed by a piecewise linear function called Chapeau function (Figure 4.24, a)

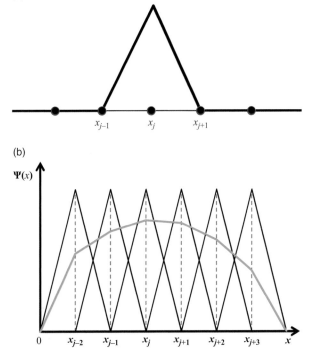

Figure 4.24   (a) Chapeau function for finite element j. (b) Linear combination of basis functions (black line) to define a piecewise (in green) linear function that approximates the solution $\Psi(x)$.

equal to unity at the jth node and zero to all other nodes. Its mathematical formulation is

$$\Phi_j(x) = \frac{x - (j-1)\,\Delta x}{\Delta x} \qquad \text{for } (j-1)\,\Delta x < x < j\Delta x \qquad (4.265)$$

$$\Phi_j(x) = \frac{(j+1)\,\Delta x - x}{\Delta x} \qquad \text{for } j\Delta x < x < (j+1)\Delta x \qquad (4.266)$$

$$\Phi_j(x) = 0 \qquad\qquad\qquad\qquad\qquad \text{elsewhere.} \qquad (4.267)$$

An approximation of the true function is provided by a linear combination of the Chapeau functions (Figure 4.24, b). Higher-order shapes (polynomial, curvilinear elements) can also be adopted as basis functions.

Contrary to the spectral method that describes the physical quantities by expansion functions over the *entire* domain, the finite-element method represents the fields for individual elements by a combination of a small number of basis functions (Figure 4.24, b). Much of the computation is thus local to a single element. In this case, the accuracy is not anymore achieved by using a large number of basis functions but by increasing the number of elements inside the entire model domain. These elements can be made smaller where a high-resolution formulation is required. The finite element technique is attractive for complex grid geometries including unstructured meshes or when the desired precision varies over the entire domain. It has been developed and often applied for aeronautic or civil engineering applications, but is not often used in chemical transport modeling. It could become an attractive approach for unstructured or adaptive grids. An application of the method to the 1-D advection equation is given in Section 7.8. The *spectral element method* (Canuto *et al.,* 1984; Patera, 1984), which is used in some modern atmospheric general circulation models (Fournier *et al.*, 2004), is a high-order finite element technique that combines the geometric flexibility of finite elements with the high accuracy of spectral methods.

4.11 Lagrangian Approaches

The methods described in the previous sections are Eulerian since the dependent variables are calculated relative to a numerical grid attached to the rotating Earth. In the Lagrangian approach, the dependent variables such as chemical concentrations are calculated following the trajectory of infinitesimally small air parcels (called *particles*) displaced by air motions. Different types of Lagrangian models have been developed (Figure 4.25) and we give here a brief overview.

4.11.1 Models for Single Trajectories

The simplest Lagrangian models are trajectory models that track the time evolution of a single particle along its trajectory (or of an ensemble of neighboring particles to

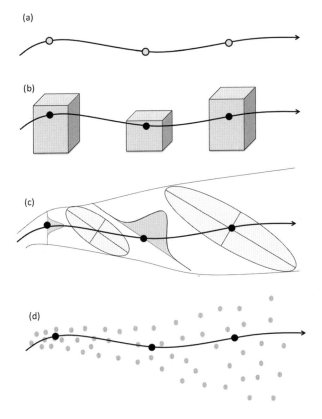

(a)

(b)

(c)

(d)

Figure 4.25 Schematic representation illustrating different Lagrangian model formulations. (a) Trajectory of a single air particle that retains its identity as it moves along a single line determined by the mean wind velocity; turbulent mixing is neglected. This formulation is appropriate when the flow is laminar. (b) Trajectories of multiple particles aggregated in a small volume (box). Each individual particle is displaced along the flow. The deformation of the box provides information on the dispersion of the particles resulting from inhomogenieties in the flow. (c) Gaussian puff advected by the wind and affected by turbulent dispersion (see Section 4.12). (d) Particle dispersion model for a large number of individual particles with random wind velocities (turbulence) generated by a stochastic (Markov) process. From Lin (2012).

account for errors in airflow). Particles are assumed to retain their identity during their displacement. They transport physical quantities such as momentum, energy (i.e., potential temperature), water vapor, or chemicals. Trajectory models are simple and useful, for example, to track the fate of a pollution plume, or to determine the sources contributing to concentrations observed at a receptor site.

The Lagrangian trajectory of a particle located at $\mathbf{r}(t)$ with three spatial components (x, y, z) is derived by integrating

$$\frac{d\mathbf{r}(t)}{dt} = \mathbf{v}(\mathbf{r}(t)) \tag{4.268}$$

in which $\mathbf{v}(u, v, w)$ is the velocity vector along the particle trajectory. If the wind field $\mathbf{v}(\mathbf{r}(t))$ and the initial position of the particle are known, the trajectory is completely determined. The "zero acceleration solution," which is computationally inexpensive but only first-order accurate, provides the position of the particle at time $t + \Delta t$ as a function of the particle at time t

$$\mathbf{r}(t + \Delta t) = \mathbf{r}(t) + \Delta t \, \mathbf{v}(\mathbf{r}(t)) \tag{4.269}$$

where Δt is the integration time step. A second-order accurate solution is obtained from the so-called "constant acceleration solution" or Pettersen's scheme

$$\mathbf{r}(t + \Delta t) = \mathbf{r}(t) + \frac{\Delta t}{2} \left[\mathbf{v}(\mathbf{r}(t)) + \mathbf{v}(\mathbf{r}(t + \Delta t)) \right] \tag{4.270}$$

This implicit expression is solved by an iterative method. Expressions with higher-order approximations can also be used. One constructs in this manner forward trajectories or, if the sign of the time interval is reversed, backward trajectories. Wind vectors used to construct the trajectory are typically interpolated from an archive of assimilated meteorological data available on a fixed grid and time interval. Errors on the calculated trajectories arise from the truncation in the finite difference scheme, the quality of the wind fields, the interpolation of the winds, poor knowledge of the unmeasured vertical wind component, and the starting position of the trajectory (Stohl, 1998). Back trajectories constructed to analyze the history of an air parcel are often useful only for a few days, beyond which the errors grow too large, especially if the backward-moving particles encounter convective situations with large unresolved vertical motions.

Local sources and sinks may affect the chemical variables transported along trajectories. The change in mixing ratio μ_i of a species i along a trajectory is obtained by integrating the Lagrangian form of the continuity equation:

$$\frac{d\mu_i}{dt} = S_i \tag{4.271}$$

Here, S_i [expressed for example in ppm s^{-1}] represents the net source rate of species i. As the particle is transported from departure point A at time t_A to arrival point B at time t_B, the change in mixing ratio is obtained by integration along the trajectory

$$\mu_i(\mathbf{r}_B, t_B) = \mu_i(\mathbf{r}_A, t_A) + \int_A^B S_i dt \tag{4.272}$$

S_i may include a source term from emission or chemical production, available as input on the same grid and time interval as the winds; and a sink from first-order loss. The source term must be interpolated in the same way as the winds, and its value along the trajectory between points A and B is often estimated at the midpoint or as the average between values at A and B. The first-order loss is applied as an exponential decay along the trajectory.

4.11.2 Stochastic Models

Lagrangian stochastic (LS) models, also called Lagrangian particle dispersion models (LPDMs), simulate the transport and dispersion of chemicals by calculating the random trajectories of an ensemble of particles in the turbulent flow. In the zeroth-order formulation, the displacement of particles is treated as a Markov chain, where the probabilities of future states do not depend on the path by which the present state was achieved. The position $\mathbf{r}$ of each particle is determined by a sequence of random increments. The change in the three components (x, y, z) of the position $\mathbf{r}$ of a fluid element is derived from the stochastic differential equations

$$dx = \alpha_x dt + \sigma_x d\xi_x \qquad dy = \alpha_y dt + \sigma_y d\xi_y \qquad dz = \alpha_z dt + \sigma_z d\xi_z \quad (4.273)$$

where the terms with α_x, α_y, and α_z represent the deterministic motions and the terms with σ_x, σ_y, and σ_z represent the stochastic component. Components $d\xi_x$, $d\xi_y$, and $d\xi_z$ are uncorrelated, and chosen so that their mean values equal zero and their variances equal dt. One can show (Boughton $et\ al.$, 1987; Sportisse, 2010) that the mean concentration obtained from the stochastic equations satisfies the Eulerian advection–diffusion equation if

$$\alpha_x = \overline{u} + \frac{\partial K_x}{\partial x} \qquad \alpha_y = \overline{v} + \frac{\partial K_y}{\partial y} \qquad \alpha_z = \overline{w} + \frac{\partial K_z}{\partial z} \quad (4.274)$$

and

$$\sigma_x = \sqrt{2K_x} \qquad \sigma_y = \sqrt{2K_y} \qquad \sigma_z = \sqrt{2K_z} \quad (4.275)$$

where $\overline{\mathbf{v}}(\overline{u}, \overline{v}, \overline{w})$ is the 3-D field of the mean (resolved) wind velocity and (K_x, K_y, K_z) are the diffusion coefficients for the three directions. Under these conditions, the particle position evolves over time step Δt as

$$x(t + \Delta t) = x(t) + \left[\overline{u} + \frac{\partial K_x}{\partial x}\right]\Delta t + \sqrt{2K_x}\Delta\xi_x$$

$$y(t + \Delta t) = y(t) + \left[\overline{v} + \frac{\partial K_y}{\partial y}\right]\Delta t + \sqrt{2K_y}\Delta\xi_y \quad (4.276)$$

$$z(t + \Delta t) = z(t) + \left[\overline{w} + \frac{\partial K_z}{\partial z}\right]\Delta t + \sqrt{2K_z}\Delta\xi_z$$

The three components $\Delta\xi_x$, $\Delta\xi_y$, $\Delta\xi_z$, known as the Wiener–Lévy process in the theory of Brownian motion, are stochastic components with mean values equal to zero and variances equal to Δt.

In the first-order method, the particle path is derived by a sequence of random increments applied to the turbulent velocity $\mathbf{v}'$ (rather than the position) of the particle. The turbulent motions are represented again by a Markov process and the variation in the wind velocity by a stochastic differential equation established by Langevin (1908) to describe Brownian motion (Thomson, 1987):

$$du = a_x dt + b_{x,x} d\xi_x + b_{x,y} d\xi_y + b_{x,z} d\xi_z$$

$$dv = a_y dt + b_{y,x} d\xi_x + b_{y,y} d\xi_y + b_{y,z} d\xi_z \quad (4.277)$$

$$dw = a_z dt + b_{z,x} d\xi_x + b_{z,y} d\xi_y + b_{z,z} d\xi_z$$

The deterministic part of the acceleration is defined by vector $\mathbf{a} = (a_x, a_y, a_z)^T$ and the random component is defined by the 3×3 matrix $\mathbf{B} = (b_{ij})$. In most cases, only the diagonal components of **B** (b_{xx}, b_{yy}, b_{zz}) are retained. In these relations, the terms $d\xi_x$, $d\xi_y$ and $d\xi_z$ denote again the spatial components of a Gaussian white noise with an average of zero and a variance of dt. Expressions for these terms are available for different types of turbulence (see Wilson and Sawford, 1996).

For a stationary and horizontally homogeneous turbulent flow with a constant mean wind $\bar{u}$ directed in the x-direction ($\bar{v} = \bar{w} = 0$) and a constant air density, Luhar (2012) writes the following Langevin equations for the three components of the velocity

$$du = -\frac{u - \bar{u}}{T_{Lx}} dt + \left(\frac{2\sigma_u^2}{T_{Lx}}\right)^{1/2} d\xi_x(t)$$

$$dv = -\frac{v}{T_{Ly}} dt + \left(\frac{2\sigma_v^2}{T_{Ly}}\right)^{1/2} d\xi_y(t) \qquad (4.278)$$

$$dw = \left[-\frac{v}{T_{Lz}} + \frac{1}{2}\left(1 + \frac{w^2}{\sigma_w^2}\right)\frac{\partial\sigma_w^2}{\partial z}\right] dt + \left(\frac{2\sigma_w^2}{T_{Lz}}\right)^{1/2} d\xi_z(t)$$

where σ_u^2, σ_v^2 and σ_w^2 denote the variances of the three spatial components of the wind velocity. The Lagrangian timescales in the three spatial directions are

$$T_{Lx} = \frac{2\sigma_u^2}{3\varepsilon} \qquad\qquad T_{Ly} = \frac{2\sigma_v^2}{3\varepsilon} \qquad\qquad T_{Lz} = \frac{2\sigma_w^2}{3\varepsilon} \qquad (4.279)$$

Here, ε denotes the turbulent kinetic energy dissipation rate. For homogeneous turbulence in the vertical direction, $\partial\sigma_w^2/\partial z = 0$. In atmospheric boundary layer problems, the mean horizontal wind velocity is usually large relative to turbulent fluctuations, and the Langevin equation is written only for the vertical wind component (Stohl, 2005).

Once the velocity components have been calculated, the position ($x(t)$, $y(t)$, $z(t)$) of the particle is obtained by integration of

$$dx = u \, dt \qquad\qquad dy = v \, dt \qquad\qquad dz = w \, dt \qquad (4.280)$$

Lagrangian models can be run forward or backward in time. In the forward mode, particles released at one or more source locations are transported by the fluid motions through the model domain. In the backward mode, particles released at a receptor point are used to determine the upwind influences at that point (Figure 4.26). A frequent application of the backward mode is to determine the surface flux footprint contributing to the atmospheric concentration at a given location.

4.11.3 Global and Regional Three-Dimensional Lagrangian Models

The Lagrangian formalism can be applied for calculating the spatial distribution and temporal evolution of chemical composition in global or regional atmospheric domains. In these models, one follows the displacement and dispersion of a large number of particles (typically 10^6 or more). The particles maintain their integrity as the model integration proceeds, so that particles of different origins in the spatial domain never mix and the chemical species that they carry are not allowed to react.

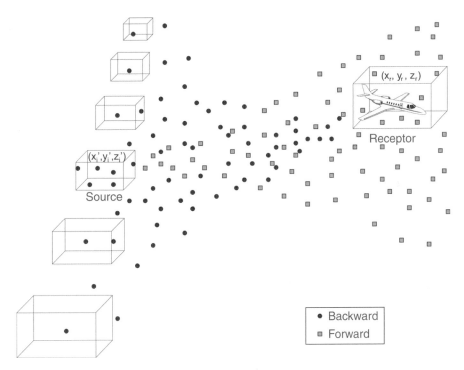

Figure 4.26 Lagrangian receptor model to determine the sources ("footprint") of chemical concentrations measured at a receptor point at a given time. A large number of Lagrangian particles are released at the receptor point and observation time, and are then tracked back in time. They diverge as a result of random turbulence, wind shear, and convection. A surface footprint for the observations can be determined from the statistics of the particles reaching the surface at all backward times. Adapted from Lin *et al.* (2003).

Accounting for chemical reactions is often done on an Eulerian grid interpolated from the Lagrangian particle information.

Lagrangian models have several advantages over their Eulerian counterparts. There is no numerical diffusion so that sharp gradients are preserved. The effective resolution in regions of particular interest can be readily enhanced by increasing the density of particles. The numerical integration of the trajectory equations is stable and the time step is therefore not limited by the Courant–Friedrichs–Lewy (CFL) criterion (see Section 7.3). The independence of calculations along individual trajectories is particularly attractive for massively parallel computing architectures.

Lagrangian formulations also have several disadvantages. The lack of mixing between air parcels generates spurious small-scale features that would be smoothed in the real atmosphere by turbulent mixing. Not accounting for mixing is problematic for simulating nonlinear chemistry and aerosol processes where mixing can greatly affect rates. Lagrangian models also generally cannot provide uniformly dense coverage of a given 3-D domain; depending on shear in the flow, particles may cluster in some regions of the domain while leaving other regions unsampled.

4.11.4 Semi-Lagrangian Models

A *semi-Lagrangian* approach is often adopted in global models. Here, the concentrations are calculated at fixed points on an Eulerian grid. The transport over a time step $\Delta t = t_{n+1} - t_n$ is calculated by initiating back trajectories over time Δt from the location of each gridpoint at time t_{n+1}. The location of the departure point, i.e., the location of the corresponding particle at the previous time t_n, is derived by integrating the trajectory backward in time. The departure point does not in general correspond to a gridpoint, and thus the mixing ratio at that departure point is derived by interpolating the values from neighboring grid points at time t_n. A great advantage of semi-Lagrangian methods is that they are computationally stable for any time step. Chemical reactions are computed on the model grid, allowing for mixing and nonlinear processes. However, interpolation of the concentration field compromises mass conservation and this requires correction at every time step. Semi-Lagrangian algorithms have been combined with finite volume Eulerian methods (Lin and Rood, 1996) to overcome Courant number limitations. More details on semi-Lagrangian methods are provided in Section 7.8.

4.12 Atmospheric Plume Models

The atmospheric dispersion and chemical evolution of a plume originating at a source point $\mathbf{r}_0$ can be represented by *Gaussian plume models*. Such models have the computational economy of trajectory models while accounting for small-scale eddy motions (turbulent diffusion) and allowing for nonlinear chemistry.

If the plume is transported in the x-direction by the prevailing wind velocity u (assumed here to be constant) and is dispersed by turbulent motions in the perpendicular (y, z) plane, the continuity equation for the density $\rho(x, y, z, t)$ [kg m^{-3}] of a given species can be expressed in an Eulerian framework by

$$\frac{\partial \rho}{\partial t} + u \frac{\partial \rho}{\partial x} = K_y \frac{\partial^2 \rho}{\partial y^2} + K_z \frac{\partial^2 \rho}{\partial z^2} \qquad (4.281)$$

where K_y and K_z [m^2 s^{-1}] are eddy mixing coefficients in the y and z directions, respectively, for which semi-empirical formulations are available (Pasquill, 1971; Seinfeld and Pandis, 2006). This simplified expression ignores chemical transformations. It assumes that the air density is uniform and that eddy mixing in the direction x of the wind can be neglected relative to the advection (this is called the *slender plume approximation*). The source rate of the species at point $\mathbf{r}_0$ is expressed as a boundary condition (see below).

An alternative approach to the above dispersion equation is to use a Lagrangian formulation. In this formulation, the density $\rho(\mathbf{r}, t)$ at point $\mathbf{r}$ and time t is determined by an ensemble of individual particles released at points $\mathbf{r}'$ and previous times t', and displaced by a randomly varying wind velocity to reach point $\mathbf{r}$ at time t. The variations in the wind velocity account for all scales of motion including the small-scale turbulent features. One can then write

$$\rho(\mathbf{r},t) = \int_0^t dt' \int_V s(\mathbf{r}',t')\, G(\mathbf{r},t;\mathbf{r}',t')\, d\mathbf{r}' \qquad (4.282)$$

where $s(\mathbf{r}',t')$ [kg m^{-3} s^{-1}] is the source term. The Green function $G(\mathbf{r},t;\mathbf{r}',t')$ defines the probability that a particle released at point $\mathbf{r}'$ at time t' reaches point $\mathbf{r}$ at time t. It is assumed here that the initial concentration is zero but the formulation can easily be generalized (see Section 4.2.8). In most applications, turbulence is considered to be stationary and homogeneous, and the probability distribution of the velocity is assumed to be Gaussian. When considering a point source, the source s is equal to zero at all points $\mathbf{r}'$ except at a single point $\mathbf{r}' = \mathbf{r}_0$.

We now consider two specific applications: the Gaussian plume model, in which the point emission is constant in time, and the puff model, in which a given mass of material is released instantaneously. The two are related in that a Gaussian plume can be viewed as a superimposition of elementary puffs released continuously in time, but the assumption of constant emission enables a straightforward analytical solution in the Gaussian plume model, while the puff model can be generalized to variable winds and emissions.

4.12.1 Gaussian Plume Models

Let's assume that the species under consideration is emitted with a constant source rate Q [kg s^{-1}] at a single point located at $\mathbf{r}_0$ with coordinates $x_0 = 0$, $y_0 = 0$ and at height $z_0 = H_e$ above the surface, so that

$$s(x,y,z) = Q\,\delta(x)\,\delta(y)\,\delta(z - H_e) \qquad (4.283)$$

Here, H_e [m] denotes the effective source height (which may account for buoyancy in the case of a heated source). The Dirac delta function $\delta(\xi)$ is equal to 1 for $\xi = 0$ and zero elsewhere, and has unit of m^{-1}. With the assumptions mentioned above and for steady-state conditions ($\partial\rho/\partial t = 0$), the Eulerian advection–diffusion form of the continuity equation takes the form

$$u\frac{\partial\rho}{\partial x} = K_y\frac{\partial^2\rho}{\partial y^2} + K_z\frac{\partial^2\rho}{\partial z^2} \qquad (4.284)$$

with the following boundary conditions

$$\rho(0,y,z) = \frac{Q}{u}\,\delta(y)\,\delta(z - H_e)$$

$$\rho(x,y,z) = 0 \quad \text{for } y \to \pm\infty \text{ and } z \to \infty \qquad (4.285)$$

An additional condition at the surface ($z = 0$) must be provided. One can assume total reflection (no uptake), or total or partial absorption (uptake). In the first case, the condition at the surface (zero flux) is written

$$K_z\frac{d\rho}{dz}(x,y,0) = 0 \qquad (4.286)$$

It requires that the vertical gradient in the concentration be zero at the surface. The analytical solution of (4.284) can be found by separation of variables

$$\rho(x,y,z) = \frac{Q}{u} \, \Phi(x,y)\Psi(x,z) \tag{4.287}$$

If, in addition, we replace the independent variable x by

$$r_y(x) = \frac{1}{u}\int_0^x K_y(x')dx' \qquad r_z(x) = \frac{1}{u}\int_0^x K_z(x')dx' \tag{4.288}$$

the PDEs for the dependent variables $\Phi(r_y, y)$ and $\Psi(r_z, z)$ are expressed by

$$\frac{\partial \Phi}{\partial r_y} = \frac{\partial^2 \Phi}{\partial y^2} \tag{4.289}$$

for the domain $0 \le r_y < \infty$ and $-\infty < y < \infty$ with the boundary conditions

$$\Phi(0,y) = \delta(y) \qquad \Phi(\infty,y) = 0 \qquad \Phi(r_y, \pm\infty) = 0 \tag{4.290}$$

and by

$$\frac{\partial \Psi}{\partial r_z} = \frac{\partial^2 \Psi}{\partial z^2} \tag{4.291}$$

for the domain $0 \le r_z < \infty$ and $0 \le z < \infty$ with the boundary conditions

$$\Psi(0,y) = \delta(z - H_e) \quad \Psi(\infty,z) = 0 \quad \Psi(r_z,\infty) = 0 \quad \text{and} \quad \frac{\partial \Psi}{\partial z}(r_z,0) = 0 \tag{4.292}$$

The corresponding solution is

$$\Phi(r_y,y) = \frac{1}{(4\pi r_y)^{1/2}} e^{-y^2/4r_y} \tag{4.293}$$

and

$$\Psi(r_z,z) = \frac{1}{(4\pi r_z)^{1/2}} \left[\exp\left(-\frac{(z-H_e)^2}{4r_z} \right) + \exp\left(-\frac{(z+H_e)^2}{4r_z} \right) \right] \tag{4.294}$$

and the spatial distribution of the concentration is therefore

$$\rho(x,y,z) = \frac{Q}{4\pi\bar{u}(r_y r_z)^{1/2}} \exp\left(-\frac{y^2}{4r_y} \right) \left[\exp\left(-\frac{(z-H_e)^2}{4r_z} \right) + \exp\left(-\frac{(z+H_e)^2}{4r_z} \right) \right] \tag{4.295}$$

Quantities r_y and r_z are provided by (4.288). If the eddy diffusion coefficients are uniform, we have

$$r_y(x) = \frac{K_y x}{u} \qquad r_z(x) = \frac{K_z x}{u} \tag{4.296}$$

The solution of the advection–diffusion equation is then expressed as a function of the eddy diffusion coefficients by

$$\rho(x,y,z) = \frac{Q}{4\pi x \left(K_y K_z\right)^{1/2}} \exp\left(-\frac{u\,y^2}{4K_y x}\right) \left[\exp\left(-\frac{u(z - H_e)^2}{4K_z x}\right)\right.$$

$$\left. + \exp\left(-\frac{u(z + H_e)^2}{4K_z x}\right)\right] \tag{4.297}$$

If we adopt a Lagrangian point of view and derive the concentration field by considering the displacement of an ensemble of fluid particles continuously emitted at point $(0, 0, H_e)$ at a constant rate Q, the application of relation (4.282) with a Gaussian probability distribution function for the particles leads to the following expression for the concentration at $t \to \infty$

$$\rho(x,y,z) = \frac{Q}{2\pi u \sigma_y \sigma_z} \exp\left(-\frac{y^2}{2\sigma_y^2}\right) \left[\exp\left(-\frac{(z - H_e)^2}{2\sigma_z^2}\right) + \exp\left(-\frac{(z + H_e)^2}{2\sigma_z^2}\right)\right] \tag{4.298}$$

where $\sigma_y(x)$ and $\sigma_z(x)$ denote the standard deviations of the particle distribution in the y and z direction, respectively (see Figure 4.27). Transport down to the surface

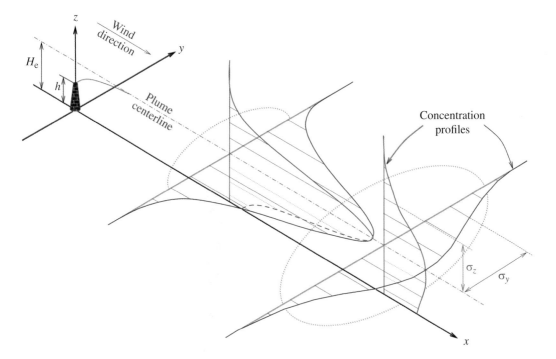

Figure 4.27 Gaussian plume released by a continuous point source (here a stack located at point $x = 0$; $y = 0$; $z = h$). The wind direction is aligned with the x-direction. The concentration distribution resulting from the dispersion in the y, z plane is shown at two downwind locations. Reproduced from Stockie (2011). Copyright © 2011 Society for Industrial and Applied Mathematics.

($\sigma_z = H_e$) occurs at a distance $x_D = u\,H_e^2/2K_z$. Beyond that distance, the effect of the surface becomes important.

The Eulerian and Lagrangian formulations lead to equivalent results if

$$\sigma_y^2(x) = 2\,r_y(x) = \frac{2}{u}\int_0^x K_y(x')dx' \qquad \text{and} \qquad \sigma_z^2(x) = 2\,r_z(x) = \frac{2}{u}\int_0^x K_z(x')dx'$$

$$(4.299)$$

If K_y and K_z are uniform,

$$\sigma_y^2(x) = \frac{2K_y x}{u} \qquad \text{and} \qquad \sigma_z^2(x) = \frac{2K_z x}{u} \qquad (4.300)$$

The standard deviations $\sigma_y(x)$ and $\sigma_z(x)$ vary therefore as the square root of the downwind distance x. Experimental studies show, however, that the exponent of this power law relationship is generally higher than 0.5. More elaborate empirical relations have therefore been established to express the standard deviations as a function of micrometeorological parameters and atmospheric stability (Pasquill, 1971).

The formulation presented here for a continuous point source can be generalized to more complex situations (Lin and Hildemann, 1996) involving, for example, multiple point sources, line sources (e.g., roads), the presence of an inversion layer aloft, absorbing rather than reflecting surfaces, and the addition of chemical and scavenging processes.

4.12.2 Puff Models

We now consider the time evolution of a puff of mass Q_P [kg] instantaneously released at time $t = 0$ and at a given point $(0, 0, H_e)$ in the atmosphere and subject to advection (constant wind speed u) in the x-direction and dispersion (constant eddy diffusion coefficients K_x, K_y, K_z) in the three spatial directions. We assume again total reflection of the material at the surface. The advection–diffusion equation to be solved is

$$\frac{\partial \rho}{\partial t} + u\frac{\partial \rho}{\partial x} = K_x\frac{\partial^2 \rho}{\partial x^2} + K_y\frac{\partial^2 \rho}{\partial y^2} + K_z\frac{\partial^2 \rho}{\partial z^2} \qquad (4.301)$$

for the source defined by

$$\rho(0, y, z, t) = \frac{Q_P}{u}\,\delta(y)\,\delta(z - H_e)\,\delta(t) \qquad (4.302)$$

The solution for the mean distribution of the concentration is

$$\rho(x, y, z, t) = \frac{Q_P}{(2\pi)^{3/2}\sigma_x\sigma_y\sigma_z}\,\exp\left(-\frac{(x - ut)^2}{2\sigma_x^2} - \frac{y^2}{2\sigma_y^2}\right)$$

$$\left[\exp\left(-\frac{(z - H_e)^2}{2\sigma_z^2}\right) + \exp\left(-\frac{(z + H_e)^2}{2\sigma_z^2}\right)\right] \qquad (4.303)$$

where the standard deviations σ_x, σ_y, and σ_z measure the dispersion of the puff relative to its center of mass located at $x = ut$. They are related to the eddy diffusion coefficients by

$$\sigma_x(x) = \sqrt{2K_x t} \qquad \sigma_y(x) = \sqrt{2K_y t} \qquad \sigma_z(x) = \sqrt{2K_z t} \qquad (4.304)$$

The solution is readily generalized to the evolution of a plume from a time-dependent point source in a variable 3-D wind. In that case we can view the plume as resulting from a continuous suite of puffs emitted sequentially. If we consider N puffs $i = 1$, N of mass $q_i \Delta t$ [kg] emitted successively at time $t_i = i\Delta t$ and advected by a 3-D wind field from source point $(0, 0, H_e)$ to position $x_i(t)$, $y_i(t)$, $z_i(t)$ at time t, the mean concentration field is given by

$$\rho(x,y,z,t) = \frac{1}{(2\pi)^{3/2}} \sum_{i=1}^{N} \frac{q_i \Delta t}{\sigma_x \sigma_y \sigma_z} \exp\left(-\frac{(x - x_i(t))^2}{2\sigma_x^2} - \frac{(y - y_i(t))^2}{2\sigma_y^2}\right)$$
$$\times \left[\exp\left(-\frac{(z - z_i(t) - H_e)^2}{2\sigma_z^2}\right) + \exp\left(-\frac{(z - z_i(t) + H_e)^2}{2\sigma_z^2}\right)\right] \qquad (4.305)$$

4.13 Statistical Models

Physical models such as the Eulerian, Lagrangian, and Gaussian plume models presented in the previous sections simulate the behavior of chemical species on the basis of conservation equations that capture the effects of chemical and transport processes in the atmosphere. *Statistical models* offer an alternative approach that does not require a physical representation of the chemical and dynamical processes involved. Such models are typically based on empirical relationships between variables established from a large number of previously observed situations. Sometimes the relationships are established from a highly detailed but computationally unaffordable physical model. Once the statistical model has been constructed, it can be applied to conditions within a certain range of validity (typically the ranges of the input variables used to construct the model). Statistical models are often used to explore relationships between an output variable and different candidate input variables, to parameterize a physical model for faster computation, or to make forecasts on the basis of observations of the present state. In the latter case, the present value of the output variable is often a useful input variable for the forecast model. We describe here two types of statistical models: multiple linear regression models and artificial neural networks.

4.13.1 Multiple Linear Regression Models

Multiple linear (or *multilinear*) regression models specify a linear relation between one dependent (output) variable noted y and P independent (input, *predictor*, or *explanatory*) variables denoted x_p ($p = 1$, P). Consider a set of N successive observations i of the dependent and independent variables (y_i and $x_{i,p}$ for $i = 1, N$). We write the linear expression

$$y_i = b_0 + b_1 x_{i,1} + b_2 x_{i,2} + \cdots + b_P x_{i,P} + e_i \tag{4.306}$$

where e_i is a residual error term, and b_i are unknown parameters to be derived from the ensemble of observations. Errors are assumed to follow a normal distribution with a zero mean value and a variance σ^2.

Equation (4.306) can be rewritten in matrix form

$$\mathbf{y} = \mathbf{X}\mathbf{b} + \mathbf{e} \tag{4.307}$$

where $\mathbf{y} = (y_1, y_2, \ldots y_N)^T$ is the N-dimensional response vector, $\mathbf{b} = (b_0, b_1, \ldots b_P)^T$ is the $P + 1$ dimensional slope vector, and $\mathbf{e} = (e_1, e_2, \ldots e_N)^T$ the N-dimensional error vector. The $N \times (P + 1)$ matrix

$$\mathbf{X} = \begin{pmatrix} 1 & x_{11} & x_{12} & \cdots & x_{1P} \\ 1 & x_{21} & x_{22} & \cdots & x_{2P} \\ \vdots & \vdots & \vdots & & \vdots \\ 1 & x_{N1} & x_{N2} \cdots & & x_{NP} \end{pmatrix}$$

is called the design matrix.

In the presence of error, design of a reliable multilinear model requires that the number of independent observations available be much larger than the number of unknown parameters b_i ($N \gg (P + 1)$). In the ordinary least squares method, optimal values $\widehat{\mathbf{b}}$ for the parameters are derived by minimizing a cost function $J(\mathbf{b})$ defined as the sum of the square differences between the observed values and their corresponding model values

$$J(\mathbf{b}) = \sum_{i=1}^{N} e_i^2 = \sum_{i=1}^{N} (y_i - b_0 - b_1 x_{i,1} - b_2 x_{i,2} - \cdots - b_P x_{i,P})^2 \tag{4.308}$$

Solving $dJ(\mathbf{b})/d\mathbf{b} = \mathbf{0}$ yields the $P + 1$ "normal equations"

$$\sum_{i=1}^{N} \sum_{p=0}^{P} x_{ij} x_{ip} \widehat{b}_p = \sum_{i=1}^{N} x_{ij} y_i \qquad j = 0, \ldots, P \tag{4.309}$$

or in matrix notation

$$(\mathbf{X}^T\mathbf{X})\widehat{\mathbf{b}} = \mathbf{X}^T\mathbf{y} \tag{4.310}$$

The solution

$$\widehat{\mathbf{b}} = (\mathbf{X}^T\mathbf{X})^{-1}\mathbf{X}^T\mathbf{y} \tag{4.311}$$

is unique, provided that the N rows of matrix $\mathbf{X}$ are linearly independent. Here $\mathbf{X}^T\mathbf{X}$ and $(\mathbf{X}^T\mathbf{X})^{-1}$ are $(P + 1) \times (P + 1)$ symmetric matrices and $\mathbf{X}^T\mathbf{y}$ a $P + 1$ dimensional vector. The fitted values of the predicted quantity are then

$$\widehat{\mathbf{y}} = \mathbf{X}\widehat{\mathbf{b}} = \mathbf{X}(\mathbf{X}^T\mathbf{X})^{-1}\mathbf{X}^T\mathbf{y} \tag{4.312}$$

The success of the fitting model is commonly measured by the *coefficient of multiple determination* R^2 (more often called *R-squared coefficient*) defined as the ratio between the variance of the fitted values $\widehat{\mathbf{y}}$ to the variance of the observed values $\mathbf{y}$. Since $\text{var}(\mathbf{y}) = \text{var}(\widehat{\mathbf{y}}) + \text{var}(\mathbf{e})$,

$$R^2 = \frac{\text{var}(\widehat{\mathbf{y}})}{\text{var}(\mathbf{y})} = 1 - \frac{\text{var}(\mathbf{e})}{\text{var}(\mathbf{y})} \tag{4.313}$$

More explicitly,

$$R^2 = 1 - \frac{\sum\limits_{i=1}^{N} (y_i - \widehat{y}_i)^2}{\sum\limits_{i=1}^{N} (y_i - \bar{y})^2} \tag{4.314}$$

where

$$\bar{y} = \frac{1}{N} \sum_{i=1}^{N} y_i \tag{4.315}$$

is the mean value of the observed data. The R^2 coefficient, whose value varies between 0 and 1, represents the fraction of the variance that is explained by the multilinear model. Some caution is needed in interpreting R^2 as the quality of the fit because R^2 will tend to increase as more predictor variables are added without actually increasing the predictive capability of the model. Adjusted coefficients of determination are used to address this problem (see Appendix E).

4.13.2 Artificial Neural Networks

Artificial neural networks are inspired by the functioning and learning ability of biological neural systems. They do not require assumptions on the relationships between input and output variables, which is an advantage over the multiple linear regression approach. The method can be applied to any smooth nonlinear relationships that exist between the variables. The artificial neural network is trained by using observational data to adjust its internal parameters until a usefully predictive input–output mapping is achieved. The predictive power can be continually improved through the ingestion of more observational data.

An artificial neural network (Figure 4.28) consists of layered interconnected nodes (or neurons) including an input layer, one or more successive "hidden layers," and an output layer. The input layer plays no computational role; it only supplies data to the first hidden layer. The output layer provides the solution. Neurons are connected to all nodes belonging to the upstream and downstream neighboring layers. The number of hidden layers is determined by the complexity of the problem. In the feed-forward network considered here, information flows only in one direction from the input to the output layer. More complex architectures include possible feedbacks between layers.

Each node is characterized by a set of numerical inputs x_i that conveys information from other upstream nodes with specific weights. The total input signal y_j to node j from all upstream nodes i is given by

$$y_j = \sum_{i=1}^{N} w_{i,j} x_i \tag{4.316}$$

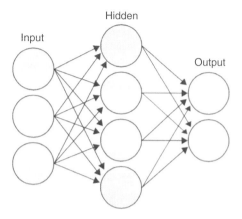

Typical artificial neural network with interconnected neurons (or nodes) including an input layer, a "hidden" layer, and an output layer. The network must be trained; from a set of M known training patterns available from observations, it learns how the input **X** represented by a $[M \times K]$ matrix relates to the output **Y** represented by a $[M \times J]$ matrix. Here, K denotes the number of input nodes (and variables) and J the number of output nodes (and variables). In the figure, $K = 3$ and $J = 2$.

where x_i is the output signal of node i and $w_{i,j}$ is the *weight* of the signal flowing from node i to node j. A second element is an activation or *transfer function F* that transforms the weighted input y_j into a numerical output x_j. The logistic function is frequently adopted:

$$x_j = F\left(y_j\right) = \frac{1}{1 + \exp\left(-y_j\right)} \tag{4.317}$$

The output is then passed to the next downstream hidden layer until one reaches eventually the output layer.

The optimal values of the individual weights $w_{i,j}$ are determined by training the system with observations. Training is the process by which one determines a combination of weights that leads to a minimum error of the output variables. This process is equivalent to the derivation of the intercept and slope coefficients in the linear regression method described in the previous section. Different mathematical algorithms are available to determine the conditions under which the error is minimal (see, e.g., Bishop, 1995). Most of them adopt some form of gradient-descent approach in which the value of the weight parameters of the models, initially small and random, are gradually modified along a steepest-gradient direction of the error function until an absolute minimum value is reached.

4.14 Operator Splitting

The continuity equations for chemical species consist of a sum of terms describing different processes for which the model provides independent formulations, as described for example by (4.10). These processes are occurring simultaneously. Ideally, the numerical algorithm used to solve the equations should account for this simultaneous coupling. This is not practical in 3-D models because of the multi-dimensionality of the problem and because the numerical algorithms best suited to treat each of the terms are very different. It is therefore often advantageous to solve for the different terms individually and sequentially over a given time step Δt. This approach is called *operator splitting*. It enables software modularity, algorithms tailored to each operator, and better performance. For example, a stiff integrator can be employed to integrate the chemical equations, while a flux scheme can be adopted to integrate the transport equations.

To describe the method, we consider a simple form of the continuity equation for a variable Ψ that represents, for example, the concentration of a chemical species. We assume that this variable is affected by two distinct atmospheric processes (such as transport and chemistry), represented by linear operators A and B. Thus we write

$$\frac{\partial \Psi}{\partial t} = A\,\Psi + B\,\Psi \qquad (4.318)$$

To apply the operator splitting method over a time interval Δt, we first update the value of Ψ from time t_n to time $t_{n+1} = t_n + \Delta t$ by calculating at each model grid point (i, j, k) an intermediate value (Ψ^*) resulting from the application of operator A over the time interval $[0, \Delta t]$:

$$\frac{\partial \Psi^*}{\partial t} = A\,\Psi^* \qquad \text{with} \qquad \Psi^*(0) = \Psi^n \qquad \text{over} \qquad [0, \Delta t] \qquad (4.319)$$

The resulting value of Ψ^* at time Δt is then used as the initial condition for the second step involving operator B:

$$\frac{\partial \Psi^{**}}{\partial t} = B\,\Psi^{**} \qquad \text{with} \qquad \Psi^{**}(0) = \Psi^*(\Delta t) \qquad \text{over} \qquad [0, \Delta t] \qquad (4.320)$$

The value of Ψ^{n+1} at time t_{n+1} is thus given by the calculated value of Ψ^{**} at time Δt.

In the scheme presented above and referred to here as an A–B scheme, the integration is initiated by applying operator A followed by operator B. If we reverse the order of the successive integration (B–A scheme), the solution will be somewhat different. Ideally the order should not matter, but in fact it does, and this represents the *operator splitting error* (Box 4.12).

When one of the two operators (say, B) is stiff, the A–B scheme described above can be slightly modified to avoid transient disturbances (Sportisse, 2000). In this case, rather than imposing an initial condition to the second substep that accounts for the calculation of the first substep, one prefers to add a source term during the second substep and replace (4.320) by

$$\frac{\partial \Psi^{**}}{\partial t} = B\,\Psi^{**} + \frac{\Psi^*(\Delta t) - \Psi^n}{\Delta t} \qquad \text{with} \qquad \Psi^{**}(0) = \Psi^n \qquad \text{on} \qquad [0, \Delta t]$$

$$(4.321)$$

This scheme represents an explicit integration with the non-stiff operator A and an implicit integration with the stiff operator B.

Assume that the continuity equation includes two linear terms representing two distinct processes

$$d\Psi/dt = \mathbf{A}_1\Psi + \mathbf{A}_2\Psi$$

where Ψ represents a vector of dimension n (e.g., species concentrations) and $\mathbf{A_1}$ and $\mathbf{A_2}$ are $n \times n$ matrices. If the solution at time t_n is known, then the analytical (true) solution at time $t_{n+1} = t_n + \Delta t$ is

$$\Psi^{true}(t_{n+1}) = \exp\left[(\mathbf{A}_1 + \mathbf{A}_2)\Delta t\right]\Psi(t_n)$$

Adopting a splitting method, the equation is solved over the splitting time interval Δt by treating the two processes $\mathbf{A}_1$ and $\mathbf{A}_2$ sequentially:

$$\frac{d\Psi^*}{dt} = \mathbf{A}_1\Psi^* \qquad \text{with } \Psi^* = \Psi(t_n) \text{ as initial condition}$$

$$\frac{d\Psi^{**}}{dt} = \mathbf{A}_2\Psi^{**} \qquad \text{with } \Psi^{**} = \Psi^*(t_{n+1}) \text{ as initial condition}$$

We find

$$\Psi^{splitting}(t_{n+1}) = \exp\left[\mathbf{A}_2\Delta t\right]\exp\left[\mathbf{A}_1\Delta t\right]\Psi(t_n)$$

The local error e (error made during one single time step Δt) due to the splitting process is

$$e(\Delta t) = \Psi^{true}(t_{n+1}) - \Psi^{splitting}(t_{n+1})$$
$$= \left\{\exp\left[(\mathbf{A}_1 + \mathbf{A}_2)\Delta t\right] - \exp\left[\mathbf{A}_2\Delta t\right]\exp\left[\mathbf{A}_1\Delta t\right]\right\}\Psi(t_n)$$

which is zero only if the two matrices commute ($\mathbf{A}_1\,\mathbf{A}_2 = \mathbf{A}_2\,\mathbf{A}_1$). If the exponentials are approximated to the second-order as

$$\exp\left[\mathbf{A}, \Delta, \mathbf{t}\right] = \mathbf{I} + \mathbf{A}\Delta t + \frac{1}{2}\mathbf{A}^2\Delta t^2 + O\left(\Delta t^3\right)$$

where $\mathbf{I}$ is the identity matrix, it is easy to show that the local splitting error is of order $O(\Delta t^2)$ if the two processes do not commute. By summing over the entire integration time, one finds that the *global splitting error* associated with the above algorithm is of the order $O(\Delta t)$, and the method is therefore said to be a first-order method.

The algorithm described above is first-order accurate (on global error, see Box 4.12) with respect to the splitting time Δt. The splitting error can be reduced by adopting a symmetric splitting scheme. One option, which is computationally expensive, is to calculate at each time step the average between the solutions derived for different possible orderings of the operators. When only two operators A and B are considered, the solution is simply the average between the values calculated by the A–B and B–A schemes:

$$\Psi^{n+1} = \frac{1}{2}\left(\Psi_{A-B}^{n+1} + \Psi_{B-A}^{n+1}\right) \tag{4.322}$$

The local error associated with the solution is fourth-order.

A more efficient approach introduced by Strang (1968) and presented here in the simple case of two operators A and B is to integrate the equations as a symmetric sequence of operators A–B–A. The integration is performed first for operator A over the time interval $[0, \Delta t/2]$, then for operator B over the full interval $[0, \Delta t]$ and finally for operator A again over the interval $[0, \Delta t/2]$. Thus

$$\frac{\partial \Psi^*}{\partial t} = A\,\Psi^* \qquad \text{with} \qquad \Psi^*(0) = \Psi^n \qquad \qquad \text{on} \qquad \left[0, \frac{\Delta t}{2}\right] \tag{4.323}$$

$$\frac{\partial \Psi^{**}}{\partial t} = B\,\Psi^{**} \qquad \text{with} \qquad \Psi^{**}(0) = \Psi^*\left(\frac{\Delta t}{2}\right) \qquad \text{on} \qquad [0, \Delta t] \tag{4.324}$$

$$\frac{\partial \Psi^*}{\partial t} = A\,\Psi^* \qquad \text{with} \qquad \Psi^*(0) = \Psi^{**}(\Delta t) \qquad \qquad \text{on} \qquad \left[0, \frac{\Delta t}{2}\right] \tag{4.325}$$

This Strang approach can be generalized to a larger number of operators. For example, advection (A), diffusion (D), and chemical (C) in the sequence $A_{\Delta t/2}D_{\Delta t/2}C_{\Delta t}D_{\Delta t/2}A_{\Delta t/2}$ where the subscript denotes the time interval over which the integration is performed.

Strang splitting leads to a second-order approximation. There is no operator splitting error if the operators commute, but this is not generally the case in atmospheric applications. For example, advection commutes with diffusion only if either the wind or diffusion fields do not vary in space. Diffusion commutes with chemistry only if the chemical source term is linear in concentration and independent of the spatial variable (Lanser and Verwer, 1998).

The computational cost of solving the 3-D transport equation on a grid of N points is generally proportional to N^3, but decreases to about $3N$ if the problem is reduced to a set of three 1-D equations by operator splitting. In the case of pure advection, where the flux divergence is written as

$$\nabla \cdot (\Psi\, v) = \frac{\partial (\Psi\, u)}{\partial x} + \frac{\partial (\Psi\, v)}{\partial y} + \frac{\partial (\Psi\, w)}{\partial z} \tag{4.326}$$

the 3-D advection equation can be solved by three sequential 1-D operators applied to the 3-D concentration field represented by vector Ψ:

$$\Psi^{(1)} = A_x(\Psi^n) \tag{4.327}$$

$$\Psi^{(2)} = A_y\left(\Psi^{(1)}\right) \tag{4.328}$$

$$\Psi^{n+1} = A_z\left(\Psi^{(2)}\right) \tag{4.329}$$

Here the advection operators A_x, A_y, A_z solve the corresponding 1-D advection equations. For A_x, for example, the advection equation is

$$\frac{\partial \Psi}{\partial t} = -u\frac{\partial \Psi}{\partial x} \tag{4.330}$$

and $A_x(\boldsymbol{\Psi}^n)$ solves this equation over the time interval $[t_n, t_{m+1}]$. Numerical methods for solving the 1-D advection equation are presented in Chapter 7. A side advantage of splitting 3-D advection into 1-D operators is that it allows the use of different time steps in different directions to address CFL criterion limitations. With stronger winds in the x-direction than in the y-direction, and weak winds in the z direction, one can use for example the following arrangement of operators:

$$\boldsymbol{\Psi}^{n+1} = A_{x,\Delta t/4}A_{y,\Delta t/2}A_{x,\Delta t/4}A_{z,\Delta t}A_{x,\Delta t/4}A_{y,\Delta t/2}A_{x,\Delta t/4}(\boldsymbol{\Psi}^n) \tag{4.331}$$

Diffusive transport over a model time step can also be decomposed into three different operators, one in each direction. Here, numerical stability requires that an implicit approach be adopted (Section 8.6). Each direction can be treated completely separately, as in the advection case. However, it is usually preferable to adopt an *alternating direction implicit (ADI) method* in which, during a third of the time step, diffusion is solved implicitly in one direction and explicitly in the other two directions. Consider the simplest case of a constant diffusion coefficient K and constant air density. The diffusion term takes the form

$$K\ \nabla^2\Psi = K\left(\frac{\partial^2\Psi}{\partial x^2} + \frac{\partial^2\Psi}{\partial y^2} + \frac{\partial^2\Psi}{\partial z^2}\right)$$

The ADI method updates the value of the 3-D gridded field $\boldsymbol{\Psi}$ from $\boldsymbol{\Psi}^n$ to $\boldsymbol{\Psi}^{n+1}$ over a time step $\Delta t = t_{n+1} - t_n$ with the sequence

$$\boldsymbol{\Psi}^{(1)} = \boldsymbol{\Psi}^n + \frac{\gamma}{3}\left[D_x\left(\boldsymbol{\Psi}^{(1)}\right) + D_y(\boldsymbol{\Psi}^n) + D_z(\boldsymbol{\Psi}^n)\right] \tag{4.332}$$

$$\boldsymbol{\Psi}^{(2)} = \boldsymbol{\Psi}^{(1)} + \frac{\gamma}{3}\left[D_x\left(\boldsymbol{\Psi}^{(1)}\right) + D_y\left(\boldsymbol{\Psi}^{(2)}\right) + D_z\left(\boldsymbol{\Psi}^{(1)}\right)\right] \tag{4.333}$$

$$\boldsymbol{\Psi}^{n+1} = \boldsymbol{\Psi}^n + \frac{\gamma}{3}\left[D_x\left(\boldsymbol{\Psi}^{(2)}\right) + D_y\left(\boldsymbol{\Psi}^{(2)}\right) + D_z\left(\boldsymbol{\Psi}^{n+1}\right)\right] \tag{4.334}$$

where

$$\gamma = K\Delta t \tag{4.335}$$

and D_x applied to grid point $\Psi_{i,j,k}$ is

$$D_x\left(\Psi_{i,j,k}\right) = \frac{\Psi_{i+1,j,k} - 2\Psi_{i,j,k} + \Psi_{i-1,j,k}}{\Delta x^2} \tag{4.336}$$

with equivalent forms for D_y and D_z. The ADI method is easily generalized to cases in which the diffusion coefficient is variable.

4.15 Filtering

Meteorological models tend to produce undesirable noise caused by dispersion errors in the numerical integration of the dynamical equations. This noise affects the atmospheric distributions of chemical species transported by the model. In order to keep model simulations numerically stable, some form of dissipation may need to be introduced (Jablonowski and Williamson, 2011). Numerical diffusion in solving the advection equation in Eulerian models (Chapter 7) often serves as an implicit filter to damp undesired noise. Filters are explicit if they are implemented by the addition of terms in the governing equations or if they are applied *a posteriori* as a correction to the calculated fields. Here we describe explicit filters that damp the noisy small-scale waves with wavelengths L of the order of $(2–4) \Delta x$ without reducing substantially the amplitude of the better-resolved scales.

As we will see, numerical filters generally involve the application of a diffusion term to the solutions of the dynamical equations in order to damp small-scale features. Aside from ensuring numerical stability, an additional purpose of this damping is to mimic turbulence-related processes that are unresolved by the model grid. The introduction of a numerical filter as diffusion operator can thus be based on physical as well as numerical considerations.

4.15.1 Diffusive Filters

A *diffusive filter* involves the addition of a diffusion term to the dynamical equations as

$$\left[\frac{\partial \Psi}{\partial t}\right]_{diff} = (-1)^{q+1} K_{2q} \nabla^{2q}\Psi \qquad q = 1, 2, 3, \ldots \qquad (4.337)$$

where q is a positive integer, $2q$ the order of the diffusion, and K_{2q} the adopted diffusion coefficient [with units $m^{2q}\ s^{-1}$]. Setting $q = 1$ corresponds to the second-order diffusion previously introduced in Section 4.2.3 to parameterize subgrid turbulent mixing. It is also often applied as an artificial sponge in the upper layers of dynamical models to avoid spurious reflection of waves at the top boundary. Second-order filters are not very scale-selective and may negatively impact the well-resolved waves produced by the model. More scale-selective hyper-diffusion schemes with higher values of q can be used. The fourth-order hyper-diffusion scheme with $q = 2$, called bi-harmonic diffusion or super-viscosity, is often adopted. The chosen value of the diffusion coefficient is somewhat arbitrary and is often regarded as a tuning parameter of the model; it has to be as small as possible to avoid dissipation of well-resolved physical waves, and large enough to ensure numerical stability of the computed solution.

4.15.2 Digital Spatial Filters

Digital spatial filters are local filters that take into account only neighboring grid points. A widely used digital filter is the linear *Shapiro filter* (Shapiro, 1970; 1975) that is based on constant-coefficient grid point operators of order m. The order of the filter determines the width of the numerical stencil (i.e., the number of neighboring grid points involved in the filtering operator). The Shapiro filter eliminates short waves from the calculated fields and thus functions equivalently to the addition of a diffusion term in the dynamical equations. In the 1-D case, we consider a function $\Psi(x)$ defined over the interval $(-\infty < x < +\infty)$ with values Ψ_i at discrete points

$x_i = i\Delta x$. The smoothed value $\{\Psi_i\}$ at point x_i using a first-order Shapiro filter is the weighted average between the unsmoothed value Ψ_i at point x_i with weight $(1 - S)$ and the two adjacent unsmoothed values at points x_{i-1} and x_{i+1} with weight $S/2$. Thus,

$$\{\Psi_i\} = (1 - S)\Psi_i + S\frac{\Psi_{i-1} + \Psi_{i+1}}{2} \qquad (4.338)$$

or equivalently,

$$\{\Psi_i\} = \Psi_i + S\frac{\Psi_{i-1} + \Psi_{i+1} - 2\Psi_i}{2} \qquad (4.339)$$

where S is the so-called smoothing element. We note three important properties of this filtering operator: (1) it is symmetric in space; (2) it involves only three values of x (the filter is local); and (3) over a large number of points, the averages of the smoothed values approach the averages of the unsmoothed values.

To derive the properties of the filter, we express at grid point $x_i = i\Delta x$ the Fourier component of functions Ψ and $\{\Psi\}$ for wavenumber k [m^{-1}] corresponding to wavelength $L = 2\pi/k$, as

$$\Psi_k(x_j) = A_k \exp[ikx_i] \qquad (4.340)$$

and

$$\{\Psi_k(x_i)\} = \{A_k\} \exp[ikx_i] \qquad (4.341)$$

The response function $g(k) = \{A_k\}/A_k$ of the filter is

$$|g(k)| = 1 - 2S \sin^2\left(\frac{k\Delta x}{2}\right) = 1 - 2S \sin^2\left(\frac{\pi\Delta x}{L}\right) \qquad (4.342)$$

The value of S must be chosen so that the response function is positive and smaller than 1. This requires $0 \le S \le \frac{1}{2}$. In most applications, the value $S = \frac{1}{2}$ is adopted, and the first-order filter operator is expressed by

$$\{\Psi_i\} = \frac{\Psi_{i-1} + 2\Psi_i + \Psi_{i+1}}{4} \qquad (4.343)$$

with the corresponding response function

$$|g(k)| = 1 - \sin^2\left(\frac{k\Delta x}{2}\right) = \cos^2\left(\frac{k\Delta x}{2}\right) \qquad (4.344)$$

The filter can be applied more than once to achieve greater smoothing of the solution.

One can also use higher-order filters whose amplitude response are provided by

$$|g(k)| = 1 - \sin^{2m}\left(\frac{k\Delta x}{2}\right) \qquad (4.345)$$

where the order of the filter m is an integer multiple of 2. Define the difference operator δ as

$$\delta\Psi_i = \Psi_{i+1/2} - \Psi_{i-1/2} \qquad (4.346)$$

with, for example,

$$\delta^2\Psi_i = \delta(\delta\Psi_i) = \delta\Psi_{i+1/2} - \delta\Psi_{i-1/2} = \Psi_{i+1} - 2\Psi_i + \Psi_{i-1} \quad (4.347)$$

The filter operators of orders 1, 2, 4, …, m are defined successively as

$$\{\Psi_i\}^{(1)} = \left[1 + \left(\frac{\delta}{2}\right)^2\right]\Psi_i \quad (4.348)$$

$$\{\Psi_i\}^{(2)} = \left[1 + \frac{\delta^2}{4}\right]\left[1 - \frac{\delta^2}{4}\right]\Psi_i \quad (4.349)$$

$$\{\Psi_i\}^{(4)} = \left[1 + \frac{\delta^4}{16}\right]\left[1 - \frac{\delta^4}{16}\right]\Psi_i \quad (4.350)$$

$$\cdots\cdots$$

$$\{\Psi_i\}^{(m)} = \left[1 + \left(\frac{\delta}{2}\right)^m\right]\left[1 - \left(\frac{\delta}{2}\right)^m\right]\Psi_i \quad (4.351)$$

For example, the filter operators of orders 1 and 2 are

$$\{\Psi_i\}^{(1)} = \frac{1}{4}[\Psi_{i-1} + 2\Psi_i + \Psi_{i-1}] \quad (4.352)$$

$$\{\Psi_i\}^{(2)} = \frac{1}{16}[-\Psi_{i-2} + 4\Psi_{i-1} + 10\Psi_i + 4\Psi_{i-1} - \Psi_{i-2}] \quad (4.353)$$

Shapiro filters totally eliminate the shortest resolvable wave corresponding to two grid cells ($L = 2\Delta x$), and damp to a lesser extent the amplitude of the other small-scale resolvable waves ($L = 3\Delta x$ and $4\Delta x$).

Figure 4.29 shows the response function of the Shapiro filter for different orders m after 1 and 1000 applications, respectively. It highlights the cumulative character of the filtering operation, specifically in the case of the low-order filters, which

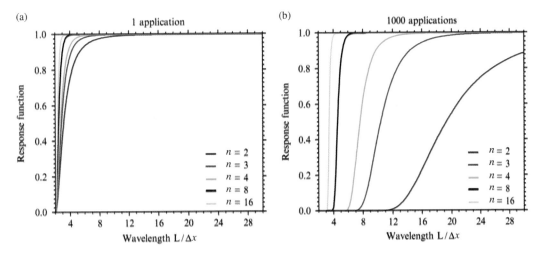

Figure 4.29 Response function for 1-D Shapiro filters of order 2, 3, 4, 8, and 16 with $S = \frac{1}{2}$ after (a) one application and (b) 1000 applications. Reproduced from Jablonowski and Williamson (2011).

strongly damp relatively large waves (up to the $12\Delta x$ wave in the case of a second-order filter with 1000 applications). The order-8 filter is often adopted since, after a large number of applications, it effectively eliminates all waves with wavelengths less than $4\Delta x$, but preserves almost entirely the waves with wavelengths larger than $6\Delta x$.

If the first-order Shapiro filter is applied twice in sequence with smoothing factors S equal to $+\frac{1}{2}$ and $-\frac{1}{2}$, respectively, we find an intermediate value

$$\{\Psi_i\} = \frac{1}{2}\Psi_i + \frac{\Psi_{i-1} + \Psi_{i+1}}{4} \tag{4.354}$$

and a final filtered value

$$\{\{\Psi_i\}\} = \frac{3}{2}\{\Psi_i\} - \frac{\{\Psi_{i-1}\} + \{\Psi_{i+1}\}}{4} \tag{4.355}$$

The resulting response function of the two-stage operator is

$$|g(k)| = \frac{1 + \cos(k\Delta x)}{2} \frac{3 - \cos(k\Delta x)}{2} \tag{4.356}$$

This particular filter, referred to as the *Shuman filter* (Shuman, 1957), is very effective for attenuating short waves while preserving large waves: the $L = 2\Delta x$ wave is totally eliminated and the amplitude of the response function is equal to 0.75 for $L = 4\Delta x$ and 0.98 for $L = 8\Delta x$. The effect of the filter is very small for wavelengths $L > 8\Delta x$.

A 2-D function $\Psi(x, y)$ can be filtered by applying the 1-D Shapiro operator successively in directions x and y (with the corresponding indices i and j). The resulting operator involves nine discrete points:

$$\{\Psi_{i,j}\}^{x,y} = \Psi_{i,j} + \frac{S}{2}(1 - S)\left[\Psi_{i-1,j} + \Psi_{i,j+1} + \Psi_{i+1,j} + \Psi_{i,j-1} - 4\Psi_{i,j}\right]$$
$$+ \frac{S^2}{4}\left[\Psi_{i-1,j+1} + \Psi_{i+1,j+1} + \Psi_{i+1,j-1} + \Psi_{i-1,j-1} - 4\Psi_{i,j}\right] \tag{4.357}$$

with a response function

$$|g(k,h)| = \left[1 - 2S\left(1 - \sin^2\left(\frac{k_x\Delta x}{2}\right)\right)\right]\left[1 - 2S\left(1 - \sin^2\left(\frac{k_y\Delta y}{2}\right)\right)\right] \tag{4.358}$$

Here k_x and k_y represent the wavenumbers in the x and y directions, respectively. An alternative is to define the 2-D smoothing function as

$$\{\Psi_{i,j}\}^{x,y} = \frac{1}{2}\left(\{\Psi_{i,j}\}^x + \{\Psi_{i,j}\}^y\right) \tag{4.359}$$

which results in a five-point operator

$$\{\Psi_{i,j}\}^{x,y} = \Psi_{i,j} + \frac{S}{4}\left[\Psi_{i-1,j} + \Psi_{i,j+1} + \Psi_{i+1,j} + \Psi_{i,j-1} - 4\Psi_{i,j}\right] \tag{4.360}$$

with a response function

$$|g(k,h)| = 1 - S\left(\sin^2\left(\frac{k\Delta t}{2}\right) + \sin^2\left(\frac{h\Delta t}{2}\right)\right) \tag{4.361}$$

In the preceding discussion, we have assumed that the spatial domain is infinite and we have therefore ignored the influence of the domain boundaries. This is not of concern if the domain is periodic, as in global model applications. For limited domains, values of the field Ψ imposed at the boundaries may strongly influence the filtered function inside the domain, especially if the filtering is repeated a large number of times. For a simple three-point operator in one direction, the influence of the boundary propagates to the interior of the domain by one grid point from both ends at each application of the filtering process. Shapiro (1970) discusses the boundary effects in detail and shows that, for example, a successful filtering procedure with periodic boundaries may result in the spurious growth of undesired waves inside the domain if fixed conditions are imposed at the boundaries. Some adjustments in the filtering procedure can be applied to limit the influence of the boundary conditions.

4.15.3 Spectral Filters

Spectral filters are commonly used in global longitude–latitude grid point models to damp noise in the vicinity of the pole where convergence of the meridians reduces the longitudinal spacing Δx between grid points. This reduced spacing can lead to violation of the CFL criterion and numerical instability (see Chapter 7). Short waves resulting from such numerical instability are damped or eliminated by applying a 1-D Fourier filter in the zonal direction. The grid data are first transformed into the spectral space via Fourier transform, and the resulting Fourier coefficients a_m corresponding to dimensionless wavenumber m are modified to become

$$\{a_m\} = F(m)\, a_m \tag{4.362}$$

with $F(m)$ being the so-called response function. In practical applications, the Fourier filter is applied only poleward of a cut-off latitude φ_c, and the strength of the filter is gradually enhanced toward the pole. This can be accomplished by increasing the number of wavenumbers m affected by the filtering process and by choosing a response function whose value decreases with latitude. Figure 4.30 shows two examples of response functions; the first one expressed by

$$F(m) = \min\left[1.0, \left(\frac{\cos\varphi}{\cos\varphi_c}\right)^{2q} \frac{1}{\sin^2(m\Delta\lambda/2)}\right] \tag{4.363}$$

corresponds to a strong filter and the second one

$$F(m) = \min\left[1.0, \left(\frac{\cos\varphi}{\cos\varphi_c}\right) \frac{1}{\sin(m\Delta\lambda/2)}\right] \tag{4.364}$$

to a weaker filter. Here, φ denotes latitude and $\Delta\lambda$ the angular longitudinal resolution. The positive integer parameter q can be chosen to modify the strength of the filter.

In the final step of the filtering process, the fields are converted back into the grid point space by an inverse Fourier transform. The advantage of the Fourier filter is that it can be made very scale-selective and dependent on latitude; the drawback is that all data along latitude rings are needed. The use of local spectral filters has been

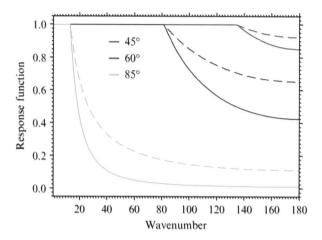

Figure 4.30 Response function of a strong Fourier filter with $q = 1$ (solid line) and a weaker (dashed line) filter represented as a function of the dimensionless zonal wavenumber at latitudes of $45°$, $60°$, and $85°$, respectively. The cut-off latitude is assumed to be $40°$. Reproduced from Jablonowski and Williamson (2011).

proposed for models that utilize local spectral methods like the discontinuous Galerkin approach or the spectral element method (Section 4.10). See Vandeven (1991) and Boyd (1998) for more details.

4.15.4 Time-Smoothing Filters

In certain cases, it is appropriate to apply a slight time smoothing to the solution of a differential equation. In the *Robert–Asselin filter* (Robert, 1969; Asselin, 1972), the solution Ψ^n at time t_n is replaced by

$$\{\Psi^n\} = \Psi^n + \nu\left(\{\Psi^{n-1}\} - 2\Psi^n + \Psi^{n+1}\right) \tag{4.365}$$

and the high frequencies in the solution are damped. Note that the value of the function Ψ at time t_{n-1} is replaced by its already filtered value. Such a filter with a parameter ν of the order of 1% is often applied when the leapfrog advection scheme is used, since it removes the odd–even checkerboard pattern generated by that numerical algorithm.

4.16 Interpolation and Remapping

Models quantify the values of variables at discrete locations (grid points in Eulerian models, location of individual particles in Lagrangian models) and discrete time steps. One often needs model information intermediate between these discrete points. If the value of a function $F(\mathbf{r})$ is known for a sequence of $N + 1$ selected values $\mathbf{r}_i$ of the independent variable $\mathbf{r}$, the process by which one estimates the value of this function for intermediate values is called *interpolation*. Interpolation methods may be used for comparing model results to observations. They may also be used to convert model variables to a different grid, a procedure known as *regridding*. Yet another application is for converting between a Lagrangian particle field and an Eulerian grid, as is done in semi-Lagrangian transport schemes (Chapter 7) or for treating nonlinear chemistry in an otherwise Lagrangian modeling framework.

The interpolation process consists of determining a function $\Psi(\mathbf{r})$ called *interpolant* whose value is strictly equal to the true function $F(\mathbf{r})$ for specified values $\mathbf{r}_i$ of the independent variable $\mathbf{r}$, and approximates this function in the intervals between the different nodes $\mathbf{r}_i$. Thus, we impose at $N + 1$ points

$$\Psi(\mathbf{r}_i) = F(\mathbf{r}_i) \equiv F_i \qquad (i = 0, N) \qquad (4.366)$$

The interpolation error $e(\mathbf{r})$ at point $\mathbf{r}$ is defined as the difference between the true function $F(\mathbf{r})$ and the interpolating function $\Psi(\mathbf{r})$

$$e(\mathbf{r}) = F(\mathbf{r}) - \Psi(\mathbf{r}) \qquad (4.367)$$

The fitting process adopted for the interpolation often consists of minimizing the mean squared error.

Interpolation schemes must model $F(\mathbf{r})$ by some plausible functional form. In 1-D problems, for example, the function $F(x)$ is approximated by polynomials passing through the nodes where the value of the function is known. In *global polynomial interpolation*, a single polynomial $\Psi^{(M)}(x)$ of degree M passing through $M + 1$ points is defined for the entire domain under consideration. In *piecewise* methods, a different polynomial function is defined in each of the M specified intervals $[x_i, x_{i+1}]$ covering the entire domain. Conditions are imposed at the breakpoints x_i between intervals to ensure continuity and smoothness of the resulting interpolant.

4.16.1 Global Polynomial Interpolation

If we know the value of a function $F(x)$ at $M + 1$ distinct points $(x_0, x_1, \ldots, x_M)$ in a specified domain $[a, b]$, we can define a single polynomial $\Psi^{(M)}(x)$ of degree M

$$\Psi^{(M)}(x) = c_0 + c_1 x + c_2 x^2 + \cdots + c_M x^M \qquad (4.368)$$

that approximates $F(x)$ and passes through all these points. The $M + 1$ coefficients c_i $(i = 0, \ldots, M)$ are determined by expressing that the polynomial verifies the known values F_i of the function at each point x_i:

$$c_0 + c_1 x_i + c_2 x_i^2 + \cdots + c_M x_i^M = F_i \qquad (4.369)$$

We solve therefore the system of $M + 1$ equations

$$\begin{pmatrix} 1 & x_0 & x_0^2 & \cdots & x_0^M \\ 1 & x_1 & x_1^2 & \cdots & x_1^M \\ \vdots & \vdots & \vdots & & \vdots \\ 1 & x_M & x_M^2 & \cdots & x_M^M \end{pmatrix} \begin{pmatrix} c_0 \\ c_1 \\ \vdots \\ c_M \end{pmatrix} = \begin{pmatrix} F_0 \\ F_1 \\ \vdots \\ F_M \end{pmatrix} \qquad (4.370)$$

where the square matrix of dimension $M + 1$ is referred to as the Vandermonde matrix. This matrix is nonsingular if all points x_i are distinct. The solution of this system can be obtained by standard techniques such as the LU decomposition (see Box 6.2). This technique, however, is computationally expensive since it requires of the order of M^3 operations. More efficient numerical methods requiring only $O(M^2)$ operations are available (Press *et al.*, 2007). Vandermonde matrices are notoriously ill-conditioned, so the method should be used only if the data points are well-spaced and the values of the function well-behaved. The Newton or Lagrange interpolations described below are often preferred.

Newton interpolation

A polynomial of degree M can be expressed by the Newton expression

$$\Psi_i^{(M)}(x) = a_0 + a_1(x - x_0) + a_2(x - x_0)(x - x_1) + \cdots \\ + a_M(x - x_0)(x - x_1)\cdots(x - x_{M-1}) \tag{4.371}$$

or equivalently

$$\Psi^{(M)}(x) = \sum_{k=0}^{M} a_k\, \varphi_{k-1}(x) \tag{4.372}$$

with

$$\varphi_j(x) = \prod_{i=0}^{j} (x - x_i) \tag{4.373}$$

and $\varphi_{-1}(x) = 1$. The coefficients a_i, called divided differences, are computed from

$$a_0 = F(x_0) \quad a_1 = F[x_1, x_0] \quad a_2 = F[x_2, x_1, x_0] \quad a_M = F[x_M, x_{M-1}, \ldots, x_1, x_0] \tag{4.374}$$

where the bracket functions are defined by

$$F[x_i, x_j] = \frac{F(x_i) - F(x_j)}{(x_i - x_j)} \tag{4.375}$$

$$F[x_i, x_j, x_k] = \frac{F[x_i, x_j] - F[x_j, x_k]}{(x_i - x_k)} \tag{4.376}$$

$$\ldots\ldots$$

$$F[x_M, x_{M-1}, \ldots, x_1, x_0] = \frac{F[x_M, x_{M-1}, \ldots, x_1] - F[x_{M-1}, \ldots, x_1, x_0]}{(x_M - x_0)} \tag{4.377}$$

An illustrative example ($M = 2$) is given by the quadratic interpolation polynomial $\Psi^{(2)}(x)$,

$$\Psi^{(2)}(x) = a_0 + a_1(x - x_0) + a_2(x - x_0)(x - x_1) \tag{4.378}$$

whose value is equal to F_0, F_1, and F_2 at the three points x_0, x_1, and x_2, respectively. By application of (4.374), one finds

$$a_0 = F_0 \quad a_1 = \frac{(F_1 - F_0)}{(x_1 - x_0)} \quad a_2 = \frac{1}{(x_2 - x_0)}\left[\frac{(F_2 - F_1)}{(x_2 - x_1)} - \frac{(F_1 - F_0)}{(x_1 - x_0)}\right] \tag{4.379}$$

An advantage of the method is that the Newton interpolation algorithm can be expressed as a recursive process since

$$\Psi^{(M+1)}(x) = \Psi^{(M)}(x) + a_{M+1}\,\varphi_M(x) \tag{4.380}$$

if

$$\varphi_M(x) = \prod_{j=0}^{M}(x - x_j) \tag{4.381}$$

and

$$a_{M+1} = \frac{\left[F(x_{M+1}) - \Psi^{(M)}(x_{M+1})\right]}{\varphi_M(x_{M+1})} \tag{4.382}$$

Thus, one can easily calculate $\Psi^{(M+1)}(x)$ without having to re-compute all coefficients if the polynomial $\Psi^{(M)}(x)$ is known.

Lagrange interpolation

In the Lagrange formulation, one expresses the interpolation function by

$$\Psi^{(M)}(x) = \sum_{i=0}^{M} L_i(x)\,F_i \tag{4.383}$$

where

$$L_i(x) = \prod_{\substack{j=0 \\ j \neq i}}^{M} \frac{(x - x_j)}{(x_i - x_j)} \tag{4.384}$$

In this expression, each of the $M+1$ terms is of degree M and is equal to zero at each node x_j except at one of them (denoted x_i), where it is equal to F_i. For example, if the value of the function F is known at three points x_0, x_1, and x_2, ($M = 2$), the functional form of the second-order polynomial given by the Lagrange formula is

$$\Psi^{(2)}(x) = \frac{(x - x_1)(x - x_2)}{(x_0 - x_1)(x_0 - x_2)}F_0 + \frac{(x - x_0)(x - x_2)}{(x_1 - x_0)(x_1 - x_2)}F_1 + \frac{(x - x_0)(x - x_1)}{(x_2 - x_0)(x_2 - x_1)}F_2 \tag{4.385}$$

The Lagrange formulation is not recursive.

As shown by Figure 4.31, the use of a single interpolating polynomial $\Psi^{(M)}(x)$ for the global domain captures broad features, but often produces excessive variations (oscillatory artifacts) in the intervals between data points and specifically in the first and last intervals of the domain (Runge's phenomenon). Very inaccurate approximations can be found if the interpolant $\Psi^{(M)}(x)$ is used to extrapolate data beyond the limits of the data point domain. High-order polynomial interpolation is often ill-conditioned as small changes in the data lead to large differences in the values derived in the interval between nodes (overfitting). Finally, the errors resulting from local outliers (e.g., measurement error at a given station) propagate to the entire polynomial domain.

Piecewise linear interpolation

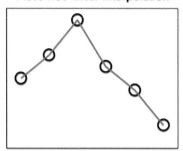

Full degree polynomial interpolation

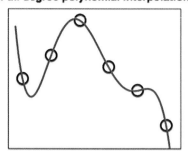

Shape-preserving Hermite interpolation

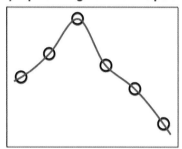

Spline interpolation

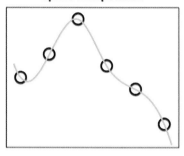

Figure 4.31 Comparison of different interpolation methods: piecewise linear, global polynomial, piecewise spline, and piecewise cubic Hermite. Reproduced From Moler (2004).

Rational function interpolation

It is often preferable to approximate the true function $F(x)$ that passes through $M + 1$ points by a rational function $R(x)$ that is the quotient of two polynomials, one of degree n and the second of degree m, with $n + m = M$:

$$\Psi(x) = \frac{p(x)}{q(x)} = \frac{p_0 + p_1 x + p_2 x^2 + \cdots + p_n x^n}{q_0 + q_1 x + q_2 x^2 + \cdots + q_m x^m} \tag{4.386}$$

The use of rational polynomials often leads to much better approximations than the use of ordinary polynomials, especially for a large number of nodes. The main drawback is that there is no control over the occurrence of poles (zero denominator) in the domain of interpolation. This problem can be avoided by using rational polynomials of higher degrees and, for example, by making the degree of the numerator and the denominator equal to M. An example is the barycentric interpolant

$$\Psi(x) = \frac{\displaystyle\sum_{i=0}^{M} \frac{w_i F_i}{(x - x_i)}}{\displaystyle\sum_{i=0}^{M} \frac{w_i}{(x - x_i)}} \tag{4.387}$$

If the weights $w_0, w_1, \ldots, w_N$ are chosen as

$$w_i = \prod_{\substack{j=0 \\ j \neq i}}^{M} \frac{1}{x_i - x_j} \tag{4.388}$$

the existence of poles is avoided. An alternative solution, that also prevents poles, is to simply specify (Press *et al.*, 2007)

$$w_i = (-1)^i \qquad (i = 0, M) \tag{4.389}$$

4.16.2 Piecewise Interpolation

Rather than defining a single interpolation function for the entire domain of interest $[x_0, x_N]$, it is often preferable to define a different interpolation function for each individual interval $[x_i, x_{i+1}]$, and express that the junction of these functions and potentially of their derivatives is continuous at each breakpoint (or partition point) x_i in the domain. To describe such piecewise interpolation algorithms, we consider again a 1-D function $F(x)$ representing a physical quantity whose values $F_i = F(x_i)$ are known at the data points x_i. We wish to estimate the value of this function at an arbitrary point x located between two of these partition points. The most desirable methods provide piecewise polynomial functions with a high degree of smoothness at the nodes where they connect.

The simplest piecewise interpolation method, called *nearest-neighbor interpolation*, is to approximate the function at point x by the value F_i corresponding to the closest point x_i. The method can be generalized in the 2-D case, leading to a mosaic of cells called a *Voronoi diagram*. In each cell i, the value of the interpolant is constant and equal to the value of the function at the data point $\mathbf{r}_i$. The resulting interpolant is a discontinuous function, which does not fulfill the requirement of smoothness at the junction between intervals.

A better method is *piecewise linear interpolation* in which the value of the function between partition points x_i and x_{i+1} is approximated (1-D case) by

$$\Psi_i^{(1)}(x) = F_i + \frac{F_{i+1} - F_i}{x_{i+1} - x_i}(x - x_i) \tag{4.390}$$

This linear interpolation is commonly applied in models of atmospheric composition and is often implemented in more than one dimension. Consider, for example, a function $F(x, y)$ of two independent variables x and y whose values $F_{i,j}$ are known at selected points (x_i, y_j). If we define the reduced variables

$$s = \frac{x - x_i}{x_{i+1} - x_i} \qquad\qquad t = \frac{y - y_j}{y_{j+1} - y_j} \tag{4.391}$$

whose values range between 0 and 1, the bilinear interpolation function $\Psi(x, y)$ at an unsampled point (x, y) is given by

$$\Psi(x, y) = (1 - s)(1 - t)F_{i,j} + s(1 - t)F_{i+1,j} + t(1 - s)F_{i,j+1} + s\,t\,F_{i+1,j+1} \tag{4.392}$$

Linear interpolation is characterized by discontinuities in the derivatives of the interpolant at the boundaries of the intervals, which is a major disadvantage of the method (see Figure 4.31 for the 1-D case) The difficulty can be addressed by considering higher-order piecewise interpolation methods, for example cubic interpolation algorithms. A frequently used algorithm based on cubic Hermite basis functions (Fritsch and Carlson, 1980) provides a monotone interpolant (no overshoots or undershoots).

In 2-D problems, a bicubic interpolation is performed by applying successively a 1-D cubic interpolation procedure in each direction. The resulting interpolated surface is smoother than the surfaces obtained by the bilinear interpolation.

The error resulting from the approximation by a polynomial function depends on the size of the intervals between nodes x_i, the location of the selected intermediate point x, and the properties of function $F(x)$. In the case of a simple linear interpolation, the error can be derived by expanding $F(x)$ in Taylor series about x_i, evaluating the first derivative of $F(x)$ at point x_i, and expressing the interpolant $\Psi^{(1)}(x)$ as a linear function taking the known values F_i and F_{i+1} at points x_i and x_{i+1}. Here, the superscript (1) refers to the order of the interpolant. The resulting error is

$$e^{(1)}(x) = F(x) - \Psi^{(1)}(x) = \frac{1}{2}(x - x_i)(x - x_{i+1})\frac{d^2F}{dx^2} \qquad (4.393)$$

where the second derivative is calculated at a point within the interval $[x_i, x_{i+1}]$. If Δx represents the length of this interval, the maximum error, which occurs at the midpoint x_m, is given by

$$\max\left|e^{(1)}(x)\right| = \frac{(\Delta x)^2}{8}\frac{d^2F}{dx^2}\bigg|_{x=x_m} \qquad (4.394)$$

For a piecewise polynomial interpolation of degree M, the error is proportional to the $(M+1)^{\text{th}}$ derivative of F

$$e^{(M)}(x) = F(x) - \Psi^{(M)}(x) = \varphi_M(x)\frac{1}{(M+1)!}\frac{d^{M+1}F}{dx^{M+1}} \qquad (4.395)$$

where φ_n is given by (4.381).

Polynomial spline interpolation

Consider a domain $[x_0, x_N]$ split into N intervals $[x_i, x_{i+1}]$. Spline piecewise interpolation is provided by a set of N polynomial pieces of degree p defined on each interval, so that the adjacent polynomial pieces and their $p-1$ derivatives are continuous a junction points x_i. In the cubic spline method, which is often adopted, we express the interpolant in each interval by a cubic polynomial

$$\Psi_i(x) = a_i + b_i(x - x_i) + c_i(x - x_i)^2 + d_i(x - x_i)^3 \qquad (i = 0, \ldots, N-1) \tag{4.396}$$

The $4N$ unknown coefficients a_i, b_i, c_i, and d_i are determined as follows (Figure 4.32). First, we require that the polynomial matches the values F_i of the data at each breakpoint x_i ($2N$ conditions)

$$\Psi_{i-1}(x_i) = \Psi_i(x_i) = F_i \qquad (i = 1, \ldots, N-1) \qquad (4.397)$$

$$\Psi_0(x_0) = F_0 \quad \text{and} \quad \Psi_{N-1}(x_N) = F_N \qquad (4.398)$$

Second, to make the interpolation as smooth as possible, we request that the first and second derivatives of $\Psi(x)$ be continuous at each breakpoint x_i ($2N - 2$ conditions):

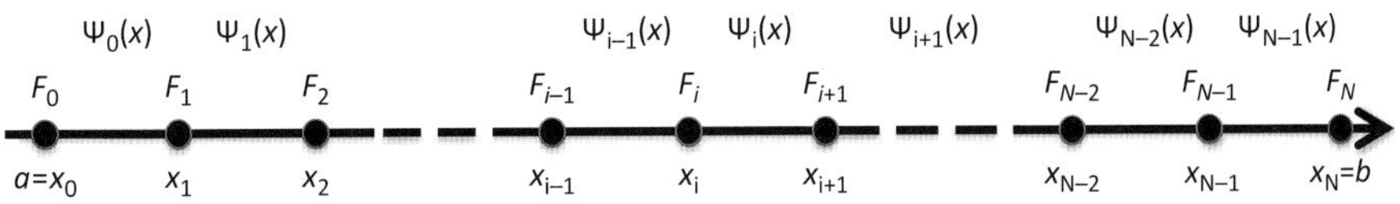

Figure 4.32 Spline functions $\Psi_i(x)$ defined in each interval $[x_i, x_{i+1}]$ of the entire domain $[x_0, x_N]$. The value of the "true" function $F(x)$ and of the spline function are equal to F_i at nodes x_i. In the cubic spline method, continuity of the first and second derivatives is also imposed at the nodes.

$$\Psi'_{i-1}(x_i) = \Psi'_i(x_i) \qquad \text{and} \qquad \Psi''_{i-1}(x_i) = \Psi''_i(x_i) \qquad (i = 1, \ldots, N-1) \quad (4.399)$$

Third, we must add two conditions required to solve the system of $4N$ unknowns. A standard choice, referred to as natural boundary conditions, is to impose that the second derivative of the interpolant is zero at both endpoints x_0 and x_N of the domain

$$\Psi''_0(x_0) = 0 \qquad \text{and} \qquad \Psi''_{N-1}(x_N) = 0 \qquad\qquad (4.400)$$

An alternative is to prescribe the slope of the spline function at each boundary if the first derivative of the original function $F(x)$ is known at both endpoints:

$$\Psi'_0(x_0) = F'(x_0) \qquad \text{and} \qquad \Psi'_{N-1}(x_N) = F'(x_N) \qquad (4.401)$$

The $4N$ coefficients a_i, b_i, c_i, and d_i are then derived by applying these conditions to polynomial (4.396) and to its first and second derivatives, respectively. The solution of the resulting system requires that a tridiagonal system of equations be solved, which is performed through only $O(N)$ operations (see Box 4.4). The cubic spline interpolation method is therefore computationally efficient. Further, the interpolation function is relatively smooth (Figure 4.31) since its first and second derivatives are continuous functions. The bicubic spline algorithm generalizes the method to two dimensions.

The spline interpolation has the advantage of capturing both broad and detailed features, but in some cases it may be smoother than wished; it also has occasionally the tendency to oscillate. Experience shows that the use of spline polynomials of a degree higher than three seldom yields any real advantage.

4.16.3 Distance-Weighted Interpolation

In *distance-weighted interpolation*, we express the interpolant by a linear combination of radial basis functions $\phi(|\mathbf{r} - \mathbf{r}_i|)$, called *influence functions*. These functions express the degree to which a data point situated at location $\mathbf{r}_i$ influences its surroundings. They are expressed as a function of the radial distance $d(\mathbf{r}, \mathbf{r}_i) = |\mathbf{r} - \mathbf{r}_i|$. We write therefore

$$\Psi(\mathbf{r}) = \sum_{i=0}^{N} w_i(\mathbf{r}) \, F_i = \frac{\sum\limits_{i=0}^{N} F_i \, \phi(d(\mathbf{r}, \mathbf{r}_i))}{\sum\limits_{i=0}^{N} \phi(d(\mathbf{r}, \mathbf{r}_i))} \qquad (4.402)$$

The weights $w_i(\mathbf{r})$ at location $\mathbf{r}$ of a given sampling point i are thus given by

$$w_i(\mathbf{r}) = \frac{\phi(d(\mathbf{r}, \mathbf{r}_i))}{\sum\limits_{i=0}^{N} \phi(d(\mathbf{r}, \mathbf{r}_i))} \tag{4.403}$$

In general, only the data points $\mathbf{r}_i$ located close to the target point $\mathbf{r}$ are taken into consideration when applying (4.402). A variety of forms are used for the radial basis function $\phi(d(\mathbf{r}, \mathbf{r}_i))$ to describe the decrease in the influence of a data point with distance d from the point. An example is the Gaussian form

$$\phi(d(\mathbf{r}, \mathbf{r}_i)) = \exp\left[-\frac{(d(\mathbf{r}, \mathbf{r}_i))^2}{c} \right] \tag{4.404}$$

where c is an adjustable parameter. In the *inverse distance weighting* (*IDW*) *method* (Shepard, 1968), the radial basis function is

$$\phi(d(\mathbf{r}, \mathbf{r}_i)) = \frac{1}{|\mathbf{r} - \mathbf{r}_i|^p} \tag{4.405}$$

where the choice of the power parameter p (typically 1–20) defines the smoothness of the solution.

Distance-weighted methods using radial basis functions are simple to implement and computationally inexpensive. A disadvantage is that errors cannot be characterized. Another issue is that the interpolant is sensitive to the sampling configuration. When adopting weighting functions that are purely radial, observations clustered in particular directions (and often providing redundant information) carry an artificially large weight. This can be corrected by multiplying the radial basis function ϕ by an anisotropy correction factor that is a function of the angles between every pair of stations relative to the unsampled point $\mathbf{r}$.

4.16.4 Kriging

The methods of interpolation discussed above are often qualified as deterministic because the variation of the physical quantity, described by a single "true" function $F(\mathbf{r})$, is approximated by a single interpolation function $\Psi(\mathbf{r})$. An alternative approach is to assume that $F(\mathbf{r})$ is a random field with several possible realizations among an ensemble of distributions that verify the known data $F_i = F(\mathbf{r}_i)$ at N points $\mathbf{r}_i$. The stochastic method introduced by South African mining engineer D. G. Krige (1951) and formalized by French mathematician G. Matheron (1962) is commonly used to interpolate spatially distributed geophysical data (Cressie, 1993). It provides the mean and variance of the ensemble of possible realizations at every point within a defined region. Rather than expressing weights as a function of the distance between sampled and unsampled data points as in the *IDW* method, the kriging approach accounts for the spatial correlation between data points.

We express again the interpolant $\Psi(\mathbf{r})$ that approximates $F(\mathbf{r})$ at any unsampled point $\mathbf{r}$ by a linear combination

$$\Psi(\mathbf{r}) = \sum_{i=1}^{N} w_i(\mathbf{r})\, F_i \qquad (4.406)$$

where the weights $w_i(\mathbf{r})$ must be determined from the assumed probabilistic behavior of the random field. Several variants of the kriging method have been developed. We present here the *ordinary kriging* method, which is the most widely used.

The method assumes that, for each pair of variables $F(\mathbf{r}_i)$ and $F(\mathbf{r}_j)$, a covariance exists that depends only on the separation vector $\mathbf{d} = \mathbf{r}_j - \mathbf{r}_i$. The difference between variables $F(\mathbf{r})$ and $F(\mathbf{r} + \mathbf{d})$ is treated as a stationary unbiased variable with mean

$$\overline{F(\mathbf{r} + \mathbf{d}) - F(\mathbf{r})} = 0 \qquad (4.407)$$

and variance

$$\mathrm{var}[F(\mathbf{r} + \mathbf{d}) - F(\mathbf{r})] = \overline{[F(\mathbf{r} + \mathbf{d}) - F(\mathbf{r})]^2} = 2\gamma(\mathbf{d}) \qquad (4.408)$$

Function $\gamma(\mathbf{d})$ is called the semi-variance function; it is determined experimentally as a function of the norm of the separation distance d between different data points in the domain by

$$\gamma(d) = \frac{1}{2N} \sum_{1}^{N} [F(r_i) - F(r_i + d)]^2 \qquad (4.409)$$

When displayed graphically (Figure 4.33), function $\gamma(d)$ constitutes a so-called *semi-variogram* or simply *variogram*. The experimental data are fitted by an analytical curve, from which values $\gamma(i, j)$ of the semi-variance for each pair of points (i, j) are determined. As shown below, the values of $\gamma(i, j)$ will be used to derive the weights needed to calculate the interpolant.

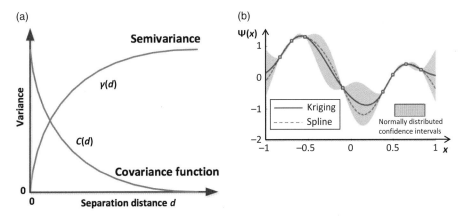

Figure 4.33 (a) Typical variogram $\gamma(d)$ and equivalent covariance function $C(d)$ (covariogram) as a function of the distance d between data points. One can show that $\gamma(d) = C(0) - C(d)$, where $C(0) = \mathrm{Var}\,(F(\mathbf{x}))$. The semivariance increases with the distance d, while the covariance function decreases. From Gentile *et al.* (2012). (b) Graphical illustration of 1-D data interpolation by kriging and spline methods. The open squares indicate the location of the data. The kriging interpolation is shown in red, and corresponds to the means of the normally distributed confidence intervals shown in gray. The spline interpolation polynomials are shown by the blue dashed line. Source: Wikimedia Commons.

The weights w_i are determined by imposing conditions. First, we express that the estimate $\Psi(\mathbf{r})$ must be unbiased (mean of the difference between the true function and the interpolant equal to zero)

$$\overline{\Psi(\mathbf{r}) - F(\mathbf{r})} = 0 \quad \text{or} \quad \sum_{i=1}^{N}\left(w_i\overline{F_i}\right) - m = \sum_{i=1}^{N}(w_i m) - m = 0 \qquad (4.410)$$

where m is the mean of $F(\mathbf{r})$. This implies that the sum of the weights is equal to 1:

$$\sum_{i=1}^{N} w_i = 1 \qquad (4.411)$$

Second, we minimize the mean square error between the true and interpolant functions, subject to the unbiased condition, by minimizing

$$\begin{aligned}
G(\mathbf{w}, \mu) &= \overline{[F(\mathbf{r}) - \Psi(\mathbf{r})]^2} - 2\,\mu\sum_{i=1}^{N}(w_i - 1) \\
&= \overline{\left[F(\mathbf{r}) - \sum_{1}^{N} w_i F_i\right]^2} - 2\,\mu\sum_{i=1}^{N}(w_i - 1)
\end{aligned} \qquad (4.412)$$

where μ is a so-called Lagrange multiplier (a parameter used in optimization theory to find the local maxima and minima of a function subject to equality constraints). This is achieved by equating to zero the partial derivative of G with respect to w_i and μ. After some algebraic manipulations, one obtains the following linear system

$$\begin{aligned}
\sum_{i=1}^{N} w_i\gamma(k, i) - \mu &= \gamma(0, k) \qquad \text{for } k = 1, \ldots, N \\
\sum_{i=1}^{N} w_i &= 1
\end{aligned} \qquad (4.413)$$

or, in matrix form,

$$\begin{pmatrix}
\gamma(1,1) & \gamma(1,2) & \cdots & \gamma(1,N) & -1 \\
\gamma(2,1) & \gamma(2,2) & \cdots & \gamma(2,N) & -1 \\
\cdots & \cdots & \cdots & \cdots & \cdots \\
\gamma(N,1) & \gamma(N,2) & \cdots & \gamma(N,N) & -1 \\
1 & 1 & \cdots & 1 & 0
\end{pmatrix} \cdot \begin{pmatrix} w_1 \\ w_2 \\ \vdots \\ w_N \\ \mu \end{pmatrix} = \begin{pmatrix} \gamma(0,1) \\ \gamma(0,2) \\ \vdots \\ \gamma(0,N) \\ 1 \end{pmatrix} \qquad (4.414)$$

Here, index 0 refers to the unsampled position $\mathbf{r}$ where we seek an estimate of the interpolant Ψ. The solution of the system provides the value for an unsampled point $\mathbf{r}$ of the optimal weights w_i and the Lagrangian multiplier μ. The variance for ordinary kriging

$$\sigma^2 = Var[F(\mathbf{r})] - \sum_{i=1}^{N} w_i(\mathbf{r})\,\gamma(0, i) + \mu \qquad (4.415)$$

is a measure of the interpolation error.

The kriging method has the advantage of providing an approximation of the true function that is based on a spatial statistical analysis of the data, and therefore automatically accounts for the possible clustering between data points. Its strength stems from the use of the semi-variance γ rather than geometric distances. It is particularly suited for situations in which data are sparse. A drawback of the method is its complexity and computational burden. Finally, the assumption of stationarity is not always valid. The ordinary kriging method is illustrated by Figure 4.33b.

4.16.5 Correction for Local Effects

Some corrections in the interpolation process may be introduced to account for local effects. We examine here two different approaches that address this issue, taking as an example the interpolation of chemical concentrations measured by a network of monitoring stations.

Innovation kriging (IK) method

The mapping of chemical concentrations by the ordinary kriging method described in the previous section accounts exclusively for observations at monitoring stations. The interpolation can potentially be improved, and specifically account for small-scale patterns such as chemical plumes, by including additional information from a chemical transport model. In the IK method (Blond *et al.*, 2003), this is performed by replacing (4.406) with

$$\Psi(\mathbf{r}) = \mathsf{M}(\mathbf{r}) + \sum_{i=1}^{N} w_i(\mathbf{r})[F_i - \mathsf{M}(\mathbf{r}_i)] \tag{4.416}$$

where F_i denotes the chemical concentration field observed at N stations located at points $\mathbf{r}_i$, and $\mathsf{M}(\mathbf{r})$ is a first guess (*prior*) of the interpolated field provided by the model (background value). The correction term applied to this prior estimate of the field

$$H_i \equiv H(\mathbf{r}_i) = F_i - \mathsf{M}(\mathbf{r}_i) \tag{4.417}$$

is called the *increment* or *innovation*. Again, the weight functions w_i need to be determined by optimization. If the spatial resolution of the model is sufficiently high, the prior estimate $\mathsf{M}(\mathbf{r})$ may exhibit detailed patterns not seen by the monitoring stations.

In the IK technique, the weighting functions w_i are determined by applying the ordinary kriging procedure to the innovation H_i rather than to the observations F_i. The system resulting from the optimization process is similar to system (4.413), but with the covariances applying to the innovation H_i rather than to the observed concentrations F_i. Figure 4.34 illustrates how the IK method can outperform the ordinary kriging technique by providing model information on small-scale features undamped by the observations.

Local empirical corrections

Another approach to account for local influences missing from the original inter-polation process is to use additional statistical information related to some identified

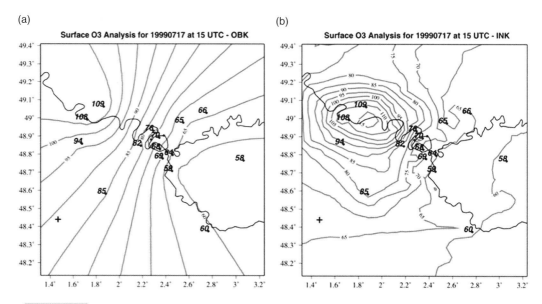

Figure 4.34 Production of a map of surface ozone mixing ratios [ppbv] in the vicinity of Paris based on surface observations at several monitoring stations on July 17, 1999. The measured mixing ratios at these stations are indicated next to the black dots. (a): Analysis by the ordinary kriging method based on the information provided only by the monitoring stations. (b): Analysis by the innovation kriging method in which data from the observing stations are combined with prior information from a chemical transport model. The model predicts the presence of an ozone plume downwind of Paris with a maximum concentration of 115 ppbv. The maps produced by the two techniques are very different, which highlights in this case the importance of the prior information provided by the model. From Blond *et al.* (2003).

forcing factor. Consider for example the mapping of surface ozone pollution in an urban area. Ozone may be locally titrated by conversion to NO_2 near emission hotspots of nitric oxide (NO) following reaction (3.18). This titration is important to characterize for population exposure but occurs at scales ~1 km that cannot be properly captured by the ozone observation network. Instead, one can establish an empirical statistical relationship between ozone concentrations and some relevant parameter for which finer mapping is available, such as land use or population density (Figure 4.35), and use this relationship to estimate smaller-scale features. The empirical relationship is first used to "detrend" the original observational data (remove the actual contribution of the small-scale forcing factors on the observed concentrations). The detrended values are then interpolated through one of the methods described above. The local influence is then reintroduced by a "retrending" procedure based on the empirical relationship. Results shown in Figure 4.35 for NO_2 and ozone highlight the importance of the local empirical correction in densely populated areas.

4.16.6 Conservative Remapping

Data available in one coordinate system must often be transformed to a different coordinate system. The transfer process is generally based on the interpolation of the physical quantities from a *source grid* to a *target* or *destination grid*, an operation

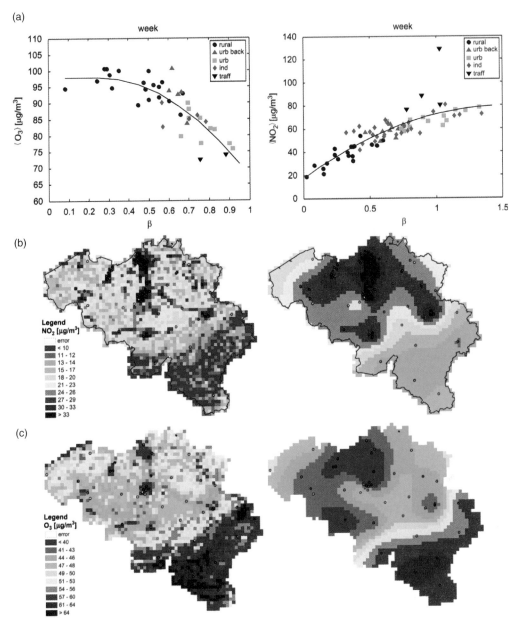

Figure 4.35 Empirical correction applied to improve air quality estimates in Belgium. (a): Trend functions for NO_2 and O_3: average maximum one-hour concentration values as a function of land-use parameter β (here for weekday summer values between 2001 and 2006). Factor β represents a weighted and normalized sum of land-use indicators including, for example, urban fabric, industrial areas, road and rail networks, arable land, agricultural areas, forests, wetlands, etc. (b): Annual mean NO_2 surface concentration for year 2006 obtained by the ordinary kriging interpolation method (right) and adjusted to account for the local effects of land use (left). The small circles on the map indicate the location of the monitoring stations. (c): Same, but for surface ozone. From Janssen *et al.* (2008).

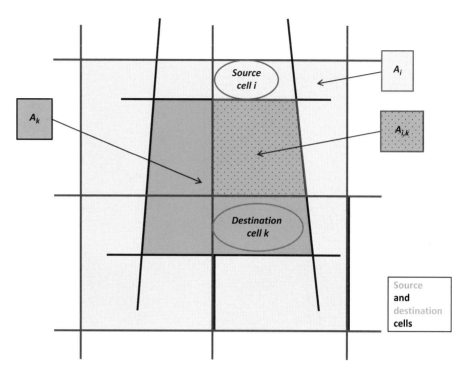

Figure 4.36 Schematic representation of remapping from a source grid (rectangular, in blue) to a destination grid (Mercator projection, in green). The stippled area $A_{i,k}$ represents the overlap between source grid cell i (area A_i) and destination cell k (area A_k).

called *remapping* or *regridding*. A simple conservative area-weighted method is to use interpolation weights provided by the fractional overlap between source and destination grid cells (Figure 4.36). We denote by A_i the area of source grid cell i, by A_k the area of target grid cell k, and by $A_{i,k}$ the overlap between the two. By definition,

$$A_k = \sum_{i=1}^{N_k} A_{i,k} \tag{4.418}$$

where N_k is the number of cells i that fall within the destination grid cell k.

If $F_i(\mathbf{r})$ is an *intensive* variable in a source grid cell i (i.e., a quantity whose value is independent of the size of the grid cell), and $\mathbf{r}$ is the spatial coordinate for the source grid, the mean value $\overline{\Psi}_k$ of the variable in the destination grid cell k is given by

$$\overline{\Psi}_k = \frac{1}{A_k} \sum_{i=1}^{N_k} \int_{A_{i,k}} F_i \, dA \tag{4.419}$$

If we assume that the mean value of F_i is the same for the entire source grid cell and for the overlapping area with destination grid cell k, so that

$$\overline{F}_i = \frac{1}{A_i} \int_{A_i} F_i \, dA = \frac{1}{A_{i,k}} \int_{A_{i,k}} F_i \, dA \tag{4.420}$$

then we can compute $\overline{\Psi}_k$ simply as

$$\overline{\Psi}_k = \frac{1}{A_k} \sum_{i=1}^{N_k} A_{i,k} \, \overline{F}_i \tag{4.421}$$

If F_i is an *extensive* variable (i.e., a quantity whose value is proportional to the grid cell area), then cumulative values must be used in the transformation and the mean quantity for the destination grid cell k is given by

$$\overline{\Psi}_k = \sum_{i=1}^{N_k} \left(\frac{A_{i,k} \, \overline{F}_i}{A_i} \right) \tag{4.422}$$

The above expressions represent a first-order area-weighted scheme that is generally good enough when a single remapping operation needs to be done. The calculation of overlapping area $A_{i,k}$ is not always straightforward when the source and destination grids use different geographic projections. This problem can be addressed by dividing the source grid cell into a large number of mini-cells with area A_m such that $A_m \ll A_k$. The mini-cells are attributed to point geographical locations and one just needs to count the number $N_{i,k}$ of mini-cells that fall within destination cell k to derive

$$A_{i,k} \approx N_{i,k} A_m \tag{4.423}$$

The first-order remapping method discussed above can be improved by expanding $F_i(\mathbf{r})$ as a Taylor series around the centroid $\mathbf{r}_i$ of source cell i

$$F_i(\mathbf{r}) = \overline{F}_i + \nabla_i F \cdot (\mathbf{r} - \mathbf{r}_i) \tag{4.424}$$

where $\nabla_i F$ represents the spatial gradient of F in cell i, and

$$\mathbf{r}_i = \frac{1}{A_i} \int_{A_i} \mathbf{r} \, dA \tag{4.425}$$

The regridded field is now second-order accurate if $\nabla_i F$ is at least a first-order approximation of the gradient, as may be derived from the difference between adjacent grid cells. We then obtain

$$\overline{\Psi}_k = \frac{1}{A_k} \sum_{i=1}^{N_k} \left(A_i \overline{F}_i + \int_{A_{ik}} \nabla_i F \cdot (\mathbf{r} - \mathbf{r}_i) \, dA \right) \tag{4.426}$$

The integral on the right-hand side can be approximated in simple ways. The second-order remapping scheme is useful to reduce error involved in repeated interconversions between a source grid and destination grid. When a large number of back-and-forth remappings between the two grids are performed, the signature of the coarse grid on the fine grid becomes visible in the case of the first-order interpolation. The second-order remapping retains better the shape of the original function.

Remapping the three components of the wind velocities between two atmospheric grids requires special attention. The components of the wind velocities are related to each other by the continuity equation for air, and errors introduced in the interpolation process may therefore lead to the violation of mass conservation. The problem is addressed by multiplying the wind velocity components on the source grid by the local air concentration (or pressure), and by applying the conservative interpolation procedure to the resulting mass flux. The interpolated field on the destination grid is then divided by the air concentration (or pressure) to derive the interpolated wind velocity.

References

Andrews, D. G., Holton, J. R. and Leong, C. B. (1987) *Middle Atmosphere Dynamics*, Academic Press, Orlando.

Andrews D. G. and McIntyre M. E. (1976) Planetary waves in horizontal and vertical shear: The generalized Eliassen–Palm relation and the mean zonal acceleration, *J. Atmos. Sci.*, **33**, 2031–2048.

Andrews D. G. and McIntyre M. E. (1978) An exact theory of nonlinear waves on a Lagrangian-mean flow, *J. Fluid Mech.*, **89**, 609–646.

Arakawa A. and Lamb V. R. (1977) Computational design of the basic dynamical processes of the UCLA general circulation model, *Methods Comput. Phys.*, **17**, 174–265.

Asselin R. (1972) Frequency filter for time integrations, *Mon. Wea. Rev.*, **100**, 487–490.

Banks P. M. and Kockarts G. (1973) *Aeronomy*, Academic Press, New York.

Bishop C. M. (1995) *Neural Networks for Pattern Recognition*, Clarendon Press, Oxford.

Blond N., Bel L., and Vautard R. (2003) Three-dimensional ozone data analysis with an air quality model over the Paris area, *J. Geophys. Res*, **108** (D23), 4744, doi:10.1029/2003JD003679.

Boughton B. A., Delarentis J. M., and Dunn W. W. (1987) A stochastic model of particle dispersion in the atmosphere, *Boundary-Layer Meteorology*, **80**, 147–163.

Bourke W. (1974) A multi-level spectral model: I.Formulation and hemispheric integrations, *Mon. Wea. Rev.*, **102**, 687–701.

Boyd J. (1976) The noninteraction of waves with the zonally averaged flow on a spherical earth and the interrelationships of eddy fluxes of energy, heat and momentum, *J. Atmos. Sci.*, **33**, 2285–2291.

Boyd J. (1998) Two comments on filtering (artificial viscosity) for Chebyshev and Legendre spectral and spectral element methods: Preserving boundary conditions and interpretation of the filter as a diffusion. *J. Comput. Phys.*, **143**, 283–288.

Brasseur G. P. and Solomon S. (2005) *Aeronomy of the Middle Atmosphere: Chemistry and Physics of the Stratosphere and Mesosphere*, 3rd edition, Springer, Amsterdam.

Canuto V. M., Goldman I., and Hubickyj O. (1984) A formula for the Shakura–Sunyaev turbulent viscosity parameter, *Astrophys. J.*, **280**, L55–L88, doi: 10.1086/184269.

Chapman S. and Cowling T. G. (1970) *The Mathematical Theory of Non-uniform Gases*, 3rd edition, Cambridge University Press, Cambridge.

Charney J. G., Fjörtoft R., and von Neumann J. (1950) Numerical integration of the barotropic vorticity equation, *Tellus*, **2**, 237–254.

Courant R. (1943) Variational methods for the solution of problems of equilibrium and vibrations, *Bull. Amer. Math. Soc.*, **49**, 1–23.

Cressie N. (1993) *Statistics for Spatial Data*, Wiley, Chichester.

Cruse H. (2006) *Neural Networks as Cybernetic Systems*, 2nd edition, Brains, Mind and Media, Bielefeld.

de Bruyns Kops S. M., Riley J. J., and Kosaly G. (2001) Direct numerical simulation of reacting scalar mixing layers, *Phys. Fluids*, **13**, 5, 1450–1465.

Dietachmayer G. S. and Droegemeier K. K. (1992) Application of continuous dynamic grid adaption techniques to meteorological modelling. Part I: Basic formulation and accuracy, *Mon. Wea. Rev.*, **120**, 1675–1706.

Durran D. R. (2010) *Numerical Methods for Fluid Dynamics*, Springer, Amsterdam.

Eliasen E., Machenhauer B., and Rasmussen E. (1970) *On a Numerical Method for Integration of the Hydrodynamical Equations with a Spectral Representation of the Horizontal Fields*, Institute of Theoretical Meteorology, University of Copenhagen, Copenhagen.

Enting I. G. (2000) Green's function methods of tracer inversion, *Geophys. Monograph.*, **114**, 19–31,

Fournier A., Taylor M. A., and Tribbia J. (2004) The spectral element atmosphere model (SEAM): High resolution parallel computation and localized resolution of regional dynamics, *Mon. Wea. Rev.*, **132**, 726–748.

Fritsch F. N. and Carlson R. E. (1980) Monotone piecewise cubic interpolation, *SIAM J. Numer. Anal.*, **17**, 238–246.

Garcia, R. and Solomon S. (1983) A numerical model of the zonally averaged dynamical and chemical structure of the middle atmosphere, *J. Geophys. Res.* **88**, 1379–1400.

Garcia-Menendez F., Yano A., Hu Y., and Odman M. T. (2010) An adaptive grid version of CMAQ for improving the resolution of plumes, *Atmos. Pollut. Res.*, **1**, 239–249.

Gentile M., Courbin F., and Meylan G. (2013) Interpolating point spread function anisotropy, *Astronomy & Astrophysics*, **549**, A1.

Hall T. M. and Plumb R. A. (1994) Age of air as a diagnostic of transport, *J. Geophys. Res.*, **99**, 1059–1070.

Haltiner G. J. and Williams R. T. (1980) *Numerical Prediction and Dynamic Meteorology*, Wiley, Chichester.

Hubbard M. E. (2002) Adaptive mesh refinement for three-dimensional off-line tracer advection over the sphere, *Int. J. Numer. Methods Fluids*, **40**, 369–377.

Jablonowski C. and Williamson D. L. (2011) The pros and cons of diffusion, filters and fixers in atmospheric general circulation models. In *Numerical Techniques for Global Atmospheric Models* (Lauritzen P. H., Jablonowski C., Taylor M. A., and Nair R. D., eds.), Springer-Verlag, Berlin.

Jacob D. J. (1999) *Introduction to Atmospheric Chemistry*, Princeton University Press, Princeton, NJ.

Janssen S., Dumont G., Fierens F., and Mensink C. (2008) Spatial interpolation of air pollution measurements using CORINE land cover data, *Atmos. Env.*, **42**, 4884–4903.

Kasahara A. (1974) Various vertical coordinate systems used for numerical weather prediction, *Mon. Wea. Rev.*, **102**, 504–522.

Krige D. G. (1951) A statistical approach to some basic mine valuation problems on the Witwatersrand, *J. Chem., Metal. Mining Soc. South Africa*, **52**, 119–139.

Langevin P. (1908) On the theory of Brownian motion, *C. R. Acad. Sci. (Paris)*, **146**, 530–533.

Lanser D. and Verwer J. G. (1998) Analysis of operator splitting for advection–diffusion–reaction problems from air pollution modeling. CWI Report MAS-R9805.

Lanser D. and Verwer J. G. (1999) Analysis of operator splitting for advection–diffusion-reaction problems from air pollution modelling, *J. Comput. Appl. Math.*, **111**, 210–216.

Laprise R. (1992) The resolution of global spectral models. *Bull. Amer. Meteor. Soc.*, **73**, 1453–1454.

Lauritzen P. H., Jablonowski C., Taylor M. A., and Nair R. D. (2011) *Numerical Techniques for Global Atmospheric Models*, Springer, Amsterdam.

Lin J. C. (2012) Lagrangian modeling of the atmosphere: An introduction. In *Lagrangian Modeling of the Atmosphere* (Lin J., Brunner D., Gerbig C., *et al.*, eds.), American Meteorological Union, Washington, DC.

Lin J. C., Gerbig C., Wofsy S. C., *et al.* (2003) A near-field tool for simulating the upstream influence of atmospheric observations: The Stochastic Time-Inverted Lagrangian Transport (STILT) model, *J. Geophys. Res.*, **108**(D16), 4493, doi:10.1029/202JD003161.

Lin J. S. and Hildemann L. (1996) Analytical solutions of the atmospheric diffusion equation with multiple sources and height-dependent wind speed and eddy diffusivities, *Atmos. Environ.*, **30**(2), 239–254.

Lin S.-J. and Rood R. B. (1996) Multidimensional flux-form semi-Lagrangian scheme, *Mon. Wea. Rev.*, **124**, 2046–2070.

Liu S. C., McAfee J. R., and Cicerone R. J. (1982) Radon-222 and tropospheric vertical transport, *J. Geophys. Res.*, **89**, 7291–7297.

Luhar A. K. (2012) Lagrangian particle modeling of dispersion in light winds. *Lagrangian Modeling of the Atmosphere* (Lin J., Brunner D., Gerbig C., *et al.*, eds.), American Meteorological Union, Washington, DC.

Matheron G. (1962) *Traité de géostatistique appliquée*. Editions Technip, Paris.

McWilliams J. (2006) *Fundamentals of Geophysical Fluid Dynamics*, Cambridge University Press, Cambridge.

Mesinger F. (1984) A blocking technique for representation of mountains in atmospheric models. *Riv. Meteor. Aeronautica*, **44**, 195–202.

Moler, C. (2004) *Numerical Computing with MATLAB*, Society for Industrial and Applied Mathematics, Philadelphia, PA.

Neufeld Z. and Hernandez-Garcia E. (2010) *Chemical and Biological Processes in Fluid Dynamics*, Imperial College Press, London.

Odman M. T., Mathur R., Alapaty K., *et al.* (1997) Nested and adaptive grids for multiscale air quality modeling. In *Next Generation Environmental Models and Computational Methods* (Delic G. and Wheeler M. F., eds.), Society for Industrial and Applied Mathematics, Philadelphia, PA.

Odman M. T., Khan M. N., Srivastava R. K., and McRae D. S. (2002) Initial application of the adaptive grid air pollution model. In *Air Pollution Modeling and Its Applications* (Borrego C. and Schayes G., eds.), Kluwer Academic/Plenum Publishers, New York.

Orszag S. A. (1970) Transform method for the calculation of vector-coupled sums: Application to the spectral form of the vorticity equation, *J. Atmos. Sci.*, **27**, 890–895.

Pasquill F. (1971) Atmospheric dispersion of pollutants, *Q. J. Roy. Meteor. Soc.*, **97**, 369–395.

Patera A. T. (1984) A spectral element method for fluid dynamics: Laminar flow in a channel expansion, *J. Compute. Phys.*, **54**, 468–488.

Phillips N. A. (1957) A coordinate system having some special advantages for numerical forecasting, *J. Meteor.* **14**, 184–185.

Prather M. J. (2007) Lifetimes and time-scales in atmospheric chemistry. *Phil. Trans. R. Soc. A*, **365**, 1705–1726, doi: 10.1098/rsta.2007.2040.

Press W. H., Teukolsky S. A., Vetterling W. T., and Flannery B. P. (2007) *Numerical Recipes: The Art of Scientific Computing*, Cambridge University Press, Cambridge.

Qaddouri A. (2008) Optimized Schwarz methods with the Yin–Yang grid for shallow water equations. In *Domain Decomposition Methods in Science and Engineering* (Langer U., Discacciati M., Keyes D. E., Widlund O., and Zulehner, W., eds.) Springer, New York.

Reynolds O. (1883) An experimental investigation of the circumstances which determine whether the motion of water shall be direct or sinuous, and of the law of resistances in parallel channels. *Phil. Trans. Roy. Soc. London*, **174**, 935–982.

Robert A. (1969) The integration of a spectral model of the atmosphere by the implicit method. *Proceedings of the WMO/IUGG Symposium on NWP.* Japan Meteorological Society, Tokyo.

Seinfeld J. H. and Pandis S. N. (2006) *Atmospheric Chemistry and Physics: From Air Pollution to Climate Change*, 2nd edition, Wiley, New York.

Shapiro R. (1970) Smoothing, filtering and boundary effects, *Rev. Geophys. Spac. Phys.*, **8**, 2, 359–387.

Shapiro R. (1975) Linear filtering, *Math. Comput.*, **29**, 1094–1097.

Shepard D. (1968) A two-dimensional interpolation function for irregularly-space data. In *Proceedings of the 1968 ACM National Conference*, doi:10.1145/800186.810616.

Shuman F. G. (1957) Numerical methods for weather prediction, II: Smoothing and filtering, *Mon. Wea. Rev.*, **85**, 357–361.

Skamarock W. C. and Klemp J. B. (1993) Adaptive grid refinement for two-dimensional and three-dimensional nonhydrostatic atmospheric flow, *Mon. Wea. Rev.*, **121**, 788–804.

Sportisse B. (2000) An analysis of operator splitting techniques in the stiff case, *J. Comput. Phys.*, **161**, 140–168, doi: 10.1006/jcph.2000.6495.

Sportisse B. (2010) *Fundamentals in Air Pollution*, Springer, Amsterdam.

Srivastava R. K., McRae D. S., and Odman M. T. (2000) An adaptive grid algorithm for air-quality modeling, *J. Comput. Phys.*, **165**, 437–472, doi: 10.1006/jcph.2000.6620.

Steyn D. G. and Rao S. T. (2010) Air pollution modeling and its application. In *Proceedings of the 30th NATO/SPS International Technical Meeting on Air Pollution Modelling and Its Application*, Springer, New York.

Stockie J. M. (2011) The mathematics of atmospheric dispersion modeling, *SIAM Rev.*, **53** (2) 349–372.

Stohl A. (1998) Computation, accuracy and applications of trajectories: A review and bibliography, *Atmos. Environ.*, **32**, 6, 947–966.

Stohl A., Forster C., Frank A., Seibert P., and Wotawa G. (2005) Technical note: The Lagrangian particle dispersion model FLEXPART version 6.2, *Atmos. Chem. Phys.*, **5**, 2461–2474.

Strang G. (1968) On the construction and comparison of difference schemes, *SIAM J. Numer. Anal.*, **5**, 3, 506–517.

Thomson D. J. (1987) Criteria for the selection of stochastic models of particle trajectories in turbulent flows, *J. Fluid. Mech.*, **180**, 529–556.

Tomlin A. S., Berzins M., Ware J., Smith J., and Pilling M. J. (1997) On the use of adaptive gridding methods for modelling chemical transport from multi-scale sources, *Atmos. Environ.*, **31**, 2945–2959.

Vandeven H (1991) Family of spectral filters for discontinuous problems. *J. Sci. Comput.*, **6**, 159–192.

Warner T. T. (2011) *Numerical Weather and Climate Prediction*, Cambridge University Press, Cambridge.

Washington W. M., Buja L., and Craig A. (2009) The computational future for climate and Earth system models: On the path to petaflop and beyond, *Phil. Trans. R. Soc. A*, **367**, 833–846.

Williamson D. L. and Laprise R. (1998) Numerical approximations for global atmospheric general circulation models. In *Numerical Modelling of the Global Atmosphere for Climate Prediction* (Mote P. and O'Neill A., eds.), Kluwer Academic Publishers, Dordrecht.

Wilson J. D. and Sawford B. L. (1996) Review of Lagrangian stochastic models for trajectories in the turbulent atmosphere, *Boundary-Layer Meteorology*, **78**, 191–210.

5 Formulations of Radiative, Chemical, and Aerosol Rates

5.1 Introduction

We saw in Chapter 4 how the continuity equations for reactive chemical species in the atmosphere include chemical production and loss terms determined by kinetic rate laws. Similarly, we saw that the continuity equations for aerosols include formation and growth terms determined by microphysical properties. Here we present the formulations of these different terms.

We begin in Section 5.2 with the equations of *radiative transfer* that govern the propagation of radiation in the atmosphere. This determines the rates of photolysis reactions, which play a particularly important role in driving atmospheric chemistry as described in Chapter 3. We go on to present the general formulations of chemical kinetics in atmospheric models including gas-phase reactions (Section 5.3), reactions in aerosol particles and clouds (Section 5.4), and the design of chemical mechanisms (Section 5.5). In Section 5.6 we describe the computation of aerosol microphysical processes as needed to model the evolution of aerosol size distributions.

5.2 Radiative Transfer

Radiative transfer describes the propagation of radiation in the atmosphere. Radiation is energy propagated by electromagnetic waves. These waves represent oscillating electric and magnetic fields traveling at the speed of light. The oscillations are characterized by their frequency ν [Hz] or wavelength λ [m]. Frequency ν and wavelength λ are related by

$$\lambda \nu = c \tag{5.1}$$

where $c = 3.00 \times 10^8$ m s^{-1} is the speed of light in vacuum. The radiation is quantized as photons with energy $h\nu$ [J], where h is the Planck constant (6.63×10^{-34} J s). The intensity of radiation that propagates through the atmosphere is affected by emission, absorption, and scattering processes. We refer to the *radiation spectrum* as the distribution of energy contributed by photons of different wavelengths. Solar radiation is mainly in the ultraviolet (UV, $\lambda < 0.4$ μm), the visible (Vis, $0.4 < \lambda < 0.7$ μm), and the shortwave infrared (SWIR, $0.7 < \lambda < 3$ μm). Radiation emitted by the Earth and its atmosphere is mainly in the 5–20 μm range, called terrestrial IR (TIR). Solar radiation is sometimes called *shortwave* and terrestrial radiation *longwave*.

We present in this section the basic theory of radiative transfer to describe the photon flux in the atmosphere as governed by emission, scattering, and absorption of radiation. We will refer to the spectral distribution of a physical quantity (such as the photon flux F) as the *spectral density* of this quantity or its *monochromatic value*, expressed by the derivative versus wavelength ($F_\lambda = dF/d\lambda$) or versus frequency ($F_\nu = dF/d\nu$). It is common practice in the spectroscopy literature to express radiative quantities as a function of wavelengths in the UV–Vis region of the spectrum, and as a function of frequencies or of wavenumbers ($1/\lambda$) in the IR. For consistency in the presentation we will express radiative quantities as a function of wavelength throughout. Conversion to frequency or wavenumber is straightforward.

5.2.1 Definitions

Radiance

The radiation field can be described by the spectral density L_λ of the *radiance* (also called the intensity). The radiance is defined as the amount of energy d^4E [J] in wavelength interval $d\lambda$ [nm] traversing horizontal surface dS [m^2] during a time interval dt [s] in solid angle $d\Omega$ [sr] inclined at an angle θ relative to the vertical (Figure 5.1). Thus, for a *pencil of light* propagating in the direction Ω defined by angle θ, the spectral density of the radiance is defined by

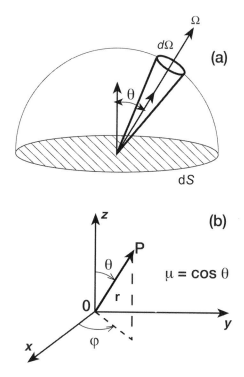

Figure 5.1 (a) Geometry of a pencil of light propagating in a solid angle $d\Omega$ and traversing a horizontal surface dS at a zenith angle θ. (b) Coordinates of point P defined as distance **r** from origin O, azimuthal angle φ, and zenith angle θ.

$$L_\lambda = \frac{d^4 E}{d\lambda \, dt \, dS \, d\Omega \cos\theta} \tag{5.2}$$

and has units of W m^{-2} sr^{-1} nm^{-1}, where wavelength λ is in nm.

Beer–Lambert law

A pencil of radiation traversing an optically active medium such as air is affected by its interaction with that medium. The Beer–Lambert law states that the attenuation of the radiance traversing an infinitesimally thin layer is proportional to the density of the medium, the thickness ds of the layer, and the radiance of the light. In the absence of radiative emission by the medium, we have

$$dL_\lambda(\lambda, s) = -\beta_{ext}(\lambda, s) \, L_\lambda(\lambda, s) \, ds \tag{5.3}$$

where $\beta_{ext}(\lambda, s)$ [m^{-1}] is the *extinction or attenuation coefficient*. $\beta_{ext}(\lambda, s)$ is a property of the medium. It can be assumed to be proportional to the mass density $\rho(s)$ [kg m^{-3}] or (for a gas) the number density $n(s)$ [molecules cm^{-3}] of the medium, the proportionality coefficient being the wavelength-dependent *mass extinction cross-section* k_{ext} [m^2 kg^{-1}] or the *molecular extinction cross-section* σ_{ext} [cm^2 molecule^{-1}]:

$$\beta_{ext}(\lambda, s) = k_{ext}(\lambda, s) \, \rho(s) = \sigma_{ext}(\lambda, s) \, n(s) \tag{5.4}$$

Integration of expression (5.3) between geometrical points s_0 and s yields

$$L_\lambda(\lambda, s) = L_\lambda(\lambda, s_0) \exp\left[-\int_{s_0}^{s} \beta_{ext}(\lambda, s') \, ds' \right] \tag{5.5}$$

The *optical depth* at wavelength λ between geometrical points s_0 and s is given by

$$\tau(\lambda, s_0, s) = \int_{s_0}^{s} \beta_{ext}(\lambda, s') \, ds' = \int_{s_0}^{s} k_{ext}(\lambda, s') \, \rho(s') ds' = \int_{s_0}^{s} \sigma_{ext}(\lambda, s') \, n(s') ds' \tag{5.6}$$

and the corresponding *transmission function* T between s_0 and s is

$$\mathsf{T}(\lambda, s_0, s) = \exp\left[-\tau(\lambda, s_0, s)\right] \tag{5.7}$$

Following standard atmospheric chemistry usage, we define the optical depth at altitude z as the extinction in the *vertical* direction

$$\tau(\lambda, z) = \int_{z}^{\infty} \beta_{ext}(\lambda, z) \, dz \tag{5.8}$$

The extinction along an inclined direction is then referred to as the *slant optical depth* or *optical path*.

Extinction includes processes of *absorption* (conversion of radiation to other forms of energy, such as heat) and *scattering* (change in the direction of the incident radiation). The extinction coefficients and optical depths are often separated into additive absorption (*abs*) and scattering (*scat*) components:

$$\beta_{ext} = \beta_{abs} + \beta_{scat} \tag{5.9}$$

and

$$\tau = \tau_{abs} + \tau_{scat} \tag{5.10}$$

The ratio between the scattering extinction and the total extinction is called the *single scattering albedo* $\omega(\lambda)$:

$$\omega(\lambda) = \frac{\beta_{scat}(\lambda)}{\beta_{scat}(\lambda) + \beta_{abs}(\lambda)} \tag{5.11}$$

Finally, in an absorbing atmosphere, it is customary to express the transmission as a function of the *path length* [kg m^{-2}] between geometric points s_0 and s

$$u(s_0, s) = \int_{s_0}^{s} \rho_{abs}(s') \, ds' \tag{5.12}$$

where ρ_{abs} [kg m^{-3}] is the mass density of the absorber. If $k_{abs}(\lambda, u)$ [m^2 kg^{-1}] is the mass absorption cross-section,

$$T(\lambda, u) = \exp\left[-\int_{u} k_{abs}(\lambda, u) \, du\right] \tag{5.13}$$

or, if we assume a homogeneous atmosphere where k_{abs} is only dependent on wavelength λ and not on pressure or temperature,

$$T(\lambda, s_0, s) = \exp\left[-k_{abs}(\lambda) \, u(s_0, s)\right] \tag{5.14}$$

Radiative transfer equation

In addition to being attenuated by its interaction with matter, the energy of a pencil of light can be strengthened as a result of local *radiative emission* by the material, or through *scattering* of radiation from all directions into that pencil of light. These two processes lead to an increase in the local radiance expressed as

$$dL_\lambda(\lambda, s) = j(\lambda, s) \, ds \tag{5.15}$$

Here, $j(\lambda, s)$ is a radiative source term from emission or scattering that can be assumed proportional to the extinction ($j(\lambda, s) \sim \beta_{ext}(\lambda, s)$) since the same processes are involved. Defining the *source function* as $J(\lambda, s) = j(\lambda, s)/\beta_{ext}(\lambda, s)$, we obtain a simple form of the radiative transfer equation

$$\frac{dL_\lambda(\lambda, s)}{\beta_{ext}(\lambda, s) \, ds} = -L_\lambda(\lambda, s) \; + \; J(\lambda, s) \tag{5.16}$$

A general expression for the radiative equation in a 3-D inhomogeneous atmospheric medium is given by Liou (2002):

$$\frac{1}{\beta_{ext}(\mathbf{r})}(\mathbf{\Omega} \bullet \nabla)L_\lambda(\lambda, \mathbf{r}, \mathbf{\Omega}) = -L_\lambda(\lambda, \mathbf{r}, \mathbf{\Omega}) + J(\lambda, \mathbf{r}, \mathbf{\Omega}) \tag{5.17}$$

where $L_\lambda(\lambda, \mathbf{r}, \mathbf{\Omega})$ and $J(\lambda, \mathbf{r}, \mathbf{\Omega})$ represent respectively the monochromatic radiance and source function in the direction defined by the vector $\mathbf{\Omega}$ (see Figure 5.1) and at the location defined by the vector **r**. In many applications, it can be assumed that radiative quantities and atmospheric parameters vary only with altitude (plane-parallel atmosphere). In this case, if the vertical optical depth rather than the geometric altitude is adopted as the independent variable

$$d\tau = -\beta_{ext}\, dz = \beta_{ext} \cos(\theta)\, ds$$

the radiative transfer equation takes the form

$$\mu \frac{dL_\lambda(\lambda, \tau, \mu, \varphi)}{d\tau} = -L_\lambda(\lambda, \tau, \mu, \varphi) + J(\lambda, \tau, \mu, \varphi) \tag{5.18}$$

where $\mu = \cos(\theta)$. The first term on the right-hand side of this equation accounts for the attenuation of light following the Beer–Lambert law, while the second term J represents the radiative source term (local emission or light scattered from other directions).

The radiance [W m^{-2} sr^{-1}] at a point **r** of the atmosphere and for a direction $\mathbf{\Omega}$ is given by the spectral integration of L_λ

$$L(\mathbf{r}, \mathbf{\Omega}) = \int_0^\infty L_\lambda(\lambda, \mathbf{r}, \mathbf{\Omega})\, d\lambda \tag{5.19}$$

The spectral density of the *irradiance* $F_\lambda(\lambda, \mathbf{r}, \mathbf{\Omega})$ [W m^{-2} nm^{-1}] at point **r** is defined as the energy flux density traversing a surface of unit area perpendicular to direction $\mathbf{\Omega}$ integrated over all directions $\mathbf{\Omega}'$ of the incoming pencils of light. It is thus provided by the integration over all directions of the normal component of the monochromatic radiance

$$F_\lambda(\lambda, \mathbf{r}, \mathbf{\Omega}) = \int_{4\pi} L_\lambda(\lambda, \mathbf{r}, \mathbf{\Omega}') \cos(\mathbf{\Omega}, \mathbf{\Omega}')\, d\mathbf{\Omega}' \tag{5.20}$$

This quantity is used to describe the exchanges of radiative energy in the atmosphere and hence to quantify its thermal budget. In a plane-parallel atmosphere with horizontal surface as reference, the spectral density of the irradiance is calculated as a function of altitude z by

$$F_\lambda(\lambda, z) = \int_0^{2\pi} d\varphi \int_{-1}^{1} \mu\, L_\lambda(\lambda, z, \mu, \varphi)\, d\mu \tag{5.21}$$

where φ is the azimuthal angle (Figure 5.1). One often defines the upward and downward fluxes, $F_\lambda\!\uparrow(\lambda, z)$ and $F_\lambda\!\downarrow(\lambda, z)$ as

$$F_\lambda^\uparrow(\lambda, z) = \int_0^{2\pi} d\varphi \int_{-1}^{0} \mu\, L_\lambda(\lambda, z, \mu, \varphi)\, d\mu \qquad (\mu > 0) \tag{5.22}$$

$$F_\lambda^\downarrow(\lambda, z) = -\int_0^{2\pi} d\varphi \int_0^{1} \mu\, L_\lambda(\lambda, z, \mu, \varphi)\, d\mu \qquad (\mu < 0) \tag{5.23}$$

with the net irradiance density being $F_\lambda(\lambda, z) = F_\lambda\!\uparrow(\lambda, z) - F_\lambda\!\downarrow(\lambda, z)$. The irradiance [W m^{-2}] is obtained by integrating F_λ over the entire electromagnetic spectrum

$$F(z) = \int_0^\infty F_\lambda(\lambda, z)\, d\lambda \tag{5.24}$$

The *diabatic heating* Q in units of $[\text{W m}^{-3}]$ resulting from the absorption of radiation (radiative energy absorbed and converted into thermal energy) is the divergence of the irradiance

$$Q(\mathbf{r}) = -\mathbf{\nabla} \bullet F \tag{5.25}$$

In a plane-parallel atmosphere, the diabatic heating rate can be expressed in units of $[\text{K s}^{-1}]$ as

$$\tilde{Q}(z) = -\frac{1}{\rho_a\, c_p}\, \frac{dF(z)}{dz} \tag{5.26}$$

In models using isobaric coordinates, the heating rate $Q(p)$ is expressed as a function of atmospheric pressure p by;

$$\tilde{Q}(p) = \frac{g}{c_p}\frac{dF(p)}{dp} \tag{5.27}$$

Here, c_p $[\text{J K}^{-1}\text{ kg}^{-1}]$ is the specific heat at constant pressure, ρ_a $[\text{kg m}^{-3}]$ the air density, and g the gravitational acceleration.

The radiance measures the photon flux from a particular direction, and the irradiance measures the photon flux through a horizontal surface. The irradiance is relevant to atmospheric heating, as expressed by (5.26). However, photolysis of molecules is determined by the flux of photons originating from all directions. This is measured by the *actinic flux*, whose spectral density or *actinic flux density* $[\text{W m}^{-2}\text{ nm}^{-1}]$ is the integral of the monochromatic radiance over all solid angles

$$\Phi_\lambda(\lambda, \mathbf{r}) = \int_{4\pi} L_\lambda(\lambda, \mathbf{r}, \mathbf{\Omega})\, d\Omega \tag{5.28}$$

The photolysis of atmospheric molecules occurs regardless of the direction of the incident photon and is therefore dependent on the actinic flux rather than the irradiance. For a plane-parallel atmosphere, the actinic flux density at altitude z is given by integration over all zenithal and azimuthal directions μ and φ in spherical coordinates:

$$\Phi_\lambda(\lambda, z) = \int_0^{2\pi} d\varphi \int_{-1}^1 L_\lambda(\lambda, z, \mu, \varphi)\, d\mu \tag{5.29}$$

The actinic flux density is commonly expressed as a photon flux density $q_\lambda(\lambda, z)$ $[\text{photons m}^{-2}\text{ s}^{-1}\text{ nm}^{-1}]$:

$$q_\lambda(\lambda, z) = \frac{\Phi_\lambda(\lambda, z)}{h\nu} = \frac{\Phi_\lambda(\lambda, z)\lambda}{hc} \tag{5.30}$$

At altitude z, the actinic flux $q(\Delta\lambda, z)$ for a wavelength interval $\Delta\lambda$ $[\text{photons m}^{-2}\text{ s}^{-1}]$ is obtained by spectral integration of the spectral density q_λ over this interval:

$$q(\Delta\lambda, z) = \int_{\Delta\lambda} q_\lambda(\lambda, z)\, d\lambda \tag{5.31}$$

Atmospheric chemistry models typically use spectrally integrated actinic fluxes with wavelength intervals $\Delta\lambda \sim 1$–10 nm to calculate photolysis frequencies.

5.2.2 Blackbody Radiation

A blackbody is an idealized physical body that absorbs all incident electromagnetic radiation. All blackbodies at a given temperature emit radiation with the same spectrum. The laws of blackbody emission are fundamental for understanding radiative transfer in the atmosphere.

Planck's law

Planck's law describes the spectral density of radiative emission for a blackbody at temperature T [K] under thermodynamic equilibrium. It assumes that photons are distributed with frequency ν according to Boltzmann statistics. Under these conditions, the spectral density B_ν [W m^{-2} Hz^{-1}] of the blackbody radiance is given by the *Planck function*

$$B_\nu(\nu, T) = \frac{2h\nu^3}{c^2} \frac{1}{e^{h\nu/kT} - 1} \tag{5.32}$$

where k is the Boltzmann constant (1.38×10^{-23} J K^{-1}). One can also express the Planck function for the spectral density B_λ [W m^{-2} nm^{-1}] as a function of wavelength:

$$B_\lambda(\lambda, T) = \frac{2hc^2}{\lambda^5} \frac{1}{e^{hc/\lambda kT} - 1} \tag{5.33}$$

Stefan–Boltzmann law

Integration of the spectral density B_λ over all wavelengths yields the total radiance

$$B(T) = \int_0^\infty B_\lambda(\lambda, T)\, d\lambda = \int_0^\infty \frac{2hc^2}{\lambda^5} \frac{1}{e^{hc/\lambda kT} - 1}\, d\lambda = bT^4 \tag{5.34}$$

where $b = 2\pi^4 k^4/(15\, c^2 h^3)$. The blackbody emission flux F [W m^{-2}] is obtained by performing a hemispheric integration of the radiance and, since the blackbody radiance is isotropic, we write

$$F(T) = B(T) \int_0^{2\pi} d\varphi \int_0^{+1} \mu d\mu = \pi B(T) = \sigma T^4$$

The flux varies therefore with the fourth power of the absolute temperature. This expression represents the *Stefan–Boltzmann law*, and the proportional factor $\sigma = 2\pi^5 k^4/(15\, c^2 h^3) = 5.67 \times 10^{-8}$ Wm^{-2} K^{-4} denotes the *Stefan–Boltzmann constant*.

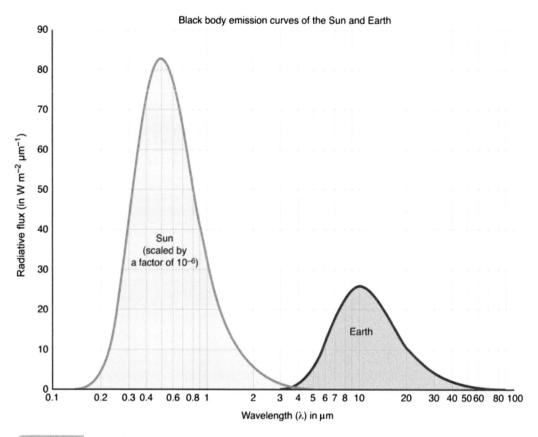

Figure 5.2 Blackbody spectra at 5800 K and 253 K, corresponding to the effective temperatures of the Sun and the Earth.

Wien's displacement law

By differentiating $B_\lambda(\lambda, T)$ with respect to λ, we find that the maximum wavelength of emission λ_{max} is inversely proportional to the blackbody temperature:

$$\lambda_{max} = \frac{hc}{5kT} = \frac{2897 \,[\text{K} \,\mu\text{m}]}{T} \qquad (5.35)$$

The mean temperature of the Sun is 5800 K, so solar radiation peaks in the visible at 0.5 μm. The mean surface temperature of the Earth is 288 K, so terrestrial emission is in the infrared peaking at approximately 10 μm. There is almost no overlap between the solar and terrestrial radiation spectra so that these two types of radiation can be treated separately (see Figure 5.2).

Kirchhoff's law

Under thermodynamic equilibrium, the *emissivity* of a body at a given wavelength (defined as the ratio of the monochromatic emitting intensity to the value given by the Planck function) is equal to its *absorptivity* (defined as the ratio of the mono-chromatic absorbed intensity to the value of the Planck function). In the case of a blackbody, the values of the emissivity and absorptivity are equal to 1. Kirchhoff's

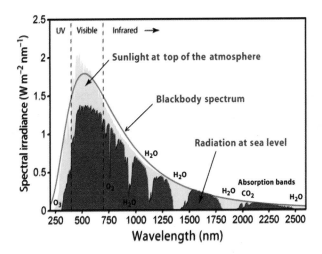

Figure 5.3 Spectral density of the solar irradiance spectrum at the top of the atmosphere and at sea level over the range 250–2500 nm. The 5800 K blackbody spectrum (thin black line) is shown for comparison. Absorption features by several radiatively active gases are indicated. Source: Robert A Rodhe, Wikimedia Commons.

law implies that we can compute the emission flux density of any object simply by knowing its surface temperature and its absorption spectrum.

5.2.3 Extra-Terrestrial Solar Spectrum

The spectrum of radiation emitted by the Sun can be approximated as that of a blackbody at a temperature $T = 5800$ K. The spectral density of the solar flux [W m^{-2} nm^{-1}] traversing a surface perpendicular to the direction of the solar beam at the top of the Earth's atmosphere can then be expressed by

$$\Phi_{\infty,\lambda}(\lambda, T) = \beta_R \pi B_\lambda(\lambda, T) \tag{5.36}$$

where the dilution factor

$$\beta_R = \left[\frac{R_{Sun}}{d}\right]^2 \tag{5.37}$$

accounts for the distance between the Sun and the Earth ($d = 1.471 \times 10^8$ km at the perihelion and 1.521×10^8 km at the aphelion). Here R_{Sun} is the solar radius (6.96×10^5 km). The total solar flux at the top of the Earth's atmosphere,

$$\Phi_\infty(T) = \beta_R \sigma T^4 \tag{5.38}$$

is equal to 1380 W m^{-2} and is called the *solar constant.*

The observed solar spectrum deviates from the theoretical blackbody curve because it includes contributions from different solar layers at different temperatures. The extra-terrestrial (top of the atmosphere) spectral density of the solar flux is shown in Figure 5.3. Also shown is the spectral irradiance at the surface, which is weaker than at the top of the atmosphere because of atmospheric scattering and absorption. Absorption by ozone, water vapor, and CO_2 is responsible for well-defined bands in the spectrum where

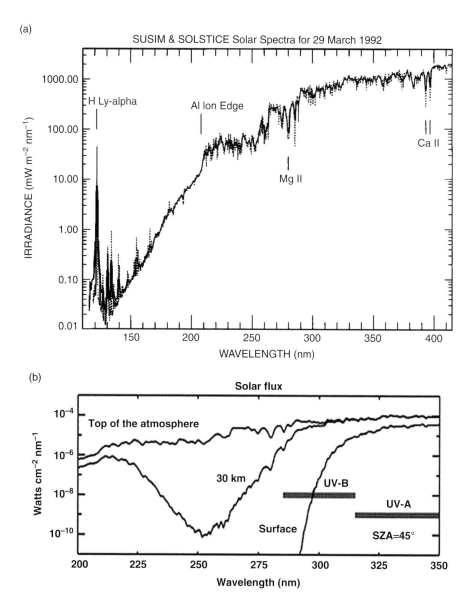

(a)

SUSIM & SOLSTICE Solar Spectra for 29 March 1992

(b)

Solar flux

Figure 5.4 Solar irradiance spectra at the top of the atmosphere at 119–420 nm (a) and at different altitudes at 200–350 nm (b). The top panel (from Woods *et al.*, 1996) shows the Lyman-alpha line at 121 nm from solar H atoms. The bottom line illustrates the strong absorption of radiation at wavelengths shorter than 300 nm by oxygen and ozone in the Earth's atmosphere.

surface radiation is strongly depleted. A more detailed solar spectrum in the UV region ($\lambda < 400$ nm), where sufficient energy is available to dissociate molecules and initiate photochemistry, is shown in Figure 5.4. Radiation in that region of the spectrum interacts strongly with the Earth's atmosphere through absorption by molecular oxygen and ozone, and through scattering by air molecules. For wavelengths shorter than 300 nm, the radiation reaching the Earth's surface is orders of magnitude lower than that at the top of the atmosphere.

5.2.4 Penetration of Solar Radiation in the Atmosphere

Transfer of radiation in the Earth's atmosphere is sensitive to absorption and scattering by atmospheric molecules, aerosol particles, clouds, and the surface. We examine these different processes here.

Absorption only

We first consider the simple case in which absorption plays the dominant role, with scattering assumed to be negligible. This assumption is often used in the middle and upper atmosphere, where scattering is weak because the atmosphere is thin and there are no clouds. The direct incoming solar beam then represents the dominant component of solar radiation. The actinic flux density $q_\lambda(\lambda; z, \theta_0)$ at altitude z and for a given solar zenith angle θ_0 is proportional to the radiance associated with that beam. From the Beer–Lambert law, we have

$$
\begin{aligned}
q_\lambda(\lambda, z, \theta_0) &= q_{\lambda,\infty}(\lambda) \exp\left[-\sigma_{abs}(\lambda) \int_s^\infty n(s')\, ds'\right] \\
&= q_{\lambda,\infty}(\lambda) \exp\left[-\mathcal{F}(z, \theta_0)\, \sigma_{abs}(\lambda) \int_z^\infty n(z')\, dz'\right] \\
&= q_{\lambda,\infty}(\lambda) \exp\left[-\mathcal{F}(z, \theta_0)\, \tau_{abs}(\lambda, z)\right]
\end{aligned}
\tag{5.39}
$$

where $q_{\lambda,\infty}(\lambda)$ is the spectral density of the solar actinic flux at the top of the atmosphere. The exponential attenuation describes the absorption of the solar beam by an absorber with number density n and wavelength-dependent absorption cross-section σ_{abs}. The *air mass factor* $\mathcal{F}(z, \theta_0)$, defined as the ratio of the slant column density to the vertical column density,

$$
\mathcal{F}(z, \theta_0) = \frac{\displaystyle\int_s^\infty n(s')\, ds'}{\displaystyle\int_z^\infty n(z')\, dz'}
\tag{5.40}
$$

accounts for the influence of the solar inclination. If we neglect the effect of the Earth's curvature and assume a plane-parallel atmosphere, the air mass factor is simply

$$
\mathcal{F}(z, \theta_0) = \sec \theta_0
\tag{5.41}
$$

This approximation is generally acceptable if the solar zenith angle θ_0 is less than 75°. Otherwise, a more complex approach must be adopted to account for the Earth's sphericity (see Smith and Smith, 1972; Brasseur and Solomon, 2005).

When several absorbers i are contributing to the attenuation of radiation, the total optical depth is the sum of the optical depth associated with the individual species:

$$
\tau_{abs}(\lambda, z) = \sum_i \tau_{abs,i}(\lambda, z)
\tag{5.42}
$$

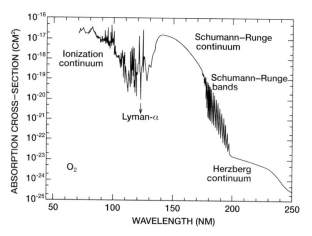

Figure 5.5 Absorption cross-section [cm^2 molecule^{-1}] of molecular oxygen between 50 and 250 nm featuring the Schumann–Runge continuum (130–170 nm), the Schumann–Runge bands (175–205 nm), and the Herzberg continuum (200–242 nm). Note the weak absorption cross-section at the wavelength corresponding to the intense solar Lyman-α line. At wavelengths shorter than 102.6 nm, absorption of radiation leads to photo-ionization of O$_2$.

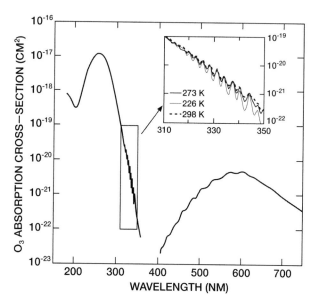

Figure 5.6 Absorption cross-section [cm molecule^{-1}] of ozone between 180 and 750 nm with the Hartley band (200–310 nm), the temperature-dependent Huggins bands (310–400 nm) and the weak Chappuis bands (beyond 400 nm).

In the middle and upper atmosphere, the optical depth is due primarily to the absorption by ozone and molecular oxygen, so that

$$\tau_{abs}(\lambda, z) = \tau_{abs,O3}(\lambda, z) + \tau_{abs,O2}(\lambda, z) \tag{5.43}$$

The spectral distributions of the absorption cross-sections for O$_2$ and O$_3$ are shown in Figures 5.5 and 5.6. Spectral regions of importance are listed in Table 5.1.

Table 5.1 Spectral regions for atmospheric absorption by O_2 and O_3	
Wavelength	Atmospheric absorbers
121.6 nm	Solar Lyman-α line, absorbed by O_2 in the mesosphere. No absorption by O_3.
130–175 nm	O_2 Schumann–Runge continuum. Absorption by O_2 in the thermosphere.
175–205 nm	O_2 Schumann–Runge bands. Absorption by O_2 in the mesosphere and upper stratosphere. Effect of O_3 can be neglected in the mesosphere, but is important in the stratosphere.
200–242 nm	O_2 Herzberg continuum. Absorption of O_2 in the stratosphere and weak absorption in the mesosphere. Absorption by the O_3 Hartley band is also important (see below).
200–310 nm	O_3 Hartley band. Absorption by O_3 in the stratosphere leading to formation of the $O(^1D)$ atom.
310–400 nm	O_3 Huggins bands. Absorption by O_3 in the stratosphere and troposphere leading to formation of the $O(^3P)$ atom.
400–850 nm	O_3 Chappuis bands. Weak absorption by O_3 in the troposphere with little attenuation all the way to the surface.

Absorption and scattering

Radiation is scattered by air molecules, aerosols, and clouds. Scattering refers to a change in the direction of incident radiation without loss of energy. The scattering properties of a medium are characterized by the efficiency with which the incoming radiation is scattered (*scattering efficiency*, defined next) and by the distribution of angles of the scattered radiation relative to the incident beam (*scattering phase function*, also defined below). The theory of Lorenz (1890) and Mie (1908) describes the interaction between a plane electromagnetic wave and a spherical particle based on the Maxwell equations. Scattering by a sphere is uniform over all azimuth angles so that the scattering phase function is characterized by a single angle Θ $[0°, 180°]$ relative to the incident beam. The scattering properties are determined by the particle size parameter

$$\alpha = \frac{\pi D_p}{\lambda}$$

introduced in Section 3.9.4 where D_p is the particle diameter, and on the *refraction index* m_r defined as the ratio of the speed of light in vacuum to that in the scattering medium. The size parameter determines the scattering regime (Figure 5.7). The refraction index is commonly expressed as a complex number to account for both scattering and absorption:

$$m = m_r - i\, m_i \tag{5.44}$$

where the imaginary component m_i is a measure of the absorption efficiency. Both m_r and m_i are wavelength-dependent.

Scattering by air molecules

Scattering of light by air molecules is described by the *Rayleigh theory*, which can be viewed as the asymptotic case of the Lorenz–Mie theory for a size parameter $\alpha \ll 1$. Under these assumptions, the scattering cross-section [cm^2 molecule^{-1}] is found to be

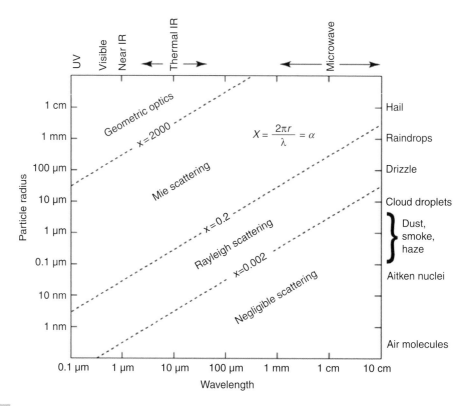

Figure 5.7 Different scattering regimes as a function of radiation wavelength and particle radius. Reproduced from Petty (2006).

$$\sigma_{scat}(\lambda) = \frac{8\pi^3}{3\lambda^4 n_a^2}(m_r - 1)^2 f(\delta) \qquad (5.45)$$

where n_a is the air number density and m_r is the real part of the refractive index of air. In this limit there is no dependence on particle size. The correction factor

$$f(\delta) = \frac{6 + 3\delta}{6 - 7\delta}$$

accounts for the anisotropy of non-spherical molecules with the anisotropy factor for air molecules being $\delta = 0.035$. The factor $m_r - 1$ is approximately proportional to the air density n_a so that the dependence of the scattering cross-section on air density largely cancels. The dominant feature of the Rayleigh scattering cross-section as described by (5.45) is the λ^{-4} dependence.

Scattering by aerosols and cloud droplets

Atmospheric scattering due to very large particles such as cloud droplets ($\alpha \gg 1$) is described by the laws of *geometric optics*, which can be regarded as another asymptotic approximation of the electromagnetic theory. In this formulation, the direction of propagation of light rays is modified by local reflection and refraction processes. The case of smaller aerosol particles is more complex as particle dimensions are typically

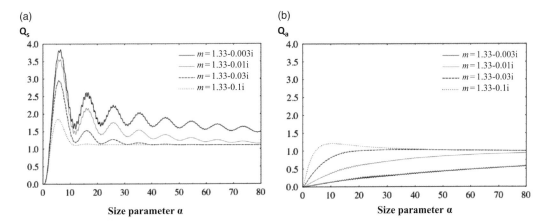

Figure 5.8 Scattering (a) and absorption (b) efficiencies as a function of the aerosol size parameter α for various amounts of absorption (imaginary part of the refraction index *m*). From Frank Evans, University of Colorado, personal communication.

of the same order of magnitude as the radiation wavelength. The *Lorenz–Mie theory* of light applied to spherical aerosol particles with diameter D_p provides general analytical expressions for the *scattering* and *absorption efficiencies*

$$Q_{scat} = \frac{4\sigma_{scat}}{\pi D_p^2} \tag{5.46}$$

and

$$Q_{abs} = \frac{4\sigma_{abs}}{\pi D_p^2} \tag{5.47}$$

defined as the ratio of the scattering (σ_{scat}) and absorption (σ_{abs}) cross-sections to the geometric cross-section $\pi D_p^2/4$, as a function of size parameter α. For example, the scattering efficiency is expressed by the following expansion

$$Q_{scat} = c_1 \alpha \left[1 + c_2 \alpha + c_3 \alpha^2 + \cdots \right] \tag{5.48}$$

where the coefficients c_i are provided by the theory as a function of the refraction index. When considering the scattering of visible light by molecules ($\alpha \sim 10^{-3}$), the dominant contribution is provided by the first-order term

$$c_1 = \frac{8}{3} \left(\frac{m_r^2 - 1}{m_r^2 + 2} \right)^2 \tag{5.49}$$

and describes Rayleigh scattering presented earlier.

For aerosol and cloud particles, the dependence of scattering efficiency on particle size becomes important, while the dependence on wavelength is less pronounced than for air molecules. The extinction efficiency Q_{ext} is defined as

$$Q_{ext} = Q_{scat} + Q_{abs} \tag{5.50}$$

Values of Q_{scat} and Q_{abs} as a function of size parameter α are presented in Figure 5.8 for different values of the refraction index. Scattering is most efficient when the particle radius is equal to the wavelength of incident radiation ($\alpha = 2\pi$). It is inefficient for very small particles ($\alpha \ll 1$). For very large particles ($\alpha \gg 1$), Q_{scat} approaches a diffraction limit. The scattering coefficient for a given aerosol size distribution is obtained by integration of the particle cross-sections over the size distribution weighted by the scattering efficiency:

$$\beta_{scat} = \frac{\pi}{4} \int_0^\infty D_p^2 \, Q_{scat}(D_p) \; n_N(D_p) \; dD_p \qquad (5.51)$$

A similar equation applies to compute β_{abs}.

Source function and phase function

When the effects of both absorption and scattering on the solar radiance are taken into account and local emissions are ignored, the source term J that appears in the radiative transfer equation ((5.18) for a plane-parallel atmosphere) must account for light scattered from the direction of the Sun (μ_0, φ_0) and from all other directions (μ', φ'). The source term J can be expressed as

$$\begin{aligned} J(\lambda, \tau, \mu, \varphi) &= \frac{\omega(\lambda)}{4\pi} P(\lambda, \mu, \varphi, \mu_0, \varphi_0) \, \Phi_\lambda(\lambda, \infty) \exp\left[-\frac{\tau(\lambda, z)}{\mu_0} \right] \\ &+ \frac{\omega(\lambda)}{4\pi} \int_0^{2\pi} d\varphi' \int_{-1}^1 P(\lambda, \mu, \varphi, \mu', \varphi') \, L_\lambda(\lambda, \tau, \mu', \varphi') \, d\mu' \end{aligned} \qquad (5.52)$$

where $\omega(\lambda) = \beta_{scat}/\beta_{ext}$ is again the *single scattering albedo*. The first term on the right-hand side of (5.52) represents the single scattering of the direct solar radiation (whose irradiance at the top of the atmosphere is $\Phi_\lambda(\lambda, \infty)$), and the second term accounts for multiple scattering. The *phase function* $P(\lambda, \mu, \varphi, \mu', \varphi')$ defines the probability density that a photon originating from direction (μ', φ') is scattered in direction (μ, φ). For spherical particles, the phase function depends only on the scattering angle Θ between the direction of the incident and the scattered radiation. It is often expressed as a function of parameter $\mu_s = \cos \Theta$, which is related to the azimuthal and zenithal directions by

$$\mu_s = \cos \Theta = \mu\mu' + \left(1 - \mu^2\right)^{1/2}\left(1 - \mu'^2\right)^{1/2} \cos(\varphi - \varphi') \qquad (5.53)$$

The phase function P is normalized so that

$$\frac{1}{4\pi} \int_0^{2\pi} d\varphi \int_0^\pi P(\Theta) \sin \Theta \, d\Theta = \frac{1}{4\pi} \int_0^{2\pi} d\varphi \int_{-1}^1 P(\mu_s) \, d\mu_s = 1 \qquad (5.54)$$

For isotropic scattering, P is constant and equal to 1. The non-isotropy of the scattering process can be expressed by the *asymmetry factor*

$$g = \int_{-1}^1 P(\mu_s) \, \mu_s \, d\mu_s \qquad (5.55)$$

Its value is equal to 1 if all the light is scattered forward, −1 if it is entirely scattered backward, and 0 if scattering is isotropic. The angular distribution of the scattered

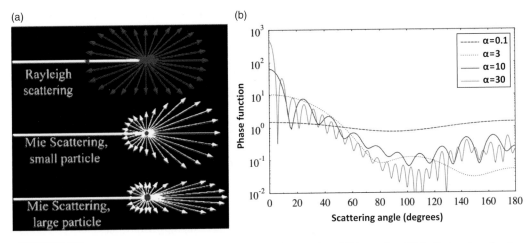

(a)   (b)

(a) Schematic representation of light scattered by particles of different sizes. (b) Scattering phase function derived from the Lorenz–Mie theory as a function of the scattering angle Θ for different values of the particle size parameter α. In the case of Rayleigh scattering ($\alpha \ll 1$), the same amount of light is scattered in the forward and backward directions. Mie scattering tends to favor scattering in the forward direction, especially in the case of large particles. From deepocean.net (a) and Frank Evans, University of Colorado, personal communication (b).

energy, and hence the asymmetry factor, can be derived from theory. In the most general case, it varies with the value of the size parameter α and with the degree of polarization of light. Light emitted by the Sun is unpolarized, but becomes partially polarized after scattering with molecules and particles in the atmosphere.

For scattering of unpolarized light by gas molecules, Rayleigh's theory applies ($\alpha \ll 1$), and the phase function is found to be

$$P(\Theta) = \frac{3}{4}\left(1 + \cos^2\Theta\right) \tag{5.56}$$

or equivalently

$$P(\mu_s) = \frac{3}{4}\left(1 + \mu_s^2\right) \tag{5.57}$$

In the presence of spherical aerosol particles ($\alpha \sim 1$), the Lorenz–Mie theory applies and the phase function takes on a complicated form. Derivation is presented in radiative transfer textbooks such as Liou (2002). The phase function is then often described for modeling purposes in terms of associated Legendre polynomials, or more simply by the asymmetry factor. For an ensemble of particles characterized by a size distribution $n_N(D_p)$, the mean phase function $P(\mu_s)$ is derived by averaging the size-dependent phase function $P(D_p, \Theta)$ or $P(D_p, \mu_s)$ and weighting it by the scattering efficiency Q_{scat}

$$P(\mu_s) = \frac{\displaystyle\int_0^\infty D_p^2\, Q_{scat}(D_p)\, P(D_p, \mu_s)\, n_N(D_p)\, dD_p}{\displaystyle\int_0^\infty D_p^2\, Q_{scat}(D_p)\, n_N(D_p)\, dD_p} \tag{5.58}$$

Sample phase functions $P(D_p, \Theta)$ calculated by the Lorenz–Mie theory are shown in Figure 5.9. Scattering by large particles is characterized by a strong forward component with a value for the asymmetry factor of about 0.8.

Solution of the radiative transfer equation

The solution of the radiative transfer equation (5.18) for the upward radiance at a vertical level corresponding to an optical depth of τ is

$$L_\lambda(\lambda, \tau, \mu, \varphi) = L_\lambda(\lambda, \tau_s, \mu, \varphi) \exp\left[-\frac{\tau_s - \tau}{\mu}\right] + \int_0^\tau J(\lambda, \tau', \mu, \varphi) \exp\left[-\frac{\tau' - \tau}{\mu}\right] \frac{d\tau'}{\mu} \quad (\mu > 0)$$

(5.59)

where τ_s is the optical depth at the surface. For the downward radiance

$$L_\lambda(\lambda, \tau, \mu, \varphi) = L_\lambda(\lambda, 0, \mu, \varphi) \exp\left[-\frac{\tau}{-\mu}\right] + \int_0^\tau J(\lambda, \tau', \mu, \varphi) \exp\left[-\frac{\tau - \tau'}{-\mu}\right] \frac{d\tau'}{-\mu} \quad (\mu < 0)$$

(5.60)

Finally, for $\mu = 0$, the horizontal radiance is

$$L_\lambda(\lambda, \tau, \mu = 0, \varphi) = J(\lambda, \tau, \mu = 0, \varphi) \tag{5.61}$$

Different numerical methods are available to obtain approximate solutions to the integro-differential radiative transfer equation in an absorbing and scattering atmosphere. These include, for example, iterative Gauss, successive order, discrete ordinate, two-stream, and Monte-Carlo methods. See Lenoble (1977) and Liou (2002) for more details.

In the *successive order method*, the solution is obtained by solving iteratively $(n = 0, 1, \ldots)$

$$\mu \frac{dL_\lambda^{(n+1)}(\tau, \mu, \varphi)}{d\tau} = -L_\lambda^{(n+1)}(\tau, \mu, \varphi) + \frac{\omega(\lambda)}{4\pi} \int_0^{2\pi} d\varphi' \int_{-1}^1 P(\lambda, \mu, \varphi, \mu', \varphi') L_\lambda^{(n)}(\lambda, \tau, \mu', \varphi') d\mu'$$

(5.62)

with the zeroth order radiance given by the Beer–Lambert law applied to the incoming direct solar flux

$$L_\lambda^{(0)}(\lambda, \tau, \mu, \varphi) = \Phi_\lambda(\lambda, \infty) \exp\left[-\frac{\tau(\lambda, z)}{\mu_0}\right] \delta(\mu - \mu_0)\, \delta(\varphi - \varphi_0) \tag{5.63}$$

Here, $\delta(x - x')$ represents the Dirac function, which is equal to one for $x = x'$ and zero otherwise. The final radiance is the sum of the different components

$$L_\lambda = \sum_0^\infty L_\lambda^{(n)} \tag{5.64}$$

In the *discrete ordinates method* (Chandrasekhar, 1950; Stamnes *et al.*, 1988), the radiance is expanded as a Fourier series about the cosine of the azimuthal angle,

$$L_\lambda(\lambda, \tau, \mu, \varphi) = \sum_{m=0}^N L_\lambda^m(\lambda, \tau, \mu) \cos(m\varphi) \tag{5.65}$$

while the phase function is expanded into associated Legendre polynomials:

$$P(\lambda, \mu, \varphi, \mu', \varphi') = \sum_{m=0}^{N} [2 - \delta(0 - m)] \left(\sum_{l=m}^{N} (2l + 1) P_l^m(\mu) P_l^m(\mu') \right) \cos [m(\varphi - \varphi')]$$

(5.66)

The radiative transfer equation (5.18) in three variables (τ, μ, φ) is replaced by $N + 1$ uncoupled integro-differential equations for $L^m(\tau, \mu)$ ($m = 0, 1, 2, \ldots N$) in two variables (τ; μ). The integration over μ is replaced by an accurate Gaussian quadrature formula at $2N + 1$ values of μ_i ($i = -N, \ldots, -1, 0, 1, \ldots, N$), chosen to be the $2N$ roots of the Legendre polynomial $P_{2N}(\mu)$.

In the *two-stream method*, the phase function (5.66) is expanded in terms of Legendre polynomials, with only the two first terms of the expansion being retained ($N = 1$). The radiative transfer equation for hemispherically averaged radiances in a plane-parallel atmosphere is approximated by replacing the integrals over the zenith angles that characterize the source function J (see (5.52)) by a Gaussian summation with only two quadrature points, corresponding to ascending and descending directions, respectively. The diffuse radiance is therefore divided into an upward-propagating and a downward-propagating component, and two coupled equations, one for each stream, must be solved. In the *Eddington method*, both the phase function and the radiance are expanded in terms of Legendre polynomials and only the two first terms of this expansion are retained, with $P(\mu, \mu') = 1 + 3g\mu\mu'$

$$L(\lambda, \tau, \mu) = L_0(\lambda, \tau) + L_1(\lambda, \tau)\mu \qquad (-1 \leq \mu \leq 1) \qquad (5.67)$$

Here g is the asymmetry factor (see 5.55). Coefficients L_0 and L_1 are derived by solving two coupled differential equations established by inserting the above expression into the azimuthally averaged transfer equation for diffuse radiation. The two-stream and Eddington approximations provide computationally efficient methods and are commonly used. They are inaccurate in the presence of clouds, where photons are repeatedly scattered by cloud droplets in a predominantly forward direction. The *delta-Eddington* approximation addresses this issue by adjusting the phase function to account for the strong forward contribution in the multiple scattering process (Joseph *et al.*, 1976). In this case, the phase function for the fraction of the scattered light that resides in the forward peak is expressed by a delta function.

Albedo

Radiation reflected by the surface must be included in radiance calculations through a lower boundary condition of the type

$$L_\lambda(\mu > 0, \varphi, z = 0) = \mathcal{A} \, L_\lambda(\mu < 0, \varphi, z = 0) \qquad (5.68)$$

where the surface albedo $\mathcal{A}$ varies with surface type (Figure 5.10), wavelength, and the incident and reflected zenith and azimuthal angles. It is frequently assumed that the albedo is isotropic (the surface is then called *Lambertian*) but this is often not precise enough for retrieval of atmospheric parameters such as aerosol optical depth

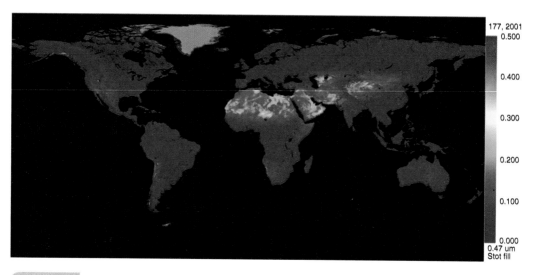

Figure 5.10 Land surface albedo at 470 nm. Data based on observations by the Moderate Resolution Imaging Spectroradiometer (MODIS) in June 2001. Areas where no data are available are shown in gray. Image from MODIS Atmosphere support group incl. E. Moody, NASA Goddard Space Flight Center and C. Schaaf, Boston University and National Aeronautics and Space Administration (NASA).

from satellite measurements of solar backscatter. In those cases one needs to describe the full angular dependence of the surface albedo, known as the *bidirectional reflectance distribution function* (*BRDF*).

5.2.5 Emission and Absorption of Terrestrial Radiation

In the infrared, at wavelengths larger than approximately 3.5 μm, the radiation field is determined primarily by radiative emission from the Earth's surface and the atmosphere. The contribution of solar radiation is small, and scattering by air molecules can be neglected. Because of the limited overlap between solar (shortwave) and terrestrial (longwave) radiation, a clear distinction can be made in the approach that is adopted to solve the radiative transfer equation. In the case of longwave radiation, the spatial distribution of the radiance is derived by integrating the radiative transfer equation (5.18) in which the source function is represented by the Planck function $B_\lambda(\lambda, T)$. This formulation requires that collisions be sufficiently frequent so that the energy levels of the molecules are populated according to the Boltzmann distribution, a condition called *local thermodynamic equilibrium* (*LTE*). Local thermodynamic equilibrium is met in the lower atmosphere but breaks down above 60–90 km altitude, depending on the spectral band. In the limit of no scattering and for LTE conditions, radiative transfer is described by *Schwarzschild's equation*

$$\mu \frac{dL_\lambda(\lambda, \tau, \mu, \varphi)}{d\tau} = L_\lambda(\lambda, \tau, \mu, \varphi) - B(\lambda, T(\tau)) \tag{5.69}$$

Assuming azimuthal symmetry (no dependence of the radiance on angle φ), and providing the boundary conditions $L_\lambda(\lambda, z_{surface}, \mu > 0) = B_\lambda(\lambda, T_s)$ at the surface and

$L_\lambda(\lambda, \infty, \mu < 0) = 0$ at the top of the atmosphere, the upward and downward components of the radiance at altitude z and for the zenith direction μ are obtained by integration of (5.69)

$$L_\lambda(\lambda, z, \mu) = B_\lambda(\lambda, T_s)\, \mathsf{T}(\lambda, z, 0, \mu) + \int_0^z B_\lambda(\lambda, z') \frac{\partial \mathsf{T}(\lambda, z', z, \mu)}{\partial z'}\, dz' \quad (\mu > 0)$$

(5.70)

and

$$L_\lambda(\lambda, z, \mu) = -\int_z^\infty B_\lambda(\lambda, z') \frac{\partial \mathsf{T}(\lambda, z', z, \mu)}{\partial z'}\, dz' \quad (\mu < 0) \qquad (5.71)$$

Here T_s represents the temperature at the Earth's surface and

$$\mathsf{T}(\lambda, z', z, \mu) = \exp\left[-\int_{z'}^z \beta_{abs}(\lambda, \zeta) \frac{d\zeta}{\mu} \right]$$

(5.72)

denotes the transmission function between altitudes z' and z for an inclination μ and wavelength λ. As stated in Section 5.2, $\beta_{abs}(\lambda, z)$ represents the absorption coefficient $[\mathrm{m}^{-1}]$ proportional to the concentration of the absorbers and to their wavelength-dependent absorption cross-sections. The transmission function T was previously introduced in equation (5.7).

Absorption of radiation by molecules in the IR involves transitions between vibrational and rotational energy levels of the molecule, in contrast to absorption in the UV which involves transitions between electronic levels. Vibrational transitions generally require $\lambda < 20$ µm, while rotational transitions require $\lambda > 20$ µm. Combined vibrational–rotational transitions create fine structure in the absorption spectrum. The radiation emitted by the Earth is mainly in the wavelength range $\lambda < 20$ µm, so that absorption by molecules generally involves vibrational transitions. Molecules absorbing in that range reduce the flux of terrestrial radiation escaping to space and are called *greenhouse gases*. A selection rule of quantum mechanics is that vibrational transitions are allowed only if they change the dipole moment of the molecule. All molecules with an asymmetric distribution of charge (H_2O, O_3, N_2O, CO, chlorofluorocarbons) or the ability to acquire a distribution of charge by stretching or flexing (CO_2, CH_4) are greenhouse gases. Homonuclear diatomic molecules and single atoms (N_2, O_2, Ar) are not greenhouse gases. A peculiarity of the Earth's atmosphere is that the dominant constituents are not greenhouse gases. Figure 5.11 shows the atmospheric absorption of terrestrial radiation in the IR from 1 to 16 µm. The strongest bands are the 15 µm and 4.3 µm CO_2 bands, the 9.6 µm ozone band, the 6.3 µm water band, the 7.66 µm methane band, the 7.78 µm and 17 µm N_2O bands, and the 4.67 µm CO band.

Calculation of the IR radiance by (5.70) and (5.71) requires quantitative knowledge of the absorption spectra for the different radiatively active trace gases. Detailed radiative transfer models calculate the radiance under different conditions and derive the corresponding atmospheric heating rates. *Line-by-line models* with very high spectral resolution can account for individual absorption lines. Absorption

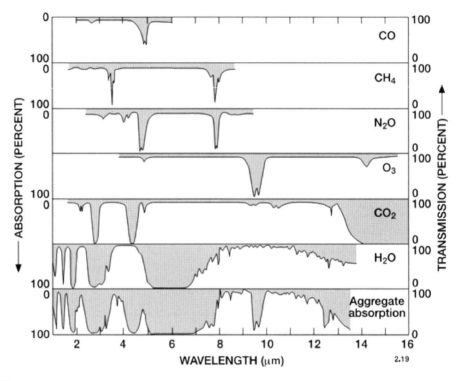

Figure 5.11 Vertical atmospheric transmission (absorption) of infrared radiation from the surface to the top of the atmosphere represented as a function of wavelength (1–16 μm) for different radiatively active gases. The different absorption bands and an aggregate spectrum are represented. Adapted from Shaw (1953).

lines have an extremely narrow natural width determined by the uncertainty principle of quantum mechanics, but are broadened in the atmosphere by thermal motions of the molecules (pressure-independent Doppler broadening) and also in the lower atmosphere by collisions between molecules (pressure-dependent collisional broadening). Line-by-line models are too computationally costly for application in atmospheric models, and parameterizations must be developed. Narrowband and broad-band models have been developed to simplify the calculation of the mean transmittance over specified spectral intervals. Some of these models assume that the same line repeats itself periodically as a function of wavenumber (e.g., regular model of Elsasser, 1938), while others assume a particular statistical distribution of the line positions and intensities within each spectral interval (e.g., random model of Goody, 1952).

The *correlated k-distribution method* (Goody *et al.*, 1989; Fu and Liou, 1992) is a computationally efficient algorithm used to calculate the average transmission $T_{\Delta v}$ over a given frequency (or wavenumber) interval Δv without having to perform a tedious spectral integration that accounts for the complexity of the rapidly varying absorption coefficient $k(v)$ inside this interval. This is accomplished by replacing the spectral integration required for the calculation of the mean transmission (expressed here for the case of a homogeneous atmosphere, see (5.13)

$$T_{\Delta v}(u) = \frac{1}{\Delta v} \int_{\Delta v} \exp\left[-k(v)\,u\right] dv \tag{5.73}$$

by an integration in the k space

$$T_{\Delta v}(u) = \int_0^\infty \exp\left[-k(v)\,u\right] f(k)\,dk \tag{5.74}$$

Here the function $f(k)dk$ represents the fraction of the interval Δv in which the mass absorption cross-section $k(v)$ has a value between k and $k + dk$. According to this expression, $f(k)$ is the inverse Laplace transform of the transmission function. If

$$g(k) = \int_0^k f(k')\,dk' \tag{5.75}$$

represents the cumulative probability distribution, which is a monotonically increasing smooth function, (5.74) can be expressed as

$$T_{\Delta v}(u) = \int_0^1 \exp\left[-k(g)\,u\right] dg \tag{5.76}$$

Since $k(g)$ is also a smoothly varying function, the mean transmission over a spectral interval Δv can be estimated by using a simple and efficient numerical quadrature (with a small number of k-intervals) to calculate the integral. Liou (2002) and Goody (1995) discuss the method in the more realistic case of a non-homogeneous atmosphere in which parameter k varies with temperature and pressure.

5.3 Gas Phase Chemistry

5.3.1 Photolysis

When a photon absorbed by a molecule exceeds in energy one of the chemical bonds of that molecule, it can cause cleavage of the bond and break the molecule into two fragments. This process is called *photolysis*. For example, molecular oxygen has a chemical bond between its O atoms corresponding to the energy of a 242 nm photon. It follows that only photons of greater energy (shorter wavelength) can drive O_2 photolysis:

$$O_2 + hv \;(\lambda \leq 242 \text{ nm}) \rightarrow O + O \tag{5.77}$$

The rate at which O_2 is photolyzed is given by

$$\frac{d[O_2]}{dt} = -J_{O2}[O_2] \tag{5.78}$$

where J_{O2} is the *photolysis frequency* [generally expressed in s^{-1}]. In the general case of photolysis of a molecule A, we have

$$\frac{d[A]}{dt} = -J_A[A] \tag{5.79}$$

The photolysis frequency of A is derived by spectral integration of the product of three quantities: the wavelength-dependent absorption cross-section $\sigma_A(\lambda)$, the *quantum yield* $\varepsilon_A(\lambda)$ and the local solar actinic flux q_λ

$$J_A(z) = \int_0^{\lambda_{max}} \varepsilon_A(\lambda)\, \sigma_A(\lambda)\, q_\lambda(z)\, d\lambda \tag{5.80}$$

where λ_{max} is the wavelength corresponding to the energy threshold for dissociation of the molecule. The absorption cross-section [cm^2 molecule^{-1}], called σ_{abs} in Section 5.2, represents the ability of a molecule to absorb a photon at a particular wavelength, and the quantum yield represents the probability that this absorption will lead to photolysis.

Numerical integration of (5.80) over spectral intervals $\Delta\lambda$ is straightforward when the absorption spectrum is a continuum. In certain cases, however, the absorption spectrum exhibits complex structures of discrete bands with many narrow spectral lines. Examples are the Schumann–Runge bands (175–205 nm) of molecular oxygen and several bands of nitric oxide (e.g., the δ-bands) in the same spectral area. In this case, rather than reducing by several orders of magnitude the size of the spectral interval used in the numerical integration, parameterizations are adopted to use effective values of the absorption cross-section averaged over large spectral intervals (see Box 5.1).

Box 5.1 — Photolysis in the Schumann–Runge Bands

The Schumann–Runge bands (175–205 nm) feature high-frequency variations in the absorption cross-sections of molecular oxygen (see Figure 5.5). These high-frequency variations complicate the calculation of photolysis rates by numerical integration over the wavelength spectrum. Accounting for the Schumann–Runge bands is important for computing photolysis frequencies in the upper stratosphere and mesosphere. Scattering is negligible at those altitudes so that the actinic flux is defined by attenuation of the direct beam by O_2 and O_3. The photolysis frequency of a molecule A at altitude z can be calculated as:

$$J_A(z, \theta_0) = \sum_k \sigma_A(\Delta\lambda_k)\, q_{k,\infty}(\Delta\lambda_k)\, \widetilde{T_{O2}}(\Delta\lambda_k, z, \theta_0)\, \widetilde{T_{O3}}(\Delta\lambda_k, z, \theta_0)$$

where $\sigma_A(\Delta\lambda_k)$ is the mean absorption cross-section over the wavelength interval $\Delta\lambda_k$, $q_{k,\infty}(\Delta\lambda_k) = \int_{\Delta\lambda_k} q_{\lambda,\infty}(\lambda)\, d\lambda$ is the mean top-of-atmosphere actinic flux over that interval, $\widetilde{T_{O2}}$ and $\widetilde{T_{O3}}$ are the *effective* O_2 and O_3 transmission functions from the top of the atmosphere averaged over $\Delta\lambda_k$, and θ_0 is the solar zenith angle. We have assumed a quantum yield of unity for simplicity of notation. The effective O_2 transmission function $\widetilde{T_{O2}}$ is defined by

$$\widetilde{T_{O2}}(\Delta\lambda_k, z, \theta_0) = \frac{\int_{\Delta\lambda_k} q_{\lambda,\infty}(\lambda)\, T_{O2}(\lambda, z, \theta_0)\, d\lambda}{\int_{\Delta\lambda_k} q_{\lambda,\infty}(\lambda)\, d\lambda}$$

where T_{O2} is the actual transmission function accounting for the fine absorption structure.

For $A \equiv O_2$, the photolysis frequency is given by

$$J_{O2}(z, \theta_0) = \sum_k \sigma_{02}^{SRB}(\Delta\lambda_k, z, \theta_0) \, q_{k,\infty}(\Delta\lambda_k) \, \widetilde{T}_{02}(\Delta\lambda_k, z, \theta_0) \, \widetilde{T}_{03}(\Delta\lambda_k, z, \theta_0)$$

where the *effective* O_2 cross-section for wavelength interval $\Delta\lambda_k$ is defined as

$$\sigma_{02}^{SRB}(\Delta\lambda_k, z, \theta_0) = \frac{\int_{\Delta\lambda_k} \sigma_{02}(\lambda) \, q_{\lambda,\infty}(\lambda) \, \widetilde{T}_{02}(\Delta\lambda_k, z, \theta_0) \, d\lambda}{\int_{\Delta\lambda_k} q_{\lambda,\infty}(\lambda) \, \widetilde{T}_{02}(\Delta\lambda_k, z, \theta_0) \, d\lambda}$$

Rather than performing a computationally expensive line-by-line calculation of the effective parameters σ_{02}^{SRB} and $\widetilde{T}_{02}$, these parameters can be fitted as a function of the O_2 slant column density

$$N = \sec\theta_0 \int_z^\infty n(O_2) \, dz$$

(see e.g., Kockarts, 1994).

Gijs *et al.* (1997) adopt the following expression

$$\ln\left[\sigma_{02}^{SRB}(N)\right] = A(N)[T(N) - 220\,\text{K}] + B(N)$$

where $T(N)$ is the temperature [K] at the altitude where the column is equal to N. Coefficients A and B are expressed as a function of N, using Chebyshev polynomial fits. The effective transmission is then derived by noting that

$$\frac{d\left(\widetilde{T}_{02}(N)\right)}{dN} = -\sigma_{02}^{SRB}(N)$$

Other methods to parameterize these effective coefficients have been developed by Fang *et al.* (1974), Minschwaner *et al.* (1993), Zhu *et al.* (1999), and others. Minschwaner and Siskind (1993) have derived fast methods to calculate the photolysis frequency of nitric oxide using a similar approach. Chabrillat and Kockarts (1997) propose a parameterization for the calculation of photolysis frequencies in the spectral range close to the Lyman (121 nm), where both the solar flux and the O_2 absorption cross-section vary rapidly over a narrow spectral interval.

5.3.2 Elementary Chemical Kinetics

Gas-phase chemical reactions can be classified for kinetic purposes as *unimolecular*, *bimolecular*, and *termolecular* (or *three-body*). A unimolecular reaction involves the dissociation of a molecule by photons (*photolysis*) or heat (*thermolysis*). It has the general form

$$A + h\nu \rightarrow C + D \quad \text{(photolysis)} \tag{5.81}$$

$$A + M \rightarrow C + D + M \quad \text{(thermolysis)} \tag{5.82}$$

where A is the reactant and C and D are its dissociation products. M is an inert molecule, such as N_2 or O_2, that transfers energy to the reactant by collision. The rate of reaction is given by

$$-\frac{d[A]}{dt} = \frac{d[C]}{dt} = \frac{d[D]}{dt} = k[A] \qquad (5.83)$$

where k is a *rate constant* (equivalently called *rate coefficient*) for the reaction, usually given in $[s^{-1}]$. For photolysis reactions, k is the photolysis frequency and the J notation of (5.79) is commonly used.

A bimolecular reaction has the general form

$$A + B \rightarrow C + D \qquad (5.84)$$

and the rate of reaction is given by

$$-\frac{d[A]}{dt} = -\frac{d[B]}{dt} = \frac{d[C]}{dt} = \frac{d[D]}{dt} = k[A][B] \qquad (5.85)$$

where the rate constant k is usually given in units of $[cm^3 \ molecule^{-1} \ s^{-1}]$. The reaction rate $k[A][B]$ is proportional to the number of collisions Z_{AB} per unit time between A and B. This collision frequency can be derived from the gas kinetics theory; it is proportional to the collision cross-section $\pi(r_A + r_B)^2$ and the thermal velocity

$$v_{th} = \left[\frac{8kT}{\pi} \left(\frac{1}{m_A} + \frac{1}{m_B} \right) \right]^{\frac{1}{2}} \qquad (5.86)$$

if r and m are the molecular radii and masses of A and B. Thus

$$Z_{AB} = \pi(r_A + r_B)^2 v_{th}[A] \ [B] \qquad (5.87)$$

Reaction of A and B involves formation of an *activated complex* AB* that breaks down either to the original reactants A and B or to the products C and D. The minimum energy needed to form the activated complex is called the *activation energy* E_a $[J \ mol^{-1}]$. Gas kinetics theory shows that the fraction of collisions with an energy larger than E_a is proportional to $\exp[-E_a/\mathcal{R}T]$, so that the reaction rate coefficient can be written

$$k(T) = \mathscr{P} \ \pi \ (r_A + r_B)^2 \left[\frac{8kT}{\pi} \left(\frac{1}{m_A} + \frac{1}{m_B} \right) \right]^{\frac{1}{2}} \exp\left[\frac{-E_a}{\mathcal{R}T} \right] \qquad (5.88)$$

where the *steric factor* $\mathscr{P}$ accounts for processes that are not included in the simple collision theory. This last equation provides a justification for the empirical Arrhenius equation (Figure 5.12)

$$k(T) = A \exp\left[\frac{-E_a}{\mathcal{R}T} \right] \qquad (5.89)$$

Here, $\mathcal{R}$ is again the gas constant equal to 8.314 J mol^{-1} K^{-1} and A is the so-called pre-exponential factor. We see from (5.88) that A varies with the square root of the temperature T, but this weak dependence is often ignored. If the enthalpy ΔH of the

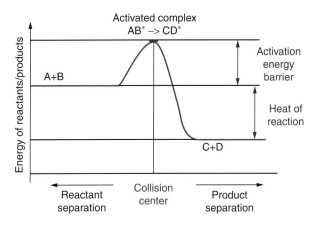

Figure 5.12 Energy transfer along a reaction path. The *activation energy* is the minimum amount of energy needed for colliding species to react. The *heat of reaction* is the potential energy difference between the reactants and products. The reaction is said to be *exothermic* if heat is released by the reaction. Otherwise, it is said to be *endothermic*, and heat must be absorbed from the environment.

reaction (difference between the enthalpy of formation of the products C and D and of the reactants A and B) is negative, the reaction is said to be *exothermic* and may proceed at a rapid rate. Otherwise the reaction is said to be *endothermic* and much less likely to occur at a rapid rate.

A three-body reaction describes the combination of two reactants A and B to form a single product AB, and requires an inert *third body* (M) to stabilize the product. The reactants collide to form a product AB* that is internally excited due to conversion of the kinetic energy of the colliding molecules:

$$(k_1); \qquad A + B \rightarrow AB^* \tag{5.90}$$

The excited product AB* either decomposes

$$(k_2); \qquad AB^* \rightarrow A + B \tag{5.91}$$

or is thermally stabilized by collision with M (in the atmosphere, M is usually N_2 or O_2):

$$(k_3); \qquad AB^* + M \rightarrow AB + M^* \tag{5.92}$$

where the asterisk describes the addition of internal energy to M. This internal energy is eventually dissipated as heat. Thus, the overall reaction is written

$$A + B + M \rightarrow AB + M \tag{5.93}$$

Although M has no net stoichiometric effect in the overall reaction, it is important to include it in the notation of the reaction because it can play a kinetic role. The rate of the overall reaction is given by

$$-\frac{d[A]}{dt} = -\frac{d[B]}{dt} = \frac{d[AB]}{dt} = k_3[AB^*][M] \tag{5.94}$$

where [M] is effectively the air density [molecules cm^{-3}]. AB* has a very short lifetime and is therefore lost as rapidly as it is produced; this defines a *quasi-steady* *state* (as opposed to true steady state, since the concentration of AB* is still changing with time). The corresponding steady-state expression is

$$k_1[\text{A}][\text{B}] = (k_2 + k_3[\text{M}])[\text{AB}^*] \tag{5.95}$$

Replacing into (5.94), we obtain

$$-\frac{d[\text{A}]}{dt} = -\frac{d[\text{B}]}{dt} = \frac{d[\text{AB}]}{dt} = \frac{k_1 k_3[\text{A}][\text{B}][\text{M}]}{k_2 + k_3[\text{M}]} = k[\text{A}][\text{B}] \tag{5.96}$$

with

$$k = \frac{k_1 k_3[\text{M}]}{k_2 + k_3[\text{M}]} \tag{5.97}$$

which is the *Lindemann–Hinshelwood* rate expression for a three-body reaction. In the *low-pressure limit* $[\text{M}] \ll k_2/k_3$, we have $k \to (k_1 k_3/k_2)[\text{M}]$ so that the rate depends linearly on the air density. In the *high-pressure limit* $[\text{M}] \gg k_2/k_3$, we have $k \to k_1$; the rate is then independent of the air density, as $[\text{M}]$ is sufficiently high to ensure that AB* stabilizes by reaction (5.92) rather than decomposes by reaction (5.91).

A standard formulation for the rate expression of a three-body reaction is

$$k = \frac{k_o[M]}{1 + \dfrac{k_o[M]}{k_\infty}} F \tag{5.98}$$

Here, $k_o = k_1 k_3/k_2$ is the low-pressure rate constant, $k_\infty = k_1$ is the high-pressure rate constant, and F is a correction factor for the transition regime between low-pressure and high-pressure limits. The Lindemann–Hinshelwood rate expression has $F = 1$. More accurate is the *Troe expression*:

$$\log F = \frac{\log F_C}{1 + \left[\log \dfrac{k_o[M]}{k_\infty} \right]^2} \tag{5.99}$$

where F_C is the *broadening factor*. Kinetic data for three-body reactions are commonly reported as F_C, k_o, and k_∞, with temperature dependences for k_o and k_∞.

5.4 Chemical Mechanisms

As we saw in Chapter 4, the chemical operator of an atmospheric model solves the chemical evolution equation

$$\frac{dn_i}{dt} = P_i - \ell_i n_i \tag{5.100}$$

for an ensemble of species. Here n_i is the number density [molecules cm^{-3}] of the ith species, P_i is the production rate [molecules cm^3 s^{-1}] representing the sum of contributions from all reactions producing i, and $\ell_i n_i$ [molecules cm^{-3} s^{-1}] is the loss rate representing the sum of contributions from all reactions consuming i. If P_i and ℓ_i are constants, then the solution to (5.100) is a simple exponential approach to the steady state P_i/ℓ_i:

$$n_i(t) = n_i(0) \exp\left[-\ell_i t\right] + \frac{P_i}{\ell_i}\left(1 - \exp\left[-\ell_i t\right]\right) \tag{5.101}$$

The problem is more complicated if P_i and ℓ_i depend on the concentrations of other species that are themselves coupled to species i. This is frequently the case in atmospheric chemistry because of catalytic cycles, reaction chains, and common dependences on oxidant concentrations. One must then solve (5.100) for the ensemble of coupled species as a system of coupled ordinary differential equations (ODEs). Computational methods for this purpose are described in Chapter 6.

Here, we discuss the general task of defining the ensemble of coupled species and reactions that need to be taken into account in a chemical transport model to address a particular problem. This collection of reactions represents the *chemical mechanism* of the model. It includes not only the species of direct interest to the problem but also the precursors and reactants for these species, which themselves may have precursors and reactants. The mechanism must represent a closed system where the concentrations of all species can be computed. Box 5.2 gives an example of a simple mechanism.

Box 5.2 | **A Simple Mechanism for Tropospheric Ozone Formation**

We describe here an simple mechanism for production of ozone from oxidation of hydrocarbons in the presence of NO_x (see Chapter 3). The mechanism includes just nine coupled reacting species: O_3, OH, HO_2, RO_2, RH, $HCHO$, CO, NO, NO_2. Emissions of NO_x and RH complete the closure. No closure is needed for species that are only products (such as H_2, $ROOH$, H_2O_2, HNO_3). Although this mechanism is considerably oversimplified relative to mechanisms used in research models, it serves to illustrate some of the ideas presented in the text. Comments on individual reactions (1–3 and 7) are listed following the mechanism.

1. $O_3 + h\nu + H_2O \rightarrow 2OH + O_2$
2. $RH + OH \xrightarrow{O_2} RO_2 + H_2O$
3. $RO_2 + NO \xrightarrow{O_2} HCHO + HO_2 + NO_2$
4. $HO_2 + NO \rightarrow OH + NO_2$
5. $NO_2 + h\nu \xrightarrow{O_2} NO + O_3$
6. $NO + O_3 \rightarrow NO_2 + O_2$
7. $HCHO + h\nu \xrightarrow{O_2} HO_2 + CO + 0.5H_2$
8. $HCHO + OH \xrightarrow{O_2} CO + HO_2 + H_2O$
9. $CO + OH \xrightarrow{O_2} CO_2 + HO_2$
10. $RO_2 + HO_2 \rightarrow ROOH + O_2$
11. $HO_2 + HO_2 \rightarrow H_2O_2 + O_2$
12. $NO_2 + OH + M \rightarrow HNO_3 + M$

1. This reaction convolves four elementary reactions: (1a) $O_3 + h\nu \rightarrow O(^1D) + O_2$, (1b) $O(^1D) + M \rightarrow O(^3P) + M$, (1c) $O(^3P) + O_2 + M \rightarrow O_3 + M$, (1d) $O(^1D) + H_2O \rightarrow 2\,OH$. $O(^3P)$ and $O(^1D)$ have lifetimes of much less than a second and can be assumed to be at steady state through the

Box 5.2 (*cont.*)

above reactions. Thus the overall rate of reaction (1) is computed in the mechanism as $-d[O_3]/dt = k_1[O_3] = (k_{1a}k_{1d}[O_3][H_2O])/(k_{1b}[M] + k_{1d}[H_2O])$.

2. RH in this reaction represents a lumped hydrocarbon accounting for the overall reactivity of hydrocarbons RH_i with elementary rate coefficients k_i for oxidation by OH. Thus $[RH] = \Sigma [RH_i]$ and $k_2 = (\Sigma k_i[RH_i])/\Sigma[RH_i]$. The RH + OH reaction produces the R radical, which immediately adds O_2 to produce the lumped organic peroxy radical RO_2. Thus, O_2 is involved in the stoichiometry of the reaction although it does not control the rate; customary practice is then to put it on top of the reaction bar.

3. This reaction is not stoichiometrically balanced and does not conserve carbon. The reaction RO_2 + NO actually produces RO and NO_2, but we assume that RO immediately adds O_2 to produce HCHO and HO_2. This is based on analogy with the fate of CH_3O_2 and CH_3O produced from methane oxidation. It is obviously a very crude treatment, as higher RO radicals may react by various pathways to produce a range of oxygenated organic compounds. However, we may not have the information needed for the RO species of interest, and/or accounting for the full suite of compounds would greatly increase the number of species in the mechanism. An important attribute of the formulation of reaction (3) is that it conserves radicals through the formation of HO_2.

7. This reaction represents the sum of two branches for HCHO photolysis, with an assumed 50:50 branching ratio: (7a) HCHO + $h\nu \rightarrow$ H + CHO and (7b) HCHO + $h\nu \rightarrow$ CO + H_2. H and CHO both rapidly add O_2 to yield HO_2 and CO.

The list of possible reactions involving atmospheric species is exceedingly large, but only a small fraction is sufficiently fast to need to be taken into account. Placing limits on the size of the chemical mechanism is essential for computational tractability of the model. The computational cost is mainly driven by the number of coupled species and the stiffness of the system. If a species plays a significant role in the chemical mechanism, but is not coupled to the others, it should be separated from the coupled system and its chemical evolution calculated independently.

A number of compilations of reactions of atmospheric relevance are available in the literature. Many of these reactions have large uncertainties in their rate coefficients and products, reflecting the difficulty of laboratory measurements of reaction kinetics. A particular challenge is the chemistry of organic species, which involves a very large number of species and a cascade of oxidation products including radicals with varying functionalities and volatility. Most of the reaction rate constants have never been actually measured except for the smallest organic compounds, and must instead be inferred by analogy with reactions of similar species. To limit the size of the mechanism as well as to reflect the limitations in our chemical knowledge, large organic compounds are typically lumped into classes of species with the same functionality or volatility, and the evolution of a particular class is represented by a single *surrogate* or *lumped species*.

The choice of species to be included in the coupled chemical mechanism must balance chemical completeness and computational feasibility. Although experience and chemical intuition are important in the construction of a chemical mechanism, one can also use objective considerations. Consider for example the construction of a chemical mechanism to compute OH radical concentrations in the troposphere. Tropospheric OH has a concentration $\sim 10^6$ molecules cm^{-3} and a lifetime ~ 1 s, so that the important reactions controlling OH concentrations must have rates $\sim 10^4$–10^6 molecules cm^{-3} s^{-1}. Species for which total production rates (P_i) and loss rates ($\ell_i n_i$) are orders of magnitude lower under the conditions of interest will not play a significant role in OH chemistry, either directly or indirectly, and can thus be decoupled or eliminated from the mechanism. Starting from a large ensemble of candidate species, it is thus possible to construct objectively a reduced mechanism. Calculations of P_i and $\ell_i n_i$ can be made locally in a chemical transport model so that reduced mechanisms adapted to the local conditions can be selected (Santillana *et al.*, 2010).

5.5 Multiphase and Heterogeneous Chemistry

So far we have discussed reactions involving the gas phase. Liquid and solid phases in the forms of aerosols and clouds enable a different type of chemistry. Chemical species partition between the gas and the condensed phase, and reactions can take place at the surface or in the bulk of the condensed phase (Figure 5.13). Standard usage is to refer to this chemistry as *multiphase* or *heterogeneous*. Some authors make a distinction between multiphase chemistry as involving reactions in the bulk condensed phase, and heterogeneous chemistry as involving surface reactions, but atmospheric chemistry literature tends to use these two terms interchangeably.

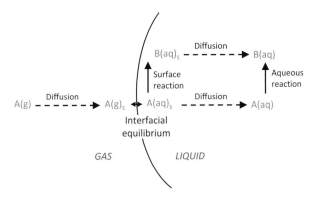

Figure 5.13 General schematic for uptake of a chemical species A by an aqueous aerosol particle with subsequent reaction to produce non-volatile species B. Chemical reactions are indicated by solid arrows and molecular diffusion by dashed arrows. The s subscript indicates surface species with properties possibly different from the bulk.

We see from Figure 5.13 that heterogeneous chemistry involves a combination of molecular diffusion, interfacial equilibrium, and chemical reaction. We examine here how these processes determine the rate of the overall reaction.

5.5.1 Gas–Particle Equilibrium

Chemical species in the atmosphere partition between the gas and particle phases in a manner determined by their free energies in each phase. Equilibrium partitioning always holds at the gas–particle interface, and extends to the bulk gas and particle phases in the absence of mass transfer limitations (Figure 5.13). Mass transfer limitations will be discussed in Section 5.5.2. For liquid particles, the timescale to achieve gas–particle bulk equilibrium is typically less than a few minutes.

Gas–particle equilibrium for a species X is described by

$$X(g) \rightleftarrows X(a) \tag{5.102}$$

where $X(g)$ and $X(a)$ denote the species in the gas and aerosol phases, respectively. The general form of the equilibrium constant is

$$K = \frac{[X(a)]}{p_x} \tag{5.103}$$

where p_x is the partial pressure of X and $[X(a)]$ is the concentration in the particle phase. Different measures of concentration are used depending on the type of particle phase.

Aqueous particles

When the particle phase is an aqueous solution (aqueous aerosol or cloud), the equilibrium expression (5.103) is called *Henry's law* and the equilibrium constant K is called the *Henry's law constant*. $[X(a)]$ (commonly written $[X(aq)]$) is the molar concentration (or *molarity*) of the species in solution. In the atmospheric chemistry literature, K is commonly given in units of [M atm^{-1}] where M denotes moles per liter of solution. The Henry's law constant varies with temperature T [K] according to the *van't Hoff law*

$$K(T) = K(T_o) \exp\left[-\frac{\Delta H}{\mathcal{R}} \left(\frac{1}{T} - \frac{1}{T_o} \right) \right] \tag{5.104}$$

where T_o is the reference temperature commonly taken as 298 K, ΔH is the enthalpy of dissolution [J mol^{-1}] at that reference temperature, and $\mathcal{R} = 8.314$ J K^{-1} mol^{-1} is the ideal gas constant. ΔH is always negative for a gas-to-aqueous transition so that K increases with decreasing temperature. Table 5.2 lists the Henry's law constants for selected species. Dependences on temperature are strong; a typical value $\Delta H/\mathcal{R} = -5900$ K implies a doubling of K for every 10 K temperature decrease.

Fast dissociation or complexation of the species in the aqueous phase can increase the actual solubility beyond the physical solubility specified by Henry's law. Consider for example the dissolution and dissociation of an acid HA:

Table 5.2 Henry's law constants K for selected species[a]			
Species	$K(298\ K)$ [M atm^{-1}]	$\Delta H/\mathcal{R}$ [K]	K^* [M atm^{-1}]
O_3	1.1 (−2)	−2400	1.8(−2)
CH_3OOH	3.1 (2)	−5200	9.5(2)
SO_2	1.2(0)	−3150	1.6(3)
CH_2O	1.7 (0)	−3200	1.4(4)
HCOOH	8.9 (3)	−6100	2.2(5)
H_2O_2	8.3 (4)	−7400	4.1(5)
NH_3	7.4(1)	−3400	9.2(8)
HNO_3	2.1 (5)	−8700	4.3(11)

[a] Read 1.1(−2) as 1.1×10^{-2}. Effective Henry's law constants K^* are calculated at $T = 280$ K and pH = 4.5, including complexation with water for dissolved CH_2O, acid dissociation for dissolved SO_2, HCOOH, and HNO_3, and protonation for NH_3. Data are from the Jacob (1986, 2000) compilations.

$$HA(g) \rightleftarrows HA(aq) \tag{5.105}$$

$$HA(aq) \rightleftarrows H^+ + A^- \tag{5.106}$$

with acid dissociation constant K_a [M]:

$$K_a = \frac{[H^+][A^-]}{[HA(aq)]} \tag{5.107}$$

To account for the dissociation of HA in the aqueous phase, we define the *effective Henry's law constant* K^* [M atm^{-1}] as

$$K^* = \frac{[HA(aq)] + [A^-]}{p_{HA}} = K\left(1 + \frac{K_a}{[H^+]}\right) \tag{5.108}$$

Similar expressions can be derived for other dissociation and complexation processes.

The dimensionless partitioning coefficient f of an atmospheric species X between the aqueous phase and the gas phase can be defined by the concentration ratio in the two phases, referenced in both cases to the volume of air. By making use of the ideal gas law we obtain:

$$f(X) = \frac{n_X(aq)}{n_X(g)} = L K \mathcal{R} T \tag{5.109}$$

Here, $n_X(aq)$ and $n_X(g)$ are the concentrations of X in the aqueous and gas phases, respectively, both in units of molecules per cm^3 of air. L is the atmospheric *liquid water content* [cm^3 liquid water per cm^3 of air]. K in (5.109) should be replaced by K^* if X(aq) dissociates or complexes. With K in units of [M atm^{-1}], the ideal gas constant is $\mathcal{R} = 0.08205$ atm M^{-1} K^{-1}. Liquid water contents are typically in the range 10^{-9}–10^{-11} for aqueous aerosol under non-cloud conditions, so that $f \ll 1$ for all species in Table 5.2 except for NH_3 (if aerosol pH is low) and HNO_3 (if aerosol pH is high). Gas–aerosol equilibrium of NH_3 and HNO_3 is discussed in Section 3.9. By contrast, application of (5.109) to a precipitating cloud with liquid water content $L \sim 1 \times 10^{-6}$ and typical pH range 4–5 yields $f \gg 1$ for HNO_3, NH_3, H_2O_2, and HCOOH, and $f \sim 1$ for CH_2O. We conclude that HNO_3, NH_3, H_2O_2, and HCOOH are efficiently scavenged by rain, CH_2O is partly scavenged, and SO_2, CH_3OOH, and O_3 are not efficiently scavenged. SO_2 can be efficiently scavenged only if it oxidizes rapidly to sulfate in the aqueous phase (see Section 3.8). In-cloud partitioning is discussed further in Section 8.8 as a driver of wet deposition.

Organic particles

Gas–particle equilibrium involving non-aqueous solutions is harder to quantify because of uncertainty in the composition of the particle phase. Formation of organic aerosol is thought to involve at least in part an equilibrium partitioning between semi-volatile organic vapors and the organic phase of the aerosol:

$$K = \frac{[X(a)]}{C_O[X(g)]} \qquad (5.110)$$

where $[X(g)]$ and $[X(a)]$ are the gas- and particle-phase concentrations of X, and C_O is the concentration of pre-existing organic aerosol, all in units of mass per volume of air. Nonlinearity in gas–particle partitioning arises because condensation of X contributes to C_O. An alternative way to express the same equilibrium is with respect to the volatility of the species, $C_X^* = 1/K$ in units of mass per volume of air. The partitioning coefficient between the particle and the gas phases is then given by $f = C_O/C_X^*$. As in the case of aqueous-phase partitioning, f can range over many orders of magnitude depending on the species; most species will be in a limiting regime of near-total fractionation in the gas phase ($f \ll 1$) or in the particle phase ($f \gg 1$), with some species switching between regimes depending on C_O. Because of the very large number and poor characterization of the species contributing to organic aerosol formation, an effective modeling approach can be to partition the ensemble of organic species into order-of-magnitude *volatility classes* C_i^* that are transported independently in the model as their total (gas + particle) concentration C_i. Gas–particle partitioning is then diagnosed locally by

$$C_O = \sum_i f_i C_i \qquad \text{with} \qquad f_i = \frac{C_O}{C_i*} \qquad (5.111)$$

The solution to C_O must be obtained iteratively. This is known as the *volatility basis set* (*VBS*) approach (Donahue *et al.*, 2006).

Solid particles

Equilibrium between the gas phase and solid particles is in general poorly understood, with the important exception of H_2SO_4–HNO_3–NH_3 aerosol for which well-established bulk thermodynamics apply (Martin, 2000). For example, in the simple case of dry ammonium nitrate (NH_4NO_3) aerosol, equilibrium can be expressed by a temperature-dependent equilibrium constant for condensation:

$$K_{NH4NO3} = p_{NH3}\,p_{HNO3} \tag{5.112}$$

If NH_3 and HNO_3 concentrations are sufficiently low that the product $p_{NH3}\,p_{HNO3}$ is less than K_{NH4NO3}, then no NH_4NO_3 aerosol forms. If $p_{NH3}\,p_{HNO3}$ is greater than K_{NH4NO3}, then NH_3 and HNO_3 are in excess and condense to produce NH_4NO_3 aerosol. This condensation decreases p_{NH3} and p_{HNO3} until equilibrium (5.112) is met, at which point the aerosol is in equilibrium with the gas phase. In this manner, knowledge of K_{NH4NO3} constrains how much NH_4NO_3 aerosol will form given initial concentrations of NH_3 and HNO_3.

Different equilibrium formulations apply in the case of a gas adsorbing heterogeneously onto a solid aerosol surface, for example HNO_3 adsorbing on dry dust. One can then describe the uptake in terms of a number of available condensation sites on the aerosol surface, with kinetic expressions to compute adsorption/desorption rates at these sites. A simple model for monolayer uptake is the *Langmuir isotherm*:

$$\theta = \frac{K'p_X}{1 + K'p_X} \tag{5.113}$$

where θ is the fraction of occupied condensation sites, and the equilibrium constant K' is the ratio of the adsorption and desorption rate constants.

5.5.2 Mass Transfer Limitations

Achievement of gas–particle thermodynamic equilibrium as described in Section 5.5.1 may be limited by mass transfer in the gas and particles phases. The uptake rate of a gas-phase species by an aerosol can be expressed in a general form by

$$\left[\frac{dn_g}{dt}\right]_{in} = -k_T A n_g \tag{5.114}$$

where n_g is the number density in the bulk gas phase [molecules per cm^3 of air], equivalent to $n(g)$ in Section 5.5.1, A is the aerosol surface area concentration [cm^2 per cm^3 of air], and k_T is a mass transfer rate coefficient [cm s^{-1}] that depends on the thermal velocity of molecules, the probability that collision will result in uptake, and any limitations from molecular diffusion. The rate of volatilization from the aerosol surface to the gas phase is proportional to the surface concentration n_s in the aerosol phase

$$\left[\frac{dn_g}{dt}\right]_{out} = k_T A \frac{n_s}{K} \tag{5.115}$$

where K is the thermodynamic equilibrium constant between the gas and particle phases, with units for n_s and K chosen so that Kn_s has units of [molecules per cm^3 of air]. The mass transfer rate constant k_T is the same for uptake and for volatilization so that steady state $[dn_g/dt]_{in} + [dn_g/dt]_{out} = 0$ yields $n_s = K\,n_g$. The net rate of transfer at the interface is

$$\frac{dn_g}{dt} = -k_T A \left[n_g - \frac{n_s}{K}\right] \tag{5.116}$$

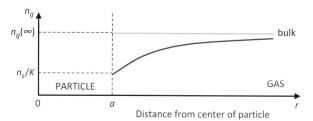

Figure 5.14   Gas-phase concentration gradient in the vicinity of an aerosol particle of radius a.

and vanishes to zero at equilibrium. The form of k_T depends on the *Knudsen number* $Kn = \lambda/a$, where a is the aerosol particle radius and λ is the *mean free path* for molecules in the gas phase. The mean free path for air is $\lambda = 0.068$ μm at standard conditions of temperature and pressure (STP: 273 K, 1 atm) and varies inversely with pressure. In the limit $Kn \gg 1$, the gas-phase concentration in the immediate vicinity of the particle is the same as in the bulk, since the particle does not interfere with the random motion of the gas molecules. This is called the *free molecular regime*. From the kinetic theory of gases, we then have

$$k_T = \frac{v\alpha}{4} \tag{5.117}$$

where $v = (8kT/\pi m)^{1/2}$ is the mean thermal velocity of molecules, α is the *mass accommodation coefficient* representing the probability that a gas molecule impacting the surface is absorbed in the bulk, k is the Boltzmann constant, and m is the molecular mass. α generally increases with the solubility of the gas and decreases with increasing temperature. It can approach unity for a highly soluble gas at low temperature, but may be several orders of magnitude lower for a gas of low solubility.

The mass transfer limitation takes on a different form in the limit $Kn \ll 1$. Under those conditions, gas molecules in the immediate vicinity of the surface undergo a large number of collisions with the surface before being able to escape the influence of the surface and migrate to the bulk. Thus, the gas concentration at the interface is controlled by local equilibrium with the aerosol, and transport between the surface and the bulk gas phase is controlled by molecular diffusion. This is called the *continuum regime*. A steady-state concentration gradient is established between the particle surface and the bulk gas phase, in contrast to the free molecular regime where there is no such gradient (Figure 5.14). Assuming a spherical aerosol, the gas-phase diffusion equation in spherical coordinates is

$$D_g \nabla^2 n_g(r) = D_g \frac{1}{r^2} \frac{d}{dr}\left(r^2 \frac{dn_g(r)}{dr}\right) = 0 \tag{5.118}$$

where D_g is the molecular diffusion coefficient and r is the distance from the center of the particle. Solving for the flux F at the gas–particle interface, we obtain

$$F = -D_g \frac{dn}{dr}\bigg|_{r=a} = \frac{D_g}{a}\left[n_g(\infty) - \frac{n_s}{K}\right]$$ (5.119)

where $n_g(\infty)$ is the bulk gas-phase concentration far away from the surface (n_g in (5.116)). Thus, in the continuum regime,

$$k_T = \frac{D_g}{a}$$ (5.120)

Mass transfer in the continuum regime is not dependent on the mass accommodation coefficient α; this is because gas molecules trapped in the immediate vicinity of the surface collide many times with the surface and thus eventually become incorporated in the bulk aerosol phase even if α is low. By contrast, mass transfer in the free molecular regime is not dependent on the particle radius a, because the particles are too small to affect the motion of molecules.

The free molecular regime generally applies to stratospheric aerosols where $a \sim$ 0.1 μm and $\lambda \sim 1$ μm. The continuum regime generally applies to tropospheric clouds where $a \sim 10$ μm and $\lambda \sim 0.1$ μm. Tropospheric aerosols are often in the *transition regime* since $a \sim \lambda \sim 0.1$–1 μm. Exact solution of mass transfer for the transition regime is complicated. Schwartz (1986) showed that the solution can be approximated to within 10% by harmonic addition of the mass transfer rate coefficients for the free molecular and continuum regimes as two conductances operating in series:

$$k_T = \left[\frac{a}{D_g} + \frac{4}{v\alpha}\right]^{-1}$$ (5.121)

Equation (5.121) can be applied in the general case to calculations of gas–aerosol mass transfer. Unless in the free molecular regime, one should integrate over the aerosol size distribution in order to resolve the dependence of k_T on the particle radius a.

The volatilization component of the gas–aerosol transfer flux was expressed in (5.115) as a function of the surface concentration n_s. This is not in general a known quantity and we would like to relate it to the bulk aerosol phase concentration, which is more easily measured or modeled. The mixing timescale for a particle of radius a is given by $\tau_{mix} = a^2/\pi^2 D_a$, where D_a is the molecular diffusion coefficient in the aerosol phase. For a liquid aqueous phase $D_a \sim 10^{-4}$ cm^2 s^{-1} and thus for a particle with $a \sim 1$ μm we have $\tau_{mix} \sim 10^{-5}$ s. This is in general sufficiently short to ensure complete mixing of the aerosol phase so that the surface concentration equals the bulk concentration. There are a few exceptions where the diffusion equation needs to be solved in the aerosol phase (Jacob, 2000).

5.5.3 Reactive Uptake Probability

Detailed treatment of heterogeneous chemistry in atmospheric models requires solution of the chemical evolution equations in the aerosol phase coupled to the gas phase through mass transfer. A simplified treatment is possible when the heterogeneous chemistry of interest can be reduced to a first-order chemical loss in the aerosol phase for a species transferred from the gas phase. Since the mass transfer

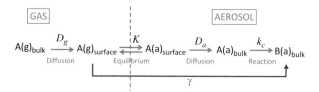

Figure 5.15 The reactive uptake probability γ convolves processes of gas–aerosol interfacial equilibrium, aerosol-phase diffusion, and reaction.

processes are themselves first-order, the combined process can be encapsulated in a first-order loss equation. For this purpose, we define the *reactive uptake coefficient* γ as the probability that a molecule impacting the particle surface will undergo irreversible reaction rather than volatilization back to the gas phase. The rate of loss for the species from the gas phase can then be represented by (5.121) but with γ replacing α in the formulation of k_T:

$$k_T = \left[\frac{a}{D_g} + \frac{4}{v\gamma} \right]^{-1} \tag{5.122}$$

The reactive uptake coefficient γ combines the processes of interfacial equilibrium, aerosol-phase diffusion, and reaction (Figure 5.15). It is a particularly helpful formulation because results from laboratory experiments can often be reported as γ. One can also relate γ to the actual chemical rate coefficient for loss in the aerosol phase. If the loss is a surface reaction, then γ simply compounds the mass accommodation α by the probability that the molecule will react on the surface [rate coefficient k_S in unit of s^{-1}] versus desorb [rate coefficient k_D in unit of s^{-1}]. Thus,

$$\gamma = \frac{\alpha\, k_S}{k_S + k_D} \tag{5.123}$$

If the loss is a first-order reaction taking place in the bulk of a liquid aerosol phase [effective rate coefficient k_C in unit of s^{-1}], then the effect of diffusion in the aerosol phase needs to be considered. Solution of the diffusion equation for a spherical particle with a zero-flux boundary condition at the particle center yields

$$\gamma = \left[\frac{1}{\alpha} + \frac{v}{4K\mathcal{R}T(D_a k_C)^{1/2} f(q)} \right]^{-1} \tag{5.124}$$

where

$$f(q) = \coth q - \frac{1}{q} \tag{5.125}$$

and $q = a(k_C/D_a)^{1/2}$ is a dimensionless number called the *diffuso-reactive parameter* (Schwartz and Freiberg, 1981). $f(q)$ represents a sphericity correction for the limitation of uptake by diffusion in the aerosol phase and is a monotonously increasing function of q. Limits are $f(q) \to q/3$ for $q \to 0$ and $f(q) \to 1$ for $q \to \infty$.

It is important to recognize the sphericity correction when applying γ values measured in the laboratory to atmospheric aerosols, because the geometry used in the laboratory is different from that in the atmosphere. The laboratory measurements are often for a bulk liquid phase with planar surface ($a \rightarrow \infty \Rightarrow q \rightarrow \infty$), so that the effective γ for an aerosol will be lower than the laboratory-reported value. Physically, this is because the small size of the aerosol particles does not allow diffusion of the dissolved gas into an infinite bulk but instead forces re-volatilization to the gas phase. A proper treatment requires that one relate γ measured in the laboratory to k_C, which is the fundamental variable, and then apply (5.124) (with integration over the aerosol size distribution) for the actual atmospheric aerosol conditions. This is generally ignored in atmospheric models under the assumption that uncertainties in k_C trump other uncertainties, so that only order-of-magnitude estimates of γ are possible in any case.

5.6 Aerosol Microphysics

The size distribution of aerosol particles evolves continuously in the atmosphere as a result of *microphysical processes* including particle nucleation, gas condensation, coagulation, activation to cloud droplets, and sedimentation. These processes are computationally challenging to represent in models because of the wide ranges of particle sizes, compositions, and morphologies that need to be resolved (Chapter 3). Processes of nucleation and aerosol–cloud interactions are also highly nonlinear, so that averaging in models can cause large errors.

Because of these difficulties, a common practice in chemical transport models is to simulate only the total mass concentrations of the different aerosol components (sulfate, nitrate, organic carbon, black carbon, dust, sea salt, etc.), integrating over all sizes or across fixed size ranges with no transfer between ranges. As pointed out in Section 4.3, the continuity equations for the aerosol components are then of the same form as for gases. The models must still assume a size distribution for the different aerosol components in order to compute radiative effects, heterogeneous chemical rates, and deposition rates. A log-normal size distribution is often assumed for the dry number size distribution function n_N (Chapter 3):

$$n_N\left(\log D_p\right) = \frac{N_0}{(2\pi)^{1/2} \log \sigma_g} \exp\left[-\frac{1}{2}\left(\frac{\log\left(D_p/D_m\right)}{\log \sigma_g} \right)^2 \right] \qquad (5.126)$$

Here D_p is the dry particle diameter, D_m is the median dry diameter, and σ_g is the *geometric standard deviation* characterizing the variance in $\log(D_p/D_m)$. Different values of D_m and σ_g are usually adopted for different chemical components of the aerosol on the basis of observations. Aerosol surface area and mass size distributions are deduced from the moments of the number size distribution function as described in Chapter 3. The aerosol size distribution is expressed in terms of dry sizes because aerosol water is highly fluctuating as a function of the local relative humidity. The contribution from aerosol water is derived by applying component-specific, multiplicative *hygroscopic growth factors* to the size distributions as a function of local relative humidity. See Martin *et al.* (2003) for an example of such an approach.

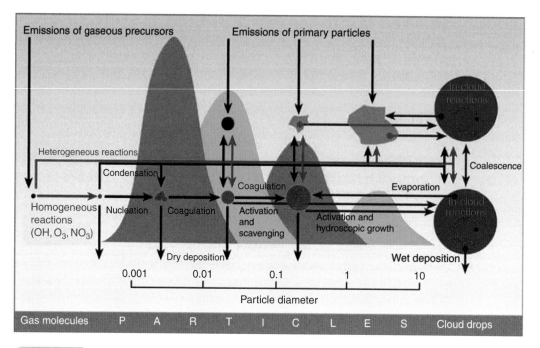

Figure 5.16 Schematic representation of the microphysical processes that determine the evolution of aerosol particles from their formation through nucleation to their activation and conversion into cloud droplets. Particle diameter is in units of μm. Major aerosol modes are highlighted, including (in order of increasing size) the nucleation, Aitken, accumulation, and coarse modes. Reproduced from Heintzenberg *et al.* (2003).

Assumption of fixed aerosol size distributions can be a large source of model error, for example in describing precipitation scavenging which is a strong function of particle size. Accurate computation of the time-evolving size distribution is also critical for addressing aerosol radiative effects, aerosol–cloud interactions, and air quality impacts. This requires solving the continuity equation for aerosols (Section 4.3) with terms describing the different aerosol microphysical processes forcing changes in the local size distribution. Here we present standard rate expressions and model formulations for this purpose.

Figure 5.16 is a schematic representation of the different processes included in models of the aerosol size distribution. It shows the multimodal distribution of the aerosol as previously described in Chapter 3 with a *nucleation mode* (particle diameter less than 0.01 μm), *Aitken nuclei mode* (0.01–0.1 μm), *accumulation mode* (0.1–1 μm), and *coarse mode* (larger than 1 μm). The nucleation and Aitken nuclei modes dominate the number density but represent a very small fraction of the mass density. They are formed by nucleation, grow by gas condensation, and are lost by coagulation. The accumulation mode is produced from the smaller modes by coagulation and gas-condensation. Growth of particles beyond 1 μm is slow because gas condensation adds little mass and Brownian diffusion decreases, slowing down coagulation. This results in "accumulation" in the 0.1–1 μm range. The accumulation mode generally dominates the total aerosol surface area and makes a major contribution to total aerosol mass. Accumulation mode particles are also efficient *cloud condensation nuclei* (*CCN*) and drive formation of cloud droplets under supersaturated conditions. Particles in the coarse mode tend to be directly emitted to the atmosphere as dust or sea salt, and are removed relatively rapidly by precipitation scavenging and sedimentation.

In the following discussion, we express the particle number distribution n_N as a function of the particle volume (V) and omit the subscript N in the notation. The size distribution is thus noted $n(V)$. We saw in Section 4.3 how the continuity equation could be applied to model the evolution of the aerosol size distribution in response to microphysical processes. We also saw how the local evolution in response to these processes could be described by separating the contributing terms:

$$\frac{\partial n(V)}{\partial t} = \left[\frac{\partial n(V)}{\partial t}\right]_{nucleation} + \left[\frac{\partial n(V)}{\partial t}\right]_{condensation/evaporation} + \left[\frac{\partial n(V)}{\partial t}\right]_{coagulation}$$

$$(5.127)$$

Here we give formulations for these individual terms and describe their practical computation in models. The formulations are taken from Seinfeld and Pandis (2006), to which the reader is referred for a detailed presentation of aerosol microphysical processes.

5.6.1 Formulation of Aerosol Processes

Nucleation

Formation of new particles in the atmosphere is driven by clustering of molecules from the gas phase. Cluster formation requires very large supersaturations and takes place in a highly localized manner (*nucleation bursts*) when such supersaturations are achieved (and then relaxed through the nucleation process). Achievement of large supersaturations requires gas mixtures with very low vapor pressure. *Binary nucleation* can take place in the atmosphere from H_2SO_4–H_2O mixtures, as H_2SO_4 has very low pressures over H_2SO_4–H_2O solutions at all relative humidities of atmospheric relevance. *Ternary nucleation* involves a third component gas, typically ammonia (H_2SO_4–NH_3–H_2O mixture), for which the H_2SO_4 vapor pressure is even lower. See Chapter 3 for H_2SO_4 production mechanisms. The possible role of organic molecules in contributing to nucleation is a topic of current research.

The critical step in nucleation is the formation of thermodynamically stable clusters that then grow rapidly by subsequent gas condensation. Clustering of molecules must overcome a nucleation barrier that can be expressed thermodynamically by the surface tension of the growing clusters, or at a molecular level in terms of the internal energy of successive clusters. In the continuity equation for aerosols, nucleation behaves as a flux boundary condition populating the bottom of the size distribution (the nucleation mode). If we assume that the size of particles produced by the nucleation process corresponds to a volume V_0, and if J_0 represents the *nucleation rate*, we write

$$\left[\frac{\partial n(V)}{\partial t}\right]_{nucleation} = J_0\,\delta(V, V_0)$$

$$(5.128)$$

See Seinfeld and Pandis (2006) for formulations of nucleation rates. The nucleation rate is a very strong function of the partial pressures of the nucleating gases, varying over orders of magnitude in response to relatively small changes in partial pressure. This nonlinear behavior is difficult to capture in models.

Condensation/evaporation

Gas condensation on existing particles causes these particles to grow. In Section 4.3 we expressed the corresponding term in the continuity equation for aerosols as

$$\left[\frac{\partial n(V)}{\partial t}\right]_{condensation/evaporation} = -\frac{\partial(I(V)n(V))}{\partial V} \tag{5.129}$$

where $I(V) = dV/dt$ is the condensation growth rate. Equation (5.129) is called the *condensation equation*. It is mathematically similar to the advection equation and numerical algorithms face the same difficulties, discussed in Chapter 7. Numerical diffusion associated with the algorithm can lead to erroneous damping of the peak values of the distribution. Numerical dispersion leads to wakes around peak values, particularly near fronts in the size distribution (Seigneur *et al.*, 1986).

The condensation growth rate for species i is proportional to the difference between the bulk vapor pressure p_i (far from the particle) and the equilibrium vapor pressure $p_{eq,i}$. It is expressed by (Seinfeld and Pandis, 2006):

$$I_i(V) = \frac{2\pi D_g(V) m_i}{\mathcal{R}T}\left(\frac{6V}{\pi}\right)^{1/3} f(Kn,\alpha)\left[p_i - p_{eq,i}\right] \tag{5.130}$$

where $D_g(V)$ [cm^2 s^{-1}] is the Brownian diffusion coefficient for the particle, and m_i [kg mol^{-1}] is the molecular mass of species i. Function $f(Kn, \alpha)$ is a correction factor for the non-continuum regime (Section 5.5.2) that depends on the Knudsen number Kn and the mass accommodation coefficient α.

Coagulation

Coagulation is the process by which two particles that collide by Brownian motion stick together to form a new, larger particle. It shifts the size distribution toward larger sizes and reduces the number of smaller particles.

The coagulation rate $J_{i,j}$ [m^{-3} s^{-1}] resulting from the collisions between two particles i and j is proportional to the number density N_i and N_j [m^{-3}] of these two particles:

$$J_{i,j} = \beta_{i,j} N_i N_j \tag{5.131}$$

where $\beta_{i,j}$ [m^3 s^{-1}] is the binary *coagulation coefficient*. In the continuum regime with $Kn \ll 1$ (particle diameter considerably larger than the mean free path), $\beta_{i,j}$ is given by

$$\beta_{i,j} = 2\pi(D_i + D_j)(D_{p,i} + D_{p,j}) \tag{5.132}$$

where $D_{p,i}$ is the diameter of particle i and D_i is the Brownian diffusion coefficient given by the Stokes–Einstein formula

$$D_i = \frac{kT}{3\pi\,\mu\,D_{p,i}} \tag{5.133}$$

where μ is the viscosity of air. Thus, in this case,

$$\beta_{i,j} = \frac{2kT}{3\mu}\frac{(D_{p,i}+D_{p,j})}{D_{p,i}D_{p,j}} \tag{5.134}$$

In the free molecular regime with $Kn \gg 1$ (very small particles with diameters considerably smaller than the mean free path), the coagulation coefficient becomes

$$\beta_{i,j} = \frac{\pi}{4}\left(D_{p,i}+D_{p,j}\right)^2\left(v_i^2+v_j^2\right)^{1/2} \tag{5.135}$$

where $v_i = (8kT/\pi m_{p,i})^{1/2}$ is the mean thermal velocity with $m_{p,i}$ the mass of particle i. In the transition regime, one generally adopts the continuum regime formula with a correction factor (Seinfeld and Pandis, 2006). Coagulation is most effective when the sizes of the two particles that collide are very different (i.e., collision between a small particle with high thermal velocity and a large particle that provides a large area for collision).

The rate of change in the aerosol size distribution resulting from coagulation is given by

$$\left[\frac{\partial n(V)}{\partial t}\right]_{coagulation} = \frac{1}{2}\int_{V_0}^{V-V_0}\beta(V-V',V')\,n(V-V')\,n(V')\,dV'$$
$$-n(V)\int_{V_0}^{\infty}\beta(V,V')\,n(V')dV' \tag{5.136}$$

where V_0 denotes the volume of the smallest particles considered in the size distribution (associated typically with nucleation, cf. (5.128)). Factor ½ is introduced to avoid double counting.

5.6.2 Representation of the Size Distribution

Computation of the microphysical terms in the continuity equation for aerosols requires that the size distribution be approximated with a limited number of parameters. This is illustrated in Figure 5.17 with discrete, spline, sectional, modal, and monodisperse approximations. The discrete representation provides a fine resolution of the size distribution function with a value for each discrete size. The spline defines a single continuous size distribution function over the whole size range. The sectional representation partitions the size distribution into discrete size intervals called "bins," with fixed values of the distribution functions within these intervals. The modal representation superimposes several continuous functions covering different ranges of the size distribution, one for each mode. The monodisperse representation assigns a single size to each mode. We elaborate here on the most popular methods.

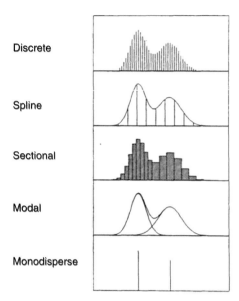

Figure 5.17 Numerical approximations of the aerosol size distribution function. From Whitby *et al.* (1991).

Sectional method

In the sectional method (Gelbard *et al.*, 1980; Adams and Seinfeld, 2002), the aerosol size domain is divided into a discrete number of bins K within which the size distribution functions are assumed to be constant. Thus the size distribution function is represented by K parameters $N_k(k = 1, K)$ representing the number concentration of particles in bin k bounded by volumes $[V_k, V_{k+1}]$:

$$N_k = \int_{V_k}^{V_{k+1}} n(V)\, dV \qquad (k = 1, K) \qquad (5.137)$$

Applying the continuity equation (5.127) to these parameters yields a system of K coupled ODEs:

$$\frac{dN_1}{dt} = -N_1 \sum_{j=1}^{K} \left(\beta_{1,j} N_j \right) - (p_1 + \gamma_1)N_1 + \gamma_2 N_2 + J$$

$$\frac{dN_k}{dt} = \frac{1}{2} \sum_{j=1}^{k-1} \left(\beta_{j,k-j} N_j N_{k-j} \right) - N_k \sum_{j=1}^{K} \left(\beta_{k,j} N_j \right) + p_{k-1} N_{k-1} \qquad (5.138)$$

$$- (p_k + \gamma_k)N_k + \gamma_{k+1} N_{k+1} \qquad (k = 2, K)$$

Nucleation (J) provides a source of particles to the smallest bin ($k = 1$). Condensation growth (p) and evaporation (γ) provide source/sink terms for adjacent bins. Coagulation couples the whole size distribution. This system of coupled ODEs is mathematically equivalent to the system representing chemical production/loss terms for different species in a chemical mechanism. Numerical algorithms for solving such systems are presented in Chapter 6.

Moments and modal

The detailed aerosol size distribution generally contains more information than required to address a specific problem. Rather than calculating its time evolution by solving the conventional continuity equation (5.127), it is often sufficient and computationally more efficient to estimate the low-order moments of the size distribution (Friedlander, 1977). The moment of order k is defined by

$$M_k = \int_0^\infty D_p^k\, n\!\left(D_p\right) dD_p \tag{5.139}$$

where $n(D_p)\, dD_p$ represents the number of particles (assumed to be spherical) per unit volume of air in the diameter size range $[D_p, D_p + dD_p]$. The evolution equations for moment M_k of the aerosol distribution (called moment dynamics equations or MDEs) are derived by starting from the aerosol continuity equation (5.127) with $n(D_p)$ as the independent variable, multiplying each term by D_p^k, and integrating each term over all particle diameters (Whitby and McMurry, 1997). The equations can be solved if all terms can be expressed with only moments as the dependent variables. This requires either an assumption on the mathematical form of the size distribution (see *modal method* below) or approximations to the terms (closure relations) to force them to be expressed in terms of moments. The latter approach describes the *method of moments* (MOM) and makes no a-priori assumptions on the form of the size distribution. However, there may be large errors associated with the closure relations.

In the *modal method* (Whitby, 1978), the aerosol size distribution is specified as the superimposition of a limited number K of functional forms, each representing a

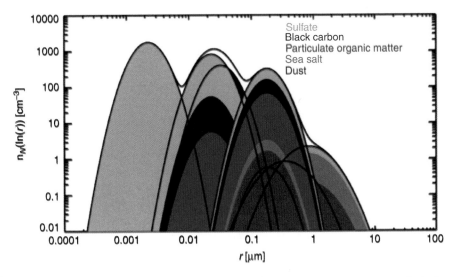

Figure 5.18 Modal distributions of different aerosol components (sulfate, black carbon, organic matter, sea salt, dust). The aerosol size distribution for each component is represented by the number concentrations in seven log-normal modes. The smallest (nucleation) mode is exclusively sulfate. *r* is particle radius. Based on the model of Stier *et al.* (2005). From Stier and Feichter, personal communication.

particular aerosol *mode*. For example, a given mode may be defined by a log-normal size distribution with imposed median diameter $D_{m,k}$ and geometric standard deviation $\sigma_{g,k}$ (5.126). The contribution of each mode to the overall size distribution is then defined by the number concentration of particles in that mode. Figure 5.18 gives an example. The microphysical terms of the continuity equation are integrated over each mode and transfer particles between modes. The evolution of the aerosol size distribution can be calculated using low-order moments by noting that these moments characterize the parameters of the distribution. For example, a log-normal distribution is fully defined by its first three moments ($k = 0, 1, 2$ in (5.139)) characterizing the total number of particles, the mean diameter, and the standard deviation of the distribution. The modal method is computationally much faster than the sectional method but relies on the suitability of decomposing the actual aerosol size distribution along specified modes.

References

Adams P. J. and Seinfeld J. H. (2002) Predicting global aerosol size distributions in general circulation models, *J. Geophys. Res.*, **107**, 4370, doi:10.1029/2001JD001010.

Brasseur G. P. and Solomon S. (2005) *Aeronomy of the Middle Atmosphere: Chemistry and Physics of the Stratosphere and Mesosphere*, 3rd edition, Springer, New York.

Chabrillat S. and Kockarts G. (1997) Simple parameterization of the absorption of the solar Lyman-α line, *Geophys. Res. Lett.*, **24** (21), 2659–2662, doi: 10.1029/97GL52690, correction: *Geophys. Res. Lett.* **25** (1), 79, doi: 10.1029/97GL03569.

Chandrasekhar S. (1950) *Radiative Transfer*, Oxford University Press, Oxford (Reprinted by Dover Publications, 1960).

Donahue N. M., Robinson A. L., Stanier C. O., and Pandis S. N. (2006) Coupled partitioning, dilution, and chemical aging of semivolatile organics, *Environ. Sci. Technol.*, **40**, 2635–2643.

Elsasser W. M. (1938) Mean absorption and equivalent absorption coefficient of a band spectrum, *Phys. Rev.*, **54**, 126–129.

Fang T. M., Wofsy S. C., and Dalgarno A. (1974) Capacity distribution functions and absorption in Schumann–Runge bands of molecular oxygen, *Planet Space Sci.*, **22**, 413–425.

Friedlander S. K. (1977) *Smoke, Dust and Haze: Fundamentals of Aerosol Behavior*, Wiley, New York.

Fu Q. and Liou K. N. (1992) On the correlated k-distribution method for radiative transfer in nonhomogeneous atmosphere, *J. Atmos. Sci.*, **49**, 2139–2156.

Gelbard F., Tambour Y., and Seinfeld J. J. (1980) Sectional representation for simulating aerosol dynamics, *J. Colloid Interface Sci.*, **76**, 357–375.

Gijs A., Koppers A., and Murtagh D. P. (1997) Model studies of the influence of O_2 photodissociation parameterizations in the Schumann–Runge bands on ozone related photolysis in the upper atmosphere, *Ann. Geophys.*, **14**, 68–79.

Goody R. M. (1952) A statistical model for water vapor absorption, *Quart. J. Roy. Met. Soc.*, **78**, 165–169.

Goody R. (1995) *Principles of Atmospheric Physics and Chemistry*, Oxford University Press, Oxford.

Goody R., West R., Chen L., and Crisp D. (1989) The correlated-k method for radiation calculations in nonhomogeneous atmosphere, *J. Quant. Spectrosc. Radiat. Transfer*, **42**, 539–550.

Heintzenberg J., Raes F., and Schwartz S. E. (2003) Tropospheric aerosols. In *Atmospheric Chemistry in a Changing World* (Brasseur G. P., Prinn R. G., and Pszenny A. P., eds), Springer, New York.

Jacob D. J. (1986) The chemistry of OH in remote clouds and its role in the production of formic acid and peroxymonosulfate, *J. Geophys. Res.*, **91**, 9807–9826.

Jacob D. J. (2000) Heterogeneous chemistry and tropospheric ozone. *Atmos. Environ.*, **34**, 2131–2159.

Joseph J. H., Wiscombe W. J., and Weinman J. A. (1976) The delta-Eddington approximation for radiative flux transfer, *J. Atmos. Sci.*, **33**, 2452–2459.

Kockarts G. (1994) Penetration of solar radiation in the Schumann–Range bands of molecular oxygen: A robust approximation, *Ann. Geophys.*, **12** (12), 1207–1217, doi: 10.1007/BF03191317.

Lenoble J. (1977) *Standard Procedures to Compute Atmospheric Radiative Transfer in a Scattering Atmosphere*, Vol. I, International Association of Meteorology and Atmospheric Physics (IAMAP), Boulder, CO.

Liou K. N. (2002) *An Introduction to Atmospheric Radiation*, Vol. 84, 2nd edition Academic Press, New York.

López-Puertas M. and Taylor F. W. (2001) *Non-LTE Radiative Transfer in the Atmosphere*, World Scientific Publishing, Singapore.

Lorenz L. V. (1890) Lysbevaegelsen i og uden for en af plane Lysbolger belyst Kugle, *Det Kongelige Danske Videnskabernes Selskabs Skrifter*, **1**, 1–62.

Martin R.V., Jacob D.J., Yantosca R.M., Chin M., and Ginoux P. (2003) Global and regional decreases in tropospheric oxidants from photochemical effects of aerosols, *J. Geophys. Res.*, **108**, 4097.

Martin S. T. (2000) Phase transitions of aqueous atmospheric particles, *Chem. Rev.*, **100**, 3403–3453.

Mie G. (1908) Beiträge zur Optik trüber Medien, speziell kolloidaler Metallösungen, *Ann. Phys.*, **330**, 377–445.

Minschwaner K. and Siskind D. E. (1993) A new calculation of nitric oxide photolysis in the stratosphere, mesosphere, and lower atmosphere, *J. Geophys. Res.*, **98** (111), 20401–20412, doi: 10.1029/93JD02007.

Minschwaner K., Salawitch R. J., and McElory M. B. (1993) Absorption of solar radiation by O_2: Implications for O3 and lifetimes of N_2O, CFCl3, and CF2Cl2, *J. Geophys. Res.*, **98**, 10543–10561, doi: 10.1029/93JD00223.

Petty G. W. (2006) *A First Course in Atmospheric Radiation*, 2nd edition, Sundog Publications, Madison, WI.

Rayleigh L. (1871) On the light from the sky, its polarization and colour, *Phil. Mag.*, **41**, 107–120.

Santillana M., Le Sager P., Jacob D. J., and Brenner M. P. (2010) An adaptive reduction algorithm for efficient chemical calculations in global atmospheric chemistry models, *Atmos. Environ.*, **44**, 4426–4431.

Schwartz S. E. (1986) Mass transport considerations pertinent to aqueous-phase reactions of gases in liquid-water clouds. In *Chemistry of Multiphase Atmospheric Systems* (Jaeschke W., ed.), Springer-Verlag, Berlin.

Schwartz S. E. and Freiberg J. E. (1981) Oxidation of SO_2 in aqueous droplets: Mass-transport limitation in laboratory studies and the ambient atmosphere, *Atmos. Environ.*, **15**, 1129–1144.

Seigneur C., Hudischewskj A. B., Seinfeld J. H., *et al.* (1986) Simulations of aerosol dynamics: A comparative review of mathematical models, *Aerosol Sci. Technol.*, **5** (2), 205–222.

Seinfeld J. H. and Pandis S. N. (2006) *Atmospheric Chemistry and Physics: From Air Pollution to Climate Change*, Wiley, New York.

Shaw J. (1953) Solar radiation, *Ohio J. Sci.*, **53**, 258.

Smith F. L. III and Smith C. (1972) Numerical evaluation of Chapman's grazing incidence integral Ch (X, χ), *J. Geophys. Res.*, **77**, 19, 3592–3597, doi: 10.1029/JA077i019p03592.

Stamnes K., Tsay S. C., Wiscombe W., and Jayawerra K. (1988) Numerically stable algorithm for discrete-ordinate-method radiative transfer in multiple scattering and emitting layered media, *Appl. Opt.*, **27**, 2502–2509.

Stier P. Feichter J., Kinne S., *et al.* (2005) The aerosol–climate model ECHAM5-HAM, *Atmos. Chem. Phys.*, **5**, 1125–1156.

Whitby K. T. (1978) The physical characteristics of sulfur aerosols, *Atmos. Environ.*, **12**, 135–159.

Whitby E. R. and McMurry P. H. (1997) Modal aerosol dynamics modeling, *Aerosol Sci.Technol.*, **27**, 673–688.

Whitby E. R., McMurry P. H., Shankar U., and Binkowski F. S. (1991) *Modal Aerosol Dynamics Modeling*, Atmospheric Research and Exposure Assessment Laboratory, Research Triangle Park, NC.

Woods T. N., Prinz D. K., Rottman G. J., *et al.* (1996) Validation of the UARS solar ultraviolet irradiances: Comparison with the ATLAS 1 and 2 measurements, *J. Geophys. Res.*, **101** (D6), 9541–9569, doi: 10.1029/96JD00225.

Zhu X., Yee J.-H., Lloyd S. A. and Storbel D. F. (1999) Numerical modelling of chemical–dynamical coupling in the upper stratosphere and mesosphere, *J. Geophys. Res.*, **104**, 23995–24011, doi: 10.1029/1999JD900476.

Numerical Methods for Chemical Systems

6.1 Introduction

Solving the 3-D continuity equations for chemical species in atmospheric models requires splitting of the transport and chemistry operators. We present in this chapter an overview of numerical methods for the chemistry operator, which solves the evolution of the system driven by chemical kinetics independently of transport. Complexity arises from the large number of coupled species in standard mechanisms for atmospheric chemistry, with time constants ranging over many orders of magnitude. The associated computational requirements are very high and this is a major challenge for the inclusion of atmospheric chemistry in Earth system models.

For a chemical mechanism involving K chemically interacting species, the task of the chemistry operator is to solve the following initial value problem over time step Δt,

$$\frac{d\mathbf{\psi}}{dt} = \mathbf{s}(\mathbf{\psi}, t) \qquad (6.1)$$

where $\mathbf{\psi} = (\psi_1, \psi_2, \ldots, \psi_K)^T$ is the vector of concentrations for the K species with known initial value $\mathbf{\psi}(t_o)$, and $\mathbf{s}$ is a vector of chemical production and loss rates. Each component of $\mathbf{s}$ is a sum of terms describing the rates of individual reactions. Equation (6.1) describes a system of coupled ordinary differential equations (ODEs) with time as the only coordinate. There is no spatial dependence since the chemical evolution is a function of local concentrations only. Although we refer to solution of (6.1) as the "chemistry operator," $\mathbf{s}$ may also include non-chemical terms such as emission, precipitation scavenging, and dry deposition rates. Any local process independent of transport (and hence with no spatial dependence) can be included in the chemistry operator. When solving for aerosol microphysics, $\mathbf{\psi}$ may represent particle concentrations of different sizes with $\mathbf{s}$ including particle formation and growth terms (Section 4.3).

Nonlinearity arises in (6.1) because the rates of bimolecular and three-body reactions involve products of concentrations. This nonlinearity can be highlighted by rewriting (6.1) as:

$$\frac{d\mathbf{\psi}}{dt} = \mathbf{A}\mathbf{\psi} + \sum_{i=1}^{K} \left(\mathbf{\psi}^T \mathbf{Q}_i \mathbf{\psi}\right) \mathbf{e}_i + \mathbf{f} \qquad (6.2)$$

Here $\mathbf{A}$ is a $(K \times K)$ diagonal matrix of unimolecular reaction rate coefficients in the mechanism, $\mathbf{Q}_i$ is a $(K \times K)$ upper triangular matrix of bimolecular and three-body rate coefficients for reactions producing or consuming species i, $\mathbf{e}_i$ is the ith column of the identity matrix of dimension K (zeros except for 1 in row i), and $\mathbf{f} = (f_1, f_2, \ldots, f_K)^T$ is an independent forcing term. Nonlinearity is introduced by $\mathbf{\Psi}^T \mathbf{Q}_i \mathbf{\Psi}$, which is a summation of terms of form $q_{ijk} \Psi_j \Psi_k$ producing or consuming species i. The form $q_{ijk} \Psi_j \Psi_k$ applies to both bimolecular and three-body reaction rates since the concentration of the "third body" in a three-body reaction is not computed from the mechanism but is instead independently specified (see Chapter 5).

The general approach for obtaining $\mathbf{\psi}(t_o + \Delta t)$ from knowledge of $\mathbf{\psi}(t_o)$ is to use a finite difference approximation of the temporal derivative in (6.1). Differences lie in the way that $\mathbf{s}(\mathbf{\psi}, t)$ is estimated. Fast ODE solvers use *explicit* methods where $\mathbf{s}$ is calculated on the basis of the known concentrations at t_o and previous time steps. In these solvers, as we will see, the time step must be smaller than the shortest time constant in the system in order to maintain stability. This is a major obstacle for atmospheric chemistry applications because radical species central to the chemical mechanisms have very short lifetimes. The systems of ODEs describing atmospheric chemistry mechanisms are *stiff* (Box 6.1), that is, the time constants for the different

---

Box 6.1   **Stiff Systems of Equations**

A system of first-order differential equations in time

$$\frac{d\psi_i}{dt} = s_i(t, \mathbf{\psi}) \quad (i = 1, 2, \ldots K) \tag{I}$$

is said to be *stiff* if the timescales for change of the dependent variables ψ_i range over many orders of magnitude. We saw in Section 4.4 that the characteristic timescales of the system are defined by the inverses of the eigenvalues λ_i of the Jacobian matrix $\mathbf{J} = \partial\mathbf{s}/\partial\mathbf{\psi}$ where $\mathbf{s} = (s_1, \ldots s_K)^T$ and $\mathbf{\psi} = (\psi_1, \ldots \psi_K)^T$. Stability of the solution requires that the real component of the eigenvalues be negative and this is always met in relevant mechanisms for atmospheric chemistry (with only transient exceptions). Thus

$$\text{Re}(\lambda_i) < 0 \quad \text{for } i = 1, 2, \ldots K \tag{II}$$

We define the *stiffness ratio* R as

$$R = \frac{\max |\text{Re}(\lambda_i)|}{\min |\text{Re}(\lambda_i)|} >> 1 \tag{III}$$

A simple example of a stiff system is given by (Press *et al.*, 2007):

$$\frac{du}{dt} = 998 \ u + 1998 \ v$$

$$\frac{dv}{dt} = -999 \ u - 1999 \ v \tag{IV}$$

with the initial conditions $u(0) = 1$ and $v(0) = 0$. The two eigenvalues of this system are -1 and -1000, corresponding to characteristic timescales of 1 and 0.001, and a stiffness ratio of 10^3. The analytical solution is:

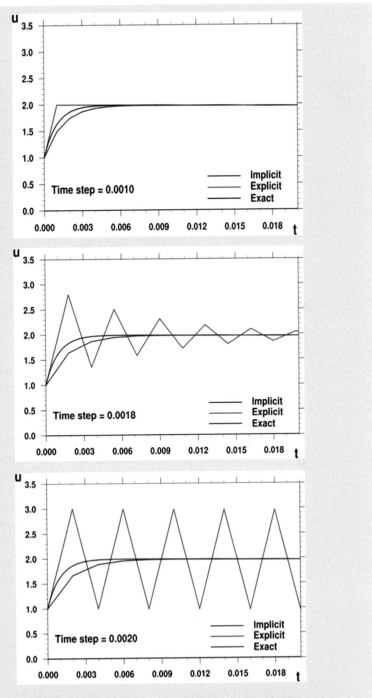

Box 6.1
Figure 1 Solution of stiff system (IV) for the range $0 < t < 0.02$ and three values of the time step Δt. The exact analytical solution for u is compared to numerical solutions by fully implicit and explicit methods and for different time steps.

Box 6.1 (*cont.*)

$$u(t) = 2e^{-t} - e^{-1000t}$$
$$v(t) = -e^{-t} + e^{-1000t} \tag{V}$$

Box 6.1 Figure 1 shows the exact solution for u for $t \in [0, 0.02]$ as well as numerical solutions by fully explicit and implicit algorithms with different time steps. u increases from 1 to 2 over the first characteristic timescale of 0.001, and then declines on the much longer timescale of 1 (not shown in the figure). The implicit algorithm is stable for all time steps and provides accurate solution of the asymptotic behavior for the first characteristic timescale, though not of the transient behavior. The explicit algorithm is stable when $\Delta t = 0.001$ but incurs oscillatory behavior at longer time steps that becomes undamped for $\Delta t = 0.002$. Even though the e^{-1000t} term becomes rapidly negligible as t increases, the explicit solution still requires a time step of less than 0.002 throughout the integration. This is burdensome if we are interested in computing the solution over long times ($t \gg 1$) in order to capture the second characteristic timescale of the system.

species coupled through the mechanism vary over many orders of magnitude. The stiffness is defined by the *stiffness ratio* $R = \tau_L/\tau_S$ where τ_L and τ_S are the longest and shortest time constants in the system, corresponding roughly to the longest and shortest lifetimes of species in the mechanism (see Section 4.4). Solution with an explicit solver requires time steps $\Delta t \sim \tau_S$, but we are generally interested in solutions over timescales $\sim \tau_L$. Thus the number of time steps required with an explicit solver is of order R. A typical mechanism for atmospheric chemistry may have stiffness $R \sim 10^8$, making for a formidable computational problem on a 3-D model grid.

An alternate approach is to use an *implicit* solver, where **s** in (6.1) is estimated on the basis of the unknown concentrations at time $t_o + \Delta t$. The system of coupled ODEs then becomes a system of K coupled algebraic equations to solve for $\psi(t_o + \Delta t)$. Implicit methods have far less severe restrictions on size of time step to remain stable. However, they require computationally expensive constructions to obtain the solution of the system of equations. High-order implicit solvers such as *Rosenbrock* and *Gear* are often used as standards of accuracy in 3-D models. Other algorithms provide a compromise between accuracy and computational performance.

The choice of a particular numerical algorithm for the chemistry operator depends on several considerations including stability, positivity, accuracy, mass conservation, and computational efficiency of the method (Zhang *et al.*, 2011). Positivity of the solution is essential for stability as otherwise the kinetics equations immediately diverge. This condition can severely restrict the size of the time step. Mass conservation is essential if quantitative tracking of chemical budgets is needed. Some slack in accuracy is often considered acceptable because the kinetics equations have stable solutions (Section 4.4) so that inaccuracies will dampen rather than grow. Computational efficiency may be of no concern for a box model but critically important for a 3-D model.

6.2 General Considerations

6.2.1 Fully Explicit Equation

The simplest method to solve (6.1) is the single step *forward Euler* or fully explicit algorithm:

$$\boldsymbol{\psi}^{n+1} = \boldsymbol{\psi}^n + \Delta t \, \mathbf{s}(t_n, \boldsymbol{\psi}^n) \tag{6.3}$$

where n is a time index, $t_n = t_0 + n \, \Delta t$ $(n = 0, 1, 2, \ldots)$ is the discretized time level, Δt is the integration time step, and $\boldsymbol{\psi}^n$ and $\boldsymbol{\psi}^{n+1}$ are approximations to the solution $\boldsymbol{\psi}(t)$ at time levels t_n and t_{n+1}, respectively. The source term $\mathbf{s}(t_n, \boldsymbol{\psi}^n)$, also noted $\mathbf{s}^n$, is evaluated at time t_n for the known approximation $\boldsymbol{\psi}^n$. This method, which is first-order accurate, is called fully explicit because the unknown $\boldsymbol{\psi}^{n+1}$ is represented strictly as a function of the known quantities at previous time t_n. It is mass-conserving but positivity is not guaranteed. Equation (6.3) is a single step algorithm because only time levels t_n and t_{n+1} are involved. Higher-order multi-step fully explicit schemes that express $\boldsymbol{\psi}^{n+1}$ as a function of the solution at previous time levels t_n, t_{n-1}, t_{n-2}, etc. are described in Section 6.2.3.

The explicit algorithm is appealingly simple but suffers from severe stability restrictions, as can be illustrated with a trivial example. Consider a single chemical species subject only to a linear loss. The corresponding differential equation is

$$\frac{d\psi}{dt} = -\ell \psi \tag{6.4}$$

with loss coefficient ℓ assumed to be constant. Applying the forward Euler algorithm, we have

$$\psi^{n+1} = \psi^n (1 - \ell \Delta t) \tag{6.5}$$

which is an approximation to the exact analytic solution

$$\psi(t_{n+1}) = \psi(t_n) \, \exp\left[-\ell \Delta t\right] \tag{6.6}$$

Stability requires that $\left| \psi^{n+1} / \psi^n \right| < 1$ as the integration proceeds. Equation (6.5) meets this stability criterion if

$$\Delta t < \frac{2}{\ell} \tag{6.7}$$

Positivity of the solution requires the more stringent criterion

$$\Delta t < \frac{1}{\ell} \tag{6.8}$$

Thus the time step for a system of several chemical species must be smaller than the chemical lifetime of the fastest reacting species.

Table 6.1 Solution of equation $d\psi/dt = -\psi$ at time $t = 2$ with $\psi(0) = 1$			
	Solution at time $t = 2$		
Time step	Exact	Fully explicit	Fully implicit
0.0001	0.1353	0.1353	0.1353
0.001	0.1353	0.1352	0.1355
0.01	0.1353	0.1340	0.1367
0.1	0.1353	0.1216	0.1486
1.0	0.1353	0.0000	0.2500
2.0	0.1353	−1.0000	0.3333

6.2.2 Fully Implicit Equation

The constraint on the time step associated with fully explicit methods can be overcome by using the *backward Euler* or fully implicit algorithm. In this case, the solution to (6.1) is approximated by

$$\boldsymbol{\psi}^{n+1} = \boldsymbol{\psi}^n + \Delta t \, \mathbf{s}\left(t_{n+1}, \boldsymbol{\psi}^{n+1}\right) \tag{6.9}$$

in which the source term $\mathbf{s}$ is evaluated at time t_{n+1} and is therefore expressed as a function of the unknown quantity $\boldsymbol{\psi}^{n+1}$. The fully implicit algorithm ensures the positivity of the solution and also conserves mass. The resulting system of algebraic equations in $\boldsymbol{\psi}^{n+1}$ requires numerical solution except in trivial cases.

In the simple linear example (6.4), the backward Euler scheme leads to

$$\psi^{n+1} = \frac{\psi^n}{(1 + \ell \, \Delta t)} \tag{6.10}$$

We have $\left|\psi^{n+1}/\psi^n\right| < 1$ for any positive value of Δt, thus the numerical scheme is unconditionally stable. The ratio tends to zero for large values of Δt, mirroring the analytic solution. The algorithm is first-order accurate; the error in the solution can be estimated by comparing the analytic solution $\exp\left[-\ell \, \Delta t\right]$ with the approximation $1/(1 + \ell \, \Delta t)$.

Table 6.1 shows the solution of (6.4) for $t = 2$, when ℓ is chosen to be 1 and the initial value is $\psi(0) = 1$. The exact (analytic) solution is compared to the approximate solution derived with the fully explicit and fully implicit algorithms for different values of the time step Δt. As expected, for both algorithms, the accuracy of the solution decreases as the time step increases. Although the implicit algorithm has the advantage of stability, that does not make it any more accurate.

A simple approach to solve implicit equation (6.9) is to linearize the source term $\mathbf{s}$ around the solution $\boldsymbol{\psi}$ at time t_n:

$$\mathbf{s}\left(t_{n+1}, \boldsymbol{\psi}^{n+1}\right) = \mathbf{s}(t_n, \boldsymbol{\psi}^n) + \mathbf{J}\left(\boldsymbol{\psi}^{n+1} - \boldsymbol{\psi}^n\right) \tag{6.11}$$

where $\mathbf{J} = \partial\mathbf{s}/\partial\boldsymbol{\psi}$ is the Jacobian matrix of partial derivatives estimated for $\boldsymbol{\psi} = \boldsymbol{\psi}^n$ with elements $J_{i,j} = \partial s_i/\partial\psi_j$. Using (6.11) in (6.9) we obtain

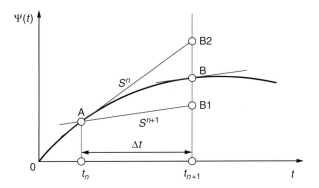

Figure 6.1 Determination of $\psi(t)$ at time level t_{n+1} from its known value at time t_n. The tangent A–B1 is proportional to the source term s at time t_{n+1}, while the tangent A–B2 is proportional to the source term at time t_n. Points B1 and B2 represent approximations to the true solution B obtained by the implicit and explicit Euler algorithms, respectively. The respective errors are defined by the distances B–B1 and B–B2.

$$\boldsymbol{\psi}^{n+1} = \boldsymbol{\psi}^n + \Delta t \left[\mathbf{I} - \mathbf{J}\Delta\mathbf{t}\right]^{-1} \mathbf{s}(t_n, \boldsymbol{\psi}^n) \tag{6.12}$$

in which $\mathbf{I}$ is the identity matrix. Applying this approach involves solving a matrix system of equations. This linearization method, when applied to an implicit equation, is called the *semi-implicit Euler method*. It is usually but not unconditionally stable. Other methods to solve implicit equations are discussed in Section 6.4.

6.2.3 Improving Accuracy

Both the Euler forward and backward methods are asymmetrical since the time derivatives are evaluated in one case at the beginning of the time interval and in the other case at the end of the interval. They are therefore only first-order accurate in Δt. Figure 6.1 illustrates the difference between the forward and backward methods, highlighting the errors incurred in the first-order approximation.

The accuracy of the solution $\boldsymbol{\psi}^{n+1}$ can be improved by making the solver more symmetric relative to time levels t_n and t_{n+1}. This can be done by taking the average of $\mathbf{s}$ between time levels t_n and t_{n+1}, which is equivalent to adopting a time-centered derivative:

$$\boldsymbol{\psi}^{n+1} = \boldsymbol{\psi}^n + \frac{\Delta t}{2} \left(\mathbf{s}(t_n, \boldsymbol{\psi}^n) + \mathbf{s}\left(t_{n+1}, \boldsymbol{\psi}^{n+1}\right)\right) \tag{6.13}$$

Equation (6.13) defines the *Crank–Nicholson scheme*. This semi-implicit algorithm is second-order accurate. Numerical solution is required as in the backward Euler fully implicit scheme. The solution is not guaranteed to be positive.

It is also possible to increase accuracy in the framework of an explicit algorithm by using *predictor-corrector* methods. In such a method, the prediction step derives a first estimate of the solution $(\mathbf{u}^{n+1})$ at time t_{n+1} from the forward Euler equation:

$$\mathbf{u}^{n+1} = \boldsymbol{\psi}^n + \mathbf{k}_1\Delta t \tag{6.14}$$

where the slope $\mathbf{k_1} = \mathbf{s}(t_n, \boldsymbol{\psi}^n)$ is calculated at time level t_n. The solution is improved by applying a correction step in which the slope is replaced by the average of $\mathbf{k_1}$ at time t_n and an estimate $\mathbf{k_2} = \mathbf{s}(t_{n+1}, \mathbf{u}^{n+1})$ at time t_{n+1}:

$$\boldsymbol{\psi}^{n+1} = \boldsymbol{\psi}^n + \frac{1}{2}(\mathbf{k}_1 + \mathbf{k}_2)\Delta t \qquad (6.15)$$

It can be shown that if the third derivative of the solution is a continuous function, this improved Euler method is a second-order scheme. The choice of the time step Δt remains constrained by the stability criteria of explicit methods.

In the *midpoint method*, the solution $\boldsymbol{\psi}^{n+1}$ is derived from a Euler formula in which $\mathbf{s}$ is estimated at an intermediate time level $t_{n+1/2} = t_n + \Delta t/2$. In this algorithm, the first step is to derive an estimate $\mathbf{u}^{n+1/2}$ of the solution at midpoint of interval Δt

$$\mathbf{u}^{n+1/2} = \boldsymbol{\psi}^n + \mathbf{k}_1 \frac{\Delta t}{2} \qquad (6.16)$$

with again $\mathbf{k}_1 = \mathbf{s}(t_n, \boldsymbol{\psi}^n)$. In the second step, the solution is computed using the entire time interval

$$\boldsymbol{\psi}^{n+1} = \boldsymbol{\psi}^n + \mathbf{k}_2 \Delta t \qquad (6.17)$$

where $\mathbf{k}_2 = \mathbf{s}(t_{n+1/2}, \mathbf{u}^{n+1/2})$ is an estimate of the source term at the midpoint between time levels t_n and t_{n+1}. Due to its symmetrical nature, the midpoint method is second-order accurate.

The accuracy of the solution can also be improved by applying an s-stage *Runge–Kutta method* defined by

$$\boldsymbol{\psi}^{n+1} = \boldsymbol{\psi}^n + \Delta t \sum_{i=1}^{s} b_i \, \mathbf{k}_i \qquad (6.18)$$

where

$$\mathbf{k}_i = \mathbf{s}\left(t + c_i \Delta t; \boldsymbol{\psi}^n + \Delta t \sum_{j=1}^{s} a_{i,j} \, \mathbf{k}_j\right) \qquad (6.19)$$

with $b_i(i = 1, \ldots s)$ and $a_{i,j}(i, j = 1, \ldots s)$ chosen to meet desired accuracy and stability conditions, and $c_i = \Sigma_j \, a_{i,j}$. If all coefficients $a_{i,j} \neq 0$, the method is fully implicit and numerically highly stable. Most applications, however, are based on the explicit Runge–Kutta method in which coefficients $a_{i,j} = 0$ for $j \geq i$. The method is less robust than the implicit version but is easier to apply. In the case of the classic explicit *fourth-order Runge–Kutta method*, for example, the solution at time t_{n+1} is provided by

$$\boldsymbol{\psi}^{n+1} = \boldsymbol{\psi}^n + \frac{\Delta t}{6}[\mathbf{k}_1 + 2\mathbf{k}_2 + 2\mathbf{k}_3 + \mathbf{k}_4] \qquad (6.20)$$

where

- $\mathbf{k}_1 = \mathbf{s}(t_n, \boldsymbol{\psi}^n)$ represents the source term at time level t_n,
- $\mathbf{k}_2 = \mathbf{s}(t_n + \Delta t/2, \boldsymbol{\psi}^n + \mathbf{k}_1 \Delta t/2)$ denotes a first estimate of the source term at the midpoint of the interval $[t_n, t_{n+1}]$,

- $k_3 = s(t_n + \Delta t/2, \psi^n + k_2 \Delta t/2)$ represents an improved estimate of the source term at the midpoint, and
- $k_4 = s(t_{n+1}, \psi^n + k_3 \Delta t)$ estimates the source term at time level t_{n+1}, using the value of k_3 calculated at the midpoint.

Explicit Runge–Kutta methods are more stable than the forward Euler algorithm. They are usually implemented with an adaptive step size procedure to meet a user-required error tolerance. As in other explicit methods, the time step is constrained by the shortest lifetime in the system and this can make implementation for atmospheric problems impractical. Implicit Runge–Kutta methods (Hairer *et al.*, 2002) are characterized by high stability, but the resulting system of equations is difficult to solve. The diagonally implicit Runge–Kutta method (*DIRK*), in which $a_{i,j} = 0$ for $j > i$, but the diagonal elements $a_{ii} \neq 0$, is simpler to implement than the fully implicit case. The *RADAU5* solver, introduced by Hairer *et al.* (1993) and implemented in some chemical models, is a one-step implicit Runge–Kutta method that is fifth-order accurate.

Linear multi-step algorithms, which retain the information from several previous time steps (t_n, t_{n-1}, t_{n-2}, etc.), also provide solutions with higher order of accuracy. They can be expressed either by an explicit expression

$$\psi^{n+1} = \psi^n + \Delta t \sum_{j=0}^{s-1} b_j \, s\left(t_{n-j}, \psi^{n-j}\right) \tag{6.21}$$

or by an implicit expression

$$\psi^{n+1} = \psi^n + \Delta t \sum_{j=-1}^{s-2} b_j \, s\left(t_{n-j}, \psi^{n-j}\right) \tag{6.22}$$

where b_j are constant coefficients and s corresponds to the order of the method. When combined, these two relations represent a multi-step predictor-corrector scheme. In this case, function $s(t_{n+1}, \psi^{n+1})$ appearing in the correction step (6.22) is estimated by using the values of ψ^{n+1} derived by the prediction step (6.21). An example is the *Adams–Bashforth–Moulton* scheme. For a value $s = 3$, for example, the prediction step is

$$\mathbf{u}^{n+1} = \psi^n + \frac{\Delta t}{12} \left[23 \, s(t_n, \psi^n) - 16 \, s\left(t_{n-1}, \psi^{n-1}\right) + 5 \, s\left(t_{n-2}, \psi^{n-2}\right)\right] \tag{6.23}$$

and the correction step is

$$\psi^{n+1} = \psi^n + \frac{\Delta t}{12} \left[5 \, s\left(t_{n+1}, \mathbf{u}^{n+1}\right) + 8 \, s(t_n, \psi^n) - s\left(t_{n-1}, \psi^{n-1}\right)\right] \tag{6.24}$$

Even though (6.22) is an implicit expression, the introduction of a predictor-corrector approach transforms the scheme into a fully explicit scheme with the associated stability requirements.

Accuracy can also be improved by applying the *extrapolation method* introduced by Lewis Fry Richardson. This method allows the construction of high-order solutions by applying the same algorithm with decreasing time steps. It is based on asymptotic expansion of the truncation error in

h-powers, where h is the time step. The approximate solution $\psi^{n+1}(h)$ at time t_{n+1} derived with some numerical algorithm L_h and time step h can be expressed as

$$L_h\{\psi^n\} = \psi^{n+1}(h) = \psi(t_{n+1}) + \mathbf{E}_m(\psi)h^m + \mathbf{O}(h^{m+1}) \qquad (6.25)$$

where $\psi(t_{n+1})$ represents the true solution at time t_{n+1}. Index m represents the order of the scheme, while $\mathbf{E}_m(\psi)h^m$ and $\mathbf{O}(h^{m+1})$ are the errors associated with algorithm L_h of order h^m and h^{m+1}, respectively. When the same algorithm is applied with smaller time steps h/k ($k = 1,2,3, \ldots$), we write similarly

$$L_{h/k}\{\psi^n\} = \psi^{n+1}(h/k) = \psi(t_{n+1}) + \mathbf{E}_m(\psi)\left[\frac{h}{k}\right]^m + \mathbf{O}(h^{m+1}) \qquad (6.26)$$

Combining (6.25) and (6.26) yields Richardson's recurrence formula that provides a higher-order approximation

$$\psi^{n+1} = \frac{k^m\,\psi^{n+1}(h/k) - \psi^{n+1}(h)}{k^m - 1} \qquad (6.27)$$

If, for example, $k = 2$ and if $\psi^{n+1}(h)$ and $\psi^{n+1}(h/2)$ are the numerical solutions obtained by a first-order algorithm ($m = 1$) with time steps h and $h/2$, respectively, the accuracy of the solution is improved by using

$$\psi^{n+1} = 2\,\psi^{n+1}(h/2) - \psi^{n+1}(h) \qquad (6.28)$$

Examples of extrapolation methods are given in Sections 6.3.1 and 6.3.3.

6.2.4 Explicit Versus Implicit Solvers

The comparison between fully explicit and fully implicit methods highlights the advantages and disadvantages of both approaches (Sandu et al., 1997b). Fully explicit equations are usually simple to solve, but stability and positivity considerations may constrain the integration time steps to prohibitively small values. Fully implicit methods are unconditionally stable and positive, so that the time step can be large, limited by accuracy requirements. However, they require solution of a system of algebraic equations at each time step, involving in general the construction and inversion of a Jacobian matrix. This can be computationally costly. Methods have been developed to reduce the stiffness of chemical systems by separating species between short-lived and long-lived and solving for each group separately, with an implicit method used for the short-lived subset only (Gong and Cho, 1993). However, separation is often difficult because atmospheric chemical mechanisms typically involve a continuum of lifetimes and the lifetimes vary with the local conditions. Adaptive separation can be done locally within a model simulation by calculating species lifetimes before applying the chemistry operator (Santillana et al., 2010), but this involves substantial computational overhead.

6.3 Explicit Solvers

We first present several explicit algorithms that use various methods to relax the requirement for short time steps while keeping the computational advantage of the explicit solution. We rewrite the system of equations to separate production and loss terms as

$$\frac{d\boldsymbol{\psi}}{dt} = \mathbf{p}(t, \boldsymbol{\psi}) - \mathbf{L}(t, \boldsymbol{\psi})\boldsymbol{\psi} \tag{6.29}$$

where the vector $\mathbf{p}$ and diagonal matrix $\mathbf{L}$ are functions of the unknown concentrations $\boldsymbol{\psi}$. We denote p_k^n as the production $p_k(t_n, \boldsymbol{\psi}^n)$ and ℓ_k^n as the loss coefficient $\ell_k(t_n, \boldsymbol{\psi}^n)$ for species k at time t_n. The loss rate of species k is generally a linear function of its concentration, hence the utility of separating out the loss coefficient.

6.3.1 Exponential Approximation

The *exponential method*, one of the earliest methods used to treat chemical processes in atmospheric models, is motivated by the form of (6.29), which has a trivial exponential solution if $\mathbf{p}$ and $\mathbf{L}$ are constant. Assuming that $\mathbf{p}$ and $\mathbf{L}$ can be approximated as constant over the time step Δt, we obtain the following explicit expression for each species k:

$$\psi_k^{n+1} = \psi_k^n \exp\left[-\ell_k^n \Delta t\right] + \left(1 - \exp\left[-\ell_k^n \Delta t\right]\right) \frac{p_k^n}{\ell_k^n} \tag{6.30}$$

The solution provided by this first-order algorithm is positive for any integration time step, and the algorithm does not require any matrix manipulation. However, the method does not conserve mass and it requires small time steps to be accurate. The accuracy can be improved by considering the implicit form of the exponential approximation

$$\psi_k^{n+1} = \psi_k^n \exp\left[-\ell_k^{n+1} \Delta t\right] + \left(1 - \exp\left[-\ell_k^{n+1} \Delta t\right]\right) \frac{p_k^{n+1}}{\ell_k^{n+1}} \tag{6.31}$$

This equation can be easily solved by an iteration procedure, starting from an initial iterate $\psi_k^{n+1}{}_{(0)} = \psi_k^n$. The number of iterations required to ensure a given level of accuracy may be different for the different chemical species within the system.

An *extrapolated* form of the exponential approximation proposed by Jay *et al.* (1995) provides a second-order accurate algorithm. Omitting index k, we compute a first estimate of the solution at time t_{n+1} following (6.30)

$$\psi_{\Delta t}^{n+1} = \psi^n \exp\left[-\ell^n \Delta t\right] + \left(1 - \exp\left[-\ell^n \Delta t\right]\right) \frac{p^n}{\ell^n} \tag{6.32}$$

A second estimate of ψ^{n+1} is derived by a two-step integration using time step $\Delta t/2$

$$\psi^{n+1/2} = \psi^n \exp\left[-\ell^n \frac{\Delta t}{2}\right] + \left(1 - \exp\left[-\ell^n \frac{\Delta t}{2}\right]\right) \frac{p^n}{\ell^n} \tag{6.33}$$

$$\psi_{\Delta t/2}^{n+1} = \psi^{n+1/2} \exp\left[-\ell^{n+1/2}\frac{\Delta t}{2}\right] + \left(1 - \exp\left[-\ell^{n+1/2}\frac{\Delta t}{2}\right]\right)\frac{p^{n+1/2}}{\ell^{n+1/2}} \qquad (6.34)$$

The solution at time t_{n+1} is then given by the extrapolation relation (6.28):

$$\psi^{n+1} = 2\psi_{\Delta t/2}^{n+1} - \psi_{\Delta t}^{n+1} \qquad (6.35)$$

6.3.2 Quasi Steady-State Approximation

The computation of exponential functions in the algorithms described in Section 6.3.1 requires substantial amounts of computer time. In a scheme proposed by Hesstvedt *et al.* (1978), called *Quasi Steady State Approximation (QSSA)*, classification of species by lifetime reduces the number of exponential functions. The species are separated according to their e-folding time (chemical lifetime $\tau_k = 1/\ell_k$), and different algorithms are applied:

- For long-lived species with $\tau_k > 100\ \Delta t$, a fully explicit Euler forward algorithm:

$$\psi_k^{n+1} = \psi_k^n + \Delta t\left(p_k^n - \ell_k^n \psi_k^n\right) \qquad (6.36)$$

- For intermediate-lived species with $0.1\ \Delta t < \tau_k < 100\ \Delta t$, an exponential approximation:

$$\psi_k^{n+1} = \psi_k^n \exp\left[-\ell^n \Delta t\right] + \left(1 - \exp\left[-\ell^n \Delta t\right]\right)\frac{p_k^n}{\ell_k^n} \qquad (6.37)$$

- For short-lived species with $\tau_k < 0.1\ \Delta t$, a steady-state value:

$$\psi_k^{n+1} = \frac{p_k^n}{\ell_k^n} \qquad (6.38)$$

This method is more efficient than the pure exponential solver. Its accuracy is highly dependent on the choice of the integration time step.

6.3.3 Extrapolation Technique (ET)

As described in Section 6.2.3, the extrapolation method combines the solutions obtained by a low-order algorithm with different time steps using Richardson's recurrence formula. In the extrapolation technique proposed by Dabdub and Seinfeld (1995), also called *ET solver*, the numerical algorithm is a predictor-corrector scheme. The predictor, which calculates a first estimate ψ_k^* of the solution at time t_{n+1}, is provided by the explicit exponential formula (6.30)

$$\psi_k^* = \psi_k^n \exp\left[-\ell^n \Delta t\right] + \left(1 - \exp\left[-\ell^n \Delta t\right]\right)\frac{p_k^n}{\ell_k^n} \qquad (6.39)$$

The formula for the corrector is chosen according to the lifetime $\tau_k = 1/\ell_k$ of chemical species k.

- For long-lived species ($\tau_k > 100\ \Delta t$), one adopts the trapezoidal rule

$$\psi_k^{n+1} = \psi_k^n + \frac{\Delta t}{2}\left(p_k^n - \ell_k^n \psi_k^n + p_k^* - \ell_k^* \psi_k^*\right) \qquad (6.40)$$

- For intermediate species ($0.1\,\Delta t < \tau_k < 100\,\Delta t$), the corrector uses an exponential form

$$\psi_k^{n+1} = Z_k^* + \left(\psi_k^* - Z_k^*\right) \exp\left[-\left(\frac{1}{\ell_k^n} + \frac{1}{\ell_k^*}\right)\frac{\Delta t}{2}\right] \tag{6.41}$$

where Z_k^* is defined as

$$Z_k^* = \frac{1}{4}\left(p_k^n + p_k^*\right)\left(\frac{1}{\ell_k^n} + \frac{1}{\ell_k^*}\right) \tag{6.42}$$

- For short-lived species ($\tau_k < 0.1\,\Delta t$)

$$\psi_k^{n+1} = Z_k^* \tag{6.43}$$

The correctors can be iterated until the relative difference between successive approximations becomes smaller than a user-imposed tolerance.

6.3.4 CHEMEQ Solver

In the CHEMEQ solver proposed by Young and Boris (1977) and as implemented by Saylor and Ford (1995), a distinction is made again between chemical species according to their lifetime $\tau_k = 1/\ell_k$. The corrector formulas are derived from the implicit trapezoidal rule, but applied in an explicit way.

- For long-lived species ($\tau_k > 5\,\Delta t$), we use:
  Predictor:

$$\psi_k^* = \psi_k^n + \Delta t\left(p_k^n - \ell_k^n\,\psi_k^n\right) \tag{6.44}$$

  Corrector:

$$\psi_k^{n+1} = \psi_k^n + \frac{\Delta t}{2}\left(p_k^n - \ell_k^n\psi_k^n + p_k^* - \ell_k^*\psi_k^*\right) \tag{6.45}$$

- For intermediate species ($0.2\,\Delta t < \tau_k < 5\,\Delta t$) we use the more accurate asymptotic integration formula:
  Predictor:

$$\psi_k^* = \frac{\left(2\tau_k^n - \Delta t\right)\psi_k^n + 2\Delta t\,p_k^n\,\tau_k^n}{2\tau_k^n + \Delta t} \tag{6.46}$$

  Corrector:

$$\psi_k^{n+1} = \frac{\left(\tau_k^n + \tau_k^* - \Delta t\right)\psi_k^n + \frac{\Delta t}{2}\left(p_k^n + p_k^*\right)\left(\tau_k^n + \tau_k^*\right)}{\tau_k^n + \tau_k^* + \Delta t} \tag{6.47}$$

- For short-lived species ($\tau_k < 0.2\,\Delta t$), steady state is assumed:

$$\psi_k^{n+1} = \frac{p_k^n}{\ell_k^n} \tag{6.48}$$

Iterations on the corrector are performed until convergence is reached.

6.3.5 TWOSTEP method

The TWOSTEP method (Verwer, 1994) is based on the second-order backward differentiation formula (see Table 4.2):

$$\frac{3\psi^{n+1} - 4\psi^n + \psi^{n-1}}{2\Delta t} = s^{n+1} \tag{6.49}$$

or

$$\psi^{n+1} = y^n + \frac{2}{3}\Delta t\, s^{n+1} \tag{6.50}$$

with

$$y^n = \frac{4}{3}\psi^n - \frac{1}{3}\psi^{n-1} \tag{6.51}$$

Again s^{n+1} represents the source term $s(t_{n+1}, \psi^{n+1})$. This algorithm is a two-step method. The solution at time t_{n+1} is expressed as a function of the solutions ψ^n and ψ^{n-1} at times t_n and t_{n-1}. When the source term is replaced by the rate of production p and the loss coefficient matrix L, the solution becomes

$$\psi^{n+1} = \left(I + \frac{2}{3}\Delta t\, L^{n+1}\right)^{-1} \left(y^n + \frac{2}{3}\Delta t\, p^{n+1}\right) \tag{6.52}$$

with I being the identity matrix. The value of ψ^{n+1} can be obtained by applying an iterative procedure provided, for example, by the Jacobi or Gauss–Seidel method (see Box 6.2).

The TWOSTEP method is second-order accurate and does not require matrix manipulation. The solution is always positive and approaches its steady-state value for large time steps. Mass is not fully conserved by the Jacobi and Gauss–Seidel iterative procedures (Box 6.2).

Box 6.2   **Solutions of Linear Algebraic Equations: LU Decomposition, Jacobi and Gauss–Seidel Iteration**

Different numerical methods are available to solve a system of N linear equations

$$Ay = b \tag{I}$$

In the *LU* decomposition method, matrix A with elements a_{ij} is decomposed into the product of two matrices L and U

$$A = L\,U \tag{II}$$

where L, the lower triangular matrix, includes non-zero elements l_{ij} only on the diagonal and below, and U, the upper triangular matrix, includes non-zero elements u_{ij} only on the diagonal and above. If A, L, and U are 3×3 matrices, equation (II) is written

$$\begin{pmatrix} a_{1,1} & a_{1,2} & a_{1,3} \\ a_{2,1} & a_{2,2} & a_{2,3} \\ a_{3,1} & a_{3,2} & a_{3,3} \end{pmatrix} = \begin{pmatrix} l_{1,1} & 0 & 0 \\ l_{2,1} & l_{2,2} & 0 \\ l_{3,1} & l_{3,2} & l_{3,3} \end{pmatrix} \begin{pmatrix} u_{1,1} & u_{1,2} & u_{1,3} \\ 0 & u_{2,2} & u_{2,3} \\ 0 & 0 & u_{3,3} \end{pmatrix}$$

System (I) becomes

$$\mathbf{Ay} = (\mathbf{LU})\mathbf{y} = \mathbf{L}\,(\mathbf{Uy}) = \mathbf{b} \tag{III}$$

Its solution is found by solving sequentially the two triangular systems

$$\mathbf{Lz} = \mathbf{b} \qquad \text{and} \qquad \mathbf{Uy} = \mathbf{z} \tag{IV}$$

first by forward substitution

$$z_1 = \frac{b_1}{l_{1,1}} \quad \text{and} \quad z_i = \frac{1}{l_{i,i}}\left[b_i - \sum_{j=1}^{i-1} l_{i,j} z_j \right] \quad (i = 2, 3, \dots, N)$$

and then by back-substitution

$$y_N = \frac{b_N}{u_{N,N}} \quad \text{and} \quad y_i = \frac{1}{u_{i,i}}\left[b_i - \sum_{j=i+1}^{N} u_{i,j} y_j \right] \quad (i = N-1, N-2, \dots, 1)$$

The decomposition of matrix $\mathbf{A}$ into triangular matrices $\mathbf{L}$ and $\mathbf{U}$ is performed by deriving the values of $l_{i,j}$ and $u_{i,j}$ from the N^2 equations of system (II). Since this system includes $N^2 + N$ unknowns, N of them can be specified: For example, the diagonal elements in one of the triangular matrices can be set equal to 1.

The Jacobi and Gauss–Seidel methods can be described as follows (Press *et al.*, 2007). We first split matrix $\mathbf{A}$ into its diagonal part $\mathbf{D}$, its lower triangle $\mathbf{L}$ part (with zeros on the diagonal) and its upper triangle $\mathbf{U}$ part (also with zeros on the diagonal). Thus, we write

$$\mathbf{A} = \mathbf{L} + \mathbf{D} + \mathbf{U} \tag{V}$$

In the *Jacobi* iteration method, we write for iteration step $(r + 1)$

$$\mathbf{Dy}_{(r+1)} = -(\mathbf{L} + \mathbf{U})\mathbf{y}_{(r)} + \mathbf{b} \tag{VI}$$

$$\mathbf{y}_{(r+1)} = \mathbf{y}_{(r)} - \mathbf{D}^{-1}\left[\mathbf{Ay}_{(r)} - \mathbf{b} \right] \tag{VII}$$

The value of $\mathbf{y}_{(r+1)}$ can easily be derived since $\mathbf{D}$ is a diagonal matrix. The method converges slowly and is most effective when matrices $\mathbf{A}$ are dominated by diagonal terms.

In the *Gauss–Seidel* method, iteration $(r + 1)$ is expressed by

$$(\mathbf{L} + \mathbf{D})\mathbf{y}_{(r+1)} = -\mathbf{U}\,\mathbf{y}_{(r)} + \mathbf{b} \tag{VII}$$

or

$$\mathbf{y}_{(r+1)} = \mathbf{y}_{(r)} - (\mathbf{L} + \mathbf{D})^{-1}\left[\mathbf{Ay}_{(r)} - \mathbf{b} \right] \tag{VIII}$$

Box 6.2 *(cont.)*

This method leads to an algorithm in which the updated values for the individual components of vector **y** are used to derive the solutions of the next components of the same vector. In the successive over-relaxation method, these iterations can be accelerated by multiplying the correction vector [**A y**$_{(r)}$ − **b**] in (VII) by an over-relaxation parameter whose value is generally chosen to be between 1 and 2. In this range of values, the method is convergent.

6.4 Implicit Solvers

We now examine a few frequently used implicit integrators. As indicated earlier, implicit solvers are robust for solving stiff systems. They require, however, computationally expensive matrix manipulations. Information on the stability of these methods is provided in Appendix E.

6.4.1 Backward Euler

In the *backward Euler method*,

$$\psi^{n+1} = \psi^n + \mathbf{s}\left(t_{n+1}, \psi^{n+1}\right)\Delta t \tag{6.53}$$

the solution ψ^{n+1} is obtained by determining the roots of the K-valued vector function

$$\mathbf{g}\left(\psi^{n+1}\right) = \psi^{n+1} - \psi^n - \mathbf{s}\left(t_{n+1}, \psi^{n+1}\right)\Delta t = 0 \tag{6.54}$$

where K is the number of species in the system. Solution can be obtained with the *Newton–Raphson* iteration method. In this case, function $\mathbf{g}(\psi^{n+1})$ at iteration $(r + 1)$ is developed as a Taylor series about a previous estimate of the solution $\psi_{(r)}^{n+1}$ at iteration (r). Thus

$$\mathbf{g}\left(\psi_{(r+1)}^{n+1}\right) = \mathbf{g}\left(\psi_{(r)}^{n+1}\right) + \mathbf{J}\left(\psi_{(r+1)}^{n+1} - \psi_{(r)}^{n+1}\right) + \cdots \tag{6.55}$$

Here **J** is the Jacobian matrix whose elements are given by $J_{i,j} = \partial g_i / \partial \psi_j$. Neglecting the higher-order terms, the value $\psi_{(r+1)}^{n+1}$ for which $\mathbf{g}(\psi_{(r+1)}^{n+1}) = 0$ is derived by

$$\psi_{(r+1)}^{n+1} = \psi_{(r)}^{n+1} - \mathbf{J}^{-1}\mathbf{g}\left(\psi_{(r)}^{n+1}\right) \tag{6.56}$$

with $\psi_{(0)}^{n+1} = \psi^n$ as the initial iterate. The iteration proceeds until convergence is reached to within a user-prescribed tolerance. The Newton–Raphson iteration conserves mass when the analytic form of the Jacobian matrix is used and recalculated for each iteration; this property may be lost, however, when approximations for the Jacobian are used.

The backward Euler method requires repeated construction and inversion of the Jacobian matrix. Inversion can be sped up by noting that the matrix is usually *sparse* (matrix with many zero elements) as many pairs of chemical species are not directly coupled. Various methods exist for computationally efficient inversion of sparse matrices (see, for example, Press *et al.*, 2007). In cases when the interactions between groups of species are weak, the Jacobian matrix can be broken into smaller matrices enabling more efficient solution (Hertel *et al.*, 1993; Sandilands and McConnell, 1997).

In another approach proposed by Shimazaki (1985), the chemical source term $\mathbf{s}$ is linearized as follows:

$$\mathbf{s} = \mathbf{p} - \mathbf{L}\boldsymbol{\psi} \tag{6.57}$$

where $\mathbf{p}$ is a vector of production rates and $\mathbf{L}$ is a diagonal matrix of loss rate coefficients. We then write

$$\boldsymbol{\psi}^{n+1} = \boldsymbol{\psi}^n + \Delta t \left[\mathbf{p}\left(t_{n+1}, \boldsymbol{\psi}^{n+1}\right) - \mathbf{L}\left(t_{n+1}, \boldsymbol{\psi}^{n+1}\right)\boldsymbol{\psi}^{n+1} \right] \tag{6.58}$$

The solution can be obtained by applying an iterative procedure:

$$\boldsymbol{\psi}^{n+1}_{(r+1)} = \left[\mathbf{I} + \Delta t\, \mathbf{L}^{n+1}_{(r)} \right]^{-1} \left[\boldsymbol{\psi}^n + \Delta t\, \mathbf{p}^{n+1}_{(r)} \right] \tag{6.59}$$

if $\mathbf{p}^{n+1} = \mathbf{p}(t_{n+1}, \boldsymbol{\psi}^{n+1})$, $\mathbf{L}^{n+1} = \mathbf{L}(t_{n+1}, \boldsymbol{\psi}^{n+1})$. $\mathbf{I}$ denotes the identity matrix, and (r) represents the iteration index ($r = 0, 1, 2, \ldots$). The initial iteration uses $\boldsymbol{\psi}^{n+1}_{(0)} = \boldsymbol{\psi}^n$. Convergence restrictions on the adopted time step Δt depend on the functional forms of vector $\mathbf{p}$ and matrix $\mathbf{L}$. Convergence may be difficult when the chemical coupling between the different species included in the system is strong. Linearization affects mass conservation but the situation is improved when the quadratic terms such as $k\,\psi_1\,\psi_2$ are linearized as $k(\psi_1^{n+1}\,\psi_2^n + \psi_1^n\,\psi_2^{n+1})/2$ and linear terms such as $k\,\psi$ are expressed as $k(\psi^{n+1} + \psi^n)/2$ (Ramaroson *et al.*, 1992).

A particularly useful feature of the backward Euler method (6.53) is its flexibility in the choice of chemical constraints applied to the system of coupled species. This makes it attractive for analysis of chemical mechanisms using box models where computational requirements are not a concern. The functions $\mathbf{g}(\boldsymbol{\psi}^{n+1})$ that are used to define the solution system do not necessarily need to be finite difference forms of the chemical kinetic equations. They can be any constraint that we choose. For example, steady-state solution of the system is obtained by using

$$\mathbf{g}\left(\boldsymbol{\psi}^{n+1}\right) = \mathbf{p}^{n+1} - \mathbf{L}^{n+1}\boldsymbol{\psi}^{n+1} = 0 \tag{6.60}$$

Individual constraints can also be applied to any particular species or groups of species. For example, we might want to impose conservation of the sum ψ_T of concentrations for a family of species $j = 1, \ldots q$:

$$g_k\left(\boldsymbol{\psi}^{n+1}\right) = \psi_T - \sum_{j=1}^{q} \psi_j^{n+1} = 0 \tag{6.61}$$

Here, (6.61) replaces the kinetic equation for one chosen member of the chemical family. This allows chemical cycling within the family while holding constant the total concentration of the family (see Box 3.1). For example, one may impose a fixed concentration of $NO_x \equiv NO + NO_2$ in the model while allowing the individual concentrations of NO and NO_2 to change. This is done by replacing the kinetic equation for NO_2 by the NO_x mass conservation equation ($[NO] + [NO_2] = [NO_x]$, where $[NO_x]$ is imposed). The chemical kinetic equation is retained for NO. Other potentially useful constraints that can be expressed by the form $g_k(\boldsymbol{\psi}^{n+1}) = 0$ include chemical equilibria between species, charge balance for aqueous-phase ion chemistry, etc.

6.4.2 Rosenbrock Solvers

The *Rosenbrock methods* (Rosenbrock, 1963), which can be regarded as a general-ization of the Runge–Kutta methods, are non-iterative implicit algorithms that are particularly well adapted to stiff systems. If we apply only one Newton–Raphson iteration to the full implicit algorithm with $\boldsymbol{\psi}^n$ being the initial iterate, we obtain

$$\boldsymbol{\psi}^{n+1} = \boldsymbol{\psi}^n + \mathbf{k}\,\Delta t \tag{6.62}$$

and solve

$$\mathbf{k} = \mathbf{s}(\boldsymbol{\psi}^n) + \mathbf{J}\,\mathbf{k}\,\Delta t \tag{6.63}$$

where $\mathbf{J}$ is the Jacobian matrix of the chemical source function $\mathbf{s}$. The idea behind the Rosenbrock methods (Hairer and Wanner, 1996) is to derive stable integration formulas that generalize expressions (6.62) and (6.63) and use s stages to achieve a high order of accuracy (i.e., high-order method). An s-stage Rosenbrock method applied to an autonomous problem $d\boldsymbol{\psi}/dt = \mathbf{s}(\boldsymbol{\psi})$ seeks a solution of the form

$$\boldsymbol{\psi}^{n+1} = \boldsymbol{\psi}^n + \Delta t \sum_{i=1}^{s} b_i\,\mathbf{k}_i \tag{6.64}$$

with s linear equations

$$\mathbf{k}_1 = [\mathbf{s}(\boldsymbol{\psi}^n) + \mathbf{J}\,\Delta t \gamma_{11}\mathbf{k}_1]$$

$$\mathbf{k}_i = \left[\mathbf{s}\left(\boldsymbol{\psi}^n + \Delta t \sum_{j=1}^{i-1}\alpha_{ij}\mathbf{k}_j\right) + \mathbf{J}\,\Delta t \sum_{j=1}^{i}\gamma_{ij}\mathbf{k}_j\right] \quad \text{for } 2 \le i \le s \tag{6.65}$$

that can be rearranged as

$$[\mathbf{I} - \Delta t\,\mathbf{J}\,\gamma_{ii}]\mathbf{k}_i = \left[\mathbf{s}\left(\boldsymbol{\psi}^n + \Delta t \sum_{j=1}^{i-1}\alpha_{ij}\mathbf{k}_j\right) + \mathbf{J}\,\Delta t \sum_{j=1}^{i-1}\gamma_{ij}\mathbf{k}_j\right] \tag{6.66}$$

where $\mathbf{I}$ is again the identity matrix. The method-specific coefficients b_i, α_{ij}, and γ_{ij} are fixed constants independent of the problem, chosen to obtain a desired order of accuracy and to ensure stability for stiff problems. Equation (6.66) can be solved successively for $\mathbf{k}_1$, $\mathbf{k}_2$, ..., $\mathbf{k}_s$, using, for example, an LU decomposition process or, when possible, by a suitable sparse matrix procedure. A comprehensive treatment of the Rosenbrock methods is provided by Hairer and Wanner (1996). See also Rosen-brock (1963) and Press *et al.* (2007).

For a non-autonomous system $d\boldsymbol{\psi}/dt = \mathbf{s}(t, \boldsymbol{\psi})$, the definition of $\mathbf{k}_i$ in expression (6.65) is changed to

$$\mathbf{k}_i = \left[\mathbf{s}\left(t_n + \alpha_i\Delta t, \boldsymbol{\psi}^n + \Delta t \sum_{j=1}^{i-1}\alpha_{ij}\mathbf{k}_j\right) + \gamma_i\mathbf{J}\,(\Delta t)^2 + \mathbf{J}\,\Delta t \sum_{j=1}^{i}\gamma_{ij}\mathbf{k}_j\right]$$

$$\alpha_i = \sum_{j=1}^{i-1}\alpha_{ij} \quad \text{and} \quad \gamma_i = \sum_{j=1}^{i}\gamma_{ij}$$

The Rosenbrock solvers are one-step algorithms. Like fully implicit methods, they conserve mass during the integration if the true analytic form of the Jacobian is used.

Positivity of the solution is not guaranteed. The Rosenbrock solvers, like the Runge–Kutta solvers, form successive results that approximate the solution at intermediate time levels. A disadvantage of the Rosenbrock solvers is that they require an evaluation of the Jacobian at each time step, several matrix vector multiplications, and the resolution of a linear system. The cost of the method, however, can be reduced (Sandu *et al.*, 1997a) by keeping the Jacobian unchanged during several time steps of the integration, and by approximating the Jacobian by a matrix of higher sparsity (this option will not preserve mass conservation).

An example of a Rosenbrock method is the second-order *ROS2 solver* ($s = 2$):

$$\psi^{n+1} = \psi^n + \frac{1}{2}\mathbf{k}_1\Delta t + \frac{1}{2}\mathbf{k}_2\Delta t \tag{6.67}$$

with

$$\begin{aligned}
\mathbf{k}_1 &= \left[\mathbf{s}(\psi^n) + \gamma\,\mathbf{J}\,\mathbf{k}_1\Delta t \right) \right] \\
\mathbf{k}_2 &= \left[\mathbf{s}(\psi^n + \mathbf{k}_1\Delta t) + \mathbf{J}\left(-2\gamma\,\mathbf{k}_1\Delta t + \gamma\,\mathbf{k}_2\Delta t \right) \right]
\end{aligned} \tag{6.68}$$

To maximize stability, the value $\gamma = 1 + 1/\sqrt{2}$ is recommended (Verwer *et al.*, 1999).

Another Rosenbrock algorithm that is accurate for stiff systems is the *RODAS3 solver* (Sandu *et al.*, 1997a) for which $s = 4$. The solution is given by

$$\psi^{n+1} = \psi^n + \frac{5}{6}\mathbf{k}_1\Delta t - \frac{1}{6}\mathbf{k}_2\Delta t - \frac{1}{6}\mathbf{k}_3\Delta t + \frac{1}{2}\mathbf{k}_4\Delta t \tag{6.69}$$

with

$$\begin{aligned}
\mathbf{k}_1 &= \left[\mathbf{s}(\psi^n) + \frac{1}{2}\mathbf{J}\,\mathbf{k}_1\Delta t \right] \\
\mathbf{k}_2 &= \left[\mathbf{s}(\psi^n) + \mathbf{J}\,\mathbf{k}_1\Delta t + \frac{1}{2}\mathbf{J}\,\mathbf{k}_2\Delta t \right] \\
\mathbf{k}_3 &= \left[\mathbf{s}(\psi^n + \mathbf{k}_1\Delta t) - \frac{1}{4}\mathbf{J}\,\mathbf{k}_1\Delta t - \frac{1}{4}\mathbf{J}\,\mathbf{k}_2\Delta t + \frac{1}{2}\mathbf{J}\,\mathbf{k}_3\Delta t \right] \\
\mathbf{k}_4 &= \left[\mathbf{s}\left(\psi^n + \frac{3}{4}\mathbf{k}_1\Delta t - \frac{1}{4}\mathbf{k}_2\Delta t + \frac{1}{2}\mathbf{k}_3\Delta t \right) + \frac{1}{12}\mathbf{J}\,\mathbf{k}_1\Delta t + \frac{1}{12}\mathbf{J}\,\mathbf{k}_2\Delta t - \frac{2}{3}\mathbf{J}\,\mathbf{k}_3\Delta t + \frac{1}{2}\mathbf{J}\,\mathbf{k}_4\Delta t \right]
\end{aligned} \tag{6.70}$$

Verwer *et al.* (1999), who compared *ROS2* and *RODAS3*, suggest that the second algorithm is less stable when using large fixed time steps, and that, in general, the first method performs with higher stability for nonlinear atmospheric kinetics problems.

6.4.3 Gear Solver

Most of the algorithms discussed in previous sections are single-step methods, in which the solution ψ^{n+1} at time t_{n+1} is calculated as a function of the solution ψ^n at time t_n. In a multi-step method, the solution ψ^{n+1} is derived as a function of the solutions ψ^n, ψ^{n-1}, ψ^{n-2}, ... at previous time levels t_n, t_{n-1}, t_{n-2}, ... A general formulation for a multi-step algorithm of order s is given by Byrne and Hindmarsh (1975):

$$\boldsymbol{\psi}^{n+1} = \sum_{k=0}^{s} \alpha_k \boldsymbol{\psi}^{n-k} + \Delta t \sum_{k=-1}^{s} \gamma_k \mathbf{s}\left(t_{n-k}, \boldsymbol{\psi}^{n-k}\right) \qquad (6.71)$$

where α_k and γ_k are method-specific constants selected to ensure stability. Single-step methods correspond to the particular case of $s = 0$ and $\alpha_0 = 1$.

Different explicit and implicit multi-step methods are available to solve ordinary differential equations, and are briefly discussed in Chapter 4. A widely used multi-step method that is particularly well adapted to stiff problems is the implicit *Gear's solver* (Gear, 1971), also called *backward differentiation formulae* (*BDF*). Here s can be as high as 6, only γ_{-1} is non-zero among the γ coefficients, and α_k is selected by stability and accuracy considerations. Thus, we write the BDF

$$\boldsymbol{\psi}^{n+1} = \sum_{k=0}^{s} \alpha_k \boldsymbol{\psi}^{n-k} + \Delta t\, \gamma\, \mathbf{s}\left(t_{n+1}, \boldsymbol{\psi}^{n+1}\right) \qquad (6.72)$$

The specific expressions used in the Gear's algorithm for different orders (1 to 6) are the following:

Order 1 : $\boldsymbol{\psi}^{n+1} = \boldsymbol{\psi}^n + \Delta t\, \mathbf{s}(t_{n+1}, \boldsymbol{\psi}^{n+1})$

Order 2 : $\boldsymbol{\psi}^{n+1} = \dfrac{4}{3} \boldsymbol{\psi}^n - \dfrac{1}{3} \boldsymbol{\psi}^{n-1} + \dfrac{2}{3} \Delta t\, \mathbf{s}\left(t_{n+1}, \boldsymbol{\psi}^{n+1}\right)$

Order 3 : $\boldsymbol{\psi}^{n+1} = \dfrac{18}{11} \boldsymbol{\psi}^n - \dfrac{9}{11} \boldsymbol{\psi}^{n-1} + \dfrac{2}{11} \boldsymbol{\psi}^{n-2} + \dfrac{6}{11} \Delta t\, \mathbf{s}\left(t_{n+1}, \boldsymbol{\psi}^{n+1}\right)$

Order 4 : $\boldsymbol{\psi}^{n+1} = \dfrac{48}{25} \boldsymbol{\psi}^n - \dfrac{36}{25} \boldsymbol{\psi}^{n-1} + \dfrac{16}{25} \boldsymbol{\psi}^{n-2} - \dfrac{3}{25} \boldsymbol{\psi}^{n-3} + \dfrac{12}{25} \Delta t\, \mathbf{s}\left(t_{n+1}, \boldsymbol{\psi}^{n+1}\right)$

Order 5 : $\boldsymbol{\psi}^{n+1} = \dfrac{300}{137} \boldsymbol{\psi}^n - \dfrac{300}{137} \boldsymbol{\psi}^{n-1} + \dfrac{200}{137} \boldsymbol{\psi}^{n-2} - \dfrac{75}{137} \boldsymbol{\psi}^{n-3} + \dfrac{12}{137} \boldsymbol{\psi}^{n-4}$
$\qquad\qquad + \dfrac{60}{137} \Delta t\, \mathbf{s}\left(t_{n+1}, \boldsymbol{\psi}^{n+1}\right)$

Order 6 : $\boldsymbol{\psi}^{n+1} = \dfrac{360}{147} \boldsymbol{\psi}^n - \dfrac{450}{147} \boldsymbol{\psi}^{n-1} + \dfrac{400}{147} \boldsymbol{\psi}^{n-2} - \dfrac{225}{147} \boldsymbol{\psi}^{n-3}$
$\qquad\qquad + \dfrac{72}{147} \boldsymbol{\psi}^{n-4} - \dfrac{10}{147} \boldsymbol{\psi}^{n-5} + \dfrac{60}{147} \Delta t\, \mathbf{s}\left(t_{n+1}, \boldsymbol{\psi}^{n+1}\right)$

This algorithm is very robust, accurate, and, in most cases, stable up to order 6 (see Section 4.8.1). It requires, however, as in the case of the backward Euler algorithm, the resolution of a nonlinear algebraic system. In the Livermore Solver for Ordinary Differential Equations (*LSODE*) (Hindmarsh, 1977), the solution to (6.72) or equivalently to

$$\mathbf{g}\left(\boldsymbol{\psi}^{n+1}\right) = \boldsymbol{\psi}^{n+1} + \sum_{k=0}^{s} \alpha_k \boldsymbol{\psi}^{n-k} + \Delta t\, \gamma\, \mathbf{s}\left(t_{n+1}, \boldsymbol{\psi}^{n+1}\right) = 0$$

is found by applying a Newton-Raphson iterative procedure

$$\mathbf{g}\left(\boldsymbol{\psi}^{n+1}_{(r+1)}\right) = \mathbf{g}\left(\boldsymbol{\psi}^{n+1}_{(r)}\right) + \left[\frac{\partial \mathbf{g}}{\partial \boldsymbol{\psi}^{n+1}}\right]_{(r)} \left(\boldsymbol{\psi}^{n+1}_{(r+1)} - \boldsymbol{\psi}^{n+1}_{(r)}\right) = 0$$

where r represents an iteration index and $\boldsymbol{\psi}^{n+1}_{(0)} = \boldsymbol{\psi}^n$. This leads to the linear algebraic system

$$(\mathbf{I} - \Delta t \, \gamma \, \mathbf{J}) \, \left(\boldsymbol{\psi}_{(r+1)}^{n+1} - \boldsymbol{\psi}_{(r)}^{n+1} \right) = -\boldsymbol{\psi}_{(r)}^{n+1} + \Delta t \, \gamma \, \mathbf{s} \left(t_{n+1}, \boldsymbol{\psi}_{(r)}^{n+1} \right) + \sum_{k=0}^{s} \alpha_k \boldsymbol{\psi}^{n-k}$$

$$(6.73)$$

that is solved by using, for example, an *LU* decomposition technique (see Box 6.2). The Jacobian matrix $\mathbf{J} = \partial \mathbf{s}/\partial \boldsymbol{\psi}$, which appears in the predictor matrix $(\mathbf{I} - \Delta t \, \gamma \, \mathbf{J})$ is usually sparse, so that computationally efficient sparse-matrix techniques can be applied to solve the system. The Jacobian should in principle be re-evaluated at each step of the iteration. In most practical applications, however, it is calculated only at the start of the iteration, or occasionally re-evaluated as the iteration proceeds. Gear's method offers a strategy to keep the solution error below a user-specified tolerance, by varying the order of the backward differentiation scheme and, when appropriate, by reducing the time step according to the intermediate results of the computation.

References

Byrne G. D. and Hindmarsh A. C. (1975) A polyalgorithm for the numerical solution of ordinary differential equations, *ACM Trans. Math. Softw.*, **1**, 71–96.

Dabdub D. and Seinfeld J. H. (1995) Extrapolation techniques used in the solution of stiff ODEs associated with chemical kinetics of air quality models, *Atmos. Environ.*, **29**, 403–410.

Gear C. W. (1971) *Numerical Initial Value Problems in Ordinary Differential Equations*, Prentice-Hall, Englewood Cliffs, NJ.

Gong W. and Cho H.-R. (1993) A numerical scheme for the integration of the gas phase chemical rate equations in a three-dimensional atmospheric models, *Atmos. Environ.*, **27A** (14), 2147–2160.

Hairer E. and Wanner G. (1996) *Solving Ordinary Differential Equations II. Stiff and Differential Algebraic Problems*, Springer, New York.

Hairer E., Norsett S. P. and Wanner G. (1993) *Solving Ordinary Differential Equations I. Nonstiff Problems*, 2nd edition, Springer, New York.

Hairer E., Lubich C., and Wanner G. (2002) *Geometric Numerical Integration, Structure Preserving Methods for Ordinary Differential Equations*, Springer Verlag, Berlin.

Hertel O., Berkowicz R., and Christensen J. (1993) Test of two numerical schemes for use in atmospheric transport-chemistry models, *Atmos. Environ.*, **27A**, 16, 2591–2611.

Hesstvedt E., Hov O., and Isacksen I. (1978) Quasi-steady-state-approximation in air pollution modelling: Comparison of two numerical schemes for oxidant prediction, *Int. J. Chem. Kinet.*, **10**, 971–994.

Hindmarsh A. C. (1977), *GEARB: Solution of Ordinary Differential Equations Having Banded Jacobian*, LLNL, Report UCID 30059, Rev. 2.

Jay L. O., Sandu A., Potra F. A., and Carmichael G. R. (1995) Improved QSSA methods for atmospheric chemistry integration. Reports on Computational Mathematics No 67/1995, Dept. Mathematics, University of Iowa.

Press W. H., Teukolsky S. A., Vetterling W. T., and Flannery B. P. (2007) *Numerical Recipes: The Art of Scientific Computing*, 3rd edition, Cambridge University Press, Cambridge.

Ramaroson R., Pirre M., and Cariolle D. (1992) A box model for on-line computations of diurnal variations in multidimensional models: Application to the one-dimensional case, *Ann. Geophys.*, **10**, 416–428.

Rosenbrock H. H. (1963) Some general implicit processes for the numerical solution of differential equations, *J. Comput.*, **5**, 329–330.

Sandilands J. and McConnell J. (1997), Evaluation of a reduced Jacobian chemical solver, *J. Geophys. Res.*, **102** (D15), 19073–19087

Sandu A., Verwer J. G., Blom J. G., *et al.* (1997a) Benchmarking stiff ODE solvers for atmospheric chemistry problems II: Rosenbrock solvers, *Atmos. Environ.*, **31** (20), 3459–3472.

Sandu A., Verwer J. G., Van Loon M., *et al.* (1997b) Benchmarking stiff ODE solvers for atmospheric chemistry problems I: Implicit vs explicit, *Atmos. Environ.*, **31** (19), 3151–3166.

Santillana M., Le Sager P., Jacob D. J., and Brenner M. P. (2010) An adaptive reduction algorithm for efficient chemical calculations in global atmospheric chemistry models, *Atmos. Environ.*, **44**, 35, 4426–4431.

Saylor R. D. and Ford G. D. (1995) On the comparison of numerical methods for the integration of kinetic equations in atmospheric chemistry and transport models, *Atmos. Environ.*, **29** (19), 2585–2593.

Shimazaki T. (1985) *Minor Constituents in the Middle Atmosphere*, D. Reidel, Norwell, MA.

Verwer J. G. (1994) Explicit methods for stiff ODEs from atmospheric chemistry. In *Numerical Mathematics Conference NUMDIFF-7*, Halle, Germany.

Verwer J. G., Spee E. J., Blom J. G., and Hundsdorfer W. (1999) A second-order Rosenbrock method applied to photochemical dispersion problems, *SIAM J. Sci. Comput.*, **20** (4), 1456–1480.

Young T. R. and Boris J. P. (1977) A numerical technique for solving stiff ordinary differential equations associated with the chemical kinetics of reactive flow problems, *J. Phys. Chem.*, **81**, 2424–2427.

Zhang H., Linford J. C., Sandu A., and Sander R. (2011) Chemical mechanism solvers in air quality models, *Atmosphere*, **2**, 510–532, doi: 10.3390/atmos2030510.

7 Numerical Methods for Advection

7.1 Introduction

The distribution of chemical species in the atmosphere is affected by air motions ranging from the global circulation down to the millimeter scale, at which point molecular diffusion takes over to dissipate kinetic energy. Air motions conserve the mixing ratios of the transported species since the air molecules are transported in the same way as the species. A plume of an inert chemical species transported in the atmosphere may stretch and filament, but it retains its initial mixing ratio until the filaments have become thin enough for molecular diffusion to dissipate gradients.

Representing this conservative transport in an atmospheric model is a major challenge because models cannot resolve the full range of spatial and temporal scales involved. Even if they could, chaotic behavior in solving the equation of motion would prevent a deterministic representation of the flow. Assimilation of meteorological observations can force model winds to approximate the real atmosphere, but only on the large scales of the observational network and at the cost of small-scale numerical noise introduced by the assimilation process.

From a model perspective, it is useful to distinguish between transport by the resolved large-scale winds, which can be simulated deterministically; and transport by the unresolved small-scale winds, which must be represented stochastically. The distinction between large-scale and small-scale is defined by the resolution of the model. Transport by resolved winds is commonly called *advection*, while transport by unresolved winds is called *eddy flow*, *turbulence*, or (in the vertical) *convection*. This chapter focuses on the numerical schemes used to solve the advection problem. Schemes for smaller-scale unresolved transport are presented in Chapter 8.

Numerical methods should preserve the properties of the continuous partial differential equations (PDEs) that they attempt to approximate. Desirable properties of numerical methods for advection are listed by Rasch and Williamson (1990a), Williamson (1992), and Lauritzen *et al.* (2011). They include:

1. *Accuracy.* The solution must be close to the true state.
2. *Stability.* The solution must not diverge away from the true state.
3. *Monotonicity.* The solution should not generate spurious maxima or minima in mixing ratios. Since initial conditions for mixing ratios are positive, monotonicity implies positivity of the solution.
4. *Conservation.* In the absence of sources and sinks, total mass must be conserved during advection. The algorithm should also conserve the second moment (variance) of the advected quantity.

(a)

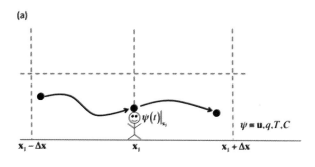

(b)

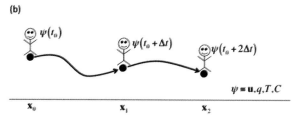

Figure 7.1 Eulerian and Lagrangian perspectives. In the Eulerian representation (a), the observer is located at fixed points (model grid points) and tracks the change in the calculated state variable ψ (e.g., mixing ratio C) as air parcels move by. In the Lagrangian representation (b), the observer tracks the change in the variable ψ in individual air parcels as they move with the flow. Reproduced from Lin (2012).

5. *Transportivity.* Transport should be downwind only.
6. *Locality.* The solution at a given point must not be controlled by the concentrations far away from that point.
7. *Correlativity.* Relationships between species in the flow must be preserved.
8. *Flexibility.* An advection scheme is most useful if it can be implemented on different grids and at different resolutions; this makes it in particular applicable for adaptive grids.
9. *Efficiency.* A more computationally efficient algorithm facilitates simulations with higher resolution, over longer periods, and/or involving a larger number of transported species.

In Chapters 1 and 4 we drew a distinction between *Eulerian* and *Lagrangian* models for atmospheric transport (Figure 7.1). A Eulerian model solves the advection equation on a fixed reference grid, while a Lagrangian model tracks particles as they are transported in the atmosphere. Both have advantages and disadvantages. A Eulerian model provides a complete solution over the atmospheric domain with regular spatial resolution, but is subject to numerical noise and to stability constraints. A Lagrangian model has no limitations from numerical diffusion or stability, but it has uneven spatial resolution and cannot easily handle nonlinear chemistry. Eulerian and Lagrangian approaches are sometimes combined to benefit from the advantages of each, as in semi-Lagrangian advection schemes.

We focus this chapter on the basic approaches to solve the advection equation, including description of some classic schemes. The schemes used in current models of atmospheric composition are based on these classic schemes, but often include refinements that we cannot present in detail.

Section 7.2 presents different forms of the advection equation. Section 7.3 reviews *finite difference* methods used to solve the advection equation in a Eulerian framework, while Section 7.4 focuses on *finite volume* methods. Flux-corrected methods introduced to preserve the monotonicity of the solution are discussed in Section 7.5. Selected advanced *Eulerian* numerical methods are presented in Section 7.6. Sections 7.7 and 7.8 describe the methods used in *Lagrangian* and *semi-Lagrangian* models.

7.2 The Advection Equation

An atmospheric species transported in a model is commonly called a *tracer*. The atmospheric advection of a tracer i is determined by its local mass density ρ_i and by the wind field $\mathbf{v}$. The corresponding mass flux $\mathbf{F}_i$ is the product

$$\mathbf{F}_i = \rho_i \mathbf{v} \tag{7.1}$$

As shown in Chapter 4, the local rate of change in the density due to advective transport is the divergence $\nabla \cdot \mathbf{F}_i$ of this flux. This defines the continuity equation for an inert tracer (no local sources or sinks):

$$\frac{\partial \rho_i}{\partial t} + \nabla \cdot (\rho_i \mathbf{v}) = 0 \tag{7.2}$$

Equation (7.2) is the Eulerian flux form of the advection equation, previously discussed in Chapter 4. Transport in a Eulerian model is often expressed in terms of mass fluxes across grid cell interfaces. This can be expressed by integration over a finite volume V_c (usually a model grid cell) of the advection equation (7.2):

$$\frac{1}{V_c} \frac{\partial}{\partial t} \int_{V_c} \rho_i dV + \frac{1}{V_c} \int_{V_c} \nabla \cdot (\rho_i \mathbf{v}) dV = 0 \tag{7.3}$$

The first term in this integral equation represents the time evolution of the mean density $\langle \rho_i \rangle$ of tracer i in the finite volume V_c. The second term can be transformed by applying the divergence (or Gauss–Ostrogradsky) theorem (see Appendix E), which states the equivalence between the volume integral over V_c, and the surface integral over the closed boundary of volume V_c. Equation (7.3) then becomes:

$$\frac{\partial \langle \rho_i \rangle}{\partial t} + \frac{1}{V_c} \int_{S_c} \mathbf{F}_i \cdot \mathbf{n} \, dS = 0 \tag{7.4}$$

where $\mathbf{F}_i$ represents the flux vector of tracer i across surface S_c of the boundary of the cell and $\mathbf{n}$ is a unit outward vector normal to the cell boundary. This expression states that the change of the mean density inside the finite cell is determining by the net flux of material in and out of the cell. Equations (7.2) and (7.4) are often called *conservative forms* of the advection equation.

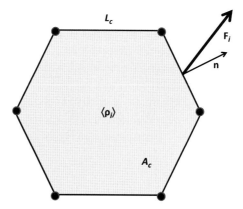

Formulation of tracer advection using (7.5) in the case of a 2-D hexagonal grid cell with surface A_c and boundaries L_c. The flux $\mathbf{F}_i$ across a particular side of the hexagon is schematically represented. Unit vector $\mathbf{n}$ is normal (outward) to the cell boundary and $\langle \rho_i \rangle$ is the average density of tracer i in the cell.

When applied to a 2-D (e.g., horizontal) model, (7.4) becomes

$$\frac{\partial \langle \rho_i \rangle}{\partial t} + \frac{1}{A_c} \int_{L_c} \mathbf{F}_i \bullet \mathbf{n} \, dl = 0 \tag{7.5}$$

where the integral is now calculated along the boundary line L_c of the 2-D cell with area A_c. Figure 7.2 illustrates the different elements needed to solve (7.5) in the case of a 2-D hexagonal grid cell. Finally, in the case of a 1-D problem (along direction x), the expression becomes

$$\frac{\partial \langle \rho_i \rangle}{\partial t} + \frac{1}{\Delta x} \left(F_i^r - F_i^l \right) = 0 \tag{7.6}$$

where Δx represents the size of the 1-D grid cell. F_i^r and F_i^l are the tracer fluxes at the right and left edges of the cell (positive rightward). As stated in Chapter 4, these expressions constitute the basis for finite volume methods; they are particularly suitable when solving the equations on complex or irregular grids.

When expressed in terms of mass mixing ratio $\mu_i = \rho_i / \rho_a$ where ρ_a is the density of air, the continuity equation becomes (advective form):

$$\frac{\partial \mu_i}{\partial t} + \mathbf{v} \bullet \nabla \mu_i = 0 \tag{7.7}$$

or

$$\frac{d\mu_i}{dt} = 0 \tag{7.8}$$

where the total derivative operator (derivative along the flow) is expressed by:

$$\frac{d}{dt} = \frac{\partial}{\partial t} + \mathbf{v} \bullet \nabla \tag{7.9}$$

These forms were previously derived in Chapter 4. Equation (7.8) specifies the invariance of the mixing ratio along flow trajectories. By contrast, tracer densities ρ_i along flow trajectories may change because air is a compressible fluid (see example below). Equation (7.7) is the Eulerian advective form of the advection equation, while (7.8) is the Lagrangian form.

Solution of the advection equation in 3-D models involves operator splitting along individual dimensions. Thus, the 1-D advection equation is solved successively in the three dimensions to obtain the 3-D solution. We will focus our discussion in the following sections on the 1-D problem as it is most relevant for model applications. In this case, the flux-form advection equation (7.2) becomes:

$$\frac{\partial \rho_i(x, t)}{\partial t} + \frac{\partial}{\partial x}[u(x, t)\rho_i(x, t)] = 0 \tag{7.10}$$

and the advective-form equation (7.7) becomes:

$$\frac{\partial \mu_i(x, t)}{\partial t} + u(x, t)\frac{\partial \mu_i(x, t)}{\partial x} = 0 \tag{7.11}$$

where $u(x, t)$ denotes the 1-D wind velocity. Rewriting the flux-form equation (7.10) as

$$\frac{\partial \rho_i(x, t)}{\partial t} + u(x, t)\frac{\partial \rho_i(x, t)}{\partial x} = -\rho_i(x, t)\frac{\partial u(x, t)}{\partial x} \tag{7.12}$$

shows that it is identical to the advective form (7.11), but with an additional term $\rho_i \, \partial u/\partial x$ that describes the compressibility of the flow. This term acts as a source of ρ_i when $u(x, t)$ decreases with x (compression) and as a sink when $u(x, t)$ increases with x (expansion). If the wind velocity is constant in space and time ($u(x, t) = c$), the flux form of the advection equation is simply

$$\frac{\partial \rho_i(x, t)}{\partial t} + c\frac{\partial \rho_i(x, t)}{\partial x} = 0 \tag{7.13}$$

Thus the advection equation applies identically to density and mixing ratio in an incompressible flow.

Figure 7.3 shows the solution of the 1-D advection equation for the two cases of a constant and a spatially variable wind velocity. For a constant velocity, the advection of both the density and the mixing ratio is represented by a simple translation (without deformation) of the initial function in the direction of the velocity. If the velocity decreases with space, the initial distribution of both quantities is distorted as the material is advected. The value of the maximum mixing ratio is unchanged, but the maximum value of the density is enhanced. Advection can thus modify extrema of tracer densities in a diverging flow.

In many applications, tracers are not only advected but are also diluted by turbulent diffusion. The 1-D advection–diffusion equation is given by:

$$\frac{\partial \rho_i(x, t)}{\partial t} + c\frac{\partial \rho_i(x, t)}{\partial x} - K\frac{\partial^2 \rho_i(x, t)}{\partial x^2} = 0 \tag{7.14}$$

where K [m^2 s^{-1}] is a diffusion coefficient. The relative importance of advection versus diffusion is measured by the dimensionless *Péclet number Pe*:

$$Pe = \frac{c\,L}{K} \tag{7.15}$$

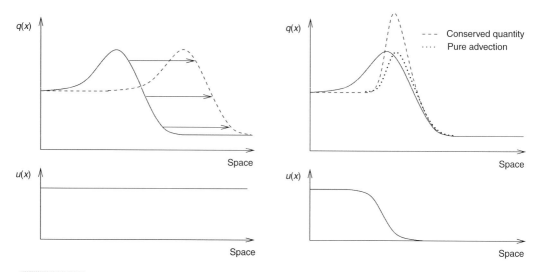

Figure 7.3 One-dimensional advection in the x-direction of a scalar function, noted here $q(x, t)$, for a constant velocity u (left) and a spatially varying velocity $u(x)$ (right). The initial distribution of the function is represented by the solid line. On the left panel, this initial function (density or mixing ratio) is translated in the direction of the constant velocity u. On the right panel, the dotted curve (labeled pure advection) corresponds to a case where the term $-q\ \partial u/\partial x$ is omitted and is therefore representative of the advection of a tracer mixing ratio. In this case, the spatial distribution of the function is modified, but the maximum value of the function is unchanged. The dashed curve is obtained by including the term $-q\ \partial u/\partial x$ and is therefore representative of the advection of a tracer density. In this case, the area under the curve is maintained during the advection process, but the maximum value of the function is not preserved. From C. P. Dullemond with permission.

where L is a characteristic length for the advection. The Péclet number can be regarded as a measure of the ratio of the diffusive timescale to the advective timescale. For conditions typical of horizontal flow with a wind velocity $c \sim 10$ m s^{-1}, a characteristic length L for long-range transport ~ 1000 km and an eddy diffusion coefficient K of 10^5 m^2 s^{-1}, the Péclet number is ~ 100 and the transport is thus dominated by advection. For vertical transport in the boundary layer with a typical vertical velocity $c \sim 0.01$ m s^{-1}, a characteristic length L of 100 m, and an eddy diffusion coefficient of 100 m^2 s^{-1}, the Péclet number is equal to 10^{-2} and diffusion becomes dominant. The solution of the transport equation for a boxcar function (shock front) subject to simultaneous advection and diffusion under a Péclet number of approximately 1 is shown schematically in Figure 7.4. Note the gradual deformation of the shape of the function under the influence of diffusion. The area under the curve, however, is conserved.

When considering a discretized form of the advection–diffusion equation, one often introduces the *numerical Péclet number* as:

$$Pe = \frac{c\ \Delta x}{K} \tag{7.16}$$

where Δx represents the grid size of the model. The numerical Péclet number measures the relative importance of advection and turbulent diffusion at the smallest spatial scale

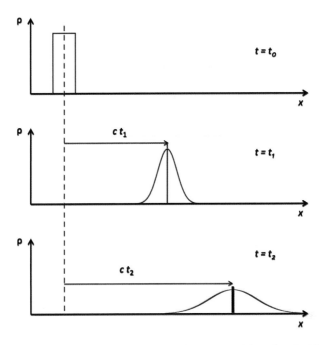

Figure 7.4   Solution of the 1-D advection–diffusion equation for a Péclet number of the order of 1. The initial function at time $t = t_0$ is represented by a boxcar function. The gradual displacement of the function is due to the advection term and deformation (spread) of the shape of the function is caused by the diffusion term. Adapted from Slingerland and Kump (2011).

resolved by the model. The advection–diffusion equation is either parabolic (diffusion-dominated) or hyperbolic (advection-dominated), depending on the Péclet number. Numerical algorithms treating atmospheric diffusion are discussed in Chapter 8.

7.3 Elementary Finite Difference Methods

We examine here the fundamental properties of several advection algorithms applied to the simple 1-D advection equation (along direction x), with a constant velocity c (taken to be positive) and fixed grid spacing Δx. Generalization to variable wind velocity and grid spacing is presented in Section 7.3.5. As in the previous chapters dealing with numerical algorithms, we represent the field of the transported quantity by the generic mathematical symbol Ψ. The advection of a non-negative scalar function Ψ is described by the first-order hyperbolic PDE:

$$\frac{\partial \Psi}{\partial t} + c \frac{\partial \Psi}{\partial x} = 0 \tag{7.17}$$

To solve this equation, initial and boundary conditions must be specified. The initial condition can be expressed as

$$\Psi(x, 0) = G(x) \tag{7.18}$$

where $G(x)$ represents the spatial distribution of the tracer distribution at $t = 0$. The resulting analytic solution is simply the translation without any change in shape of function $G(x)$ in the x-direction at a velocity c. Thus,

$$\Psi(x, t) = G(x - ct) \tag{7.19}$$

When the spatial domain, rather than being infinite, extends from $x = a$ to $x = b$, a condition must be specified at one boundary of the domain. An advection problem (hyperbolic equation) is well-posed if a boundary condition on the value of Ψ (Dirichlet condition) is imposed at the inflow boundary. If $c > 0$, the condition must be expressed at $x = a$:

$$\Psi(a, t) = H_a(t) \tag{7.20}$$

while if $c < 0$ the condition must be applied at $x = b$:

$$\Psi(b, t) = H_b(t) \tag{7.21}$$

A periodic boundary condition, such as on a sphere, can be imposed as

$$\Psi(b, t) = \Psi(a, t) \tag{7.22}$$

In this case, the mass that flows out of the domain at boundary $x = b$ flows back into the domain at boundary $x = a$.

Function $\Psi(x, t)$ can be represented in an unbounded or periodic domain as a Fourier series with components (harmonics) characterized by their wavenumbers k [m^{-1}] corresponding to wavelengths $L = 2\pi/k$:

$$\Psi(x, t) = \sum_{k=-\infty}^{\infty} A_k(t) e^{ikx} \tag{7.23}$$

In order to analyze the fundamental properties of different elementary algorithms, we assume that the problem is periodic in space and consider that a single harmonic (k) of function $\Psi(x, t)$ at grid point $x_j = j\,\Delta x$ and time t takes the value

$$\Psi_k(x_j, t) = A_k(t) e^{ikj\Delta x} \tag{7.24}$$

Here, Δx represents the grid spacing, assumed to be uniform, and j is the grid index. The advection equation becomes:

$$\frac{\partial \Psi_k}{\partial t} + c\,i\,k\,\Psi_k = 0 \tag{7.25}$$

In this idealized case, the advection velocity c can be interpreted as the phase speed of the wave defined by Ψ_k. Because c is constant, the phase speed is independent of wavenumber k. Therefore all waves propagate at the same speed. Accurate numerical algorithms have to preserve this property as much as possible. In this particular case, the group speed c_g, which is indicative of the rate at which the energy propagates, is equal to c and is independent of the wavenumber k. This analytic form of the advection equation will later be compared to the forms provided by numerical analogs.

7.3.1 Methods Using Centered Space Differences

When adopting a uniform grid spacing Δx, the space derivative $\partial \Psi/\partial x$ can be approximated by a *centered space difference* to yield the second-order accurate expression:

$$\frac{\partial \Psi}{\partial x} = \frac{\Psi_{j+1} - \Psi_{j-1}}{2\Delta x} + O\left(\Delta x^2\right) \tag{7.26}$$

where $O(\Delta x^2)$ is the truncation error. For a wave with wavenumber k, the finite difference expression can be written as

$$\frac{\Psi_{j+1} - \Psi_{j-1}}{2\Delta x} = \frac{ik \sin (k\,\Delta x)}{k\,\Delta x} A e^{ik\,j\Delta x} \tag{7.27}$$

Under these conditions, the approximate form of the advection equation

$$\frac{\partial \Psi}{\partial t} + \left[\frac{c\,ik \sin (k\,\Delta x)}{k\,\Delta x}\right]\Psi = 0 \tag{7.28}$$

can be compared with the exact analytic equation (7.25). When centered finite differences are used to approximate space derivatives, the phase velocity c^* associated with the numerical solution

$$c^* = c\left[\frac{\sin (k\Delta x)}{k\Delta x}\right] \tag{7.29}$$

varies with wavenumber k, while the true phase velocity c is independent of k (see Figure 7.5). Thus, even though all wavenumber components that characterize function Ψ should move at exactly the same speed, the shorter wavelengths are trailing the longer waves. As a result, the different Fourier components of the advected function are displaced along axis x at different velocities and the numerical solution is distorted. This property that arises from the space differencing is named *numerical dispersion* and leads to phase errors. In space-centered approximations, the shortest wavelength that can be resolved by the model ($L = 2\Delta x$ or $k\Delta x = \pi$) does not move at all since its phase speed is zero (see Figure 7.5). For long waves (small values of $k\Delta x$), the phase speed c^* provided by the numerical scheme approaches the true value of c. Figure 7.5 also shows the variation of the group velocity c_g^*,

$$c_g^* = \frac{d(k\,c)}{dk} = c \cos (k\Delta x) \tag{7.30}$$

for the centered difference scheme. For wavelengths between $4\Delta x$ and $2\Delta x$, the group velocity c_g^* is negative. This means that energy can propagate upstream, which is an undesirable property of the scheme.

When the space derivative is approximated by a fourth-order scheme over a grid with constant spacing,

$$\frac{\partial \Psi}{\partial x} = \frac{\Psi_{j-2} - 8\Psi_{j-1} + 8\Psi_{j+1} - \Psi_{j+2}}{12\,\Delta x} + O\left(\Delta x^4\right) \tag{7.31}$$

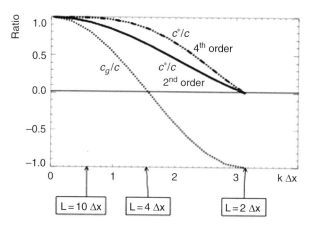

Figure 7.5 Ratios between the *phase velocity* c^* associated with the numerical solution and the true phase velocity c as a function of $k\Delta x$ for the second-order and fourth-order space derivatives. Here k refers to the wavenumber of the different Fourier components of the signal and Δx to the grid spacing. The corresponding wavelengths L for three particular waves are indicated. The graph highlights the lag in the advection of the waves relative to the advective motion. This effect is more pronounced for the shortest wavelengths. The ratio between the group velocity c_g resulting from the algorithm and the true group velocity c is also shown in the case of the second-order derivative.

the phase velocity c^* becomes:

$$c^* = c\left[\frac{4}{3}\frac{\sin(k\Delta x)}{k\Delta x} - \frac{1}{3}\frac{\sin(2k\Delta x)}{(2k\Delta x)}\right] \tag{7.32}$$

This improves over the second-order scheme, as shown in Figure 7.5, but still fails for wavelengths close to $L = 2\Delta x$.

Euler Forward Scheme (FTCS)

A simple numerical method to solve (7.17) is the *Euler forward scheme*, which approximates the time derivative by a forward difference and the space derivative by a centered difference:

$$\frac{\Psi_j^{n+1} - \Psi_j^n}{\Delta t} = -c\frac{\Psi_{j+1}^n - \Psi_{j-1}^n}{2\Delta x} \tag{7.33}$$

where Δt is the time step and Δx the grid spacing (both assumed to be constant), and n and j are the indices referring to time and space, respectively, with

$$t_n = n\,\Delta t \quad \text{and} \quad x_j = j\,\Delta x$$

This algorithm, which is also referred to as the FTCS method (**f**orward-in-**t**ime, **c**entered-in-**s**pace), is first-order accurate in time and second-order accurate in space. Its solution,

$$\Psi_j^{n+1} = \Psi_j^n - \frac{\alpha}{2}\left(\Psi_{j+1}^n - \Psi_{j-1}^n\right) \tag{7.34}$$

where

$$\alpha = c\frac{\Delta t}{\Delta x} \tag{7.35}$$

is the Courant number, can be easily computed because the algorithm is *explicit* (the solution Ψ_j^{n+1} at time t_{n+1} for each point x_j is derived directly from quantities that are already known at time t_n). The method is also *one-step* because only one calculation is required to advance the integration from time level t_n to the new time level t_{n+1}. It is a *two-level* scheme since only two time levels (t_n and t_{n+1}) are involved in the calculation.

An important consideration when assessing an algorithm is its *stability*. To address this, we apply the *von Neumann's stability analysis* presented in Box 7.1. For the FTCS algorithm, the *amplification coefficient* $g(k)$ as a function of wavenumber k is:

$$g(k) = 1 - i\alpha \sin(k\Delta x) \tag{7.36}$$

The *amplification factor*,

$$|g(k)| = \left[1 + \alpha^2 \sin^2(k\Delta x)\right]^{\frac{1}{2}} \tag{7.37}$$

is greater than one for all values of the wavenumber k. As a result, any numerical error produced by the algorithm grows exponentially with time. As highlighted in Box 7.1, the Euler FTCS algorithm is thus *unconditionally unstable* and must be rejected.

However, the FTCS method can become conditionally stable if a numerical diffusion term is added to the right-hand side of the advective equation, with diffusion coefficient K:

$$K\frac{\partial^2 \Psi}{\partial x^2} \approx \frac{K}{\Delta x^2}\left(\Psi_{j+1}^n - 2\Psi_j^n + \Psi_{j-1}^n\right) \tag{7.38}$$

In this case, the discretized equation becomes:

$$\Psi_{j+1}^n = \Psi_j^n - \frac{\alpha}{2}\left(\Psi_{j+1}^n - \Psi_{j-1}^n\right) + \beta\left(\Psi_{j+1}^n - 2\Psi_j^n + \Psi_{j-1}^n\right) \tag{7.39}$$

Here α is again the Courant number, and

$$\beta = K\frac{\Delta t}{\Delta x^2} \tag{7.40}$$

is the so-called Fourier number. The amplification coefficient for wavenumber k derived from the von Neumann's analysis becomes

$$g(k) = 1 + 2\beta[\cos(k\Delta x) - 1] - i\alpha \sin(k\Delta x) \tag{7.41}$$

with the corresponding amplitude (Figure 7.6)

$$|g(k)| = \left[(1 - 2\beta[1 - \cos(k\Delta x)])^2 + \alpha^2 \sin^2(k\Delta x)\right]^{\frac{1}{2}} \tag{7.42}$$

and phase

$$\tan\Phi(k) = \frac{-\alpha \sin(k\Delta x)}{1 - 2\beta[1 - \cos(k\Delta x)]} \tag{7.43}$$

Box 7.1   The von Neumann Stability Analysis

The von Neumann analysis provides a methodology for assessing the stability of numerical algorithms. It applies to linear PDEs with periodic boundary conditions. We first consider the analytic solution of the advection equation by noting that any function Ψ can be expressed as the superposition of an infinite number of waves. We perform therefore a discrete Fourier transform of function Ψ, which is advected in the x-direction with a positive and constant velocity c. Consider the advection of a single wave harmonic with wavenumber k. At time t, this harmonic is expressed by:

$$\Psi_k(x, t) = a\, e^{ik(x-ct)}$$

After a time interval Δt corresponding to a displacement of the wave over a distance $c\Delta t$, function Ψ takes the form:

$$\Psi_k(x, t + \Delta t) = a\, e^{ik[x-c(t+\Delta t)]} = \Psi_k(x, t)e^{-ikc\Delta t}$$

The *amplification coefficient* or *gain* $g(k)$ for harmonic k is the complex function defined as the ratio between function Ψ_k after and before the advection step. Thus,

$$g(k) = \frac{\Psi_k(t + \Delta t)}{\Psi_k(t)} = e^{-ikc\Delta t}$$

with a modulus (also called amplitude and here *amplification factor*)

$$|g(k)| = [g(k)\cdot g^*(k)]^{1/2} = 1$$

and a *phase*

$$\varphi(k) = -kc\Delta t$$

Here g^* is the complex conjugate of g. The amplification factor $|g(k)|$ represents the relative change in the amplitude of the harmonic of wavenumber k after one computational time step.

These relations highlight two properties that should be reproduced as closely as possible when numerical approximations to the exact solution are sought: (1) the amplitude of all wave harmonics is unchanged during an advection process, and (2) the phase of harmonics k varies according to $\varphi = -\alpha k \Delta x$, where α is the Courant number and Δx the grid spacing of the model. If, for example, the value of $g(k)$ resulting from the use of a numerical approximation is less than 1, the amplitude of the wave is reduced by advection. Conversely, if it is larger than 1, the amplitude of the wave grows and the method becomes rapidly *unstable*. Thus, the numerical stability of an algorithm for advection requires that $|g(k)| \leq 1$ for all waves resolved by the model. Similarly, errors on the phase cause waves of a spectrum to lag the displacement of other waves, leading to numerical dispersion.

To apply the von Neumann analysis to a numerical algorithm, let us assume that the advection equation (7.17) is approximated by the FTCS algorithm (centered difference scheme for the space derivative):

$$\Psi_j^{n+1} = \Psi_j^n - \frac{\alpha}{2}\left(\Psi_{j+1}^n - \Psi_{j-1}^n\right)$$

where α is the Courant number. We apply a discrete Fourier transform and consider a single harmonic with wavenumber k:

$$\Psi_j^n(k) = e^{ikx_j}$$

It results from the above FTCS formulation that:

$$\Psi_j^{n+1} = e^{ikx_j} - \frac{\alpha}{2}\left[e^{ik\left(x_j+\Delta x\right)} - e^{ik\left(x_j-\Delta x\right)}\right] = e^{ikx_j}\left[1 - \frac{\alpha}{2}\left(e^{ik\Delta x} - e^{-ik\Delta x}\right)\right]$$
$$= \Psi_j^n[1 - i\alpha\sin(k\Delta x)]$$

The transfer function based on the FTCS algorithm is therefore:

$$g(k) = \frac{\Psi_j^{n+1}}{\Psi_j^n} = 1 - i\,\alpha\sin(k\Delta x)$$

Its amplitude is given by

$$|g(k)| = \left[(\text{Re}[g(k)])^2 + (\text{Im}[g(k)])^2\right]^{\frac{1}{2}} = \left[1 + \alpha^2\sin^2(k\Delta x)\right]^{\frac{1}{2}}$$

and the phase $\Phi(k)$ associated with this particular algorithm is derived from:

$$\tan\Phi(k) = \frac{\text{Im}[g(k)]}{\text{Re}[g(k)]} = -\alpha\sin(k\Delta x)$$

A comparison of these expressions with the values derived from the analytic solution shows that the FTCS scheme is *unconditionally unstable* since the modulus of the transfer function is larger than one for all values of wavenumber k, even for very small time steps. The phase error $E(k)$ is given by:

$$E(k) = \Phi(k) - \varphi(k) = \Phi(k) + \alpha\,k\Delta x$$

The von Neumann analysis is adopted in this chapter to assess the stability conditions of different numerical schemes. It is a necessary and sufficient condition for stability in the case of linear finite difference equations with constant coefficients. It is a necessary but not sufficient stability condition for nonlinear equations.

Necessary and sufficient conditions for the stability of this scheme are

$$0 \leq \beta \leq \frac{1}{2} \quad \text{and} \quad \alpha^2 \leq 2\beta \tag{7.44}$$

Thus, the FTCS method can be used if a small diffusion term is added to the advection equation and the time step is sufficiently short.

Lax Scheme

In the *Lax method* (Lax, 1954), the term Ψ_j^n used in the FTCS scheme is replaced by ½ $\left(\Psi_{j+1}^n + \Psi_{j-1}^n\right)$, and the approximate form of the advection equation becomes:

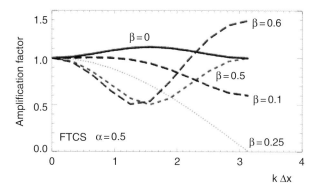

Amplification factor $|g(k)|$ as a function of $k\Delta x$ when the advection equation is approximated by the FTCS algorithm for a constant velocity c and the Courant number equal to 0.5. Parameter $\beta = k\Delta t/\Delta x^2$ represents the effect of added diffusion to the advection equation. The case with $\beta = 0$ corresponds to pure advection and is unconditionally unstable (amplification factor > 1). When diffusion is added, the solution is stable ($|g(k)| < 1$) for $\beta < 0.5$, but becomes unstable for larger values of β.

$$\Psi_j^{n+1} = \frac{1}{2}\left(\Psi_{j+1}^n + \Psi_{j-1}^n\right) - \alpha\frac{\Psi_{j+1}^n - \Psi_{j-1}^n}{2} \tag{7.45}$$

When applying the von Neumann stability analysis, we obtain the amplification coefficient:

$$g(k) = \cos(k\Delta x) - i\,\alpha\sin(k\Delta x) \tag{7.46}$$

whose amplitude

$$|g(k)| = \left[\cos^2(k\Delta x) + \alpha^2\sin^2(k\Delta x)\right]^{\frac{1}{2}} \tag{7.47}$$

remains less than or equal to unity for Courant numbers $\alpha \leq 1$.

This stability condition $\alpha \leq 1$, which applies to many Eulerian schemes, is called the *Courant–Friedrichs–Lewy* (*CFL*) *criterion*. Over a time step Δt, the displacement of tracer should never exceed a distance larger than the grid spacing Δx. When a longitude–latitude grid is adopted in the model, this condition imposes severe limitations in the vicinity of the pole where the grid spacing in the longitudinal direction becomes very small. This is often circumvented by the application of numerical filters (see Section 7.10) or by the use of a reduced grid (see Section 4.7.3). The CFL condition also requires that increases in the spatial resolution of a model be accompanied by a proportional decrease in the value of the time step.

The stabilization of the solution in the Lax scheme can be understood by rearranging (7.45) as

$$\frac{\Psi_j^{n+1} - \Psi_j^n}{\Delta t} = -c\frac{\Psi_{j+1}^n - \Psi_{j-1}^n}{2\Delta x} + \frac{\Delta x^2}{2\Delta t}\left[\frac{\Psi_{j+1}^n - 2\Psi_j^n + \Psi_{j-1}^n}{\Delta x^2}\right] \tag{7.48}$$

which is the FTCS form of equation

$$\frac{\partial\Psi}{\partial t} = -c\frac{\partial\Psi}{\partial x} + \frac{\Delta x^2}{2\Delta t}\frac{\partial^2\Psi}{\partial x^2} \tag{7.49}$$

Thus, the Lax scheme stabilizes the FTCS solution by adding a numerical diffusion term to the advection equation with diffusion coefficient $\Delta x^2/2\Delta t$. Because of this numerical diffusion, a disturbance at grid point j propagates not only to downwind grid point $(j+1)$, but also to upwind point $(j-1)$. There results a transportivity error.

Lax–Wendroff Scheme

The *Lax–Wendroff scheme* (Lax and Wendroff, 1960, 1964), also called the Leith (1965) or Crowley (1968) scheme, is designed to provide the minimum amount of added numerical diffusion required to provide stability to the FTCS solution. The value of advected function Ψ at time level $n + 1$ and at grid point j is derived from:

$$\Psi_j^{n+1} = \Psi_j^n - \alpha\left(\Psi_{j+1/2}^{n+1/2} - \Psi_{j-1/2}^{n+1/2}\right) \tag{7.50}$$

where the value of Ψ at half time level $(n + \frac{1}{2})$ and at half grid point $(j + \frac{1}{2})$ is estimated using the Lax scheme:

$$\Psi_{j+1/2}^{n+1/2} = \frac{1}{2}\left(\Psi_{j+1}^n + \Psi_j^n\right) - \frac{\alpha}{2}\left(\Psi_{j+1}^n - \Psi_j^n\right) \tag{7.51}$$

The resulting approximation to the advection equation:

$$\Psi_j^{n+1} = \Psi_j^n - \frac{\alpha}{2}\left(\Psi_{j+1}^n - \Psi_{j-1}^n\right) + \frac{\alpha^2}{2}\left(\Psi_{j+1}^n - 2\Psi_j^n + \Psi_{j-1}^n\right) \tag{7.52}$$

is second-order accurate in time even though it involves only two time levels. Here again, the third term in the right-hand side of the equation can be viewed as a diffusion term added to the FTCS scheme. In contrast to the Lax method, the effective diffusion coefficient $c^2\Delta t/2$ is proportional to the time step Δt, so that numerical diffusion can be reduced by adopting smaller time steps.

The amplification coefficient is:

$$g(k) = 1 - i\,\alpha\sin(k\Delta x) - \alpha^2[1 - \cos(k\Delta x)] \tag{7.53}$$

with amplification factor

$$|g(k)| = \left\{1 - 4\alpha^2\left[1 - \alpha^2\right]\sin^4(k\Delta x/2)\right\}^{1/2} \tag{7.54}$$

The stability criterion is again satisfied for $\alpha \leq 1$. Some amplitude dampening occurs in the solution for all $\alpha < 1$, but it is weak for wavelengths that are large compared to the grid spacing Δx. For the smallest wave that can be resolved by the grid ($L = 2\Delta x$ or equivalently $k\Delta x = \pi$), the amplification coefficient is as low as zero for $\alpha^2 = 0.5$ so the wave disappears. However, for a wave with $L = 4\,\Delta x$, the dissipation is already considerably smaller; in this case the minimum amplification coefficient is 0.8 for $\alpha^2 = 0.5$.

The phase $\Phi(k)$ associated with wavenumber k (see Box 7.2) is deduced from

$$\tan\Phi(k) = \frac{-\alpha\sin(k\Delta x)}{1 - \alpha^2(1 - \cos(k\Delta x))} \tag{7.55}$$

An important consideration is the nature of the phase errors produced by second-order algorithms such as the Lax–Wendroff scheme. If one applies a Taylor expansion to the trigonometric functions appearing in (7.55), one can show that the phase error for wavenumber k is given by

$$E(k) = - \arctan \left[\frac{\alpha \sin{(k\Delta x)}}{1 - \alpha^2 (1 - \cos{(k\Delta x)})} \right] + \alpha(k\Delta x) \approx \alpha(k\Delta x)^3 \frac{\alpha}{6} (1 - \alpha) \quad (7.56)$$

where the trigonometric functions have been approximated using Taylor expansions. The resulting lag per unit time in the displacement $\delta x(k) = E(k)/k$ of the waves increases with the square of their wavenumber k since

$$\frac{d}{dt} \delta x(k) \approx \frac{E(k)}{k\Delta t} = c \frac{E(k)}{\alpha(k\Delta x)} \infty (k\Delta x)^2 \quad (7.57)$$

The propagation of the waves is therefore fastest for the shortest wavelengths (and thus for wavelengths that approach the grid size Δx), which generates ripples in the advected signal. This type of behavior, shown here in the case of the Lax–Wendroff algorithm, is common to all second-order schemes and, as stated by *Godunov's theorem*, the monotone behavior of a numerical solution cannot be assured for linear finite difference methods with more than first-order accuracy. This theorem introduces a major limitation in the development of numerical schemes that treat advection.

Implicit Schemes

The implicit or Euler backward-in-time, centered-in-space (BTCS)

$$\frac{\Psi_j^{n+1} - \Psi_j^n}{\Delta t} = -c \frac{\Psi_{j+1}^{n+1} - \Psi_{j-1}^{n+1}}{2\Delta x} \quad (7.58)$$

with the recursive expression

$$\Psi_j^{n+1} = \Psi_j^n - \frac{\alpha}{2} \left(\Psi_{j+1}^{n+1} - \Psi_{j-1}^{n+1} \right) \quad (7.59)$$

is first-order accurate in time and second-order in space. The amplification coefficient derived from the von Neumann analysis is:

$$g(k) = \frac{1}{1 + i\alpha \sin{(k\Delta x)}} \quad (7.60)$$

The resulting amplification factor

$$|g(k)| = \left(\frac{1}{1 + \alpha^2 \sin^2{(k\Delta x)}} \right)^{\frac{1}{2}} \quad (7.61)$$

is smaller than unity for any value of the Courant number. The method is therefore *unconditionally stable*, allowing for the adoption of any arbitrary time step Δt, which is a great advantage. The solution, however, cannot be retrieved as easily as in the case of explicit schemes. In the implicit case, a system of J algebraic equations (if J is the number of grid points that are not associated with boundary conditions) must be solved, which is computationally expensive. In the 1-D case, the system of equations is tridiagonal and can be solved efficiently with the Thomas algorithm (see Box 4.4). Another limitation of the method is that it has limited accuracy, with the shortest wavelengths being more rapidly attenuated that the longer wavelengths.

Accuracy of the solution can be improved by combining the FTCS and BTCS approaches. The Crank–Nicholson algorithm, written as

$$\frac{\Psi_j^{n+1} - \Psi_j^n}{\Delta t} = -c \frac{\Psi_{j+1}^{n+1} - \Psi_{j-1}^{n+1} + \Psi_{j+1}^n - \Psi_{j-1}^n}{4\Delta x} \tag{7.62}$$

is second-order accurate in time and space. It is implicit because it includes terms evaluated at time t_{n+1} on the right-hand side. The amplification coefficient is given by

$$g(k) = \frac{1 + i\alpha \sin(k\Delta x/2)}{1 - i\alpha \sin(k\Delta x/2)} \tag{7.63}$$

and the amplification factor is equal to 1 for all wavenumbers and all Courant numbers:

$$|g(k)| = 1 \tag{7.64}$$

The algorithm is thus unconditionally stable.

Matsuno Scheme

The Matsuno scheme is a two-step explicit–implicit algorithm that is first-order accurate in time and second-order accurate in space. The first step is to predict an intermediate value Ψ_j^* of the transported quantity at time level $n + 1$ by using a simple FTCS (Euler forward) approach:

$$\frac{\Psi_j^* - \Psi_j^n}{\Delta t} = -c \frac{\Psi_{j+1}^n - \Psi_{j-1}^n}{2\Delta x} \tag{7.65}$$

The predicted values are then substituted into the space derivative term, and a correction step is applied:

$$\frac{\Psi_j^{n+1} - \Psi_j^n}{\Delta t} = -c \frac{\Psi_{j+1}^* - \Psi_{j-1}^*}{2\Delta x} \tag{7.66}$$

By eliminating the intermediate terms Ψ^*, one derives after some manipulations:

$$\Psi_j^{n+1} = \Psi_j^n - \frac{\alpha}{2} \left(\Psi_{j+1}^n - \Psi_{j-1}^n \right) + \left(\frac{\alpha}{2} \right)^2 \left(\Psi_{j+2}^n - 2\Psi_j^n + \Psi_{j-2}^n \right) \tag{7.67}$$

This explicit expression approximates an advection equation with an additional diffusion term that approaches zero for very small Δt:

$$\frac{\partial \Psi}{\partial t} + c \frac{\partial \Psi}{\partial x} - \frac{c^2 \Delta t}{4} \frac{\partial^2 \Psi}{\partial x^2} = 0 \tag{7.68}$$

The amplification coefficient is:

$$g(k) = 1 - i\,\alpha \sin(k\Delta x) - \alpha^2 \sin^2(k\Delta x) \tag{7.69}$$

with corresponding amplification factor:

$$|g(k)| = \left[1 - \alpha^2 \sin^2(k\Delta x) + \alpha^4 \sin^4(k\Delta x) \right]^{\frac{1}{2}} \tag{7.70}$$

Even though this scheme bears some resemblance to implicit schemes, it is stable for the usual Courant condition ($\alpha \leq 1$) of the explicit method rather than the condition that applies to implicit methods (unconditional stability).

If, in the Matsuno scheme, the correction step is replaced by

$$\frac{\Psi_j^{n+1} - \Psi_j^n}{\Delta t} = -c\,\frac{\Psi_{j+1}^* - \Psi_{j-1}^* + \Psi_{j+1}^n - \Psi_{j-1}^n}{4\Delta x} \tag{7.71}$$

in which the space derivative term is calculated as the average between the intermediate estimates and the estimates at time t_n, we obtain the *Heun* scheme (see Table 7.1), which is second-order accurate in space and time like the Crank–Nicholson algorithm described earlier. The method is unconditionally unstable unless a small diffusion term is artificially added to the advection equation. In this case, the scheme becomes conditionally stable. If, in the Heun scheme, the predictor step is a leapfrog algorithm (see Section 7.3.3), we obtain the method proposed by Kurihara (1965), which is second-order accurate in space and time, stable for the Courant condition, and free of numerical diffusion. Unlike the leapfrog method, it is not subject to drift, but it does not provide the exact solution for $\alpha = 1$.

Figures 7.7 and 7.8 show the amplification factors $|g|$ as a function of parameters $k\,\Delta x$ and $c\,\Delta t/\Delta x$ for several of the algorithms described above.

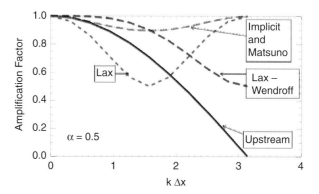

Figure 7.7 Amplification factor for different numerical methods as a function of parameter $k\,\Delta x$ for a Courant number $\alpha = 0.5$.

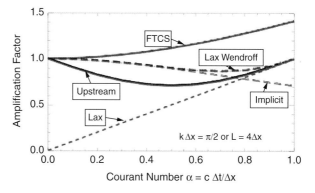

Figure 7.8 Amplification factor for different numerical methods as a function of the Courant number α for parameter $k\Delta x = \pi/2$ (corresponding to wavelength $L = 4\Delta x$).

Fourth-Order in Space Method

The algorithms discussed so far use low-order explicit or implicit forms of the finite difference equations. These algorithms can be extended to higher-order formulations. For example, the forward-in-time implicit form of the fourth-order approximation

$$\frac{\Psi_j^{n+1} - \Psi_j^n}{\Delta t} = -c \frac{\Psi_{j-2}^{n+1} - 8 \ \Psi_{j-1}^{n+1} + 8 \ \Psi_{j+1}^{n+1} - \Psi_{j+2}^{n+1}}{12\Delta x} \tag{7.72}$$

is unconditionally stable. The matrix corresponding to this system is a banded matrix with five terms that can be inverted with a fast method (Press *et al.*, 2007).

7.3.2 Methods Using Space-Uncentered Differences

In the algorithms discussed previously, the space derivative $\partial \Psi / \partial x$ is approximated by a second-order accurate centered difference. An alternative approach is to adopt a first-order accurate *backward-in-space finite difference*,

$$\frac{\partial \Psi}{\partial x} = \frac{\Psi_j - \Psi_{j-1}}{\Delta x} + O(\Delta x) \tag{7.73}$$

When introduced in the 1-D advection equation (7.17) together with a *forward in time derivative*, one obtains the *upstream* (or *upwind differencing*) *method* (Courant *et al.*, 1952; Godunov, 1959). Consistent with physical considerations, this algorithm provides a solution that depends on the behavior of Ψ in the direction from which the flow emanates, and not from the function downstream. Thus, for $c > 0$, we write a forward-in-time, backward-in-space (FTBS) expression:

$$\frac{\Psi_j^{n+1} - \Psi_j^n}{\Delta t} = -c \frac{\Psi_j^n - \Psi_{j-1}^n}{\Delta x} \quad \text{for } c > 0 \tag{7.74}$$

or equivalently:

$$\Psi_j^{n+1} = (1 - \alpha)\Psi_j^n + \alpha\Psi_{j-1}^n \quad \text{for } \alpha > 0 \tag{7.75}$$

For $c < 0$, the advection equation is approximated by a forward-in-time, forward-in-space (FTFS) expression:

$$\frac{\Psi_j^{n+1} - \Psi_j^n}{\Delta t} = -c \frac{\Psi_{j+1}^n - \Psi_j^n}{\Delta x} \quad \text{for } c < 0 \tag{7.76}$$

or

$$\Psi_j^{n+1} = (1 + \alpha)\Psi_j^n - \alpha\Psi_{j+1}^n \quad \text{for } \alpha < 0 \tag{7.77}$$

The amplification coefficient (for $c > 0$) is:

$$g(k) = 1 - \alpha[1 - \cos(k\Delta x)] - i\alpha \sin(k\Delta x) \tag{7.78}$$

with amplitude

$$|g(k)| = (1 + 2\alpha(\alpha - 1)[1 - \cos(k\Delta x)])^{\frac{1}{2}} \tag{7.79}$$

The amplitude remains below unity as long as the Courant condition ($\alpha \leq 1$) is verified. The phase $\Phi(k)$ is given by:

$$\tan \Phi = \frac{-\alpha \sin (k\Delta x)}{1 - \alpha(1 - \cos (k\Delta x))} \tag{7.80}$$

At the stability limit, when $\alpha = 1$, the amplitude $|g(k)| = 1$ and the phase $\Phi = -k\,\Delta x$. In this case, the solution provided by the upstream scheme is exact. In the general case with $\alpha < 1$, the solution is dampened (numerical diffusion) with the highest wavenumbers (or smallest wavelengths) more rapidly attenuated than the lower wavenumbers. This explains why the sharp corners of the square waves in Figures 7.10 and 7.12 are rounded by the upstream method.

The upstream method is monotonic and sign-preserving, but it is only first-order accurate in space and time and suffers therefore from numerical diffusion. This point can be intuitively understood by noting that the algorithm expressions (7.74) and (7.76) approximate to second-order in Δx and Δt the advection–diffusion equation:

$$\frac{\partial \Psi}{\partial t} + c\frac{\partial \Psi}{\partial x} = \frac{\partial}{\partial x}\left[K \frac{\partial \Psi}{\partial x}\right] \tag{7.81}$$

with diffusion coefficient $K = 0.5(c\,\Delta x - c^2 \Delta t) = 0.5\,c\,\Delta x(1 - \alpha)$.

Uncentered methods other than the upstream scheme have been proposed to reduce excessive numerical diffusion. For example, the approximation proposed by Warming and Beam (1976)

$$\Psi_j^{n+1} = \Psi_j^n - \alpha\left(\Psi_j^n - \Psi_{j-1}^n\right) - \frac{\alpha}{2}(1 - \alpha)\left(\Psi_j^n - 2\Psi_{j-1}^n + \Psi_{j-2}^n\right) \tag{7.82}$$

is second-order accurate in time and space and is stable for $0 \leq \alpha \leq 2$. It is equivalent to a Lax–Wendroff scheme in which the centered space differences are replaced by backward differences.

The Quadratic Upstream Interpolation for Convective Kinematics (QUICK) scheme of Leonard (1979) employs four points to approximate the first-order space derivative. For a constant wind velocity $c \geq 0$ and grid spacing Δx, the advection equation (7.17) is first discretized as a centered-in-space, time-forward explicit scheme:

$$\Psi_j^{n+1} = \Psi_j^n - \alpha\left(\Psi_{j+1/2}^n - \Psi_{j-1/2}^n\right) \tag{7.83}$$

where the values of the advected quantity at the left ($j - 1/2$) and right ($j + 1/2$) edges of cell j are determined by a quadratic interpolation. One derives, for example:

$$\Psi_{j+1/2} = \frac{1}{2}\left[\Psi_j + \Psi_{j+1}\right] - \frac{1}{8}\left[\Psi_{j-1} - 2\Psi_j + \Psi_{j+1}\right] \tag{7.84}$$

so that

$$\Psi_j^{n+1} = \Psi_j^n - \frac{\alpha}{8}\left(\Psi_{j-2}^n - 7\Psi_{j-1}^n + 3\Psi_j^n + 3\Psi_{j+1}^n\right) \tag{7.85}$$

The scheme is second-order accurate in space, but it is unstable unless some dissipation is added to the advection equation. Other formulations of the QUICK scheme (i.e., explicit, implicit, or semi-implicit approaches) are available (Chen and Falconer, 1992). A more elaborate algorithm, called QUICKEST, also proposed by Leonard (1979), is third-order accurate in time and space, and is stable for pure advection if $\alpha \leq 1$. In that scheme, the value of the function at the right edge is given by:

$$\Psi_{j+1/2} = \frac{1}{2}\left[\Psi_j + \Psi_{j+1}\right] - \frac{\alpha}{2}\left[\Psi_{j+1} - \Psi_j\right] - \frac{1}{8}\left(1 - \alpha^2\right)\left[\Psi_{j-1} - 2\Psi_j + \Psi_{j+1}\right] \quad (7.86)$$

The QUICK and QUICKEST schemes often generate overshoots and undershoots. They can therefore produce negative solutions. This problem is addressable by imposing flux-limiters in the integration scheme (see Section 7.5).

Finally, the algorithm proposed by Farrow and Stevens (1994), which can be regarded as an adaptation of the QUICK scheme, is expressed as a predictor-corrector integration scheme

$$\Psi_j^{n+1/2} = \Psi_j^n - \frac{\alpha}{4}\left(\Psi_{j+1}^n - \Psi_{j-1}^n\right) \quad (7.87)$$

$$\Psi_j^{n+1} = \Psi_j^n - \frac{\alpha}{2}\left[\Psi_{j+1}^{n+1/2} - \Psi_{j-1}^{n+1/2} - \frac{1}{4}\left(\Psi_{j+1}^{n+1/2} - 3\Psi_j^{n+1/2} + 3\Psi_{j-1}^{n+1/2} - \Psi_{j-2}^{n+1/2}\right)\right] \quad (7.88)$$

It is third-order accurate in space and second-order in time. A von Neumann stability analysis indicates that it is stable for Courant numbers smaller than approximately 0.6.

7.3.3 Multilevel Algorithms

In the numerical schemes discussed in previous sections, the time derivatives are approximated by a two-level forward difference. We now consider methods in which information from several earlier time levels are used to calculate the value of function Ψ at time t_{n+1}.

The *regular leapfrog method* (Courant *et al.*, 1928), which is second-order accurate in time, is based on a centered-in-time and centered-in-space (CTCS) approximation of the advection equation:

$$\frac{\Psi_j^{n+1} - \Psi_j^{n-1}}{2\Delta t} = -c\frac{\Psi_{j+1}^n - \Psi_{j-1}^n}{2\Delta x} \quad (7.89)$$

or

$$\Psi_j^{n+1} = \Psi_j^{n-1} - \alpha\left(\Psi_{j+1}^n - \Psi_{j-1}^n\right) \quad (7.90)$$

In this *three-level* algorithm, the solution "leapfrogs" from time level $(n-1)$ to time level $(n+1)$ over the time level (n) at which the space derivative term is computed.

The von Neumann stability analysis provides a quadratic equation for the amplification coefficient, whose two roots are:

$$g(k) = \pm\left[1 - \alpha^2 \sin^2(k\Delta x)\right]^{1/2} - i\,\alpha\sin(k\Delta x) \quad (7.91)$$

If $|\alpha \sin (k\,\Delta x)| > 1$, the square root term is completely imaginary, and the modulus $|g(k)|$ for one of the two roots is larger than 1, indicating instability. If $|\alpha \sin (k\,\Delta x)| \leq 1|$, which is verified for all wavenumbers when $|\alpha| \leq 1$ (CFL condition), the modulus is unity:

$$|g(k)| = \left\{ \left[1 - \alpha^2 \sin^2 (k\Delta x)\right] + \left[\alpha\ \sin (k\Delta x)\right]^2 \right\}^{1/2} = 1 \qquad (7.92)$$

and the phase shifts for the two roots ($\pm$) are respectively

$$\Phi_+ = -\sin^{-1}\left(\alpha \sin (k\Delta x)\right) \quad \text{and} \quad \Phi_- = \pi + \sin^{-1}\left(\alpha \sin (k\Delta x)\right) \qquad (7.93)$$

For Courant stable conditions, the amplitude of all waves is preserved, not dissipated. This represents the major advantage of the method. When $\alpha = 1$, the method provides the exact solution (correct amplitude and phase). If $\alpha < 1$, computational dispersion occurs as phase errors, particularly for short waves, and leads to some spurious numerical oscillations.

Expression (7.91) with a $\pm$ sign shows that the leapfrog algorithm generates two solutions with different amplification functions. One of them, called the *physical mode*, represents the meaningful solution, while the second one, referred to as the *computational mode*, is a mathematical artifact without any physical reality. This second solution propagates in the direction opposite to the flow and changes sign for every time step; it generates therefore noise that needs to be filtered with an appropriate method (see Section 4.15.4). The effect of the computational mode is visible in Figure 7.9, which shows the advection of a cosine-shaped function. Undesired oscillations with negative values of the function are produced upwind from large spatial gradients. The computational mode is most strongly excited when the initial conditions are characterized by sharp gradients.

The leapfrog algorithm tends to decouple odd and even grid points. Although, in principle, the solutions at these two types of grid point should not diverge, in practice they often do so as time progresses, causing *checkerboarding* of the solution. The problem can be addressed by adding a small dissipative term, by discarding

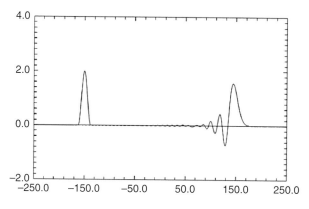

Figure 7.9   Advection by the leapfrog scheme of a cosine-shaped function with a half-width resolution of $12\Delta x$. The uniform grid is composed of 500 cells. The Courant number adopted in this example is 0.5. The solution is shown after 1600 time steps Δt. From Smolarkiewicz (2006).

the solutions at one of the two types of grid points, or by switching occasionally to an alternate advection scheme for just one time step.

Different improved leapfrog schemes have been proposed (Kim, 2003). The *upwind leapfrog scheme* introduced by Iserles (1986),

$$\frac{\left(\Psi_j^{n+1} - \Psi_j^n\right) + \left(\Psi_{j-1}^n - \Psi_{j-1}^{n-1}\right)}{2\Delta t} = -c\frac{\left(\Psi_j^n - \Psi_{j-1}^n\right)}{\Delta x} \tag{7.94}$$

or

$$\Psi_j^{n+1} = \Psi_{j-1}^{n-1} + (1 - 2\alpha)\left(\Psi_j^n - \Psi_{j-1}^n\right) \tag{7.95}$$

is characterized by a considerably lower phase error than the regular leapfrog scheme. Accuracy can be increased by adopting a fourth-order accurate spatial discretization:

$$\frac{\Psi_j^{n+1} - \Psi_j^{n-1}}{2\Delta t} = -c\frac{\Psi_{j-2}^n - 8\,\Psi_{j-1}^n + 8\,\Psi_{j+1}^n - \Psi_{j+2}^n}{12\Delta x} \tag{7.96}$$

for which the von Neumann stability analysis leads to:

$$g(k) = -i\frac{\alpha}{6}\left[8\sin\left(k\Delta x\right) - \sin\left(2k\Delta x\right)\right]$$

$$\pm\left\{1 - \left[\frac{\alpha}{6}\left[8\sin\left(k\Delta x\right) - \sin\left(2k\Delta x\right)\right]\right]^2\right\}^{1/2} \tag{7.97}$$

One derives easily that the scheme is stable if

$$\alpha < \frac{6}{8\sin\left(k\Delta x\right) - \sin\left(2k\Delta x\right)} \tag{7.98}$$

For $\alpha < 0.73$, the scheme is stable for all harmonics. The use of higher orders for the calculation of the time derivative also improves the accuracy of the solution. For example, the *four-level* algorithm:

$$\Psi_j^{n+1} = \Psi_{j-1}^{n-2} + 2(1 - 3\alpha)\left(\Psi_j^n - \Psi_{j-1}^{n-1}\right) + \frac{(1 - 2\alpha)(1 - 3\alpha)}{1 + \alpha}\left(\Psi_{j-1}^n - \Psi_j^{n-1}\right) \tag{7.99}$$

is very accurate and leads to exact solutions when $\alpha = 1/2$ or $\alpha = 1/3$. However, it is unstable for $\alpha > 1/2$.

Higher-order multi-stage methods are more accurate, but have the disadvantage of generating a larger number of computational modes. An interesting case is the third-order Adams–Bashforth scheme (see 4.197)):

$$\Psi_j^{n+1} = \Psi_j^n - \frac{\alpha}{24}\left[23\left(\Psi_{j+1}^n - \Psi_{j-1}^n\right) - 16\left(\Psi_{j+1}^{n-1} - \Psi_{j-1}^{n-1}\right) + 5\left(\Psi_{j+1}^{n-1} - \Psi_{j-1}^{n-2}\right)\right] \tag{7.100}$$

because the two undesired computational modes that are produced in this case are strongly damped if $|\alpha| < 0.72$. No filtering is required in most applications and this makes the algorithm particularly attractive, even though the solution is not positive definite. For higher values of $|\alpha|$, the amplitude of one of the computational modes becomes larger than 1, and the scheme becomes unstable.

7.3.4 Performance of Elementary Finite Difference Algorithms

Table 7.1 summarizes the properties of the different algorithms presented previously. Figure 7.10 shows a comparison for the 1-D advection of an initial square function with constant wind speed. The instability of the Euler forward algorithm is manifested in the large oscillations. The upstream algorithm preserves sign and is free of

Table 7.1 Elementary algorithms for solving the 1-D advection equation

Method	Algorithm	Stability	Accuracy	Remarks
Euler Forward	$\Psi_j^{n+1} = \Psi_j^n - \alpha/2\left(\Psi_{j+1}^n - \Psi_{j-1}^n\right)$	Unconditionally unstable	$\Delta t, \Delta x^2$	
Lax	$\Psi_j^{n+1} = 1/2\left(\Psi_{j+1}^n + \Psi_{j-1}^n\right) - \alpha/2\left(\Psi_{j+1}^n - \Psi_{j-1}^n\right)$	Stable for $\alpha < 1$	$\Delta t, \Delta x^2$	Diffusive
Leapfrog	$\Psi_j^{n+1} = \Psi_j^{n-1} - \alpha\left(\Psi_{j+1}^n - \Psi_{j-1}^n\right)$	Stable for $\alpha < 1$	$\Delta t^2, \Delta x^2$	Dispersive
Lax–Wendroff	$\Psi_j^{n+1} = \Psi_j^n - \alpha/2\left(\Psi_{j+1}^n - \Psi_{j-1}^n\right)$ $+\alpha^2/2\left(\Psi_{j+1}^n - 2\Psi_j^n + \Psi_{j-1}^n\right)$	Stable for $\alpha < 1$	$\Delta t^2, \Delta x^2$	
Implicit	$\Psi_j^{n+1} = \Psi_j^n - \alpha/2\left(\Psi_{j+1}^{n+1} - \Psi_{j-1}^{n+1}\right)$	Unconditionally stable	$\Delta t, \Delta x^2$	
Crank–Nicholson	$\Psi_j^{n+1} = \Psi_j^n - \alpha/4\left[\left(\Psi_{j+1}^n - \Psi_{j-1}^n\right) +\left(\Psi_{j+1}^{n+1} - \Psi_{j-1}^{n+1}\right)\right]$	Unconditionally stable	$\Delta t^2, \Delta x^2$	
Matsuno	$\Psi_j^{n+1} = \Psi_j^n - \alpha/2\left(\Psi_{j+1}^n - \Psi_{j-1}^n\right)$ $+\alpha^2/4\left(\Psi_{j+2}^n - 2\Psi_j^n + \Psi_{j-2}^n\right)$	Stable for $\alpha < 1$	$\Delta t, \Delta x^2$	Diffusive
Heun	$\Psi_j^{n+1} = \Psi_j^n - \alpha/2\left(\Psi_{j+1}^n - \Psi_{j-1}^n\right)$ $+\alpha^2/8\left(\Psi_{j+2}^n - 2\Psi_j^n + \Psi_{j-2}^n\right)$	Unconditionally unstable	$\Delta t^2, \Delta x^2$	
Kurihara	$\Psi_j^{n+1} = \Psi_j^n - \alpha/4[\left(\Psi_{j+1}^{n-1} - \Psi_{j-1}^{n-1}\right) +\left(\Psi_{j+1}^n - \Psi_{j-1}^n\right)]$ $+\alpha^2/4\left(\Psi_{j+2}^n - 2\Psi_j^n - \Psi_{j-2}^n\right)$	Stable for $\alpha < 1$	$\Delta t^2, \Delta x^2$	Not diffusive
Fourth-order (implicit)	$\Psi_j^{n+1} = \Psi_j^n - \alpha/12[\Psi_{j-2}^{n+1} - 8\Psi_{j-1}^{n+1}$ $+8\Psi_{j+1}^{n+1} - \Psi_{j+2}^{n+1}]$	Unconditionally stable	$\Delta t, \Delta x^4$	
Upstream ($\alpha > 0$)	$\Psi_j^{n+1} = \Psi_j^n - \alpha\left(\Psi_j^n - \Psi_{j-1}^n\right)$	Stable for $\alpha < 1$	$\Delta t, \Delta x$	Monotonic diffusive
Upstream ($\alpha < 0$)	$\Psi_j^{n+1} = \Psi_j^n - \alpha\left(\Psi_{j+1}^n - \Psi_j^n\right)$	Stable for $\alpha < 1$	$\Delta t, \Delta x$	Monotonic diffusive

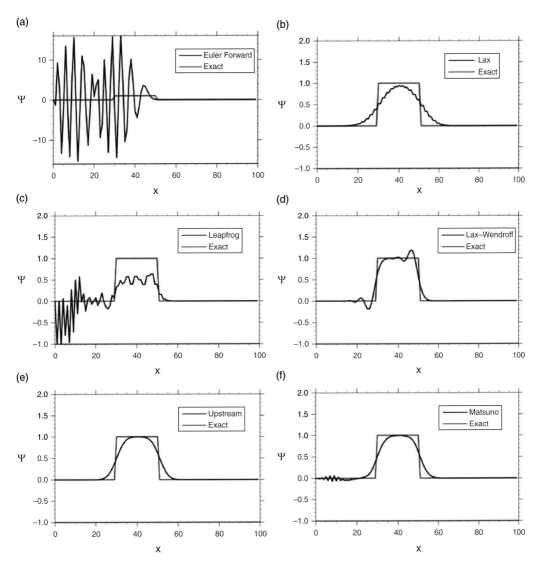

Figure 7.10   Comparison between exact analytic (red) and numerical (black) solutions of the 1-D advection equation for a square function. The velocity c is constant. The different numerical algorithms are labeled in the panels. The original square function is centered at $x = 20$ and the adopted Courant number is equal to 0.5. The periodic boundary condition for the advected field is zero at $x = 0$ and $x = 100$. The results are shown after 40 time steps. The Euler forward algorithm is unstable (note the different scale used for the y-axis).

oscillations (negligible phase lag), but it is very diffusive. The Lax and Matsuno algorithms are also very diffusive. The leapfrog algorithm is not diffusive and conserves the concentration variance but it produces oscillations with undesirable negative values. The Lax–Wendroff method is slightly diffusive and produces small unwanted oscillations with negative values. Filters are generally applied to avoid unphysical negative values, but such filters may destroy the conservation properties

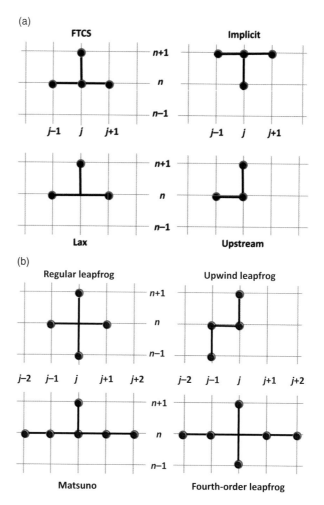

Figure 7.11 Stencils describing several numerical algorithms for the approximate solution of the 1-D advection equation.

of the numerical algorithms. The stencils associated with some of the algorithms are shown in Figure 7.11. Finally, Figure 7.12 shows the performance of different algorithms in the case of the diagonal advection of a square wave in a 2-D domain. Again, one notes the strong numerical diffusion associated with the upwind scheme and the presence of large oscillations in the case of the leapfrog scheme. The multidimensional definite advection transport algorithm (MPDATA) (Smolarkiewicz, 1984; see Section 7.6) provides positive definite solutions, but with large overshoots. The Lax–Wendroff scheme with flux limiters (see Section 7.5) performs best and the QUICKEST algorithms exhibit significant oscillations (Gross et al., 1999).

In summary, first-order methods such as the upstream algorithm are characterized by numerical diffusion in the solution and, as a result, tend to reduce the amplitude of peaks and to smooth spatial gradients that are present in the initial tracer distributions. High wavenumber components are also eliminated. Dispersion, which is common to the simple high-order methods described above, tends to distort the solution since it causes all waves, and specifically the small waves, to lag the true

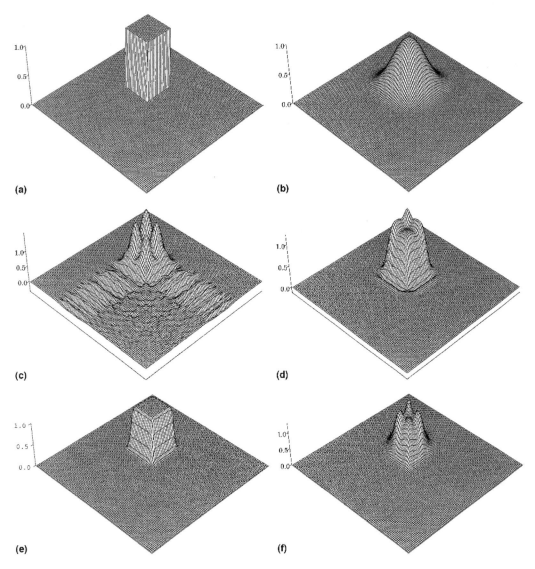

(a)

(b)

(c)

(d)

(e)

(f)

Figure 7.12 Two-dimensional advection of a square wave of width $20\Delta x$ and initial concentration $= 1$ advected over 50 grid cells in each coordinate direction for a Courant number of 0.25. (a) Exact solution; (b) first-order upstream; (c) leapfrog; (d) QUICKEST; (e) Lax–Wendroff with flux limiters, (f) MPDATA. Reproduced from Gross *et al.* (1999) with permission from the American Society of Civil Engineers (ASCE).

solution. Numerical diffusion is generally viewed as a lesser evil because it merely causes loss of information, while dispersion generates spurious information.

7.3.5 Generalization to Variable Wind Speed and Grid Size

The previous discussion has highlighted some fundamental properties of different Eulerian algorithms. In practical applications, the wind in the x-direction may not be uniform (c is replaced by $u(x, t)$), and the discretization interval Δx_j may vary along the spatial dimension x. In that case, the space derivative in the conservative 1-D flux-form equation:

$$\frac{\partial \Psi}{\partial t} + \frac{\partial(u\Psi)}{\partial x} = 0 \qquad (7.101)$$

is replaced by its second-order centered finite-difference approximation:

$$\frac{\partial(u\Psi)}{\partial x} = A_{j-1}(u\Psi)_{j-1} + B_j(u\Psi)_j + C_{j+1}(u\Psi)_{j+1} \qquad (7.102)$$

where

$$A_j = \frac{\Delta x_{j+1} - 2\Delta x_j}{\Delta x_j(\Delta x_j + \Delta x_{j+1})} \quad B_j = \frac{\Delta x_{j+1} - \Delta x_j}{\Delta x_j \Delta x_{j+1}} \quad C_j = \frac{\Delta x_j}{\Delta x_j(\Delta x_j + \Delta x_{j+1})}$$

and $\Delta x_j = x_j - x_{j-1}, \Delta x_{j+1} = x_{j+1} - x_j$. This center-difference scheme can be applied in the case of explicit, implicit, or Crank–Nicholson algorithms.

7.3.6 Mass Conservation

If we integrate the 1-D advection equation (7.17) over the spatial interval [A, B] and between time levels t_n and t_{n+1}, we find the conservation expression

$$\left[\int_A^B \Psi dx\right]^{n+1} = \left[\int_A^B \Psi dx\right]^n + c[\Psi_A - \Psi_B] \qquad (7.103)$$

where we have again assumed c to be fixed. Condition (7.103) must be met for tracer mass to be conserved with Ψ_A and Ψ_B representing boundary conditions. In the simple case of the FCTS scheme, it is easy to evaluate the left-hand side of this integral relation:

$$\Delta x\left[\sum_{j=j_A}^{j_B} \Psi_j^{n+1}\right] = \Delta x \sum_{j=j_A}^{j_B}\left(\Psi_j^n - c\frac{\Delta t}{2\Delta x}\left(\Psi_{j+1}^n - \Psi_{j-1}^n\right)\right) \qquad (7.104)$$

or

$$\Delta x\left[\sum_{j=j_A}^{j_B} \Psi_j^{n+1}\right] = \Delta x\left[\sum_{j=j_A}^{j_B} \Psi_j^n\right] + c\Delta t\left(\Psi_{j_A-1/2}^n - \Psi_{j_B+1/2}^n\right) \qquad (7.105)$$

if

$$\Psi_{j_A-1/2}^n = \left(\Psi_{j_A-1}^n + \Psi_{j_A}^n\right)/2 \quad \text{and} \quad \Psi_{j_B+1/2}^n = \left(\Psi_{j_B+1}^n + \Psi_{j_B}^n\right)/2$$

Thus, under the conditions adopted here, the accumulation of the conservative tracer Ψ in the domain [A, B] is proportional to the net flux ($c\ \Psi$) at the boundaries A and B. The finite-difference analog has preserved the integral expressed by the continuum equation (7.103). If the fluxes across the external boundaries are zero or if the domain is periodic with $\Psi_A = \Psi_B$, mass is fully conserved in the domain. If the constant velocity c is replaced by a velocity $u(x, t)$ that varies with space and time, tracer conservation will be generally obtained if one considers the finite difference analog of the flux-form equation (7.2) but not its advective form (7.7).

7.3.7 Multidimensional Cases

The 1-D advection problem can be generalized to multiple spatial dimensions. In a 2-D Cartesian space (x, y), the flux-conservative form of the advection equation is expressed by

$$\frac{\partial \Psi}{\partial t} + \frac{\partial(u\Psi)}{\partial x} + \frac{\partial(v\Psi)}{\partial y} = 0 \tag{7.106}$$

where u and v are the velocities of the wind components in the x and y directions, respectively. If we assume constant grid spacing, the discretization leads to the following expression

$$\frac{\Psi_{i,j}^{n+1} - \Psi_{i,j}^{n}}{\Delta t} + \frac{(u\Psi)_{i+1,j} - (u\Psi)_{i-1,j}}{2\Delta x} + \frac{(v\Psi)_{i,j+1} - (v\Psi)_{i,j-1}}{2\Delta y} = 0 \tag{7.107}$$

where Δx and Δy are the grid spacings in the x and y directions, and the indices i and j refer to these two directions respectively. If the differences in space are estimated at time t_n (explicit method), the algorithm is unconditionally unstable, but it can be made stable by adding numerical diffusion as in the 1-D case. If estimated at time t_{n+1} (implicit method), the algorithm is unconditionally stable. Expression (7.107) can be extended to three dimensions and solved for variable grid spacing. A Crank–Nicholson form of (7.107) can also be easily derived. The matrix of the system becomes rapidly very large and is not banded as in the 1-D case. Three-dimensional advection in practical applications is usually performed by operator splitting, with successive numerical solution of the 1-D advection equation over each dimension for individual time steps.

7.3.8 Boundary Conditions

The resolution of hyperbolic equations such as the advection equation applied to a limited spatial domain requires that a condition be imposed at the boundary through which material flows into the domain. In the 1-D case with a domain $[a, b]$, the condition must be specified at point $x = a$ if the velocity is positive $(c > 0)$ and at $x = b$ in the opposite situation $(c < 0)$. In some cases, the numerical algorithm requires that an additional condition be provided at the outflow boundary. By applying such a condition without precaution, the problem becomes ill-posed, and the algorithm may provide unstable solutions. This is the case when centered differences are used to represent space derivatives. Consider a 1-D flow on a domain $[a, b]$ with a positive constant velocity c and an inflow boundary condition $\Psi(a, t) = H(t)$ at location $x = a$. We approximate the spatial derivative by the leapfrog (CTCS) scheme $(j = 1, J)$:

$$\Psi_j^{n+1} = \Psi_j^{n-1} - \alpha\left(\Psi_{j+1}^n - \Psi_{j-1}^n\right) \tag{7.108}$$

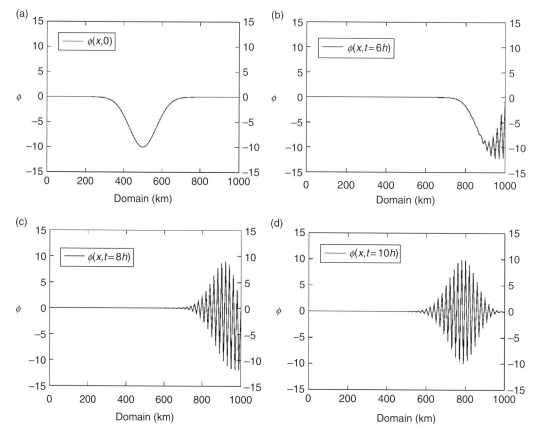

Figure 7.13 Advection of a bell-shaped function (arbitrary units) by a leapfrog scheme over a domain of 1000 km with a Courant number of 0.1 ($c = 20$ m s^{-1}, $\Delta t = 50$ s, $\Delta x = 10$ km) and boundary conditions of zero at the edges of the domain. Initial condition (a), solution after six hours (b), after eight hours (c) and after ten hours (d). From P. Termonia, Royal Meteorological Institute of Belgium.

At grid point $j = 1$, the solution is easily calculated:

$$\Psi_1^{n+1} = \Psi_1^{n-1} - \alpha\left(\Psi_2^n - \Psi_0^n\right) \tag{7.109}$$

since $\Psi_0^n = H(t_n)$ is specified. The calculation of the solution at grid point $j = J$,

$$\Psi_J^{n+1} = \Psi_J^{n-1} - \alpha\left(\Psi_{J+1}^n - \Psi_{J-1}^n\right) \tag{7.110}$$

is not straightforward because, in a well-posed problem, no value should be imposed at point $J + 1$. If a value is nevertheless imposed at this outflow boundary, e.g., $\Psi_{J+1}^n = \Psi_J^n$ or $\Psi_{J+1}^n = 0$, the scheme will produce unrealistic (unphysical) reflections that propagate upstream. An illustration is provided by Figure 7.13 that shows the advection of a bell-shaped function by a leapfrog scheme with $\Psi = 0$ at both limits of the domain. As the signal reaches the downwind boundary, spurious wave reflections (saw-toothed artifacts) are produced and propagate upstream. This is generalizable to multidimensional problems: Spurious reflections often occur at the lateral boundaries of a limited-domain nested model driven by boundary conditions from a larger-domain model. Even if the imposed boundary conditions verify the analytic solution of the advection equation, some reflections are to be expected since the numerical solution is slightly different.

In practical atmospheric applications, the sign of the wind velocity and hence the direction of the flow at the boundary of the domain frequently change as the model simulation proceeds. It is therefore difficult to identify the boundary at which a condition should be specified. This problem is usually addressed by imposing some conditions along the entire boundary of the model domain (ill-posed condition), while adding in the equations a relaxation term that damps the high-frequency signal produced at the downwind boundaries. In this case, the original advection equation is modified as (Davies, 1983):

$$\frac{\partial \Psi}{\partial t} + c\frac{\partial \Psi}{\partial x} = -\lambda(x)\left(\Psi - \widetilde{\Psi}\right) \tag{7.111}$$

where the relaxation coefficient $\lambda(x)$ is different from zero only in the boundary zones (a few grid cells near the inflow and outflow boundaries, called *buffer zones*) and $\widetilde{\Psi}$ is an externally specified field chosen to be close to the expected solution. The value of $\lambda(x)$ and the width of the relaxation zone (typically $2\Delta x$ to $6\Delta x$) need to be optimized to avoid the reflection of waves while minimizing perturbation to the solution. To ensure stability, the relaxation term should be estimated at time t_{n+1}.

An alternative damping scheme is to add a diffusion term to the advection equation:

$$\frac{\partial \Psi}{\partial t} + c\frac{\partial \Psi}{\partial x} = \frac{\partial}{\partial x}\left(K(x)\frac{\partial \Psi}{\partial x}\right) \tag{7.112}$$

where the diffusion coefficient $K(x)$ is non-zero only near the boundary zones. Davies (1983) discusses the stability conditions for this approach.

7.4 Elementary Finite Volume Methods

In *finite volume* approaches, rather than considering the values of function Ψ at specified points of a model grid, one calculates the average of this function over defined grid cells. The grid points are now viewed as the centers of grid cells, often called gridboxes. The cell boundaries are called cell *edges*, *walls*, or *interfaces*.

7.4.1 One-Dimensional Formulation

In the 1-D problem (x-direction), the location of the cell center is noted x_j, while the locations of the cell interfaces are noted $x_{j-1/2}$ (left side) and $x_{j+1/2}$ (right side). For each grid cell (j) (whose size is assumed here to be constant and equal to Δx):

$$x_{j+1/2} = \frac{1}{2}\left(x_j + x_{j+1}\right) \tag{7.113}$$

except at the left ($j = 1$) and right ($j = N$) boundaries of the model, where we adopt

$$x_{1/2} = x_1 - \frac{(x_2 - x_1)}{2} \qquad x_{N+1/2} = x_N + \frac{(x_N - x_{N-1})}{2} \tag{7.114}$$

The average value Ψ_j of the variable distribution $\psi(x, t)$ inside the cell is

$$\Psi_j = \frac{1}{\Delta x} \int_{x_{j-1/2}}^{x_{j+1/2}} \psi(x,t)\, dx \tag{7.115}$$

If $\psi(x, t)$ is a tracer concentration, then Ψ_j is the mean tracer concentration and $\Psi_j \Delta x$ the tracer mass in grid cell (j).

The exact (analytic) solution of the 1-D advection equation with fixed wind velocity c, when integrated over a time period Δt, is

$$\psi(x, t + \Delta t) = \psi(x - c\Delta t, \ t) \tag{7.116}$$

or, when the integral form is considered instead,

$$\frac{1}{\Delta x} \int_{x_{j-1/2}}^{x_{j+1/2}} \psi(x, \ t + \Delta t) \ dx = \frac{1}{\Delta x} \int_{x_{j-1/2}}^{x_{j+1/2}} \psi(x - c\Delta t, \ t) \ dx \tag{7.117}$$

Recognizing that the first term in (7.117) is equal to Ψ_j^{n+1}, and defining $x' = x - c\Delta t$, one can write:

$$\Psi_j^{n+1} = \frac{1}{\Delta x} \int_{x_{j-1/2}-c\Delta t}^{x_{j+1/2}-c\Delta t} \psi(x', t) \ dx' \tag{7.118}$$

Splitting this integral into different contributing parts, one writes equivalently:

$$\Psi_j^{n+1} = \frac{1}{\Delta x} \int_{x_{j-1/2}}^{x_{j+1/2}} \psi(x', t) \ dx' + \frac{1}{\Delta x} \int_{x_{j-1/2}-c\Delta t}^{x_{j-1/2}} \psi(x', t) \ dx' - \frac{1}{\Delta x} \int_{x_{j+1/2}-c\Delta t}^{x_{j+1/2}} \psi(x', t) \ dx' \tag{7.119}$$

or

$$\Psi_j^{n+1} = \Psi_j^n + \frac{1}{\Delta x} \int_{x_{j-1/2}-c\Delta t}^{x_{j-1/2}} \psi(x', t) \ dx' - \frac{1}{\Delta x} \int_{x_{j+1/2}-c\Delta t}^{x_{j+1/2}} \psi(x', t) \ dx' \tag{7.120}$$

The mean tracer concentration in grid cell j at time t_{n+1} is thus obtained by adding the mean value of the tracer concentration that enters grid cell (j) to the existing mean value in that cell at time t_n and removing the mean value that is transported downstream from cell (j) to cell $(j + 1)$. In this expression, the mass leaving the upwind donor cell equals the mass entering the neighboring downwind receptor cell. The finite volume method is therefore perfectly mass-conserving, which is its main advantage.

If $F_{j-1/2}^{n+1/2}$ and $F_{j+1/2}^{n+1/2}$ represent the mean fluxes through the left and right interfaces of grid cell j, respectively, averaged over time $\Delta t = t_{n+1} - t_n$, we write equivalently to (7.120):

$$\Psi_j^{n+1} = \Psi_j^n - \frac{\Delta t}{\Delta x} \left(F_{j+1/2}^{n+1/2} - F_{j-1/2}^{n+1/2} \right) \tag{7.121}$$

For the period Δt during which the subgrid function $\psi(x)$ is assumed to remain unchanged, donor cell $(j - 1)$ transfers a mass to receptor cell (j):

$$F_{j-1/2}^{n+1/2} \Delta t = \int_{x_{j-1/2}-c\Delta t}^{x_{j-1/2}} \psi(x) \ dx \tag{7.122}$$

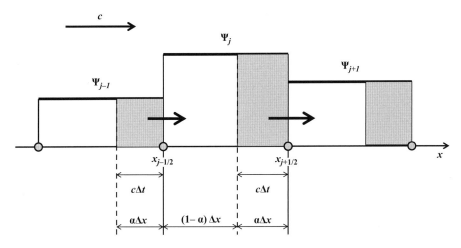

Representation of the *donor-cell scheme* in one dimension x. During time step Δt, the shaded area in cell j is transported in the x-direction to cell $j + 1$.

where the interval $[x_{j-1/2} - c\Delta t, x_{j-1/2}]$ corresponds to the shaded area in cell $(j-1)$ (Figure 7.14). Similarly, the mass transferred from donor grid cell j to receptor grid cell $(j+1)$ is

$$F_{j+1/2}^{n+1/2} \Delta t = \int_{x_{j+1/2}-c\Delta t}^{x_{j+1/2}} \psi(x)\ dx \qquad (7.123)$$

Different implementations of the finite volume method (i.e., different assumptions for the subgrid distribution of $\psi(x)$ inside each cell) lead to different estimates of the fluxes at the interfaces of the grid cells. The simplest assumption is that $\psi(x)$ is uniform inside each cell, so that the state Ψ_j at the grid center is identical to the state everywhere in the grid. This is the *donor cell* method. If the subgrid function $\psi(x)$ is assumed to vary linearly with position x inside each grid cell, the algorithm is called *piecewise linear*. If $\psi(x)$ is a second-order polynomial the algorithm is called *quadratic* or *piecewise parabolic*; see examples in Figure 7.15.

Donor-cell algorithm

In the simple donor-cell scheme with $c > 0$, the subgrid function $\psi(x)$ is uniform inside each cell, and the mass $F_{j-1/2} \Delta t$ advected from cell $(j-1)$ to cell (j) over a time period Δt is equal to $\Psi_{j-1} c \Delta t = \Psi_{j-1} \alpha \Delta x$, Simultaneously, a mass equal to $F_{j+1/2} \Delta t = \Psi_j c \Delta t = \Psi_j \alpha \Delta x$ is displaced from cell (j) to cell $(j+1)$. From (7.121), we find:

$$\Psi_j^{n+1} = \Psi_j^n - \alpha\left(\Psi_j^n - \Psi_{j-1}^n\right) = \Psi_{j-1}^n \alpha + \Psi_j^n(1 - \alpha) \qquad (7.124)$$

Similarly, if $c < 0$, we have

$$\Psi_j^{n+1} = \Psi_j^n - \alpha\left(\Psi_{j+1}^n - \Psi_j^n\right) = \Psi_j^n(1 - \alpha) - \Psi_{j+1}^n \alpha \qquad (7.125)$$

As shown by Figure 7.14 displayed for $c > 0$, term $\Psi_{j-1}^n \alpha \Delta x$ accounts for the mass transferred from the donor grid cell during the time step Δt, while term $\Psi_j(1 - \alpha) \Delta x$

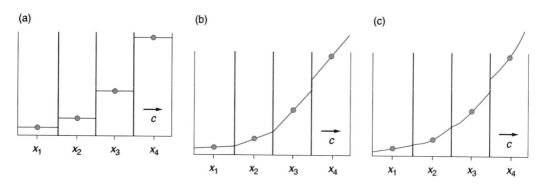

Figure 7.15 Representation of the spatial distribution of a tracer within four grid cells: zeroth-, first-, and second-order polynomials.

represents the mass that remains in cell j during this time period. The addition of these two terms represents the resulting mass in gridbox (j) at time t_{n+1}. This expression, derived in the simple case where $\psi(x)$ is assumed uniform, is identical to the upstream formula (7.74). The method is first-order accurate and is therefore characterized by large numerical diffusion.

Piecewise linear algorithm

In the case of the piecewise linear approach, we write for $x_{j-1/2} < x < x_{j+1/2}$:

$$\psi(x) = \Psi_j + b_j(x - x_j) \tag{7.126}$$

where Ψ_j denotes the value of linear function $\psi(x)$ at the center of the cell (also the mean value of $\psi(x)$ in the cell), and b_j is the *slope* of the function inside the cell. For a fixed velocity $c > 0$ and an equally spaced grid, the flux at the left cell interface is:

$$F_{j-1/2}^{n+1/2} = \frac{c}{\Delta t} \int_{t_n}^{t_{n+1}} \psi(x_{j-1/2}, t) \, dt = \frac{c}{\Delta t} \int_{t_n}^{t_{n+1}} \Psi_{j-1}^n + b_{j-1}^n \left(x_{j-1/2} - x_{j-1} - c(t - t_n)\right) \, dt \tag{7.127}$$

or

$$F_{j-1/2}^{n+1/2} = c\left[\Psi_{j-1}^n + \frac{1}{2}b_{j-1}^n(\Delta x - c\Delta t)\right] \tag{7.128}$$

Similarly, for $c < 0$, one finds

$$F_{j-1/2}^{n+1/2} = c\left[\Psi_j^n - \frac{1}{2}b_j^n(\Delta x + c\Delta t)\right] \tag{7.129}$$

From expression (7.119), one can easily deduce that the average value of subgrid function $\psi(x)$ in cell (j) at time t_{n+1} is given by:

$$\Psi_j^{n+1} = \Psi_j^n - \alpha\left(\Psi_j^n - \Psi_{j-1}^n\right) - \frac{\alpha}{2}\left(b_j^n - b_{j-1}^n\right)(1 - \alpha)\Delta x \tag{7.130}$$

if $c > 0$, and

$$\Psi_j^{n+1} = \Psi_j^n - \alpha\left(\Psi_{j+1}^n - \Psi_j^n\right) + \frac{\alpha}{2}\left(b_{j+1}^n - b_j^n\right)(1+\alpha)\Delta x \qquad (7.131)$$

if $c < 0$. This algorithm can be viewed as an extension of the donor-cell scheme with a correction term that disappears if the slope of function $\psi(x)$ is equal to zero.

The value of b_j is expressed as a function of the values of function Ψ at the center of adjacent cells. Different options are: (1) centered slope or Fromm method, (2) upwind slope or Beam–Warming method, and (3) downwind slope (equivalent to the Lax–Wendroff algorithm). The values of the b_j coefficients are respectively

$$b_j = \frac{\Psi_{j+1} - \Psi_{j-1}}{2\Delta x}, \qquad b_j = \frac{\Psi_j - \Psi_{j-1}}{\Delta x}, \qquad b_j = \frac{\Psi_{j+1} - \Psi_j}{\Delta x}$$

The resulting algorithm for $c > 0$ is in the case of the Fromm scheme (which is upwind biased)

$$\Psi_j^{n+1} = \Psi_j^n - \frac{\alpha}{4}\left(\Psi_{j+1}^n + 3\Psi_j^n - 5\Psi_{j-1}^n + \Psi_{j-2}^n\right) + \frac{\alpha^2}{4}\left(\Psi_{j+1}^n - \Psi_j^n - \Psi_{j-1}^n + \Psi_{j-2}^n\right) \quad (7.132)$$

In the case of the Beam–Warming scheme (also upwind biased), it is (see also expression 7.82)

$$\Psi_j^{n+1} = \Psi_j^n - \frac{\alpha}{2}\left(3\Psi_j^n - 4\Psi_{j-1}^n + \Psi_{j-2}^n\right) + \frac{\alpha^2}{2}\left(\Psi_j^n - 2\Psi_{j-1}^n + \Psi_{j-2}^n\right) \qquad (7.133)$$

and, in the case of the Lax–Wendroff scheme, we find the three-point stencil (spatially centered) expression that is identical to (7.52)

$$\Psi_j^{n+1} = \Psi_j^n - \frac{\alpha}{2}\left(\Psi_{j+1}^n - \Psi_{j-1}^n\right) + \frac{\alpha^2}{2}\left(\Psi_{j+1}^n - 2\Psi_j^n + \Psi_{j-1}^n\right) \qquad (7.134)$$

Figure 7.19 in Section 7.5 shows the numerical solution for advection of a step function obtained with the second-order accurate Fromm and Beam–Warming methods. The solution from the Lax–Wendroff algorithm was previously shown in Figure 7.10. In all three cases, the solution is not monotonic.

Other implementations of the finite volume method with specific subgrid distributions of function $\psi(x)$ (e.g., the algorithms of Russell and Lerner, 1981; Colella and Woodward, 1984; Prather, 1986) are discussed in Section 7.5.

7.4.2 Two-Dimensional Formulation

When extended to two dimensions (see Figure 7.16), the finite volume algorithm is expressed as:

$$\Psi_{i,j}^{n+1} = \Psi_{i,j}^n - \frac{\Delta t}{\Delta x}\left(F_{i+1/2,j}^{n+1/2} - F_{i-1/2,j}^{n+1/2}\right) - \frac{\Delta t}{\Delta y}\left(G_{i,j+1/2}^{n+1/2} - G_{i,j-1/2}^{n+1/2}\right) \qquad (7.135)$$

where F and G represent the mean fluxes in the x- and y-direction respectively. The challenge in defining accurate algorithms is to properly formulate the fluxes at the interfaces as a function of the dependent variables in the neighboring cells.

Another approach is to solve the discretized form (7.6) for the mean density $\langle\rho\rangle_j$ inside grid cell j (Dukowicz and Baumgardner, 2000; Lipscomb and Ringler, 2005; Miura, 2007; Skamarock and Menchaca, 2010):

$$\langle\rho\rangle_j^{n+1} = \langle\rho\rangle_j^n - \frac{\Delta t}{A_j}\sum L_c\left(\mathbf{F}_j^{n+1/2}\mathbf{n}\right) \qquad (7.136)$$

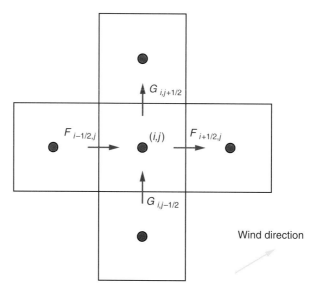

Figure 7.16 Representation of orthogonal flux components F (in the x-direction) and G (in the y-direction) across cell interfaces in two dimensions. The flux form adopted for the algorithm ensures mass conservation.

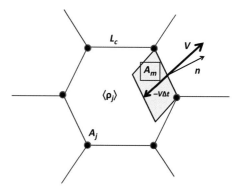

Figure 7.17 Schematic representation of the 2-D remapping algorithm of Miura (2007) in the case of a hexagonal cell. The shaded region represents the mass advected through the cell boundary over a time step Δt. Redrawn from Skamarock and Menchaca (2010).

Here, A_j is the area of the cell, $\mathbf{F}_j$ the mass flux across the interfaces of the cell, $\mathbf{n}$ a unit vector perpendicular to the cell boundaries, and L_c the length of each cell edge. The sum applies to all cell edges. The flux can be estimated through a remapping algorithm, as depicted in Figure 7.17 for a hexagonal cell where the fluid velocity $\mathbf{v}$ at one point of each cell boundary (e.g., center of the cell edge) is projected backward to define the upstream flux-area A_m (shaded parallelogram). The mean density in area A_m is derived by a polynomial fit using the mean densities in the neighboring cells at time level t_n. The flux $\mathbf{F}_j$ is then derived from the mass contained in area A_m that is displaced across the cell edge over a time interval Δt with velocity $\mathbf{v}$. The mass originating from all neighborhood cells is remapped onto cell j and provides the mean density in this cell at time level t_{n+1}. The accuracy of the scheme depends on the order of the polynomial that is adopted. In the incremental remapping of Dukowicz and Baumgardner (2000) and of Lipscomb and Ringler (2005), all endpoints on the grid are tracked back, so that the upstream flux area is a polygon. The method bears some similarities with semi-Lagrangian schemes discussed in Section 7.8.

7.5 Preserving Monotonicity: Flux-Corrected Transport

As stated in Section 7.4, the solutions provided by high-order accurate algorithms are not monotonic. Preserving monotonicity in the solution of the advection equation is an important requirement for chemical transport models. The generation of new extrema or "ripples" in the vicinity of steep gradients (including shocks and discontinuities of the solution) is unacceptable in most applications. Correction techniques have therefore been proposed to eliminate these unphysical maxima or minima caused by numerical dispersion in high-order algorithms.

One-dimensional flux-corrected advection algorithms are based on the finite volume approximation equation (7.121)

$$\Psi_j^{n+1} = \Psi_j^n - \frac{\Delta t}{\Delta x} \left(F_{j+1/2}^{n+1/2} - F_{j-1/2}^{n+1/2} \right)$$

in which $F_{j-1/2}^{n+1/2}$ and $F_{j+1/2}^{n+1/2}$ are again the flux averaged over the adopted time step at the edge of grid cell (j). The presence of spurious oscillations in the solution is avoided by preventing the total variation (TV) in the discrete representation of the solution

$$TV = \sum_j \left| \Psi_j - \Psi_{j-1} \right|$$

from increasing as the integration proceeds. This is accomplished by limiting the amplitude of the upstream and downstream fluxes, so that the following condition:

$$\sum_j \left| \Psi_j^{n+1} - \Psi_{j-1}^{n+1} \right| \leq \sum_j \left| \Psi_j^n - \Psi_{j-1}^n \right| \tag{7.137}$$

is fulfilled (total variation diminishing or TVD condition). An increase in the total variation (TV) is a measure of the formation of oscillations in the solution.

Fluxes can be limited in the discrete form of the advection equation by specifying the fluxes at each edge of the finite volume cells [here for ($j - 1/2$)] as

$$F_{j-1/2} = F_{j-1/2}^L - \Phi(r_{j-1/2}) \left[F_{j-1/2}^L - F_{j-1/2}^H \right] \tag{7.138}$$

where $F_{j-1/2}^L$ and $F_{j-1/2}^H$ represent the fluxes calculated by a low-order and a high-order method, respectively. The flux limiter functions $\Phi(r_{j-1/2})$ are expressed as a function of parameter $r_{j-1/2}$ defined as

$$r_{j-1/2} = \frac{\Psi_{j-1} - \Psi_{j-2}}{\Psi_j - \Psi_{j-1}} \qquad \text{for } c > 0 \tag{7.139}$$

$$r_{j-1/2} = \frac{\Psi_{j+1} - \Psi_j}{\Psi_j - \Psi_{j-1}} \qquad \text{for } c < 0 \tag{7.140}$$

Similar expressions can be established for the flux at the other edge $(j + 1/2)$ of the cell. One can show that, to fulfill the TVD condition, the flux limiter function must be chosen such

$$r \leq \Phi(r) \leq 2r \qquad \text{for} \qquad 0 \leq r \leq 1$$
$$1 \leq \Phi(r) \leq r \qquad \text{for} \qquad 1 \leq r \leq 2$$
$$1 \leq \Phi(r) \leq 2 \qquad \text{for} \qquad r > 2$$

In practical terms, the solution can be made monotonic by adopting for the $\Phi(r)$ a value close to zero in the vicinity of sharp gradients (low-order method) and a value close to 1–2 (higher-order method) in regions where the solution is expected to be smooth.

To illustrate the flux limiter method, we consider the linear piecewise scheme discussed in Section 7.4. In this particular case, the flux at the left interface of a grid cell (j) can be expressed by (7.128) and (7.129) or

$$F_{j-1/2}^{n+1/2} = c \left[\Psi_{j-1}^n + \frac{1}{2}(1 - \alpha) b_{j-1}^n \Delta x \right] \qquad \text{for} \quad c > 0 \tag{7.141}$$

$$F_{j-1/2}^{n+1/2} = c \left[\Psi_j^n - \frac{1}{2}(1 + \alpha) b_j^n \Delta x \right] \qquad \text{for} \quad c < 0 \tag{7.142}$$

A flux limiter required to preserve monotonicity is introduced by adjusting the second term in expression (7.142). This is accomplished by replacing $b_{j-1} \Delta x$ and $b_j \Delta x$ in equations (7.141–7.142) by $\Phi(r_{j-1/2}) (\Psi_j - \Psi_{j-1})$ The concept of flux limiter is therefore similar to the concept of *slope limiter*, which is sometimes used to characterize the flux-corrected transport. The resulting "corrected flux" is therefore

$$F_{j-1/2}^{n+1/2} = c \left[\Psi_{j-1}^n + \frac{1}{2}(1 - \alpha) \Phi\left(r_{j-1/2}^n\right)\left(\Psi_j^n - \Psi_{j-1}^n\right) \right] \qquad \text{for} \quad c > 0 \tag{7.143}$$

$$F_{j-1/2}^{n+1/2} = c \left[\Psi_j^n - \frac{1}{2}(1 + \alpha) \Phi\left(r_{j-1/2}^n\right)\left(\Psi_j^n - \Psi_{j-1}^n\right) \right] \qquad \text{for} \quad c < 0 \tag{7.144}$$

with $r_{j-1/2}$ defined by expressions (7.139) or (7.140), depending on the sign of the wind velocity c. By adjusting all indices in the above expression, one finds the value of the corrected flux at the right edge of cell (j), and the solution Ψ_j^{n+1} at time t_{n+1} is derived by applying (7.121) with an appropriate choice for the limiter $\Phi(r)$. Note that the choice of $\Phi(r) = 0$ and $\Phi(r) = 1$ for all values of r, corresponds the donor cell and Lax–Wendroff algorithms, respectively. Similarly, the choice $\Phi(r) = r$ and $\Phi(r) = (1 + r)/2$ corresponds respectively to the Beam–Warming and the Fromm schemes discussed in Section 7.4. None of these four limiters satisfy the TVD condition, and as a result the corresponding schemes do not provide monotonic solutions.

Several formulations for flux limiters that satisfy the TVD condition and hence lead to monotonic solutions have been proposed (Roe, 1986). The following expressions

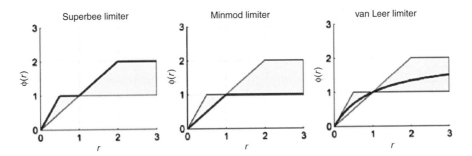

Figure 7.18
Flux/slope limiter functions (blue curve) for the superbee, minmod and van Leer algorithms superimposed on the regions (shaded) in which the TVD condition is met. Courtesy of Graham W. Griffiths.

$$\Phi(r) = \max \left[0, \min\left(1, r\right)\right]$$

define the *minmod* method,

$$\Phi(r) = \max \left[0, \min\left(2r, 1\right), \min\left(r, 2\right)\right]$$

the *superbee* method, and

$$\Phi(r) = \frac{r + |r|}{1 + |r|}$$

the *van Leer* algorithm. Figure 7.18 shows as a function of parameter r the values of the three limiters $\Phi(r)$, as well as the domain in which the TVD condition is met. Figure 7.19 shows how the application of a minmod and superbee flux correction improves the solution. Several other limiters have been proposed to enforce monotonicity of the solution.

An interesting numerical method that overcomes the excessive diffusion of the upstream algorithm and provides monotonic solutions is the flux-corrected scheme developed by van Leer (1977, 1979). To describe this algorithm, we start again from the finite volume expression

$$\Psi_j^{n+1} = \Psi_j^n - \frac{\Delta t}{\Delta x}\left(F_{j+1/2} - F_{j-1/2}\right) \tag{7.145}$$

where Ψ_j applies to the center of cell j and $F_{j+1/2}$ and $F_{j-1/2}$ are the time-averaged flux across boundaries $j + \frac{1}{2}$ and $j - \frac{1}{2}$, respectively. We assume here that the wind field is not uniform, and $F_{j+1/2}$ is therefore computed as the product of the velocity $u_{j+1/2}$ by an estimate of Ψ at the grid cell boundary. This estimate is obtained from a Taylor's series expansion on the gridded Ψ field:

$$F_{j-1/2}^{n+1/2} = u_{j-1/2}^n \left[\Psi_{j-1}^n + \frac{1}{2}\left(1 - u_{j-1/2}^n \frac{\Delta t}{\Delta x}\right)\Delta_{j-1}^n\Psi\right] \qquad \text{for} \quad u_{j-1/2}^n > 0 \tag{7.146}$$

$$F_{j-1/2}^{n+1/2} = u_{j-1/2}^n \left[\Psi_j^n - \frac{1}{2}\left(1 + u_{j-1/2}^n \frac{\Delta t}{\Delta x}\right)\Delta_j^n\Psi\right] \qquad \text{for} \quad u_{j-1/2}^n < 0 \tag{7.147}$$

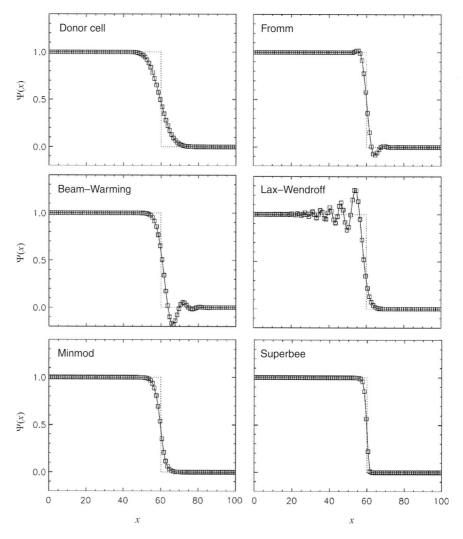

Figure 7.19 Advection of a sharp discontinuity (step function) using six different algorithms: the first-order diffusive *donor-cell* scheme; the second-order non-monotonic *Fromm* and *Beam–Warming* algorithms; and the flux-corrected *minmod* and *superbee* methods. The results are obtained after 300 time steps with $\Delta t = 0.1$ over a grid of 100 points with a spacing $\Delta x = 1$. From C. P. Dullemond with permission.

where $\Delta_j^n \Psi / \Delta x$ corresponds to the slope b_j of the subgrid function $\psi(x)$. Van Leer proposes for $\Delta_j^n \Psi$ an expression that limits the flux at the edge of the grid cells (see also Allen *et al.*, 1991):

$$\Delta_j^n \Psi = 2 \frac{\left(\Psi_j^n - \Psi_{j-1}^n\right)\left(\Psi_{j+1}^n - \Psi_j^n\right)}{\left(\Psi_{j+1}^n - \Psi_{j-1}^n\right)} \tag{7.148}$$

if $(\Psi_j - \Psi_{j-1})(\Psi_{j+1} - \Psi_j) > 0$, and by $\Delta_j^n \Psi = 0$ otherwise. The algorithm is easily extended to two dimensions by using expression (7.135) rather than (7.145) as the initial step. Although considerably less diffusive than the upstream scheme, this algorithm still contains scale-dependent diffusion. Once the spatial distribution of the transported quantity has diffused to a preferred shape, the diffusion decreases considerably.

7.6 Advanced Eulerian Methods

Several advanced schemes for solving the advection equation have been developed with the purpose of avoiding the spurious oscillations found in high-order methods and the excessive numerical diffusion characteristic of low-order methods. They can be viewed as an extension of some of the fundamental methods discussed in the previous sections.

The MPDATA Scheme of Smolarkiewicz

The *Multidimensional Definite Advection Transport Algorithm* (MPDATA) proposed by Smolarkiewicz (1983, 1984) focuses on compensating the first-order error of the upstream scheme by reducing the implicit numerical diffusion. Starting from the upstream scheme:

$$\Psi_j^* = \Psi_j^n - \alpha\left(\Psi_j^n - \Psi_{j-1}\right) \tag{7.149}$$

which provides a first guess Ψ_j^* for the solution at time t_{n+1}, the algorithm uses a second step in which the velocity c is replaced by a compensatory "anti-diffusion velocity" u^A defined as

$$u^A = \frac{K_{visc}}{\Psi}\frac{\partial\Psi}{\partial x} \quad \text{for } \Psi > 0 \quad \text{and} \quad u^A = 0 \text{ for } \Psi = 0 \tag{7.150}$$

with $K_{visc} = 0.5(c\Delta x - c^2\Delta t) = 0.5\,c\,\Delta x(1 - \alpha)$. Here, the ratio $(1/\Psi)\,\partial\Psi/\partial x$ is calculated iteratively using the latest estimate Ψ_j^{n+1} of the solution (Ψ_j^* at the first iterative step). The value of the anti-diffusion velocity at half mesh point $j + \frac{1}{2}$ is therefore

$$u^A = \frac{K_{visc}}{\Psi}\frac{\partial\Psi}{\partial x} = \frac{2K_{visc}}{\Delta x}\left(\frac{\Psi_{j+1}^{n+1} - \Psi_j^{n+1}}{\Psi_{j+1}^{n+1} + \Psi_j^{n+1} + \varepsilon}\right) \tag{7.151}$$

where ε is a small value that ensures that u^A is equal to zero when Ψ_j^* and Ψ_{j+1}^* are equal to zero. The "anti-diffusion" step becomes:

$$
\begin{aligned}
\Psi_j^{n+1} = \Psi_j^* &- \frac{\Delta t}{2\Delta x}\left[\left(u_{j+1/2}^A + |u_{j+1/2}^A|\right)\Psi_j^* + \left(u_{j+1/2}^A - |u_{j+1/2}^A|\right)\Psi_{j+1}^*\right] \\
&+ \frac{\Delta t}{2\Delta x}\left[\left(u_{j+1/2}^A + |u_{j+1/2}^A|\right)\Psi_j^* + \left(u_{j+1/2}^A - |u_{j+1/2}^A|\right)\Psi_{j-1}^*\right]
\end{aligned} \tag{7.152}
$$

Several iterations can be performed to improve accuracy.

This simple and computationally efficient algorithm is positive definite (if the initial condition is positive) with considerably less implicit diffusion than in the

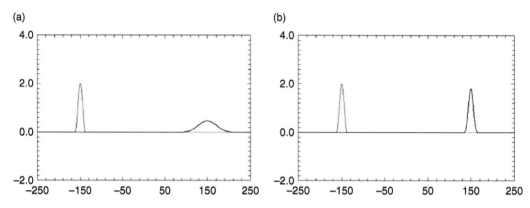

Figure 7.20 Comparison between two advection algorithms applied to a cosine-shaped function (resolved with 12 intervals). (a) first-order upwind scheme; (b) second-order accurate Smolarkiewicz scheme in which the numerical diffusion that characterizes the upwind scheme is compensated by the introduction of an "anti-diffusion" velocity. As in Figure 7.9, the adopted grid is uniform with 500 cells. The Courant number is 0.5 and the solution is shown after 1600 time steps. From Smolarkiewicz (2006).

upstream method (Figure 7.20). It is stable under the CFL condition. It does not preserve monotonicity of the transported quantities and, in general, the solutions are not free from small oscillations. The algorithm can easily be extended to multiple dimensions, with an anti-diffusion pseudo velocity defined in each direction. Smolarkiewicz (2006) expanded his MPDATA algorithm to arbitrary finite volume frameworks.

The SHASTA Scheme of Boris and Book

The *Sharp and Smooth Transport Algorithm* (SHASTA) proposed by Boris and Book (1973) is an Eulerian finite difference algorithm that makes use of the flux-corrected transport (FCT) technique described in Section 7.5. It ensures monotonicity of the solution, conserves mass, and handles steep gradients and shocks particularly well. The scheme includes an advection step followed by a corrective step that reduces the effect of the diffusion produced by the first step. We consider here the 1-D case and assume a variable velocity $u(x)$.

Advection step. The SHASTA algorithm first defines fluid elements formed by connecting linearly adjacent values (Ψ_j and Ψ_{j+1} in Figure 7.21). Each resulting trapezoidal element is displaced by the distance $u \, \Delta t$. Since the wind velocity $u(x)$ is variable in space, the advection is not a simple translation of the initial element; contraction or dilatation along x can take place. We assume a Courant number less than 0.5, so that the function at grid point j can never be advected further than the grid cell boundaries. After the displacement of the function is completed, the displaced elements are interpolated back onto the original Eulerian grid (see Figure 7.21).

In their algorithm, Boris and Book (1973) prescribe the wind velocities at the grid points x_{j-1}, x_j, x_{j+1}, and at the intermediate time level $t_{n+1/2}$. They deduce at each grid point j a first approximate value for the function Ψ and time level t_{n+1}:

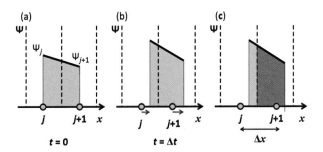

Figure 7.21 Advection of a fluid element. (a) Initial condition. (b) Location and shape of the fluid element after the advection step. During a time step Δt the two boundaries of the fluid element at location j and $j + 1$ are displaced by a distance $u_j \Delta t$ and $u_{j+1} \Delta t$, respectively. Here, the wind velocities are provided at the intermediate time $t_{n+1/2}$. At the end of the advection step, the fluid element is deformed if the velocity $u(x)$ is not uniform. (c) Interpolation of the fluid element onto the grid. The light orange fraction remains in cell j while the darker orange fraction goes into cell $j + 1$. Adapted from Boris and Book (1973).

$$\Psi_j^* = \frac{1}{2} Q_-^2 \left(\Psi_{j-1}^n - \Psi_j^n \right) + \frac{1}{2} Q_+^2 \left(\Psi_{j+1}^n - \Psi_j^n \right) + (Q_- + Q_+) \Psi_j^n \tag{7.153}$$

where

$$Q_- = \frac{\frac{1}{2} + u_j^{n+1/2} \frac{\Delta t}{\Delta x}}{1 - \frac{\Delta t}{\Delta x} \left(u_{j-1}^{n+1/2} - u_j^{n+1/2} \right)} \tag{7.154}$$

and

$$Q_+ = \frac{\frac{1}{2} - u_j^{n+1/2} \frac{\Delta t}{\Delta x}}{1 + \frac{\Delta t}{\Delta x} \left(u_{j+1}^{n+1/2} - u_j^{n+1/2} \right)} \tag{7.155}$$

For a uniform velocity c, the displacement of the trapezoid corresponds to a translation without deformation, and expression (7.153) becomes

$$\Psi_j^* = \Psi_j^n - \frac{\alpha}{2} \left(\Psi_{j+1}^n - \Psi_{j-1}^n \right) + \left(\frac{1}{8} + \frac{\alpha^2}{2} \right) \left(\Psi_{j+1}^n - 2\Psi_j^n + \Psi_{j-1}^n \right) \tag{7.156}$$

with $\alpha = c \, \Delta t/\Delta x$. This expression includes a two-sided differencing expression that approximates the advection, and a diffusion approximation where the diffusion coefficient is the sum of an independent term $(1/8)$ and a velocity-dependent term $(\alpha^2/2)$. This second term is smaller since $c \, \Delta t/\Delta x$ is chosen to be less than 0.5. Without the velocity-independent diffusivity, (7.156) is identical to the Lax–Wendroff algorithm. Following the von Neumann analysis, the amplification coefficient associated with the advective step is

$$g(k) = 1 + \left(\frac{1}{4} + \alpha^2 \right) [1 - \cos(k\Delta x)] - i\alpha \sin(k\Delta x) \tag{7.157}$$

and the corresponding amplification factor is

$$|g(k)| = \left\{ \left(1 - \frac{1}{4}(1 - \cos{(k\Delta x)}) \right)^2 - \frac{\alpha^2}{2}(1 - 2\alpha^2)\left((1 - \cos{(k\Delta x)})^2 \right) \right\}^{1/2}$$

(7.158)

The value of this factor is smaller than one for all wave harmonics if the Courant number is less than 0.5.

Correction (anti-diffusion) step. Assuming that the diffusivity in (7.156) is only weakly velocity-dependent ($\alpha \ll 0.5$), a second step is applied to remove the excessive diffusion produced by the advection step. Thus, we write:

$$\Psi_j^{n+1} = \Psi_j^* - \frac{1}{8}\left(\Psi_{j+1}^* - 2\Psi_j^* + \Psi_{j-1}^* \right)$$

(7.159)

where Ψ_j^* is the approximation for the transported function derived by the first (advective) step. This can be rewritten as

$$\Psi_j^{n+1} = \Psi_j^* - \left(f_{j+1/2} - f_{j-1/2} \right)$$

(7.160)

where

$$f_{j\pm1/2} = \pm\frac{1}{8}\left(\Psi_{j\pm1}^* - \Psi_j^* \right)$$

(7.161)

represents the amount of material ("flux") crossing the boundaries of grid cell j during the time step Δt. The amplification coefficient associated with the anti-diffusion step

$$g(k) = 1 + \frac{1}{4}(1 - \cos{(k\Delta x)})$$

(7.162)

is real, so that the anti-diffusion step does not affect the phase properties of the solution.

The overall amplification factor for the two consecutive steps is:

$$|g(k)| = \left\{ \left(1 - \frac{1}{16}(1 - \cos{(k\Delta x)})^2 \right)^2 \right.$$

$$\left. - \frac{\alpha^2}{2}(1 - 2\alpha^2)(1 - \cos{(k\Delta x)})^2\left(1 + \frac{1}{4}(1 - \cos{(k\Delta x)}) \right)^2 \right\}^{1/2}$$

(7.163)

in which the velocity-dependent (or α-dependent) term is generally small. Again, the method is stable when the amplification factor is less than or equal to 1.

The anti-diffusion correction step can introduce spurious extrema and negative values, which can again be avoided by applying an FCT constraint. The monotonicity of the solution is indeed preserved if the anti-diffusion flux f never produces values for Ψ at any grid point j that are larger than the values at the neighboring points. This is achieved if, rather than using (7.161), the value of the flux is replaced by the following FCT condition:

$$f_{j+1/2} = \text{sign}(\Delta_{j+1/2})\max\left\{ 0, \min\left[\Delta_{j-1/2}\text{sign}(\Delta_{j+1/2}), \frac{1}{8}|\Delta_{j+1/2}|, \Delta_{j+3/2}\text{sign}(\Delta_{j+1/2}) \right] \right\}$$

(7.164)

where

$$\Delta_{j+1/2} = \Psi_{j-1} - \Psi_j \tag{7.165}$$

The FCT step can be improved by replacing the factor 1/8 in (7.164) with a factor that accounts for the velocity dependence of the diffusivity. Further improvements to the SHASTA method that lead to more accurate solutions have been introduced by Boris and Book (1976).

The Piecewise Parabolic Method

In their piecewise parabolic method (PPM), Colella and Woodward (1984) assume that the subgrid distribution $\psi(x)$ of the tracer concentration inside cell j can be represented by a quadratic function:

$$\psi(x) = \psi_{j-1/2} + y(x)\left[\psi_{j+1/2} - \psi_{j-1/2} + d_j(1 - y(x))\right] \tag{7.166}$$

where $y = (x - x_{j-1/2})/\Delta x$. Coefficients $\psi_{j-1/2}$ and $\psi_{j+1/2}$ are the values of $\psi(x)$ at the boundaries $x_{j-1/2}$ and $x_{j+1/2}$ of cell j, and

$$d_j = 6\left[\Psi_j^n - \frac{1}{2}\left(\psi_{j+1/2} + \psi_{j-1/2}\right)\right] \tag{7.167}$$

where Ψ_j^n is the mean concentration in grid cell j at time t_n. For constant spacing Δx, it can be shown through interpolation from Ψ_j^n in nearby zones that the value of $\psi(x)$ at for example $x_{j+1/2}$ can usually be expressed as:

$$\psi_{j+1/2} = \frac{7}{12}\left(\Psi_j^n + \Psi_{j+1}^n\right) - \frac{1}{12}\left(\Psi_{j+2}^n + \Psi_{j-1}^n\right) \tag{7.168}$$

When the grid cells are unequally spaced, the expressions for $\psi_{j+1/2}$ and $\psi_{j-1/2}$ are more complicated. Colella and Woodward (1984) propose a slightly modified interpolation procedure in the presence of sharp discontinuities (shocks) to ensure that these discontinuities remain sharp during the advection step. The effect of this "steepening" process (Carpenter et al., 1990) is shown in Figure 7.22. Large discontinuities can still produce oscillations in the post-shock flow. In this case, it is advised to introduce some dissipation in the neighborhood of the shock. This can be achieved, for example, by flattening the interpolation profile in the vicinity of the discontinuity, which is equivalent to reducing locally the order of the method. In this case, coefficient $\psi_{j+1/2}$ can be replaced, for example, by

$$\psi_{j+1/2}^{flat} = \Psi_j f_j + \psi_{j+1/2}\left(1 - f_j\right) \tag{7.169}$$

where $f_j \in [0, 1]$ is an adjustable factor. Far away from discontinuities or if the shock profile is sufficiently broad, coefficient f_j should be set to zero.

The solution at time t_{n+1} is obtained by:

$$\Psi_j^{n+1} = \Psi_j^n - \alpha\left(a_{j+1/2}^n - a_{j-1/2}^n\right) \tag{7.170}$$

where, for example,

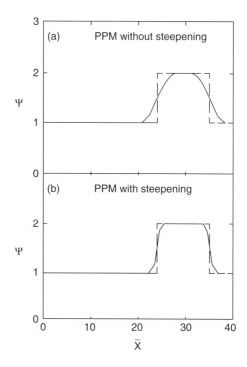

Advection of a square function using the piecewise parabolic method (PPM) (Carpenter *et al.*, 1990) without steepening (a) and with steepening (b). The spatial domain extends over 40 Δx with cyclic boundary conditions. The Courant number $\alpha = 0.5$. The numerical solution is shown after 1000 time steps (12.5 revolutions). From Müller (1992). Copyright © American Meteorological Society, used with permission.

$$a_{j+1/2}^n = \psi_{j+1/2} - \frac{\alpha}{2}\left[\psi_{j+1/2} - \psi_{j-1/2} - \left(1 - \frac{2\alpha}{3}\right)d_j\right] \qquad (7.171)$$

Other algorithms for rendering the PPM shape-preserving are presented by Colella and Sekora (2008).

The Crowley–Tremback–Bott Scheme

To improve the accuracy of the first-order upstream method, Crowley (1968), Tremback *et al.* (1987), and Bott (1989a, 1989b) have also proposed to represent the transported quantity within each grid cell j by a polynomial $\psi_{j,\ell}$ of order ℓ (with ℓ assumed to be an even integer number). Thus, at time level n, we write

$$\psi_{j,\ell}^n(y) = \sum_{k=0}^{\ell} a_{j,k}^n y^k \qquad (7.172)$$

where $y = (x - x_j)/\Delta x$ is a dimensionless variable such that $-\frac{1}{2} \leq y \leq \frac{1}{2}$. We assume again that the grid spacing Δx is uniform. Coefficients $a_{j,k}^n$ are determined from the requirement that the value of $\psi_{j,\ell}^n(y)$ agree with the value of Ψ_j^n at grid points

Table 7.2 Coefficients $a_{j,k}$ for the $\ell = 2$ and $\ell = 4$ versions of the Bott's area preserving flux form algorithm (after Bott, 1989a and Chlond, 1994)

	$\ell = 2$	$\ell = 4$
$a_{j,0}$	$-\dfrac{1}{24}\left[\Psi_{j+1} - 26\Psi_j + \Psi_{j-1}\right]$	$\dfrac{1}{1920}\left[9\Psi_{j+2} - 116\Psi_{j+1} + 2134\Psi_j - 116\Psi_{j-1} + 9\Psi_{j-2}\right]$
$a_{j,1}$	$\dfrac{1}{2}\left[\Psi_{j+1} - \Psi_{j-1}\right]$	$\dfrac{1}{48}\left[-5\Psi_{j+2} + 34\Psi_{j+1} - 34\Psi_{j-1} + 5\Psi_{j-2}\right]$
$a_{j,2}$	$\dfrac{1}{2}\left[\Psi_{j+1} - 2\Psi_j + \Psi_{j-1}\right]$	$\dfrac{1}{48}\left[-3\Psi_{j+2} + 36\Psi_{j+1} - 66\Psi_j + 36\Psi_{j-1} - 3\Psi_{j-2}\right]$
$a_{j,3}$	–	$\dfrac{1}{12}\left[\Psi_{j+2} - 2\Psi_{j+1} + 2\Psi_{j-1} - \Psi_{j-2}\right]$
$a_{j,4}$	–	$\dfrac{1}{12}\left[\Psi_{j+2} - 4\Psi_{j+1} + 6\Psi_j - 4\Psi_{j-1} + \Psi_{j-2}\right]$

($i = j - \ell/2, \ldots, j, \ldots j + \ell/2$), and that the area covered by $\psi_{j,\ell}^n(y)$ in grid cell j equals $\Psi_j^n \, \Delta x$. Thus, the coefficients $a_{j,k}^n$ are expressed as a function of the values of $\Psi_j, \Psi_{j+1}, \ldots$ at the $(\ell + 1)$ neighboring points. Table 7.2 provides the values of the coefficients derived for second-order and fourth-order polynomials, while Tremback *et al.* (1987) also considers higher order schemes.

As before, the solution for grid cell j at time t_{n+1} is provided by the finite-volume approximation:

$$\Psi_j^{n+1} = \Psi_j^n - \frac{\Delta t}{\Delta x}\left(F_{j+1/2} - F_{j-1/2}\right)$$

In the Crowley–Tremback–Bott scheme, the fluxes $F_{j+1/2}$ and $F_{j-1/2}$ at the right and left boundaries of grid cell j are estimated from (7.122) and (7.123) in which $\psi_j(y, t)$ is replaced by its polynomial approximation (7.172) of order ℓ.

In the more general case where the velocity u in the x-direction is spatially variable, (7.123) is replaced by (see Bott, 1989a, 1989b; Chlond, 1994)

$$F_{j+1/2} = \frac{\Delta t}{\Delta x}\left(I_{j+1/2}^+ - I_{j+1/2}^-\right) \tag{7.173}$$

where

$$I_{j+1/2}^+ = \int_{1/2-\alpha^+}^{1/2} \psi_j(z, t)\,dz \quad \text{and} \quad I_{j+1/2}^- = \int_{-1/2}^{1/2-\alpha^+} \psi_j(z, t)\,dz$$

are area integrals in which

$$z = (y - x_j)/\Delta x$$

$$\alpha^+ = \alpha_{j+1/2}^+ = \max\left(0, u_{j+1/2}^n \Delta t/\Delta x\right) \quad \text{and} \quad \alpha^- = \alpha_{j+1/2}^- = \max\left(0, u_{j+1/2}^n \Delta t/\Delta x\right)$$

Again, function $\psi_j(z, t)$ can be approximated by an area-preserving polynomial (7.172). Substitution of this polynomial into the above expressions yields the following expressions for integrals $I_{j+1/2}^+$ and $I_{j+1/2}^-$:

$$I_{j+1/2,\ell}^+ = \sum_{k=0}^{\ell} a_{j,k} \frac{1 - (1 - 2\alpha^+)^{k+1}}{(k+1)2^{k+1}} \qquad (7.174)$$

and

$$I_{j+1/2,\ell}^- = \sum_{k=0}^{\ell} a_{j,k}(-1)^k \frac{1 - (1 - 2\alpha^-)^{k+1}}{(k+1)2^{k+1}} \qquad (7.175)$$

This polynomial fitting method is not exempt from localized unphysical oscillations near sharp spatial gradients. To address the problem, Bott (1989a, 1989b) introduces nonlinear flux limiters and imposes that the total amount of outflow from gridbox j during a time step Δt be limited to $\Psi_j^n \Delta x/\Delta t$. In addition, the flux $F_{j+1/2}$ is set to zero if it does not have the same sign as the velocity $u_{j+1/2}$. These two conditions are fulfilled if, rather than using (7.173), the flux $F_{j+1/2}$ is expressed as

$$F_{j+1/2} = \frac{\Delta x}{\Delta t}\left(\beta_{j+1/2}\tilde{I}_{j+1/2,\ell}^+ - \beta_{j+3/2}\tilde{I}_{j+1/2,\ell}^-\right) \qquad (7.176)$$

where

$$\tilde{I}_{j+1/2,\ell}^+ = \max\left(I_{j+1/2,\ell}^+, 0\right), \qquad \tilde{I}_{j+1/2,\ell}^- = \max\left(I_{j+1/2,\ell}^-, 0\right)$$

and

$$\beta_{j+1/2} = \min\left\{1, \frac{\Psi_j^n}{\max\left(\tilde{I}_{j+1/2,\ell}^+ + \tilde{I}_{j+1/2,\ell}^-, \varepsilon\right)}\right\}$$

Here ε is a small value added to avoid numerical unstable situations if $\tilde{I}_{j+1/2,\ell}^+ + \tilde{I}_{j+1/2,\ell}^- = 0$.

Spatial functional distributions other than polynomials have been used to interpolate the dependent variables between grid points. Spalding (1972), for example, uses an exponential fitting technique, which prevents the spurious oscillations associated with the Bott scheme near sharp gradients, and ensures positivity of the solution. The method, however, is diffusive and computationally expensive. Chlond (1994) uses a hybrid scheme in which the polynomial and exponential interpolation methods are combined. A switch is used so that the polynomial scheme is applied in regions where the distribution of the transported quantity is smooth, and the exponential fitting technique is applied near sharp gradients.

The Prather Scheme

The algorithm presented by Prather (1986) is an extension of the Russell and Lerner scheme in which the advected function inside a grid cell is represented by a second-order polynomial. In each grid cell, we represent the 3-D (x, y, z) distribution of the tracer mixing ratio ψ by

$$\psi(x, y, z) = a_0 + a_x x + a_{xx} x^2 + a_y y + a_{yy} y^2 + a_z z + a_{zz} z^2 + a_{xy} xy + a_{yz} xy + a_{xz} xz \tag{7.177}$$

within a rectangular gridbox of volume $V = \Delta x\, \Delta y\, \Delta z$ with $0 \leq x \leq \Delta x$, $0 \leq y \leq \Delta y$, and $0 \leq z \leq \Delta z$. The same function can also be expressed by a linear combination of orthogonal second-order polynomials K_k:

$$\psi(x, y, z) = m_0 K_0 + m_x K_x + m_y K_y + m_{yy} K_{yy} + m_z K_z + m_{zz} K_{zz}$$
$$+ m_{xy} K_{xy} + m_{yz} K_{yz} + m_{xz} K_{xz} \tag{7.178}$$

where m_k are moment coefficients. By definition, the orthogonal functions satisfy the conditions:

$$\int_V K_i K_j \, dV = 1 \qquad (i \neq j) \tag{7.179}$$

where $dV = dx\, dy\, dz$. Prather (1986) provides ten orthogonal polynomials that apply to the algorithm in three dimensions:

$$K_0 = 1$$

$$K_x(x) = x - \frac{\Delta x}{2} \quad K_{xx}(x) = x^2 - x\Delta x + \frac{(\Delta x)^2}{6}$$

$$K_y(y) = y - \frac{\Delta y}{2} \quad K_{yy}(y) = y^2 - y\Delta y + \frac{(\Delta y)^2}{6}$$

$$K_z(z) = z - \frac{\Delta z}{2} \quad K_{zz}(z) = z^2 - z\Delta z + \frac{(\Delta z)^2}{6}$$

$$K_{xy}(x, y) = \left(x - \frac{\Delta x}{2}\right)\left(y - \frac{\Delta y}{2}\right)$$

$$K_{yz}(y, z) = \left(y - \frac{\Delta y}{2}\right)\left(z - \frac{\Delta z}{2}\right)$$

$$K_{xz}(x, z) = \left(x - \frac{\Delta x}{2}\right)\left(z - \frac{\Delta z}{2}\right)$$

with appropriate normalization factors. He also provides linear expressions that relate coefficients a_k and m_k. An upstream method is used to transport simultaneously the zeroth-order (mass), first-order (slope), and second-order (curvature) moments of the tracer distribution in each grid cell. The moments S_i are defined by

$$S_0 = \int_V \psi(x, y, z) K_0 dV = m_0 V \tag{7.180}$$

$$S_x = \frac{6}{\Delta x} \int_V \psi(x, y, z) K_x(x) dV = \frac{m_x V \Delta x}{2} \tag{7.181}$$

$$S_{xx} = \frac{30}{(\Delta x)^2} \int_V \psi(x, y, z) K_{xx}(x) dV = \frac{m_{xx} V (\Delta x)^2}{6} \tag{7.182}$$

$$S_{xy} = \frac{36}{\Delta x \Delta y} \int_V \psi(x, y, z) K_{xy}(x, y) dV = \frac{m_{xy} V \Delta x \Delta y}{4} \tag{7.183}$$

Parallel expressions are derived for S_y, S_{yy}, S_{xy}, S_z, S_{zz}, and S_{xz}.

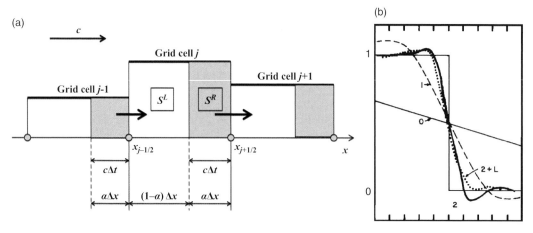

Figure 7.23 (a) Advection along direction x of moments during a time step Δt from grid cell j to $j + 1$ after decomposition of moment S into sub-moments S^L, which remains in gridbox j, and S^R, which is transferred from gridbox j to adjacent downwind gridbox $j + 1$. Coefficient $\alpha = V^R/(V^R + V^L) = c\Delta t/\Delta x$ is the Courant number. (b) Comparison of the distribution of a tracer advected across 200 gridboxes by the basic upstream scheme (0), the first-order moments method (1) and the second-order moments method (2). Note the presence of overshoots and undershoots in the second-order scheme. Positivity is ensured by placing limits on the high-order moments (curve $2 + L$). From Prather (1986).

To illustrate the method, we consider a 2-D problem and assume that the velocity $c > 0$ of the flow is uniform and directed in the x-direction. The distribution of the tracer mixing ratio inside a 2-D gridbox is expressed by

$$\psi(x, y) = m_0 K_0 + m_x K_x(x) + m_{xx} K_{xx}(x) + m_y K_y(y) + m_{yy} K_{yy}(y) + m_{xy} K_{xy}(x, y) \tag{7.184}$$

Within a cell j of volume V, we define the sub-volume V^R of the fluid that will be removed from this cell and added to the neighboring cell $j + 1$ over a time step Δt (Figure 7.23):

$$V^R = c\Delta t \Delta y \Delta z \tag{7.185}$$

The volume V^L of the fluid remaining in the original cell j is

$$V^L = (\Delta x - c\Delta t)\Delta y \Delta z \tag{7.186}$$

The method involves two consecutive steps:

First step: *Decomposition of the moments S_k for each grid cell into sub-moments S_K^R and S_K^L associated with the fraction of the tracer that is advected to the downwind grid cell and the fraction of the tracer that remains in the grid cell during time step Δt.* We have

$$S_0^R = \alpha[S_0 + (1 - \alpha)S_x + (1 - \alpha)(1 - 2\alpha)S_{xx}]$$

$$S_x^R = \alpha^2[S_x + 3(1 - \alpha)S_{xx}] \quad S_y^R = \alpha[S_y + (1 - \alpha)S_{xy}]$$

$$S_{xx}^R = \alpha^3 S_{xx} \qquad S_{yy}^R = \alpha S_{yy} \quad S_{xy}^R = \alpha^2 S_{xy}$$

where the Courant number $\alpha = c\Delta t/\Delta x = V^R/V$ is assumed to be smaller than 1. The fraction of the tracer remaining in the cell j during time step Δt is located in the sub-box of volume V^L extending from bounds $x_{j-1/2}$ to $x_{j+1/2} - c\Delta t$. The corresponding sub-moments S_K^L are:

$$S_0^L = (1 - \alpha)[S_0 - \alpha S_x - \alpha(1 - 2\alpha)S_{xx}]$$
$$S_x^L = (1 - \alpha)^2[S_x - 3\alpha S_{xx}] \quad S_y^L = (1 - \alpha)[S_y - \alpha S_{xy}]$$
$$S_{xx}^L = (1 - \alpha)^3 S_{xx} \quad S_{yy}^L = (1 - \alpha)S_{yy} \quad S_{xy}^L = (1 - \alpha)^2 S_{xy}$$

Second step: *Advection step and reconstruction of the moments for the complete grid cell.* The advection is performed by transporting moments S_K^R from grid cell j to adjacent cell $j + 1$, while maintaining moments S_K^L in their original box. For time t_{n+1}, the moments for the entire gridbox j can be reconstructed by calculating the updated moments:

$$S_0 = S_0^R + S_0^L$$
$$S_x = \alpha S_x^R + (1 - \alpha)S_x^L + 3[(1 - \alpha)S_0^R - \alpha S_0^L]$$
$$S_{xx} = \alpha^2 S_{xx}^R + (1 - \alpha)^2 S_{xx}^L + 5\{\alpha(1 - \alpha)(S_x^R - S_x^L) + (1 - 2\alpha)[(1 - \alpha)S_0^R - \alpha S_0^L]\}$$
$$S_y = S_y^R + S_y^L \quad S_{yy} = S_{yy}^R + S_{yy}^L$$
$$S_{xy} = \alpha S_{xy}^R + (1 - \alpha)S_{xy}^L + 3[(1 - \alpha)S_y^R - \alpha S_y^L]$$

From these new moments derived on the full grid cell, one derives the coefficients m_k using (7.180)–(7.183), and from there the spatial distribution of the tracer mixing ratio inside the grid cell.

The Prather method is less diffusive than the slope scheme (Figure 7.23b), but it adds to the computational and memory requirements because at each time step ten moments must be computed and stored for each grid cell in the 3-D case. As with other high-order methods, the scheme produces overshoots and undershoots that can lead to negative solutions. Placing a limit on the magnitude of the higher-order moments can ensure positivity of the solution (Figure 7.23b). The scheme is absolutely stable for Courant numbers ranging from 0.2764 to 0.7236 (Prather, 1986), but is marginally unstable over the rest of the domain [0, 1]. Phase errors are extremely small.

7.7 Lagrangian Methods

Lagrangian advection methods divide the atmosphere into a large number of air parcels and follow the displacement of their centroids as a function of time. Tracer mixing ratios are conserved in these displacements. The trajectory of the centroid is determined from the wind velocity at the centroid location. Because the wind velocities are generally provided at the discrete points where observations are performed or at the grid points of an Eulerian meteorological model, their values must be interpolated at the location of each centroid. Although the motion of the air parcels does not follow any grid, model results can still be provided at regularly spaced grid points by interpolation from the randomly located air parcels situated in the vicinity of these grid points.

In the general case of a multidimensional model with variable wind velocity $\mathbf{v}(\mathbf{r}, t)$, where $\mathbf{r}$ is the parcel location, the parcel trajectories are calculated from:

$$\frac{d\mathbf{r}(t)}{dt} = \mathbf{v}(\mathbf{r}(t), t) \tag{7.187}$$

A second-order accurate solution of this differential equation is obtained by solving, for example, the implicit expression (4.270):

$$\mathbf{r}(t + \Delta t) = \mathbf{r}(t) + \frac{1}{2}\Delta t[\mathbf{v}(\mathbf{r}(t), t) + \mathbf{v}(\mathbf{r}(t + \Delta t), t + \Delta t)] \tag{7.188}$$

where $\mathbf{r}(t)$ is the location of the air parcel at time t (departure point) and $\mathbf{r}(t + \Delta t)$ is the position of the parcel at time $t + \Delta t$ (arrival point). This implicit equation can be solved by an iteration and interpolation procedure (Kida, 1983; Stohl, 1998).

The use of computationally expensive iterative implicit methods can be avoided if the position vector $\mathbf{r}(t)$ is expanded by a Taylor series in which terms of order higher than $(\Delta t)^2$ are neglected:

$$\mathbf{r}(t + \Delta t) = \mathbf{r}(t) + \left(\frac{d\mathbf{r}}{dt}\right)_t \Delta t + \frac{1}{2}\left(\frac{d^2\mathbf{r}}{dt^2}\right)_t (\Delta t)^2 + O\left[(\Delta t)^3\right] \approx \mathbf{r}(t) + \mathbf{v}\Delta t + \frac{1}{2}\boldsymbol{\gamma}(\Delta t)^2 \tag{7.189}$$

Here, the wind velocity $\mathbf{v} = (d\mathbf{r}/dt)_t$ and the acceleration $\boldsymbol{\gamma} = (d\mathbf{v}/dt)_t = (d^2\mathbf{r}/dt^2)_t$ are calculated at the departure point (time level t). The acceleration $\boldsymbol{\gamma}$ is easily obtained from the spatial variation of the velocity field if the local rate of change in the velocity $\Delta\mathbf{v}/\Delta t$ can be neglected over the time interval Δt. Thus:

$$\boldsymbol{\gamma} = \frac{d\mathbf{v}}{dt} = \frac{\partial\mathbf{v}}{\partial t} + \mathbf{v}\cdot\boldsymbol{\nabla}\mathbf{v} \approx \mathbf{v}\cdot\boldsymbol{\nabla}\mathbf{v} \tag{7.190}$$

In the above equations, a random wind velocity $\mathbf{v}'$ is often added to the mean wind velocity $\mathbf{v}$ to account for small-scale turbulent motions. An example of a simulation produced over a period of three weeks by a Lagrangian particle dispersion model following a volcanic eruption is shown in Figure 7.24.

In global Lagrangian models, the position of an air parcel is often defined by its longitude λ, latitude φ, and altitude z (or pressure p). Thus, after a time step Δt, an air parcel originally located at point $[\lambda(t), \varphi(t), z(t)]$ is displaced to a new point whose position $[\lambda(t + \Delta t), \varphi(t + \Delta t), z(t + \Delta t)]$ is derived from

$$\lambda(t + \Delta t) = \lambda(t) + \frac{\Delta x \Delta y}{a^2[\sin(\varphi + \Delta\varphi) - \sin(\varphi)]} \tag{7.191}$$

$$\varphi(t + \Delta t) = \varphi(t) + \frac{\Delta y}{a} \tag{7.192}$$

$$z(t + \Delta t) = z(t) + \Delta z \tag{7.193}$$

where a is the Earth's radius. The geometric displacements Δx, Δy, and Δz are expressed as a function of the components (u, v, w) of the wind velocity $\mathbf{v}$ and the components $(\gamma_x, \gamma_y, \gamma_z)$ of the wind acceleration $\boldsymbol{\gamma}$ by

$$\Delta x = u\Delta t + \frac{1}{2}\gamma_x(\Delta t)^2 \quad \Delta y = v\Delta t + \frac{1}{2}\gamma_y(\Delta t)^2 \quad \Delta z = w\Delta t + \frac{1}{2}\gamma_z(\Delta t)^2 \tag{7.194}$$

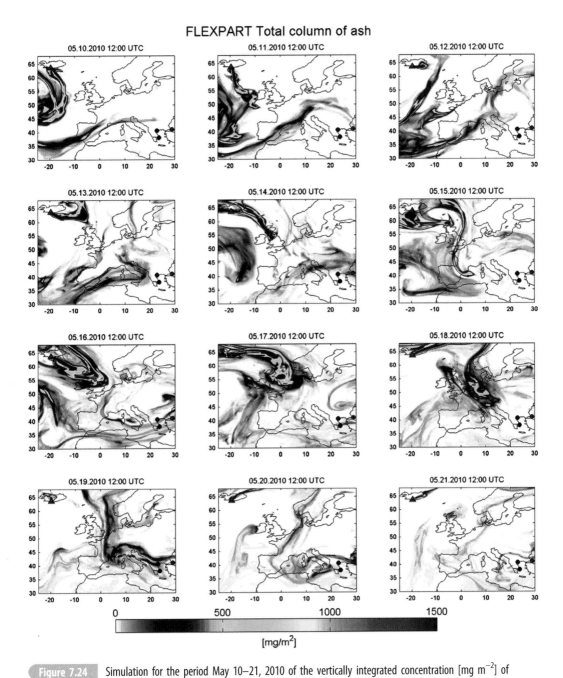

Figure 7.24 Simulation for the period May 10–21, 2010 of the vertically integrated concentration [mg m^{-2}] of atmospheric ash resulting from the eruption of the Eyjafjalljokull volcano in southern Iceland. Transport was simulated using the Lagrangian particle dispersion model FLEXPART (Stohl *et al.*, 1998), which traces the displacement of a large number of particles by the mean winds to which random motions representing turbulence and convection are superimposed. The model is driven by assimilated meteorological data on a $0.18° \times 0.18°$ grid with 91 levels in the vertical. Reproduced from Papayannis *et al.* (2012).

From (7.190), the three components of the acceleration (γ_x, γ_y, γ_z) are expressed by

$$\gamma_x = \frac{u}{a\cos\varphi}\frac{\partial u}{\partial\lambda} + \frac{v}{a}\frac{\partial u}{\partial\varphi} + w\frac{\partial u}{\partial z} \tag{7.195}$$

$$\gamma_y = \frac{u}{a\cos\varphi}\frac{\partial v}{\partial\lambda} + \frac{v}{a}\frac{\partial v}{\partial\varphi} + w\frac{\partial v}{\partial z} \tag{7.196}$$

$$\gamma_z = \frac{u}{a\cos\varphi}\frac{\partial w}{\partial\lambda} + \frac{v}{a}\frac{\partial w}{\partial\varphi} + w\frac{\partial w}{\partial z} \tag{7.197}$$

Lagrangian methods have several advantages over Eulerian methods. First, since all tracers follow the same trajectory, a single calculation of the air parcel displacement can be used to immediately infer the transport of all tracers. Second, the stability of the algorithm is not constrained by the value of the Courant number as in the explicit Eulerian methods, so the adopted time step is limited by accuracy rather than by stability considerations. Other desirable requirements are met: during the parcel displacement, mass is conserved and the sign of the transported function is maintained. Thus, unless interpolation procedures are not carefully performed, the method also guarantees monotonicity, transportivity and locality of the solution.

The Lagrangian methods also have several disadvantages. First, inaccuracies in the interpolation of the wind velocities lead to errors in the calculation of the parcel trajectories, and these errors can accumulate as the time integration proceeds. Second, the initially uniform distribution of air parcels may become highly irregular over time as a result of errors in the wind interpolation. As a result, the tracer concentration may become under-determined in certain parts of the domain while being over-determined in others. Third, Lagrangian transport does not allow for mixing between air parcels even when they are closely located. As a result, contrary to the Eulerian algorithms that often produce excessive diffusion, Lagrangian methods require that some diffusive mixing be added to account for interactions between air parcels. This is critical in particular for the treatment of nonlinear chemistry and aerosol microphysics.

7.8 Semi-Lagrangian Methods

Semi-Lagrangian transport (SLT) methods combine important advantages of the Lagrangian and Eulerian methods. The upstream SLT method (Figure 7.25) consists of using Lagrangian back-trajectories to relate concentrations on a regular Eulerian grid at the end of a model time step to the concentrations at the beginning of the time step. We first consider a version of the SLT scheme in which *points* in the atmosphere with given tracer mixing ratios are displaced with the flow during a model time step to reach locations coincident with the model grid points. We then consider a finite volume version of the SLT scheme in which *volumes* of air are displaced with the flow (with no mass transfer through their boundaries) to reach at the end of the model time step a volume of air that is coincident with a model grid cell.

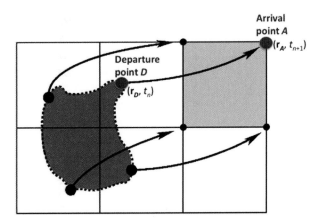

Figure 7.25   Schematic representation of the semi-Lagrangian method in two dimensions. A parcel located at the arrival point A at time level t_{n+1} was located at the departure point D at time level t_n. The value of a conserved quantity such as the mixing ratio of a passive tracer is unchanged as the parcel is displaced from point D to point A during the time step. The value of the quantity at the departure point D is derived by interpolation from the neighboring grid points at time t_n. In the finite volume version of the SLT scheme, one considers the displacement of a given volume of air (a surface in two dimensions) from its departure position (blue area) to its arrival position (gray area) coincident with a Eulerian grid cell of the model. Reproduced with permission from Peter Hjort Lauritzen (personal communication), National Center for Atmospheric Research.

7.8.1 Grid Point Based SLT Schemes

During a time step $\Delta t = t_{n+1} - t_n$, one determines the backward trajectory of the atmospheric points that reach the different Eulerian grid points of the model at time level t_{n+1}. The trajectories are calculated using interpolated gridded velocities. Consider an arrival point A at time t_{n+1} on the Eulerian grid (Figure 7.25). The location $\mathbf{r}_D$ of the upwind departure point D at time t_n is determined by backward integration of (7.187). For example, we write the equation

$$\mathbf{r}_D = \mathbf{r}_A - \frac{\Delta t}{2}\left[\mathbf{v}_D + \mathbf{v}_A\right] \qquad (7.198)$$

which has to be solved iteratively since the velocity $\mathbf{v}_D$ depends on the location of the departure point, which is not known a priori. In general, the departure points do not coincide with model grid points. The tracer mixing ratio at departure point D and time level t_n is determined by the interpolation of the surrounding values of the mixing ratio at the closest regular Eulerian grid points. For an inert tracer, the mixing ratio μ remains constant along the trajectory between D and A. Thus:

$$\mu(\mathbf{r}_A, t_{n+1}) = \mu(\mathbf{r}_D, t_n) \qquad (7.199)$$

The starting point of the backward trajectory can be derived by replacing the 3-D advection problem with three successive 1-D problems (Seibert and Morariu, 1991). For the advection along the x-axis, the trajectory is determined by integrating

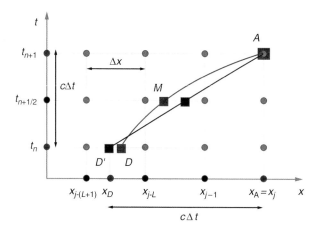

Figure 7.26 Representation of the semi-Lagrangian scheme in the 1-D case for a variable wind speed $c(x, t)$. The trajectory is represented by the curve DA (in red). The arrival point A at time t_{n+1} is coincident with grid point x_j of the Eulerian model grid. The location of the departure point D is derived from a back-trajectory calculation and is not coincident with a model grid point. It is located between grid points $x_{j-(L+1)}$ and x_{j-L}. (In this figure, $L = 2$, but it could be larger for longer time steps). By approximating the variable wind velocity by its value at the midpoint M (or by the average between the wind speeds at the departure and arrival points), the curve DA (in red) can be approximated by the straight line $D'A$ (in blue), and the departure point D by point D'. The approximation can be improved by iteration.

$$\frac{dx}{dt} = u(x) \tag{7.200}$$

between the departure point x_D and the arrival point (coincident with grid point $x_A = x_j$; see Figure 7.26). Thus,

$$\int_{x_D}^{x_j} \frac{dx}{u(x)} = \int_{t_n}^{t_n+\Delta t} dt \tag{7.201}$$

Approximating the wind along the trajectory by

$$u(x) = u(x_j) + (x - x_j)\frac{\partial u}{\partial x} \tag{7.202}$$

(7.201) can be solved analytically and the departure point x_D is found to be

$$x_D = x_j - \left[1 - \exp\left(-\Delta t\frac{\partial u}{\partial x}\right)\right] u(x_j)\left(\frac{\partial u}{\partial x}\right)^{-1} \tag{7.203}$$

where the partial derivative $\partial u/\partial x$ is numerically calculated as

$$\frac{\partial u}{\partial x} = \frac{(u_{j+1} - u_j)}{(x_{j+1} - x)} \quad \text{for } u_j \leq 0 \quad \text{and} \quad \frac{\partial u}{\partial x} = \frac{(u_j - u_{j-1})}{(x_j - x_{j-1})} \quad \text{for } u_j > 0 \tag{7.204}$$

In the simple 1-D case in which the wind velocity $u = c$ is constant and positive (see Figure 7.26), the departure point is given by

$$x_D = x_A - c\Delta t \tag{7.205}$$

and the mixing ratio at the departure point D is provided for example by a linear interpolation between the values at the closest grid points (indices $m - 1 = j - (L + 1)$ and $m = j - L$ in Figure 7.26)

$$\mu(x_D, t_n) = \mu\left(x_{j-(L+1)}, t_n\right) + \frac{\left(x_D - x_{j-(L+1)}\right)}{\Delta x}\left[\mu\left(x_{j-L}, t_n\right) - \mu\left(x_{j-(L+1)}, t_n\right)\right] \tag{7.206}$$

When the Courant number $\alpha = c\Delta t/\Delta x$ is smaller than 1, the departure point D is located between grid point x_{j-1} and the arrival point x_j (A), and it is straightforward to show that

$$\mu\left(x_j, t_{n+1}\right) = \mu\left(x_{j-1}, t_n\right) + \frac{\left(\Delta x - c\Delta t\right)}{\Delta x}\left[\mu\left(x_j, t_n\right) - \mu\left(x_{j-1}, t_n\right)\right] \tag{7.207}$$

because the mixing ratio remains constant during the displacement of the parcel. Adopting the more classic notation, we write

$$\mu_j^{n+1} = \alpha\,\mu_{j-1}^n + (1 - \alpha)\,\mu_j^n \tag{7.208}$$

More generally, if the Courant number α has a *ceiling* (smallest larger integer) of L, such that $\alpha' = L - \alpha$ is positive and smaller than unity, then the departure point D is located in the grid cell $[j - L, j - L + 1]$. The interpolation formula (7.208) then becomes:

$$\mu_j^{n+1} = \alpha'\,\mu_{j-L}^n + (1 - \alpha')\,\mu_{j-L+1}^n \tag{7.209}$$

Numerical approximation (7.209) is equivalent to expression (7.75) that describes the notoriously dissipative Eulerian upstream method. Thus, if one uses a linear interpolation as implemented here, the semi-Lagrangian method is excessively diffusive. By using higher-order interpolation schemes, the intensity of the diffusion can be considerably reduced. Cubic spline functions (Bermejo, 1990) or biquadratic polynomials (Lauritzen *et al.*, 2010) are often adopted. Williamson and Rasch (1989) and Rasch and Williamson (1990b) examine several possible interpolators and assess their ability to preserve the shape of the advected fields. Accurate interpolation schemes add to the computational costs of the method.

Non-interpolating semi-Lagrangian schemes have also been developed (see, e.g., Ritchie, 1986). In this case, the vector that defines the back-trajectory is decomposed in the sum of a vector that reaches the grid point G closest to the departure point D and a residual vector pointing from this grid point to the departure point D. To determine the value of the transported function at grid point G, no interpolation is needed. The transport along the second vector is performed using a classic Eulerian method, for which the Courant number is always smaller than 1. The overall advection for a given time step is thus always stable, but the advection for the second substep has the dispersive/diffusive properties of the Eulerian scheme that is adopted.

As in the case of the pure Lagrangian methods, SLT algorithms are stable for relatively large time steps. The stability condition is provided by the Lipschitz criterion (trajectories may not cross each other), which is considerably less severe than the CFL condition. To illustrate the performance of the semi-Lagrangian

(a)

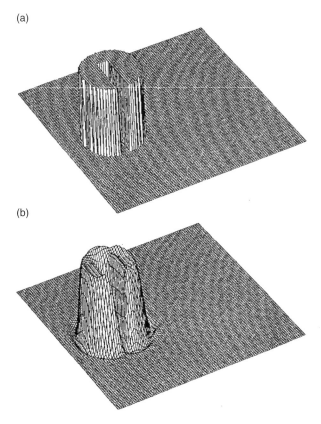

(b)

Advection (solid-body rotation) of a "slotted" cylinder using a semi-Lagrangian method with a constant uniform angular velocity about the center of the domain. The adopted Courant number is 4.2. A cubic-spline interpolator is adopted: (a) shows the initial condition and (b) the shape of the cylinder after six revolutions. From Staniforth and Côté (1991). Copyright © American Meteorological Society, used with permission.

method, Figure 7.27 shows the evolution of a "slotted" cylinder after six revolutions of solid-body rotation at uniform angular velocity about the center of the domain (Staniforth and Côté, 1991). The adopted Courant number of 4.2 is considerably larger than required for the stability of Eulerian schemes. For the same accuracy, the SLT method is more computationally efficient than Eulerian methods.

A major disadvantage of SLT grid point schemes is that they do not conserve mass. Numerical adjustment is necessary and different methods can be used for this purpose (Rasch and Williamson, 1990b; see also Section 7.9). The simplest is to compare the total mass $M_i(t_{n+1})$ of tracer i at time t_{n+1} over the model domain to the total mass $M_i(t_n)$ at time t_n, and apply a uniform multiplicative correction $M_i(t_n)/M_i(t_{n+1})$. to the mixing ratios at t_{n+1}.

7.8.2 Finite Volume Based SLT Schemes

To avoid the artificial mass correction process required by SLT grid point schemes, conservative finite volume SLT schemes have been developed (Nair and

Machenhauer, 2002; Nair *et al.* 2003; Zerroukat *et al.*, 2002, 2007; Lauritzen *et al.*, 2006). Rather than transporting the tracer mixing ratio at specific points in the model domain, these schemes advect variable finite volume elements that contain a specified mass of the tracer. By this process, the total mass (or the averaged mass density) of the tracer in the "departure" volume is equal to its mass in the "arrival" volume, ensuring mass conservation. In a 2-D formulation, the conservation of the total mass inside a Lagrangian grid cell of area A that moves and is distorted with the fluid motion is expressed as:

$$\frac{\partial}{\partial t} \int_{A(t)} \rho \, dA = 0 \quad \text{or equivalently} \quad \int_{A(t)} \rho \, dA = \int_{A(t+\Delta t)} \rho \, dA \qquad (7.210)$$

After discretization in time, we have

$$\left\langle \rho_j^{n+1} \right\rangle A_j^{n+1} = \left\langle \rho_j^n \right\rangle A_j^n \qquad (7.211)$$

where A_j^n and A_j^{n+1} are the surface areas of the cell j at the departure and arrival time levels, respectively, and

$$\left\langle \rho_j \right\rangle = \frac{1}{A_j} \int_{A_j} \rho_j \, dA \qquad (7.212)$$

is the mean density in grid cell j. The surface A_j^{n+1} at time t_{n+1} coincides with a Eulerian grid cell of the model. The value of the mean density $\langle \rho_j^n \rangle$ in the departure area A_j^n is derived by interpolation of the solution at time t_n. The solution at time t_{n+1} is obtained by a remapping process. The method is illustrated in Figure 7.28 in a 1-D configuration. In this simple case, one defines the mean density of a tracer in a grid cell of size Δx:

$$\left\langle \rho_j \right\rangle = \frac{1}{\Delta x_j} \int_{\Delta x_j} \rho_j(x) \, dx \qquad (7.213)$$

The solution at the arrival time t_{n+1} is

$$\left\langle \rho_j^{n+1} \right\rangle = \frac{\left\langle \rho_j^n \right\rangle \Delta x_j^D}{\Delta x_j} \qquad (7.214)$$

If the wind velocity u varies as a function of space and time, the size of the gridbox (noted Δx at the arrival location) varies during the back-trajectory step and becomes $\Delta x^D \neq \Delta x$ at the departure location.

Flux-form finite volume SLT schemes are conservative for tracer transport if the winds originate from a general circulation model (GCM) where the exact same scheme was applied to solve the Navier–Stokes equation for momentum. This requirement can be achieved in online GCM simulations of chemical tracers, but is generally not achievable in offline chemical transport models (CTMs) (Jöckel *et al.*, 2001; Horowitz *et al.*, 2003). This problem in offline CTMs often results from inconsistencies between the advection scheme and the surface pressure tendency provided by the dynamical model. A mass fixer is often applied to alleviate this

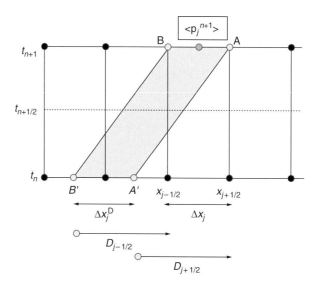

Schematic representation of the 1-D finite volume mass conserving Lagrangian method of Lin and Rood (1996). The "volume" Δx at the arrival time t_{n+1} is represented by the grid cell AB, and its value Δx^D at departure time t_n is represented by the distance $A'B'$. If the wind velocity u is not constant in space, the displacements at the cell interfaces $D_{j+1/2}$ and $D_{j-1/2}$ for points A' and B' are not equal and Δx is different from Δx^D (Lin and Rood, 1996). Copyright © American Meteorological Society, used with permission.

problem (Horowitz *et al.*, 2003), but may violate monotonicity requirements and generate non-physical transport.

7.9 Spectral, Finite Element, and Spectral Element Methods

The *spectral method* is widely used to solve the dynamical equations in GCMs. In this approach, the fields such as the temperature or the winds are represented by a series of continuous basis functions such as spherical harmonics in the horizontal plane (see Chapter 4) and by a finite difference formulation in the vertical direction. Spectral methods have also been used in these models to represent the advection of moisture. To illustrate the methodology, we consider again the 1-D case in the x-direction and we approximate the solution of the advection equation by the expansion

$$\psi(x, t) \approx \sum_{k=1}^{K} a_k(t) \Phi_k(x) \qquad (7.215)$$

in which $\Phi_k(x)$ represents a set of orthogonal functions (e.g., elementary trigonometric functions) and a_k are unknown coefficients that depend on time t. Expression (7.215) is introduced in the advection equation (7.17), leading to a system of K differential equations for the coefficients $a_k(t)$ that can be solved by standard methods. When spherical geometry is used, the solution is instead expressed as a function of spherical harmonics (see equation (4.252)).

With the proper choice of parameters, the spectral method provides accurate and stable results and can be mass-conservative. However, it is not shape-preserving (monotonic, positive definite); overshoots and undershoots can be produced (see Box 4.7). Negative values of the transported fields can be eliminated by using appropriate filtering and filling schemes, but these corrupt the correlations between tracers which is critical for nonlinear chemistry. In addition, the algorithm does not satisfy the criterion of locality. Because of these limitations, spectral methods are generally not well suited for chemical applications.

In the *finite element method*, the solution $\psi(x, t)$ is also provided by expansion (7.215), but with the basis functions $\Phi_k(x)$ defined only on a small region of space $[A, B]$ called finite elements (see Section 4.10). In the Galerkin approach, the coefficients a_k are derived by requiring that the error arising from representing function $\psi(x, t)$ by expansion (7.215),

$$e_K = \frac{\partial}{\partial t} \sum_{k=1}^{K} a_k(t)\, \Phi_k(x) + c \frac{\partial}{\partial x} \sum_{k=1}^{K} a_k(t)\, \Phi_k(x) \qquad (7.216)$$

be orthogonal to the basis functions. This condition is expressed by the integral over the domain $[A, B]$:

$$\int_{A}^{B} e_K\, \Phi_i(x)\, dx = 0 \qquad (7.217)$$

for all values of $i \in [0, K]$. If in (7.217) one replaces the error e_K by expression (7.216), one obtains the system of K-coupled ODEs:

$$\sum_{k=1}^{K} \frac{\partial a_k(t)}{\partial t} \int_{A}^{B} \Phi_i(x)\, \Phi_k(x)\, dx + c \sum_{k=1}^{K} a_k \int_{A}^{B} \Phi_i(x) \frac{\partial \Phi_k(x)}{\partial x}\, dx = 0 \quad (i = 1, \dots K)$$

$$(7.218)$$

which is solved to obtain the coefficients $a_k(t)$. For this purpose, the time derivatives are usually approximated by finite differences.

The *spectral element method* (Patera, 1984) is a finite element technique in which a high-degree spectral method is applied within each element. As discussed by Nair *et al.* (2011), the spectral element method combines the geometric flexibility of the traditional finite element methods with the high accuracy, rapid convergence, and weak numerical dispersion and dissipation of the classical spectral methods. It is not inherently conservative, but can be engineered to ensure a user-required level of mass conservation.

Although finite element and spectral element methods have so far mainly been used for engineering applications, they are now emerging as promising methods for atmospheric problems. Their local domain decomposition property makes them particularly well suited for massively parallel computer architectures, and they can be easily applied when the model domain is geometrically complex or when grid refinement is needed in specified atmospheric regions.

7.10 Numerical Fixers and Filters

An important requirement for the numerical method applied to solve the advection equation is that the solution be positive and that mass be conserved. This is not always the case, and if so one can apply a-posteriori corrections to restore these desired properties. Jablonowski and Williamson (2011) provide an extensive review of the use of fixers and filters in atmospheric models.

Fixers. Negative values in the transported variables (e.g., tracer mixing ratio) can be eliminated by introducing "*numerical fixers*" that borrow mass from surrounding (or downstream) grid points. Rasch and Williamson (1990a) describe a possible implementation of local and global fixers. When a negative value is encountered during a point-by-point scan of the calculated quantities on each horizontal surface, the immediate neighboring points are examined, and if sufficient mass is available to "fill the hole," the negative value is set to zero and the values at the neighboring points are reduced proportionally. If there is not sufficient mass available, no action is taken. The application of this local *filling method* does conserve mass, but may not eliminate all negative values. In a second step, a global filter is applied in which the remaining negative values are set to zero; this violates mass conservation but a renormalization can be applied by scaling to the global masses in the domain, as described in Section 7.8.1 to enforce mass conservation in SLT schemes. This filling process produces diffusion and does not ensure monotonicity of the corrected fields; it can also be computationally expensive.

Filters. In several of the algorithms described previously, short waves may grow excessively in the solution of the advection equation, producing undesired noise and even catastrophic instability. These waves can be eliminated by appropriate *smoothing* or *filtering*. One option is to add a small diffusive term in the advection equation, which will smooth the solution and suppress high wavenumbers. In spectral models, numerical noise can easily be suppressed by omitting wavenumbers larger than a specified value and highlighting only the scales of interest. Spectral filtering can also be used in grid point models by applying a Fourier transform to the solution, which eliminates the high frequencies in the signal, and applying an inverse Fourier transform. Such filtering is often applied in polar regions where the meridians converge and a longitudinal grid point spacing becomes so small that, without filtering, the solution would become unstable. Finally, numerical filters such as the linear Shapiro filter presented in Section 4.15.2 are often applied to eliminate two-grid interval waves completely while having little effect on longer waves (Shapiro, 1971). Other high-frequency filters have been developed by Asselin (1972) and Forester (1977).

7.11 Concluding Remarks

In this chapter we have reviewed different numerical algorithms used to approximate the solution of the linear advection equation. No existing method fully addresses modelers' requirements. The examination of simple numerical schemes reveals that high-order algorithms are generally not monotonic and occasionally produce undesired negative values. Low-order algorithms such as the upstream method preserve the sign of the solution, but are excessively diffusive. Thus, practical applications must adopt more elaborate schemes that address some of the drawbacks that characterize the simple methods. Modern schemes are often upstream-based Eulerian finite volume methods that are mass conservative, positive definite, and possess good phase-error characteristics. They may use adaptive time steps to meet CFL stability requirements or semi-Lagrangian options to get around these requirements. Specific, often complex nonlinear algorithms are developed to reduce the numerical diffusion that characterizes upstream methods. These complex schemes can yield significant improvement in accuracy, but often with enhanced computational costs.

Finite volume Eulerian methods, in which a subgrid distribution of the transported quantity is specified, provide highly accurate solutions and are popular in global CTMs. In many respects, they are superior to classic grid point methods. Computational cost depends on the user tolerance for numerical diffusion. The Prather scheme is regarded as a reference among Eulerian models. It produces accurate solutions with little diffusion. However, it has large computing time and storage requirements. A van Leer scheme may enable higher grid resolution, compensating for the lower accuracy.

Lagrangian methods are popular for source-oriented and receptor-oriented transport problems in which one is concerned with transport from a point source or transport contributing to concentrations at a receptor point. However, they do not provide the regular full-domain solution achievable by Eulerian methods and cannot properly represent nonlinear chemistry or aerosol microphysics. Semi-Lagrangian methods are very popular in global CTMs because their numerical stability is not as severely constrained by choice of time step as in the case of Eulerian schemes. They are sometimes used as a back-up scheme in cases where the regular Eulerian solver violates the CFL criterion.

In summary, there is no single advection scheme that is universally best. The choice of scheme depends on the type of problem being solved, the tolerance for different kinds of error, and the computational demands. For any scheme, it is important to verify that basic criteria of stability and mass conservation are met. The material in this chapter should enable readers to understand the issues associated with different advection schemes and to make informed choices in selecting appropriate schemes for their applications.

References

Allen D. J., Douglass A. R., Rood R. B., and Guthrie P. D. (1991) Application of a monotonic upstream-biased transport scheme to three-dimensional constituent transport calculations, *Mon. Wea. Rev.*, **119**, 2456–2464.

Asselin R. (1972) Frequency filter for time integrations, *Mon. Wea. Rev.*, **100**, 487–490.

Bermejo R. (1990) Notes and correspondence on the equivalence of semi-Lagrangian schemes and particle-in-cell finite element methods, *Mon. Wea. Rev.*, **118**, 979–987.

Boris J. P. and Book D. L. (1973) Flux-corrected transport. I. SHASTA, a fluid transport algorithm that works, *J. Comput. Phys.*, **11**, 38–69.

Boris J. P. and Book D. L. (1976) Flux-corrected transport. III. Minimal-error FCT algorithms, *J. Comput. Phys.*, **20**, 397–431.

Bott A. (1989a) A positive definite advection scheme obtained by non-linear renormalization of the advective fluxes, *Mon. Wea. Rev.*, **117**, 1006–1015.

Bott A. (1989b) Reply, *Mon. Wea. Rev.*, **117**, 2633–2636.

Carpenter R. L., Droegemeier K. K., Woodward P. R., and Hane C. E. (1990) Application of the piecewise parabolic method (PPM) to meteorological modeling. *Mon. Wea. Rev.*, **118**, 586–612.

Chen Y. and Falconer R. A. (1992) Advection–diffusion modelling using the modified QUICK scheme, *Int. J. Numer. Meth. Fluids*, **15**, 1171–1196.

Chlond A. (1994) Locally modified version of Bott's advection scheme, *Mon. Wea. Rev.*, **122**, 111–125.

Colella P. and Sekora M. D. (2008) A limiter for PPM that preserves accuracy at smooth extrema, *J. Comput. Phys.*, **227** (15), 7069–7076.

Colella P. and Woodward P. R. (1984) The piecewise parabolic method (PPM) for gasdynamical simulations, *J. Comput. Phys.*, **54**, 174–201.

Courant R., Friedrichs K., and Lewy H. (1928) Über die partiellen Differenzengleichungen der mathematischen Physik. *Math. Ann.*, **100**, 32–74.

Courant R., Isaacson E., and Rees M. (1952) On the solution of nonlinear hyperbolic differential equations by finite difference, *Commun. Pure Appl. Math.*, **5**, 243–255.

Crowley W. P. (1968) Numerical advection experiments, *Mon. Wea. Rev.*, **96**, 1–11.

Davies H. C. (1983) Limitations of some common lateral boundary schemes used in regional NWP models, *Mon. Wea. Rev.*, **111**, 1002–1012.

Dukowicz J. K. and Baumgardner J. R. (2000) Incremental remapping as a transport/advection algorithm, *J. Comput. Phys.*, **160**, 318–335.

Dullemond C. P. (2009) *Numerical Fluid Dynamics*, Lecture Notes, Zentrum für Astronomie, Ruprecht-Karls Universität, Heidelberg.

Farrow D. E. and Stevens D. P. (1994) A new tracer advection scheme for Bryan Cox type ocean general circulation models, *J. Phys. Oceanogr.*, **25**, 1731–1741.

Forester C. K. (1977) Higher order monotonic convective difference schemes, *J. Comput. Phys.*, **23**, 1–22.

Godunov S. K. (1959) A finite difference method for the computation of discontinuous solutions of the equations of fluid dynamics, *Mat. Sb.*, **47**, 357–393.

Gross E. S., Koseff J. R., and Monismith S. G. (1999) Evaluation of advective schemes for estuarine salinity simulations, *J. Hydr. Engrg., ASCE*, **125** (1) 32–46.

Horowitz L. W., Walters S., Mauzerall D. L., *et al.* (2003) A global simulation of tropospheric ozone and related tracers: Description and evaluation of MOZART, version 2, *J. Geophys. Res.*, **108** (D24), 4784, doi: 10.1029/2002JD002853.

Iserles A. (1986) Generalised leapfrog methods, *IMA J. Numer. Anal.*, **6**, 381–392.

Jablonowski C. and Williamson D. L. (2011) The pros and cons of diffusion, filters and fixers in atmospheric general circulation models, In *Numerical Techniques for*

Global Atmospheric Models, Lecture Notes in Computational Science and Engineering (Lauritzen P. H., Jablonowski C., Taylor M. A., and Nair R. D., eds.), Springer, Berlin.

Jöckel P., von Kuhlmann R., Lawrence M. G., *et al.* (2001) On a fundamental problem in implementing flux-form advection schemes for tracer transport in 3-dimensional general circulation and chemistry transport models, *Q. J. R. Meteorol. Soc.*, **127**, 1035–1052.

Kida H. (1983) General circulation of air parcels and transport characteristics derived from a hemispheric GCM: Part 1. A determination of advective mass flow in the lower stratosphere, *J. Meteor. Soc. Japan*, **61**, 171–187.

Kim C. (2003) Accurate multi-level schemes for advection, *Int. J. Numer. Meth. Fluids*, **41**, 471–494, doi: 10.1002/fld.443.

Kurihara Y. (1965) On the use of implicit and iterative methods for the time integration of the wave equation, *Mon. Wea. Rev.*, **93**, 33–46.

Lauritzen P. H., Kaas E., and Machenhauer B. (2006) A mass-conservative semi-implicit semi-Lagrangian limited-area shallow-water model on a sphere, *Mon. Wea. Rev.*, **134**, 2588–2606.

Lauritzen P. H., Nair R. D., and Ullrich P. A. (2010) A conservative semi-Lagrangian multi-tracer transport scheme (CSLAM) on the cubed-sphere grid, *J. Comput. Phys.*, **229** (5), doi: 10.1016/j.jcp.2009.10.036

Lauritzen P. H., Ullrich P. A., and Nair R. D. (2011) Atmospheric transport schemes: Desirable properties and a semi-Lagrangian view on finite-volume discretizations. In *Numerical Techniques for Global Atmospheric Models, Lecture Notes in Computational Science and Engineering* (Lauritzen P. H., Jablonowski C., Taylor M. A., and Nair R. D., eds.), Springer, Berlin.

Lax P. D. (1954) Weak solutions of nonlinear equations and their numerical computation, *Comm. Pure Appl. Math.*, **7**, 159–193.

Lax P. D. and Wendroff B. (1960) Systems of conservation laws, *Comm. Pure and Appl. Math.*, **13**, 217–237.

Lax P. D. and Wendroff B. (1964) Difference schemes for hyperbolic equations with high order of accuracy, *Comm. Pure Appl. Math.*, **17** (3), 281–398.

Leith C. E. (1965) Numerical simulations of the earth's atmosphere, *Meth. Comput. Phys.*, **4**, 1–28.

Leonard B. P. (1979) A stable and accurate convective modelling procedure based on quadratic upstream interpolation, *Comput. Methods in Appl. Mech. Eng.*, **19**, 59–98.

Lin J. C. (2012) Lagrangian modeling of the atmosphere: An introduction. In *Lagrangian Modeling of the Atmosphere* (Lin J., Brunner D., Gerbig C., *et al.*, eds.), American Meteorological Union, Washington, DC.

Lin S. J. and Rood R. B. (1996) Multidimensional flux-form semi-Lagrangian transport schemes, *Mon. Wea. Rev.*, **124**, 2046–2070.

Lipscomb W. H. and Ringler T. D. (2005) An incremental remapping transport scheme on a spherical geodesic grid, *Mon. Wea. Rev.*, **133**, 2335–2350.

Miura H. (2007) An upwind-biased conservative advection scheme for spherical hexagonal–pentagonal grids, *Mon. Wea. Rev.*, **135**, 4038–4044.

Müller R. (1992), The performance of classical versus modern finite-volume advection schemes for atmospheric modeling in a one-dimensional test-bed, *Mon. Wea. Rev.*, **120**, 1407–1415.

Nair R. D., and Machenhauer B. (2002) The mass-conservative cell-integrated semi-Lagrangian advection scheme on the sphere, *Mon. Wea. Rev.*, **130**, 647–667.

Nair R. D., Scroggs J. S., and Semazzi F. H. M. (2003) A forward-trajectory global semi-Lagrangian transport scheme, *J. Comput. Phys.*, **193**, 275–294.

Nair R. D., Levy M., and Lauritzen P. H. (2011) Emerging methods for conservation laws. In *Numerical Techniques for Global Atmospheric Models, Lecture Notes in Computational Science and Engineering* (Lauritzen P. H., Jablonowski C., Taylor M. A., and Nair R. D., eds.), Springer, Berlin.

Papayannis A., Mamouri R. E., Amiridis V., *et al.* (2012) Optical properties and vertical extension of aged ash layers over the Eastern Mediterranean as observed by Raman lidars during the Eyjafjallajökull eruption in May 2010, *Atmos. Env.*, **48**, 56–65.

Patera A. T. (1984) A spectral element method for fluid dynamics: Laminar flow in a channel expansion, *J. Comput. Phys.*, **54**, 468–488.

Prather M. J. (1986) Numerical advection by conservation of second order moments, *J. Geophys. Res.*, **91**, 6671–6681.

Press W. H., Teukolsky S. A., Vetterling W. T., and Flannery B. P. (2007) *Numerical Recipes: The Art of Scientific Computing*, 3rd edition, Cambridge University Press, Cambridge.

Rasch P. J. and Williamson D. L. (1990a) Computational aspects of moisture transport in global models of the atmosphere, *Quart. J. Roy. Meteor. Soc.*, **116**, 1071–1090.

Rasch P. J. and Williamson D. L. (1990b) On shape-preserving interpolation and semi-Lagrangian transport, *SIAM J. Sci. Stat. Comput.*, **11**, 656–687.

Ritchie H. (1986) Eliminating the interpolation associated with the semi-Lagrangian scheme, *Mon. Wea. Rev.*, **114**, 135–146.

Roe P. L. (1986) Characteristic-based schemes for the Euler equations, *Ann. Rev. Fluid Mech.*, **18**, 337–365.

Russell G. L. and Lerner J. A. (1981) A new finite-differencing scheme for the tracer transport equation, *J. Appl. Meteor.*, **20**, 1483–1498.

Seibert P. and Morariu B. (1991) Improvements of upstream, semi-Lagrangian numerical advection schemes, *J. Appl. Meteor.*, **30**, 117–125.

Shapiro R. (1971) The use of linear filtering as a parameterization of atmospheric motion, *J. Atmos. Sci.*, **28**, 523–531.

Skamarock W. and Menchaca M. (2010) Conservative transport schemes for spherical geodesic grids: High-order reconstruction for forward-in-time schemes, *Mon. Wea. Rev.*, **138**, 4497–4508.

Slingerland R. and Kump L. (2011) *Mathematical Modeling of Earth's Dynamical Systems: A Primer*, Princeton University Press, Princeton, NJ.

Smolarkiewicz P. K. (1983) A simple positive definite advection scheme with small implicit diffusion, *Mon. Wea. Rev.*, **111**, 479–486.

Smolarkiewicz P. K. (1984) A fully multidimensional positive definite advection transport algorithm with small implicit diffusion, *J. Comput. Phys.*, **54**, 325–362.

Smolarkiewicz P. K. (2006) Multidimensional positive definite advection transport algorithm: An overview, *Int. J. Numer. Meth. Fluids*, **50** (10), 1123–1144.

Spalding D. B. (1972) A novel finite difference formulation for differential expressions involving both first and second derivatives, *Int. J. Numer. Meth. Eng.*, **4**, 551–559.

Staniforth A. and Côté J. (1991) Semi-Lagrangian integration schemes for atmospheric models: A review, *Mon. Wea. Rev.*, **119**, 2206–2223.

Stohl A. (1998) Computation, accuracy and applications of trajectories: A review and bibliography, *Atmos. Environ.*, **32** (6), 947–966.

Stohl A., Hittenberger M., and Wotawa G. (1998) Validation of the Lagrangian particle dispersion model FLEXPART against large scale tracer experiment data, *Atmos. Environ.*, **32**, 4245–4262.

Tremback C. J., Powell J., Cotton W. R., and Pielke R. A. (1987), The forward-in-time upstream advection scheme: Extension to higher order, *Mon. Wea. Rev.*, **115**, 894–902.

van Leer B. (1977) Toward the ultimate conservative difference scheme IV: A new approach to numerical convection, *J. Comp. Phys.*, **23**, 276–299.

van Leer B. (1979) Toward the ultimate conservative difference scheme V: A second order sequel to Godunov's method, *J. Comp. Phys.*, **32**, 101–136.

Warming R. F. and Beam R. M. (1976) Upwind second-order difference schemes and applications in aerodynamic flows, *AIAA Journal*, **14** (9), 1241–1249.

Williamson D. L. (1992) Review of the numerical approaches for modeling global transport. In *Air Pollution Modelling and its Applications* (Dop H. V. and Kallow G., eds.), Plenum Press, New York.

Williamson D. L. and Rasch P. J. (1989) Two-dimensional semi-Lagrangian transport with shape preserving interpolation, *Mon. Wea. Rev.*, **117**, 102–129.

Zerroukat M., Wood N., and Staniforth A. (2002) SLICE: A semi-Lagrangian inherently conserving and efficient scheme for transport problems, *Q. J. R. Meteorol. Soc.*, **128**, 2801–2820.

Zerroukat M., Wood N., and Staniforth A. (2007) Application of the Parabolic Spline Method (PSM) to a multi-dimensional conservative transport scheme (SLICE), *J. Comput. Phys.*, **225**, 935–948.

Parameterization of Subgrid-Scale Processes

8.1 Introduction

Meteorological variables affecting chemical concentrations vary on all scales down to the millimeter Kolmogorov scale, below which turbulence dissipates by molecular diffusion (Kolmogorov, 1941a, 1941b). The smallest scales cannot be represented deterministically in atmospheric models and must be parameterized in some way.

Processes that usually require parameterization include near-surface turbulence driving surface fluxes, turbulent eddies in the planetary boundary layer (PBL), and deep convective transport in updrafts and downdrafts (Figure 8.1). Clouds offer a vivid illustration of the variability of atmospheric motions on small scales and their link to large-scale effects (Figure 8.2). Parameterization of cloud processes is of importance in chemical transport models not only because of their dynamical role, but also because of their effect on radiation, precipitation scavenging, aqueous-phase chemistry, and aerosol modification.

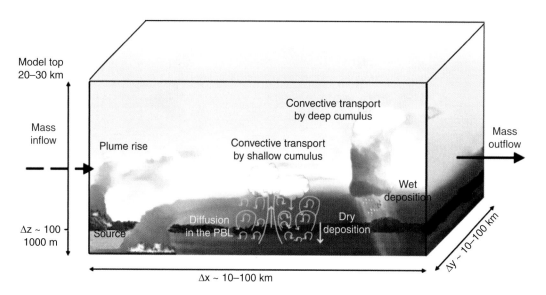

Figure 8.1 Schematic representation of subgrid physical processes affecting atmospheric transport. Courtesy of S. Freitas, INPE, CPTEC, Brazil.

Observation of clouds from aircraft provides a vivid illustration of the wide range of scales of atmospheric variability. This photograph shows cirrus clouds in the upper troposphere, the anvil of a cumulonimbus associated with deep convection, and shallow cumulus clouds at the top of the boundary layer.

Direct numerical simulations (*DNS*) provide a deterministic representation of chemical behavior in turbulent flows, with no need for parameterization, by solving the Navier–Stokes (momentum) and continuity equations on an extremely fine grid down to the dissipative Kolmogorov scales. This allows explicit simulation of small eddies but the simulation domains are very limited. The computational burden can be reduced by applying a low-pass filter to the Navier–Stokes and continuity equations, retaining only the larger eddies. In the *large-eddy simulation* (*LES*) method, introduced in 1963 by Joseph Smagorinsky for atmospheric flows, the effects of the unresolved smaller scales on large eddies are parameterized by a subgrid model. This provides a practical approach for studying processes on the scale of the PBL with sub-kilometer horizontal resolution. Another method, called *Reynolds decomposition*, expresses model variables as the sums of time-averaged and fluctuating values, and solves the corresponding Navier–Stokes and continuity equations for the time-averaged values resolved by the model. Covariance between the fluctuating values appears in the resulting equations and must be parameterized to provide "closure." In *statistical methods*, momentum and chemical concentrations are treated as random variables defined by coupled probability density functions (PDFs). The momentum PDFs may be sampled stochastically at individual time steps, and the resulting concentration PDFs are constructed from a large ensemble of random realizations of the flow.

Here we refer to *subgrid-scale processes* (or simply *subgrid processes*) as the ensemble of processes driven by transport on scales smaller than the spatial/temporal resolution of the model. Although "grid" has an Eulerian connotation, consideration of subgrid processes equally applies in Lagrangian models for scales that are not explicitly resolved. These subgrid processes affect the larger-scale composition of the atmosphere and must therefore be parameterized. The parameterizations may be based on empirical information from atmospheric observations or on the analysis of results from finer-scale models. The type of parameterization adopted depends on the resolution of the model, as this defines the scales that are explicitly resolved. As model resolution increases, the subgrid parameterizations must usually be revised.

This chapter presents general approaches to parameterization of subgrid processes in chemical transport models. Section 8.2 presents the Reynolds decomposition and averaging procedure. We then discuss methods to describe chemical covariances (Section 8.3), and present closure relations that relate eddy terms to mean quantities, including turbulent diffusion formulations (Section 8.4). Stochastic statistical approaches are discussed in Section 8.5. Numerical algorithms to solve the diffusion equation are presented in Section 8.6. We apply these concepts to the PBL (Section 8.7) and further discuss parameterizations of deep convection (Section 8.8), wet scavenging (Section 8.9), lightning (Section 8.10), gravity wave breaking (Section 8.11), mass transfer through dynamical barriers (Section 8.12), and long-lived free tropospheric plumes (Section 8.13). Gaussian plume models for boundary layer point sources were presented in Section 4.12. Near-surface turbulence driving mass transfer between the atmosphere and the surface (dry deposition, two-way exchange) is covered in Chapter 9.

8.2 Reynolds Decomposition: Mean and Eddy Components

Fluid motions are categorized as either smooth, steady *laminar* flows or irregular, fluctuating *turbulent* flows (Figure 8.3). The regime associated with fluid motions is characterized by the dimensionless *Reynolds number Re*, defined as the ratio between the nonlinear field acceleration $(\mathbf{v} \cdot \nabla)\mathbf{v}$ that generates turbulence in the Navier–Stokes equation (see Section 4.5.2 and Box 4.2) and the viscosity term $\nu \nabla^2 \mathbf{v}$ that tends to suppress it. Here ν denotes the kinematic viscosity of the fluid (1.3×10^{-5} m^2 s^{-1} for dry air at 1 atm and 273 K) and $\mathbf{v}$ the velocity of the flow. Thus, we write from dimensional considerations

$$Re = \frac{U^2/L}{\nu U/L^2} = \frac{UL}{\nu} \tag{8.1}$$

where U and L represent characteristic velocity and length scales of the flow, respectively. The characteristic velocity scale can be taken as a mean or typical wind

(a)   Laminar   (b)   Turbulent

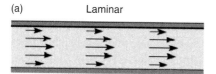
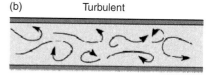

Figure 8.3   Schematic representation of laminar (a) and turbulent (b) flows between two fixed boundaries. Reproduced from W. Aeschbach-Hertig, Inst. Fuer Umweltphysik, University of Heidelberg.

speed, and the characteristic length scale can be taken as the distance of interest over which that wind speed varies. The transition between laminar and turbulent flow gradually takes place for Reynolds numbers of the order of 10^3 to 10^4. For a wind velocity of 10 m s^{-1} and a length scale of 1000 m the Reynolds number is of the order of 10^9, well into the turbulent regime. This can be generalized to any atmospherically relevant values of U and L: atmospheric flow is turbulent under all conditions, even when it is dynamically stable.

Turbulence can be generated mechanically by wind shear, and amplified or suppressed by buoyancy. The dimensionless *gradient Richardson number*, which expresses the ratio between the buoyant suppression and mechanical generation of turbulence, provides an indicator of the dynamical stability of the flow:

$$Ri = \frac{(g/\overline{\theta})\partial\overline{\theta}/\partial z}{(\partial\overline{u}/\partial z)^2} \tag{8.2}$$

Here $\overline{u}$ is the mean wind speed, $\overline{\theta}$ the mean potential temperature, g the acceleration of gravity, and z the height. The sign of the gradient Richardson number is determined by the atmospheric lapse rate. A positive value of Ri, associated with a positive vertical gradient in the mean potential temperature, corresponds to a stably stratified atmosphere. A negative value of Ri characterizes an unstable layer with the presence of convective motions. The case for which $Ri = 0$ corresponds to neutral stratification. Theory shows that, even in the presence of a mean wind shear, turbulence is suppressed if Ri exceeds a critical value Rc of about 0.25.

In Reynolds decomposition, resolved (or mean) and unresolved (or turbulent) processes are separated by expressing the model variables (such as temperature, wind velocity, humidity, chemical concentrations) as the sum of a slowly varying mean value $\overline{\Psi}$ and a rapid fluctuation Ψ' around this mean. Thus,

$$\Psi = \overline{\Psi} + \Psi' \tag{8.3}$$

Component Ψ', called the *eddy* term, is associated with the irregular and stochastic nature of the motions around a mean state (Figure 8.4). By definition, its mean value is equal to zero ($\overline{\Psi'} = 0$).

The mean term represents for example an average at a given point of the atmosphere over a time period T:

$$\overline{\Psi}(T) = \frac{1}{T}\int_T \Psi(t)\ dt \tag{8.4}$$

The deviation $\Psi'(t)$ captures the rapid fluctuations around the mean value $\overline{\Psi}$. These fluctuations are characterized by various timescales or frequencies v. The corresponding spectrum $\tilde{\Psi}(v)$ can be derived by a Fourier transform

$$\tilde{\Psi}(v) = \int\limits_{-\infty}^{+\infty} \Psi'(t)e^{-2\pi ivt}\ dt \tag{8.5}$$

Similarly, we can define the mean value of function $\Psi(x)$ over a given spatial length L and derive a spectrum of the turbulent component $\Psi'(x)$ as a function

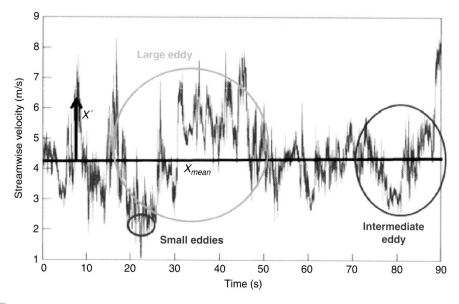

Figure 8.4 Time series of wind velocity measured at a fixed location in a turbulent boundary layer for 90 seconds. The Reynolds number was 2×10^7. Fluctuations occur over a range of timescales. Data recorded by B. Dhruva of Yale University and reproduced from Ecke (2005).

of wavenumber $k = 2\pi/\lambda$, where λ is the wavelength in radians on the spherical Earth:

$$\tilde{\Psi}(k) = \int_{-\infty}^{+\infty} \Psi'(x)e^{-ikx} \; dx \tag{8.6}$$

Turbulent motions are often described by the mean *turbulent kinetic energy* of the flow per unit mass [TKE in m^2 s^{-2}]:

$$TKE = \frac{1}{2}\left(\overline{u'^2} + \overline{v'^2} + \overline{w'^2}\right) \tag{8.7}$$

where u', v', and w' represent the fluctuations of the three wind components (u, v, w).

The spectrum of atmospheric variability to be described by eddy terms may range from mesoscale weather patterns not resolved by global models down to the millimeter scale resolved only by the DNS approach. The largest turbulent elements receive their energy from the mean flow. Through a cascade process (Richardson, 1922), this energy propagates to smaller scales and is eventually dissipated by viscosity as heat. An example of the spectral distribution $E(k)$ of the turbulent energy covering the scales from 10 m to 0.1 mm is shown in Figure 8.5. In the so-called inertial subrange, characterized by isotropic turbulence (no dominant direction), the energy decreases with increasing wavenumber according to Kolmogorov–Obukhov's $-5/3$ law (Kolmogorov, 1941a, 1941b;

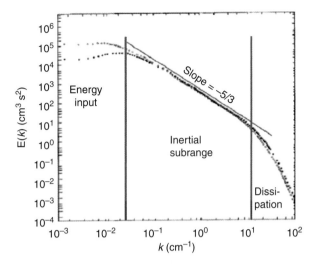

Figure 8.5 Turbulence energy spectrum $E(k)$ showing the cascade of energy from larger to smaller scales (smaller to larger wavenumbers k) with different spectral regions characterizing energy input, energy cascade, and dissipation. In a large portion of the spectrum, the turbulent kinetic energy spectrum varies as $k^{-5/3}$ (Kolmogorov, 1941a,1941b). Reproduced from W. Aeschbach-Hertig. Institute für Umweltphysik, University of Heidelberg, Germany.

Obukhov, 1941). The total specific turbulent kinetic energy is derived by integrating $E(k)$ over all wavenumbers k:

$$TKE = \int_0^\infty E(k) \; dk \tag{8.8}$$

The flux of quantity Ψ associated with the atmospheric flow of velocity $\mathbf{v}$ is expressed as

$$\mathbf{v}\Psi = (\bar{\mathbf{v}} + \mathbf{v}')(\bar{\Psi} + \Psi') \tag{8.9}$$

or

$$\mathbf{v}\Psi = \bar{\mathbf{v}}\bar{\Psi} + \mathbf{v}'\bar{\Psi} + \bar{\mathbf{v}}\Psi' + \mathbf{v}'\Psi' \tag{8.10}$$

Its average, for example, over a model grid cell, is given by

$$\overline{\mathbf{v}\Psi} = \bar{\mathbf{v}}\bar{\Psi} + \overline{\mathbf{v}'\Psi'} \tag{8.11}$$

since by definition $\overline{\mathbf{v}'} = 0$ and $\overline{\Psi'} = 0$. The first term on the right-hand side of (8.11) is the flux associated with the resolved circulation (product of the mean values). The second term is the *eddy flux* arising from the covariance between the fluctuations of $\mathbf{v}$ and Ψ.

When the Eulerian form of the continuity equation for the density ρ_i of chemical species i is considered (Section 4.2),

$$\frac{\partial \rho_i}{\partial t} + \boldsymbol{\nabla} \cdot (\rho_i \mathbf{v}) = s_i \tag{8.12}$$

the separation between spatially resolved and unresolved transport is expressed by:

$$\frac{\partial(\bar{\rho}_i + \rho_i')}{\partial t} + \nabla \cdot \left[(\bar{\rho}_i + \rho_i')(\bar{\mathbf{v}} + \mathbf{v}') \right] = \bar{s}_i + s_i' \tag{8.13}$$

By taking the average of each term and noting that $\overline{\rho_i'} = 0$, $\overline{s_i'} = 0$, and $\overline{\mathbf{v}'} = 0$, the continuity equation for the mean concentration $\bar{\rho}_i$ becomes

$$\frac{\partial\bar{\rho}_i}{\partial t} + \nabla \cdot (\bar{\rho}_i \bar{\mathbf{v}}) + \nabla \cdot \left(\overline{\rho_i' \mathbf{v}'} \right) = \bar{s}_i \tag{8.14}$$

The transport terms that appear in the continuity equation account for contributions by the mean (resolved) circulation and by the eddy (unresolved) motions. The eddy transport term must be parameterized in some way (Section 8.4). In the vertical direction, the eddy flux is generally much larger than the resolved flux because mechanical and buoyant turbulence dominate the motion. In the horizontal direction, the eddy flux is generally less important. In 2-D altitude–latitude models with no zonal resolution, often used for the stratosphere, the eddy terms account for the meridional transport associated with large-scale planetary wave disturbances, and the contributions of the mean and eddy transport terms tend to be equally important.

When expressed in terms of mass mixing ratio $\mu_i = \rho_i/\rho_a$, for species i, the continuity equation (8.14) becomes (see (4.8)):

$$\frac{\partial(\rho_a \mu_i)}{\partial t} + \nabla \cdot (\rho_a \mu_i \mathbf{v}) = s_i \tag{8.15}$$

Here again ρ_a represents the air density [kg m^{-3}]. If we apply the Reynolds decomposition for variables ρ_a, μ_i, $\mathbf{v}$, and s_i, equation (8.15) becomes

$$\frac{\partial\left[(\bar{\rho}_a + \rho_a')(\bar{\mu}_i + \mu_i')\right]}{\partial t} + \nabla \cdot \left[(\bar{\rho}_a + \rho_a')(\bar{\mu}_i + \mu_i')(\bar{\mathbf{v}} + \mathbf{v}') \right] = \bar{s}_i + s_i' \tag{8.16}$$

By taking the average of each term, we obtain:

$$\frac{\partial(\bar{\rho}_a \bar{\mu}_i)}{\partial t} + \frac{\partial\left(\overline{\rho_a' \mu_i'}\right)}{\partial t} + \nabla \cdot \left[\bar{\rho}_a\left(\bar{\mu}_i\bar{\mathbf{v}} + \overline{\mu_i'\mathbf{v}'}\right) + \overline{\rho_a'\mu_i'}\,\bar{\mathbf{v}} + \overline{\rho_a'\mathbf{v}'}\,\bar{\mu}_i + \overline{\rho_a'\mu_i'\mathbf{v}'} \right] = \bar{s}_i \tag{8.17}$$

Ignoring the fluctuations of the air density ($\rho' = 0$), (8.17) becomes:

$$\frac{\partial(\bar{\rho}_a \bar{\mu}_i)}{\partial t} + \nabla \cdot \left[\bar{\rho}_a\left(\bar{\mathbf{v}}\bar{\mu}_i + \overline{\mathbf{v}'\mu_i'}\right) \right] = \bar{s}_i \tag{8.18}$$

Making use of the continuity equation for the mean air density, one deduces that

$$\frac{\partial\bar{\mu}_i}{\partial t} + \bar{\mathbf{v}}\nabla\bar{\mu}_i + \frac{1}{\bar{\rho}_a}\nabla \cdot \left[\bar{\rho}_a\overline{\mathbf{v}'\mu_i'} \right] = \frac{\bar{s}_i}{\bar{\rho}_a} \equiv \bar{S}_i \tag{8.19}$$

If the air density can be assumed to be constant over the entire spatial domain under consideration, as often assumed in boundary layer problems, (8.19) becomes

$$\frac{\partial\bar{\mu}_i}{\partial t} + \bar{\mathbf{v}}\nabla\bar{\mu}_i + \nabla \cdot \left(\overline{\mu_i'\mathbf{v}'} \right) = \bar{S}_i \tag{8.20}$$

which shows that, in the absence of chemical reactions, the local change of mixing ratio is affected by the divergence of the eddy flux density $\overline{\mu_i' v'}$. When chemical reactions are taken into account, the mean source term $\bar{s}$ includes covariance terms that measure the correlation between the fluctuations in the concentrations of the different reactive species, as further discussed in Section 8.3.

The assumption of constant air density is often inadequate. To address this problem, the Reynolds decomposition procedure can be replaced by the so-called Favre decomposition (Hesselberg, 1926; Favre, 1958a, 1958b; Van Mieghem, 1973):

$$\Psi = \hat{\Psi} + \Psi'' \tag{8.21}$$

in which $\hat{\Psi}$ represents the density-weighted average over a volume V:

$$\hat{\Psi} = \frac{\int_V \rho_a\, \Psi\, dV}{\int_V \rho_a\, dV} = \frac{\overline{\rho_a \Psi}}{\overline{\rho_a}} \tag{8.22}$$

and Ψ'' is the departure from this average. Note that $\hat{\Psi}'' = 0$ but $\overline{\Psi''} \neq 0$. Applying a Favre averaging procedure, the continuity equation takes the exact form (Kramm and Meixner, 2000)

$$\frac{\partial \hat{\mu}_i}{\partial t} + \hat{\mathbf{v}} \boldsymbol{\nabla} \hat{\mu}_i + \frac{1}{\overline{\rho}_a} \boldsymbol{\nabla} \cdot \left[\overline{\rho_a \mathbf{v}'' \mu''} \right] = \overline{S}_i \tag{8.23}$$

This equation avoids assumptions about the fluctuation of air density. Like the averaging procedure based on the Reynolds decomposition, the Favre decomposition term representing the eddy flux must be parameterized. In what follows we will use the standard notation for the Reynolds decomposition but the equations can also be applied to the Favre averaging procedure.

8.3 Chemical Covariance

Subgrid variability affects the local chemical terms in the continuity equation through covariances between concentrations of reacting species. This can be important for modeling the evolution of chemical and aerosol plumes. It can be addressed by application of Reynolds decomposition to the chemical variables. For the simple case of a single reaction $A + B \rightarrow C$, the mean chemical production rate $\bar{s}$ for species C is

$$\bar{s} = \overline{k_{AB}\, \rho_A\, \rho_B} = \overline{k}_{AB}\left(\overline{\rho}_A \overline{\rho}_B + \overline{\rho_A' \rho_B'} \right) + \overline{\rho}_A \overline{k_{AB}' \rho_B'} + \overline{\rho}_B \overline{k_{AB}' \rho_A'} + \overline{k_{AB}' \rho_A' \rho_B'} \tag{8.24}$$

where ρ_A and ρ_B represent the density of chemical species A and B, respectively. The variation in the reaction rate constant k_{AB} generally results from fluctuations in temperature and can then be expressed by

$$k_{AB}' = \frac{\partial k_{AB}}{\partial T} T' \tag{8.25}$$

The term $\overline{\rho'_A \rho'_B}$ represents the *chemical covariance* between the concentrations of A and B, while the terms $\overline{k'_{AB}\rho'_B}$ and $\overline{k'_{AB}\rho'_A}$ account for the covariance between species concentration and temperature. The last term $\overline{k'_{AB}\rho'_A \rho'_B}$ is the third-order moment of the fluctuations in concentrations and temperature.

If we ignore the variability in the rate constant k_{AB} that results from eddy variations in the temperature ($k'_{AB} = 0$), the mean source term is expressed by the simpler relation

$$\overline{s} = \overline{k_{AB}\rho_A \rho_B} = \overline{k}_{AB}\left(\overline{\rho}_A \overline{\rho}_B + \overline{\rho'_A \rho'_B}\right) \tag{8.26}$$

The mean chemical source rate is therefore the sum of a *resolved source term* that can be explicitly calculated from the mean concentrations and an *unresolved chemical covariance term* whose value must be parameterized in some way. If chemical species A and B have a common origin the covariance term is usually positive. If they have different origins it is often negative.

The *segregation ratio* (or *intensity of segregation*) provides an estimate of the degree of mixing for A and B (Brodkey, 1981):

$$I_{AB} = \frac{\overline{\rho'_A \ \rho'_B}}{\overline{\rho}_A \ \overline{\rho}_B} \tag{8.27}$$

A value of I_{AB} equal to zero implies that the reactants are well-mixed so that chemical evolution can be computed from the grid mean concentrations. A value of -1 (anti-correlation) indicates that the reactants are fully segregated and the mean source term $\overline{s}$ is then equal to 0. From (8.26) and (8.27) one finds

$$\overline{s} = k_{AB} \ \overline{\rho_A \ \rho_B} = k_{AB}(1 + I_{AB}) \ \overline{\rho}_A \ \overline{\rho}_B \tag{8.28}$$

This expression suggests that, in theory, the effects of turbulence on chemical reactions can be accounted for by replacing the rate constants k_{AB} by effective rate constants $k_{AB,eff} = k_{AB} (1 + I_{AB})$ (Vinuesa and Vilà-Guerau de Arellano, 2005). However, inferring the value of I_{AB} and its variability is not straightforward, so this approach cannot be easily implemented.

The effect of turbulent fluctuations on a chemical reaction rate can be described by the *Damköhler number* (Damköhler, 1940, 1947):

$$Da = \frac{\tau_{turb}}{\tau_{chem}} \tag{8.29}$$

which represents the ratio between the timescale τ_{turb} associated with turbulence in the flow and the timescale τ_{chem} associated with chemical evolution. Different formulations are available to estimate the turbulence timescale. For example, in the convective atmospheric boundary layer, the time constant for transport in buoyant eddies can be expressed as the ratio between the mixed layer depth h and the *convective velocity scale* w^* defined as

$$w^* = \left[\frac{gh\left(\overline{\theta'_v w'}\right)_{surf}}{\overline{\theta}_v}\right]^{1/3} \tag{8.30}$$

For typical daytime values $h = 1000$ m and $w^* = 1\text{–}2$ m s^{-1} (Stull, 1988), the turbulent timescale is of the order of 10–15 minutes. It is considerably longer in a stable atmosphere such as at night. In the *slow chemistry limit*, in which the chemical timescale is long in comparison with the turbulent timescale ($D_a \ll 1$), the chemical species are well mixed and the concentrations can be approximated by their mean values. The chemical covariance terms can be ignored. In the opposite situation, referred to as *fast chemistry limit* ($D_a \gg 1$), chemical transformation is limited by the rate at which turbulence brings reactants together. In this case, the covariance between the fluctuating components of the concentrations is large in comparison to the product of the mean concentrations and must be estimated from closure relations.

8.4 Closure Relations

In order to solve the continuity equations including eddy contributions that arise from Reynolds decomposition, *closure relations* that relate the eddy flux and covariance terms to mean quantities must be formulated. These closure relations are effectively parameterizations. Different formulations are possible. *Local closure* schemes express unknown eddy quantities at a given model grid point as a function of known mean quantities or their derivatives at that grid point. *Non-local closure* schemes relate the unknown eddy quantities at a grid point to known mean quantities at other grid points. In *first-order closure*, the mean quantities are the only dependent variables solved by the continuity equations. In *higher-order closure*, additional equations for the higher moments are solved together with the equations for the mean quantities. For example, in second-order closure schemes, prognostic equations are expressed for covariance terms such as $\overline{\Psi'_m \Psi'_n}$ and closure formulations must then be adopted for the third moments $\overline{\Psi'_k \Psi'_m \Psi'_n}$.

8.4.1 First-Order Closure

A simple first-order closure relation assumes that the eddy flux $\overline{\mu'_i \mathbf{v}'}$ of species i is proportional to the gradient of the mean mixing ratio $\overline{\mu}_i$. This amounts to assuming analogy of turbulent mixing and molecular diffusion (Fick's law):

$$\overline{\mu'_i \mathbf{v}'} = -\mathbf{K} \nabla \overline{\mu}_i \tag{8.31}$$

Such parameterization of turbulent mixing as molecular diffusion is grounded in the observed near-Gaussian dilution of plumes emanating from point sources. It is a good assumption when eddy scales are small relative to the model grid scale. The *turbulent* (or *eddy*) *diffusion* matrix $\mathbf{K}$ has elements K_{ij} that describe turbulent diffusion in the three spatial directions (x, y, z). These elements K_{ij} are called *turbulent diffusion coefficients* and are derived from empirical relations. One generally ignores the off-diagonal terms K_{xy}, K_{yz}, etc. that allow for the existence of counter-gradient fluxes, and retain only the diagonal terms K_{xx}, K_{yy}, and K_{zz}.

We pointed out above that turbulent mixing is generally most important in the vertical direction where mean winds are slow. If only the vertical direction is considered, the eddy flux of species i is written

$$\overline{\mu_i' w'} = -K_z \frac{\partial \overline{\mu}_i}{\partial z} \tag{8.32}$$

If we neglect inhomogeneities in air density ($\rho_a' = 0$), we have

$$\overline{\rho_i' w'} = -K_z \; \overline{\rho}_a \frac{\partial}{\partial z} \left(\frac{\overline{\rho}_i}{\overline{\rho}_a} \right) \tag{8.33}$$

Here K_z is the vertical turbulent (or eddy) diffusion coefficient. The same value of K_z is assumed to apply to all chemical species, and is often derived from turbulent diffusion of momentum or specific heat: This is the so-called *similarity assumption*. K_z depends on the intensity of turbulence. Standard semi-empirical formulations of K_z for the PBL are presented in Section 8.7. The K_z formalism is also used in conceptual 1-D (vertical) models of the global atmosphere, in which case K_z values are chosen to fit observed gradients of atmospheric tracers (Liu *et al.*, 1984). Values of K_z in the boundary layer are of the order of 100 m^2 s^{-1} in the daytime (unstable atmosphere) and 0.1 m^2 s^{-1} at night (stable atmosphere). Values in the free troposphere are of the order of 10 m^2 s^{-1} and values in the stratosphere are of the order of 0.1–1 m^2 s^{-1}. The time constant that describes diffusive transport over a length scale L is $L^2/2K_z$, by analogy with Einstein's equation for molecular diffusion. For example, a 1-km thick daytime boundary layer with $K_z = 100$ m^2 s^{-1} mixes vertically on a timescale of 5×10^3 s or 1.5 hours. This timescale is much longer than the transport time $h/w*$ in buoyant updrafts introduced in Section 8.3. Thus species emitted at the surface can be injected rapidly to the top of the boundary layer in buoyant updrafts, but thorough vertical mixing of the boundary layer takes a longer time.

8.4.2 Higher-Order Closures

The first-order closure formalism has the advantage of being computationally expedient. It is not suited to strongly convective environments, where the transport is mostly accomplished by the largest eddies (Wyngaard, 1982; Vilà-Guerau de Arellano, 1992) instead of the small eddies assumed in the turbulent diffusion closure. One can address this problem by using higher-order closure approaches. This adds other equations (Stull, 1988; Garratt, 1994; Stensrud, 2007) that describe the evolution of higher order moments (e.g., eddy fluxes, covariances, turbulent kinetic energy). The equations for the mean and turbulent components are established by applying Reynolds decomposition to the dependent variables in the different prognostic equations (continuity, momentum, energy). Equations for the mean quantities are obtained by averaging all terms in the equations. Equations for the turbulent components are obtained by subtracting the equations for the mean quantities from the governing equations. From there, predictive equations can be established for the different eddy fluxes and covariances. Box 8.1 gives an example.

Box 8.1	Second-Order Closure Equations for the Turbulent Flow, Eddy Fluxes, and Covariances of Chemical Species

The governing equations that describe the interactions between chemistry and turbulent mixing of species include second moments such as eddy fluxes $\overline{w'\mu'}$ or chemical covariance $\overline{\mu'_m\mu'_n}$. The derivation of the equations for such quantities requires long algebraic manipulations. We consider here a simplified case assuming horizontal homogeneity (no derivative along horizontal directions) and no air subsidence (mean vertical wind component equals zero). We start by writing the vertical projection of the momentum equation in which we assume that the friction term is proportional to the Laplacian of the velocity

$$\frac{\partial w}{\partial t} + w\frac{\partial w}{\partial z} = -g - \frac{1}{\rho_a}\frac{\partial p}{\partial z} + \nu\frac{\partial^2 w}{\partial z^2}$$

Here ν stands for the kinematic viscosity coefficient. We now apply the Reynolds decomposition with

$$\rho_a = \overline{\rho}_a + \rho'_a \qquad w = \overline{w} + w' \qquad p = \overline{p} + p'$$

If we assume that the atmospheric mean state follows hydrostatic equilibrium conditions and if we further neglect density variations $\left(\rho'_a/\overline{\rho}_a \ll 1\right)$ in the inertia term $\partial w/\partial t$ but retain it in the gravity term (Boussinesq approximation), we find the turbulent momentum equation

$$\frac{\partial w'}{\partial t} + w\frac{\partial w'}{\partial z} = \left(\frac{\theta'_v}{\overline{\theta}_v}\right)g - \frac{1}{\overline{\rho}_a}\frac{\partial p'}{\partial z} + \nu\frac{\partial^2 w'}{\partial z^2} - \frac{\partial\overline{w'w'}}{\partial z}$$

In this equation, we have replaced the density variations by virtual potential temperature variations $\left(\rho'_a/\overline{\rho}_a = \theta'_v/\overline{\theta}_v\right)$ as deduced from the equation of state. The virtual potential temperature (potential temperature that accounts for the buoyancy effects related to humidity – see Section 2.4) is related to the value of the potential temperature θ by $\theta_v = \theta\,(1 + 0.61\,r_w)$ if r_w represents the water vapor mixing ratio by mass.

We now consider the simplified continuity equation for the mixing ratio μ that includes a molecular diffusion term

$$\frac{\partial\mu}{\partial t} + w\frac{\partial\mu}{\partial z} = D\frac{\partial^2\mu}{\partial z^2} + S$$

where D represents the diffusion coefficient. If we apply again a Reynolds decomposition with

$$\mu = \overline{\mu} + \mu' \qquad w = \overline{w} + w' \qquad S = \overline{S} + S'$$

we obtain the following equation

$$\frac{\partial\overline{\mu}}{\partial t} + \frac{\partial\mu'}{\partial t} + w'\frac{\partial\overline{\mu}}{\partial z} + w'\frac{\partial\mu'}{\partial z} = D\frac{\partial^2\overline{\mu}}{\partial z^2} + D\frac{\partial^2\mu'}{\partial z^2} + \overline{S} + S'$$

in which subsidence has been ignored and the mean vertical velocity $\overline{w} = 0$

Applying the averaging operator to each term of the equation, noting that, by continuity, $\partial w'/\partial z = 0$, we find the equation for the mean mixing ratio

Box 8.1 (*cont.*)

$$\frac{\partial \overline{\mu}}{\partial t} + \frac{\partial \overline{\mu' w'}}{\partial z} = D\frac{\partial^2 \overline{\mu}}{\partial z^2} + \overline{s}$$

where the second term reflects the effect of turbulence on the vertical distribution of the mean mixing ratio. In this expression, molecular diffusion is often neglected because it is much smaller than the eddy flux term. By subtraction, we obtain the equation for the eddy component of the mixing ratio

$$\frac{\partial \mu'}{\partial t} + w'\frac{\partial \overline{\mu}}{\partial z} + w'\frac{\partial \mu'}{\partial z} = D\frac{\partial^2 \mu'}{\partial z^2} + \frac{\partial \overline{w'\mu'}}{\partial z} + s'$$

We then derive the equation for the mean vertical eddy flux $\overline{w'\mu'}$ of the tracer mixing ratio by multiplying the momentum perturbation equation by μ' and the tracer perturbation equation by w'. We then take the Reynolds average of both equations and add them together. After some manipulations that include the transformation of the turbulent flux divergence term into its flux form, we derive the equation for vertical turbulent tracer flux:

$$\frac{\partial \overline{w'\mu'}}{\partial t} = -\overline{w'^2}\frac{\partial \overline{\mu}}{\partial z} - \frac{\partial \overline{w'^2 \mu'}}{\partial z} + \overline{\mu'\theta'}_v\frac{g}{\overline{\theta}_v} - \frac{\overline{\mu'}}{\overline{p}_a}\frac{\partial p'}{\partial z} - F + \overline{R}_{w'\mu'}$$

The terms on the right side of the equation represent the flux source/sink terms associated with the vertical gradient in the mean mixing ratio, the vertical turbulent transport of the flux, the buoyant production, the pressure covariance, tracer flux dissipation, and chemical transformations. The dissipation term F must be parameterized. For a chemical scheme that includes N photolysis reactions and M second-order reactions, the chemical term can be expressed as

$$\overline{R}_{w'\mu} = \text{sign}\sum_{i=1}^{N} J_i\,\overline{w'\mu'_i} + \text{sign}\sum_{n,m}^{M} k_{nm}\left[\overline{\mu}_m\left(\overline{w'\mu'_n}\right) + \overline{\mu}_n\left(\overline{w'\mu'_m}\right) + \overline{w'\mu'_m\mu'_n}\right]$$

with $m < n$ in the second summation. Factor sign is equal to $+1$ if the reaction constitutes a production and to -1 if it is a loss. J and k represent photolysis coefficients and reaction rate constants, respectively.

The chemical covariance $\overline{\mu'_m\mu'}_n$ that appears in the source term of the continuity equation for reactive species can be derived from a covariance budget equation (Garratt, 1994; Verver *et al.*, 1997)

$$\frac{\partial \overline{\mu'_m\mu'}_n}{\partial t} = -\overline{\mu'_m w'}\frac{\partial \overline{\mu}_n}{\partial z} - \overline{\mu'_n w'}\frac{\partial \overline{\mu}_m}{\partial z} - \frac{\partial \overline{\mu'_m\mu'_n w'}}{\partial z} - 2D\sum_{k=1}^{3}\overline{\left(\frac{\partial \mu'_m}{\partial x_k}\right)\left(\frac{\partial \mu'_n}{\partial x_k}\right)} + \overline{R}_{mn}$$

Here, the first two terms on the right-hand side of the equation represent the production of chemical covariance by the concentration gradients, the third term accounts for the vertical turbulent transport of the second moment, and the fourth term represents the dissipation by molecular diffusion. The last term accounts for the chemical influence on the covariances. For example, if k denotes the rate constant of a reaction between two species m and n, the corresponding loss term is

$$\overline{R}_{mn} = -k\left[\overline{\mu}_m\overline{\mu'_m\mu'_n} + \overline{\mu}_n\overline{\mu'_m\mu'_n} + \overline{\mu}_m\overline{\mu'^2_n} + \overline{\mu}_n\overline{\mu'^2_m} + \overline{\mu'^2_m\mu'_n} + \overline{\mu'_m\mu'^2_n}\right]$$

Additional moments including the wind variance $\overline{w'^2}$, the concentration variance $\overline{\mu'^2_i}$, and the covariance $\overline{\mu'_i\theta'_v}$ between species concentrations and the virtual temperature are provided by Stull (1988) and Garratt (1994). The triple correlation terms that appear in the second-order equations must be determined by higher-order closure equations or empirical closure expressions (Verver et al., 1997).

Second-order closure formulations include predictive equations for covariances between wind components, wind and temperature, wind and humidity, and wind and chemical concentrations. Triple correlation terms appear in the equations and lead to a new closure problem. In principle, additional differential equations can be written to describe the evolution of third moments, but in this case fourth-order moments will appear. In most practical applications, the system is closed either by neglecting these high-order moments or by introducing diagnostic expressions that include adjustable empirical parameters.

8.5 Stochastic Representation of Turbulent Reacting Flows

An alternative approach to treat turbulence is through stochastic methods (Pope, 2000). In this case, the flow velocity $\mathbf{v}(\mathbf{r}, t) = (v_1, v_2, v_3)^T$ and the vector of concentrations for N interacting chemical species $\mathbf{\Psi}(\mathbf{r}, t) = (\Psi_1, \ldots, \Psi_N)^T$ at location $\mathbf{r}$ are viewed as random variables with respect to time t. Their dynamical behavior is fully described by a *joint velocity-composition PDF*, denoted here $p_{\mathbf{v},\mathbf{\Psi}}$, that describes the likelihood for the continuous random variables $\mathbf{v}$ and $\mathbf{\Psi}$ to take given values. Specifically, the probability that the random velocity $\mathbf{v}$ *and* the random chemical concentration $\mathbf{\Psi}$ fall into the infinitesimal intervals $[\mathbf{u}, \mathbf{u} + d\mathbf{u}]$ and $[\boldsymbol{\phi}, \boldsymbol{\phi} + d\boldsymbol{\phi}]$, respectively is given by

$$\Pr[(\mathbf{u} < \mathbf{v}(\mathbf{r}, t) < \mathbf{u} + d\mathbf{u}) \text{ and } (\boldsymbol{\phi} < \mathbf{\Psi}(\mathbf{r}, t) < \boldsymbol{\phi} + d\boldsymbol{\phi})] = p_{\mathbf{v},\mathbf{\Psi}}(\mathbf{u}, \boldsymbol{\phi}; \mathbf{r}, t)\, d\mathbf{u}\, d\boldsymbol{\phi} \tag{8.34}$$

In this section the expected (mean) value of a random quantity $X(\mathbf{v}, \mathbf{\Psi})$ is denoted by brackets:

$$\langle X(\mathbf{v}, \mathbf{\Psi})\rangle = \int\limits_{-\infty}^{+\infty} d\mathbf{u} \int\limits_{-\infty}^{+\infty} X\, p_{\mathbf{v},\mathbf{\Psi}}(\mathbf{u}, \boldsymbol{\phi}; \mathbf{r}, t)\, d\boldsymbol{\phi} \tag{8.35}$$

and the fluctuating component is denoted by a prime: $X' = X - \langle X\rangle$.

The evolution of the turbulent field at a fixed point of the flow is simulated by solving a transport equation for the joint PDF. Fox (2003) provides details on the derivation of this equation. In short, the equation can be established by equating two independent expressions for the expected value of the total derivative of an arbitrary scalar function $F(\mathbf{v}, \mathbf{\Psi})$:

$$\left\langle \frac{dF}{dt} \right\rangle = \frac{\partial \langle F \rangle}{\partial t} + \sum_{i=1}^{3} \frac{\partial \langle v_i F \rangle}{\partial x_i} \tag{8.36}$$

and

$$\left\langle \frac{dF}{dt} \right\rangle = \sum_{i=1}^{3} \left\langle \frac{\partial F}{\partial v_i} \frac{dv_i}{dt} \right\rangle + \sum_{n=1}^{N} \left\langle \frac{\partial F}{\partial \Psi_n} \frac{d\Psi_n}{dt} \right\rangle \tag{8.37}$$

The total derivatives of the fluid velocity components ($i = 1, 3$) and the concentration components ($n = 1, N$) that appear in (8.37) are substituted by expressions derived from the Navier–Stokes and continuity equations, respectively

$$\frac{dv_i}{dt} = \frac{\partial v_i}{\partial t} + \sum_{j=1}^{3} v_j \frac{\partial v_i}{\partial x_j} = \nu \sum_{j=1}^{3} \frac{\partial^2 v_i}{\partial x_j^2} - \frac{1}{\rho_a} \frac{\partial p}{\partial x_i} + g_i \quad (i = 1, 3) \tag{8.38}$$

$$\frac{d\Psi_n}{dt} = \frac{\partial \Psi_n}{\partial t} + \sum_{j=1}^{3} v_j \frac{\partial \Psi_n}{\partial x_j} = D_n \sum_{j=1}^{3} \frac{\partial^2 \Psi_n}{\partial x_j^2} + S_n(\mathbf{\Psi}) \quad (n = 1, N) \tag{8.39}$$

where ν [m^2 s^{-1}] is the kinematic viscosity coefficient and D_n [m^2 s^{-1}] is a molecular diffusion coefficient. Quantity g_i denotes the gravitational acceleration and ρ_a the air density. The expected values of the different terms appearing in expressions (8.38) and (8.39) are expressed as a function of the joint PDF, applying (8.35) for the different variables. Noting that the equality between (8.36) and (8.37) must hold for any arbitrary choice of scalar function F, one derives the following transport equation (Fox, 2003; Cassiani *et al.*, 2010)

$$\frac{\partial p_{\mathbf{v}, \mathbf{\Psi}}}{\partial t} + \sum_{i=1}^{3} u_i \frac{\partial p_{\mathbf{v}, \mathbf{\Psi}}}{\partial x_i} = -\sum_{i=1}^{3} \frac{\partial}{\partial u_i} \left[p_{\mathbf{v}, \mathbf{\Psi}} \langle A_i | \mathbf{u}, \mathbf{\phi} \rangle \right] - \sum_{n=1}^{N} \frac{\partial}{\partial \phi_n} \left[p_{\mathbf{v}, \mathbf{\Psi}} \langle B_i | \mathbf{u}, \mathbf{\phi} \rangle \right] \tag{8.40}$$

where the conditional fluxes are

$$\langle A_i | \mathbf{u}, \mathbf{\phi} \rangle = \left\langle \left(\nu \sum_{j=1}^{3} \frac{\partial^2 v_i}{\partial x_j^2} - \frac{1}{\rho_a} \frac{\partial p'}{\partial x_i} \right) \middle| \mathbf{v}(\mathbf{r}, t) = \mathbf{u}, \mathbf{\Psi}(\mathbf{r}, t) = \mathbf{\phi} \right\rangle - \frac{1}{\rho_a} \frac{\partial \langle p \rangle}{\partial x_i} + g_i \tag{8.41}$$

and

$$\langle B_i | \mathbf{u}, \mathbf{\phi} \rangle = \left\langle D_n \sum_{j=1}^{3} \frac{\partial^2 \Psi_n}{\partial x_j^2} \middle| \mathbf{v}(\mathbf{r}, t) = \mathbf{u}, \mathbf{\Psi}(\mathbf{r}, t) = \mathbf{\phi} \right\rangle + S_n(\mathbf{\phi}) \tag{8.42}$$

The first terms in (8.40) refer successively to the local rate of change of the joint PDF and to its transport in the geometric space (x_i). The first term on the right-hand side of the equation accounts for the effects of viscous stress, pressure fluctuations and gravity, and the second term for the transport by molecular fluxes in the composition space and for chemical sources. Terms that require multiple-point information (gradient and Laplacian terms) are not closed, and closure expressions based on empirical information or DNS data must be added to the system.

The numerical solution of the transport equation (8.40) provides the value of the joint PDF at a specified point $\mathbf{r}$ and time t, from which mean physical quantities can be derived by integration over velocities and concentrations following (8.35). An advantage of the stochastic method over the classic Reynolds decomposition method is that it provides not only the mean values of the velocity and concentrations, but also the full PDF from which to compute higher moments of these quantities, including the covariances between fluctuating quantities. In particular, no closure assumption is needed to treat the chemical source term involving products of concentrations.

Solution of (8.40) is computationally burdensome and more economical methods are usually adopted. A common approach is to consider a simpler *composition PDF* obtained by integrating the joint velocity-composition PDF over the entire velocity phase space:

$$p_{\mathbf{\Psi}}(\boldsymbol{\phi}; \mathbf{r}, t) = \int\limits_{-\infty}^{+\infty} p_{\mathbf{v}, \mathbf{\Psi}}(\mathbf{u}, \boldsymbol{\phi}; \mathbf{r}, t) \, d\mathbf{u} \tag{8.43}$$

In this case, one assumes that the stochastic wind fields are provided by an external turbulence model. The transport equation for the composition PDF is derived by integrating (8.40) over the entire velocity phase space (see Fox, 2003 for a complete derivation). One finds:

$$\frac{\partial p_{\mathbf{\Psi}}}{\partial t} + \sum_{i=1}^{3} \langle v_i \rangle \frac{\partial p_{\mathbf{\Psi}}}{\partial x_i} = -\sum_{i=1}^{3} \frac{\partial}{\partial x_i} \left[p_{\mathbf{\Psi}} \langle v_i' | \boldsymbol{\phi} \rangle \right] - \sum_{n=1}^{N} \frac{\partial}{\partial \phi_n} \left[p_{\mathbf{\Psi}} \langle (D_n \nabla^2 \Psi_n') | \boldsymbol{\phi} \rangle \right]$$
$$- \sum_{n=1}^{N} \frac{\partial}{\partial \phi_n} \left[p_{\mathbf{\Psi}} \left(D_n \nabla^2 \langle \Psi_n \rangle + S_n(\boldsymbol{\phi}) \right) \right] \tag{8.44}$$

The terms on the left-hand side of (8.44) represent again the time evolution of the PDF and its advection by the mean wind field $\langle v_i \rangle$ in the spatial space x_i (sometimes referred to as macro-mixing). The terms on the right-hand side include the effect of unresolved concentration-conditioned velocity fluctuations (meso-mixing) and the transport in the composition space due to molecular mixing (micro-mixing). Again, the chemical term involves only single-point information, and is described in a closed form. One usually assumes that the term related to the concentration-conditioned velocity $\langle v_i' | \boldsymbol{\phi} \rangle$ can be parameterized by

$$p_{\mathbf{\Psi}} \langle v_i' | \mathbf{\Psi}(\mathbf{r}, t) = \boldsymbol{\phi} \rangle = -K \frac{\partial p_{\mathbf{\Psi}}}{\partial x_i} \tag{8.45}$$

where K is an eddy diffusion coefficient, supposed here to be isotropic. Similarly, the conditional term $\langle (D_n \nabla^2 \Psi_n') | \boldsymbol{\phi} \rangle$ that accounts for the dissipation of the concentration fluctuations by molecular diffusion is also unclosed, and can be parameterized by a linear dissipation toward the mean concentration state $\langle \Psi_n \rangle$.

It has been shown (Valiño, 1998) that the solution of the transport equation for the composition PDF (8.44) is equivalent to the solution provided by an ensemble of M stochastic partial differential equations (PDEs)

$$
\begin{aligned}
\frac{d\Psi_n^{(m)}}{dt} = & -\sum_{i=1}^{3} \langle v_i \rangle \frac{\partial \Psi_n^{(m)}}{\partial x_i} + \sum_{i=1}^{3} \frac{\partial}{\partial x_i} \left(K \frac{\partial \Psi_n^{(m)}}{\partial x_i} \right) \\
& + (2K)^{1/2} \sum_{i=1}^{3} \frac{\partial \Psi_n^{(m)}}{\partial x_i} \frac{d\xi_i^{(m)}}{dt} + S_n^{(m)}(\boldsymbol{\Psi}) - \frac{\left(\Psi_n^{(m)} - \langle \Psi_n \rangle \right)}{T_{mix}}
\end{aligned}
\tag{8.46}
$$

where m denotes one of the M possible realizations and T_{mix} a mixing timescale. The two first terms on the right-hand side of this equation represent the effects of the large-scale advection and of mixing by eddy diffusion. The third term introduces the effects of random motions with $d\xi_i$ denoting the random increment of an uncorrelated Wiener process with a zero average and a variance equal to dt (see Section 4.11.2). The fourth term evaluates the chemical sources and sinks; it implicitly accounts for the effects of fluctuations (chemical eddies) on the reaction rates. The last term introduces a relaxation toward the mean concentration. The average of the concentration field as well as the covariances are obtained from the ensemble of solutions in the M different realizations.

Stochastic model modules, if inserted into a coarser chemical transport modeling system, provide an approach to simulate the influence of chemical and dynamical subgrid processes on large-scale dynamics and chemistry, including, for example, the dispersion of reactive plumes in the PBL and the effects of subgrid heterogeneities in surface emissions. The formulation discussed above is Eulerian because the PDFs are calculated at fixed points in space. The evolution of fluid particle properties in turbulent reacting flows can also be simulated by Lagrangian stochastic models (see Section 4.11.2) in which the flow is represented by a large number of particles, each being characterized by its own position, velocity, and chemical composition. These different properties evolve according to stochastic Lagrangian model equations, and each sampled particle is assumed to be representative of a different realization of the flow.

8.6 Numerical Solution of the Diffusion Equation

Diffusion equations are often used in atmospheric models to parameterize the effects of small-scale turbulence. They must be solved numerically except in idealized cases (see Box 8.2). Here, we first consider numerical methods adopted to solve the 1-D diffusion equation

$$
\frac{\partial \Psi}{\partial t} = \frac{\partial}{\partial x} \left[K \frac{\partial \Psi}{\partial x} \right]
\tag{8.47}
$$

which is frequently encountered in chemical transport models.

Box 8.2	Analytic Solution for the 1-D Diffusion Equation

Consider the 1-D diffusion equation for an inert tracer concentration $\Psi(x, t)$

$$\partial \Psi / \partial t = K \partial^2 \Psi / \partial x^2$$

where K is a constant diffusion coefficient. We assume at the initial time ($t = 0$) that the entire mass of the tracer is concentrated at the spatial origin ($x = 0$). Thus, the initial value of the concentration is $\Psi(x, 0) = S \, \delta(x)$ where $\delta(x)$ is the Dirac function (this function equals zero everywhere except at $x = 0$ and its integral value is 1). For an infinite space with $\Psi = 0$ for $x \rightarrow \pm \infty$, the analytic solution is the Gaussian expression

$$\Psi(x, t) = \left\{ S / (4\pi K t)^{1/2} \right\} \exp \left\{ - \left[x^2 / (4 K t) \right] \right\}$$

This expression provides the basis for Gaussian plume models. The plume dilutes with a variance $\sigma^2 = 2 K t$ that increases linearly with time t. See Box 8.2 Figure 1 and Section 4.12.1.

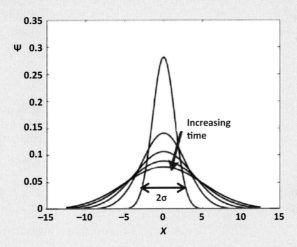

Box 8.2 Figure 1	Dilution of a plume.

8.6.1 Explicit Schemes for the 1-D Diffusion Equation

If the diffusion coefficient K is constant, the partial differential parabolic equation written now as

$$\frac{\partial \Psi}{\partial t} = K \frac{\partial^2 \Psi}{\partial x^2} \tag{8.48}$$

can be approximated by the forward-in-time, centered-in-space (FTCS) scheme:

$$\frac{\Psi_j^{n+1} - \Psi_j^n}{\Delta t} = \frac{K}{\Delta x^2} \left(\Psi_{j-1}^n - 2\Psi_j^n + \Psi_{j+1}^n \right) \tag{8.49}$$

where, as in previous chapters, Δx and Δt represent the grid interval and the time step, respectively. This expression is easy to solve due to its explicit nature. It is, however, only first-order accurate in time. Following von Neumann's stability analysis considered in Chapter 7, in which the evolution of a wave with wavenumber k is examined, the amplification factor $|g(k)|$ associated with the FTCS method is

$$|g(k)| = 1 - 2\beta(1 - \cos(k\Delta x)) \tag{8.50}$$

For the method to be stable, this quantity must be less than 1 for any value of k. This is achieved if the dimensionless parameter

$$\beta = K\frac{\Delta t}{\Delta x^2} \tag{8.51}$$

is less than ½. Thus, to ensure stability of the solution, the integration time step Δt must be smaller than $\Delta x^2/2K$. This timescale characterizes the diffusion of air parcels across the grid cell width Δx. However, even though this criterion guarantees stability, it may not be sufficient to ensure a correct simulation of the shortest resolved waves, and specifically the $2\Delta x$ wave mode. Durran (2010) shows that a more appropriate condition is

$$0 \le \beta \le \frac{1}{4} \tag{8.52}$$

In the explicit *Richardson* scheme, the time derivative in the FTCS scheme is replaced by a centered difference:

$$\frac{\Psi_j^{n+1} - \Psi_j^{n-1}}{2\Delta t} = \frac{K}{\Delta x^2}\left[\Psi_{j-1}^n - 2\Psi_j^n + \Psi_{j+1}^n\right] \tag{8.53}$$

This scheme is second-order accurate in time but is unconditionally unstable. Unconditional stability, however, can be achieved by splitting the $2\Psi_j^n$ term into $\Psi_j^{n+1} + \Psi_j^{n-1}$. The resulting *DuFort–Frankel* algorithm is:

$$\frac{\Psi_j^{n+1} - \Psi_j^{n-1}}{2\Delta t} = \frac{K}{\Delta x^2}\left[\Psi_{j-1}^n - \left(\Psi_j^{n+1} + \Psi_j^{n-1}\right) + \Psi_{j+1}^n\right] \tag{8.54}$$

Although the scheme appears implicit, the solution Ψ_j^{n+1} is easily derived if the dependent variables are known at the two previous time steps t_n and t_{n-1}:

$$\Psi_j^{n+1} = \frac{2\beta}{1 + 2\beta}\left(\Psi_{j-1}^n + \Psi_{j+1}^n\right) + \frac{1 - 2\beta}{1 + 2\beta}\Psi_j^{n-1} \tag{8.55}$$

The amplification factor is:

$$|g(k)| = \frac{2\beta\cos(k\Delta x) \pm \left[1 - 4\beta^2\sin^2(k\Delta x)\right]^{0.5}}{1 + 2\beta} \tag{8.56}$$

In spite of its unconditional stability, the DuFort–Frankel scheme does not correctly damp the short-wavelength components of the solution when implemented with an excessively large time step. It is therefore recommended to adopt a value of β smaller than ½.

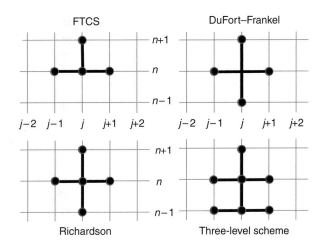

Stencils for different explicit schemes to approximate the 1-D diffusion equation. Indices j and n refer to the spatial and temporal dimensions, respectively.

Accuracy of the solution can be improved by considering a weighted time derivative over three time levels t_{n-1}, t_n, and t_{n+1}, and a more elaborate formulation of the space derivative:

$$(1+\gamma)\frac{\Psi_j^{n+1}-\Psi_j^n}{\Delta t}-\gamma\frac{\Psi_j^n-\Psi_j^{n-1}}{\Delta t}$$
$$=\frac{K}{\Delta x^2}\left[(1-\theta)\left(\Psi_{j-1}^n-2\Psi_j^n+\Psi_{j+1}^n\right)+\theta\left(\Psi_{j-1}^{n-1}-2\Psi_j^{n-1}+\Psi_{j+1}^{n-1}\right)\right] \quad (8.57)$$

where γ and θ are adjustable parameters. For $\theta = 0$, this expression corresponds to the explicit FTCS scheme if $\gamma = 0$, to the Richardson scheme if $\gamma = \frac{1}{2}$, and to the DuFort–Frankel scheme if $\gamma = -\frac{1}{2} + \beta$. Stability conditions vary with the value of parameter γ: the stability limit for β varies from about 0.35 to 0.5 when γ increases from 0 to 6. The stencils corresponding to different explicit algorithms are shown in Figure 8.6.

If, as is often the case in atmospheric applications, the diffusion coefficient K is not a constant parameter, but depends on the spatial variable x, a more general form of the diffusion equation (8.47) must be considered. In this case, the explicit FTCS algorithm, for example, takes the form

$$\frac{\Psi_j^{n+1}-\Psi_j^n}{\Delta t}=\frac{1}{\Delta x}\left\{\left[K(x)\frac{\partial\Psi}{\partial x}\right]_{j+1/2}^n-\left[K(x)\frac{\partial\Psi}{\partial x}\right]_{j-1/2}^n\right\} \quad (8.58)$$

or

$$\frac{\Psi_j^{n+1}-\Psi_j^n}{\Delta t}=\frac{1}{\Delta x^2}\left[K_{j+1/2}\left(\Psi_{j+1}^n-\Psi_j^n\right)-K_{j-1/2}\left(\Psi_j^n-\Psi_{j-1}^n\right)\right] \quad (8.59)$$

Index $j + \frac{1}{2}$ corresponds to the point located at equal distance between grid-points x_j and x_{j+1}. If the value of K is not known at such intermediate locations but only at the grid points, one can use a simple interpolation as $K_{j+1/2} = (K_j + K_{j+1})/2$. This scheme is stable only if

$$\Delta t \le \min\left(\frac{\Delta x^2}{2K_{j+1/2}}\right) \tag{8.60}$$

for all values of j. However, as indicated above, a step twice as small should be preferred to ensure a proper treatment of the shortest resolved waves. If both the diffusion coefficient and the grid spacing vary in space, we approximate the diffusion term by a second-order finite difference expression:

$$\frac{\partial}{\partial x}\left[K(x)\frac{\partial\Psi}{\partial x}\right]_j^n \approx \frac{2}{\Delta x_{j+1} + \Delta x_j}\left[K_{j+1/2}\left(\frac{\Psi_{j+1} - \Psi_j}{\Delta x_{j+1}}\right) - K_{j-1/2}\left(\frac{\Psi_j - \Psi_{j-1}}{\Delta x_j}\right)\right]^n \tag{8.61}$$

or

$$\frac{\partial}{\partial x}\left[K\frac{\partial\Psi}{\partial x}\right]_j^n \approx A_j\Psi_{j-1}^n + B_j\Psi_{j-1}^n + C_j\Psi_{j-1}^n \tag{8.62}$$

where

$$A_j = 2\frac{K_{j-1/2}}{\Delta x_j(\Delta x_j + \Delta x_{j+1})} \qquad B_j = -2\frac{\frac{K_{j-1/2}}{\Delta x_j} + \frac{K_{j+1/2}}{\Delta x_{j+1}}}{\Delta x_j + \Delta x_{j+1}} \qquad C_j = 2\frac{K_{j+1/2}}{\Delta x_{j+1}(\Delta x_j + \Delta x_{j+1})}$$

with

$$\Delta x_j = x_j - x_{j-1}, \qquad \Delta x_{j+1} = x_{j+1} - x_j \quad \text{and} \quad \Delta x_j + \Delta x_{j+1} = x_{j+1} - x_{j-1}.$$

8.6.2 Implicit Schemes for the 1-D Diffusion Equation

The implicit form of the FTCS scheme

$$\frac{\Psi_j^{n+1} - \Psi_j^n}{\Delta t} = \frac{K}{\Delta x^2}\left(\Psi_{j-1}^{n+1} - 2\Psi_j^{n+1} + \Psi_{j+1}^{n+1}\right) \tag{8.63}$$

provides solutions that are unconditionally stable because the amplification factor

$$|g(k)| = \frac{1}{1 + 2\beta(1 - \cos(k\Delta x))} \tag{8.64}$$

is less than 1 for all values of β. The errors on the solution can, however, be substantial for large values of the time step because this algorithm is only first-order accurate in time.

Expression (8.63) represents a tridiagonal system of $J - 1$ linear equations:

$$-\beta\Psi_{j-1}^{n+1} + (1 + 2\beta)\Psi_j^{n+1} - \beta\Psi_{j+1}^{n+1} = \Psi_j^n \quad (j = 1, J - 1) \tag{8.65}$$

The solution requires that boundary conditions be specified for $j = 0$ and $j = J$. The Thomas algorithm, often adopted to solve tridiagonal systems, is given in Box 4.4.

The accuracy of the fully implicit FTCS algorithm can be improved by using a combination of spatial derivatives at time t_n and t_{n+1}:

$$\frac{\Psi_j^{n+1} - \Psi_j^n}{\Delta t} = \frac{K}{2\Delta x^2}\left[\left(\Psi_{j-1}^{n+1} - 2\Psi_j^{n+1} + \Psi_{j+1}^{n+1}\right) + \left(\Psi_{j-1}^n - 2\Psi_j^n + \Psi_{j+1}^n\right)\right] \tag{8.66}$$

This algorithm, called the *Crank–Nicholson scheme,* is unconditionally stable since the associated amplification factor

$$|g(k)| = \frac{1 - 2\beta \sin^2(k\Delta x/2)}{1 + 2\beta \sin^2(k\Delta x/2)} \tag{8.67}$$

is always less than or equal to 1. It is second-order accurate in time and space. As in the fully implicit method, a system of linear equations involving a tridiagonal matrix must be solved.

A general approach is to consider a weighted time differencing over three time levels together with a combination of implicit and explicit approximations for the space derivatives:

$$(1 + \gamma) \frac{\Psi_j^{n+1} - \Psi_j^n}{\Delta t} - \gamma \frac{\Psi_j^n - \Psi_j^{n-1}}{\Delta t}$$
$$= \frac{K}{\Delta x^2} \left[\theta \left(\Psi_{j-1}^{n+1} - 2\Psi_j^{n+1} + \Psi_{j+1}^{n+1} \right) + (1 - \theta) \left(\Psi_{j-1}^n - 2\Psi_j^n + \Psi_{j+1}^n \right) \right] \tag{8.68}$$

where γ and θ are again adjustable parameters. The Crank–Nicholson scheme corresponds to the case where $\gamma = 0$ and $\theta = 1/2$. For $\gamma = 1/2$ and $\theta = 1$, one obtains the following *three-level scheme*:

$$\frac{3}{2} \frac{\Psi_j^{n+1} - \Psi_j^n}{\Delta t} - \frac{1}{2} \frac{\Psi_j^n - \Psi_j^{n-1}}{\Delta t} = \frac{K}{\Delta x^2} \left[\Psi_{j-1}^{n+1} - 2\Psi_j^{n+1} + \Psi_{j+1}^{n+1} \right] \tag{8.69}$$

which is second-order accurate in space and time, and is unconditionally stable. Figure 8.7 gives schematic representations of the different implicit schemes.

If the diffusion coefficient K varies in space, expressions similar to (8.58) can easily be established for the fully implicit and for the Crank–Nicholson schemes; in both cases, they lead to a tridiagonal system of linear equations. These algorithms are usually stable for any value of time step Δt. More elaborate and accurate forms can be adopted for the discretization of the second derivative in space (see Table 4.2). In

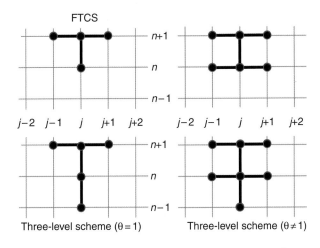

Figure 8.7 Stencils for different implicit schemes to approximate the 1-D diffusion equation. Indices j and n refer to the spatial and temporal dimensions, respectively.

this case, the matrix associated with the system of linear equations may not be tridiagonal anymore. Techniques to solve this system, such as the Gauss elimination algorithm, are often prohibitively expensive.

8.6.3 Numerical Algorithms for the Multidimensional Diffusion Equation

The explicit FTCS method described earlier can easily be generalized to more than one dimension. However, because of the constraints on the adopted time step, implicit methods are often preferred. Again, a system of coupled algebraic equations can easily be established, but the matrix of this system, although sparse, is no longer tridiagonal.

   An alternative to the fully implicit approach is to apply operator splitting and treat the problem as a succession of 1-D problems. In the popular *alternating-direction implicit (ADI) method*, the time step is divided into sub-steps at which a different dimension is treated implicitly (Figure 8.8). For example, in the 2-D case, the diffusion equation expressed as

$$\frac{\partial \Psi}{\partial t} = \frac{\partial}{\partial x}\left[K_x \frac{\partial \Psi}{\partial x}\right] + \frac{\partial}{\partial y}\left[K_y \frac{\partial \Psi}{\partial y}\right] \tag{8.70}$$

is integrated from time t_n to t_{n+1} by considering the following sequence

$$\frac{\Psi^* - \Psi^n}{\Delta t} = \frac{1}{2}\left\{\frac{\partial}{\partial x}\left[K_x \frac{\partial \Psi}{\partial x}\right]\right\}^* + \frac{1}{2}\left\{\frac{\partial}{\partial y}\left[K_y \frac{\partial \Psi}{\partial y}\right]\right\}^n \tag{8.71}$$

$$\frac{\Psi^{n+1} - \Psi^*}{\Delta t} = \frac{1}{2}\left\{\frac{\partial}{\partial x}\left[K_x \frac{\partial \Psi}{\partial x}\right]\right\}^* + \frac{1}{2}\left\{\frac{\partial}{\partial y}\left[K_y \frac{\partial \Psi}{\partial y}\right]\right\}^{n+1} \tag{8.72}$$

where Ψ^* is an intermediate value for function Ψ. Assuming constant diffusion coefficients and defining

$$\beta_x = K_x \frac{\Delta t}{\Delta x^2} \tag{8.73}$$

and

$$\beta_y = K_y \frac{\Delta t}{\Delta y^2} \tag{8.74}$$

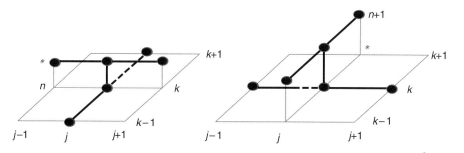

Figure 8.8   Stencil representing the two successive steps in the 2-D alternating direction method.

the resulting algebraic equations are:

$$-\frac{\beta_x}{2}\Psi^*_{j-1,k} + (1+\beta_x)\Psi^*_{j,k} - \frac{\beta_x}{2}\Psi^*_{j+1,k} = \frac{\beta_y}{2}\Psi^n_{j,k-1} + \left(1-\beta_y\right)\Psi^n_{j,k} + \frac{\beta_y}{2}\Psi^n_{j,k+1}$$
$$(j=1,J-1)$$
$$(8.75)$$

$$-\frac{\beta_y}{2}\Psi^{n+1}_{j,k-1} + \left(1+\beta_y\right)\Psi^{n+1}_{j,k} - \frac{\beta_y}{2}\Psi^{n+1}_{j,k+1} = \frac{\beta_x}{2}\Psi^*_{j-1,k} + (1-\beta_x)\Psi^*_{j,k} + \frac{\beta_x}{2}\Psi^*_{j+1,k}$$
$$(k=1,K-1)$$
$$(8.76)$$

where j and k are the indices referring to the spatial discretization in the x- and y-direction, respectively. Each step requires solving an implicit equation in one dimension; the first step is solved for parameter k fixed and the second step for parameter j fixed. This approach requires that at each time step two tridiagonal systems be solved; it is thus computationally efficient. The von Neumann stability analysis can be applied for each step and for the steps combined to yield:

$$|g(k)| = \left[\frac{1-2\beta_x\sin^2(k\Delta x/2)}{1+2\beta_y\sin^2(k\Delta y/2)}\right]\left[\frac{1-2\beta_y\sin^2(k\Delta y/2)}{1+2\beta_x\sin^2(k\Delta x/2)}\right] \qquad (8.77)$$

This method is unconditionally stable because the amplification factor is always less than or equal to 1. It is second-order accurate in space and time. We close by noting that diffusive schemes, as those discussed above, are not necessarily used to

---

Box 8.3   **Numerical Diffusion in Atmospheric Models**

In addition to the diffusion parameterization used to describe atmospheric turbulence, spurious *numerical diffusion* may be produced in Eulerian models by the adopted advection schemes (see Chapter 7) and by the use of numerical filters, which are often applied to avoid numerical instability when solving the transport or dynamical equations. This additional (unphysical) diffusion leads to excessive mixing and can be reduced by using a higher-order advection scheme and a finer grid to resolve spatial gradients. Numerical diffusion is a particular problem at dynamical barriers through which physical transport is restricted (see Section 8.12). In Lagrangian models, on the other hand, air parcels moving with the flow are considered to be isolated from other air parcels, and no mixing occurs. Without an appropriate parameterization to account for mixing processes between neighboring air parcels, Lagrangian models may overestimate spatial gradients and produce unrealistic small-scale structures (Collins *et al.*, 1997; McKenna *et al.*, 2002). To account for small-scale mixing processes in a Lagrangian model, an expression can be added that brings the mixing ratio μ_i of a given air parcel closer to the background mixing ratio $\bar{\mu}_i$. Collins *et al.* (1997) use a relaxation term $d(\bar{\mu}_i - \mu_i)$ where the degree of exchange d is considerably faster in the troposphere than in the stratosphere. The background mixing ratio in a given area is defined as the average for all air parcels located within that area. In an alternate approach, McKenna *et al.* (2002) entirely mix two air parcels when their spatial separation is smaller than a given threshold value.

represent physical processes, but rather to stabilize algorithms adopted for solving the advection equation. Further, unwanted numerical diffusion can be produced by low-order advection schemes (see Box 8.3).

8.7 Planetary Boundary Layer Processes

Section 2.10 presented an overview of the factors controlling the structure and vertical mixing of the PBL. The PBL is defined as the lower part of the atmosphere, typically 1–3 km in depth, that exchanges air with the surface on a daily basis. As discussed in Section 2.10, one can distinguish between three regimes to describe boundary layer dynamics, depending on the sensible heat flux at the surface: (1) a *convectively unstable* regime where heating of the surface drives strong buoyant motions, typical of land during daytime; (2) a *convectively stable* PBL where cooling of the surface suppresses buoyant motions, typical of land at night, and characterized by stratification of the atmosphere; (3) a *neutral* regime with little sensible heating of the surface, typical of marine conditions. Different formulations of turbulence must be considered for these three regimes.

Cumulus or stratus clouds are often present in the upper part of the PBL. Their formation is determined by evaporation of water from the surface (latent heat flux) and eventual condensation as air parcels rise. Cloud formation affects the dynamics of the PBL through both latent heat release and radiative effects. Clouds at night decrease nighttime stability both by suppressing radiative cooling of the surface and by radiative cooling at cloud top.

In this section we present approaches of varying complexity to describe vertical turbulence in the PBL and its implications for the concentrations and transport of chemical species. The simplest approach, often sufficient for rough estimates, is to use a box model for the convectively unstable or neutral mixed layer assuming vertically uniform concentrations through the mixing depth. Consider a horizontally uniform atmosphere with a time-varying mixing depth $h(t)$ through which the atmosphere is assumed to be vertically well mixed. The budget of a chemical species i is defined by the mass conservation equation

$$\frac{d\rho_i}{dt} = \frac{F_i(0) - F_i(h)}{h} + s_i \tag{8.78}$$

where ρ_i is the mass concentration, $F_i(0)$ is the net surface flux from emission and/or deposition, $F_i(h)$ is the entrainment flux at the top of the mixed layer, and s_i is the net chemical source term. Here and elsewhere, we define vertical fluxes as positive upward. As the mixed layer grows in the morning, the entrainment flux is given by

$$F_i(h) = (\rho_i - \rho_b)\frac{dh}{dt} \tag{8.79}$$

where ρ_b is the background concentration entrained from above the mixed layer, which can represent free tropospheric air or residual boundary layer air from the previous day. When the mixing depth decreases, as in the evening, air is removed from the box and $F_i(h) = 0$.

Box 8.4 — Vertical Flux Gradients in the Mixed Layer

Consider a chemical species i in the well-mixed convective boundary layer (mixed layer) with no *in-situ* production or loss, no horizontal gradient, and a constant surface flux. Assume that the mixed layer extends to the PBL top capped by a subsidence inversion, as under clear-sky daytime conditions, so that the vertical flux across the mixed layer top is zero. Further, assume that the change in air density with altitude can be neglected. The 1-D (vertical) continuity equation in Eulerian form for the chemical species is

$$\frac{\partial \rho_i}{\partial t} = -\frac{\partial F_z}{\partial z}$$

where ρ_i is the density and F_z is the vertical flux. Since the surface flux is constant and the atmosphere is well-mixed, the mass mixing ratio μ_i must change at the same rate at all altitudes: $\partial \mu_i / \partial t = \alpha$ where α is a constant. If the air density ρ_a is fixed, then $\partial \rho_i / \partial t = \alpha \rho_a$ is a constant too. It follows that $\partial F_z / \partial z$ is a constant and hence that the magnitude of the chemical flux varies linearly with altitude with a boundary condition $F_z = 0$ at the top of the mixed layer.

This result is somewhat counter-intuitive, as one might have expected the flux to be uniform with altitude in a well-mixed layer. However, that would hold only if the concentrations were constant, which cannot be the case since the flux at the top of the mixed layer is zero. This linear variation of the flux with altitude is important for interpreting vertical flux measurements from aircraft (Box 8.4 Figure 1), as these measurements will underestimate the surface flux in a predictable manner. Meteorologists refer to the *surface layer* as the lowest part of the atmosphere where the vertical fluxes are within 10% of their surface values. We see that this surface layer extends to 10% of the mixing depth; for a typical 1-km daytime mixing depth the surface layer is 100 m deep.

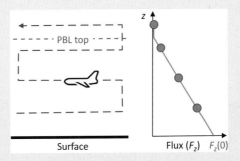

Box 8.4 Figure 1 Surface flux measurements from aircraft. The aircraft makes vertical eddy flux measurements (Section 10.2.4) on successive horizontal flight legs in the mixed layer at different altitudes, under conditions when the mixed layer extends to the PBL top (daytime, clear-sky). The line fitted to the flux measurements at different altitudes is extrapolated to the surface to derive the surface flux $F_z(0)$.

From a more fundamental perspective, the governing equations for the mean winds, temperature, and species concentrations in the PBL can be obtained by applying the Reynolds decomposition described in Section 8.2 to the momentum, energy, and continuity equations (Stull, 1988; Garratt, 1994). The continuity equation (8.19) for the mean mixing ratio of species i in Cartesian coordinates is:

$$\frac{\partial \overline{\mu}_i}{\partial t} + \overline{u}\frac{\partial \overline{\mu}_i}{\partial x} + \overline{v}\frac{\partial \overline{\mu}_i}{\partial y} + \overline{w}\frac{\partial \overline{\mu}_i}{\partial z} = \overline{s}_i - \frac{\partial}{\partial x}\overline{\mu'_i u'} - \frac{\partial}{\partial y}\overline{\mu'_i v'} - \frac{\partial}{\partial z}\overline{\mu'_i w'} \qquad (8.80)$$

Again, the eddy terms must be parameterized by assuming, for example, that the local turbulent flux is proportional to the local gradient in the mean mixing ratio (see Section 8.4.1). This assumption requires that the scale of the turbulence be smaller than the characteristic spatial scale of the flow, a condition that is not met when the size of the eddies is of the same order as the vertical extent of the boundary layer. In LES models, the largest eddies that contribute most of turbulent transport are explicitly resolved, while the dissipative processes produced by the smaller eddies are parameterized. When applied to the atmospheric boundary layer, these models have a horizontal resolution of typically a few hundred meters.

Box 8.5 illustrates LES modeling with an application to marine boundary layer chemistry. These models can provide information on the chemical segregation of species within the mixed layer and the implications for chemical reaction rates. As discussed in Section 8.3, the fate of an atmospheric species emitted at the surface and reacting in the mixed layer is characterized by its Damköhler number, which is the ratio between the integral timescale of turbulence (mixing) and the timescale for chemical loss. Under convectively unstable conditions, a chemical species released at the surface may have much higher concentrations in buoyant rising plumes (updrafts) than outside (see Box 8.5). While in an updraft, the species does not mix with other species present outside the updraft. As a result of this segregation, reaction rates may be very different than in a well-mixed atmosphere (Section 8.3). A classic example for the continental mixed layer is that of isoprene, a major biogenic hydrocarbon emitted by vegetation with a mean lifetime of less than one hour against oxidation by the OH radical. The mean lifetime of isoprene is shorter than the typical timescale for overturning of the mixed layer, so that one would expect a strong vertical gradient of concentrations. In fact, isoprene in updrafts may be sufficiently concentrated to deplete OH and thus reach the top of the mixed layer with minimal chemical loss.

Box 8.5   **LES Simulation of the Marine Boundary Layer**

An LES performed by Kazil *et al.* (2011) illustrates the complexity of PBL processes with the example of sulfate particle formation from oceanic dimethylsulfide (DMS). The model accounts for the coupling between dynamical, chemical, aerosol, and cloud processes (Grell *et al.*, 2005). Its resolution is 300 m in the horizontal and 30 m in the vertical. The figure shows an instantaneous cross-section of a 60-km South Pacific domain with three cloudy regions: a decaying convective zone in the west, a broad active convective cell in the center, and a localized convective updraft in the east. DMS is uniformly emitted from the surface and has a lifetime of hours against oxidation by OH to produce SO_2. It is rapidly transported in the localized updrafts of the convective cells. OH concentrations are particularly high at cloud tops because of scattered radiation, resulting in fast DMS oxidation and SO_2 production. SO_2 produced near cloud top is oxidized by OH in the gas phase to produce gaseous sulfuric acid [$H_2SO_4(g)$] and in clouds to produce aqueous sulfate. $H_2SO_4(g)$ initiates nucleation of new particles as shown in the upper part of the PBL around cloud tops.

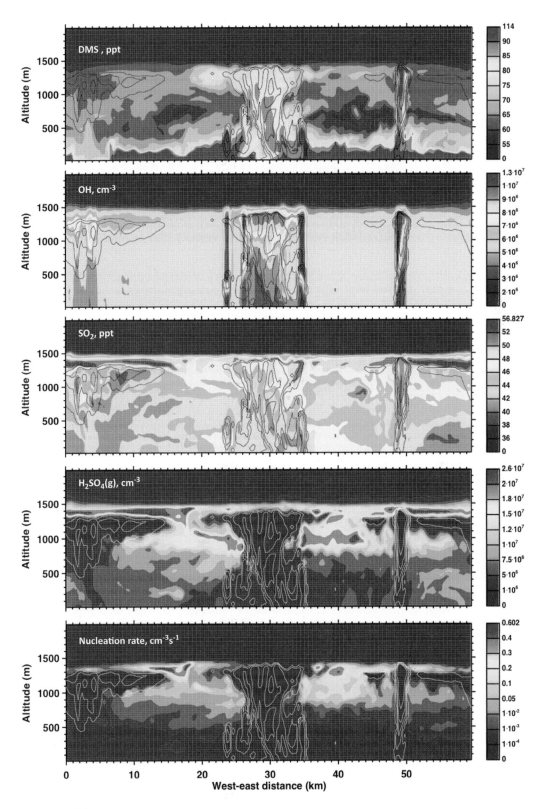

Box 8.5 Figure 1   Large-scale eddy simulation in the ocean boundary layer.

Lagrangian stochastic models are often used to simulate the transport and dispersion of trace species in the PBL (Thomson and Wilson, 2012). In the simplest of these models, the path of air particles is calculated by a sequence of random increments in position (random walk). In more sophisticated formulations that draw on the idea of Brownian motions and represent turbulent motions by a Markov chain process, the trajectory of air particles is obtained by integrating a sequence of random increments in velocity (see Section 4.11.2 for more details). Stochastic time-inverted Lagrangian transport (STILT) models are used to derive surface sources and sinks of trace species from atmospheric concentration data at a receptor location (Lin *et al.*, 2003). Lagrangian approaches are particularly effective at dealing with such receptor-oriented problems.

8.7.1 Mean Atmospheric Wind Velocity and Temperature

The solution to the continuity equation (8.80) requires that the advection terms be either specified from a meteorological analysis or calculated from the Navier–Stokes equations. The mean wind velocity $\overline{\mathbf{v}}(\overline{u}, \overline{v}, \overline{w})$ is derived from the momentum equation to which the Reynolds decomposition and averaging is applied:

$$\frac{\partial \overline{u}}{\partial t} + \overline{u}\frac{\partial \overline{u}}{\partial x} + \overline{v}\frac{\partial \overline{u}}{\partial y} + \overline{w}\frac{\partial \overline{u}}{\partial z} = -f\left(\overline{v}_g - \overline{v}\right) - \frac{\partial}{\partial x}\overline{u'u'} - \frac{\partial}{\partial y}\overline{u'v'} - \frac{\partial}{\partial z}\overline{u'w'} + F_x \quad (8.81)$$

$$\frac{\partial \overline{v}}{\partial t} + \overline{u}\frac{\partial \overline{v}}{\partial x} + \overline{v}\frac{\partial \overline{v}}{\partial y} + \overline{w}\frac{\partial \overline{v}}{\partial z} = +f\left(\overline{u}_g - \overline{u}\right) - \frac{\partial}{\partial x}\overline{u'v'} - \frac{\partial}{\partial y}\overline{v'v'} - \frac{\partial}{\partial z}\overline{v'w'} + F_y \quad (8.82)$$

Here F_x and F_y represent the influence of viscous stress and can generally be neglected in comparison to other terms in the equations. The geostrophic components of the winds $\overline{u}_g$ and $\overline{v}_g$ are related to the pressure gradients by

$$\overline{v}_g = \frac{1}{f\,\rho_a}\frac{\partial \overline{p}}{\partial x} \quad (8.83)$$

$$\overline{u}_g = -\frac{1}{f\,\rho_a}\frac{\partial \overline{p}}{\partial y} \quad (8.84)$$

Here, f denotes the Coriolis parameter $[\mathrm{s}^{-1}]$. Assuming steady-state and horizontal homogeneity with no significant subsidence in (8.81) and (8.82), the deviation of the wind from its geostrophic value in the extra-tropical boundary layer is proportional to the turbulent momentum flux divergence (Holton, 1992):

$$f\left(\overline{v} - \overline{v}_g\right) = \frac{\partial \overline{u'w'}}{\partial z} \quad (8.85)$$

$$f\left(\overline{u} - \overline{u}_g\right) = -\frac{\partial \overline{v'w'}}{\partial z} \quad (8.86)$$

The Reynolds decomposition can also be applied to the energy conservation equation to derive the mean potential temperature $\overline{\theta}$:

$$\frac{\partial \overline{\theta}}{\partial t} + \overline{u}\frac{\partial \overline{\theta}}{\partial x} + \overline{v}\frac{\partial \overline{\theta}}{\partial y} + \overline{w}\frac{\partial \overline{\theta}}{\partial z} = \frac{1}{\rho C_p}\left[Q - L_v E\right] - \frac{\partial}{\partial x}\overline{u'\theta'} - \frac{\partial}{\partial y}\overline{v'\theta'} - \frac{\partial}{\partial z}\overline{w'\theta'} + F_\theta \quad (8.87)$$

where Q represents the net radiative heating rate, L_v (2.5×10^6 J kg^{-1} at 273 K) the latent heat associated with gas–liquid water phase change, E the mass of water vapor produced by evaporation by unit volume and unit time, $C_p = 1004.67$ J kg^{-1} K^{-1} the specific heat for moist air at constant pressure, and F_θ the effect of thermal diffusivity (often neglected).

8.7.2 Boundary Layer Turbulence Closure

The closure relation required to solve the continuity equation for chemical species i in the boundary layer is often represented by a first-order local diffusion formulation:

$$\overline{\mu_i' w'} = -K_z \frac{\partial \overline{\mu}_i}{\partial z} \tag{8.88}$$

where K_z is an eddy diffusion coefficient. The resulting equation becomes

$$\frac{\partial \overline{\mu}_i}{\partial t} + \overline{u} \frac{\partial \overline{\mu}_i}{\partial x} + \overline{v} \frac{\partial \overline{\mu}_i}{\partial y} = \overline{S}_i + \left(K_z \frac{\partial \overline{\mu}_i}{\partial z} \right) \tag{8.89}$$

Similar local closure relations for the turbulent momentum and heat fluxes are expressed by

$$\overline{u'w'} = -K_m \frac{\partial \overline{u}}{\partial z} \qquad \overline{v'w'} = -K_m \frac{\partial \overline{v}}{\partial z} \qquad \overline{\theta'w'} = -K_\theta \frac{\partial \overline{\theta}}{\partial z} \tag{8.90}$$

where K_m is the *eddy viscosity coefficient* and K_θ is the *eddy diffusivity of heat*. Simple formulations for K_m and K_θ have been developed as a function of the wind shear, atmospheric stability, and PBL height (see e.g., Holtslag and Boville, 1993).

Prandtl (1925) introduced the concept of mixing length l in a neutral buoyancy environment by considering an air parcel that moves upwards by a distance z' from a reference level z in a field where the mean mixing ratio and wind velocity increase linearly with height. If no mixing occurs, the mixing ratio in the air parcel is conserved during its motion and differs from its value in the surrounding environment by a value

$$\mu' = -\left(\frac{\partial \overline{\mu}}{\partial z} \right) z' \tag{8.91}$$

Similarly, we write

$$u' = -\left(\frac{\partial \overline{u}}{\partial z} \right) z' \tag{8.92}$$

For the parcel to move upwards, it must have a turbulent velocity w'. We assume that w' is proportional to the horizontal fluctuation u' so that

$$w' = c \left| \frac{\partial \overline{u}}{\partial z} \right| z' \tag{8.93}$$

where c is a proportionality constant. The resulting mean eddy flux is

$$\overline{\mu'w'} = -c \, \overline{z'^2} \left| \frac{\partial \overline{u}}{\partial z} \right| \left(\frac{\partial \overline{\mu}}{\partial z} \right) \tag{8.94}$$

Let us define the *mixing length* l [m] as

$$l^2 = c \overline{z'^2} \tag{8.95}$$

where $\overline{z'^2}$ is the variance of the displacement distance. The mixing length measures the ability of the turbulence to mix air parcels. Its value varies with the size of the eddies, so that the overall (mean) mixing length should be provided by an integration over the spectrum of all eddy sizes. Typical values of the mean mixing length in the boundary layer range between 500 m and 1 km (Stull, 1988). It is often assumed that the mixing length varies with altitude as $l = k\,z$, where k is the von Karman constant taken to be 0.35. From (8.94) and (8.95) together with (8.88), one defines an eddy diffusion coefficient [$m^2\ s^{-1}$]:

$$K_z = l^2 \left| \frac{\partial \overline{u}}{\partial z} \right| \tag{8.96}$$

In this formulation, the value of K_z increases with the vertical wind shear (a measure of the intensity of the turbulence) and the mixing length (a measure of the mixing produced by turbulence), but is not a function of the static stability of the layer. Other formulations include a dependence of the eddy diffusion coefficient on the gradient Richardson number Ri as given by (8.2). For example (Blackadar, 1979, Stull, 1988):

$$
\begin{aligned}
K_z &= 1.1 \left[\frac{Rc - Ri}{Ri} \right] l^2 \left| \frac{\partial \overline{u}}{\partial z} \right| \quad \text{for} \quad \frac{\partial \overline{\theta}_v}{\partial z} > 0 \\
K_z &= \left[1 - 18 Ri \right]^{-1/2} l^2 \left| \frac{\partial \overline{u}}{\partial z} \right| \quad \text{for} \quad \frac{\partial \overline{\theta}_v}{\partial z} < 0
\end{aligned}
\tag{8.97}
$$

with

$$l = kz \ \text{ for } \ z < 200\,\text{m} \ \text{ and } \ l = 70\text{m} \ \text{ for } \ z > 200\text{m}$$

It is assumed that, in the convectively stable situation in which the vertical gradient of $\overline{\theta}_v$ is positive, turbulence is generated only if Ri becomes smaller than a critical value Rc equal to about 0.25. Idealized vertical profiles of the mean wind components in the boundary layer can be derived by introducing the empirical closure relations in the simplified momentum equations (8.85) and (8.86), as shown in Box 8.6. The local approach is best suited when eddies are smaller than the length scale for turbulence, as can be usually assumed for neutral or stable conditions. It underestimates vertical transport under convectively unstable conditions when large eddies of the dimension of the mixed layer become important. In this case, vertical transport has a *non-local* character and the eddy diffusion parameterization is inadequate. This can be corrected with a non-local term added to the eddy diffusion parameterization (Deardorff, 1966, 1972; Troen and Mahrt, 1986; Holtslag and Boville, 1993):

$$\overline{\mu_i' w'} = -K_z \left\{ \frac{\partial \overline{\mu}_i}{\partial z} - \gamma_c \right\} \tag{8.98}$$

where γ_c reflects the contribution of non-local turbulent transport. To better account for the entrainment of air from the free troposphere into the mixed layer, Hong *et al.* (2006) include an additional flux component as:

$$\overline{\mu_i' w'} = -K_z \left\{ \frac{\partial \overline{\mu}_i}{\partial z} - \gamma_c \right\} + \left(\overline{w' \mu_i'} \right)_h \left(\frac{z}{h} \right)^3 \tag{8.99}$$

Mean Horizontal Wind in the Boundary Layer: The Ekman Spiral

We consider the simplified momentum equations (8.85) and (8.86), and assume that the geostrophic wind is directed along the x-axis ($v_g = 0$). We express the eddy fluxes of momentum by the first-order closure relations (8.90) in which K_m is assumed to be constant

$$f\,\overline{v} = -K_m \frac{\partial^2 \overline{u}}{\partial z^2} \quad \text{and} \quad f\left(\overline{u} - \overline{u}_g\right) = K_m \frac{\partial^2 \overline{v}}{\partial z^2}$$

with the adopted boundary conditions

$$\overline{u} = 0 \text{ and } \overline{v} = 0 \text{ at } z = 0, \quad \overline{u} = \overline{u}_g \text{ and } \overline{v} = 0 \text{ as } z \to \infty.$$

The solution is given by (Ekman, 1905):

$$\overline{u} = \overline{u}_g[1 - \exp(-\gamma z) \cos(\gamma z)]$$
$$\overline{v} = \overline{u}_g \exp(-\gamma z) \sin(\gamma z)$$

where $\gamma = [f/(2K_m)]^{1/2}$. The wind vector turns with height as a spiral, diminishing in amplitude toward the surface where it is directed 45 degrees to the left of the geostrophic wind vector aloft. Whereas the geostrophic wind follows isobars, the wind in the boundary layer tilts toward the direction of low pressure. It becomes parallel to the geostrophic wind at the altitude $h = \pi/\gamma$. This height is often used as an estimate of the depth of the boundary layer.

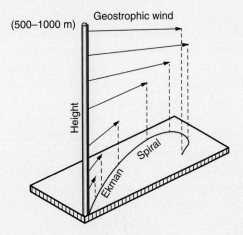

Idealized vertical structure of the horizontal wind velocity (Ekman spiral) in the atmospheric boundary layer.

where h is the depth of the mixed layer (mixing depth) and $\overline{-(w'\mu_i')}_h$ is the *entrainment flux* at level $z = h$. The entrainment flux is commonly computed as $w_e\,\Delta\overline{\mu}_i$, where $\Delta\overline{\mu}_i$ is the mixing ratio difference across the top of the mixed layer (between the mixed layer and the free troposphere above), and w_e is an *entrainment velocity* derived from observations for a species or other variable such as heat for which the flux is known. A value of 0.5 cm s^{-1} is typical for w_e.

Different formulations have been introduced to express parameters K_z and γ_c as a function of other atmospheric quantities. Holtslag and Boville (1993) and Hong *et al.* (2006) express the eddy diffusion in the mixed layer as

$$K_z = kzw_s\left(1 - \frac{z}{h}\right)^2 \tag{8.100}$$

where k is the von Karman constant, z the geometric height above the surface, h the mixing depth, and w_s the mixed layer velocity scale dependent on stability (Stull, 1988). K_z is of the order of 100 m^2 s^{-1} in the daytime convective mixed layer, 0.1–1 m^2 s^{-1} under stable nighttime conditions, and 10 m s^{-2} under neutral conditions such as over the oceans for both day and night. The correction γ_c to the local gradient is given by

$$\gamma_c = b\frac{F(0)}{hw_s} \tag{8.101}$$

where $F(0)$ is the surface flux. The value of the dimensionless proportionality factor b is about 6.5 (Troen and Mahrt, 1986).

8.7.3 Surface Layer

The lowest part of the mixed layer is called the *surface layer*. It is commonly defined as the vertical extent of the atmosphere for which vertical fluxes of conserved quantities are within 10% of their surface values (Box 8.4). It is typically 10–100 m in depth. Eddy sizes in the surface layer are constrained by the proximity to the surface. Mechanical eddies driven by surface roughness are typically more important than buoyant eddies driven by surface heating. Understanding the dynamics of the surface layer is of critical importance as it determines the rate at which chemicals are removed by dry deposition (see Chapter 9).

Parameterizations of atmospheric turbulence in the surface layer are generally based on the *similarity theory* developed by Monin and Obukhov (1954), which uses scaling arguments to provide relationships between dimensionless quantities. A central parameter is the *friction velocity* u_* [m s^{-1}] that characterizes the surface momentum flux and is defined by

$$u_*^2 = \left[\left(\overline{u'w'}\right)^2_{surf} + \left(\overline{v'w'}\right)^2_{surf}\right]^{1/2} \tag{8.102}$$

Its value is typically ~10% of the 10-m wind speed and increases with surface roughness. One also defines a potential temperature scale θ_* and mixing ratio scale μ_* as

$$\overline{w'\theta'_v} = -u_*\theta_* \text{ and } \overline{w'\mu'} = -u_*\mu_* \tag{8.103}$$

Another key parameter in similarity theory is the *Monin–Obukhov length L* [m], defined as

$$L = -\frac{u_*^3}{kg}\frac{\overline{\theta}_v}{\left(\overline{w'\theta'_v}\right)_{surf}} \tag{8.104}$$

where g is the gravitational acceleration, k is the von Karman constant, $\overline{\theta}_v$ is the mean virtual potential temperature, and $\left(\overline{w'\theta'_v}\right)_{surf}$[K m s^{-1}] the sensible heat flux at the surface. $|L|$ represents the height above the surface at which buoyant production/ suppression of turbulence by surface heating/cooling equals mechanical production of turbulence by wind shear. For a neutral boundary layer (as over the ocean), $|L| \to \infty$ because the sensible heat flux is negligible and all turbulence is generated mechanically. In a nighttime stable atmosphere over land, L is positive (typically ~100 m) as turbulence is generated mechanically and suppressed by buoyancy. In a daytime unstable atmosphere, L is negative (typically ~−100 m). For $z < |L|$, turbulence is mostly mechanical; for $z > |L|$ it is mostly determined by buoyancy.

If we assume neutral stability of the surface layer and adopt a coordinate system in which the wind is aligned with the x-direction, one deduces from a dimensional analysis that the vertical wind shear $\partial\overline{u}/\partial z$ is proportional to the friction velocity (Stull, 1988). Thus,

$$\frac{\partial\overline{u}}{\partial z} = \frac{u_*}{kz} \tag{8.105}$$

By integration over height z, one obtains a logarithmic relation for the vertical profile of the mean wind velocity:

$$\overline{u}(z) = \frac{u_*}{k}\ln\left[\frac{z}{z_{0,m}}\right] \tag{8.106}$$

Parameter $z_{0,m}$ [m], the *aerodynamic roughness length*, is the height at which the mean wind vanishes. Its value varies with the height of the physical elements that generate the surface drag (trees, ocean waves, etc.). It is typically ~3% of the height of these elements and ranges from ~10^{-5} m for smooth ice surfaces to ~1 m for a tall forest canopy.

Under buoyant conditions where neutral stability cannot be assumed, one replaces (8.105) by an expression written in term of a dimensionless wind shear:

$$\Phi_m = \frac{kz}{u_*}\frac{\partial\overline{u}}{\partial z} \tag{8.107}$$

Empirical values of Φ_m commonly used in models and going back to Businger *et al.* (1971) and Dyer (1974) are

$$\Phi_m = 1 + \beta_m\frac{z}{L} \qquad \text{for } \frac{z}{L} > 0 \ (\text{stable})$$

$$\Phi_m = 1 \qquad \text{for } \frac{z}{L} = 0 \ (\text{neutral}) \tag{8.108}$$

$$\Phi_m = \left(1 - \gamma_m\frac{z}{L}\right)^{-1/4} \qquad \text{for } \frac{z}{L} < 0 \ (\text{unstable})$$

with $\beta_m = 4.7$ and $\gamma_m = 15.0$.

Expression (8.107) can be integrated to yield the vertical profile of the mean wind velocity in the surface layer:

$$\bar{u}(z) = \frac{u_*}{k}\left[\ln\left(\frac{z}{z_{0,m}}\right) - \Psi_m\right] \tag{8.109}$$

where the correction term Ψ_m is

$$\Psi_m(z) = \int_{z_{0,m}}^{z} [1 - \Phi_m(z')]\frac{dz'}{z'} \tag{8.110}$$

When adopting the empirical expressions (8.108), the correction term Ψ_m for the mean wind velocity is

$$\Psi_m = -\beta_m\frac{(z - z_{0,m})}{L} \qquad\qquad\qquad \text{for } \frac{z}{L} > 0 \text{ (stable)}$$

$$\Psi_m = 0 \qquad\qquad\qquad\qquad\qquad\qquad \text{for } \frac{z}{L} = 0 \text{ (neutral)}$$

$$\Psi_m = 2\ln\left[\frac{1+x}{1+x_0}\right] + \ln\left[\frac{1+x^2}{1+x_0^2}\right] - 2\tan^{-1}(x) + 2\tan^{-1}(x_0) \text{ for } \frac{z}{L} < 0 \text{ (unstable)}$$

$$\tag{8.111}$$

where

$$x = \frac{1}{\Phi_m(z)} = \left[1 - \gamma_m\frac{z}{L}\right]^{1/4} \quad \text{and} \quad x_0 = \frac{1}{\Phi_m(z_0)} = \left[1 - \gamma_m\frac{z_{0,m}}{L}\right]^{1/4} \tag{8.112}$$

Under this formalism, the correction factor Ψ_m and the wind velocity (Figure 8.9) are equal to zero at $z = z_{0,m}$ in all situations. In the stable case, $\Phi_m > 1$ and $\Psi_m < 0$, while in the unstable case, $0 < \Phi_m < 1$ and $\Psi_m > 0$. When $z \gg z_{0,m}$ one can assume $x_0 \approx 1$ in the above expressions.

Finally, by combining relation (8.102) written in the 1-D case (x-direction) with equation (8.107) and the eddy diffusion formulation for the momentum flux (see equation (8.90))

$$\overline{u'w'} = -K_m\frac{\partial\bar{u}}{\partial z} \tag{8.113}$$

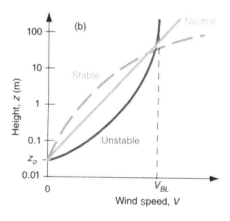

Figure 8.9 Variation with height of the mean wind velocity in the surface layer for different static stability conditions. The wind scale is linear and the altitude scale is logarithmic. Reproduced from Wallace and Hobbs (2006), based on Ahrens (2000) and Stull (1988).

one derives an expression for the momentum eddy diffusion coefficient K_m:

$$K_m(z) = \frac{kzu_*}{\Phi_m} \tag{8.114}$$

Relationships similar to (8.107) can be obtained for the virtual potential temperature and mixing ratio gradients, respectively:

$$\Phi_h = \frac{kz}{\theta_*}\frac{\partial\overline{\theta}_v}{\partial z} \quad \text{and} \quad \Phi_\mu = \frac{kz}{\mu_*}\frac{\partial\overline{\mu}}{\partial z} \tag{8.115}$$

Empirical expressions for Φ_h are available from Businger *et al.* (1971):

$$\Phi_h = Pr + \beta_h\frac{z}{L} \qquad \text{for } \frac{z}{L} > 0 \quad \text{(stable)}$$

$$\Phi_h = Pr \qquad\qquad\quad \text{for } \frac{z}{L} = 0 \quad \text{(neutral)} \tag{8.116}$$

$$\Phi_h = Pr\left(1 - \gamma_h\frac{z}{L}\right)^{-1/2} \quad \text{for } \frac{z}{L} < 0 \quad \text{(unstable)}$$

where the Prandtl number $Pr = K_m/K_h$ represents the ratio between the eddy diffusion coefficients for momentum and heat. Businger *et al.* (1971) estimate $\beta_h = 4.7$, $\gamma_h = 9.0$ and $Pr \approx 0.74$ for a von Karman constant $k = 0.35$. Hogstrom (1988) derives $\beta_h = 7.8$, $\gamma_h = 11.6$ and $Pr \approx 0.95$ for $k = 0.4$. It is generally assumed that the dimensionless gradients for the virtual potential temperature and chemical mixing ratios are equal, i.e., $\Phi_\mu \approx \Phi_h$. The vertical profiles of the potential virtual temperature and species mixing ratio in the surface layer are obtained by integration of (8.115) with (8.116):

$$\overline{\theta}_v(z) = \overline{\theta}_v(z_{0,h}) + Pr\frac{\theta_*}{k}\left[\ln\left(\frac{z}{z_{0,h}}\right) - \Psi_h\right] \tag{8.117}$$

$$\overline{\mu}(z) = \overline{\mu}(z_{0,\mu}) + Pr\frac{\mu_*}{k}\left[\ln\left(\frac{z}{z_{0,\mu}}\right) - \Psi_\mu\right] \tag{8.118}$$

where

$$\Psi_h(z) = \int_{z_{0,h}}^{z}[1 - \Phi_h(z')]\frac{dz'}{z'} \quad \text{and} \quad \Psi_\mu(z) = \int_{z_{0,\mu}}^{z}[1 - \Phi_\mu(z')]\frac{dz'}{z'} \tag{8.119}$$

Here, $z_{0,h}$, and $z_{0,\mu}$ are the roughness lengths for the virtual potential temperature and species mixing ratio, and are generally much smaller than $z_{0,m}$.

The vertical eddy diffusion coefficient for heat and chemical species is given by

$$K_h = \frac{\overline{w'\theta'_v}}{\partial\overline{\theta}/\partial z} = \frac{u_*\theta_*}{\partial\overline{\theta}/\partial z} = \frac{kzu_*}{\Phi_h} \approx K_\mu = \frac{kzu_*}{\Phi_\mu} \tag{8.120}$$

Over regions where roughness elements are packed together, (e.g., forest canopies, urban centers), the altitude above the ground at which the mean wind vanishes is shifted by a *displacement height d*, and expressions (8.109), (8.117), and (8.118) are replaced by:

$$\overline{u}(z) = \frac{u_*}{k}\left[\ln\left(\frac{z-d}{z_0}\right) - \Psi_m\left(\frac{z-d}{L}\right)\right] \tag{8.121}$$

$$\overline{\theta}_v(z) = \overline{\theta}_v(z_0) + Pr\frac{\theta_*}{k}\left[\ln\left(\frac{z-d}{z_0}\right) - \Psi_h\left(\frac{z-d}{L}\right)\right] \qquad (8.122)$$

$$\overline{\mu}(z) = \overline{\mu}(z_0) + Pr\frac{\mu_*}{k}\left[\ln\left(\frac{z-d}{z_0}\right) - \Psi_\mu\left(\frac{z-d}{L}\right)\right] \qquad (8.123)$$

Values of d are typically of the order of 65–75% of the height of the roughness elements and define the effective surface.

8.8 Deep Convection

Deep convective motions occur in the troposphere when surface heating and latent heat release are sufficiently strong that rising air parcels can pierce through the top of the planetary boundary layer (Chapter 2). Continued latent heat release as water condenses then produces intense updrafts. Vertical winds typically reach 10 m s^{-1}. The updrafts form large cumulonimbus clouds (Figure 8.10) with intense rainfall. Eventually they encounter a sufficiently stable layer (which could be the tropopause) to stop their ascent. They can also be weakened by entraining free tropospheric air. Outflow from convective updrafts forms an anvil where air is released to the surrounding atmosphere.

Deep convection is particularly important as a mechanism for vertical transport in the tropics and mid-latitudes during summer, when strong surface heating occurs. Air parcels in convective updrafts are transported from the boundary layer to the upper troposphere in less than one hour. Air can be *entrained* into the updraft at all levels in the convective column, broadening and diluting the updraft. *Detrainment* (outflow) from the updraft mostly takes place in the anvil near the top of the cloud. Rapid *downdrafts* can take place as precipitation evaporates to cool sinking air parcels, bringing free tropospheric air down to the ground. Outside the convective cells, a slow downward flow (subsidence) compensates for the net upward flow occurring inside the cells. Water-soluble gases and aerosols are efficiently scavenged in the precipitating updrafts, suppressing their release in the outflow. Scavenging is considered in Section 8.9. Particularly strong updrafts generate lightning due to separation of electrical charge between the cloud and the ground. The resulting production of nitrogen oxides (NO$_x$) is discussed in Section 8.10.

Convective motions can be simulated explicitly in *cloud-resolving models* that use a LES with horizontal resolution of less than 1 km. In coarser-resolution models they must be parameterized. The parameterization must recognize the organized nature of deep convection across levels in the model horizontal grid. An eddy diffusion parameterization would be physically incorrect because it assumes that turbulence involves scales much smaller than the model vertical grid, whereas deep convection involves rapid unidirectional upward transport across a number of vertical grid levels. Observed vertical profiles of chemical mixing ratios near convective

(a)

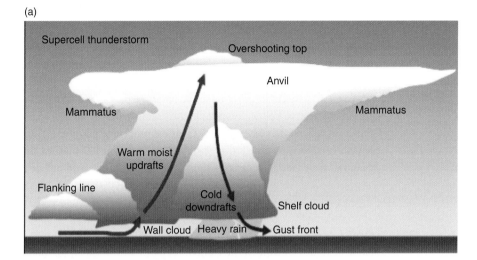

(b)

Figure 8.10 (a) Schematic representation of a deep convective system. Courtesy of Cameron Douglas Craig. (b) Photograph of a thunderstorm cell with an updraft reaching the upper troposphere.

cells often feature a "C-shape" for species originating in the boundary layer and discharged in the convective outflow, bypassing the intermediate levels. This cannot be reproduced by an eddy diffusion parameterization, which can only produce a monotonic down-gradient change of concentrations with altitude.

Parameterization of convection is a major area of research in atmospheric dynamics and many different schemes are used in meteorological models. Online chemical transport models apply the same convective transport equations to chemical and meteorological variables, including scavenging for water-soluble species for which analogy with scavenging of condensed water is commonly used. Offline chemical transport models must have their own convective transport module to replicate the convective motions from the parent meteorological model. This is preferably done by using archived convective mass fluxes provided by the meteorological model.

A standard assumption in convective parameterizations is that the net updraft in the subgrid convective cell is compensated by subsidence in the non-convective fraction of the grid cell, so that there is no net vertical motion of air on the grid scale. The non-convective subsiding fraction is assumed to represent the bulk of the grid cell. The change of the *average* mass density of species i inside a grid cell is given by:

$$\left[\frac{\partial \rho_i}{\partial t}\right]_{conv} = \frac{\partial(\rho_a \overline{\mu}_i)}{\partial t} = -\frac{\partial}{\partial z}\left[M_u\left(\mu_{i,u} - \overline{\mu}_i\right) + M_d\left(\mu_{i,d} - \overline{\mu}_i\right)\right] \tag{8.124}$$

where $\rho_a(z)$ denotes the mean air density, which varies with altitude z, while $M_u(z)$ and $M_d(z)$ represent the vertical fluxes [$\mathrm{kg\,m^{-2}\,s^{-1}}$] of air in the updrafts and downdrafts summed over the gridbox. These two fluxes are assumed to be positive when directed upwards; thus M_u is positive and M_d is negative. The differences $\mu_{i,u}(z) - \overline{\mu}_i(z)$ and $\mu_{i,d}(z) - \overline{\mu}_i(z)$ are the excess mass mixing ratios of chemical i inside the drafts relative to the atmospheric background (grid cell mean) mixing ratio $\overline{\mu}_i$. An equivalent form of (8.124) is

$$\frac{\partial(\rho_a \overline{\mu}_i)}{\partial t} = -\frac{\partial}{\partial z}\left[M_u \mu_{i,u} + M_d \mu_{i,d} + M_e \overline{\mu}_i\right] \tag{8.125}$$

where $M_e = -(M_u + M_d)$ represents the subsidence flux on the grid scale.

If $E_u(z)$ and $D_u(z)$ [$\mathrm{kg\,m^{-3}\,s^{-1}}$] represent the rates of entrainment into and detrainment from the updrafts in the convective system, the mass flux of air M_u varies with height according to

$$\frac{\partial M_u(z)}{\partial z} = E_u(z) - D_u(z) \tag{8.126}$$

with $M_u = M_b$ at the base (altitude z_b) and $M_u = 0$ at the top of the convective cloud system. Similarly for the downdraft, we write

$$\frac{\partial M_d(z)}{\partial z} = E_d(z) - D_d(z) \tag{8.127}$$

with $M_d = 0$ at the top of the convective cloud system. Figure 8.11 shows an example of entrainment and detrainment of air derived by a simple convective parameterization (Mari *et al.*, 2000). There is large entrainment at cloud base and large detrainment from the updraft at the top of the convective column.

In the absence of chemical or physical transformations occurring in the clouds, the mixing ratio of species i in the updraft (u) and in the downdraft (d) are given by

$$\frac{\partial\left(M_u\,\mu_{i,u}\right)}{\partial z} = E_u(z)\,\overline{\mu}_i(z) - D_u(z)\,\mu_{i,u}(z) \tag{8.128}$$

$$\frac{\partial\left(M_d\,\mu_{i,d}\right)}{\partial z} = E_d(z)\,\overline{\mu}_i(z) - D_d(z)\,\mu_{i,d}(z) \tag{8.129}$$

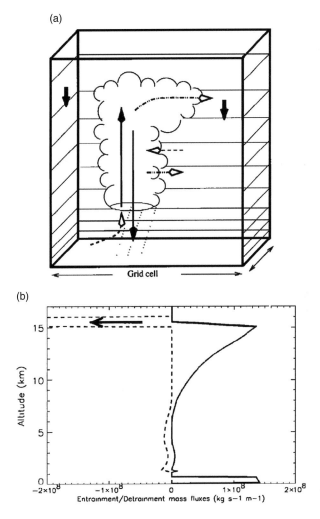

Figure 8.11 (a) Schematic representation of updraft, downdraft, and compensating subsidence in a model grid column. (b) Entrainment (solid curve) and detrainment (dashed curve) fluxes derived from a mesoscale model. The arrow shows the detrainment in the anvil of the cloud. From Mari *et al.* (2000).

We thus see that knowledge of the vertical distribution of entrainment and detrainment rates, as well as of the mean mixing ratio $\bar{\mu}_i(z)$ in the grid cells, allows us to calculate the mixing ratios $\mu_{i,u}(z)$ and $\mu_{i,d}(z)$ inside the updrafts and downdrafts, and from there to calculate the tendency in $\bar{\mu}_i(z)$ associated with convection for the whole grid column.

From a numerical perspective, the clouds inside a model grid column are treated as a 1-D (vertical) system and the above equations are discretized as a function of altitude (Figure 8.12). Updraft and downdraft pipes extend from the bottom to the top of the convective system. They are isolated from the gridboxes through which they extend vertically, and exchange mass with those gridboxes only through

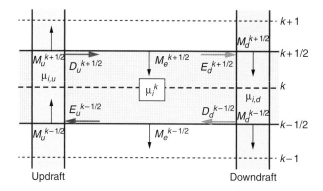

Figure 8.12 Discretization of the tendency equation that accounts for the effect of convection on the budget of species i and on the mean mixing ratio noted here μ_i^k at level k of a model. Processes include the updraft M_u, the downdraft M_d, the compensating subsidence flux M_e, and the entrainment E and detrainment D associated with the updrafts (u) and downdrafts (d), respectively.

entrainment and detrainment. There can be several updraft and downdraft pipes within a grid column, representing convective systems of different vertical extent. A downdraft pipe may be associated with each updraft pipe and represents in general a small fraction of the corresponding updraft flux. Chemical mass fluxes through each pipe are calculated as a balance between updraft (or downdraft) and entrainment/detrainment terms. The fluxes at the base of each updraft pipe, and at the top of each downdraft pipe, are initialized by entraining air from the grid cell. Mean concentrations in individual grid cells are modified by convection through entrainment, detrainment, and large-scale subsidence as follows:

$$
\left[\frac{\partial \overline{\rho}_i(z)}{\partial t}\right]^k_{conv} = -E_u^{k-1/2}\overline{\mu}_i^{k-1/2} + D_u^{k+1/2}\mu_{i,u}^{k+1/2} - E_d^{k+1/2}\overline{\mu}_i^{k+1/2} + D_d^{k-1/2}\mu_{i,d}^{k-1/2}
$$

$$
-\frac{1}{\Delta z}\left[(M_e\overline{\mu}_i)^{k+1/2} - (M_e\overline{\mu}_i)^{k-1/2}\right] \tag{8.130}
$$

where the fluxes (E, D, and M) are defined at the edges of the grid cell ($k - \frac{1}{2}$, $k + \frac{1}{2}$, etc.), while the mean mixing ratio is calculated at its center ($k - 1$, k, $k + 1$). Using an upstream scheme,

$$
(M_e \ \overline{\mu}_i)^{k+1/2} = M_e^{k+1/2}\overline{\mu}_i^{k+1} \quad \text{if} \quad M_e^{k+1/2} < 0
$$

$$
(M_e \ \overline{\mu}_i)^{k+1/2} = M_e^{k+1/2}\overline{\mu}_i^k \quad \text{if} \quad M_e^{k+1/2} > 0 \tag{8.131}
$$

the flux divergence associated with the subsidence flow is:

$$
(M_e \ \overline{\mu}_i)^{k+1/2} - (M_e \ \overline{\mu}_i)^{k-1/2} = M_e^{k+1/2} \ \overline{\mu}_i^{k+1} - M_e^{k-1/2} \ \overline{\mu}_i^k \quad \text{if } M_e < 0 \tag{8.132}
$$

Scavenging of water-soluble species is done by adding a loss term to the updraft mass balance equation (8.128):

$$
\frac{\partial(M_u \ \mu_{i,u})}{\partial z} = E_u(z) \ \overline{\mu}_i \ (z) - D_u(z) \mu_{i,u}(z) - \frac{k_i M_u \ \mu_{i,u}}{w} \tag{8.133}
$$

where w is the updraft velocity and k_i is a first-order loss constant $[s^{-1}]$ describing the scavenging. Further discussion of scavenging is presented in the next section.

8.9 Wet Deposition

Wet deposition is a general term to describe the removal of gases and particles by precipitation. *Scavenging* refers to the process by which wet deposition takes place. One refers to *large-scale scavenging* or *convective scavenging* depending on whether the precipitation in the meteorological model results from grid-scale motion of water vapor or from a subgrid convective parameterization. The distinction is important because convective scavenging must be applied to the subgrid convective updrafts, thus requiring coupling of convective transport and scavenging (8.133). By contrast, large-scale scavenging can be decoupled from grid-scale transport through operator splitting. Note that large-scale precipitation is still often subgrid (even though it is diagnosed from grid-scale motion), in which case its scavenging affects only the precipitating fraction of the grid cell.

It is also important to distinguish between *in-cloud scavenging* followed by precipitation (a process called *rainout*), and *below-cloud scavenging* by precipitation (called *washout*). Rainout and washout involve different physical processes. In-cloud scavenging followed by evaporation of precipitation below the cloud can release species at lower altitudes, resulting in downward motion rather than actual deposition. Cirrus precipitation is an important example of this effective downward motion (Lawrence and Crutzen, 1998).

The reader is referred to Seinfeld and Pandis (2006) for a detailed description of scavenging mechanisms for aerosols and gases. Here we limit our attention to the practical implementation of scavenging in chemical transport models. This typically involves consideration of three processes: (1) scavenging in convective updrafts associated with convective precipitation, (2) rainout and washout applied to large-scale precipitation and to convective anvils (convective precipitation outside the updraft), and (3) partial or total release below cloud as precipitation evaporates. Items (2) and (3) can be treated within the same algorithm, but (1) requires a separate algorithm.

8.9.1 Scavenging in Convective Updrafts

Scavenging in convective updrafts must be computed as part of the convective transport algorithm to prevent soluble species from being detrained at the top of the cloud without having experienced scavenging. It must be applied as air rises in the updraft from one model vertical level to the next in order to properly account for entrainment and detrainment fluxes at different levels. It must allow for different phases of cloud condensate, often ranging from 100% liquid at the bottom (*warm cloud*) to 100% solid at the top (*cold cloud*). There is typically an intermediate stage, with temperatures ranging from about –10 °C to –40 °C, at which the cloud contains both liquid and solid condensate (*mixed cloud*). In the mixed cloud, liquid cloud droplets will freeze upon contact with ice crystals (*riming*), and the resulting production of large ice crystals drives precipitation formation.

Let us consider the most general case of a mixed cloud, where both liquid and ice condensate are present. Warm and cold clouds can be treated as limiting cases. The scavenging rate constant k_i [s^{-1}] for species i in the updraft (8.133) is expressed as

$$k_i = \left(\varepsilon_i f_{i,L} + f_{i,I} \right) k \tag{8.134}$$

where k [s^{-1}] is the rate constant for conversion of cloudwater to precipitation and is typically $10^{-3} - 10^{-2}$ s^{-1} (Kain and Fritsch, 1990); $f_{i,L}$ and $f_{i,I}$ are the fractions of the species in the air parcel present in the liquid and ice water respectively; and $\varepsilon_i \leq 1$ is the retention efficiency of the species as liquid water is converted to precipitation. For a warm cloud, $\varepsilon_i = 1$ because freezing does not take place. For a mixed cloud, ε_i accounts for exclusion from the ice matrix as droplets freeze, and is highly dependent on species type as well as on the freezing mechanism and rate (Stuart and Jacobson, 2006). As air is lifted in the updraft from one model level to the next, the fraction F_i of species i scavenged from the updraft is computed as:

$$F_i = 1 - \exp \left(-\frac{k_i \Delta z}{w} \right) \tag{8.135}$$

where Δz is the distance between level midpoints and w is the updraft velocity, which may be provided by the meteorological model or need to be assumed (typically 5–10 m s^{-1}).

In the case of aerosol particles, the fraction incorporated in the condensed phase depends on aerosol and cloud microphysics in a complex way. It is sometimes assumed that $k_i \approx k$, meaning that all aerosol is in the condensed phase. A distinction is often made between hydrophobic and hygroscopic aerosol, and more sophisticated treatments can be used in models that resolve the aerosol size distribution (Seinfeld and Pandis, 2006).

In the case of gases and for a warm cloud, the fraction $f_{i,L}$ present in the liquid phase is determined by the effective Henry's law constant K_i^* [M atm^{-1}] (see Section 5.5.1):

$$f_{i,L} = \frac{K_i^* L \mathcal{R} T}{1 + K_i^* L \mathcal{R} T} \tag{8.136}$$

where L is the cloud liquid water content [m^3 water per m^3 air], T is the temperature [K], and $\mathcal{R} = 0.08205$ atm M^{-1} K^{-1} is the ideal gas constant. For a typical precipitating cloud with $L \sim 1 \times 10^{-6}$, gases with $K_i^* \gg 10^4$ M atm^{-1} are efficiently scavenged ($f_{i,L} \rightarrow 1$) while gases with $K_i^* \ll 10^4$ M atm^{-1} are not scavenged ($f_{i,L} \rightarrow 0$). Gases can also be taken up by ice crystals in mixed or cold clouds. The fraction $f_{i,I}$ present in the ice phase can be estimated from a surface coverage model (such as a Langmuir isotherm) or a co-condensation model (Mari *et al.*, 2000).

8.9.2 Rainout and Washout

Scavenging outside convective updrafts, including in-cloud scavenging (rainout) and below-cloud scavenging (washout), can be treated as a first-order loss process in the precipitating column. For each grid column the meteorological model must provide information on the vertical distribution of precipitation rates. Let P_j represent the precipitation rate [cm water s^{-1}] through the bottom of model level j. One must apply rainout to the in-cloud levels where new precipitation forms ($P_j > P_j + 1$) and washout to the below-cloud levels where precipitation evaporates ($P_j \leq P_{j+1}$). We start the scavenging calculation at the top of each precipitating column and progress downward level by level, applying rainout or washout/reevaporation as appropriate.

Rainout can be computed similarly to scavenging in the convective updrafts described in Section 8.9.1. The fraction F_i of species i scavenged from a grid cell over a time step Δt is given by

$$F_i = f_A(1 - \exp[-k_i\Delta t]) \tag{8.137}$$

where f_A is the areal fraction of the grid cell experiencing precipitation and k_i [s^{-1}] is a first-order rainout rate constant. From knowledge of the rainout rate constant k of the condensed water in the cloud, we can calculate k_i in the same manner as in the case of convective updrafts (8.134). Values of f_A and k may be supplied as part of the hydrological information from the driving meteorological model. If not, they need to be estimated, and classic parameterizations for this purpose are available from Giorgi and Chameides (1986) and Balkanski *et al.* (1993). Accounting for $f_A < 1$ when precipitation is subgrid in scale is important because scavenging of water-soluble species from a precipitating column is highly efficient. Assuming $f_A = 1$ would lead to an overestimate of scavenging.

Washout involves the below-cloud uptake of aerosol particles and gases by hydrometeors (raindrops or ice crystals). For aerosol particles and highly soluble gases, washout is a kinetic process limited by mass transfer (collision rates for particles, molecular diffusion for soluble gases). The scavenged fraction F_i of species i for a grid cell experiencing washout over a time step Δt is given by:

$$F_i = f_A\left(1 - \exp\left[-k_i'\frac{P_j}{f_A}\Delta t\right]\right) \tag{8.138}$$

where k_i' [cm^{-1}] is a first-order washout rate constant, typically ~1 cm^{-1} for aerosol particles and highly soluble gases. Parameterizations for k_i' are given for example by Feng (2007, 2009) for scavenging of aerosol particles by rain and snowfall, and by Levine and Schwartz (1982) for scavenging of highly soluble gases by rain. Evaporation of precipitation below cloud will not release these species to the surrounding air if the hydrometeors simply shrink; evaporation must be complete. In the case of partial evaporation, one must make an assumption about the fraction of hydrometeors that shrink and the fraction that evaporate entirely. A 50/50 assumption is often made.

For moderately soluble gases where washout is not limited by mass transfer, we can assume equilibrium between the hydrometeors and the surrounding air within the precipitating fraction of the grid cell. In that case, the fraction of species i that is incorporated in the liquid or ice can be calculated using $f_{i,L}$ and $f_{i,I}$ as given in Section 8.9.1 for the case of convective updrafts. In particular, $f_{i,L}$ is given by:

$$f_{i,L} = \frac{K_i^* L_P \mathcal{R} T}{1 + K_i^* L_P \mathcal{R} T} \tag{8.139}$$

where L_P is the rainwater content of the precipitating fraction of the grid cell defined as the volume of precipitation to which a unit volume of air is exposed over timestep Δt:

$$L_P = \frac{P_j \Delta t}{f_A \Delta z} \tag{8.140}$$

This equilibrium treatment allows for evaporative release of gases below cloud level to respond directly to the downward decrease of P_j from level to level. Let m_j represent the mass of a species i in grid cell j and Δm_{j+1} represent the mass scavenged into the grid cell through the top over time Δt by precipitation overhead. The mass Δm_j transported out through the bottom of the grid cell by precipitation over time Δt is then given by:

$$\Delta m_j = f_i \left(f_A m_j + \Delta m_{j+1} \right) \tag{8.141}$$

8.10 Lightning and NO$_x$ Production

Deep convective storms with strong updrafts generate lightning as a result of electrical charge separation between the cloud and the surface. Heating in the lightning bolt produces a plasma with temperatures exceeding 10^6 K. This leads to the production of nitric oxide (NO) from air molecules, initiated by thermolysis of O_2 (*Zel'dovich mechanism*):

$$
\begin{aligned}
O_2 &\rightleftarrows O + O \\
O + N_2 &\rightleftarrows NO + N \\
\underline{N + O_2} &\underline{\rightleftarrows NO + O} \\
Net: N_2 + O_2 &\rightleftarrows 2NO
\end{aligned}
\tag{8.142}
$$

Estimates for the global source of NO$_x$ from lightning range from 1 to 20 Tg N a^{-1} (Schumann and Huntrieser, 2007), which can be compared to a global NO$_x$ source from fossil fuel combustion of about 30 Tg N a^{-1}. Lightning releases NO$_x$ in the upper troposphere, where it is particularly efficient for producing ozone and OH. As such, it plays a major role in determining tropospheric oxidant levels.

Figure 8.13 shows the global climatological distribution of lightning observed from space. Lightning mainly occurs over land where intense heating of the ground leads to strong convective updrafts. Lightning NO$_x$ is mostly released in the detrainment zone at the top of the updraft (Ott *et al.*, 2010). Active nonlinear chemistry producing ozone takes place in this detrainment ozone as lightning NO$_x$ interacts with water vapor and VOCs injected in the updraft.

The standard way to represent the lightning source of NO$_x$ in models is as part of the parameterization for convective transport, thus ensuring that the association between lightning NO$_x$ and convective outflow driving nonlinear chemistry is captured. The lightning flash rate in the convective column is parameterized on the basis of the strength and/or depth of the convection, and a NO$_x$ yield per flash is assumed that may vary depending on the energy of the flash. Some models distinguish between cloud-to-ground and intra-cloud flashes in that regard.

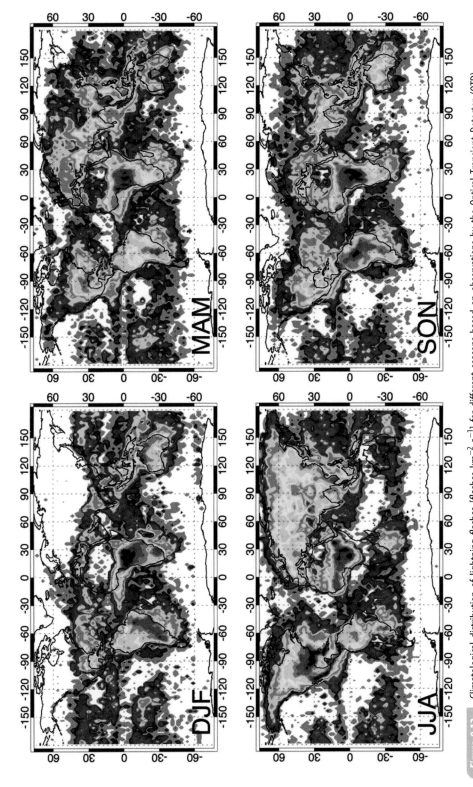

Figure 8.13 Climatological distribution of lightning flashes [flashes km^{-2} a^{-1}] for different seasons based on observations by the Optical Transient Detector (OTD). From Christian *et al.* (2003).

Different model parameterizations have been proposed to compute lightning flash rates in deep convection. Cloud-resolving models use the convective updraft velocity as the best predictor variable, but this variable is generally not available in coarser-resolution models. A common parameterization for global models is that of Price and Rind (1992), which expresses the flash frequency F [flashes per minute] as a steep function of the cloud top height H [km], with separate expressions for continental (c) and maritime (m) convection:

$$F_c = 3.44 \times 10^{-5} H^{4.9} \qquad F_m = 6.4 \times 10^{-4} H^{1.73} \qquad (8.143)$$

Other parameterizations relate lightning flash frequency to deep convective mass fluxes (Allen et al., 2000) or to convective precipitation (Meijer et al., 2001). None of these parameterizations have much success in reproducing lightning observations (Murray et al., 2012). Models for the present-day atmosphere can constrain the distribution of lightning flash frequencies with satellite and ground-based observations, and apply these locally or regionally to the deep convective updrafts simulated by the model (Sauvage et al., 2007; Murray et al., 2012). This offers a more realistic representation of present-day lightning but it cannot be used to simulate past or future climates.

Estimates of NO_x yields from lightning flashes span a wide range from 30 to 1000 moles per flash (Price et al., 1997; Théry et al., 2000; Schumann and Huntrieser, 2007). Yields depend on the energy of the flash, but this is very poorly constrained. The general practice in atmospheric models is to adjust the global lightning NO_x source to a value that is compatible with observed atmospheric concentrations of reactive nitrogen oxides (NO_y, including NO_x and its oxidation products) and tropospheric ozone. This leads to a global source in the range 2–8 Tg N a^{-1} (Solomon et al., 2007). The implied NO_x yields per flash are in the range 200–500 moles, consistent with atmospheric observations (Murray et al., 2012).

8.11 Gravity Waves

Gravity waves are oscillations that develop in stably stratified air when air parcels are displaced vertically, for example by mountain ranges or by neighboring thunderstorms. Their horizontal wavelengths are typically ~10–100 km. The vertical propagation of these waves depends on the vertical profile of the mean horizontal wind speed; waves are absorbed at a critical level where the phase speed c of the wave is equal to the mean wind speed u. As the waves propagate vertically, their amplitude increases as the inverse of the air density. The perturbation of temperature in the mesosphere or lower thermosphere becomes so large that the air becomes convectively unstable and the waves break (Figure 8.14, a). The momentum transported by the wave from lower atmospheric levels is transferred to the mean flow, which leads to an attenuation of the zonal flow and triggers a meridional circulation directed from the summer to the winter hemisphere (Figure 8.14, b). Vertical mixing also takes place.

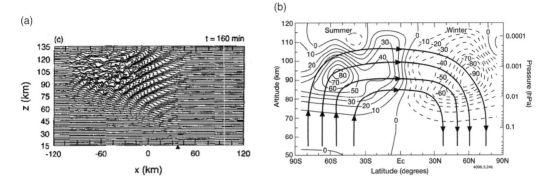

Figure 8.14 (a) Two-dimensional model of the potential temperature field (isentropes) perturbed by a prescribed gravity wave source located at the triangle along the x-axis. The figure shows the propagation of the gravity wave after 160 minutes of model integration, i.e., at a mature stage of wave breaking. Reproduced from Prusa *et al.* (1996). Copyright © American Meteorological Society, used with permission. (b) Global distribution of wind acceleration [m s^{-1} d^{-1}] in the upper atmosphere as driven by gravity wave breaking above 70 km altitude. The resulting global mean meridional circulation, schematically represented by arrows, is characterized by upward (downward) motions in the summer (winter) hemisphere. Reproduced from Brasseur and Solomon (2005).

Gravity waves cannot be explicitly represented at the grid resolution of global models and their effects must therefore be parameterized. Different formulations have been proposed (Lindzen, 1981; Medvedev and Klaassen, 1995; Hines, 1997a, 1997b). In a very simple approach, Lindzen (1981) derives the gravity wave drag (G) and vertical eddy diffusion coefficient (K_z) from the following expressions:

$$G = \frac{-k\,(u-c)^3}{2HN} \tag{8.144}$$

$$K_z = \frac{k\,(u-c)}{2HN^3} \tag{8.145}$$

where k is the horizontal wavenumber of the wave, H is the atmospheric scale height, and N is the Brunt–Väisälä frequency ($N^2 = g\,\partial\ln\theta/\partial z$ if θ is the potential temperature and z the altitude). In most parameterizations, rather than considering the propagation of a single gravity wave, a spectrum of waves with different phase velocities is considered.

8.12 Dynamical Barriers

As discussed in Chapter 2, dynamical barriers in the atmosphere limit the rate at which mass is exchanged between different atmospheric regions. Transport across these barriers often involves subgrid processes. An important case is the tropopause, where a strong inversion severely restricts transport. Long-lived species penetrate the stratosphere in regions of tropical upwelling. Water vapor is trapped during this transport by condensation and precipitation at the cold temperatures of the tropopause. Downward transport across the tropopause occurs mostly as small-scale tongues of air (*tropopause folds*) that form in connection with meteorological disturbances. These tongues may eventually mix in the troposphere. Diffusive numerical transport schemes can lead to spurious fluxes across the tropopause because vertical gradients are particularly large.

Dynamical barriers also restrict meridional transport in the stratosphere. The subtropical barrier isolates tropical rising air (the *tropical pipe*) from mid-latitude influences. A second barrier arises from the two polar vortices that isolate polar regions from lower latitudes. Transport through these barriers involves dynamical disturbances at scales that are generally unresolved by models. For example, narrow filaments are stripped away from the polar vortex in response to planetary wave breaking events. These very thin structures are stretched around the vortex before they mix with the surrounding air masses.

The need to resolve dynamical barriers has motivated the development of Lagrangian models of stratospheric transport (Fairlie *et al.*, 1999). These Lagrangian models describe the deformation and dissipation of the filaments on the basis of the flow divergence (McKenna *et al.* 2002). Comparisons to observations show that the Lagrangian models are far more effective than their Eulerian counterparts in generating and preserving the filamentary structures.

8.13 Free Tropospheric Plumes

The free troposphere, ranging from the top of the boundary layer (~2 km) to the tropopause, is on average a convectively stable environment. Observations show that chemical plumes injected into the free troposphere by convection, volcanoes, or stratospheric intrusions can retain their identity as well-defined layers for a week or more as they are transported on intercontinental scales. Vertical soundings of the free troposphere often reveal the presence of distinct chemical layers, typically ~1 km thick and stretching horizontally in filaments spread over ~1,000 km (Thouret *et al.*, 2000; Heald *et al.*, 2003). Global Eulerian models have great difficulty in reproducing such layered structures in the free troposphere. The plumes dissipate much too quickly, even when they are sufficiently thick that they should be resolved at the model grid scale. This problem is very different in nature from the turbulent diffusion of boundary layer plumes emitted from point sources, typically simulated with a Gaussian plume or puff model (Section 4.12). Boundary layer plumes dissipate on a timescale of hours, but free tropospheric layers persist for considerably longer because of the convectively stable environment.

Figure 8.15 from Rastigejev *et al.* (2010) illustrates the problem. It shows the transport over nine days of an inert chemical in a 2-D (horizontal) model of the free troposphere over the Pacific. The model has $2° \times 2.5°$ horizontal resolution. The chemical is released uniformly at $t = 0$ over a $12° \times 15°$ domain, resolved by 6×6 grid squares. Transport is solely by advection, computed with a second-order

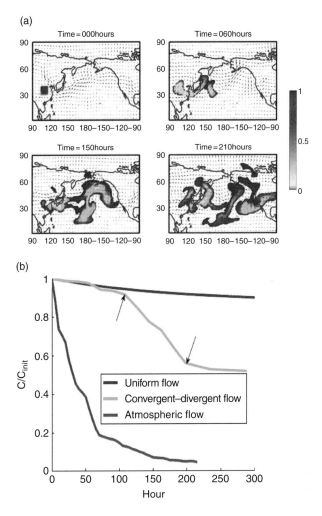

(a)

(b)

Figure 8.15 Free tropospheric advection of a chemically inert plume in a 2-D version of the GEOS-Chem chemical transport model with $2° \times 2.5°$ horizontal resolution. The plume is released at time $t = 0$ as a uniform layer over a $12° \times 15°$ domain. (a) Shows the evolution of plume mixing ratios over nine days in a variable atmospheric flow at 4 km altitude from the NASA GEOS meteorological data assimilation. (b) Shows the decay of peak mixing ratios in the plume for the atmospheric flow of the left panel, for a uniform flow, and for a uniform flow with convergent–divergent perturbation applied between 125 and 200 h. From Rastigejev *et al.* (2010).

accurate piecewise parabolic scheme. The advection equation should perfectly conserve the mixing ratio in the plume, but the model plume is instead rapidly dissipated. The bottom panel of Figure 8.15 shows the decay of the maximum mixing ratio in the plume with time. After two days the maximum mixing ratio has dropped to 40% of the initial value; after one week it is less than 10%. As shown in Figure 8.15, this fast numerical decay of the plume is caused by the strong variability of the atmospheric flow. A simulation using the same model with uniform flow shows only 10% dissipation in two weeks, reflecting the high-order accuracy of the advection scheme. Introducing a convergent–divergent pattern in this uniform flow causes a sharp increase in plume dissipation.

Plume dissipation in variable flow as illustrated by Figure 8.15 is due to divergence of the wind, causing stretching of the plume. The divergence is measured by the *Lyapunov exponent* λ, defined as the exponential rate at which nearby trajectories diverge from each other: $\lambda = \partial u / \partial x$ where u is the wind speed in the direction x of the flow. In the absence of molecular diffusion, the continuity equation prescribes that the chemical concentration within the plume should remain constant in time even with stretching. However, numerical diffusion in the advection algorithm causes the plume to decay rapidly when stretched. We represent the numerical diffusion by a diffusivity D normal to the flow. In this conceptual 2-D example the diffusion is taken to be horizontal; but the same argument applies to diffusion in the vertical. To estimate the rate of decay, we need to know the characteristic length scale over which the concentration decays at the edge of the plume. This length scale r_b is determined by a balance between diffusion and stretching. Intuitively, if the plume is very thick, stretching dominates and the plume filaments; conversely, if the plume is very thin, diffusion dominates and the plume thickens. There is an equilibrium thickness for which diffusion and stretching are in balance. The rate constant for diffusion is $\sim D/r_b^2$, while the rate constant for stretching is λ. Balance between diffusion and stretching thus implies

$$r_b = \sqrt{\frac{D}{\lambda}} \tag{8.146}$$

If we assume that the mixing ratio μ of the chemical species is uniform in the plume and zero in the surrounding background, then the rate of decay of μ is given by the diffusive outflux through the boundary, namely

$$V\frac{d\mu}{dt} = -DS\frac{\mu}{r_b} \tag{8.147}$$

where V and S are the volume and surface area of the plume. Now $V/S = W$, the width of the plume in the direction perpendicular to the stretching direction of the flow. Hence the mixing ratio in the plume decays exponentially as $\mu \sim \exp[-\alpha t]$, with

$$\alpha = \frac{\sqrt{D\lambda}}{W} \tag{8.148}$$

This implies the following physical picture for the decay of an initially thick plume in stretched flow. The thickness of the plume decreases in time due to the stretching of the flow until $W = r_b$, at which point stretching and diffusion precisely match so that the plume thickness does not decrease further but the mixing ratio continues to decay. Replacing $W = r_b$ in (8.148) implies $\alpha = \lambda$, so that the decay rate is equal to the Lyapunov exponent of the flow (Chella and Ottino, 1984; Balkovsky and Fouxon, 1999). The decay rate of a stretched plume thus approaches a limit that is independent of the numerical diffusion.

We can now understand the numerical decay of the plume shown in Figure 8.15. Under a uniform flow the characteristic timescale for plume decay is

$$\tau_u = \frac{W^2}{D} \tag{8.149}$$

where W is the width of the plume. On the other hand, in a divergent flow this decay timescale is

$$\tau_d = \frac{W}{\sqrt{D\lambda}} \tag{8.150}$$

where λ is the Lyapunov exponent of the flow. We see that $\tau_u > \tau_d$ when $W > (D/\lambda)^{1/2} = r_b$. A plume thicker than r_b decays faster than a simple estimate from numerical diffusion would suggest. Ultimately the plume decays at a rate that is determined by the Lyapunov exponent.

Let us consider the consequences for the sensitivity of plume decay to grid resolution. A straightforward analysis demonstrates that the numerical diffusivity near sharp boundaries is $D \sim u\Delta x$ where Δx is the grid spacing (Rastigejev *et al.*, 2010). This is the case even with a higher-order advection algorithm, as the higher order of accuracy is contingent on adequate resolution of gradients on the grid scale, which fails when the boundaries are sharp (i.e., when the plume is resolved by only a few grid cells).

With $D \sim u\Delta x$ we find

$$r_b \sim \sqrt{u\Delta x/\lambda} = \sqrt{u\Delta x/\nabla u} \tag{8.151}$$

Thus the length scale r_b is roughly the geometric mean of the grid spacing Δx and the length scale $u/\nabla u$ over which the velocity field varies. The length scale below which numerical diffusion is important is not the grid resolution but a much larger (flow-dependent) length scale. For the flow field in Figure 8.15, which varies over $\sim 10^4$ km and with horizontal grid resolution ~ 100 km, this crossover scale is ~ 1000 km.

These arguments imply that the decay rate of a plume with initial width $W > r_b$ is initially set by numerical diffusion. Ultimately, the plume will be stretched so that $W = r_b$, at which point the decay rate approaches the Lyapunov exponent of the flow. Increasing the grid resolution of the model delays the attainment of this regime, but only moderately so as $r_b \sim \Delta x^{1/2}$. Improving resolution of plumes by a factor of 2 would require a factor of 4 increase in grid resolution. The situation is in fact worse because stretching of the flow increases as the grid resolution increases and smaller eddies are resolved (Wild and Prather, 2006). Numerical tests by Rastigejev *et al.* (2010) indicate that $r_b \sim \Delta x^{1/4}$ because of this effect. Increasing the resolution of plumes by a factor of 2 would thus require a factor of 8 increase in grid resolution.

Lagrangian models perform much better than Eulerian models in preserving plumes during long-range transport. As discussed in Section 8.12, Lagrangian models have been used in the stratosphere to improve the simulation of transport across dynamical barriers. However, Eulerian models are generally preferred in global applications for several reasons, including better representation of area sources, ability to describe nonlinear chemistry and aerosol evolution, and completeness and smoothness of the solution. One possible approach is to use embedded

Lagrangian plumes within the Eulerian framework, as is sometimes done in regional air quality models to describe Gaussian plumes originating from point sources (Section 4.12). Another approach is to use an adaptive grid model where localized increases in grid resolution are triggered by strong concentration gradients (Box 4.6). These approaches have yet to be implemented in global models.

References

Ahrens C. D. (2000) *Essentials of Meteorology: An Invitation to the Atmosphere*, 3rd edition, Thomson Brooks/Cole, Belmont, CA.

Allen D., Pickering K., Stenchikov G., Thompson A., and Kondo Y. (2000) A three-dimensional total odd nitrogen (NO_y) simulation during SONEX using a stretched-grid chemical transport model, *J. Geophys. Res.*, **105**: doi: 10.1029/1999JD901029.

Balkanski Y. J., Jacob D. J., Gardner G. M., Graustein W. C., and Turekian K. K. (1993) Transport and residence times of tropospheric aerosols inferred from a global three-dimensional simulation of ^{210}Pb, *J. Geophys. Res.*, **98**(D11), 20573–20586, doi:10.1029/93JD02456.

Balkovsky E. and Fouxon A. (1999) Universal long-time properties of Lagrangian statistics in the Batchelor regime and their application to the passive scalar problem, *Phys. Rev. E*, **60**, 4164.

Blackadar A. K. (1979) *High Resolution Models of the Planetary Boundary Layer, Advances in Environmental and Scientific Engineering*, Vol. **I**, Gordon and Breach, New York.

Brasseur G. P. and Solomon S. (2005) *Aeronomy of the Middle Atmosphere: Chemistry and Physics of the Stratosphere and Mesosphere*, 3rd edition, Springer, New York.

Brodkey R. S. (1981) Fundamentals of turbulent motions, mixing and kinetics, *Chem. Eng. Comm.*, **8**, 1–23.

Businger J. A., Wyngaard J. C., Izumi Y., and Bradley E. F. (1971) Flux profile relationships in the atmospheric surface layer, *J. Atmos. Sci.*, **28**, 181–189.

Cassiani M., Vinuesa J. F. Galmarini S. and Denby D. (2010) Stochastic fields methods for sub-grid scale emission heterogeneity in mesoscale atmospheric dispersion flows, *Atm. Chem. Phys.*, **10**, 267–277.

Chella R. and Ottino J. M. (1984) Conversion and selectivity modifications due to mixing in unpremixed reactors, *Chem. Eng. Sci.*, **39**, 551.

Christian H. J., Blakeslees, R., Boccippio, D., *et al.* (2003) Global frequency and distribution of lightning as observed from space by the Optical Transient Detector, *J. Geophys. Res.*, **108**(D1), 4005, doi:10.1029/2002JD002347.

Collins W. J., Stevenson D. S., Johnson C. E., and Derwent R. G. (1997) Tropospheric ozone in a global-scale three-dimensional Lagrangian model and its response to NO_x emission controls, *J. Atmos. Chem.*, **26**, 223–274.

Damköhler G. (1940) Influence of turbulence on the velocity of flames in gas mixtures, *Z. Elektrochem*, **46**, 601–626.

Damköhler G. (1947) The effect of turbulence on the flame velocity in gas mixtures, Technical report NACA TM 1112.

Deardorff J. W. (1966) The counter gradient heat flux in the lower atmosphere and in the laboratory, *J. Atmos. Sci.*, **23**, 503–506.

Deardorff J. W. (1972) Theoretical expression for the countergradient vertical heat flux, *J. Geophys. Res.*, **77** (30), 5900–5904.

Durran D. R. (2010) *Numerical Methods for Fluid Dynamics: with Applications to Geophysics*, 2nd edition, Springer, New York.

Dyer A. J. (1974) A review of flux-profile relations, *Bound. Layer Meteor.*, **1**, 363–372.

Ecke R. (2005) The turbulence problem: An experimentalist's perspective, *Los Alamos Sci.*, **29**, 124–141.

Ekman V. W. (1905) On the influence of the earth's rotation on ocean currents, *Ark. Mat. Astron. Fys.*, **2**, 11, 1–52.

Fairlie T. D., Pierce R. B., Al-Saadi J. A., *et al.* (1999) The contribution of mixing in Lagrangian photochemical predictions of polar ozone loss over the Arctic in summer 1997, *J. Geophys. Res.*, **104**, 26597–26609.

Favre A. (1958a) Equations statistiques des gaz turbulents: Masse, quantité de movement, *C. R. Acad. Sci. Paris*, **246**, 2576–2579.

Favre A. (1958b) Equations statistiques des gaz turbulents: Énergie totale, énergie interne. *C. R. Acad. Sci. Paris*, **246**, 2723–2725

Feng J. (2007) A 3-mode parameterization of below-cloud scavenging of aerosols for use in atmospheric dispersion models, *Atmos. Environ.*, **41**, 6808–6822,.

Feng J. (2009), A size-resolved model for below-cloud scavenging of aerosols by snowfall, *J. Geophys. Res.*, **114**, D08203, doi:10.1029/2008JD011012.

Fox R. O. (2003) *Computational Models for Turbulent Reacting Flows*, Cambridge University Press, Cambridge.

Garratt J. R. (1994) *The Atmospheric Boundary Layer*, Cambridge University Press, Cambridge.

Giorgi F. and Chameides W. L. (1986) Rainout lifetimes of highly soluble aerosols and gases as inferred from simulations with a general circulation model, *J. Geophys. Res.*, **91**, 14367–14376.

Grell G. A., Peckham S. E., Schmitz R., *et al.* (2005), Fully coupled online chemistry within the WRF model, *Atmos. Environ.*, **39**, 6957–6975, doi: 10.106/j.atmosenv.2005.04.027.

Heald C. L., Jacob, D., Fiore, A., *et al.* (2003) Asian outflow and transpacific transport of carbon monoxide and ozone pollution: An integerated satellite aircraft and model perspective, *J. Geophys. Res.*, **108**, 4804.

Hesselberg T. (1926) Die Gesetze des ausgegleichenen atmosphaerischen Bewegungen. *Beitr. Physik freien Atmosphaere.*, **12**, 141–160.

Hines C. O. (1997a) Doppler-spread parameterization of gravity wave momentum deposition in the middle atmosphere: Part 1. Basic formulation, *J. Atmos. Solar. Terr. Phys.*, **59**, 371–386.

Hines C. O. (1997b) Doppler-spread parameterization of gravity wave momentum deposition in the middle atmosphere: Part 2. Broad and quasi monochromatic spectra, and implementation, *J. Atmos. Solar. Terr. Phys.*, **59**, 387–400.

Hogstrom U. (1988) Non-dimensional wind and temperature profiles in the atmospheric surface layer: A re-evaluation, *Bound. Layer Meteor.*, **42**, 55–78.

Holton J. R. (1992) *An Introduction to Dynamical Meteorology,* 3rd edition, Academic Press, New York.

Holtslag A. A. M. and Boville B. (1993) Local versus nonlocal boundary-layer diffusion in a global climate model, *J. Climate*, **6**, 1825–1842.

Hong S. Y., Noh Y., and Dudhia J. (2006) A new vertical diffusion package with an explicit treatment of entrainment processes, *Mon. Wea. Rev.*, **134**, 2318–2341.

Kain J. S. and Fritsch J. M. (1990) A one-dimensional entraining/detraining plume model and its application in convective parameterization, *J. Atmos. Sci.*, **47**, 2784–2802.

Kazil J., Wang H., Feingold G., *et al.* (2011) Modeling chemical and aerosol processes in the transition from closed to open cells during VOCALs-Rex, *Atmos. Chem. Phys.*, **11**, 7491–7514, doi: 10.5194/acp-11-7491-2011.

Kolmogorov A. N. (1941a) The local structure of turbulence in incompressible viscous fluid for very large Reynolds number, *Dokl. Akad. Nauk SSSR*, **30**, 301–303. Reprinted in *Proc. R. Soc. Lond., Ser: A*, **434**, 9–13, 1991

Kolmogorov A. N. (1941b) Dissipation of energy in the locally isotropic turbulence, *Dokl. Akad. Nauk SSSR*, **31**, 19–21. Reprinted in *Proc. R. Soc. Lond., Ser: A*, **434**, 15–17, 1991.

Kramm G. and Meixner F. X. (2000) On the dispersionof trace species in the atmospheric boundary layer: A reformulation of the governing equations for the turbulent flow of the compressible atmosphere, *Tellus*, **52A**, 500–522.

Lawrence M. G. and Crutzen P. J. (1998) The impact of cloud particle gravitational settling on soluble trace gas distribution, *Tellus*, **50B**, 263–289.

Levine S. Z. and Schwartz S. E. (1982) In-cloud and below-cloud scavenging of nitric acid vapor, *Atmos. Environ.*, **16** (7), 1725–1734.

Lin J. C., Gerbig C., Wofsy S. C., *et al.* (2003) A near-field tool for simulating the upstream influence of atmospheric observations: The Stochastic Time-Inverted Lagrangian Transport (STILT) model, *J. Geophys. Res.*, **108**, 4493, doi:10.1029/2002JD003161.

Lindzen R. S. (1981) Turbulence and stress due to gravity wave and tidal breakdown, *J. Geophys. Res.*, **86**, 9707–9714.

Liu S. C., McAfee J. R., and Cicerone R.J. (1984) Radon 222 and tropospheric vertical transport, *J. Geophys. Res.*, **89**, 7291–7297.

Mari C., Jacob D. J., and Bechtold P. (2000) Transport and scavenging of soluble gases in a deep convective cloud, *J. Geophys. Res.*, **105**, 22255–22267.

McKenna D. S., Grooss J. U., Gunther G., *et al.* (2002) A new Chemical Lagrangian Model of the Stratosphere (CLaMS): 2. Formulation of chemistry scheme and initialization, *J. Geophys. Res.*, **107**, 4309.

Medvedev A. S. and Klaassen G. P. (1995) Vertical evolution of gravity wave spectra and the parameterization of associated wave drag, *J. Geophys. Res.*, **110** (D12), 25841–25853, doi: 10.1029/95JD02533.

Meijer E. W., van Velthoven P. E. J., Brunner D. W., Huntrieser H., and Kedler H. (2001) Improvement and evaluation of the parameterisation of nitrogen oxide production by lightning, *Phys. Chem. Earth. (C)*, **26**, 577–583.

Monin A. S. and Obukhov A. M. (1954) Basic laws of turbulent mixing in the atmosphere near the ground, *Tr. Akad. Nauk., SSSR Geophyz Inst.*, No 24 (151), 1963–1987.

Murray L. T., Jacob D. J., Logan J. A., *et al.*, (2012) Optimized regional and interannual variability of lightning in a global chemical transport model constrained by LIS/OTD satellite data, *J. Geophys. Res.*, **117**, D20307, doi:10.1029/2012JD017934

Obukhov, A. M. (1941) On the distribution of energy in the spectrum of turbulent flow, *Dokl. Akad. Nauk SSSR*, **32**, 22–24.

Ott L. E., Pickering K. E., Stenchikov G. L., *et al.* (2010) Production of lightning NO_x and its vertical distribution calculated from three-dimensional cloud-scale chemical transport model simulations, *J. Geophys. Res.*, **115**, D04301, doi: 10.1029/2009JD011880.

Pope S. B. (2000) *Turbulent Flows*, Cambridge University Press, Cambridge.

Prandtl L. (1925) Über die ausgebildete Turbulenz, *Z. Angew. Math. Mech.*, **5**, 136–138.

Price C. and Rind D. (1992) A simple parameterization for calculating global lightning distribution, *J. Geophys. Res.*, **97**, 9919–9933.

Price C., Penner J., and Prather M. (1997) NO_x from lightning: 1. Global distribution based on lightning physics, *J. Geophys. Res.*, **102**, 5929–5941.

Prusa J. M., Smolarkiewicz P. K., and Garcia R. R. (1996) On the propagation and breaking at high altitudes of gravity waves excited by tropospheric forcing, *J. Atmos. Sci.*, **53** (15), 2186–2216.

Rastigejev Y., Park R., Brenner M. P., and Jacob D. J. (2010) Resolving intercontinental pollution plumes in global models of atmospheric transport, *J. Geophys. Res.*, **115**, D02302.

Richardson L. F. (1922) *Weather Prediction by Numerical Process.* Cambridge University Press, Cambridge, reprinted 1965.

Sauvage B., Martin R. V., van Donkelaar A., *et al.* (2007) Remote sensed and in situ constraints on processes affecting tropical tropospheric ozone, *Atmos. Chem. Phys.*, **7**, 815–838.

Schumann U. and Huntrieser H. (2007) The global lightning-induced nitrogen oxides source, *Atmos. Chem. Phys.*, **7**, 3823–3907.

Seinfeld J. H. and Pandis S. N. (1996) *Atmospheric Chemistry and Physics: From Air Pollution to Climate Change*, Wiley, New York.

Seinfeld J. H. and Pandis S. N. (2006) *Atmospheric Chemistry and Physics: From Air Pollution to Climate Change*, Wiley, Chichester.

Solomon S., Qin D., Manning M., *et al.* (eds.) (2007) *Contribution of Working Group I to the Fourth Assessment Report of the Intergovernmental Panel on Climate Change*, Cambridge University Press, Cambridge.

Stensrud D. J. (2007) *Parameterization Schemes: Keys to Understanding Numerical Weather Prediction Models*, Cambridge University Press, Cambridge.

Stuart A. L. and Jacobson M. Z. (2006). A numerical model of the partitioning of trace chemical solutes during drop freezing, *J. Atmos. Chem.*, **53**(1), 13–42.

Stull R. B. (1988) *An Introduction to Boundary Layer Meteorology*, Kluwer, Dordrecht.

Théry C., Laroche P., and Blanchet P. (2000) EULINOX: The European lightning nitrogen oxides experiment. In *EULINOX Final Report* (Höller, H. and Schumann, U., eds.), Deutches Zentrum für Luft- und Raumfahrt, Köln.

Thomson D. J. and Wilson J. D. (2012) History of Lagrangian stochastic models for turbulent dispersion. In *Lagrangian Modeling of the Atmosphere* (Lin J., Brunner D., Gerbig C., *et al.*, eds.), American Geophysical Union, Washington, DC.

Thouret V., Cho J., Newell R., Marenco A., and Smit H. (2000) General characteristics of tropospheric trace constituent layers observed in the MOZAIC program, *J. Geophys. Res.*, **105**, 17379–17392.

Troen L. and Mahrt L. (1986) A simple model of the atmospheric boundary layer: Sensitivity to surface evaporation, *Bound. Layer Meteorol.*, **37**, 129–148.

Valiño L. (1998) A field Monte-Carlo formulation for calculating the probability density function of a single scalar in a turbulent flow, *Flow, Turbulence and Combustion*, 60, 157. doi:10.1023/A:1009968902446

Van Mieghem J. (1973) *Atmospheric Energetics*, Oxford University Press, Oxford.

Verver G. H. L., van Dop H., and Holtslag A. A. M. (1997) Turbulent mixing of reactive gases in the convective boundary layer, *Bound. Layer Meteorol.*, **85**, 197–222.

Vilà-Guerau de Arellano J. (1992) A review of turbulent flow studies relating to the atmosphere, *ACTA Chimica Hungarica-Models in Chemistry*, **129** (6), 889–902.

Vinuesa J.-F. and Vilà-Guerau de Arellano J. (2005) Introducing effective reaction rates to account for the inefficient mixing of the convective boundary layer, *Atmos. Environ.*, **39**, 445–461.

Wallace J. M. and Hobbs P. V. (2006) *Atmospheric Science: An Introductory Survey*, Academic Press, New York.

Wild O. and Prather M. J. (2006) Global tropospheric ozone modeling: Quantifying errors due to grid resolution, *J. Geophys. Res.*, **111**, D11305, doi: 10.1029/2005JD006605.

Wyngaard J. C. (1982) Planetary boundary layer modeling. In *Atmospheric Turbulence and Air Pollution Modelling*, (Nieuwstadt F. T. M. and Van Dop H., eds.), Reidel, Norwell, MA.

9 Surface Fluxes

9.1 Introduction

Solving the continuity equation for the concentrations of atmospheric species requires boundary conditions at the Earth's surface. The surface can act either as a source or a sink. The boundary condition can be expressed as a vertical flux or as a concentration. The surface flux is called an *emission* when upward and a *deposition* when downward. Direct deposition of gas molecules and aerosol particles to the surface is called *dry deposition*, to distinguish it from *wet deposition* driven by precipitation scavenging (Chapter 8). Many species can be both emitted and dry deposited, and the difference between the two represents the *net surface flux*.

Emission processes include volatilization of gases from the surface, mechanical lifting of particles by wind action, and forced injection of volatile and particulate material from combustion and volcanoes. Injection may take place at significant altitudes above the surface (smokestacks, volcanoes, large fires, aircraft) and this is implemented in atmospheric models as sources at the corresponding vertical model levels.

Dry deposition of gases may involve absorption by liquid surfaces or adsorption to solid surfaces. Dry deposition of particles involves sticking to surfaces by diffusion, interception, and impaction. Very large particles are also removed by gravitational settling. *Fog deposition* is a special case of particle deposition in which fog droplets containing dissolved gases and particles are removed by settling or impaction on surfaces.

Emission and dry deposition may be coupled through surface processes. In the simplest such case, deposited gases and particles may be temporarily stocked at the surface and then re-emitted. There may also be biogeochemical, transport, and other processes that take place within the surface reservoir, causing the re-emitted species to be different from that deposited or to be re-emitted in a new location. Accounting for these processes requires that the atmospheric model be coupled to a model for the surface reservoir that tracks the material deposited, its transformations and transport, and the eventual emission. Coupled models for the atmosphere and surface reservoirs are called *Earth system models*, *global biogeochemical models*, or *multimedia models*, with the preferred terminology depending on their level of detail. Earth system models couple atmosphere and surface reservoirs in a global 3-D dynamical framework; global biogeochemical models generally have simpler (or absent) representations of transport; and multimedia models are often regional in scale and empirically based. Coupled models must still use a surface boundary condition for their atmospheric component, and the computations of emission and dry deposition follow the same approaches as atmosphere-only models.

In this chapter we first discuss the different processes emitting material to the atmosphere and their representation in models (Section 9.2). We then discuss the representation of dry deposition as a one-way uptake by the surface (Section 9.3). Finally, we discuss two-way exchange in which the surface provides both a source and a sink (Section 9.4).

9.2 Emission

Emissions in atmospheric models are usually provided by *bottom-up emission inventories* that calculate emissions from knowledge of the underlying processes. In these inventories, the emission flux E_i of species i is computed in general form as:

$$E_i = A \times F_i \times S_i \tag{9.1}$$

where A is the *activity rate* for the process driving the emission, F_i is an *emission factor* for species i that measures the amount of emission per unit of activity, and S_i is a scaling factor to account for local meteorological variables, surface properties, and other effects not included in the specifications of A and F_i. For example, emission of SO_2 from coal combustion may be computed as the product of an annual coal combustion rate (A), the amount of SO_2 emitted per unit mass of coal burned (F_i), and a seasonal scaling factor to account for changing power demand (S_i). Emission of ammonia from livestock manure may be computed as the product of the number of heads of livestock (A), a mean rate of ammonia emitted per head (F_i), and a temperature-dependent scaling factor (S_i). Scaling factors are often calculated within the atmospheric model at individual time steps to yield time-dependent emissions consistent with the local model environment.

The bottom-up approach provides a consistent framework to quantify emissions guided by our best knowledge of the driving processes. A given process may emit many different species, but the activity rate A is common to all. Information on A is obtained from socioeconomic, ecological, or other geographical databases. Emission factors F_i for the different species emitted by a given activity are typically estimated from field or laboratory experiments. Scaling factors S_i adjust the emissions to account for information that is not resolved in the activity rate databases, or for conditions in which emission factors differ from the base case F_i.

Bottom-up emission inventories give the total emission of a species as the sum of contributions from different activities. This enables atmospheric chemistry models to determine the contributions of different source types to atmospheric concentrations and to make future projections. For example, a bottom-up inventory for NO_x emissions with sector information for power plants and vehicles can be used to separate the contributions of these two source types to ozone pollution. Projections of future activity rates from a socioeconomic model can be used through the bottom-up approach to project future emissions and from there future atmospheric concentrations.

A defining feature of a bottom-up emission inventory is that it is not directly constrained by observed atmospheric concentrations. As a result, an atmospheric

Species	Terrestrial biogenic	Open fires	Ocean biogenic	Anthropogenic	Volcanic	Lightning	Mechanical	Total
NO_x (as N)	11	7	–	32	–	5	–	55
CO	80	460	20	610	–	–	–	1170
Methane	190	50	–	290	–	–	–	530
Isoprene	520	–	–	–	–	–	–	520
SO_2 (as S)	–	1	–	57	10	–	–	68
Ammonia	3	6	8	45	–	–	–	62
Black carbon (as C)	–	11	–	7	–	–	–	18
Dust	–	–	–	–	–	–	1500	1500
Sea salt	–	–	–	–	–	–	5000	5000

Table 9.1 Global emissions to the atmosphere (Tg a^{-1})

Typical estimates for circa 2015. Dash indicates a zero or negligible source.

model simulation driven by bottom-up emissions may simulate atmospheric concentrations that disagree with observations. Analysis of this disagreement may point to errors in the bottom-up emission estimates and the need to improve these estimates. We refer to atmospheric observations as providing *top-down constraints* on emissions.

Top-down constraints from atmospheric observations can be used to optimize emissions in two different ways. The first is to use observed surface air concentrations as boundary conditions for the atmospheric model. This completely ignores bottom-up information on emissions, and is often done in the case of long-lived gases for which atmospheric concentrations are known better than emissions. The emissions can then be diagnosed from the atmospheric model implicitly by mass balance (i.e., to balance the loss computed by the model). The second, more general way to use top-down information from atmospheric observations is to apply correction factors to the bottom-up emissions in order to match the observations. This can be done by statistical optimization using various *inverse modeling* methods (see Chapter 11). Top-down correction factors applied to the bottom-up inventories improve by design the simulation of observed atmospheric concentrations, but can be difficult to interpret in terms of the underlying processes because they are statistical fits with no intrinsic physical meaning. Ultimately, the best use of top-down constraints is to guide improvements in the bottom-up inventories.

We describe here standard methods to produce bottom-up emission inventories for different processes. Table 9.1 gives global emission estimates for selected species, with contributing processes broadly classified as terrestrial biogenic, open fires, oceanic, anthropogenic, volcanic, lightning, and mechanical. This classification follows standard practice in the atmospheric chemistry literature, but there are ambiguities and inconsistencies that need to be recognized. For example, terrestrial biogenic emissions associated with agriculture are classified as anthropogenic, but those affected by inadvertent human influence (such as nitrogen deposition) are generally not. Open fires are generally not classified as anthropogenic, although most are set by humans. Oceanic emissions are often biogenic, but are separated from terrestrial biogenic emissions because they are derived by different bottom-up methods. The following subsections cover terrestrial biogenic, open fire, volcanic, anthropogenic, and mechanical emissions in order. Oceanic emissions are generally computed as a two-way exchange process (Section 9.4). Lightning is coupled to deep convection and was covered in Section 8.10.

9.2.1 Terrestrial Biogenic Emissions

Biological organisms emit a wide range of volatile compounds through growth, metabolism, and decay. Photosynthesis and respiration are dominant processes. Photosynthesis converts CO_2 to molecular oxygen and releases volatile organic by-products. Respiration is either aerobic, in which molecular oxygen is converted to CO_2, or anaerobic, in which another oxidant such as nitrate or sulfate is used to oxidize organic carbon. Biogeochemical carbon models provide estimates of photo-synthesis and respiration rates, as well as related quantities such as net primary productivity (NPP). They also differentiate between *autotrophic* respiration by green plants and *heterotrophic* respiration by decomposers. Box 9.1 gives a summary of the major processes and rates.

Carbon fluxes are often used as activity rates in (9.1) to estimate the terrestrial biogenic emissions of other species with emission factors. Deriving emission factors for individual species requires field or laboratory measurements that must then be extrapolated to produce regional and global estimates. Meteorological variables such as light, temperature, and soil moisture often have a large effect on emissions and are applied in the model as local scaling factors.

We present here three basic algorithms to compute the terrestrial biogenic emis-sions of methane, nonmethane volatile organic compounds (NMVOCs), and NO_x in atmospheric models. Emissions of other species generally follow algorithms of similar structure.

Methane. The main natural source of methane is wetlands, where bacteria reduce organic carbon to methane under anaerobic conditions. Some of that methane is oxidized as it rises to the surface and encounters aerobic waters, while the rest escapes to the atmosphere. A simple formulation (Kaplan, 2002) expresses the methane emission rate [E, gCH_4 m^{-2} d^{-1}] in a given model grid square as a function of wetland fractional extent [W, m^2 m^{-2}], heterotrophic carbon respiration [R, gC m^{-2} d^{-1}], and an emission factor [F, gCH_4 gC^{-1}] dependent on temperature (T) and the depth of the water table (D):

$$E = W \times R \times F(T,D) \tag{9.2}$$

This formulation can be applied in models using gridded wetland and water table data available from satellites (Bloom *et al.*, 2010). More advanced formulations derive methane emissions from a full biogeochemical model (Figure 9.1; Riley *et al.*, 2011) or account for seasonal variation in the pool of organic carbon reducible to methane (Bloom *et al.*, 2012).

Box 9.1 — Terrestrial Carbon Cycle

Biogeochemical models of the terrestrial carbon cycle describe the flow of carbon as it is captured from the atmosphere by photosynthesis, transferred through different ecosystem pools, and eventually respired back to the atmosphere. The rate of photosynthesis by green plants is called the *gross primary productivity* (*GPP*). Some of the carbon fixed by green plants is respired by the plants themselves; this is called *autotrophic* respiration. The rest is transferred to other ecosystem pools through litter fall and plant mortality. The amount of carbon that is fixed by green plants and not autotrophically respired is the NPP. It represents the net source of carbon to the ecosystem from green plants. Most of that carbon is eventually consumed by *decomposers* (bacteria and other biota) through *heterotrophic* respiration. The net amount of carbon delivered to the ecosystem by green plants (NPP) and not consumed by decomposers is called the *net ecosystem productivity* (*NEP*). It represents the net accumulation of carbon in the undisturbed ecosystem. Disturbances such as fires, erosion, and harvest provide an additional sink for that carbon. The net accumulation of carbon after all these disturbances have been taken into account is called the *net biome production* (*NBP*).

Box 9.1 Figure 1 gives current global estimates of these different carbon fluxes. Half of the carbon fixed by green plants (GPP) is transferred to other ecosystem pools (NPP) while the rest is autotrophically respired. Eighty percent of the transferred carbon is respired by decomposers and the remaining 20% accumulates in the undisturbed ecosystem (NEP). Fires, erosion, and harvest balance most of the NEP. The residual NBP is 1.4 Pg C a^{-1}, just 1% of the GPP. The NBP represents the global build-up of terrestrial carbon, so that the terrestrial biosphere is not in steady state. This is of fundamental importance for our understanding of anthropogenic perturbation to the carbon cycle because it balances a significant part of the current fossil fuel source of CO_2 (6.4 Pg C a^{-1}).

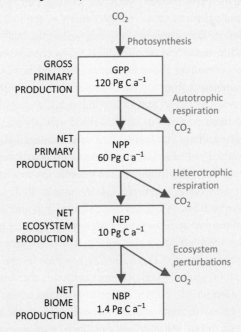

Box 9.1 Figure 1   Global flows in the terrestrial carbon cycle.

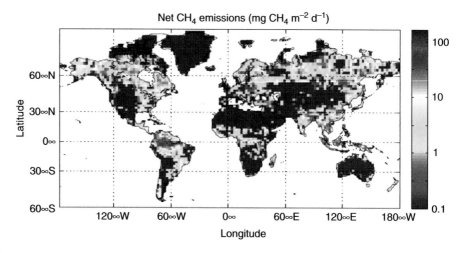

Annual emission of methane from wetlands (Riley *et al.*, 2011).

Nonmethane volatile organic compounds. Terrestrial plants are the largest global source of NMVOCs (Guenther *et al.*, 2006). Major species emitted by plants include isoprene, terpenes, sesquiterpenes, alkenes, carbonyls, and alcohols. They may be emitted as by-products of photosynthesis, as responses to injury, and from metabolism and decay. Emission fluxes depend on plant type, life stage (*phenology*), and foliage density; on radiative and meteorological variables within the canopy; and on external perturbations such as cutting, air pollution, and insect infestation. A standard measure of foliage density is the *leaf area index* (*LAI;* m^2 leaf per m^2 of land surface, counting only one side of the leaf). NMVOC flux measurements can be made at the leaf or plant level using chamber devices, at the canopy level from towers extending above the canopy top, and at the landscape level from aircraft. Flux measurements from towers and aircraft are generally made by *eddy correlation*, i.e., $E = \overline{w'C'}$ where w' and C' are the turbulent (residual) components from fast collocated measurements of vertical wind velocity and atmospheric concentrations. The measured fluxes, including their environmental dependences, are then extrapolated to produce regional and global emission inventories. The extrapolation is done with varying detail depending on the information available. It generally resolves different *plant functional types* (*PFTs*). A simple PFT classification might resolve only deciduous trees, evergreen trees, shrubs, and grasses. A more elaborate classification might resolve deciduous trees into tropical and temperate, broadleaf and fineleaf, etc.

The most advanced bottom-up emission inventories have been developed for isoprene ($CH_2=C(CH_3)-CH=CH_2$), which is the dominant NMVOC emitted by vegetation globally and accounts alone for about half of the global NMVOC source (Guenther *et al.*, 2006). Isoprene is produced in the chloroplasts of plants and is released to the atmosphere through leaf stomata. Emission only takes place in daytime when the stomata are open. Canopy emission fluxes depend on plant species, foliage density, leaf age, temperature, photosynthetically active radiation (PAR), and water stress. This is commonly represented in bottom-up emission models by multiplying base emissions E_o tabulated for each PFT under standard conditions with an ensemble of scaling factors describing the sensitivity to local environmental variables. In the MEGAN emission model (Guenther *et al.*, 2012), the canopy emission flux E for a given PFT is given as

$$E = E_o \times C_{CE} \times \Lambda \times \gamma_{PAR} \times \gamma_T \times \gamma_{AGE} \times \gamma_{SM} \tag{9.3}$$

where E_o is the base emission per unit area of Earth surface under standard conditions, Λ is the LAI, and the dimensionless scaling factors γ describe the sensitivity to above-canopy radiation (*PAR*), surface air temperature (*T*), leaf age distribution (*AGE*), and soil moisture (*SM*). The coefficient C_{CE} enforces $E = E_o$ under standard conditions, which for MEGAN are defined as $T = 303$ K, PAR = 1000 μmol photons m^{-2} s^{-1}, $\Lambda = 5$ m^2 m^{-2}, a leaf age distribution of 80% mature, 10% growing, and 10% senescing, and a volumetric soil moisture of 0.3 m^3 m^{-3}. The total isoprene emission flux for a given model gridsquare is obtained by summing the contributions from all PFTs in that gridsquare. Bottom-up emission inventories for other species generally follow the same kind of algorithms as for isoprene, but with less sophistication.

Figure 9.2 shows the MEGAN base emissions E_o under standard conditions for Europe and for Central/North America. Values are high for tropical forests, the southeastern USA, and boreal forests, reflecting PFTs with strong potential for isoprene emission. Values are low in the US Midwest where crops are poor isoprene emitters. Figure 9.3 shows as examples of scaling factors the dependences of isoprene emission on air temperature (γ_T) and LAI ($\Lambda \times \gamma_{PAR}$), taken from Guenther *et al.* (2006). Emission depends both on the instantaneous temperature and on the temperature for the past ten days. The dependence on LAI would be linear were it not for canopy light extinction measured by γ_{PAR}. Because of this extinction, there is a saturation effect limited by the penetration of light in the canopy.

Combining base emissions and scaling factors through (9.3) yields the global mean distributions of isoprene emission in Figure 9.4. Emissions are highest in tropical forests because of elevated temperature, LAI, and PAR. Emissions at northern mid-latitudes show strong seasonality driven by phenology and temperature.

Nitrogen oxides. Nitrogen is essential to life and has an active biogeochemical cycle in terrestrial ecosystems. Specialized bacteria present in all ecosystems convert atmospheric nitrogen (N_2) to ammonia, a process called *biofixation*, and the resulting *fixed nitrogen* then cycles through the ecosystem. Fixed nitrogen can also be directly delivered to the ecosystem by fertilizer application or by deposition of atmospheric ammonia and nitrate. Biological processes that cycle nitrogen within the ecosystem include *assimilation* (conversion of inorganic nitrogen to biological material), *mineralization* (conversion of organic nitrogen to inorganic forms), *nitrification* (aerobic microbial oxidation of ammonium to nitrite and on to nitrate), and *denitrification* (anaerobic microbial reduction of nitrate to N_2). Volatile N_2O and NO are generated as by-products of nitrification and denitrification.

Emission fluxes of ammonia, N_2O, and NO from the terrestrial biosphere are of great interest for atmospheric chemistry. They are determined by the biogeochemical cycling of nitrogen. In turn, the deposition of ammonia and nitrate is an important source of nitrogen to the terrestrial biosphere. Ideally, the emissions would be computed in a coupled atmosphere–land model tracking the chemical cycling of nitrogen in the atmospheric and terrestrial reservoirs. Simpler parameterizations are generally used in atmospheric models in which emission is computed as a function of

(a)

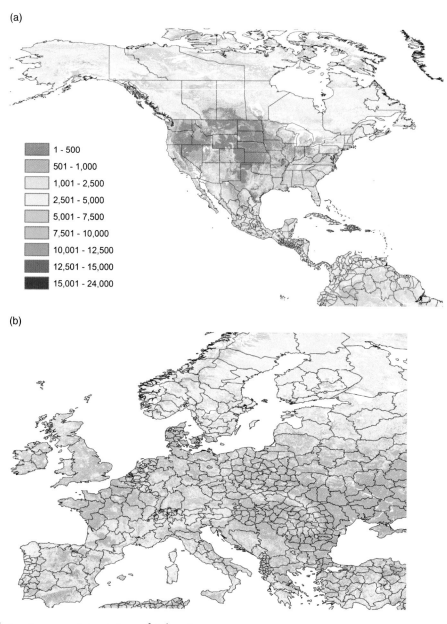

(b)

Figure 9.2 Base isoprene emission E_o [µg m^{-2} h^{-1}] under standard conditions in (a) Central/North America and (b) Europe. From A. Guenther and C. Wiedinmeyer, NCAR, personal communication.

soil nitrogen availability, temperature, and soil moisture. For example, Hudman *et al.* (2012) parameterize the soil emission E_i of NO for different biomes i as the product of functions describing respectively the dependences on soil nitrogen enrichment (N), temperature (T), soil moisture measured by the fraction of water-filled pore space (θ), and time since the last precipitation event (l):

$$E_i = f_{1,i}(N) \times f_2(T) \times f_3(\theta) \times f_4(l) \qquad (9.4)$$

(a)

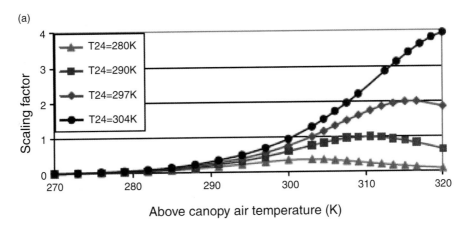

(b)

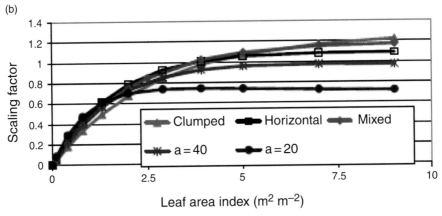

Figure 9.3 Scaling factors γ_T (a) and $\Lambda\gamma_{PAR}$ (b) in the MEGAN bottom-up isoprene emission inventory (9.3). Here, T_{24} is the air temperature for the past 1–10 days (assumed constant), and the calculation of γ_{PAR} is for two solar zenith angles (20° and 40°) and three leaf angular distributions (clumped, horizontal, and mixed). The figure shows how isoprene emission saturates as LAI exceeds 2 and light penetration inside the canopy becomes the limiting factor. Adapted from Guenther *et al.* (2006).

Here, N includes contributions from fertilizer input and atmospheric deposition (thus coupling emissions to deposition). The temperature function is a measure of microbial activity. The soil moisture function peaks for θ = 0.2–0.3; at lower values of θ bacterial activity is limited by water availability, while at higher values the clogging of soil pores leads to anaerobic conditions where emission of N_2O and N_2 dominates over emission of NO. The pulsing function f_4 describes the observed surge of emissions upon precipitation after an extended dry period (dry season), when water-stressed bacteria reactivate to mobilize excess nitrogen.

Figure 9.5 shows the global annual soil emissions of NO computed by Hudman *et al.* (2012) from (9.4). Emissions are high in agricultural areas of northern mid-latitudes, reflecting the heavy use of fertilizer. Dry grasslands in South America and Africa also have high emissions, largely driven by the pulsing at the end of the dry season. Some of the soil emissions of NO may be oxidized to NO_2 within the

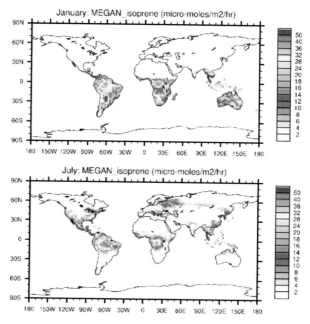

Figure 9.4 Global distribution of isoprene emission in January and July. From Guenther *et al.* (2012).

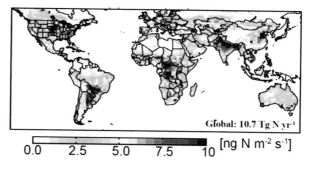

Figure 9.5 Annual emission of NO from soils. From Hudman *et al.* (2012).

canopy, and this NO_2 may then deposit to leaves, thus limiting export to the above-canopy atmosphere. However, the leaves may also be a source of NO_2. These canopy effects are very poorly understood and often not included in models.

9.2.2 Open Fires

Open fire emissions include contributions from wildfires, prescribed fires, land clearing, and agricultural management. These emissions are often labeled in the literature as *biomass burning*, but that leaves ambiguity as to whether biofuels are included. Most fires are set by humans, although some wildfires are triggered by lightning. Even when set by humans, fires are not generally classified as "anthropogenic" in emission inventories because they may have happened anyway even without human intervention. In fact, human intervention may be to suppress wildfires.

Chemical species	Savanna and grassland	Tropical forest	Extra-tropical forest	Crop residue	Pasture maintenance
Table 9.2 Emission factors for open fires					
CO_2	1686	1643	1509	1585	1548
CO	63	93	122	102	135
CH_4	1.9	5.1	5.68	5.82	8.71
NMVOCs	12.4	26	27	25.7	44.8
H_2	1.7	3.36	2.03	2.59	–
NO_x	3.9	2.55	1.12	3.11	0.75
N_2O	–	–	0.38	–	–
Organic aerosol	2.62	4.71	9.1	2.3	9.64
Black carbon	0.37	0.52	0.56	0.75	0.91

Emission factors [g kg^{-1}] for species emitted from combustion of different types of biomass. NO_x is given as NO. From the review of Akagi *et al.* (2011).

Fires emit mostly CO_2, CO, and H_2O, but also many other trace species. Emissions depend on the type of vegetation, the vegetation density, and the fire intensity. Fire information is usually available from satellites and ground surveys as area burned a_k over a time period Δt for a vegetation type k. From there, one can compute the emission rate $E_{i,k}$ of species i from the fire as the product of the area burned per unit time da_k/dt, the *fuel load* Γ_k [kg biomass m^{-2}], the fraction of fuel combusted or *burning efficiency* β_k, and an emission factor $F_{i,k}$ [g species emitted per kg fuel burned]:

$$E_{i,k} = F_{i,k} \times \beta_k \times \Gamma_k \times \frac{da_k}{dt} \qquad (9.5)$$

The burning efficiency depends on the fire intensity and on meteorological conditions, and also varies between different ecosystem components. Emission factors are determined from laboratory fire experiments or from sampling of fire plumes, generally using CO_2 as the normalization factor. They can be very different between successive *flaming* and *smoldering* stages of a fire. For example, NO_x emission factors are much higher in the flaming stage while CO emission factors are much higher in the smoldering stage. The different stages of a fire are generally not resolved in models because of lack of detailed temporal information. In most cases, models use mean emission factors compiled from data for different vegetation types (Table 9.2).

Figure 9.6 shows an inventory of CO emissions from open fires in September 2000. September is the end of the dry season in the southern tropics, and fire activity is particularly intense there. Most tropical fires are from agricultural management, in particular savanna burning. There is also a contribution from land clearing. Fires at northern mid-latitudes include contributions from wildfires (as in Siberia and Canada), prescribed burning (e.g., the southeast USA), and agricultural waste burning (e.g., West Asia).

Plumes from large fires are buoyant due to the heat released by combustion and can thus be lofted to the free troposphere above the PBL. This lofting is important to recognize in models because it affects the subsequent transport and chemistry of the fire plumes, and allows smoke particles to reach the free troposphere without being

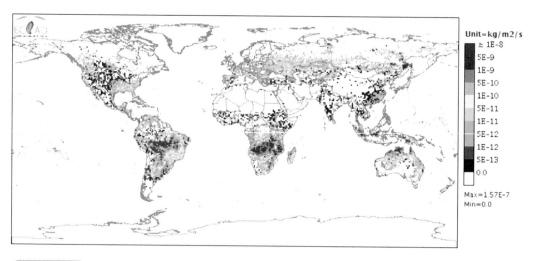

Figure 9.6   Emission of CO from open fires in September 2000. Source: ECCAD database (Granier *et al.*, 2011; Lamarque *et al.*, 2010).

scavenged by precipitation. The height reached by the plume is determined by the fire size and intensity, and by the thermodynamic stability of the background atmosphere. Latent heat release can increase the height reached by the plume and lead to the formation of deep convective clouds, a process called *pyroconvection*. A standard plume-rise formulation used in atmospheric models is that of Freitas *et al.* (2007). Figure 9.7 illustrates its application to a boreal fire in central Canada (large fire size and intensity) and a grassland fire in Texas (small fire size and intensity). The boreal fire plume rises to 3 km altitude while the grassland fire plume remains in the PBL.

9.2.3 Volcanoes

Volcanoes play a fundamental role in the cycling of elements on geologic timescales by transferring material from the lithosphere to the atmosphere. On the shorter perspective of atmospheric sources and sinks, volcanoes are of most interest as sources of ash and sulfur gases (mainly SO_2 and H_2S). Volcanoes often release material in the free troposphere. Large volcanic eruptions inject material into the lower stratosphere, and the resulting long-lived sulfate aerosol has important implications for climate and for stratospheric ozone.

Volcanic emissions can be non-eruptive or eruptive. Non-eruptive emissions are released at the volcano mouth while eruptive emissions are injected to higher altitude. Eruptive emissions are usually brief and variable, although some volcanoes can be in continuous eruption for many years. Worldwide databases of volcanic eruptions are available with eruption dates and strengths measured by the logarithmic *volcanic explosivity index* (*VEI*). The VEI is an integer measure that ranges from 0 (non-explosive) to 8 (colossal). Volcanic emissions and injection heights are commonly assigned in models as a function of VEI or using direct observations. Satellite observations have greatly increased the ability to map volcanic SO_2 emissions (Schnetzler *et al.*, 2007; Figure 9.8).

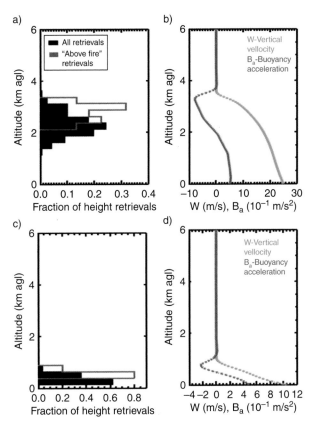

Figure 9.7 Plume rise from a large boreal forest fire in Canada (a and b) and from a small grassland fire in Texas (c and d). The left panels show the plume rise inferred from aerosol retrievals by the MISR satellite instrument. The right panels show results from the 1-D plume rise model of Freitas *et al.* (2007), where the point of zero vertical velocity marks the top of the plume. From Val Martin *et al.* (2012).

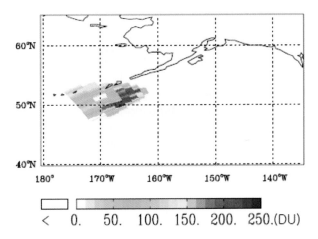

Figure 9.8 SO₂ plume from the Kasatochi volcanic eruption in the Aleutians observed by the OMI satellite instrument on August 8, 2008. From Wang *et al.* (2013).

9.2.4 Anthropogenic Emissions

Anthropogenic emissions span a wide range of processes of which combustion, industrial leaks, and agricultural activities are the most important. The "anthropogenic" label in the literature can be ambiguous and inconsistent. For example, some anthropogenic inventories include prescribed and agricultural fires while others do not. Anthropogenic inventories typically include emissions of ammonia from agricultural fertilizer, but may not include emissions of NO_x from the same process. Regional inventories may include emissions from aircraft in airports but not in the air. They may include ship emissions in ports but not at sea. Because of definitional problems such as these, care is needed when using anthropogenic emission inventories. It is important to ascertain which sources are included.

Anthropogenic emissions are usually better quantified than other emissions because activity rates are available as economic data and emission factors are documented for air quality management purposes. Emission inventories commonly distinguish between *area sources* and *point sources.* Area sources include vehicles and other individually small sources for which emissions are distributed over the activity area with best estimates of emission factors. Point sources are concentrated discharges from localized sources such as power plants. These emissions are often released by smokestacks hundreds of meters above the surface, and height information may be provided in the inventory. Large point sources may have continuous emission monitoring devices installed in their stacks to comply with air quality regulations, in which case the emissions are particularly well quantified.

Anthropogenic emission inventories are produced by various groups and agencies to serve air quality management and climate modeling needs. They may cover the whole world or limited geographical domains. Regulatory models used for air quality management typically construct their own highly detailed emission inventories over regional domains separating individual sources and with temporal resolution as fine as hourly. At the other end, many inventories are available only as gridded annual totals. In such cases, temporal information on emissions (diurnal, weekday/weekend, seasonal, interannual) needs to be independently provided using scaling factors.

Figure 9.9 shows as an example a global inventory of NO_x anthropogenic emissions in 2008. Emissions are mainly from fossil fuel combustion and peak in the densely populated regions of developed countries. Emissions over the oceans are from ships and aircraft.

9.2.5 Mechanical Emissions: Sea Salt and Dust

Wind stress on the Earth's surface causes mechanical emission of aerosol particles including sea salt, mineral dust, pollen, and plant debris. Sea salt and dust are dominant components of the coarse-mode (supermicron) aerosol over ocean and land, respectively, and generally make important contributions to total aerosol mass concentrations and optical depth. Pollen and plant debris have more localized influences.

Sea Salt Aerosol

Emission of sea salt particles is mostly driven by the entrainment of air into seawater by wave-breaking. The resulting air bubbles rise and burst at the sea surface,

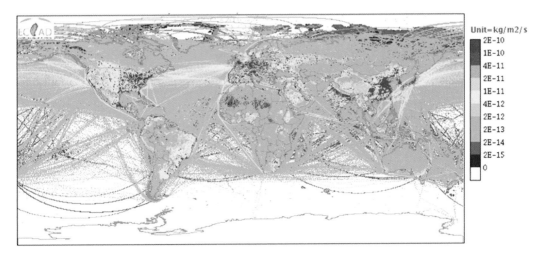

Figure 9.9 Anthropogenic NO$_x$ emissions from the MACCity inventory at 0.5° x 0.5° resolution for the year 2008. Courtesy: C. Granier, Centre National de la Recherche Scientifique (CNRS).

injecting particles into the air. The emission flux is a strong function of wind speed. A commonly used emission parameterization is that of Monahan *et al.* (1986), modified by Gong (2003) to better fit observations, and by Jaeglé *et al.* (2011) to include dependence on sea surface temperature (SST):

$$\frac{dE}{dr} = 1.373 \ u_{10}^{3.41} r^{-A} \left(1 + 0.057 r^{3.45}\right) \times 10^{1.607 \exp\left[-B^2\right]} \times g(T_C) \qquad (9.6)$$

with

$$
\begin{aligned}
A &= 4.7(1 + 30r)^{-0.017 r^{-1.44}} \\
B &= 1 - 2.31 \log r \\
g(T_C) &= 0.3 + 0.1 T_C + 0.0076 T_C^2 + 0.00021 T_C^3
\end{aligned}
\qquad (9.7)
$$

Here, dE/dr is the emission flux size distribution function [particles m^{-2} s^{-1} μm^{-1}] at 80% relative humidity (RH), r is the particle radius [μm] at 80% RH (about twice the dry radius), u_{10} [m s^{-1}] is the wind speed at 10 m above the surface, and T_C [°C] is the SST. Figure 9.10 shows the resulting number size distribution of the emitted particles, featuring a peak at 0.1 μm consistent with observations (Gong, 2003). The dependence of emissions on SST reflects the strong sensitivity of seawater viscosity to temperature: Warmer waters are less viscous, allowing for faster rise of small bubbles and hence a larger particle source.

Figure 9.11 shows the global mass flux of sea salt aerosol computed from (9.6) using assimilated meteorological data. Emission is highest at southern mid-latitudes where winds are strongest, though this maximum is mitigated by cold SSTs. Warm waters of the tropics have higher emission than would be computed solely from a wind speed dependence.

Mineral Dust

Mineral dust is emitted by sandblasting of soils, a process called *saltation*. Wind lifts large *sand* particles (diameter $D_s > 50$ μm) that travel over only short horizontal distances before falling back to the surface by gravity. As the sand particles fall, they

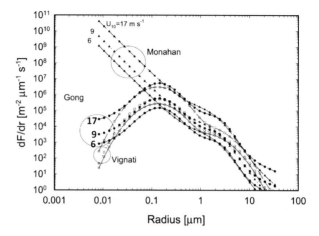

Emission flux size distribution function computed with the Gong (2003) parameterization for three different wind speeds (6, 9, 17 m s^{-1}) and compared to the parameterizations of Monahan *et al.* (1986) and Vignati *et al.* (2001). The original Monahan *et al.* (1986) parameterization features an increase in emission with decreasing radius below 0.1 μm that is inconsistent with observations. Adapted from Gong (2003).

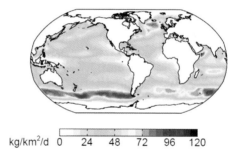

Annual mean mass emission flux of sea salt aerosol. From Jaeglé *et al.* (2011).

eject dust particles of diameter D_d small enough to be transported over long distances in the atmosphere. These fine particles are classified as clay ($D_d < 2$ μm) and silt ($2 < D_d < 50$ μm).

Experimental data show that the dust emission flux is proportional to the horizontal saltation flux from the transport of sand particles (Gillette, 1979). Over bare soils, the saltation flux Q [kg m^{-1} s^{-1}] can be expressed as a function of the friction velocity u_*[m s^{-1}] (defined in Chapter 8) and a threshold friction velocity u_{*_t} by

$$Q(D_s) = \frac{c \rho_a u_*^3}{g} \left(1 + \frac{u_{*_t}(D_s)}{u_*}\right) \left(1 - \frac{u_{*_t}^2(D_s)}{u_*^2}\right) S \qquad (9.8)$$

Here ρ_a is the air density [kg m^{-3}], g is the acceleration of gravity [m s^{-2}], S a preference source term, and c a constant of proportionality derived from wind-tunnel experiments and typically taken to be $c = 2.61$. The preference source term S accounts for accumulated erodible sediments in a given grid square due, for example, to topography or run-off areas; Ginoux et al. (2001) assume that large amounts of sediments accumulate primarily in valleys and depressions and adopt the empirical formulation

$$S = \left(\frac{z_{\max} - z}{z_{\max} - z_{\min}} \right)^5 \tag{9.9}$$

where z denotes the mean altitude of the model grid cell under consideration, while z_{min} and z_{max} represent the maximum and minimum elevations in the surrounding $10° \times 10°$ area (typical size of a hydrological basin). The threshold friction velocity u_{*t} is a function of the sand particle diameter D_s, and represents the capacity of the soil to resist wind erosion. Its value is determined by factors such as soil texture, soil moisture, and the presence of vegetation and other roughness elements (Xi and Sokolik, 2015). In the case of dry soils, u_{*t} has a value of about 0.2 m s^{-1} for $D_s = 100$ μm.

The emission flux of *dust* particles resulting from the bombardment of *saltating* particles (sand grains) of size D_s is calculated by assuming that the flux of dust [kg m^{-2} s^{-1}] corresponding to a particle size bin i ($i = 1, I$) of increment ΔD_i and mean diameter D_i, is given by

$$\hat{E}(D_i, \Delta D_i, D_s) = \alpha(D_i, \Delta D_i, D_s)Q(D_s) \tag{9.10}$$

The sandblast efficiency α [m^{-1}] can be derived from theoretical considerations (Shao, 2004) or from wind tunnel experiments. The dust emission from size bin D_i is then given by integrating $\hat{E}(D_i, \Delta D_i, D_s)$ between the lower and upper limits d_1 and d_2 of the size of the saltating particles

$$E(D_i, \Delta D_i) = \int_{d_1}^{d_2} \hat{E}(D_i, \Delta D_i, D_s)p(D_s)dD_s$$

Here $p(D_s)$ is the size distribution of the sand particles (often assumed to be a composite of log-normal distributions). The total emission rate of dust is obtained by summing the emission for all I bins:

$$E = \sum_{i=1}^{I} E(D_i, \Delta D_i)$$

Darmenova et al. (2009) review different physical parameterizations adopted in dust emission models.

The above formulation requires detailed information on soil characteristics that may not be available. Simpler formulations are used in global models (Ginoux *et al.*, 2001; Zender *et al.*, 2003). Ginoux *et al.* (2001, 2012) compute the dust emission flux as:

$$\begin{aligned} E &= S f_A u_{10}^2(u_{10} - u_{10,t}) \quad \text{for } u_{10} > u_{10,t} \\ &= 0 \quad \text{for } u_{10} < u_{10,t} \end{aligned} \tag{9.11}$$

Here, u_{10} denotes the 10 m wind speed, $u_{10,t}$ is a threshold, f_A is the fractional area of land suitable for saltation, and S is an adjustable global scaling factor to match dust observations. Dust emission in this formulation has a cubic dependence on wind speed, and is therefore controlled by gusty conditions that are poorly resolved in atmospheric models. The scaling factor S is intended to correct for this effect and varies with the model grid resolution. Global models typically choose S to yield a global dust emission of about 1500 Tg a^{-1} as this is found to provide a good fit to observations.

Figure 9.12 shows the global distribution of natural and anthropogenic dust emissions estimated by Ginoux *et al.* (2012). Natural emission is dominated globally by the Sahara and also has substantial contributions from the Middle East,

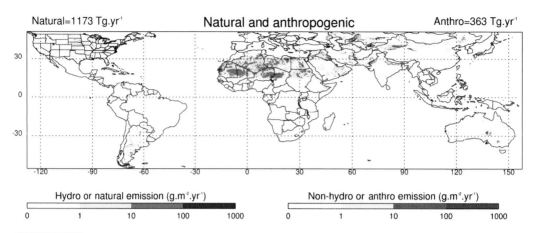

Figure 9.12   Annual mean dust emission from natural and anthropogenic sources. From Ginoux *et al.* (2012).

the Gobi desert, and the North American West. There are large anthropogenic dust emissions from dry and eroded agricultural areas.

9.3 One-Way Dry Deposition

Dry deposition or *surface uptake* is the process by which gases and particles are transferred from the atmosphere to the Earth's surface. It is a major sink for many atmospheric species. Except for very large particles, it does not take place by gravity, which is negligibly slow. It takes place instead by turbulent transfer to the surface followed by surface uptake. *One-way deposition* as described here assumes that the deposition is irreversible so that the surface is a terminal sink. Generalization to *two-way exchange* is presented in Section 9.4.

9.3.1 Dry Deposition Velocity

The dry deposition sink for a species i is computed as the *dry deposition flux* $F_{D,i}$ [molecules cm^{-2} s^{-1}] applied to the lowest altitude z_1 resolved by the model (lowest model grid point). Proper physical description requires that the dry deposition flux computed at z_1 represent the flux at the actual surface. This holds if z_1 is within the surface layer (Section 8.7.3), typically 50–100 m deep, where vertical fluxes can be assumed uniform. $F_{D,i}$ depends on the number density $n_i(z_1)$ at altitude z_1, the efficiency of vertical transfer from altitude z_1 to the surface, and the efficiency of loss at the surface. If the loss rate at the surface has a first-order dependence on the surface number density $n_i(0)$, as is usually the case, then the deposition flux has a first-order dependence on $n_i(z_1)$:

$$F_{D,i} = -w_{D,i}(z_1)n_i(z_1) \tag{9.12}$$

Here, $w_{D,i}(z_1)$ is the *dry deposition velocity* [cm s^{-1}] of species i at altitude z_1. It is called a "velocity" because of its units, but it describes in fact a turbulent process and not a simple one-way flow. One-way *gravitational settling* is important only for very large aerosol particles and is covered in Section 9.3.8. The flux is defined as positive when upward, thus the dry deposition flux in (9.12) is negative.

Conservation of the vertical flux in the air column below z_1 is an important assumption in the computation of dry deposition using (9.12). Aside from z_1 being in the surface layer, it requires that the atmospheric lifetime of the depositing species against chemical loss be long relative to the timescale for turbulent transfer from z_1 to the surface. The latter timescale is of the order of minutes for z_1 in the range 10–100 m. Shorter-lived species require finer vertical resolution near the surface to compute dry deposition, although one might be able to assume in those cases that dry deposition is negligible relative to chemical loss.

9.3.2 Momentum Deposition to a Flat Rough Surface

Insight into the deposition of chemical species can be gained from similarity to deposition of momentum, Consider the simple case of momentum deposition to a flat rough surface (Figure 9.13). Momentum is transported to the surface by turbulence. Turbulent eddies in the surface layer are sufficiently small that an eddy diffusion parameterization is adequate (Section 8.7.3). Let $\rho_a u$ be the mean scalar horizontal momentum where ρ_a is the air density and u is the mean horizontal wind speed. The momentum deposition flux F_m is related to the vertical gradient of the horizontal momentum by:

$$F_m = -K_z \frac{d\rho_a u}{dz} \approx -K_z \rho_a \frac{du}{dz} \qquad (9.13)$$

where we neglect the small variation of ρ_a with altitude. The eddy diffusion coefficient K_z has units $[\text{cm}^2 \text{ s}^{-1}]$ and needs to be empirically specified. Dimensionality considerations are helpful here. K_z can be viewed as the product of a length scale [cm] and a velocity scale $[\text{cm s}^{-1}]$. We expect K_z to increase with distance from the surface as eddies become less restricted by the surface boundary. Thus z is an appropriate length scale. We also expect K_z to increase as the momentum deposition flux increases, and this can be expressed in terms of the friction velocity $u_* = (|F_m|/\rho_a)^{1/2}$ introduced in Section 8.7.3. Therefore:

$$K_z = ku_* z \qquad (9.14)$$

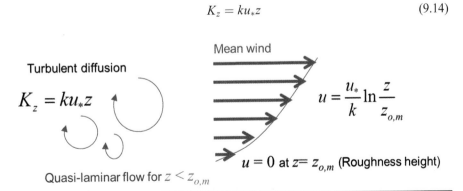

FLAT ROUGH SURFACE

Figure 9.13   Log law for the horizontal wind speed over a flat rough surface.

where $k = 0.35$ is the von Karman constant. Replacing (9.14) and the definition of the friction velocity into (9.13), we get:

$$du = \frac{u_*}{k}\frac{dz}{z} \qquad (9.15)$$

and by integration,

$$u = \frac{u_*}{k}\ln z + c \qquad (9.16)$$

where c is an integration constant. We see from the form of (9.16) that the mean wind speed must die out ($u = 0$) at some distance above the surface called the *roughness height for momentum* $z_{0,m}$. Applying this boundary condition to (9.16) we obtain the *log law* for the wind (equation (8.106)):

$$u = \frac{u_*}{k}\ln\frac{z}{z_{0,m}} \qquad (9.17)$$

Field observations show that this relationship is generally well obeyed. Plots of $\ln z$ vs. u from experimental data can be fitted to a straight line, and the values of u_* and $z_{0,m}$ can be derived from the slope and intercept. The thin layer $[0, z_{0,m}]$ close to the surface is viewed as a *quasi-laminar boundary layer* in which molecular diffusion plays an important role.

9.3.3 Big-Leaf Model for Dry Deposition

The formulation of momentum deposition to a flat rough surface (Section 9.3.2) provides a simple basis for parameterizing deposition of chemical species to a complex canopy. This parameterization is called the *big-leaf model* or *resistance-in-series model* (Hicks *et al.*, 1987). Figure 9.14 gives a schematic. From the atmospheric perspective, the canopy is modeled as a flat rough surface based at a displacement height d above the ground (Section 8.7.3). Depositing species are delivered to the surface by turbulent and quasi-laminar transfer (Section 9.3.2) and penetrate into the surface medium, where they are eventually removed. Think of the "big leaf" as a porous medium above which the airflow follows atmospheric

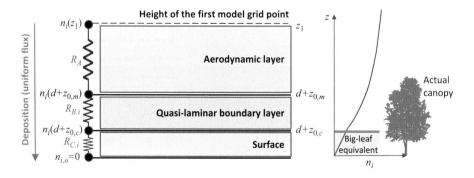

Figure 9.14 Schematic of the big-leaf model for one-way dry deposition. Vertical axis is not to scale.

dynamics for deposition to a flat rough surface (Section 9.3.2), and below which some combination of surface processes leads to the actual uptake. Vertical transport to the big-leaf surface takes place by atmospheric turbulence down to altitude $d + z_{0,m}$, and final transport to the surface through the quasi-laminar boundary layer is facilitated by molecular diffusion. The log law for the wind (9.17) needs to be adjusted for the displacement height (Section 8.7.3):

$$u = \frac{u_*}{k} \ln\left(\frac{z - d}{z_{0,m}}\right) \tag{9.18}$$

where the altitude z is relative to the actual Earth surface below the canopy. Typically d is about 2/3 of the canopy height, $z_{0,m}$ is about 1/30 of the canopy height, and u_* is about 1/10 of the wind speed. Assuming similarity between turbulent transport of chemicals and momentum, a similar log law applies to the vertical concentration profile of species i, but with a non-zero concentration as boundary condition at altitude $d + z_{0,c}$:

$$n_i(z) - n_i(d + z_{0,c}) = \frac{u_*}{k} \ln\left(\frac{z - d}{z_{0,c}}\right) \tag{9.19}$$

Here, $z_{0,c}$ is the roughness height for depositing species (assumed to be the same for all species) and $d + z_{0,c}$ is the effective height of the big-leaf surface. The quasi-laminar boundary layer is thus defined as the layer $[d + z_{0,m}, d + z_{0,c}]$, from the point where the wind dies out down to the effective surface. In one-way deposition the surface is a terminal sink for the depositing species, and this is enforced by a boundary condition $n_{i,o} = 0$ within the big-leaf medium (Figure 9.14).

Downward vertical transfer in the $[z_1, d + z_{0,m}]$ column takes place by turbulence; thus we write for that column:

$$F_{D,i} = -K_z(z)\, n_a(z)\, \frac{dC_i(z)}{dz} \approx -K_z(z)\frac{dn_i(z)}{dz} \tag{9.20}$$

where C_i is the mixing ratio of species i. The turbulent flux is proportional to the mixing ratio gradient in the eddy diffusion formulation, but we can neglect the vertical dependence of the air density n_a within the surface layer and write the flux as proportional to the number density gradient. Integration of equation (9.20) yields

$$F_{D,i} = -\frac{n_i(z_1) - n_i(d + z_{0,m})}{\displaystyle\int_{d+z_{0,m}}^{z_1} \frac{dz}{K_z(z)}} = -\frac{n_i(z_1) - n_i(d + z_{0,m})}{R_A} \tag{9.21}$$

where R_A [s cm^{-1}] is the *aerodynamic resistance* to deposition:

$$R_A = \int_{d+z_{0,m}}^{z_1} \frac{dz}{K_z(z)} \tag{9.22}$$

The term "resistance" reflects the analogy with electrical circuits, taking $n_i(z_1) - n_i(d + z_{0,m})$ as the analog of a difference in potential and $F_{D,i}$ as the analog of a current intensity.

Following on the analogy with electrical circuits, we can define a quasi-laminar boundary layer resistance $R_{B,i}$ [s cm^{-1}] (commonly called *boundary resistance*) to describe vertical transport through the quasi-laminar boundary layer:

$$F_{D,i} = -\frac{n_i(d + z_{0,m}) - n_i(d + z_{0,c})}{R_{B,i}} \tag{9.23}$$

and a surface resistance $R_{C,i}$ [s cm^{-1}] to describe the uptake at the surface:

$$F_{D,i} = -\frac{n_i(d + z_{0,c})}{R_{C,i}} \tag{9.24}$$

where $n_i(d + z_{0,c})$ is the concentration in contact with the surface. We combine (9.21), (9.23), and (9.24) to eliminate $n_i(d + z_{0,m})$ and $n_i(d + z_{0,c})$, and obtain:

$$F_{D,i} = -\frac{n_i(z_1)}{R_A + R_{B,i} + R_{C,i}} = -\frac{n_i(z_1)}{R_i} \tag{9.25}$$

where $R_i = R_A + R_{B,i} + R_{C,i}$ [s cm^{-1}] is the *total resistance* to dry deposition and is the inverse of the dry deposition velocity (9.12). Thus:

$$w_{D,i} = \frac{1}{R_i} = \frac{1}{R_A + R_{B,i} + R_{C,i}} \tag{9.26}$$

We see by analogy to Ohm's law that R_i is the sum of three resistances in series describing resistance to turbulent transport through the surface layer (R_A), resistance to diffusion through the quasi-laminar boundary layer ($R_{B,i}$), and resistance to surface uptake ($R_{C,i}$), as illustrated in Figure 9.14. By calculating the individual resistances we can derive the dry deposition velocity, and by comparing the magnitudes of the individual resistances we can determine the process limiting dry deposition. In the following subsections we describe the calculation of the individual resistances.

9.3.4 Aerodynamic Resistance

Equation (9.22) expresses R_A as a function of the eddy diffusion coefficient K_z. For flow over a flat rough surface, we have $K_z = ku_*z$ (Section 9.3.2), and correcting for the displacement height yields $K_z = ku_*(z - d)$. Replacing into (9.22):

$$R_A = \int_{d+z_{0,m}}^{z_1} \frac{dz}{ku_*(z - d)} = \frac{1}{ku_*} \ln\left(\frac{z_1 - d}{z_{0,m}}\right) \tag{9.27}$$

This expression applies for neutral buoyancy conditions when the log law for the wind holds. The atmosphere can be assumed neutral when mechanical turbulence dominates over buoyant turbulence, that is when $z_1 \ll |L|$ where L is the Monin–Obukhov length (Section 8.7.3). When this condition is not satisfied, a stability correction factor Ψ_m must be introduced in the formulation of the vertical wind profile as given in Section 8.7.3 (see also expression (9.18)):

$$u = \frac{u_*}{k}\left[\ln\left(\frac{z - d}{z_{0,m}}\right) - \Psi_m\left(\frac{z - d}{L}\right)\right] \tag{9.28}$$

and the expression for the aerodynamic resistance becomes:

$$R_A = \frac{1}{k\,u_*} \left[\ln\left(\frac{z_1 - d}{z_{0,m}}\right) - \Psi_m\left(\frac{(z_1 - d)}{L}\right) \right] \tag{9.29}$$

Correction formulas are generally applicable up to $z \approx |L|$. At higher altitudes, the parameterization of turbulence becomes more complicated as buoyant plumes dominate and the surface layer assumption of uniformity of vertical fluxes may not be valid. Values of $|L|$ generally exceed 100 m so that a lowest model level $z_1 < 100$ m is adequate. Very unstable conditions can have smaller values of $|L|$, but the aerodynamic resistance in the $[|L|, z_1]$ column is then negligibly small and R_A in (9.29) can be calculated by replacing z_1 with $|L|$. Very stable conditions at night can lead to ground-based inversions and very small positive values of L. In that case the aerodynamic resistance computed at $z_1 > L$ is very large, deposition is restricted to the shallow layer $[0, L]$, and the concentration at z_1 is decoupled from that in surface air. It may be best from the model perspective to ignore deposition under such conditions as it operates only on a small atmospheric mass. One should not expect then for the model to be able to reproduce surface observations.

9.3.5 Quasi-Laminar Boundary Layer Resistance

The quasi-laminar boundary layer resistance (boundary resistance) $R_{B,i}$ in the big-leaf model measures the resistance to transfer from the zero-momentum point at altitude $d + z_{0,m}$ to the big-leaf surface at altitude $d + z_{0,c}$. Even though turbulence technically dies out at $d + z_{0,m}$ in the eddy diffusion parameterization for momentum, there is in reality still some turbulence to carry species down to the surface. A first estimate of $R_{B,i}$ can thus be made from (9.27):

$$R_{B,i} = \int_{d+z_{0,c}}^{d+z_{0,m}} \frac{dz}{ku_*(z - d)} = \frac{1}{ku_*} \ln\left(\frac{z_{0,m}}{z_{0,c}}\right) \tag{9.30}$$

Molecular diffusion also plays a significant role in the thin quasi-laminar boundary layer, and the corresponding rate depends on the molecular diffusion coefficient D_i. This can be accounted for by the semi-empirical correction of Hicks *et al.* (1987):

$$R_{B,i} = \frac{1}{ku_*} \ln\left(\frac{z_{0,m}}{z_{0,c}}\right) \left(\frac{Sc_i}{Pr}\right)^{2/3} \tag{9.31}$$

where the *Schmidt number* $Sc_i = \nu/D_i$ is the ratio between the kinematic viscosity of air ($\nu = 0.15$ cm^2 s^{-1} at standard temperature and pressure) and the molecular diffusion coefficient D_i. The *Prandtl number* Pr is the ratio of the kinematic viscosity to the thermal diffusivity of air ($Pr = 0.72$ at standard temperature and pressure). The term $\ln(z_{0,m}/z_{0,c})$ is roughly 2 for vegetated canopies and 1 for bare surfaces and water. The boundary resistance computed in this manner is a very crude approximation, but is of little importance for computing the deposition velocity for gases since comparison of (9.31) to (9.27) indicates that $R_A \gg R_{B,i}$.

In the case of aerosol particles, the molecular diffusion coefficient must be replaced by the *Brownian diffusion coefficient* describing the random motion of particles. The Brownian diffusion coefficient is inversely dependent on particle size. For particles larger than ~0.1 μm, Brownian diffusion is very small, but transfer to the surface is then facilitated by *interception* (when particles carried by the airflow hit the surface) and *inertial impaction* (when particles deviate from the airflow as it curves around surface elements). There is detailed theory for these aerosol processes (Slinn, 1982; Seinfeld and Pandis, 2006) though practical application is limited by complexity of the canopy. Interception and impaction are generally the limiting factors for dry deposition of >0.1 μm aerosol particles, but the corresponding resistance is generally referred to as surface resistance in the literature. Thus for aerosol particles of diameter D_p the deposition velocity is typically computed as $w_D(D_p) = 1/(R_A + R_C(D_p))$ where the surface resistance $R_C(D_p)$ accounts for Brownian diffusion, interception, and impaction. Seinfeld and Pandis (2006) give a detailed discussion of these processes. The parameterization for $R_C(D_p)$ by Zhang *et al.* (2001) is frequently used in models.

9.3.6 Surface Resistance

The surface resistance $R_{C,i}$ in the big-leaf model describes the physical and chemical uptake taking place on the ensemble of canopy surfaces. For aerosol particles, collision with surfaces takes place by Brownian diffusion, interception, and impaction (Section 9.3.5). For gases, the uptake involves surface adsorption or absorption followed by chemical reaction. Deposition of gases can take place to the *stomata* (open pores) of leaves, within which gases diffuse to eventually react in the leaf *mesophyll*. It can also take place to the waxy surfaces of leaves, called *cuticles*, and to the ground and other surfaces.

The overall surface resistance is commonly decomposed into processes representing uptake by different canopy elements and parameterized as an ensemble of resistances in parallel and in series. Figure 9.15 from Wesely and Hicks (2000) shows a standard scheme. In that scheme, uptake by the canopy takes place in parallel to the canopy leaves, the lower canopy, and the ground. Uptake by canopy leaves takes place in parallel to the stomata and to the cuticles, and uptake by the stomata is described by two resistances in series representing diffusion through the stomata and reaction at the mesophyll. Uptake to the lower canopy and to the ground involves aerodynamic resistance to transfer through the canopy. The overall surface resistance $R_{C,i}$ is computed from this network of resistances by adding resistances in series, and adding conductances (inverses of resistances) in parallel, as one would do for electrical resistances. Wesely and Hicks (2000) and other literature provide estimates of each resistance in Figure 9.15 for different canopy types, gas chemical properties (e.g., effective Henry's law constant, oxidant potential), and meteorological variables.

Table 9.3 gives surface resistances for SO_2 and ozone computed with the Wesely and Hicks (2000) model for a deciduous forest in summer (full canopy) and winter (no leaf canopy), during day and night. SO_2 and ozone are commonly used as reference species for dry deposition because relatively large observational databases

Table 9.3 Surface resistances $R_{C,i}$ (s cm^{-1}) for a deciduous forest canopy

Species		Day	Night
SO$_2$	summer	1.3	10
	winter	9.8	10
O$_3$	summer	1.1	9.5
	winter	6.1	30

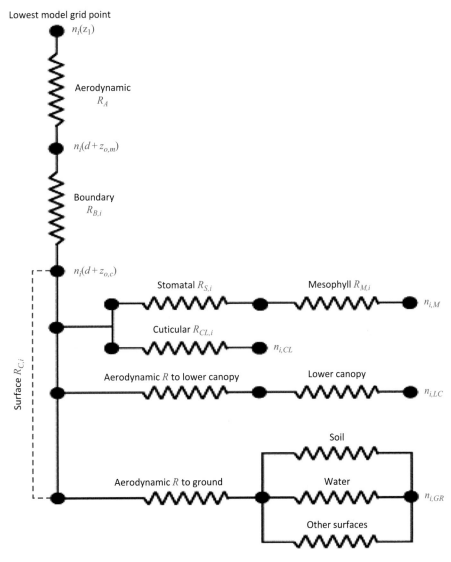

Figure 9.15 Surface resistance model from Wesely and Hicks (2000) separating contributions from vegetation, lower canopy, and the ground. The leaf mesophyll, lower canopy vegetation, and ground all have their own internal concentrations ($n_{i,M}$, $n_{i,LG}$ $n_{i,GR}$) as boundary conditions instead of the single concentration $n_{i,o}$ given in Figure 9.14. For one-way deposition these boundary conditions are all set to zero. Adapted from Wesely and Hicks (2000).

are available for both. Uptake of SO_2 is driven by its effective water solubility while uptake of ozone is driven by its reactivity as an oxidant. Uptake is particularly efficient at the leaf stomata, where water-soluble gases dissolve in the leaf water and oxidants react with unsaturated organic compounds. The overall surface resistances for SO_2 and ozone are much smaller in summer than in winter and much smaller in daytime than at night, reflecting the importance of the stomata, which are open only during daytime.

An illustrative back-of-the-envelope estimate of the surface resistance can be made for deposition of a highly water-soluble or reactive species to a leafy canopy in the daytime. In that case, leaves account for most of the total depositing surface in the canopy and reaction at the mesophyll is fast ($R_{M,i} \approx 0$ in Figure 9.15). Thus $R_{C,i} \approx R_{S,i}$, where $R_{S,i}$ is the stomatal resistance for the whole leaf canopy. Measurements of the stomatal resistance $R_{s,w}$ for water vapor exchange per unit area of leaf indicate a typical value of 2 s cm^{-1}. The corresponding stomatal resistance $R_{s,i}$ for species i per unit area of leaf scales to that of water by the inverse ratio of molecular diffusion coefficients, and the molecular diffusion coefficients are in turn inversely proportional to the square root of the molecular weights. Thus we have:

$$R_{C,i} \approx \frac{R_{s,w}}{\Lambda} \left(\frac{M_i}{M_w} \right)^{1/2} \tag{9.32}$$

where Λ is the LAI introduced in Section 9.2.1, and M_i and M_w are the molecular weights of species i and water vapor, respectively. Taking ozone as an example and a typical mid-latitudes forest LAI of 3, we obtain $R_{C,O3} = 1.1$ s cm^{-1}, which is the value in Table 9.3.

9.3.7 Factors Controlling the Dry Deposition Velocity

The deposition velocity of a gas as described by the resistance-in-series model (9.26) can be limited by aerodynamic transfer if $R_A \gg R_{C,i}$ or by surface uptake if $R_A \ll R_{C,i}$. It is never limited by transfer in the quasi-laminar boundary layer because $R_A \gg R_{B,i}$ in all cases. Whether aerodynamic transfer or surface uptake is limiting depends on species properties, canopy properties, and atmospheric stability. Gases with weak surface reactivity have low deposition velocities generally limited by the surface resistance. At the other extreme, strong acids like HNO_3 have zero surface resistance and their deposition velocity is always limited by aerodynamic transfer. Aerosol particles have highly variable deposition velocities depending on their size and on the canopy structure. Deposition velocities are smallest for particles in the 0.1–1 μm range. Smaller particles are efficiently removed by Brownian diffusion, while larger particles are efficiently removed by inertial impaction.

Deposition velocities over land vary strongly between day and night as driven both by atmospheric stability and by surface resistance (Table 9.3). This is illustrated in Figure 9.16 with measured ozone deposition velocities above a forest canopy. The deposition velocity is low at night when the atmosphere is stable and the stomata are closed. It increases rapidly at sunrise when the stomata start to open, and peaks in midday when the stomata are most open and the atmosphere is unstable. A mean afternoon decline in dry deposition is often observed due to increasing cloudiness

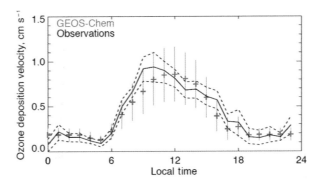

Figure 9.16 Diurnal variation of the ozone dry deposition velocity over a pine forest in North Carolina, April 15 to May 15, 1996. Mean observations and standard errors from Finkelstein *et al.* (2000) are compared to mean values from the GEOS-Chem model using a resistance-in-series parameterization. Model standard deviations describing day-to-day variability are also shown. From Katherine Travis, Harvard, personal communication.

resulting in partial stomatal closure. Also shown in Figure 9.16 are model values computed with a standard big-leaf resistance-in-series scheme and including variability driven by temperature and solar radiation.

Figure 9.17 illustrates the geographical and seasonal variations in ozone dry deposition velocity as calculated from a global model. Values are much lower over ocean than over land because ozone is poorly soluble in water. Values are also much lower in winter than in summer due to the absence of a leaf canopy and the suppression of deposition by snow.

9.3.8 Gravitational Settling

Gravitational settling is an important contributor to the deposition velocity in surface air only for aerosol particles larger than about 10 μm. The gravitational settling velocity near the Earth's surface is of the order of 1 cm s^{-1} for a 10 μm particle and 0.01 cm s^{-1} for a 1 μm particle. Gravitational settling is more important in the free troposphere and stratosphere, where vertical motions are otherwise slow and the settling velocity is higher than in surface air because of lower atmospheric pressure. Because of this, it is important to add gravitational settling as a term in the continuity equation for particles larger than about 1 μm (Chapter 4). Here we present equations for the gravitational settling velocity of particles that are applicable both for computing deposition at the surface and vertical motion through the atmosphere.

The settling velocity w of a particle of mass m_p is determined by equilibrium between gravity and drag:

$$m_p \frac{dw}{dt} = m_p g - F_{drag} \qquad (9.33)$$

The drag is given by:

$$F_{drag} = \frac{1}{2} C_D \, a_p \, \rho_a \, w^2 \qquad (9.34)$$

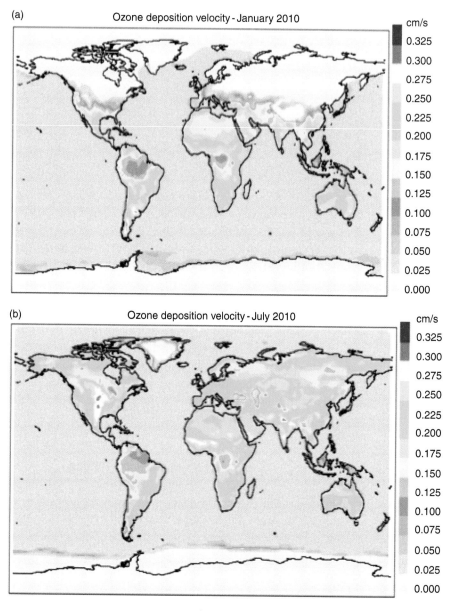

Figure 9.17 Monthly mean ozone deposition velocity [cm s^{-1}] in January and July calculated in the model of Lamarque *et al.* (2010).

where a_p is the projected area of the particle normal to the flow, ρ_a is the air density, and C_D is an empirical drag coefficient. For spherical aerosol particles ($a_p = \pi\, D_p^2/4$):

$$F_{drag} = \frac{1}{8}\, \pi\, C_D\, \rho_a\, D_p^2\, w^2 \tag{9.35}$$

where D_p is the particle diameter. For particles that are not very large relative to the mean free path of air molecules ($\lambda = 0.065$ μm at 298 K and 1 atm, but $\lambda = 0.42$ μm at

215 K and 100 hPa in the lower stratosphere), a dimensionless *slip-correction* factor C_c must be introduced to account for non-continuum effects:

$$F_{drag} = \frac{\pi \, C_D \, \rho_a \, D_p^2 \, w^2}{8 \, C_c} \tag{9.36}$$

with

$$C_c(D_p) = 1 + \frac{2\lambda}{D_p} \left[1.257 + 0.4 \exp\left(-0.55 \frac{D_p}{\lambda} \right) \right] \tag{9.37}$$

This correction factor decreases the drag and therefore increases the settling velocity. The mean free path is computed as

$$\lambda = \frac{2\mu}{p(8M_a/\pi \mathcal{R} T)^{1/2}} \tag{9.38}$$

where p is the atmospheric pressure, M_a is the molecular weight of air, $\mathcal{R}$ is the ideal gas constant, T is the absolute temperature [K], and μ is the dynamic viscosity [kg m^{-1} s^{-1}] given by

$$\mu = \mu_o \left(\frac{T_o + 120}{T + 120} \right) \left(\frac{T}{T_o} \right)^{3/2} \tag{9.39}$$

where $T_o = 298$ K and $\mu_o = 1.8 \times 10^{-5}$ kg m^{-1} s^{-1}.

The *terminal settling velocity* w_s of a particle, obtained from equilibrium between gravity and drag ($dw/dt = 0$ in (9.33) is given by

$$w_s = \left[\frac{4}{3} \frac{g \, \rho_p C_c D_p}{C_D \, \rho_a} \right]^{\frac{1}{2}} \tag{9.40}$$

where ρ_p is the mass density of the particle. The drag is a function of the Reynolds number Re:

$$Re = \frac{w \, D_p}{v} = \frac{w \, D_p}{\mu/\rho_a} \tag{9.41}$$

For near-surface conditions (1 atm and 298 K), the Reynolds number is less than 0.1 for D_p smaller than 20 μm. Under low Reynolds numbers ($Re < 0.1$), the drag coefficient can be expressed as $C_D = 24/Re$, and w_s is then given by

$$w_s = \frac{\rho_p \, D_p^2 \, g \, C_c}{18\mu} \tag{9.42}$$

At higher Reynolds numbers the following equations for the drag apply:

$$
\begin{aligned}
C_D &= \frac{24}{Re} \left[1 + \frac{3}{16} Re + \frac{9}{160} Re^2 \ln (2Re) \right] \qquad 0.1 < Re < 2 \\
C_D &= \frac{24}{Re} \left[1 + 0.15 Re^{0.687} \right] \qquad\qquad\qquad\quad 2 < Re < 500
\end{aligned}
\tag{9.43}
$$

The terminal settling velocity must then be calculated iteratively using (9.40) and (9.43) to account for the dependence of C_D on Re, and hence w_s.

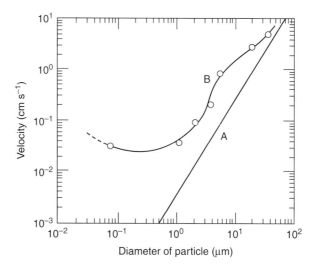

Diameter of particle (μm)

Curve A: Gravitational settling velocity of aerosol particles as a function of particle diameter.
Curve B: Deposition velocity of aerosol particles onto a grass surface. From Hobbs (2000).

Figure 9.18 shows typical gravitational settling velocities as a function of particle size and compares these velocities to measurements of particle dry deposition velocity to a grass surface. Gravitational settling accounts for 10% of the overall deposition velocity for 1 μm particles and 20% for 10 μm particles (typical of fog). Even though the gravitational settling velocity increases by two orders of magnitude from 1 to 10 μm, removal by inertial impaction also becomes more efficient. Gravitational settling dominates deposition for particles larger than 30 μm (for context, the diameter of a small raindrop is 100 μm).

9.4 Two-Way Surface Flux

The one-way deposition model described in Section 9.3 assumes that the surface is a terminal sink for depositing species. This assumption is expressed in the big-leaf model by the boundary condition of a zero concentration in the surface reservoir (Figure 9.14 and equation (9.24)). Consider instead as boundary concentration a non-zero concentration $n_{i,o}$ in the surface reservoir. Equation (9.24) then becomes

$$F_{D,i} = -\frac{n_i(d + z_{0,c}) - n_{i,o}}{R_{C,i}} \tag{9.44}$$

Combining (9.21), (9.23), and (9.44) yields:

$$F_{D,i} = -w_{D,i}[n_i(z_1) - n_{i,o}] \tag{9.45}$$

We see that a non-zero concentration within the surface reservoir (called a *compensation point*) implies a surface emission flux $w_{D,i}\, n_{i,o}$ offsetting the deposition flux –

$w_{D,i}\, n_i(z_1)$. The emission is subject to the same resistances to transfer as deposition. $w_{D,i}$ is then called a *transfer velocity, exchange velocity,* or *piston velocity* rather than a deposition velocity.

Proper representation of the non-zero compensation point in an atmospheric model depends on the nature of the source that maintains this compensation point. If the source is atmospheric deposition, this means that reaction within the surface reservoir is not sufficiently fast for the surface to be a terminal sink; re-emission to the atmosphere is a competing pathway. In that case, $n_{i,o}$ is dependent on $n_i(z_1)$, and a relationship between the two must be specified. This may be as simple as assuming a fixed proportionality, e.g., $n_{i,o} = sKn_i(z_1)$ where K is an equilibrium constant between the surface reservoir and the atmosphere (such as Henry's law for an air–water interface) and s is a saturation ratio. Or it may be as complex as a full biogeochemical model for the surface reservoir in which the gross deposition flux $-w_{D,i}n_i(z_1)$ is an input and the surface emission flux $w_{D,i}n_{i,o}$ is an output. The atmospheric model must then be coupled to the biogeochemical model.

Frequently, however, the compensation point can be considered to be independent of atmospheric deposition. This occurs when production within the surface reservoir dominates over the supply from atmospheric deposition. In such cases, the gross deposition flux $-w_{D,i}\, n_i(z_1)$ and the surface emission flux $w_{D,i}\, n_{i,o}$ are decoupled: the gross deposition flux is determined by the atmospheric concentration while the emission flux is not, so they are best computed and diagnosed as separate quantities. The surface concentration $n_{i,o}$ may be specified from observations or computed with a biogeochemical model for the surface reservoir. The gross deposition flux is the relevant sink to the surface from the perspective of the atmospheric budget, and the surface emission flux is the relevant source.

In the calculation of two-way exchange by (9.45), the same exchange velocity $w_{D,i}$ is used to compute gross deposition $-w_{D,i}n_i(z_1)$ and surface emission $w_{D,i}n_{i,o}$. This reflects the conservation of the vertical flux between z_1 and the point in the surface reservoir where the concentration $n_{i,o}$ is specified. The resistance-in-series model described in Section 9.3 for one-way deposition can thus be adapted to two-way exchange simply by specifying a non-zero $n_{i,o}$ at the surface reservoir endpoint. For example, the formulation of $R_{C,i}$ in Figure 9.15 includes three surface reservoir endpoints: inside the leaf mesophyll, at the lower canopy surface, and at the ground surface. In the one-way deposition model, concentrations at these endpoints are taken to be zero. Two-way exchange can be simulated by substituting non-zero values. A non-zero concentration is often specified in the leaf mesophyll to represent emission from leaves.

A common application of two-way surface exchange is the *two-film model* for the air–sea interface (Liss, 1973). In this case, the two-way exchange problem is relatively well posed. A single endpoint concentration $n_{i,o}$ in the bulk near-surface seawater can be specified from ship observations or from an ocean biogeochemistry model. The air–sea equilibrium is characterized by Henry's law. Transfer across the air–sea interface can be characterized by two resistances in series, one for the gas phase and one for the water phase. The two-film model as generally formulated in the literature follows standard conventions from the oceanography community. Thus vertical transfer in the gas and water phases is measured by *conductances* $k_{G,l}$

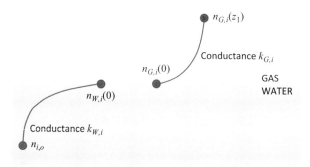

$n_{G,i}(z_1)$

Conductance $k_{G,i}$

$n_{G,i}(0)$

GAS
WATER

$n_{W,i}(0)$

Conductance $k_{W,i}$

$n_{i,o}$

Figure 9.19   Two-film model for air–sea exchange.

and $k_{W,i}$ that are the inverse of resistances, and the Henry's law equilibrium constant H_i is defined as the dimensionless ratio of air to water concentrations (in contrast, atmospheric chemists generally define the Henry's law constant as the ratio of water to air concentrations in units of M atm^{-1}).

Figure 9.19 is a schematic of the two-film model. Conservation of the vertical flux F_i is assumed between the lowest atmospheric model point z_1 and the bulk water phase where a concentration $n_{i,o}$ is specified:

$$F_i = k_{G,i}[n_{G,i}(0) - n_{G,i}(z_1)] = k_{W,i}[n_{i,o} - n_{W,i}(0)] \tag{9.46}$$

where $n_{G,i}$ and $n_{W,i}$ refer to the concentrations in the gas and water phases respectively. Application of the effective Henry's law constant $H_i = n_{G,i}(0)/n_{W,i}(0)$ at the air–sea interface allows us to express the flux in terms of bulk concentrations only:

$$F_i = K_i[H_i\, n_{i,o} - n_{G,i}(z_1)] \tag{9.47}$$

where K_i [cm s^{-1}] is the air–sea exchange velocity obtained by adding the gas-phase and water-phase conductances in parallel:

$$\frac{1}{K_i} = \frac{1}{k_{G,i}} + \frac{H_i}{k_{W,i}} \tag{9.48}$$

The marine atmosphere has near-neutral stability with a roughness height determined by wind-driven waves. It follows that turbulent mass transfer can be parameterized as a function of wind speed only, and the wind at 10-m height (u_{10}) is used for that purpose. Molecular diffusion at the interface depends on the Schmidt number Sc_i, which is different in the air and water phases. Johnson (2010) gives a detailed review of different parameterizations for $k_{G,I}$ and $k_{W,i}$, shown in Figure 9.20 as a function of wind speed. A simple expression for $k_{G,I}$ is that of Duce *et al.* (1991):

$$k_{G,i} = \frac{u_{10}}{770 + 45M_i^{1/3}} \tag{9.49}$$

where $k_{G,I}$ and u_{10} have the same units [m s^{-1}] and M_i is the molecular weight in [g mol^{-1}]. On the water side, the parameterization of Nightingale *et al.* (2000) is often used:

$$k_{W,i} = \left(0.222\, u_{10}^2 + 0.333\, u_{10}\right)\left[\frac{Sc_{W,i}}{600}\right]^{0.5} \tag{9.50}$$

Table 9.4 Freshwater Henry's law constants expressed as dimensionless gas/water concentration ratios

Species	Henry's law constant H_i (dimensionless)
O_2	3.2×10^1
CO_2	1.2×10^0
Dimethylsulfide	8.2×10^{-2}
Acetone	1.4×10^{-3}
H_2O_2	4.6×10^{-6}

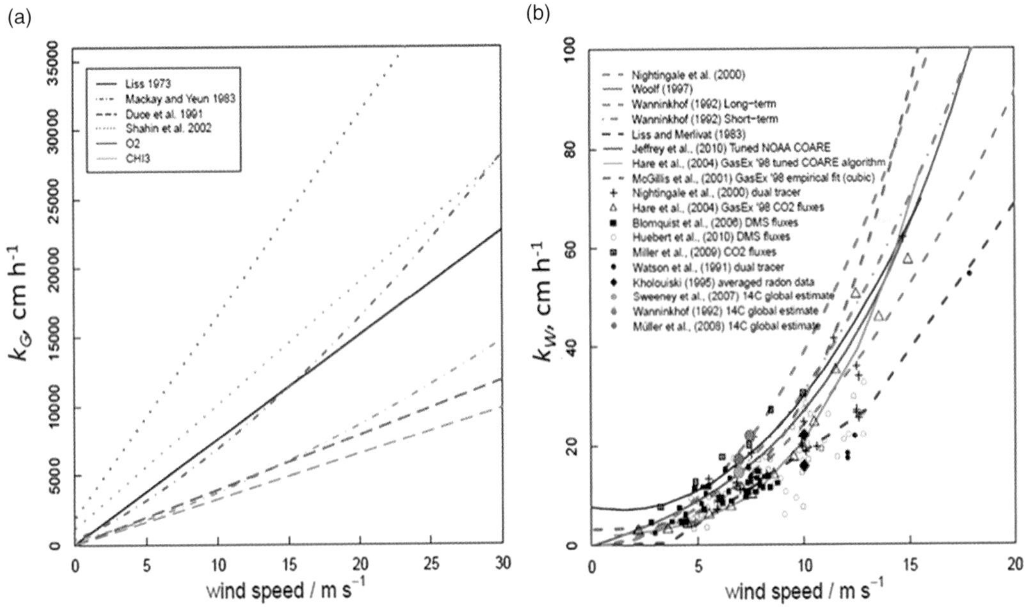

(a)   (b)

Figure 9.20 Gas-phase and water-phase conductances in the two-film model for air–sea exchange as a function of 10-m wind speed. (a) Different parameterizations of $k_{G,i}$ for air–sea exchange of O_2 and CHI_3. (b) Different parameterizations of $k_{W,i}$ for a species with Schmidt number in water $Sc_{W,I} = 660$. Different parameterizations can differ by more than a factor of 2 for a given wind speed and this reflects current uncertainty. Adapted from Johnson (2010).

where $k_{W,i}$ is in units of $[\text{cm h}^{-1}]$, u_{10} is in units of $[\text{m s}^{-1}]$, and the Schmidt number in water ($Sc_{W,I}$) has been normalized to that of CO_2 ($Sc_{W,CO2} = 600$).

We see from (9.48) that the overall exchange velocity K_i can be limited by transfer either in the gas or in the water phase depending on the relative magnitudes of H_i and $k_{W,i}/k_{G,i}$. Table 9.4 gives values of H_i for a few species in pure water at 298 K; Johnson (2010) gives an exhaustive list. Values for seawater are typically 20% lower than for pure water. Assuming as an example a typical wind speed $u_{10} = 5$ m s^{-1} and the molecular diffusion properties of CO_2, we derive from (9.49) and (9.50) $k_{G,i} = 0.6$ cm s^{-1} and $k_{W,i} = 0.002$ cm s^{-1}. For highly water-soluble species with $H_i < 10^{-3}$, such as H_2O_2, the exchange

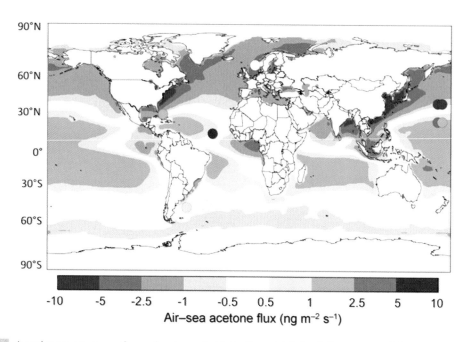

Figure 9.21 Annual mean net air–sea fluxes of acetone calculated with a global chemical transport model assuming a fixed surface ocean acetone concentration of 15 nM. Circles indicate ship observations. From Fischer *et al.* (2012).

velocity is limited by transfer in the gas phase and $K_i \approx k_{G,i}$. For sparingly water-soluble species with $H_i > 10$ such as CO_2, the exchange velocity is limited by transfer in the water phase and $K_i \approx k_{W,i}/H_i$. Gases of intermediate solubility such as methanol or acetone are in a transition regime where exchange is limited by transfer in both the gas and water phases. H_i increases with temperature for all gases, so that the ocean may be a net sink for gases of intermediate solubility at low temperatures and a net source at high temperatures.

Figure 9.21 from Fischer *et al.* (2012) illustrates the two-film model with the net air–sea flux of acetone computed with a global atmospheric model. A fixed seawater acetone concentration of 15 nM is assumed. Acetone in the atmosphere has continental sources (anthropogenic, terrestrial biogenic) and atmospheric sinks (photolysis, oxidation). Net acetone air–sea fluxes are downward at northern mid-latitudes due to relatively high atmospheric concentrations and cold ocean temperatures. They are upward in the tropics due to warm ocean temperatures. They are close to zero at southern mid-latitudes where atmospheric acetone is mostly controlled by a balance between oceanic emission and deposition. On a global scale, there is a close balance in that model between emission of acetone from the ocean (80 Tg a^{-1}) and deposition to the ocean (82 Tg a^{-1}). Even though the ocean is a net sink for acetone, ocean emission accounts for about half of the global acetone source of $150 \, \text{Tg} \, \text{a}^{-1}$ and thus plays an important role in controlling atmospheric concentrations.

References

Akagi S. K., Yokelson R. J., Wiedinmyer C., *et al.* (2011) Emission factors for open and domestic biomass burning for use in atmospheric models, *Atmos. Chem. Phys.*, **11**, 4039–4072.

Bloom A. A., Palmer P. I., Fraser A., Reay D. S., and Frankenberg C. (2010) Large-scale controls of methanogenesis inferred from methane and gravity spaceborne data, *Science*, **327**, 322–325.

Bloom A., Palmer P., Fraser A., and Reay D. (2012) Seasonal variability of tropical wetland CH_4 emissions: The role of the methanogen-available carbon pool, *Biogeosciences*, **9**, 2821–2830.

Darmenova K., Sokolik I. N., Shao Y., Marticorena B., and Bergametti G. (2009) Development of a physically based dust emission module within the Weather Research and Forecasting (WRF) model: Assessment of dust emission parameterizations and input parameters for source regions in Central and East Asia, *J. Geophys. Res.*, **114**, D14201, doi:10.1029/2008JD011236.

Duce R.A., Liss P. S., Merrill J. T., *et al.* (1991) The atmospheric input of trace species to the world ocean, *Global Biogeochem. Cycles*, **5**, 193–259.

Finkelstein P.L., Ellestad T. G., Clarke J. F., *et al.* (2000) Ozone and sulfur dioxide dry deposition to forests: Observations and model evaluation, *J. Geophys. Res.*, **105**, 15365–15377.

Fischer E. V., Jacob D. J., Millet D. B., Yantosca R. M., and Mao J. (2012) The role of the ocean in the global atmospheric budget of acetone, *Geophys. Res. Lett.*, **39**, L01807.

Freitas S. R., Longo K. M., Chatfield R., *et al.* (2007) Including the sub-grid scale plume rise of vegetation fires in low resolution atmospheric transport models, *Atmos. Chem. Phys.*, **7**, 3385–3398.

Gillette D. A. (1979) Environmental factors affecting dust emission by wind erosion. In *Sahara Dust* (Morales C., ed.), Wiley, Chichester.

Ginoux P., Chin M., Tegen I., *et al.* (2001) Sources and distributions of dust aerosols simulated with the GOCART model, *J. Geophys. Res.*, **106**(D17), 20255–20273.

Ginoux P., Clarisse L., Clerbaux C., *et al.* (2012) Mixing of dust and NH_3 observed globally over anthropogenic dust sources, *Atmos. Chem. Phys.*, **12**, 7351–7363, doi: 10.5194/acp-12-7351-2012.

Gong S. L. (2003) A parameterization of sea-salt aerosol source function for sub- and super-micron particles, *Global Biogeochem. Cycles*, **17**, 1097.

Granier C., Bessagnet B., Bond T., *et al.* (2011) Evolution of anthropogenic and biomass burning emissions at global and regional scales during the 1980–2010 period, *Climatic Change*, doi 10.1007/s10584-011-0154-1.

Guenther A., Karl T., Harley P., *et al.* (2006) Estimates of global terrestrial isoprene emissions using MEGAN (Model of Emissions of Gases and Aerosols from Nature), *Atmos. Chem. Phys.*, **6**, 3181–3210.

Guenther A. B., Jiang X., Heald C. L., *et al.* (2012) The Model of Emissions of Gases and Aerosols from Nature Version 2.1 (MEGAN 2.1): An extended and updated framework for modeling biogenic emissions, *Geosci. Model Dev.*, **5**, 1471–1492.

Hicks B. B., Baldocchi D. D., Meyers T. P., Hosker Jr. R. P., and Matt D. R. (1987) A preliminary multiple resistance routine for deriving dry deposition velocities from measured quantities, *Water, Air and Soil Pollution*, **36**, 311–330.

Hobbs P. V. (2000) *Introduction to Atmospheric Chemistry*, Cambridge University Press, Cambridge.

Hudman R. C., Moore N. E., Mebust A. K., *et al.* (2012) Steps towards a mechanistic model of global soil nitric oxide emissions: Implementation and space-based constraints, *Atmos. Chem. Phys.*, **12**, 7779–7795.

Jaeglé L., Quinn P. K., Bates T. S., Alexander B., and Lin J. T. (2011) Global distribution of sea salt aerosols: New constraints from in situ and remote sensing observations, *Atmos. Chem. Phys.*, **11**, 3137–3157.

Johnson M. T. (2010) A numerical scheme to calculate temperature and salinity dependent air–water transfer velocities for any gas, *Ocean Sci.*, **6**, 913–932.

Kaplan J. O. (2002) Wetlands at the Last Glacial Maximum: Distribution and methane emissions, *Geophys. Res. Lett.*, **29**(6), 1079.

Lamarque J.-F., Bond T. C., Eyring V., *et al.* (2010) Historical (1850–2000) gridded anthropogenic and biomass burning emissions of reactive gases and aerosols: Methodology and application, *Atmos. Chem. Phys.*, **10**, 7017–7039, doi: 10.5194/acp-10-7017-2010.

Liss P. (1973) Processes of gas exchange across an air–water interface, *Deep Sea Res.*, **20**, 221–238.

Monahan E. C., Spiel D. E., and Davidson K. L. (1986) A model of marine aerosol generation via whitecaps and wave disruption. In *Oceanic Whitecaps* (Monahan E. and Niocaill G. M.), D. Reidel, Norwell, MA.

Nightingale P. D., Malin G., Law C. S., *et al.* (2000) In situ evaluation of air–sea gas exchange parameterization using novel conservative and volatile tracers, *Global. Biogeochem. Cycles*, **14**, 373–387.

Riley W. G., Subin Z. M., Lawrence D. M., *et al.* (2011) Barriers to predicting change in global terrestrial methane fluxes: Analyses using CLM4Me, a methane biogeochemistry model integrated in CESM, *Biogeosciences*, **8**, 1925–1953.

Schnetzler C. C., Bluth G. J. S., Krueger A. J., and Walter L. S. (2007) A proposed volcanic sulfur dioxide index (VSI), *J. Geophys. Res.*, **102**, 20087–20091.

Seinfeld J. H. and Pandis S. N. (2006) *Atmospheric Chemistry and Physics: From Air Pollution to Climate Change*, 2nd edition, Wiley, New York.

Shao, Yaping (2004) Simplification of a dust emission scheme and comparison with data, *J. Geophys. Res.*, D10202, doi:10.1029/2003JD004372.

Slinn W. G. N. (1982) Predictions for particle deposition to vegetative canopies, *Atmos. Env.*, **16**, 1785–1794, doi:10.1016/0004-6981(82)90271-2.

Val Martin M., Kahn R. A., Logan J. A., *et al.* (2012) Space-based observational constraints for 1-D fire smoke plume-rise models, *J. Geophys. Res.*, **117**, D22204.

Vignati E., de Leeuw G., and Berkowicz R. (2001) Modeling coastal aerosol transport and effects of surf-produced aerosols on processes in the marine atmospheric boundary layer, *J. Geophys. Res.*, **106**, 20225–20238.

Wang J., Park S., Zeng J., *et al.* (2013) Modeling of 2008 Kasatochi volcanic sulfate direct radiative forcing: Assimilation of OMI SO_2 plume height data and comparison with MODIS and CALIOP observations, *Atmos. Chem. Phys.*, **13**, 1895–1912.

Wesely M. L. and Hicks B. B. (2000) A review of the current status of knowledge on dry-deposition, *Atmos. Environ.*, **34**, 2261–2282.

Xi X. and Sokolik I. N. (2015) Seasonal dynamics of threshold friction velocity and dust emission in Central Asia, *J. Geophys. Res.*, **120**, 1536–1564, doi:10.1002/2014JD022471.

Zender C. S., Bian H., and Newman D. (2003) Mineral Dust Entrainment and Deposition (DEAD) model: Description and 1990s dust climatology, *J. Geophys. Res.*, **108**(D14), 4416, doi: 10.1029/2002JD002775.

Zhang L., Gong S., Padro J., and Barrie L. (2001) A size-segregated particle dry deposition scheme for an atmospheric aerosol module, *Atmos. Environ.*, **35**, 549–560.

10 Atmospheric Observations and Model Evaluation

10.1 Introduction

Atmospheric chemistry models try to provide a physically based approximation to real-world behavior that serves to understand the real world and from there to predict future changes. The approximation comes with some error – by definition, a model is not perfect. As the saying goes, "All models are wrong, but some are useful." To make a model useful, it is critical to quantify its error. From there we may find that the error is acceptably small for the application of interest. Alternatively we may find that the error is too large and this then provides motivation for improving the model and often advancing scientific knowledge.

Quantifying model error requires reference to truth. Truth is elusive. Observations of atmospheric composition are our best resource. But they are sparse and have their own errors. Model error can never be fully characterized, but it can be estimated through statistical comparisons to observations. This chapter reviews simple metrics for this purpose, and also discusses the use of models as tools to interpret atmospheric observations in terms of processes. Formal approaches for error characterization and model optimization are presented in Chapter 11.

Different terms are used in the literature to describe the testing of models by comparison to observations. The word *validation* is often used but implies an exercise in legitimation to demonstrate that the model is true (*valid*) within certain error bounds. This may be appropriate terminology for regulatory models, where conclusions from the model have to hold in a court of law, but less so for research models. The term *verification* is sometimes used for operational applications (such as to *verify* a model forecast), but is inadequate for research applications where we may be more interested in *falsifying* the model, i.e., find out where the model is wrong so that we may improve it. We prefer here to adopt the term *evaluation*, which implies a broad assessment of model results, considering possible positive and negative outcomes, to understand the *value* of the model. Model evaluation offers the possibility of identifying unexplained behavior and from there advancing knowledge.

There are four types of model error. The first is error in our understanding of the physics as expressed by the model equations. The second is error in model *parameters* such as reaction rate constants or emissions that are input to the model equations. The third is numerical error in our approximate methods for solving the equations. The fourth is error in model implementation due to incorrect coding (bugs!). From an atmospheric chemist's perspective, the first two errors are the most interesting because addressing them deepens our understanding of the physical system. But the other two are important to recognize. Numerical error can be estimated by conducting simulations for different grid resolutions and time steps, by using different numerical solvers, or by comparing to analytical solutions for simple ideal cases (see Chapters 6 and 7). Bugs should of course be hunted down, and are often revealed by comparisons to other models. A complex 3-D model is probably never bug-free, but over time we can hope that the bugs that remain have little impact (and are therefore hard to detect!).

Consider a situation where the model departs from observations more than we deem acceptable, and we have established that this is not due to numerical, implementation, or measurement errors. We are then left with the task of improving the model physics or improving model parameters. Usually the first reach (because it is easiest) is to adjust the model parameters. These parameters have error ranges that can generally be estimated from the literature, such as uncertainty in rate constants. Adjusting model parameters within their error ranges is a perfectly legitimate exercise, and in fact the optimization of selected model parameters (called *state variables*) is the objective of *inverse modeling*, described in Chapter 11.

Adjustment of model parameters is often done in a simple way by constraining the model to match observations. This is called *model calibration* or *tuning*. A danger is that by ascribing all model error to the choice of some parameters we may be missing the opportunity to diagnose error in other parameters or in model physics – the familiar story of the drunk at night who looks for his missing keys under the lamppost because that's where the light is. Model tuning may lead to the model getting the right result for the wrong reasons. To avoid this situation it is important to evaluate the model for a wide range of species, conditions, and statistics. Ad hoc model tuning of multiple parameters by trial and error to fit a limited number of observations is poor practice and may lead to the model behaving like a house of cards – precariously fitting the observations available ("don't change a thing!") but ready to collapse when new observations or objective improvements to model parameters are brought in.

This brings up the importance of using a large ensemble of observations for model evaluation. Using observations taken in a wide range of dynamical and chemical environments can test model behavior over different conditions, building confidence in the capability of the model to simulate changes and make predictions. Using observations of chemically coupled species is particularly useful for revealing errors in the model chemistry. For example, a model that simulates sulfate aerosol with no bias but overestimates the precursor SO_2 may be producing sulfate with incorrect kinetics. Examining the relationship between two species with common emissions can help to separate emission errors from dynamical errors, as the latter will tend to affect both species similarly.

Research models used in atmospheric chemistry are generally versatile – they are intended to be applicable to a wide range of problems. The choice of application dictates such things as model domain and resolution, chemical mechanism, emission inventories, etc. It also defines the *error tolerance*. For some applications, we may be satisfied with a factor of 2 uncertainty; for other applications the tolerance may be much less. It is important to establish the error tolerance as it will affect the conclusions to be drawn from model evaluation. It is also important to identify what ensemble of observations can best evaluate the model for the particular application. These observations may not have been taken yet, which then calls for an experimental program as companion to the model study. The experimental program may take the form of a field campaign targeted at providing the observations needed for model evaluation. Such field campaigns involve tight partnership between experimenters and modelers, including, for example, the use of model forecasts to guide the day-to-day collection of observations in a way that can best test the model.

The concept of partnership between model and observations can be expanded by viewing the model as an integral part of the *atmospheric observing system* needed to answer a particular question. The observing system may include measurements from diverse platforms including ground-based sites, aircraft, and satellites. The model provides a common platform to integrate information from instruments measuring different species and operating on platforms with different measurement locations and schedules. Model evaluation with the ensemble of observations provides a check on the consistency of observations and enables constraints from multiple platforms. This can be formally done through *data assimilation*, as discussed in Chapter 11.

This chapter presents basic elements for carrying out model evaluation. Section 10.2 gives a primer on experimental methods and platforms. Error characterization for measurements and models is presented in Section 10.3, followed by general approaches to model evaluation in Section 10.4. Section 10.5 gives elementary statistical metrics. Statistical significance of differences is covered in Section 10.6. Section 10.7 discusses the use of models as tools to interpret atmospheric observations.

10.2 Atmospheric Observations

Measurements of atmospheric concentrations and fluxes are the main sources of data used to evaluate atmospheric chemistry models. Measurements are made *in situ*, when the instrument probes air from its vicinity, or *remotely*, when the instrument records a spectroscopic signal integrated over an atmospheric line of sight. Measurements are made routinely as part of long-term monitoring programs or intensively as part of field campaigns. Long-term monitoring programs may involve surface networks, sondes, commercial aircraft, or satellites. They are typically for a limited suite of species and provide information on short-term variability (events), long-term trends, and spatial patterns. They are particularly useful for long-term statistics and can be compared to the corresponding model statistics. Field campaigns typically provide a broader array of measurements deployed at specific locations of interest and for limited time. They generally focus on improving understanding of specific processes and are often geared to test model simulations of these processes. In such cases the models play a critical role in designing the field campaign and in interpreting the observations.

General methods for measuring concentrations include spectroscopy, mass spectrometry, chromatography, wet chemistry, and filters. Spectroscopic methods observe the interaction of atmospheric gases or particles with electromagnetic radiation. This radiation may be generated with a laser (*active methods*) or originate naturally from solar or terrestrial emission (*passive methods*). Mass spectrometry involves the ionization of an atmospheric sample followed by deflection of the ions in an imposed electromagnetic field. The angle of deflection is determined by the ratio of the electric charge to the mass of the ion. Chromatography involves the flow of an atmospheric sample through a narrow *retention column* in which individual species are separated by their different flow rates. Individual species are identified by their retention time in the column and their concentrations are measured by a detector at the exit of the column. Wet chemistry methods involve the capture of atmospheric gases and particles in a liquid sample, either by bubbling or spraying, followed by chemical analysis of the sample. Filter methods collect atmospheric samples through a porous filter, sometimes chemically treated. The filter is then analyzed by optical methods, gravimetric methods, or liquid-phase extraction followed by wet chemistry methods. Table 10.1 gives an overview of widely used measurement methods for different atmospheric species, and the following subsections provide additional information on specific methods and measurement platforms. More detailed information can be found, for example, in Finlayson-Pitts and Pitts (2000), Baron and Willeke (2005), Farmer and Jimenez (2010) and Burrows *et al.* (2011).

10.2.1 *In-Situ* Observations of Gases

Mass spectrometry (MS). In this method, the chemical species present in air samples are ionized, then an electromagnetic field is applied that separates ions according to their charge-to-mass ratios. The detection of specific lines in the mass spectrum provides quantitative information on the chemical composition of the air injected in the instrument. If the chemical species to be measured is selectively ionized by charge transfer of injected positive or negative ions, the instrument is called a chemical ionization mass spectrometer (CIMS). For example, acids such as H_2SO_4, HNO_3, or HCl can be ionized by charge transfer of reagent SF_6^-. Charge transfer from positive water ions (H^+H_2O) is called proton transfer reaction–mass spectrometry (PTR-MS) and provides a method to measure the atmospheric abundance of a wide range of organic species.

Gas chromatography (GC). In this method, air samples are injected in a narrow tube (GC column), and the chemical species are separated as they flow through the column and interact differently with the material in the column (Figure 10.1). The species are identified by their retention time in the column, and their concentrations are determined by a detector in the output stream (such as mass spectrometer, flame ionization, electron capture, thermo-ionic detectors). This technique is commonly used to measure organic species. If the detector is a mass spectrometer, the method is referred to as GC-MS.

Electrochemical ozonesondes. These small, lightweight balloon-borne instruments are routinely used to measure the vertical profile of ozone. The device contains electrodes immersed in an aqueous solution of potassium iodide (KI). When ozone enters the sensor, iodine molecules (I_2) are formed:

Table 10.1 *In-situ* and remote sensing methods for measurements of atmospheric composition

Species	*In-situ* methods	Remote sensing methods
H_2O	Frost point hygrometer Lyman alpha absorption Tunable diode laser	IR spectroscopy Microwave spectroscopy Raman lidar Filter radiometry
CO_2	Gas chromatography IR gas correlation	IR spectroscopy Filter radiometry
CO	Gas correlation Chemical conversion Differential absorption	IR spectroscopy Gas correlation radiometry
CH_4	Gas chromatography Tunable diode laser Differential absorption Gas correlation	IR spectroscopy Filter radiometry
VOCs	Gas chromatography PTR-MS Chemical ionization mass spectrometry	IR spectroscopy
O_3	UV absorption Chemiluminescence Electrochemical sondes	UV/Vis spectroscopy IR spectroscopy Microwave spectroscopy Lidar
N_2O	Gas chromatography Tunable diode laser Differential absorption	IR spectroscopy Radiometry
NO	Chemiluminescence	IR spectroscopy
NO_2	Photolysis and chemiluminescence Laser-induced fluorescence	UV/Vis spectroscopy IR spectroscopy
HNO_3	Tunable diode laser Ion chromatography Filter and wet chemistry	IR spectroscopy Filter radiometry
N_2O_5	Cavity ringdown	IR spectroscopy
HCl, HF	Tunable diode laser	IR spectroscopy
Cl, ClO	Resonance fluorescence	Microwave spectroscopy
OCS	Tunable diode laser	IR spectroscopy
SO_2	Ion chromatography Chemiluminescence	UV spectroscopy IR spectroscopy
DMS, CS_2, H_2S	Gas chromatography	
OH	Resonance fluorescence Laser-induced fluorescence Chemical ionization mass spectrometry Radioisotope chemistry	Lidar UV spectroscopy DOAS Far-IR spectroscopy
HO_2, RO_2	Radical amplifier Laser-induced fluorescence	Far-IR spectroscopy
CH_2O	Gas chromatography Tunable diode laser Wet chemical methods Laser-induced fluorescence	UV and IR spectroscopy

Species	*In-situ* methods	Remote sensing methods
Table 10.1 (*cont.*)		
H$_2$O$_2$	High-performance liquid chromatography	Far-IR spectroscopy
	Chemical ionization mass spectrometry	
O$_2$, N$_2$, H$_2$, Ar, Ne, He	Mass spectroscopy	
Aerosol	Filters	Lidar
	Optical particle counters	UV/Vis/IR spectroscopy
	Condensation nuclei counters	
	Cascade impactors	
	Differential mobility analyzers	
	Mass spectrometry	
	Electron microscopy	

Modified from Mankin *et al.* 1999 and James Crawford (personal communication).

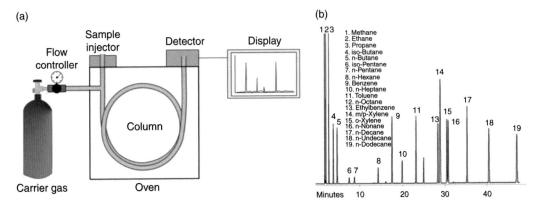

(a)

(b)

Figure 10.1 (a) Schematic representation of a gas chromatograph. Reproduced with permission from Lagzi *et al.* (2013). (b) Example of a chromatogram for a mixture of hydrocarbons as a function of the respective chemical species retention time [minutes] in the GC column, showing the intensity of peaks and their chemical identification.

$$2\,KI + O_3 + H_2O \rightarrow I_2 + O_2 + 2\,KOH \qquad (10.1)$$

The conversion of iodine into iodide at the cathode of the instrument produces a weak electrical current proportional to the mass flow rate of ozone through the cell. By measuring this current and the rate at which air enters the cell, ozone concentrations can be derived. The Brewer–Mast (BM) ozonesonde operates with a single electrochemical cell that includes electrodes between which a small electrical potential is applied to prevent current flow in the cell unless free iodine is present. The ozonesonde referred to as electrochemical concentration cell (ECC) is made of two separate cells each containing slightly different concentrations of a KI solution, and connected by an ion bridge. No external electrical power is required since the driving electromagnetic force is provided by the difference in the KI concentrations. The instrument must be calibrated and temperature corrections must be applied.

Tunable-diode laser spectroscopy. Many methods to derive the concentrations of chemical species, either in the laboratory or in the atmosphere, are based on the analysis of the spectral signature resulting from the interaction of the species with radiation. In tunable-diode laser (TDL) spectroscopy, the light source is provided by a TDL whose emission wavelength is adjusted to match a characteristic absorption line of the target gas in the path of the laser beam. The gas concentration is derived from the measurement of the light intensity detected, for example, by a photodiode.

Resonance fluorescence. This method is based on the measurement of resonance radiation scattered by gas molecules that are irradiated in an instrument at a wavelength corresponding to a particular electronic transition for the gas. The induced fluorescence spectrum is analyzed to deduce the gas concentration. The concentration of OH, for example, can be derived from the laser-induced fluorescence (LIF) associated with the $A^2\Sigma^+$ ($v'=0$) $\rightarrow$ $X^2\Pi$ ($v''=0$) electronic transition near 308 nm.

Chemiluminescence. The detection of nitric oxide (NO) is based on reaction with ozone present in excess in a reaction chamber:

$$NO + O_3 \rightarrow NO_2 + O_2 + hv \tag{10.2}$$

The radiative emission produced by this reaction covers a broad spectrum (600–3000 nm with a maximum intensity around 1200 nm) and can be detected by a sensitive photoelectric device. Its intensity is proportional to the concentration of NO. In the presence of a mixture of NO and NO_2, a measurement of the total NO_x concentration can be achieved by converting NO_2 to NO on a catalytic surface upstream of the reaction chamber. Similarly, a measurement of total reactive nitrogen oxides (NO_y, including NO_x and its oxidation products) can be made by catalytic reduction to NO followed by measurement of the NO concentration by chemiluminescence.

10.2.2 *In-Situ* Observations of Aerosols

In-situ sampling of aerosols is generally performed through an inlet that transports the particles to a collector or detector. In the ideal case, inlets should draw the totality of the particles in a specified size range. In reality this is not the case because particles deviate from the inlet airflow. Much effort has been devoted to the design of efficient *isokinetic* inlets that minimize this effect. The problem is particularly difficult in the case of aircraft sampling because of the fast and complex airflow surrounding the fuselage.

Total aerosol concentration. The total aerosol mass concentration can be determined by collecting particles on a filter and weighing the filter under controlled temperature and humidity conditions. Another technique is the β-*gauge* that measures the attenuation of β-radiation through a particle-laden filter. The attenuation, caused by electron scattering in the filter media, is proportional to the total number of atomic electrons, and provides therefore information on the total mass density of the sample. Aerosol number concentrations can be measured by growing particles by condensation in a supersaturated environment until they are large enough to be detected optically. This is the approach used in *condensation nuclei (CN) counters*.

Aerosol size distribution. The most common instrument used for counting and sizing particles is the *optical particle counter*. This instrument measures the amount of light scattered by individual particles as they flow through a tightly focused beam of light. The scattered light is directed to a photodetector. The size of the particles can be derived from the resulting electrical signal on the basis of a calibration curve. *Differential mobility analyzers* use an electric field to separate particles according to their mobility, which is a function of particle size. In the *cascade impactor*, the aerosol size distribution is measured by injecting air into a device containing a cascade of plates (impactors) around which the airflow is deviated. The largest particles do not follow the curvilinear air streamlines passing around the first impactor and, instead, hit the detection plate. This process, which involves increasingly narrowing nozzles, is repeated several times to extract from the beam the gradually smaller particles (see Figure 10.2).

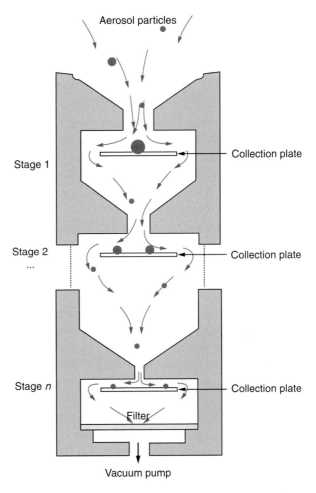

Figure 10.2   Schematic representation of a cascade impactor device. The airflow is accelerated as the gas passes through several gradually narrowing nozzles. Smaller particles remain in the flow, while larger particle with higher inertia hit the collection plates. Increasingly smaller particles are trapped by subsequent collection plates. Reproduced with permission from Lagzi *et al.* (2013).

Aerosol composition. The chemical composition of the aerosol can be determined by collecting particles on filters and subsequently analyzing the filter substrate. This analysis can be done by aqueous or organic extraction, in which the chemical species of interest are dissolved. The composition of the liquid sample is then determined by various techniques. Particles on the filter can also be heated and volatilized, and the resulting gases analyzed by gas chromatography or other methods (see Section 10.2.1).

Electron microscopy provides information on particle morphology and elemental composition. In this method, particles collected on a filter are irradiated by electrons under vacuum conditions. The X-ray energy spectrum produced by the interactions of electrons with the particles provides information on the elemental composition of the particles. A limitation is that this method will evaporate any aerosol water and other semi-volatile species.

Mass spectrometry is increasingly used to determine the chemical composition of individual particles. This enables high-frequency measurements of aerosol composition and provides size distribution information for particles with different chemical signatures.

10.2.3 Remote Sensing

Remote sensing instruments are based on the collection of spectroscopic data along a selected atmospheric line of sight. The spectra are interpreted in terms of the species concentration integrated over the line of sight, with different levels of spatial resolution depending on the instrument and the species observed. Remote sensing can be performed from the ground, aircraft, and satellites, and can use either passive or active data collection methods.

Passive remote sensing. In passive methods, the radiation source is external to the instrument and is provided by the Sun, another star, the Moon, or the Earth and its atmosphere (infrared). A detector such as a spectrometer or a radiometer captures the electromagnetic radiation from the radiation source after it has propagated through the atmosphere along an optical path. The intensity, spectral distribution, and polarization of the measured radiation provide information on atmospheric concentrations over the optical path. *Dobson spectrophotometers* (Figure 10.3) were developed in the mid-1920s by G. M. B. Dobson to investigate atmospheric circulation by measuring changes in atmospheric ozone, they are now deployed globally to verify observations from satellites. The instrument derives total column and vertical profiles of ozone by measuring the direct UV radiation from the Sun, the Moon, or the zenith sky for different wavelength pairs. The total ozone column is derived from the contrast in atmospheric absorption between 305.5 nm (strong ozone absorption) and 325.4 nm (minimal ozone absorption). The vertical profile is measured using the 311.4 and 332.4 nm wavelength pair at high solar zenith angles. The measurement is based on the *Umkehr effect* (Götz *et al.*, 1934), which describes the reversal (Umkehr in German) of the curve that represents the log-intensity ratio of the scattered light as a function of the solar zenith angle. This reversal, which is observed for a solar zenith angle of approximately 88 degrees when the wavelengths are 311.4 and 332.4 nm, results from the existence of an ozone maximum in the stratosphere.

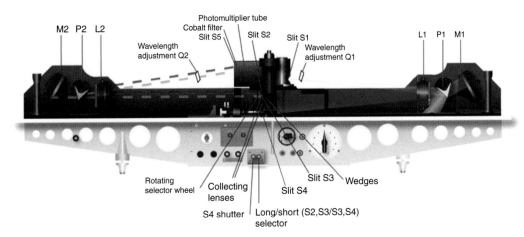

Figure 10.3 Schematic representation of the Dobson ultraviolet spectrophotometer for the measurement of ozone. Entering radiation from the Sun, the Moon, or the zenith sky is reflected by a right-angle prism and falls on slit S1. The beam is then decomposed by a first spectroscope (lens L1, prism P1, and mirror M1), which reflects the radiation back to the focal plane of the instrument. Fixed slits S2, S3, and S4 isolate the different nearby wavelengths. A second spectroscope with reversed dispersion (lens L1 and mirror M2 and prism P2) recombines the light onto a photomultiplier. A chopper alternatively allows radiation at the two wavelengths to reach the detector. The ozone column is determined from the measurement at two or more pairs of wavelengths. Reproduced from Komhyr and Evans (2008).

Instruments based on passive *differential optical absorption spectroscopy* (DOAS) measure the concentrations of gases along the light path by application of the Beer–Lambert law (see Section 5.2.4). When the Sun is used as the light source and the broad spectral signal associated with atmospheric scattering is removed from the observed spectrum, the remaining signal has spectral signatures representing absorption lines of atmospheric molecules. Difference between online and offline wavelengths measures the concentration of the absorber. Active DOAS systems using their own light source can measure the integrated concentrations of chemical species along the light path between the instrument and a reflector that may be located several hundreds of meters away. The multi-axis differential optical absorption spectroscopy (Max-DOAS) measurement technique (Hönninger *et al.*, 2004) retrieves vertical profile information by combining measurements of scattered sunlight from multiple viewing directions. This retrieval requires a detailed radiative transfer model. Ground-based Max-DOAS instruments are highly sensitive to absorbers in the lowest few kilometers of the atmosphere.

Filter radiometers, often used for spacecraft observations, measure the radiative emission of the atmosphere or the transmitted solar radiation within a particular spectral band determined by a wavelength selection device (filter). The detectors are often cooled to limit the interferences from the radiation emitted by the instrument itself. Figure 10.4 shows a cross-section of the ozone mixing ratio in the upper troposphere and lower stratosphere measured by a spaceborne multi-channel limb scanning infrared radiometer. In the *gas-filter correlation radiometry* (GFCR)

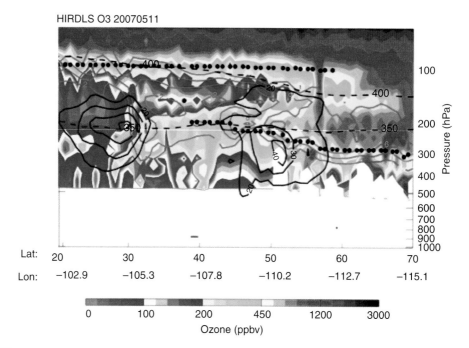

HIRDLS O3 20070511

Ozone (ppbv)

0   100   200   450   1200   3000

Figure 10.4 Cross-section of the ozone mixing ratio [ppbv] measured by the 26-channel High Resolution Dynamics Limb Sounder (HIRDLS) at 20°–70°N on November 5, 2007. The data show the intrusion of ozone-rich air masses in the vicinity of the jet stream (thin full lines) and gradual dilution as the air penetrates further into the troposphere. This ozone pattern is associated with the presence of a double tropopause (black dots) at mid-latitudes. Courtesy of W. Randel and J. Gille, NCAR.

method, the incoming radiation passes through a so-called correlation cell that is filled by the target gas and acts as a spectral filter. The difference between the signal recorded from a broadband detector and the signal emerging from the correlation cell characterizes the amount of the target gas in the atmosphere.

Fourier transform infrared (**FTIR**) *spectroscopy* measures the thermal IR radiation emitted by the Earth's surface and atmosphere with a *Michelson interferometer*, in which the incoming radiation is split into two beams by a half-transparent mirror. The first beam is directed to and reflected by a fixed flat mirror, while the second one is reflected by another flat mirror that is continuously moving along the axis of the incoming beam. As the two beams recombine, their phase shifts produce interference patterns. The resulting signal, called an interferogram, recorded as a function of the position of the moving mirror, represents the Fourier transform of the atmospheric spectrum. From there, the spectrum can be derived with high resolution. Almost all molecules have an IR spectrum from vibrational–rotational transitions and FTIR spectroscopy is therefore a versatile tool to detect a wide range of species (Table 10.1). The main limitation is interference from high-concentration species such as H_2O and CO_2. In satellite applications, limited vertical profile information can be obtained in *nadir* (downlooking) observations by exploiting known vertical gradients in temperature, and further vertical resolution can be obtained in *limb* observations at different angles.

Microwave instruments. Sensors operating in the microwave wavelength range of 0.1 to 10 cm (3–300 GHz) measure thermal emission from molecules to derive information on atmospheric parameters and chemical composition, particularly in the upper atmosphere (Kunzi *et al.*, 2011). In this method, the observed shape of spectral lines emitted by the chemical species is fitted with the shape calculated for a specified vertical distribution of the emitter.

Active remote sensing. The most common active remote sensing technique is the *light detection and ranging* instrument, known as *lidar* (Figure 10.5), which emits a coherent light beam (often pulses at a given wavelength) to the atmosphere. A small fraction of the light is scattered by atmospheric molecules or aerosol particles back to the receiver, and the vertical distribution of the scatter can be derived by timing the return. This allows for much higher vertical resolution than passive methods. Different lidar devices account for different types of scattering processes: Rayleigh, Mie, or Raman (see Section 5.2.4). The first two processes do not change the frequency of the incident photon except by a possible Doppler shift. In the third process (Raman lidar), the wavelength of the scattered radiation is slightly shifted as a result of energy exchanges between the incident radiation and atmospheric molecules. Measurement of the spectral shift allows identification of the scattering species and determination of its vertical profile.

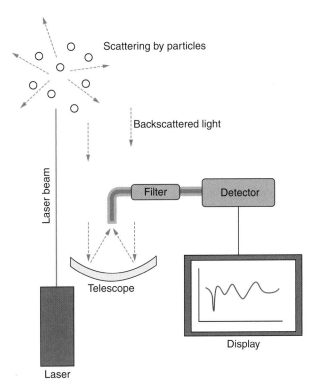

Figure 10.5 Schematic representation of a lidar system. The light produced by a laser beam directed upward is scattered by atmospheric molecules or particles. Backscattered photons are collected by a telescope and the intensity measured by a detector. Reproduced with permission from Lagzi *et al.* (2013).

If two pulses at different wavelengths are produced, e.g., by frequency multiplication of the laser output, the intensities of the two return signals are differently affected by the absorption of atmospheric species. The method, called *differential absorption lidar* (DIAL), allows the measurement of the vertical profile of ozone. Measurements made by an airborne DIAL instrument shown in Figure 10.6 highlight the complex distribution of ozone and aerosols in the troposphere, including small-scale features.

10.2.4 Measurement of Surface Fluxes

The vertical surface flux F_z of a chemical species can be determined directly by the *eddy correlation* method, in which the species number density n and the vertical wind velocity w are measured concurrently at the same location:

$$F_z = \overline{wn} \qquad (10.3)$$

Here, the measurements need to be made within the surface layer (lowest ~50 m of the atmosphere) for the flux to be representative of the surface (see Box 8.4). The usual platform is a tower extending ~10 m above the surface or canopy top. A useful flux measurement must temporally average the instantaneous flux wn over a representative collection of turbulent eddies, as represented by the averaging overbar in (10.3). The averaging time is typically about one hour. w and n can be decomposed as the sums of their time-average and fluctuating components (Section 8.2):

$$F_z = \overline{(\overline{w} + w')(\overline{n} + n')} = \overline{w}\,\overline{n} + \overline{w'n'} \qquad (10.4)$$

The first term $\overline{w}\,\overline{n}$ on the right-hand side is the *mean advective flux* representing the contribution from the mean vertical wind, and the second term $\overline{w'n'}$ is the *eddy correlation flux* representing the contribution from turbulent eddies. The mean vertical wind close to the surface is near zero, so that the mean advective flux is generally much smaller than the eddy correlation flux.

Eddy correlation flux measurements must resolve eddies of all sizes, making a significant contribution to the mean flux. This requires fast instrumentation with a measurement frequency of 1–10 Hz. Such instrumentation is often not available. If a fast measurement of the vertical velocity is available, an alternative is to use the *eddy accumulation method*. In this method, air is collected in two different storage reservoirs, one for upward flow and one for downward flow. The collected air is then analyzed and the flux is computed from the difference in mass between the reservoirs.

Another approach to estimate the vertical flux that does not require high-speed instrumentation is the *flux-gradient method*. As shown in Section 9.3, vertical transport in the surface layer can usually be parameterized as an eddy diffusion process in which the vertical flux $F_z = -K_z \partial \overline{n}/\partial z$ is proportional to the mean vertical gradient in concentration and the eddy diffusion coefficient is $K_z = ku^*z$. The friction velocity u^* can be inferred from the slope of a $\ln z$ vs. u plot by assuming the log law for the wind (Section 9.3.2). Alternatively, if the surface flux

(a)

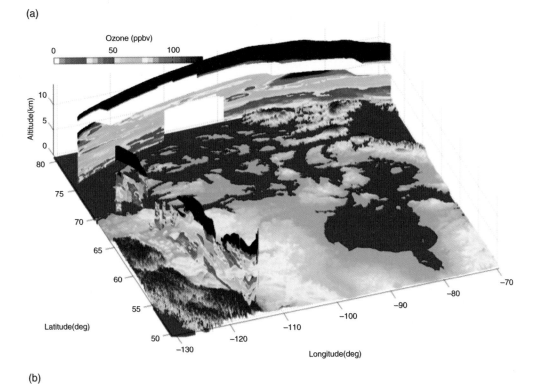

(b)

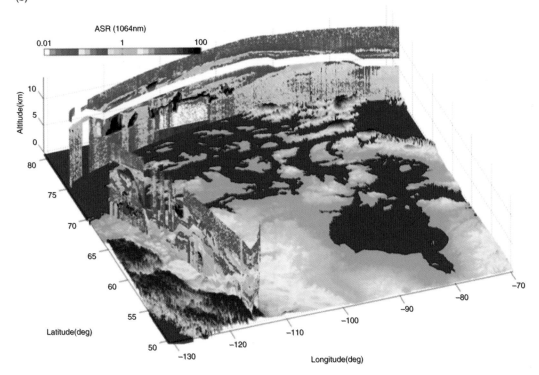

Figure 10.6 DIAL measurements of ozone mixing ratios [ppbv] (a) and aerosol scattering ratio at 591 nm (b) along a NASA DC-8 flight on July 8, 2008 from Cold Lake, Alberta, Canada to Thule, Greenland. The measurements were made during the ARCTAS field campaign. Data from J. W. Hair, NASA Lidar Applications Group, NASA Langley Research Center. Source: NASA (www.science.larc.nasa.gov).

$F_{z,\Psi}$ of a reference variable Ψ is known, one can use the similarity assumption to infer the flux of any species i by comparing the mean vertical gradients:

$$F_{z,i} = F_{z,\Psi} \frac{\overline{\partial n / \partial z}}{\overline{\partial \Psi / \partial z}} \tag{10.5}$$

Here, Ψ may represent sensible heat, water vapor, or any chemical variable for which the surface flux can be measured by eddy correlation or is otherwise known. The similarity assumption operates in both directions, so that it is possible to infer the flux of a species for which the surface is a sink ($F_{z,i} < 0$) from the flux of a variable for which the surface is a source ($F_{z,\Psi} > 0$) or vice versa.

10.2.5 Observation Platforms

Atmospheric measurements are conducted from a wide range of platforms including ground-based stations, vehicles, ships, balloons, aircraft, and satellites (Figure 10.7). These different platforms have advantages and disadvantages that often make them complementary (Table 10.2). Some can carry extensive payloads to measure a wide range of species while others are more limited. Addressing a particular scientific problem may call for a carefully designed *observing system* involving an ensemble of platforms each with a different role to play. Such an observing system must

(a)

(b)

(c)

(d)

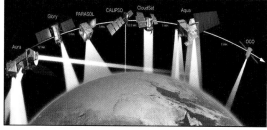

Figure 10.7   Platforms for measuring atmospheric composition. (a) Surface station installed for a field campaign in Texas (University of Houston); (b) NASA unmanned Global Hawk aircraft during the Airborne Tropical Tropopause Experiment (ATTREX) in California (2014); (c) instrumentation aboard a C-130 aircraft during the Front Range Air Pollution & Photochemistry Experiment in Colorado (FRAPPE) (2014); (d) constellation of satellites called the A-Train flying in formation (NASA).

Table 10.2 Advantages and disadvantages of observation platforms for atmospheric composition

	Surface sites	Vehicles, ships	Balloons	Aircraft	Satellites
Temporal coverage	Good	Limited	Limited	Limited	Good
Horizontal coverage	Limited	Good	Poor	Good	Good
Vertical resolution	Poor	Poor	Good	Good	Limited
Payload	Good	Good	Limited	Good	Limited

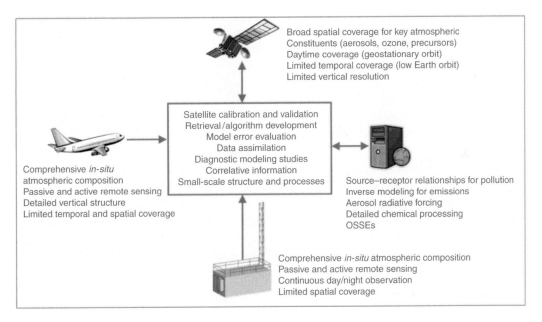

Broad spatial coverage for key atmospheric
Constituents (aerosols, ozone, precursors)
Daytime coverage (geostationary orbit)
Limited temporal coverage (low Earth orbit)
Limited vertical resolution

Satellite calibration and validation
Retrieval /algorithm development
Model error evaluation
Data assimilation
Diagnostic modeling studies
Correlative information
Small-scale structure and processes

Comprehensive *in-situ*
atmospheric composition
Passive and active remote sensing
Detailed vertical structure
Limited temporal and spatial coverage

Source–receptor relationships for pollution
Inverse modeling for emissions
Aerosol radiative forcing
Detailed chemical processing
OSSEs

Comprehensive *in-situ* atmospheric composition
Passive and active remote sensing
Continuous day/night observation
Limited spatial coverage

Figure 10.8 Observing system for atmospheric composition illustrating some applications of such a system and the role of different observing system components (ground based, aircraft, satellites, models).

generally include models to place into context the measurements taken from different platforms with different payloads, schedules, and locations. Figure 10.8 gives some general considerations for the design of such an observing system. We elaborate below on the roles of surface sites, aircraft, and satellites.

Surface sites provide local data, generally with very high accuracy and extended temporal coverage. They provide the basis for analyzing long-term trends in atmospheric chemistry as well as interannual, seasonal, and diurnal variations. Some stations record the concentrations of an ensemble of species, and the observed relationships between species can then provide constraints on their sources and chemical evolution. Other measurements that may be taken at surface sites include total columns (such as from a Dobson spectrophotometer or FTIR instrument), vertical profiles (lidar, ozonesondes), and surface fluxes (eddy correlation). Although one generally regards surface sites as serving monitoring purposes (often involving networks of similarly configured sites), they are also often used in field campaigns and provide temporal continuity.

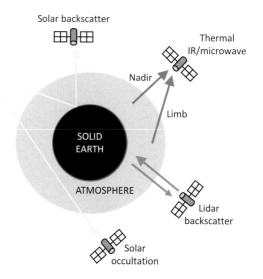

Figure 10.9   Observing strategies for atmospheric composition from low Earth orbit. Distances are not to scale.

Aircraft provide vertical coverage, and horizontal coverage beyond what ground-based stations allow. Large research aircraft with comprehensive payloads allow detailed measurements of atmospheric composition with great flexibility in operations. Observations by research aircraft provide information for relatively short periods of time (typically 1–2 months, the practical length of a field campaign). A few commercial long-range aircraft with automated instrumentation provide routine data along their flight routes at cruising altitude (upper troposphere/lower stratosphere) as well as vertical profiles during take-off and landing. Remotely piloted aircraft offer the possibility of long-endurance flights with small payloads. Vertical ranges of most aircraft do not extend above 12 km altitude but some specialized aircraft can operate up to 20 km altitude. *In-situ* measurements at higher altitudes require balloons.

Satellites provide global continuous coverage to varying degrees depending on their observation schedule, orbit track, cross-track sampling, and viewing geometry. Typical horizontal pixel resolution is of the order of 10 km for nadir view. Figure 10.9 illustrates different viewing strategies for satellites in low Earth orbit (LEO), 500–2000 km above the surface and with an orbital period of 1–2 hours. *Solar backscatter* instruments detect solar radiation backscattered by the Earth surface and its atmosphere. They generally provide information on total atmospheric columns with little or no vertical resolution. *Thermal IR/microwave* instruments detect radiation emitted by the Earth's surface and its atmosphere, and can operate either in nadir or limb mode. Nadir viewing affords better horizontal resolution and vertical penetration, but detection of the lower troposphere is limited by the need for thermal contrast with the surface. Limb viewing can achieve vertical resolution of order 1 km with horizontal resolution of order 100 km, but has little sensitivity below the upper troposphere due to interference by clouds and water vapor in the line of sight. *Lidar* instruments can achieve high vertical resolution but have no cross-track capability so their horizontal coverage is very

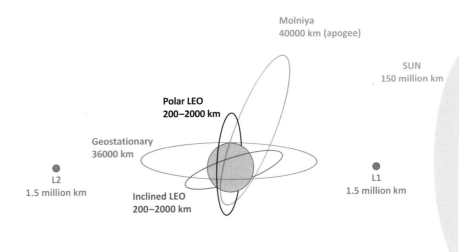

Satellite orbits and their distances from earth. Distances are not to scale.

limited. *Solar occultation* instruments detect the direct radiation from the Sun passing through the atmosphere as the satellite experiences sunrise and sunset over its orbital period. The strong signal from the Sun enables detection of species for which other methods would not achieve sufficient signal, with high vertical resolution down to cloud level. However, the measurements are sparse (twice per orbit) and the geographical coverage is limited.

Figure 10.10 shows different possible orbits for satellite observations. Observations of atmospheric composition have so far mainly been from LEO. The *polar Sun-synchronous orbit* is the most common and provides global observations at the same local time of day everywhere. Cross-track viewing can achieve global daily coverage. *Inclined orbits* provide a higher frequency of observations at low latitudes but sacrifice high latitudes. *Geostationary orbits*, where the satellite is in an equatorial plane 36 000 km away from the Earth with a 24-hour orbiting period, provide continuous data over a limited geographical domain (up to 1/3 of Earth's surface, though smaller domains are typically used); spatial resolution is limited poleward of 60° latitude. Other orbits that have been proposed for measurements of atmospheric composition include the *Molniya orbit* (high-latitude observations several times per day), the *Lagrange L1 orbit* (continuous global view of the sunlit Earth), and the *Lagrange L2 orbit* (continuous solar occultation).

The determination of atmospheric concentrations from space is considerably more complex than for *in-situ* observations. Retrievals of concentrations from the radiance spectra must account for interferences from the surface and clouds. In the case of gas retrievals, they must also account for interferences from aerosols and from other gases. In the case of aerosol retrievals, they must also account for variations in aerosol optical properties. The retrieval is generally underconstrained, which means that external *prior* information on atmospheric composition must be assumed. The prior information often comes from models, which may lead to an incestuous relationship between the satellite observations and the models that they are supposed to evaluate. This will be discussed further in Chapter 11 in the context of inverse modeling and data assimilation.

10.3 Characterization of Errors

10.3.1 Errors in Observations

Observations are characterized by systematic and random errors (Taylor, 1996; Hughes and Hase, 2010). *Systematic errors* are consistent biases that repeat themselves every time the measurement is made by the same instrument under identical conditions. They cause the measured quantity to be shifted away from its true value due to factors that reproducibly affect the measurement, such as inaccurate calibration of the instrument or, when the measurement is indirect, inaccuracy in the retrieval model. The magnitude of the systematic error determines the *accuracy* of a measurement. *Random errors* are caused by factors that affect the measurements erratically, such as photon counting noise. They determine the *precision* of the measurement. The best estimate of a measured value is the mean of individual measurements, and the random error is the distribution around this mean. When the random error distribution is near Gaussian it can be characterized by its standard deviation (see Appendix E).

The derivation of atmospheric quantities from remote sensing observations requires that a retrieval calculation be performed (see example in Box 11.5). The retrieval involves inversion of a radiative transfer model to infer the atmospheric concentrations of interest from the radiances measured by the instrument. The model represents the physics of the measurement and often requires prior information to provide the best statistical fit to the observed radiances, accounting for errors in the measurements and the model. The instrument sensitivity may have a dependence on altitude, so that the retrieved concentration profiles reflect different altitude weightings and dependences on the prior information. Satellite observations have complicated error budgets with contributions from the measured radiances and the retrieval model, and including smoothing as well as random errors. Errors on vertical profiles are usually correlated across different altitudes, so that the error statistics must be represented by an error covariance matrix (see Chapter 11).

10.3.2 Errors in Models

Error in complex models is difficult to characterize. Chemical transport models provide a continuous 3-D simulation of atmospheric composition evolving with time, but accurate observations to evaluate the model are sparse and do not cover the full range of conditions over which the model is to be applied. Statistical metrics for comparing models to observations, and from there estimating model error, are presented in Section 10.5. Model error characterization for purposes of inverse modeling can be done with the residual error variance method that lumps model and instrument error into an overall observational error (Box 11.2). We discuss here some more specific approaches for estimating model error.

Model errors originate from four sources: (1) the model equations and underlying scientific understanding; (2) the model parameters input to these equations; (3) the numerical approximations in solving the equations; and (4) coding errors. Coding errors are generally revealed by computational failure of the model under some conditions or by unexpected model behavior. Detailed output diagnostics are important for detecting coding errors, including statistical distributions that allow detection of anomalies. Benchmarking of successive model versions is essential to detect coding errors introduced during a version update. Model output should always make sense to the modeler in terms of the underlying processes. If it does not, then a bug is probably lurking and must be chased without complacency.

Errors in the numerical approximations used to solve the equations are discussed in Chapters 6 and 7. Ideally, the numerical approximations should be benchmarked against exact analytical solutions, but such analytical solutions are available only in idealized cases that are generally not relevant to the atmosphere. For example, numerical advection of a given shape in a uniform flow can be compared to the shape-preserving analytical solution, but this greatly underestimates the error in the divergent flow typical of the atmosphere (Section 8.13). In the absence of an exact calibration standard, one can still estimate numerical errors by conducting model sensitivity simulations using objectively better, higher-order numerical methods. For example, one can compare a fast chemical solver of relatively low accuracy used in standard simulations to a more accurate solver applied to the same chemical mechanism.

Errors caused by model grid resolution can be estimated by conducting sensitivity simulations at finer resolution. Extrapolation is possible with Aitken's convergence method (Wild and Prather, 2006). Let C_o be the exact solution at infinitely fine resolution for a model variable of interest and $C(h)$ the solution computed with the model at grid resolution h. We can write

$$C_o = C(h) + [C(h/2) - C(h)] + [C(h/4) - C(h/2)] + [C(h/8) - C(h/4)] + \cdots$$

$$(10.6)$$

Assume now that the model converges geometrically to the exact solution as the grid resolution increases, with a scale-independent geometric convergence factor $k < 1$ such that

$$k = \frac{C(h/4) - C(h/2)}{C(h/2) - C(h)} = \frac{C(h/8) - C(h/4)}{C(h/4) - C(h/2)} = \cdots \qquad (10.7)$$

Replacing into (10.6) we obtain the error estimate

$$C(h) - C_o = [C(h) - C(h/2)](1 + k + k^2 + \cdots) = \frac{[C(h) - C(h/2)]}{1 - k} \qquad (10.8)$$

We can calculate a value for k from (10.7) by conducting simulations at three different resolutions h, $h/2$, $h/4$, and we can further check the quality of the geometric convergence assumption by conducting a fourth simulation at resolution $h/8$. The assumption of a scale-independent geometric factor is unlikely to hold down to infinitely small resolutions. However, if k is sufficiently small, the first terms in the series may provide a good approximation of the error.

Errors due to model input parameters can be estimated in principle by conducting an ensemble of simulations where these different parameters are varied over their ranges of uncertainty, using for example a Monte Carlo method. This can be practically done for chemical mechanisms where error estimates for individual chemical rate constants are available from compilations of kinetic data. Errors in other model parameters, such as winds or emissions, are not as well quantified, and a Monte Carlo analysis over the full range of parameter space in a 3-D model would be impractical anyway. Errors can be estimated in a limited way by conducting sensitivity simulations with different meteorological fields, emission inventories, etc. Model error may be dominated by a small number of parameters, and it is part of the modeler's skill to recognize which model parameters are important and to focus error characterization accordingly.

Errors in the model formulations of chemical and physical processes can be estimated to the extent that there are objectively better or equally valid formulations to apply. For example, we can estimate the error associated with using a reduced chemical mechanism for faster computation by conducting a sensitivity simulation with the complete mechanism. Errors associated with subgrid parameterizations of processes (Chapter 8) can be estimated by comparing different choices of parameterizations, or by conducting a test simulation at high resolution where subgrid parameterization is not needed. For example, it is useful to assess the sensitivity of model results to the choice of boundary layer mixing and convective transport parameterizations.

We discussed in Chapter 4 the noise in climate models caused by chaos in the solution to the Navier–Stokes equation for momentum. There is no such chaos in the solution to the chemical continuity equations under practical atmospheric conditions. In fact, numerical errors in the solutions to chemical systems tend to dissipate with time following Le Chatelier's principle. Offline chemical transport models driven by input meteorological variables thus do not show the chaotic behavior found in climate models. However, in the case of an online chemical transport model built within a free-running climate model, the chemical concentrations develop noise driven by the noise in meteorological variables. This noise has physical basis in the internal variability of climate and needs to be characterized in the model for comparison to observations. This can be done by conducting a number of simulation years and/or by repeating the model simulations a number of times with slightly different initial meteorological conditions or physical parameters. From this ensemble of realizations we can construct probability density functions (PDFs) of concentrations to compare to the corresponding multi-year PDFs in long-term observations. Such a statistical comparison is called a *climatological* evaluation, with the PDFs representing the *climatologies* of the model and of the observations.

Intercomparisons of different models provide yet another way of estimating model error. This is done regularly in *community assessments* to determine how well different state-of-science models can reproduce specific aspects of atmospheric composition, or to estimate errors associated with future projections. It involves the comparison of simulations conducted with different models for the same conditions. For example, Figure 10.11 shows predicted surface concentrations of particle matter (PM$_{10}$) forecast for the same day by five different regional air quality models using the same chemical and meteorological initial and boundary conditions. On this particular day, several models show the formation of a dust layer over the Sahara and two of them predict an intense transport of dust particles toward eastern

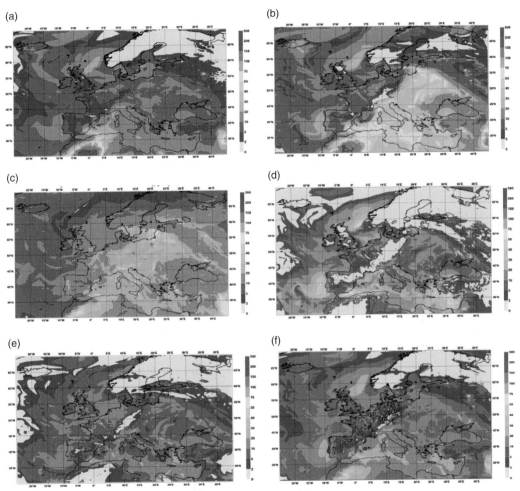

Figure 10.11 Simulated surface concentration [µg m^{-3}] of particulate matter (PM$_{10}$) on September 1, 2015, 20 UTC by five regional models (a–e) contributing to a multi-model ensemble prediction system for regional air pollution in Europe. Two of these models clearly show an intrusion of dust-rich air from the Sahara toward Southern and Eastern Europe with high dust concentrations extending from the Baltic to the Black Sea. Other models do not reproduce such a strong intrusion. The average of the six models involved in this air quality prediction is shown in (f) and is compared with observations (small color dots). The ensemble simulation is in rather good agreement with the data in the western and northern parts of Europe, but with the lack of measurements in Eastern Europe no conclusion can be drawn regarding the intensity of the Saharan dust intrusion. The color scale is identical for all graphs. From the Copernicus Atmosphere Monitoring Service (CAMS) coordinated by ECMWF and supported by the European Commission.

Europe. Other models do not capture this event and predict low concentrations of particles in most areas of Europe. The differences between these projections show the impact of the choices made in these different model formulations. In the absence of better information, a "wisdom of crowds" assumption is often made that the average of the different models (also shown) is better than any single model.

10.4 General Considerations for Model Evaluation

10.4.1 Selection of Observations

Model evaluation generally relies on the one-to-one comparison of observations to model values sampled at the same location and time. It is important to recognize that simulated and observed fields may not be exactly comparable. The model may simulate a spatial average over a grid cell while the observations are from a particular location that may not reflect the grid cell average. This is called *representation error* and is discussed in Chapter 11 in the context of inverse modeling. Spatial interpolation of the observations (see Section 4.16) may help to reduce representation error, but the error is often not random. For example, sites from surface pollution networks are often concentrated in urban areas or in the vicinity of point sources, introducing bias when comparing to a coarse-resolution model that simulates the broader regional atmosphere. It may be necessary to exclude such sites from the comparison as non-representative.

Representation error applies to temporal variability as well. Time series measured at surface sites or from aircraft often show high-frequency anomalies such as spikes driven by concentrated plumes or local meteorological conditions. The model may not be designed to capture these anomalies, either because of grid averaging or because of temporal averaging of the input data. In addition, small transport errors may cause the model to slightly misplace plumes in a way that may not be relevant for general model evaluation but weighs heavily in model comparison statistics. Such statistical outliers in the distribution of observations can be illuminating in terms of understanding processes, and often deserve attention on a case-by-case basis. However, they should be excluded from a general model evaluation data set.

Surface air observations over land often show a large diurnal cycle driven by suppressed vertical mixing in the shallow stratified surface layer at night. Nighttime concentrations may thus be very low for species taken up by the surface, and very high for species emitted at the surface. Coarse-resolution models typically cannot capture this nighttime stratification, which may not be relevant for broader model evaluation since it affects only a small volume of atmosphere and may be viewed again as a representation error. In such cases, the nighttime values must be excluded from the statistical data used for model evaluation and the focus must be on simulating daytime values, when the surface measurements are more representative of a deep mixed layer that can be captured by the model.

10.4.2 Use of Satellite Observations

Satellites provide observations with coarse spatial resolution in the vertical (nadir view) or in the horizontal (limb view) and the model fields must be correspondingly averaged for comparison. A difficulty is that the satellite retrievals make assumptions about the atmosphere (*prior information*) that may be inconsistent with the model atmosphere. In that case, a straight comparison between model and observed fields can be very misleading. It is essential to re-process the model or observed fields to simulate what the satellite would see if it was observing the model atmosphere, rather than the true atmosphere with the assumed prior information. In the case of an *optimal estimate* satellite retrieval for gases (Rodgers, 2000; Chapter 11), the satellite reports vertical concentration profiles as

$$\widehat{\mathbf{x}} = \mathbf{A}\mathbf{x} + (\mathbf{I} - \mathbf{A})\mathbf{x_A} \tag{10.9}$$

where the vector $\widehat{\mathbf{x}}$ of dimension n is the retrieved profile consisting of concentrations at n vertical levels, $\mathbf{x}$ is the true vertical profile, $\mathbf{x_A}$ is the prior estimate, $\mathbf{A}$ is the averaging kernel matrix (see Chapter 11), and $\mathbf{I}$ is the identity matrix. The satellite data set provides not only $\widehat{\mathbf{x}}$ but also $\mathbf{A}$ and $\mathbf{x_A}$. One can then compare the observations $\widehat{\mathbf{x}}$ to the corresponding model fields $\widehat{\mathbf{x}}_{\mathbf{M}}$ computed as

$$\widehat{\mathbf{x}}_{\mathbf{M}} = \mathbf{A}\mathbf{x_M} + (\mathbf{I} - \mathbf{A})\mathbf{x_A} \tag{10.10}$$

where $\mathbf{x_M}$ is the actual model vertical profile (which would be the true profile if the model were perfect). It is important to recognize that the prior term $(\mathbf{I} - \mathbf{A})\mathbf{x_A}$ is common to $\widehat{\mathbf{x}}_{\mathbf{M}}$ and $\widehat{\mathbf{x}}$, and may give the illusion of better agreement between model and observations than is actually the case. See Zhang *et al.* (2010) for methods to address this issue.

 As another example, the column concentration Ω of a gas reported by a solar backscatter instrument is often retrieved as

$$\Omega = \frac{\Omega_s}{\mathscr{F}} \tag{10.11}$$

where Ω_s is the *slant column* measured by the satellite along its line of sight, and $\mathscr{F}$ is the air mass factor (AMF) that converts the slant column to the actual vertical column. The AMF was introduced in Section 5.2.4 for a non-scattering atmosphere. In that case, it was a simple geometric conversion factor. For the actual case of a scattering atmosphere, the AMF must be computed with a radiative transfer model that accounts for the scattering properties of the surface and the atmosphere, and for the assumed relative vertical concentration profile (*shape factor*) of the gas being measured (Palmer *et al.*, 2001). The shape factor assumed in the retrieval may be inconsistent with that in the model, and this then biases the comparison of model and observed Ω. The satellite data set generally includes not only Ω but also the corresponding Ω_s (or AMF, from which Ω_s can be obtained). For the purpose of model evaluation one must discard the reported Ω, recompute the AMF by using the local shape factor from the model, and apply it to the measured slant column Ω_s. See González Abad *et al.* (2015) for simple methods to do this.

As yet another example, satellite data for aerosol optical depth (AOD) are generally retrieved from nadir measurements of the top-of-atmosphere reflectance from the Earth's surface and its atmosphere. The aerosol contribution to this reflectance, from which the AOD is derived, is obtained with a radiative transfer model including assumed aerosol size distributions and refractive indices. These assumed aerosol characteristics are generally different from those simulated locally in the chemical transport model and used to compute the model AOD. One-to-one comparison of model to observed AODs is still valid inasmuch as the AOD is a physical diagnostic quantity. The comparison is difficult to interpret, however, because differences in AODs may be attributable to model errors in either aerosol mass concentrations or aerosol optical properties, and the assumed aerosol optical properties in the satellite retrieval are also subject to error. This is a problem in particular for data assimilation, as there are multiple ways to correct a model-observation difference in AODs.

10.4.3 Preliminary Evaluation and Temporal Scales

Section 10.5 presents different statistical metrics for evaluating the ability of a model to fit large observational data sets. A first step in model evaluation should be to visually inspect the simulated and observed fields for any prominent features that need to be better understood. This visual inspection should encompass as many of the variables as possible, for different spatial domains and temporal scales, as the different perspectives can provide unique information in the driving processes and the ability of the model to simulate them. For example, examination of mean vertical profiles in an aircraft data set offers quick information on the ability of the model to simulate boundary layer mixing, planetary boundary layer (PBL) depth, ventilation to the free troposphere, and any large-scale free tropospheric bias. A large contrast in observations over land and ocean may point to the need for separate statistical evaluation of both. An inability of the model to pick up this contrast may call into question the simulation of transport or chemical loss. For time series of large data sets it can be insightful to identify dominant patterns in the observations using *empirical orthogonal functions* (*EOFs*) and diagnose the ability of the model to reproduce these patterns. Calculation of EOFs is described in Appendix E.

We elaborate here on the consideration of different timescales when comparing model to observations. These timescales can be usefully separated as intra-day (diurnal), day-to-day (synoptic), seasonal, and interannual (or long-term trends). Concentrations at surface sites often show large diurnal variations due to mixed layer growth and decay, surface sources and sinks, and photochemistry. Comparison of mean diurnal variations between model and observations can test the model representation of these processes. An example is given in Figure 10.12 for the marine boundary layer.

Model evaluation on a day-to-day scale is useful to assess the capability of the model to account for synoptic-scale variations in chemistry, boundary layer dynamics, and the advection of different air masses. Figure 10.13 shows the complexity of the day-to-day variation of species concentrations at the surface. In this particular example, which compares calculated and measured concentrations of carbon monoxide (CO) and ozone, the model slightly underestimates the mixing ratio of both species (mean bias) as well as the amplitude of the fluctuations. Further analysis would quantify these differences and assess the overall skill of the model.

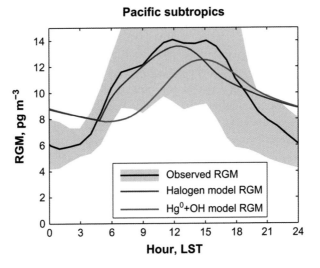

Pacific subtropics

Figure 10.12 Mean diurnal cycle of reactive gaseous mercury (RGM) in surface air over the Pacific. Observations from ship cruises (black lines, interquartile range in shading) are compared to model results with two different mechanisms for photochemical oxidation of elemental mercury (Hg⁰) to RGM. The model with halogen oxidants (red line) features a steeper morning rise than the model with oxidation by OH (blue line) and is more consistent with observations. Reproduced with permission from Holmes *et al.* (2009).

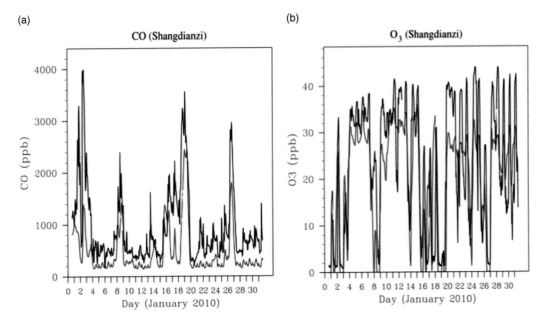

Figure 10.13 Time evolution of the surface mixing ratios [ppbv] of carbon monoxide (a) and ozone (b) in Shangdianzi, close to Beijing, China in January 2010. The values (red line) provided by the coupled meteorological and chemical regional model (WRF-Chem) are compared with surface measurements (black line). The comparison suggests that the model captures most high-pollution events (high CO concentrations) when the direction of the winds favors transport from pollution sources in the urban and industrial regions of China. During these events, ozone concentrations are generally low, presumably as a result of ozone titration by high concentrations of nitrogen oxides (not shown). During periods characterized by clean air, the model underestimates background carbon monoxide and ozone. Variations associated with diurnal variations in the height of the boundary layer are clearly visible in the ozone signal. Results provided by Idir Bouarar, Max Planck Institute for Meteorology (MPI-M). Measurements are from the Global Atmospheric Watch (GAW).

Seasonal variations provide information on influences from different climatological regimes, photochemistry, and emissions. Plotting simulated and observed
mean seasonal cycles can provide a quick revealing analysis of simulation bias.
Seasonal amplitude is generally much larger than interannual amplitude for species
with lifetimes less than a few months, so that one can usefully compare seasonal
cycles in models and observations from different years. As an example, Figure 10.14
compares the seasonal variation of simulated ozone with ozonesonde data for
different latitudes and altitudes.

Finally, comparison of observed and simulated interannual variability and long-
term trends in species concentrations indicates how well the model accounts for
climate modes and trends in emissions. As an example, Figure 10.15 compares
simulated and observed multi-year records of NO_2 column in Europe and east China,
testing the ability of emission inventories used in models to reproduce the trend of
NO_x emissions in each region.

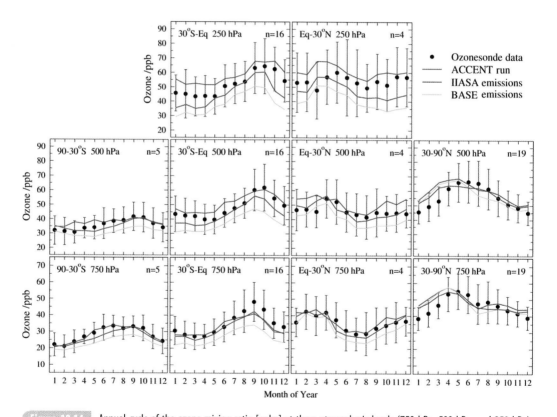

Figure 10.14 Annual cycle of the ozone mixing ratio [ppbv] at three atmospheric levels (750 hPa, 500 hPa, and 250 hPa)
averaged over four latitude bands (90–30° S, 30° S-eq, eq-30° N, 30–90° N). Comparison of multi-year
climatological ozonesonde measurements (Logan, 1999; Thompson *et al.*, 2003) with three model
simulations by the 3-D chemical transport model of Wild (2007). The differences between the BASE and the
IIASA cases result from differences in the emissions of ozone precursors; the differences between the IIASA
and ACCENT cases reflect differences in meteorology, model resolution, and the lightning source of NO_x.
Reproduced with permission from Wild (2007).

(a)

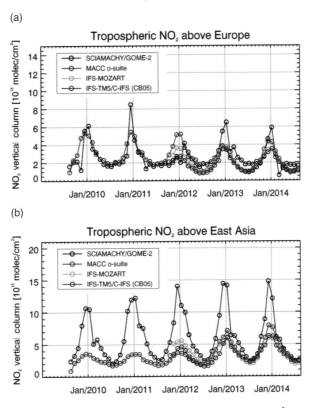

(b)

Figure 10.15 Comparison of calculated and observed seasonal evolution of the NO_2 column [cm^{-2}] for Europe (a) and East Asia (b). The black line represents retrievals from the GOME-2 and SCIAMACHY instruments. The numerical simulations are provided by two different models (TM5 and MOZART) with no data assimilation (blue and yellow lines) and by the ECMWF weather forecasting system with coupled chemistry and data assimilation (red line). There is good agreement between model results and observations in Europe, but not in East Asia, specifically during wintertime. Reproduced with permission from Eskes *et al.* (2015).

10.4.4 Aerosol Metrics

Aerosol concentrations are characterized in observations by a range of metrics including total mass concentrations for different species, total number concentrations, condensation nuclei (CN) and cloud condensation nuclei (CCN) number concentrations, size distributions (sometimes including speciation), hygroscopicity, aerosol optical depth (AOD), and absorbing aerosol optical depth (AAOD). Single-particle measurements provide additional information on particle phase and on the degree of internal mixing of different aerosol species. All of these measurements are relevant for model evaluation and provide different perspectives on aerosol sources and properties.

Many chemical transport models do not resolve aerosol microphysics and simulate only the speciated aerosol mass concentrations, treating individual aerosol species as chemicals in the model equations and ignoring the microphysical

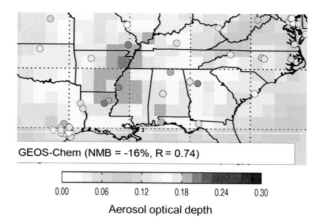

Figure 10.16 Aerosol optical depths over the Southeast USA in August–September 2013. The figure compares a mass-only aerosol simulation with the GEOS-Chem global model (background grid) to observations from the ground-based AERONET network (circles). Observations are highest in the western part of the region, which the model attributes to a dominant biogenic organic aerosol source. From Kim *et al.* (2015).

terms. These "mass-only" models can be compared directly to measurements of aerosol mass concentrations to evaluate the simulation of aerosol sources, chemistry, and loss by scavenging. They often assume fixed aerosol size distributions and optical properties for the purpose of simulating heterogeneous chemistry, aerosol radiative effects, and scavenging efficiencies. These can be compared to observations as part of the evaluation of model parameters. The simulation of radiative effects can be evaluated with measurements of AOD and AAOD, as illustrated in Figure 10.16.

Models including aerosol microphysics predict the number and size distributions of different aerosol species in addition to their mass. They can simulate the degree of mixing between different aerosol species and interactions with clouds. Such models can be evaluated with the full range of aerosol observations listed above to lend insight into particle nucleation, aerosol optical properties, chemical processes, and cloud effects. Figure 10.17 gives an example of model evaluation with observed size distributions.

10.4.5 Scatterplots

Comparison of model results to a large ensemble of observations requires statistical metrics to diagnose the significance and extent of discrepancies. A first quantitative evaluation can be made by plotting the N values simulated by the model and denoted $M_i(i = 1, N)$ as a function of the observed values $O_i(i = 1, N)$ at corresponding locations and times. The resulting *scatterplot* (Box 10.1) is characterized by a cloud of points, from which a regression line

$$\widehat{M} = a + bO \qquad (10.12)$$

| Box 10.1 | Construction of Scatterplots |

A scatterplot is a diagram that displays in Cartesian coordinates a collection of N points that represent pairs of data (x_i, y_i). It allows easy examination of different features in the (x, y) relationship including correlations, curvature in the relationship, clustering of points, presence of outliers, etc. A statistical relationship between the two quantities can be established using a best-fit procedure. Details are provided in Appendix E. This relationship is commonly expressed by a linear function:

$$\tilde{y} = a + bx$$

whose intercept a and slope b are determined by applying a linear regression method (method of least squares). The reduced major axis (RMA) regression is most appropriate to account for errors in both variables (see Appendix E).

The strength and direction of the linear relationship between the two variables are expressed by the *Pearson correlation coefficient* (r) defined by (10.27) and further discussed in Appendix E. When one quantity increases together with the other quantity, the correlation is positive and $r > 0$. In the opposite case, it is negative with $r < 0$. The value of $r \in [-1, +1]$ measures the dispersion of the data points around the regression line. If all the points fall exactly on the line then $|r| = 1$. When the values are not linked at all and the data points are fully dispersed in the graph, r is close to zero and the data are not significantly correlated. The fraction of the variance in y that is explained by the statistical model is called the *coefficient of determination* R^2 (see Appendix E). In the case of a linear regression, $R^2 = r^2$. Box 10.1 Figure 1 shows an example of a scatterplot with the corresponding regression line and coefficient of determination. The statistical significance of the correlation coefficient is discussed in Box 10.2.

Scatterplots are useful to analyze relationships between different chemical species in the atmosphere. Box 10.1 Figure 2, for example, shows the relationship between the observed concentrations of methane and CO_2 measured over southwestern Pennsylvania, color-coded by SO_2 concentrations. The diagram differentiates between air masses that are representative of the boundary layer in an urban environment and those that are characteristic of the free troposphere. Some air parcels with high methane concentrations originate from an area where extraction of natural gas from shale rock layers ("fracking") is taking place.

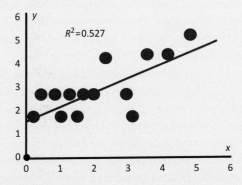

| Box 10.1 Figure 1 | Scatterplot between two arbitrary variables x and y with the regression line and the corresponding coefficient of determination R^2 deduced from the different data points. |

Box 10.1 (*cont.*)

When quantity y depends on several variables $x_1, x_2, \ldots$ the simple regression model must be replaced by a *multiple regression approach*. A linear relationship between three variables, for example, is represented by a regression plane in a 3-D space. Finally, in some situations, the linear regression model does not provide an adequate statistical model to represent the relationship between variables, and a nonlinear regression method must be applied to derive, for example, the regression coefficients of a degree-n polynomial or of other mathematical expressions.

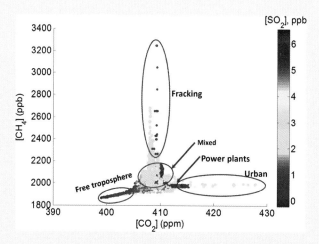

Box 10.1 Figure 2
Relationship between methane and CO_2 concentrations measured by aircraft over southwest Pennsylvania and color-coded by SO_2 concentrations. Figure provided by X. Ren and R. Dickerson, University of Maryland.

can be derived. Here $\widehat{M}$ represents a "statistical model" with coefficients a and b derived from the known values of M_i and O_i by an ordinary regression method or, if both variables are subject to uncertainties, by a reduced major axis (RMA) regression method (see Appendix E for more details). The departure of a from zero provides information on model offset (absolute bias) and the departure of b from unity provides information on relative bias, assuming that the linear model is correct. Linear regression software packages provide standard errors on a and b, but these generally assume that the linear model is correct and thus will underestimate actual errors. A better estimate of errors on a and b is obtained by *jackknife* or *bootstrap resampling* where regression coefficients are computed for different subsets of the data to yield a spread of a and b values from which the errors can be characterized.

The use of scatterplots for analyzing model results is illustrated in Figure 10.18. The panels compare the ensemble predictions of seven regional models for surface NO_2 and ozone concentrations for two different days in Europe with concentrations measured at different monitoring stations. An inspection of the data suggests that the model produces a weak summertime ozone episode with concentration values (typically 100–150 $\mu g\ m^{-3}$) fairly consistent with the observed values. Discrepancies with observations are worse for NO_2, which is strongly influenced by local pollution

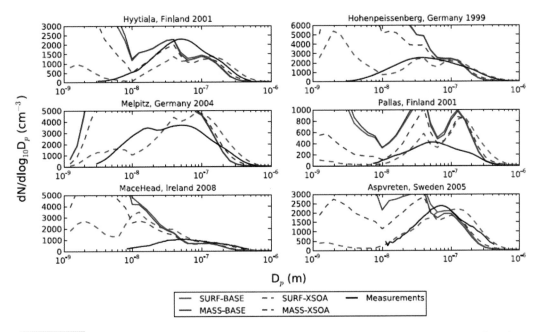

Figure 10.17 Comparison of the GEOS-Chem global model simulation with full aerosol microphysics to aerosol number size distributions measured at sites in Europe. The focus of comparison is to evaluate different model treatments of secondary organic aerosol (SOA). The BASE simulations assume SOA to be mainly biogenic, while the XSOA simulations include an additional anthropogenic source. The SURF simulations assume SOA formation to be kinetically limited by uptake to aerosol surfaces (irreversible uptake), while the MASS simulations assume SOA to be thermodynamically partitioned between the gas and pre-existing aerosol (reversible uptake). Results show that the best simulation is generally achieved for irreversible uptake including additional anthropogenic SOA. That simulation avoids in particular the overestimate in ultrafine aerosol concentrations, as the additional SOA promotes condensational growth of ultrafine aerosol to larger sizes. Reproduced with permission from D'Andrea *et al.* (2013).

sources. The scatterplots show that the NO_2 concentration values provided by the model ensemble are lower than the concentrations observed at the surface. In addition, the dispersion of the points around the 1:1 line is substantial and hence the correlation coefficient is low. In the case of ozone, the agreement between model and observations is considerably better, and the points are relatively close to the 1:1 line. The model ensemble, however, slightly overestimates the ozone concentrations in areas where the measured values are low and underestimates them where the observed values are high. More elaborate measures of model skill are discussed in the next sections.

10.5 Measures of Model Skill

Model skill is generally measured by the ability of the model to match observations of relevance to the problem of interest. In the case of operational models, such as

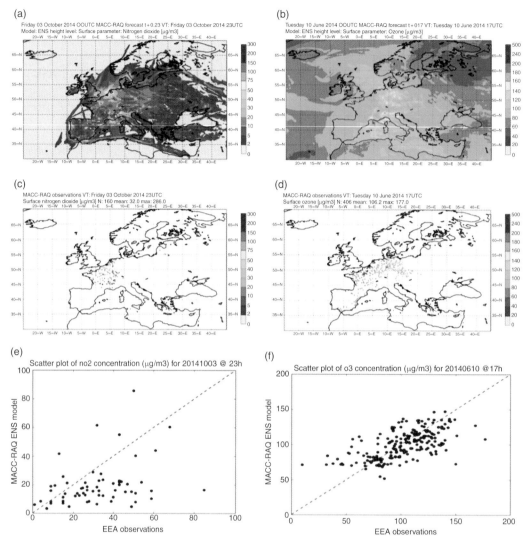

Figure 10.18 (a–b) Prediction of the NO$_2$ concentration on October 3, 2014 (a) and the ozone (b) concentrations on June 10, 2014 [µg m^{-3}] from the regional daily forecast for Europe based on the median ensemble of seven European chemistry transport models produced in the frame of the EU-MACC projects (Monitoring Atmospheric Composition and Climate, 2009–2015). (c–d) Measurements of NO$_2$ (c) and ozone (d) concentrations at European monitoring stations. (e–f) Scatterplots representing the calculated concentrations as a function of the measured values. Courtesy of Virginie Marecal, Meteo-France.

used for air quality forecasts, the ability to match observations is the ultimate goal and is often measured as a *model score*.

10.5.1 Basic Metrics

Different statistical metrics for paired comparisons of model to observed values are used to test the skill of chemical transport models. Metrics allow a general assessment of the confidence in model analyses and forecasts. Consider N paired model and observed quantities M_i and O_i, whose averages are respectively:

$$\overline{M} = \frac{1}{N}\sum_{i=1}^{N} M_i \qquad \text{and} \qquad \overline{O} = \frac{1}{N}\sum_{i=1}^{N} O_i \qquad (10.13)$$

and standard deviations are:

$$\sigma_M = \left[\frac{1}{N}\sum_{i=1}^{N}\left(M_i - \overline{M}\right)^2\right]^{\frac{1}{2}} \qquad \text{and} \qquad \sigma_O = \left[\frac{1}{N}\sum_{i=1}^{N}\left(O_i - \overline{O}\right)^2\right]^{\frac{1}{2}} \qquad (10.14)$$

The ensemble N includes data collected at different sites and/or at different times. Individual points included in the ensemble should be independent so that each paired comparison provides independent information. Observations from nearby sites may be strongly correlated and thus not provide independent information. Spatial and temporal scales over which the observations are significantly correlated can be determined with an *autocorrelogram* that plots the correlation coefficient vs. the spatial or temporal separation between data points.

As pointed out in Section 10.4, data outliers require careful consideration because they may weigh heavily in the statistical results. Model fields are generally well-behaved, and any unexplained outliers should be scrutinized as they may reveal basic model errors. Observational outliers are common and may reflect instrumental error or unusual conditions that the model is not intended to represent; if so, they should be screened from the comparison. A standard way to detect outliers is to plot the data against the normal (or log-normal) distribution, revealing extrema that depart from the distribution. These extrema should be discarded if it can be reasonably ascertained that they are not part of the population for which model evaluation is desired or feasible.

Two basic measures of model skill are often adopted: *the mean bias* (*BIAS*)

$$BIAS = \frac{1}{N}\sum_{i=1}^{N}(M_i - O_i) = \overline{M} - \overline{O} \qquad (10.15)$$

that represents the difference between the mean model and observed quantities, and the *root mean squared error* (*RMSE*):

$$RMSE = \left[\frac{1}{N}\sum_{i=1}^{N}(M_i - O_i)^2\right]^{\frac{1}{2}} \qquad (10.16)$$

that represents the spread of the individual errors. These two metrics are expressed in the same units as M_i and O_i. The *systematic root mean square error* (*RMSE$_S$*) describes the bias between observed data points O_i and the linear least square fit to the observations $\widehat{M}_i = a + bO_i$ (see (10.12) and Box 10.1). In a scatterplot, it is determined by the square of the distance between the linear regression line and the 1:1 line:

$$RMSE_S = \left[\frac{1}{N} \sum_{i=1}^{N} \left(\widehat{M}_i - O_i \right)^2 \right]^{\frac{1}{2}} \tag{10.17}$$

The *unsystematic root mean square error* ($RMSE_U$):

$$RMSE_U = \left[\frac{1}{N} \sum_{i=1}^{N} \left(M_i - \widehat{M}_i \right)^2 \right]^{\frac{1}{2}} \tag{10.18}$$

is derived as a function of the distance between the data points M_i and the linear regression line $\widehat{M}_i$. It represents the scatter of the data about the best-fit line, and can thus be regarded as a measure of model precision. A successful model is characterized by a low value of $RMSE_S$ and a value of $RMSE_U$ close to $RMSE$ because:

$$RMSE^2 = RMSE_S^2 + RMSE_U^2 \tag{10.19}$$

Measures such as the *mean absolute error* (MAE):

$$MAE = \frac{1}{N} \sum_{i=1}^{N} |M_i - O_i| \tag{10.20}$$

and the *mean absolute deviation* (MAD):

$$MAD = \frac{1}{N} \sum_{i=1}^{N} |O_i - \overline{O}| \tag{10.21}$$

involving absolute values of the differences are sometimes preferred to measures based on squared differences because they are less sensitive to high values. The *mean normalized bias* (MNB):

$$MNB = \frac{1}{N} \sum_{i=1}^{N} \left(\frac{M_i - O_i}{O_i} \right) \tag{10.22}$$

and the *mean normalized absolute error* ($MNAE$):

$$MNAE = \frac{1}{N} \sum_{i=1}^{N} \left(\frac{|M_i - O_i|}{O_i} \right) \tag{10.23}$$

may be appropriate scoring measures when the data cover an extended range of values to avoid overemphasizing the high tail of the distribution. The MNB has the disadvantage, however, of being asymmetric with respect to under- and overestimation. For example, when the model overestimates the measured value, the MNB can increase to values much larger than unity; however, when it underestimates the observation its value is limited to −1. This issue is addressed by introducing the *mean fractional bias* (MFB):

$$MFB = \frac{2}{N} \sum_{i=1}^{N} \left[\frac{M_i - O_i}{M_i + O_i} \right] \tag{10.24}$$

which varies in the range [–2, +2] and is complemented by the *mean fractional error* (*MFE*):

$$MFE = \frac{2}{N} \sum_{i=1}^{N} \left[\frac{|M_i - O_i|}{M_i + O_i} \right] \qquad (10.25)$$

A problem with the last four metrics is that they tend to overemphasize low values where relative errors may be large but of little relevance to the problem of interest. This can be corrected by taking the ratios of the sums, instead of the sums of the ratios. For example, instead of the MNB, one can use the *normalized mean bias (NMB)* for more robust statistics:

$$NMB = \frac{\sum_{i=1}^{N} (M_i - O_i)}{\sum_{i=1}^{N} O_i} \qquad (10.26)$$

Other dimensionless indices of agreement have been proposed for measuring model skill and are listed in Table 10.3. Willmott *et al.* (2012) and Chai and Draxler (2014) discuss the advantages and disadvantages of different indices.

The *Pearson correlation coefficient*

$$r = \frac{\sum_{i=1}^{N} (M_i - \overline{M})(O_i - \overline{O})}{\left[\sum_{i=1}^{N} (M_i - \overline{M})^2 \right]^{\frac{1}{2}} \left[\sum_{i=1}^{N} (O_i - \overline{O})^2 \right]^{\frac{1}{2}}} \qquad (10.27)$$

is the covariance between model and observed values normalized to the variances. It provides different information from the above metrics in that it characterizes the extent to which *patterns* in the observations are matched by the patterns in the model. Its value may range from –1 to +1. A value of +1 indicates a perfect match. A positive value indicates the level of skill with a *statistical significance* that depends on sample size (Box 10.2 and Section 10.6). Values of r near zero imply that the variability in the observations is controlled by processes that the model does not capture. A model that is able to capture the observed means but not the observed variability (non-significant r) may be getting the mean right for the wrong reasons. Negative values of r imply large model errors in the simulation of processes.

Binary prediction of atmospheric events. A model prediction of a specific event such as an air pollution episode at a given location (e.g., concentration of pollutants exceeding a regulatory threshold) can be evaluated as a binary variable by distinguishing four possible situations: (1) the event is predicted and observed; (2) the event is not predicted and not observed; (3) the event is predicted but not observed; and (4) the event is not predicted but is observed. Cases (1) and (2) are successful predictions (hits), while cases (3) and (4) are failures (misses). Consider a sample of N predictions covering a certain period of time and with each prediction having an outcome yy, nn, yn, or ny. The first letter is the prediction of whether the event occurs (y for yes, n for no) and the second letter indicates whether the event is observed. We have $N = yy + nn + yn + ny$. The skill of the model for binary prediction (event or no event) can be measured by the fraction of correct predictions (PC):

Table 10.3 Indices of agreement proposed by different authors to assess model performance

Author	Index of agreement
Willmott (1981)	$d_1 = 1 - \dfrac{\sum\limits_{i=1}^{N}(M_i - O_i)^2}{\sum\limits_{i=1}^{N}\left(|M_i - \overline{O}| + |O_i - \overline{O}|\right)^2}$
Willmott et al. (1985)	$d_2 = 1 - \dfrac{\sum\limits_{i=1}^{N}|M_i - O_i|}{\sum\limits_{i=1}^{N}\left(|M_i - \overline{O}| + |O_i - \overline{O}|\right)}$
Willmott et al. (2012)	$d_3 = 1 - \dfrac{\sum\limits_{i=1}^{N}|M_i - O_i|}{2\sum\limits_{i=1}^{N}|O_i - \overline{O}|}$ $if \sum\limits_{i=1}^{N}|M_i - O_i| \leq 2\sum\limits_{i=1}^{N}|O_i - \overline{O}|$
	$d_3 = \dfrac{2\sum\limits_{i=1}^{N}|O_i - \overline{O}|}{\sum\limits_{i=1}^{N}|M_i - O_i|_i} - 1$ $if \sum\limits_{i=1}^{N}|M_i - O_i|_i > 2\sum\limits_{i=1}^{N}|O_i - \overline{O}|$
Nash and Sutcliffe (1970)	$d_4 = 1 - \dfrac{\sum\limits_{i=1}^{N}(M_i - O_i)^2}{\sum\limits_{i=1}^{N}(O_i - \overline{O})^2}$
Legates and McCabe (1999)	$d_6 = 1 - \dfrac{\sum\limits_{i=1}^{N}|M_i - O_i|}{\sum\limits_{i=1}^{N}|O_i - \overline{O}|}$
Watterson (1996)	$d_7 = \dfrac{2}{\pi}\sin^{-1}\left[1 - \dfrac{(RMSE)^2}{\sigma_M^2 + \sigma_O^2 + (\overline{M} - \overline{O})^2}\right]$
Mielke and Berry (2001)	$d_8 = 1 - \dfrac{MAE}{\frac{1}{N^2}\sum\limits_{i=1}^{N}\sum\limits_{j=1}^{N}|M_j - O_i|}$
Douglass et al. (1999)	$d_9 = 1 - \dfrac{1}{3}\dfrac{|\overline{M} - \overline{O}|}{\sigma_O}$

$$PC = \frac{yy + nn}{N} \qquad (10.28)$$

This is a lenient metric because the PC will be high when events are rare even if the model has no skill at predicting the events. The fraction of observed events that were correctly predicted (called *probability of detection* or *POD*) is

Box 10.2   **Statistical Significance of the Correlation Coefficient**

The Pearson correlation coefficient $r(x, y)$ characterizes the strength of a linear relationship between two variables x and y. Whether the correlation is meaningful or not depends on the size of the sample that is being examined. A statistical test must determine the *significance* of the correlation coefficient for a given probability level.

The correlation of a population is said to be significant if the correlation coefficient $\rho(x, y)$ of the *entire population* is different from zero. This is different from the correlation $r(x, y)$ that can be determined for a *sample* of that population. We apply here the "null hypothesis" in which we assume that there is no correlation between x and y [$\rho(x, y) = 0$], and that, if the value $r(x, y)$ measured from a sample of limited size n is different from zero, it is due to sampling errors (the size of the sample is too small). If this hypothesis is verified for a given probability, the correlation is not significant. If, however, the null hypothesis is rejected by the statistical test, the correlation is regarded as significant at a certain *level of confidence* defined by the probability that the population is correlated. The *risk factor* is the complementary probability that the population is in fact not correlated

Significance of a correlation at a certain level of confidence can be diagnosed by *Student's t-test*. For a given correlation coefficient r and sample size n we compute

$$t = r\sqrt{\frac{n-2}{1-r^2}}$$

and compare it to the corresponding critical value from a statistical table (Appendix E). The correlation is significant if t exceeds the critical value. A confidence level of 95% corresponding to a risk factor of 5% ($p < 0.05$) is commonly used and Box 10.2 Figure 1 gives the corresponding threshold value of r for different sample sizes.

As an illustration, consider a sample of ten data points with a calculated $r = 0.6$. We derive from the above formula a value of $t = 2.12$. For a risk factor of 5% (confidence level of 95%), the table of Appendix E provides for t a critical value of 2.31. In this case, the null hypothesis of zero

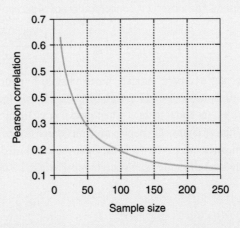

Box 10.2
Figure 1   Minimum absolute value of the Pearson correlation coefficient for significant correlation at the 95% confidence level, as a function of sample size. Reproduced from Wikipedia.

Box 10.2 (*cont.*)

correlation cannot be rejected, and the correlation coefficient derived from the sample is therefore not significant. If, however, we consider 52 data points with a measured r of only 0.3, the calculated value of t is 2.24, which is above the critical value of 2.01 found in the table, and the correlation is significant.

$$POD = \frac{yy}{yy + ny} \qquad (10.29)$$

The fraction of predicted events that did not occur (called *false alarm ratio or FAR*) is

$$FAR = \frac{yn}{yy + yn} \qquad (10.30)$$

The *critical success index*, also called the *threat score* (*TS*) is defined by

$$TS = \frac{yy}{yy + yn + ny} \qquad (10.31)$$

Some of the predictions may be successful by chance. To correct for this, one defines the *equitable threat score* (*ETS*)

$$ETS = \frac{yy - \alpha}{yy + yn + ny - \alpha} \qquad (10.32)$$

in which α is the number of events that would be predicted by chance,

$$\alpha = (yy + yn)(yy + ny) \qquad (10.33)$$

and is subtracted from the number of hits. The *ETS* is always lower than 1 and is negative if the prediction by chance is better than the actual prediction.

Grading models. When comparing different models, it is useful to attribute to each of them a grade for their ability to correctly simulate atmospheric observations. In the approach adopted by Douglass *et al.* (1999), the grade $g_{m,j}$ by which a model m represents the observed concentration of a particular species j is given by

$$g_{m,j} = 1 - \frac{1}{n_g} \frac{\left| \overline{M}_{m,j} - \overline{O}_j \right|}{\sigma_j} \qquad (10.34)$$

Here, the overbars indicate mean values, σ_j denotes the standard deviation in the observations, and n_g is a scaling factor. If this factor is taken to be equal to 3, the grade $g_{m,j}$ is equal to zero when model m is $3\sigma_j$ apart from the mean observed value.

When N chemical species are considered for evaluating model skill, and hence N diagnostics are performed, they can be combined to derive for a given model m a single overall grade G_m (Waugh and Eyring, 2008):

$$G_m = \frac{\sum_{j=1}^{N} w_j\, g_{m,j}}{\sum_{j=1}^{N} w_j} \qquad (10.35)$$

where w_j denotes a weight assigned to the diagnostic for species j as a measure of its importance.

When an ensemble of K models are used to calculate a given field j, a best estimate of the predicted concentration can be derived as:

$$\widehat{M}_j = \frac{\sum\limits_{m=1}^{K} G_m M_{m,j}}{\sum\limits_{m=1}^{K} G_m} \tag{10.36}$$

where G_m acts as a weight that favors the solutions obtained by the models with the best grades. The variance weighted by the model scores is given by:

$$\widehat{\sigma}_j^2 = \frac{\sum\limits_{m=1}^{K} G_m}{\left(\sum\limits_{m=1}^{K} G_m\right)^2 - \sum\limits_{m=1}^{K} G_m^2} \sum\limits_{m=1}^{K} G_m \left(M_{m,j} - \widehat{M}_j\right)^2 \tag{10.37}$$

10.5.2 The Taylor Diagram

Taylor (2001) proposed a concise graphical method for representing on a single figure several statistical indicators that describe the degree of agreement between model results and observations. The Taylor diagram indicates how model and observed patterns compare in terms of their correlation, their RMS differences, and the ratio of their variances.

Pattern similarities between the calculated and observed fields M_i and O_i can be quantified by the Pearson correlation coefficient r (10.27). However, this does not provide information about the relative amplitude of the two quantities. It is therefore useful to introduce the centered root mean square error ($CRMSE$):

$$CRMSE = \left\{\frac{1}{N}\sum\limits_{i=1}^{N} \left[\left(M_i - \overline{M}\right) - \left(O_i - \overline{O}\right)\right]^2\right\}^{1/2} \tag{10.38}$$

This quantity tends to zero when the patterns of the two fields *and* the associated amplitudes are similar. However, the $CRMSE$ does not indicate if the error is due to a difference in the phase or in the amplitude of the signals. An additional comparison of the two fields is their standard deviations σ_O and σ_M defined by (10.14). The Taylor diagram is constructed by recognizing that the four statistical quantities (r, $CRMSE$, σ_M and σ_O) are related by

$$(CRMSE)^2 = \sigma_M^2 + \sigma_O^2 - 2\,\sigma_M\,\sigma_O\,r \tag{10.39}$$

Figure 10.19 illustrates the Taylor diagram. The red arrows are a geometric representation of (10.39). The polar graph provides a rapid quantification of the four statistical parameters for any point on the diagram. The standard deviations of the observed field O and the model field M are represented by the radial distances from the origin; the point representative of the observations, called *reference point*, is

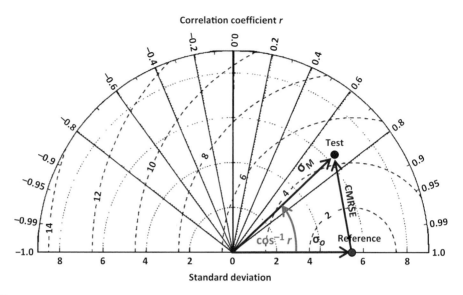

Figure 10.19 Taylor diagram summarizing the statistical comparison of a test data set (model M) to a reference data set (observations O). The observed standard deviation σ_O is plotted on the right horizontal axis. The model standard deviation σ_M is plotted as the dotted lines with values given on the left horizontal axis. The CRMSE is given by the dashed lines and the correlation coefficient r by the solid lines. The statistical fit between model and observations is given by the test point on the diagram. The reference point on the diagram indicates a perfect model. Knowledge of σ_M and r is sufficient to define the location of the test point, and from there the CRMSE is determined by the distance between the reference point and the test point.

located on the x-axis, with the abscissa equal to the corresponding standard deviation σ_O. The radial distance between the origin and the location of the *test point*, which characterizes the simulated field, is equal to the standard deviation σ_M of the calculated field. The correlation coefficient r between the observed and calculated fields is shown by the azimuthal position on the diagram. The distance between the reference and test points is the *CRMSE*.

An alternate form of the Taylor diagram is often used in which *CRMSE* and σ_M are normalized to the standard deviation σ_O of the observed field. This allows the representation of multiple data sets having different concentrations and/or units. Figure 10.20 gives an example in which comparison statistics for multiple species are shown on a single diagram. The normalized reference point is 1 on the x-axis.

Taylor (2001) and Brunner *et al.* (2003) propose different expressions to quantify the skill score of a model. These expressions assume that for a given variance, the score increases monotonically with increasing correlation. Further, for a given correlation, the score increases as the variance produced by the model approaches the variance associated with the observation. The resultant expressions for the model skill S take the form

$$S = \frac{4(1+r)^n}{\left(\dfrac{\sigma_M}{\sigma_O}+\dfrac{\sigma_O}{\sigma_M}\right)^2 + (1+r_0)^n} \qquad (10.40)$$

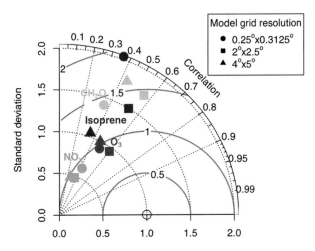

Figure 10.20 Taylor diagram for a comparison of chemical transport model results to aircraft observations of ozone, NO_x, isoprene, and formaldehyde (CH_2O) in the Southeast USA mixed layer in summer 2013. The different symbols describe model simulations at different horizontal grid resolutions ($0.25° \times 0.3125°$, $2° \times 2.5°$, $4° \times 5°$). The radial coordinate is the normalized standard deviation σ_M/σ_O. The angular coordinate is the Pearson correlation coefficient. The open circle represents the observations (reference point of Figure 10.19). The normalized *CRMSE* is shown as solid lines. The figure shows that the best model simulation is for ozone. Correlation improves when the resolution increases from $4° \times 5°$ to $2° \times 2.5°$ but then decreases at $0.25° \times 0.3125°$ because fine-scale features are more difficult to capture by the model than broad synoptic-scale features. From Yu *et al.* (2016).

where r_0 is the maximum attainable correlation. This model-dependent parameter, which must be estimated, accounts for the fact that the model is not expected to reproduce the details of the noise in the data (unforced variability). The value of exponent n is chosen according to the weight given on a good correlation versus a small RMS error. Isolines for skill scores based on such expressions can be represented on the Taylor diagram.

10.5.3 The Target Diagram

The Taylor diagram does not provide information on the mean bias (*BIAS*) between model and observed quantities. The *Target diagram* (Jolliff *et al.*, 2009; Thunis *et al.*, 2012; Figure 10.21) provides this missing information in addition to summary information about the pattern statistics, thus yielding a broader overview of their relative contribution to the total *RMSE*. Again, the values of the statistical indicators are normalized to the standard deviation of the observations σ_O. Using Cartesian coordinates, the value of $CRMSE/\sigma_O$ is displayed on the x-axis and the value of $BIAS/\sigma_O$ on the y-axis. One can show that:

$$(RMSE)^2 = (BIAS)^2 + (CRMSE)^2 \tag{10.41}$$

so that the distance between the origin and any data point displayed on the diagram represents the total *RMSE* normalized by σ_O and is therefore viewed as the *target indicator*.

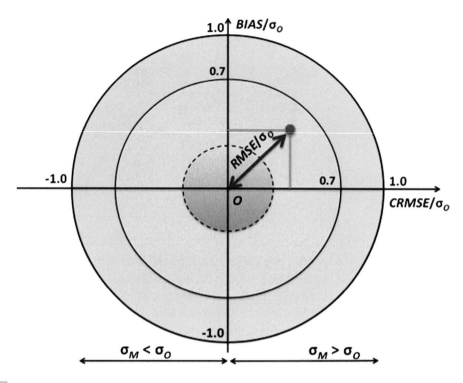

Figure 10.21 Schematic representation of the Target diagram.

Markers may be added within the diagram to better evaluate the model results. In Figure 10.21, the outermost circle corresponds to $RMSE/\sigma_O = 1$, so for all points inside this contour, the model data are positively correlated with observational data. A second contour corresponds to a higher performance, here $RMSE/\sigma_O = 0.7$. All points that represent successful model calculations are expected to appear inside this second contour. A third contour (dashed line) can be added to characterize the threshold of observational uncertainties; no meaningful improvement in the model-data agreement is obtained as the points displayed inside this circle approach the origin (target) of the diagram. Finally, the x-axis of the Target diagram is used to provide information on standard deviations: If the model standard deviation is larger than the observed one, the points are plotted on the left side of the diagram (negative abscissa); in the opposite case, they are plotted on the right side (positive abscissa). A weakness of the Target diagram is that it does not provide explicit information about the correlation coefficient.

10.6 Significance in the Difference Between Two Data Sets

An important question in the comparison of two data sets (such as model vs. observations) is whether differences between the two data sets are real or the result

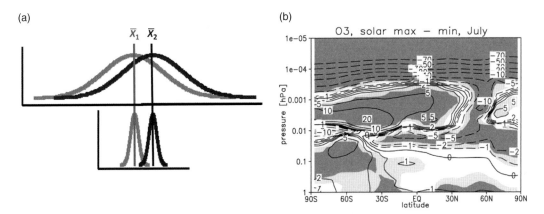

Figure 10.22   (a) Example of two distributions for random variables $\overline{X}_1$ (green) and $\overline{X}_2$ (blue). The mean values are identical for the cases characterized by high (top) and low (bottom) variability, but the overlap between the distributions is very different in the two cases. The significance of the difference between the averages of $\overline{X}_1$ and $\overline{X}_2$ is highest in the low-variability case (bottom panel with little overlap between distributions). (b) Application of Student's t-test to derive the significance of the differences in the mesospheric ozone concentrations calculated by a chemistry-climate model for high and low solar activity, respectively (July conditions). Statistical significance larger than 90% (99%) is indicated by light (dark) gray shading. Reproduced from Schmidt *et al.* (2006). Copyright © American Meteorological Society, used with permission.

of random noise. The *statistical significance* of a difference expresses the likelihood that it is real as opposed to random. Consider here the comparison between two sets of sampled data (X_1, X_2) of sizes (n_1, n_2) with distributions defined by the population means $(\overline{X}_1, \overline{X}_2)$ and unbiased estimators of their variance (σ_1, σ_2) (see Appendix E). The *Student's t-test* provides a statistical method to estimate the likelihood that the difference between the means of two distributions is significant (Figure 10.22). It assumes that the distribution of the populations is normal (Gaussian). In the student t-test, the t-variable is given by

$$t = \frac{\overline{X}_1 - \overline{X}_2}{\sigma_T \left[\dfrac{1}{n_1} + \dfrac{1}{n_2}\right]^{1/2}} \tag{10.42}$$

where σ_T is the pooled standard deviation of the two samples

$$\sigma_T^2 = \frac{(n_1 - 1)\sigma_1^2 + (n_2 - 1)\sigma_2^2}{n_1 + n_2 - 2} \tag{10.43}$$

The value of t computed from (10.42) is compared to the critical value t_c provided by a statistical table for a given value of the number of degrees of freedom $(n_1 + n_2 - 2)$ and for a user-specified risk factor p (Appendix E). If t exceeds t_c, the difference between the averages of the two distributions is considered to be significant at the specified risk level p. If p is chosen to be equal to 5%, the confidence level that the samples differ from each other is equal to 95%. Atmospheric maps showing the spatial distribution of differences between two fields usually highlight the areas where the results are statistically significant at a specified confidence level, and where they are not. Figure 10.22 gives an example. It is conventional in the literature to qualify a statistically significant result as confident, highly confident, and very highly confident if the adopted level of confidence is equal to 95%, 99%, and 99.9%, respectively.

10.7 Using Models to Interpret Observations

The model evaluation metrics described in Section 10.5 are intended to summarize the ability of a model to reproduce large ensembles of observations. They should be supplemented by more ad-hoc comparisons of temporal and spatial patterns, including relationships between species and with meteorological variables, as described in Section 10.4. Combination of these procedures is essential for establishing confidence in the model as a tool to interpret present-day atmospheric behavior and to make future projections. It provides the foundation for using the model to derive chemical budgets, conduct source–receptor analyses, infer source attribution from sensitivity simulations, etc. In this evaluation perspective, the observations are a given, and the task of the model is to reproduce them within a certain error tolerance. Here, we briefly discuss a different use of the model as a tool to *explain* the variability in the observations and from there to understand the processes that drive this variability. This involves a somewhat different perspective in model evaluation.

The general scientific approach for understanding the behavior of a complex system is to observe its variability and interpret it in terms of the driving variables. This interpretation requires a model as simplification of the system. The model may be very simple and/or qualitative, and indeed such simple models (even mental models) are often presented in observational papers as a first analysis of the data. However, simple models may be flawed by omission of important processes that are not always apparent. For complex problems in atmospheric chemistry, such as those coupling chemistry and transport, access to a 3-D model is usually required for successful interpretation. Here the purpose of the model is to distill the phenomena driving the observed variability through sensitivity simulations and/or through model simplifications to highlight the essential variables. The focus is on interpreting observations to gain scientific understanding, and the model is a tool for addressing that objective.

Interpreting observed correlations between species is an important example. These correlations can point to common sources or source regions, as in the methane vs. CO_2 relationships shown in Box 10.1 Figure 2. Changes in the relationships between different air masses provide insights into atmospheric processes or cause-to-effect connections. Quantitatively interpreting the correlations in terms of constraints on processes is, however, fraught with pitfalls because the factors driving the correlations are often not intuitive or easy to isolate. Simulation of the relationships with a 3-D model including a comprehensive treatment of processes can illuminate the interpretation of the observed relationships and in this manner advance knowledge.

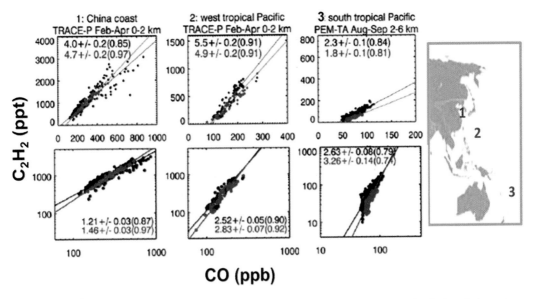

Figure 10.23 Relationships between acetylene (C$_2$H$_2$) and CO concentrations over the western Pacific. Aircraft observations for different regions (in black) are compared to results from the GEOS-Chem global 3-D chemical transport model (in red). The top row shows linear relationships and the bottom row shows log–log relationships. Reduced-major-axis (RMA) regression lines are shown with coefficient of determination (R^2) in parentheses. Errors on the regression lines are determined with the bootstrap method. Note the differences in scales between panels. Adapted from Xiao *et al.* (2007). Observations are from D. R. Blake (University of California – Irvine) and G. W. Sachse (NASA).

We illustrate this point here with the interpretation of observed correlations of acetylene (C$_2$H$_2$) with CO, as presented by Xiao *et al.* (2007; Figure 10.23). Both C$_2$H$_2$ and CO are emitted almost exclusively by combustion, and both are removed from the atmosphere by oxidation by OH with mean lifetimes of ten days and two months, respectively. Observations taken from aircraft campaigns around the world consistently show strong correlations between C$_2$H$_2$ and CO, from source regions to the most remote air masses. We would like to extract the constraints that these correlations provide for improving our understanding of emissions, atmospheric transport, and OH concentrations.

Figure 10.23 shows aircraft observations of the C$_2$H$_2$–CO relationship over the Pacific just off the China coast (boundary layer outflow), in the more remote west tropical Pacific, and in the very remote south tropical Pacific. The top panel shows the linear relationships ([C$_2$H$_2$] vs. [CO]) and the bottom panel shows the log–log relationships (log[C$_2$H$_2$] vs. log[CO]). Also shown in the figure are the correlations simulated by a global 3-D model. We see that the model reproduces the correlations but there is significant bias in the slopes. We can then use the model to understand the meaning of the correlations and the factors driving the slopes.

Let us first examine the linear correlations (top row). The correlation in the fresh Chinese outflow (top left panel) reflects the dilution of polluted Chinese air masses with background air. The transport time since emission is much shorter than the lifetimes of either C$_2$H$_2$ or CO. The C$_2$H$_2$:CO slope therefore reflects the Chinese emission ratio, providing a useful test of emission inventories. However, we find that the model slope of 4.7 is lower than the Chinese emission ratio used in the model (6.2), because the dilution takes place with non-zero background air. Thus, one cannot interpret the observed slope (4.0) as the emission ratio without accounting for this background correction. The marine air in which the outflow is diluting may further be different from the continental background air where the initial dilution took place. The best estimate of the emission ratio can be made by adjusting the emissions in the model to reproduce the observed slope.

The importance of characterizing the background is even more apparent in the correlations over the more remote west tropical Pacific (top row, middle panel). Here, the correlations are just as strong as in the fresh Chinese outflow, and the slopes are larger than in the fresh outflow, both in the observations and the model. This is counter-intuitive since one would expect chemical loss of C_2H_2 to decrease the slope, and we can turn to the model to explain this result. We find in the model that the higher slope is because the dilution is now taking place with tropical background air containing very low C_2H_2. In that case the C_2H_2–CO correlation is determined by the mixing between mid-latitude and tropical air masses, and provides little information on either emissions or chemistry. The south tropical Pacific (top row, right panel) shows lower slopes because all air masses in that case have experienced considerable chemical aging – note the differences in scale between panels. One could use the correlations over the south tropical Pacific to provide constraints on chemistry (and hence on OH concentrations), but separating chemical influence from emissions and transport is not straightforward.

Correlating the logarithms of concentrations offers a means to remove the influence of emissions. McKeen *et al.* (1996) proposed a simple Lagrangian mixing model for this purpose. Consider two species i and j in an aging air parcel receiving no fresh emission inputs and diluting at a constant rate in a uniform background. The evolution of the mixing ratio C_i of species i in that air parcel is given by

$$\frac{dC_i}{dt} = -L_i C_i - K_d(C_i - C_{i,b}) \tag{10.44}$$

Here, $L_i = k_i[\text{OH}]$ is the first-order chemical loss frequency $[\text{s}^{-1}]$ where k_i is the rate constant for reaction with OH, $K_d \, [\text{s}^{-1}]$ is a dilution rate constant, and $C_{i,b}$ is the background mixing ratio. The chemical lifetime of species i is $\tau_i = 1/L_i$. A similar equation holds for species j. Let us assume that [OH], K_d, $C_{i,b}$, and $C_{j,b}$ are constant, and let $\beta = d\ln C_j/d\ln C_i$ denote the slope of the log–log relationship. Simple analytical solutions for β are available in three limiting cases:

1. $\beta \approx \dfrac{L_i}{L_j} = \dfrac{\tau_j}{\tau_i}$ chemical loss fast relative to dilution

2. $\beta \approx \dfrac{K_d + L_i}{K_d + L_j}$ negligibly low background   (10.45)

3. $\beta \approx 1$ fast dilution relative to chemical loss

These limiting expressions are useful for interpreting correlations when the proper conditions apply. The case of the C_2H_2–CO correlation is problematic because both species have relatively long chemical lifetimes and non-negligible backgrounds.

Ehhalt *et al.* (1998) proposed an alternate simple Eulerian model in which dilution with background air is represented as an eddy diffusive process. Assuming the diffusion to take place in one dimension x with eddy diffusion coefficient K_x, we have

$$\frac{\partial C_i}{\partial t} = -L_i C_i + K_x \frac{\partial^2 C_i}{\partial x^2} \tag{10.46}$$

The steady-state solution $C_i(x)$ subject to boundary conditions $C_i(0)$ at the point of origin and $C_i(\infty) \to 0$ is given by

$$C_i(x) = C_i(0) \exp\left[-\frac{x}{\sqrt{K_x/L_i}}\right] \tag{10.47}$$

and one then finds $\beta = \sqrt{\tau_j/\tau_i}$. Ehhalt *et al.* (1998) show that this model is most realistic when chemical loss and dilution take place at comparable rates, whereas a situation where chemical loss is faster than dilution will tend toward the first limiting case of the Lagrangian mixing model $\beta \approx \tau_j/\tau_i$.

Figure 10.23 (bottom panel) applies these ideas to the C_2H_2–CO relationship by examining the log–log correlation in the observations and the model. In this case, $\tau_{CO}/\tau_{C2H2} = 3$–3.5 and $\sqrt{\tau_{CO}/\tau_{C2H2}} = 1.7 - 1.9$, where the variability reflects the temperature dependence of the rate constants. We find $\beta \approx 1$ in the fresh Asian outflow off the China coast, indicating that the correlation is driven by dilution; this confirms that the slope of the linear relationship provides a measure of the emission ratio. For the more remote regions, we find that the slope exceeds $\sqrt{\tau_{CO}/\tau_{C2H2}}$, indicating that the correlation mostly reflects chemical loss. This means then that simulation of the log–log slope provides a test of model [OH]. Sensitivity simulations presented by Xiao *et al.* (2007) to fit the observed slopes imply that tropical OH concentrations in the model are 50% too high.

This relatively simple example illustrates how observed correlations between species can provide constraints on emissions, mixing, chemistry, and other processes, but interpretation is often not obvious and mistakes can easily be made. Simulation of observed relationships with a 3-D model, complemented by model sensitivity studies, can thus be a powerful tool to advance knowledge. It is particularly satisfying if knowledge gained from the complex 3-D model can be distilled into a simpler model illuminating the fundamental processes driving the observed relationships. As the saying goes, "no model should be more complicated than it needs to be." But starting from a complicated model can provide the best guide to judicious simplification.

References

Baron, P. A. and Willeke, K. (2005) *Aerosol Measurement: Principles, Techniques, and Applications*, 2nd edition, Wiley, Chichester.

Brunner D., Staehelin J., Rogers H. L., *et al.* (2003) An evaluation of the performance of chemistry transport models by comparison with research aircraft observations: Part 1. Concepts and overall model performance, *Atmos. Chem. Phys.*, **3**, 1609–1631.

Burrows, J. P., Platt U., and Borrell P. (eds.) (2011) *The Remote Sensing of Tropospheric Composition from Space*, Springer, New York.

Chai T. and Draxler R. R. (2014) Root mean square error (RMSE) or mean absolute error (MAE)? Arguments against avoiding RMSE in the literature, *Geosci. Mod. Dev.*, **7**, 1247–1250. doi:10.5194/gmd-7-1247-2014.

D'Andrea S. D., Häkkinen S. A. K., Westervelt D. M. *et al.* (2013) Understanding global secondary organic aerosol amount and size-resolved condensational behaviour, *Atmos. Chem. Phys.*, **13**, 11519–11534.

Douglass A. R., Prather M. J., Hall T. M., *et al.* (1999) Choosing meteorological input for the global modeling initiative assessment of high-speed aircraft, *J. Geophys. Res.*, **104**, 27545–27564.

Ehhalt D. H., Rohrer F., Wahner A., Prather M. J., and Blake D. R. (1998) On the use of hydrocarbons for the determination of tropospheric OH concentrations, *J. Geophys. Res.*, **103**, 18981–18997.

Eskes H., Huijnen V., Arola A., *et al.* (2015) Validation of reactive gases and aerosols in the MACC global analysis and forecast system, *Geosci. Mod. Dev. Discuss.*, **8**, 1117–1169, doi:10.5194/gmdd-8-1117-2015.

Farmer D. K. and Jimenez J. L. (2010) Real-time atmospheric chemistry field instrumentation. *Anal.Chem.*, **82**, 7879–7884, doi:10.1021/ac1010603.

Finlayson-Pitts B. and Pitts, Jr. J. N. (2000) *Chemistry of the Upper and Lower Atmosphere*, Academic Press, New York.

González Abad G., Liu X., Chance K., *et al.* (2015) Updated Smithsonian Astrophysical Observatory Ozone Monitoring Instrument (SAO OMI) formaldehyde retrieval, *Atmos. Meas. Tech.*, **8**, 19–32.

Götz F. W. P., Meetham A. R., and Dobson G. M. B. (1934) The vertical distribution of ozone in the atmosphere, *Proc. Roy. Soc. A*, **145**, 416.

Holmes C. D., Jacob D. J., Mason R. P., and Jaffe D. A. (2009) Sources and deposition of reactive gaseous mercury in the marine atmosphere, *Atmos. Environ.*, **43**, 2278–2285.

Hönninger G., von Friedeburg C., and Platt U. (2004) Atmospheric Chemistry and Physics Multi axis differential optical absorption spectroscopy (MAX-DOAS), *Atmos. Chem. Phys.*, **4**, 231–254.

Hughes I. G. and Hase T. P. A. (2010) *Measurements and their Uncertainties: A Practical Guide to Modern Error Analysis*, Oxford University Press, Oxford.

Jolliff J. K., Kindle J. C., Shulman I., *et al.* (2009) Summary diagrams for coupled hydrodynamic-ecosystem model skill assessment, *J. Marine Syst.*, **74**, 64–82, doi:10.1016/j.jmarsys.2008.05.014

Kim P. S., Jacob D. J., Fisher J. A., *et al.*, (2015) Sources, seasonality, and trends of Southeast US aerosol: An integrated analysis of surface, aircraft, and satellite observations with the GEOS-Chem model, *Atmos. Chem. Phys.*, **15**, 10411–10433.

Komhyr W. and Evans R. (2008) *Operations Handbook: Ozone Observations with a Dobson Spectrophotometer*, World Meteorological Organization, Geneva.

Kunzi K., Bauer P., Eresmaa R., *et al.* (2011) Microwave absorption, emission and scattering: Trace gas and meteorological parameters. In *The Remote Sensing of Tropospheric Composition from Space* (Burrows, J. P., Platt U., and Borrell P., eds.), Springer-Verlag, Berlin.

Lagzi I., Meszaros R., Gelybo G., and Leelossy A. (2013) *Atmospheric Chemistry*, Eötvös Loránd University.

Legates D. R. and McCabe G. J. Jr. (1999) Evaluating the use of "goodness-of-fit" measures in hydrologic and hydroclimatic model validation, *Water Resources Res.*, **35**, 233–241.

Logan J. A. (1999) An analysis of ozonesonde data for the troposphere: Recommendations for testing 3-D models, and development of a gridded climatology for tropospheric ozone, *J. Geophys. Res.*, **104**, 16115–16149.

Logan J. A., Megretskaia, A. J., Miller, G. C., *et al.* (1999) Trends in the vertical distribution of ozone: A comparison of two analyses of ozonesonde data, *J. Geophys. Res.*, **104**, 26373–26399.

Mankin W., Atlas E., Cantrell C., Eisele E., and Fried A. (1999) Observational methods: Instruments and platforms. In *Atmospheric Chemistry and Global Change* (Brasseur, G. P., Orlando, J. J., and Tyndall, G. S., eds.), Oxford University Press, Oxford.

McKeen, S. A., Liu S., Hsie X., *et al.* (1996) Hydrocarbon ratios during PEM WEST A: A model perspective, *J. Geophys. Res.*, **101**, 2087–2109.

Mielke P. W. Jr. and Berry K. J. (2001) *Permutation Methods: A Distance Function Approach*, Springer, New York.

Nash J. E. and Sutcliffe J. V. (1970) River flow forecasting throughout conceptual models: Part I. A discussion of principles, *J. Hydro.*, **10**, 282–290.

Palmer P. I., Jacob D. J., Chance K., *et al.* (2001) Air mass factor formulation for spectroscopic measurements from satellites: Application to formaldehyde retrievals from GOME, *J. Geophys. Res.*, **106** (14), 14539–14550.

Rodgers C. D. (2000) *Inverse Methods for Atmospheric Sounding*, World Sci., Tokyo.

Schmidt H., Brasseur G. P., Charron M., *et al.* (2006) The HAMMONIA chemistry climate model: Sensitivity of the mesopause region to the 11-year solar cycle and CO_2 doubling, *J. Climate*, **19**, 3903–3931.

Taylor K. E. (2001) Summarizing multiple aspects of model performance in a single diagram, *J. Geophys. Res.*, **106**, 7183–7192.

Taylor J. R. (1996) *An Introduction to Error Analysis: The Study of Uncertainties in Physical Measurements*, University Science Books, Sausalito, CA.

Thompson A. M., Witte J. C., McPeters R. D., *et al.* (2003) Southern hemisphere additional ozonesondes (SHADOZ) 1998–2000 tropical ozone climatology 1: Comparison with Total Ozone Mapping Spectrometer (TOMS) and ground-based measurements, *J. Geophys. Res.*, **108**, 8238, doi:10.1029/2001JD000967.

Thunis P., Georgieva E., and Pederzoli A. (2012) A tool for evaluating air quality model performances in regulatory applications. *Env. Mod. Soft.*, **38**, 220–230.

Watterson I. G. (1996) Non-dimensional measures of climate model performance, *Int. J. Climatol.*, **16**, 379–391.

Waugh D. W. and Eyring V. (2008) Quantitative performance metrics for stratospheric-resolving climate-chemistry models, *Atmos. Chem. Phys.*, **8**, 5699–5713.

Wild O. (2007) Modelling the global tropospheric ozone budget: Exploring the variability in current models, *Atmos. Chem. Phys.*, **7**, 2643–2660.

Wild O. and Prather M. J. (2006) Global tropospheric ozone modeling: Quantifying errors due to grid resolution, *J. Geophys. Res.*, **111**, D11305.

Willmott C. J. (1981) On the validation of models, *Phys. Geogr.*, **2**, 184–194.

Willmott C. J., Ackelson S. G., Davis R. E., *et al.* (1985) Statistics for the evaluation of model performance, *J. Geophys. Res.*, **90**, 8995–9005.

Willmott C. J., Robeson S. M., and Matsuura K. (2012) A refined index of model performance, *Int. J. Climatol.*, **32**, 2088–2094.

Xiao Y. P., Jacob D. J., and Turquety S. (2007) Atmospheric acetylene and its relationship with CO as an indicator of air mass age, *J. Geophys. Res.*, **112**, D12305.

Yu K., Jacob D., Fisher J., *et al.* (2016) Sensitivity to grid resolution in the ability of a chemical transport model to simulate observed oxidant chemistry under high-isoprene conditions, *Atmos. Chem. Phys.*, **16**, 4369–4378.

Zhang L., Jacob D. J., Liu X., *et al.* (2010) Intercomparison methods for satellite measurements of atmospheric composition: Application to tropospheric ozone from TES and OMI, *Atmos. Chem. Phys.*, **10**, 4725–4739.

11 Inverse Modeling for Atmospheric Chemistry

11.1 Introduction

Inverse modeling is a formal approach for using observations of a physical system to better quantify the variables driving that system. This is generally done by statistically *optimizing* the estimates of the variables given all the observational and other information at hand. We call the variables that we wish to optimize the *state variables* and assemble them into a *state vector* $\mathbf{x}$. We similarly assemble the observations into an *observation vector* $\mathbf{y}$. Our understanding of the relationship between $\mathbf{x}$ and $\mathbf{y}$ is described by a model $\mathbf{F}$ of the physical system called the *forward model*:

$$\mathbf{y} = \mathbf{F}(\mathbf{x}, \mathbf{p}) + \boldsymbol{\varepsilon_O} \tag{11.1}$$

Here, $\mathbf{p}$ is a *parameter vector* including all model variables that we do not seek to optimize as part of the inversion, and $\boldsymbol{\varepsilon_O}$ is an *observational error* vector including contributions from errors in the measurements, in the forward model, and in the model parameters. The forward model predicts the effect ($\mathbf{y}$) as a function of the cause ($\mathbf{x}$), usually through equations describing the physics of the system. By *inversion* of the model we can quantify the cause ($\mathbf{x}$) from observations of the effect ($\mathbf{y}$). In the presence of error ($\boldsymbol{\varepsilon_O} \neq 0$), the solution is a best estimate of $\mathbf{x}$ with some statistical error. This solution for $\mathbf{x}$ is called the *optimal estimate*, the *posterior estimate*, or the *retrieval*. The choice of state vector (that is, which model variables to include in $\mathbf{x}$ versus in $\mathbf{p}$) is totally up to us. It depends on which variables we wish to optimize, what information is contained in the observations, and what computational costs are associated with the inversion.

Because of the uncertainty in deriving $\mathbf{x}$ from $\mathbf{y}$, we have to consider other constraints on the value of $\mathbf{x}$ that may help to reduce the error on the optimal estimate. These constraints are called the *prior* information. A standard constraint is the *prior estimate* $\mathbf{x_A}$, representing our best estimate of $\mathbf{x}$ *before* the observations are made. It has some error $\boldsymbol{\varepsilon_A}$. The optimal estimate must then weigh the relative information from the observations $\mathbf{y}$ and the prior estimate $\mathbf{x_A}$, and this is done by considering the error statistics of $\boldsymbol{\varepsilon_O}$ and $\boldsymbol{\varepsilon_A}$. Inverse modeling allows a formal analysis of the relative importance of the observations versus the prior information in determining the optimal estimate. As such, it informs us whether an observing system is effective for constraining $\mathbf{x}$.

Inverse modeling has three main applications in atmospheric chemistry, summarized in Table 11.1:

1. **Remote sensing of atmospheric composition**. Here we use radiance spectra measured by remote sensing to retrieve vertical concentration profiles. The

Table 11.1 Applications of inverse modeling in atmospheric chemistry

Application	State vector	Observations	Forward model	Prior estimate
Remote sensing	Vertical concentration profile	Radiance spectra	Radiative transfer model	Climatological profile
Top-down constraints	Surface fluxes	Atmospheric concentrations	Chemical transport model	Bottom-up inventory
Data assimilation	Gridded concentration field	Atmospheric concentrations	Mapping operator	Forecast

measured radiances at different wavelengths represent the observation vector $\mathbf{y}$, and the concentrations on a vertical grid represent the state vector $\mathbf{x}$. The forward model $\mathbf{F}$ is a radiative transfer model (Chapter 5) that calculates $\mathbf{y}$ as a function of $\mathbf{x}$ and of additional parameters $\mathbf{p}$ that may include surface emissivity, temperatures, spectroscopic data, etc. The prior estimate $\mathbf{x_A}$ is provided by previous observations of the same or similar scenes, by knowledge of climatological mean concentrations, or by a chemical transport model.

2. **Top-down constraints on surface fluxes**. Here we use measured atmospheric concentrations (observation vector $\mathbf{y}$) to constrain surface fluxes (state vector $\mathbf{x}$). The forward model $\mathbf{F}$ is a chemical transport model (CTM) that solves the chemical continuity equations to calculate $\mathbf{y}$ as a function of $\mathbf{x}$. The parameter vector $\mathbf{p}$ includes meteorological variables, chemical variables such as rate coefficients, and any characteristics of the surface flux such as diurnal variability that are simulated in the CTM but not optimized as part of the state vector. The information on $\mathbf{x}$ from the observations is called a *top-down* constraint on the surface fluxes. The prior estimate $\mathbf{x_A}$ is an inventory based on our knowledge of the processes determining the surface fluxes (such as fuel combustion statistics, land cover data bases, etc.) and is called a *bottom-up* constraint. See Section 9.2 for discussion of bottom-up and top-down constraints on surface fluxes.

3. **Chemical data assimilation**. Here we construct a gridded 3-D field of concentrations $\mathbf{x}$, usually time-dependent, on the basis of measurements $\mathbf{y}$ of these concentrations or related quantities at various locations and times. Such a construction may be useful to initialize chemical forecasts, to assess the consistency of measurements from different platforms, or to map the concentrations of non-measured species on the basis of measurements of related species. We refer to this class of inverse modeling as *data assimilation*. The corresponding state vectors are usually very large. In the time-dependent problem, the prior estimate is an atmospheric forecast model that evolves $\mathbf{x}(t)$ from a previously optimized state at time t_o to a *forecast state* at the next assimilation time step $t_o + h$. The forecast model is usually a weather prediction model including simulation of the chemical variables to be assimilated. The forward model $\mathbf{F}$ can be a simple mapping operator of observations at time $t_o + h$ to the model grid, a chemical model relating the observed variables to the state variables, or the forecasting model itself.

Proper consideration of errors is crucial in inverse modeling. Let us examine what happens if we ignore errors. We linearize the forward model $\mathbf{y} = \mathbf{F}(\mathbf{x}, \mathbf{p})$ around the prior estimate $\mathbf{x_A}$ taken as best guess:

$$\mathbf{y} = \mathbf{F}(\mathbf{x_A}, \mathbf{p}) + \mathbf{K}(\mathbf{x} - \mathbf{x_A}) + \mathbf{O}\left((\mathbf{x} - \mathbf{x_A})^2\right) \qquad (11.2)$$

where $\mathbf{K} = \nabla_{\mathbf{x}}\mathbf{F} = \partial \mathbf{y}/\partial \mathbf{x}$ is the Jacobian matrix of the forward model with elements $k_{ij} = \partial y_i/\partial x_j$ evaluated at $\mathbf{x} = \mathbf{x_A}$. The notation $\mathbf{O}((\mathbf{x} - \mathbf{x_A})^2)$ groups higher-order terms (quadratic and above) taken to be negligibly small. Let n and m represent the dimensions of $\mathbf{x}$ and $\mathbf{y}$, respectively. Assume that the observations are independent such that $m = n$ observations constrain $\mathbf{x}$ uniquely. The Jacobian matrix is then an $n \times n$ matrix of full rank and hence invertible. We obtain for $\mathbf{x}$:

$$\mathbf{x} = \mathbf{x_A} + \mathbf{K}^{-1}(\mathbf{y} - \mathbf{F}(\mathbf{x_A}, \mathbf{p})) \qquad (11.3)$$

If $\mathbf{F}$ is nonlinear, the solution (11.3) must be iterated with recalculation of the Jacobian at successive guesses for $\mathbf{x}$ until satisfactory convergence is achieved.

Now what happens if we make additional observations, such that $m > n$? In the absence of error these observations must necessarily be redundant. But we know from experience that strong constraints on an atmospheric system typically require a very large number of measurements, $m \gg n$. This is due to errors in the measurements and in the forward model, described by the observational error vector $\boldsymbol{\varepsilon_0}$ in (11.1). Thus (11.3) is not applicable in practice; successful inversion requires adequate characterization of the observational error $\boldsymbol{\varepsilon_O}$ and consideration of prior information. A standard approach to do this is to use Bayes' theorem, described in Section 11.2.

The chapter is organized as follows. Section 11.2 presents Bayes' theorem and shows how it provides a basis for inverse modeling. Section 11.3 applies Bayes' theorem to a simple scalar optimization problem in order to build intuition for the rest of the chapter. Section 11.4 introduces important vector-matrix tools for inverse modeling, including error covariance matrices, probability density functions (PDFs) for vectors, Jacobian matrices, and adjoints. Section 11.5 presents the fundamental analytical method for solving the inverse problem, Section 11.6 presents the adjoint-based method, Section 11.7 presents Markov Chain Monte Carlo (MCMC) methods, and Section 11.8 presents other optimization methods. Section 11.9 discusses means to enforce positivity in the solution to the inverse problem. Section 11.10 gives an overview of variational methods used in chemical data assimilation. Observation system simulation experiments (OSSEs) to evaluate the merits of a proposed observing system are described in Section 11.11. Inverse modeling has applications across many areas of the natural and social sciences, and a major source of confusion in the literature is the use of different terminologies and notations reflecting this diverse heritage. Here we will follow to a large extent the terminology and notation of Rodgers (2000), which we consider to be a model of elegance.

11.2 Bayes' Theorem

Bayes' theorem is the general foundation of inverse modeling. Consider a pair of vectors $\mathbf{x}$ and $\mathbf{y}$. Let $P(\mathbf{x})$, $P(\mathbf{y})$, $P(\mathbf{x}, \mathbf{y})$ represent the corresponding PDFs, so that the probability of $\mathbf{x}$ being in the range $[\mathbf{x}, \mathbf{x} + d\mathbf{x}]$ is $P(\mathbf{x})\, d\mathbf{x}$, the probability of $\mathbf{y}$ being in the range $[\mathbf{y}, \mathbf{y} + d\mathbf{y}]$ is $P(\mathbf{y})\, d\mathbf{y}$, and the probability of $(\mathbf{x}, \mathbf{y})$ being in the range $([\mathbf{x}, \mathbf{x} + d\mathbf{x}], [\mathbf{y}, \mathbf{y} + d\mathbf{y}])$ is $P(\mathbf{x}, \mathbf{y})\, d\mathbf{x}\, d\mathbf{y}$. Let $P(\mathbf{y}|\mathbf{x})$ represent the *conditional PDF* of $\mathbf{y}$ when $\mathbf{x}$ has a known value. We can write $P(\mathbf{x}, \mathbf{y})\, d\mathbf{x}\, d\mathbf{y}$ equivalently as

$$P(\mathbf{x}, \mathbf{y})\, d\mathbf{x}\, d\mathbf{y} = P(\mathbf{x})\, d\mathbf{x}\, P(\mathbf{y}|\mathbf{x})\, d\mathbf{y} \tag{11.4}$$

or

$$P(\mathbf{x}, \mathbf{y})\, d\mathbf{x}\, d\mathbf{y} = P(\mathbf{y})\, d\mathbf{y}\, P(\mathbf{x}|\mathbf{y})\, d\mathbf{x} \tag{11.5}$$

Eliminating $P(\mathbf{x}, \mathbf{y})$, we obtain *Bayes' theorem*:

$$P(\mathbf{x}|\mathbf{y}) = \frac{P(\mathbf{y}|\mathbf{x})P(\mathbf{x})}{P(\mathbf{y})} \tag{11.6}$$

This theorem formalizes the inverse problem posed in Section 11.1. Here:

- $P(\mathbf{x}|\mathbf{y})$ is the posterior PDF for the state vector $\mathbf{x}$ given the observations $\mathbf{y}$, and defines the solution to the inverse problem.
- $P(\mathbf{x})$ is the prior PDF of the state vector $\mathbf{x}$ before the measurements are made, i.e., the PDF of $\mathbf{x_A}$ defined by the error statistics for $\boldsymbol{\varepsilon_A}$.
- $P(\mathbf{y}|\mathbf{x})$ is the PDF of the observation vector $\mathbf{y}$ given the true value of $\mathbf{x}$ and accounting for errors in the measurements and in the forward model, as defined by the error statistics for $\boldsymbol{\varepsilon_O}$ (11.1).
- $P(\mathbf{y})$ is the PDF of $\mathbf{y}$ for all possible values of $\mathbf{x}$.

The optimal estimate for $\mathbf{x}$ is defined by the maximum of $P(\mathbf{x}|\mathbf{y})$, corresponding to

$$\boldsymbol{\nabla}_{\mathbf{x}} P(\mathbf{x}|\mathbf{y}) = 0 \tag{11.7}$$

where $\boldsymbol{\nabla}_{\mathbf{x}}$ is the gradient operator in the state vector space operating on all state vector elements. From (11.6) we have

$$\boldsymbol{\nabla}_{\mathbf{x}} P(\mathbf{x}|\mathbf{y}) = \frac{1}{P(\mathbf{y})} \boldsymbol{\nabla}_{\mathbf{x}} [P(\mathbf{y}|\mathbf{x})\, P(\mathbf{x})] \tag{11.8}$$

since $P(\mathbf{y})$ is independent of $\mathbf{x}$. It follows that the optimal estimation given by (11.7) can be rewritten as

$$\boldsymbol{\nabla}_{\mathbf{x}} [P(\mathbf{y}|\mathbf{x})P(\mathbf{x})] = 0 \tag{11.9}$$

This defines the *Bayesian optimal estimate solution* to the inverse problem. $P(\mathbf{y})$ does not contribute to the solution and does not appear in (11.9). Indeed, $P(\mathbf{y})$ can be viewed simply as a normalizing factor in equation (11.6) to ensure that the integral of $P(\mathbf{x}|\mathbf{y})$ over all possible values of $\mathbf{x}$ is unity. We ignore it in what follows.

Inverse modeling using Bayes' theorem as described here is an example of a *regularization method* where prior information is used to constrain the fitting of x to y. Here the constraint is the prior PDF of x. This is the most commonly used constraint in inverse modeling, but others may also be used. Some methods ignore prior information and fit x to match y using least-squares or some other norm minimization. Other regularization methods enforce different prior constraints on the solution, such as positivity, smoothness, or patterns. Examples are *Tikhonov regularizations* and *geostatistical methods.* These are covered in Section 11.7.

11.3 A Simple Scalar Example

Here we apply Bayes' theorem to solve a simple inverse problem using scalars. This allows us to introduce concepts, terminology, and equations that will be useful for understanding the solution of the more general problem involving vectors.

Consider in this example a single source releasing a species X to the atmosphere with an emission rate x (Figure 11.1). We have a prior estimate $x_A \pm \sigma_A$ for x, where σ_A^2 is the prior error variance defined by

$$\sigma_A{}^2 = E\left[(\varepsilon_A - E[\varepsilon_A])^2\right] = E\left[\varepsilon_A^2\right] \tag{11.10}$$

Here ε_A is the error on our prior estimate, and $E[\ ]$ is the *expected value operator* representing the expected mean value of the bracketed quantity for an infinitely large number of realizations. $E[\varepsilon_A]$ is the mean value of the error, called the *mean bias.* The prior estimate x_A is unbiased by definition so $E[\varepsilon_A] = 0$. This is an important point. You may think of the prior estimate as "biased" because it differs from the true value; however, before we make observations, it is equally likely to be too high or too low.

We now make a single measurement of the concentration of X downwind from the source. The measured concentration is $y = y_T + \varepsilon_I$, where y_T is the true concentration and ε_I is the instrument error. We use a CTM as forward model F to relate y_T to x:

$$y_T = F(x) + \varepsilon_M + \varepsilon_R \tag{11.11}$$

Here, ε_M describes the forward model error in reproducing the true concentration y_T given the true emission rate x. This error includes contributions from model parameters such as winds, model physics, and model numerics. There is in addition a *representation error* ε_R reflecting the mismatch between the model resolution and the measurement location (Figure 11.2). Representation error is caused by the numerical

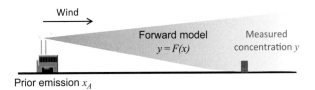

Figure 11.1 Simple example of inverse modeling. A point source emits a species X with an estimated prior emission rate x_A. We seek to improve this estimate by measuring the concentration y of X at a point downwind, and using a CTM as forward model to relate x to y.

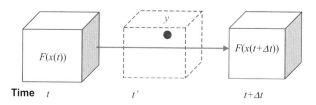

Figure 11.2 Representation error. The forward model computes concentrations as gridbox averages over discrete time steps Δt. The measurement is at a specific location within the gridbox and at an intermediate time t'. Interpolation in space and time is necessary to compare the measurement to the model, and the associated error is called the representation error.

discretization of the model equations so that the model provides simulated concentrations only on a discrete spatial grid and at discrete time steps, which may not correspond to the exact location of the measurement. Thus the model is not *representative* of the measurement, and even with interpolation some error is incurred. Representation error is not intrinsically a forward model error because it could in principle be corrected by adjusting the location and timing of the measurement.

Summing the errors, the measured concentration y is related to the true value of x by

$$y = F(x) + \varepsilon_I + \varepsilon_R + \varepsilon_M = F(x) + \varepsilon_O \qquad (11.12)$$

where $\varepsilon_O = \varepsilon_I + \varepsilon_R + \varepsilon_M$ is the *observational error* which includes instrument, representation, and forward model errors. This terminology might at first seem strange, as we are used to opposing observations to models. A very important conceptual point in inverse modeling is that instrument and model errors are inherently coupled when attempting to estimate x from y. Having a very precise instrument is useless if the model is poor or mismatched; in turn, having a very precise model is useless if the instrument is poor. Instrument and model must be viewed as inseparable partners of the observing system by which we seek to gain knowledge of x.

The instrument, representation, and model errors are uncorrelated so that their variances are additive:

$$\sigma_O^2 = \sigma_I^2 + \sigma_R^2 + \sigma_M^2 \qquad (11.13)$$

Let us assume for now that the observational error is unbiased so that $b_O = E[\varepsilon_O] = 0$; we will examine the implications of $b_O \neq 0$ later. Let us further assume that the prior and observational errors are normally distributed. Finally, let us assume that the forward model is linear so that $F(x) = kx$ where k is the model parameter; again, we will examine the implications of nonlinearity later.

We now have all the elements needed for application of Bayes' theorem to obtain an optimal estimate $\widehat{x}$ of x given y. The prior PDF for x is given by

$$P(x) = \frac{1}{\sigma_A \sqrt{2\pi}} \exp\left[-\frac{(x - x_A)^2}{2\sigma_A^2} \right] \qquad (11.14)$$

and the conditional PDF for the observation y given the true value of x is given by

$$P(y|x) = \frac{1}{\sigma_O \sqrt{2\pi}} \exp\left[-\frac{(y - kx)^2}{2\sigma_O^2} \right] \qquad (11.15)$$

Applying Bayes' theorem (11.6) and ignoring the normalizing terms that are independent of x, we obtain:

$$P(x|y) \propto \exp\left[-\frac{(x - x_A)^2}{2\sigma_A^2} - \frac{(y - kx)^2}{2\sigma_O^2}\right] \quad (11.16)$$

where $\propto$ is the proportionality symbol. Finding the maximum value for $P(x|y)$ is equivalent to finding the minimum for the *cost function J(x)*:

$$J(x) = \frac{(x - x_A)^2}{\sigma_A^2} + \frac{(y - kx)^2}{\sigma_O^2} \quad (11.17)$$

which is a least-squares sum weighted by error variances and is called a χ^2 *cost function.*

The optimal estimate $\hat{x}$ is the solution to $\partial J/\partial x = 0$:

$$\frac{dJ}{dx} = 2\frac{(x - x_A)}{\sigma_A^2} + 2k\frac{(kx - y)}{\sigma_O^2} = 0 \quad (11.18)$$

This yields

$$\hat{x} = x_A + g(y - kx_A) \quad (11.19)$$

where g is a *gain factor* given by

$$g = \frac{k\sigma_A^2}{k^2\sigma_A^2 + \sigma_O^2} \quad (11.20)$$

In (11.19), the second term on the right-hand side represents the correction to the prior estimate on the basis of the measurement y. The gain factor is the sensitivity of the optimal estimate to the observation: $g = \partial\hat{x}/\partial y$. We see from (11.20) that the gain factor depends on the relative magnitudes of σ_A and σ_O/k. If $\sigma_A \ll \sigma_O/k$, then $g \to 0$ and $\hat{x} \to x_A$; the measurement is useless because the observational error is too large. If by contrast $\sigma_A \gg \sigma_O/k$, then $g \to 1/k$ and $\hat{x} \to y/k$; the measurement is so precise that it constrains the solution without recourse to prior information.

We can also express the optimal estimate $\hat{x}$ in terms of its proximity to the true solution x. Replacing (11.12) with $F(x) = kx$ into (11.19) we obtain

$$\hat{x} = ax + (1 - a)x_A + g\varepsilon_O \quad (11.21)$$

or equivalently

$$\hat{x} = x + (1 - a)(x_A - x) + g\varepsilon_O \quad (11.22)$$

where a is the *averaging kernel* defined as

$$a = gk = \frac{\sigma_A^2}{\sigma_A^2 + (\sigma_O/k)^2} \quad (11.23)$$

The averaging kernel describes the relative weights of the prior estimate x_A and the true value x in contributing to the optimal estimate. It represents the sensitivity of the optimal estimate to the true state: $a = \partial\hat{x}/\partial x$. The gain factor is now applied to the observational error in the third term on the right-hand side.

We see that the averaging kernel simply weighs the error variances in state space σ_A^2 and $(\sigma_O/k)^2$. In the limit $\sigma_A \gg \sigma_O/k$, we have $a \to 1$ and the prior estimate does not contribute to the solution. However, our ability to approach the true solution is still limited by the term $g\varepsilon_O$ with variance $(g\sigma_O)^2$. We call $(1 - a)(x_A - x)$ the *smoothing error* since it regularizes the solution by limiting our ability to depart from the prior estimate, and we call $g\varepsilon_O$ the *observational error in state space*.

Manipulating the PDF of the optimal estimate as given by (11.16) and replacing y using (11.19) yields a Gaussian form $P(x|y) \propto \exp\left[-(x - \widehat{x})^2/2\widehat{\sigma}^2\right]$ where $\widehat{\sigma}^2$ is the harmonic sum of σ_A^2 and $(\sigma_O/k)^2$:

$$\frac{1}{\widehat{\sigma}^2} = \frac{1}{\sigma_A^2} + \frac{1}{(\sigma_O/k)^2} \tag{11.24}$$

Here $\widehat{\sigma}^2$ is the error variance on the optimal estimate, called the *posterior error variance*. It is always less than the prior and observational error variances, and tends toward one of the two in the limiting cases that we described.

Before the measurement the error variance on x was σ_A^2; after the measurement it is $\widehat{\sigma}^2$. The amount of information from the measurement can be quantified as the *relative error variance reduction* $\left(\sigma_A^2 - \widehat{\sigma}^2\right)/\sigma_A^2$. We find from (11.23) and (11.24) that this quantity is equal to the averaging kernel:

$$\frac{\sigma_A^2 - \widehat{\sigma}^2}{\sigma_A^2} = a \tag{11.25}$$

The role of the prior estimate in obtaining the optimal solution deserves some discussion. Sometimes an inverse method will be described as "not needing prior information." But that means either that any prior information is very poor compared to what can be achieved from the observing system, or that the method is suboptimal. In our example, not using prior information will yield as solution y/k with error variance σ_O^2; but since $\sigma_O^2 > \widehat{\sigma}^2$ this is not as good a solution as $\widehat{x}$. Using prior information can lead to confusion about the actual contribution of the measurement to the reported solution $\widehat{x}$. Knowledge of averaging kernels is important to avoid such confusion.

We have assumed in the above a linear forward model $y = F(x) = kx$. If the forward model is not linear, we can still calculate an optimal estimate $\widehat{x}$ as the minimum of the cost function (11.17), where we replace kx by the nonlinear form $F(x)$. We then have

$$J(x) = \frac{(x - x_A)^2}{\sigma_A^2} + \frac{(y - F(x))^2}{\sigma_O^2} \tag{11.26}$$

and the optimal estimate is given by solving $dJ/dx = 0$:

$$\frac{dJ}{dx} = 2\frac{(x - x_A)}{\sigma_A^2} + 2\frac{\partial F}{\partial x}\frac{(F(x) - y)}{\sigma_O^2} = 0 \tag{11.27}$$

The error on this optimal estimate is not Gaussian though, so (11.24) does not apply. And although we can still define an averaging kernel $a = \partial\widehat{x}/\partial x$, this averaging kernel cannot be expressed analytically anymore as a ratio of error variances; it may instead need to be calculated numerically. Obtaining error statistics on the optimal estimate is thus far more difficult. An alternative is to linearize the forward model around x_A as $k_o = \partial F / \partial x|_{x_A}$ and solve for the corresponding minimum of the cost function as in the linear case (11.18):

$$\frac{dJ}{dx} = 2\frac{(x - x_A)}{\sigma_A^2} + 2k_o\frac{(k_o x - y)^2}{\sigma_O^2} = 0 \tag{11.28}$$

This yields an initial guess x_1 for $\hat{x}$ on which we iterate by recalculating $k_1 = \partial F / \partial x|_{x_1}$, solving (11.28) using k_1, obtaining a next guess x_2, and so on until convergence. This preserves the ability for analytical characterization of observing system errors.

We have assumed in our analysis that the errors are unbiased. The prior error ε_A is indeed unbiased because x_A is our best prior estimate of x; even though x_A is biased its error is not. Another way of stating this is that we don't know that x_A is biased until after making the measurement. However, the observational error could be biased if the instrument is inaccurate or if there are systematic errors in some aspect of the forward model. In that case we must rewrite (11.12) as

$$y = F(x) + b_O + \varepsilon_O' \tag{11.29}$$

where $b_O = E[\varepsilon_O]$ is the observation bias and ε_O' is the residual random error such that $E[\varepsilon_O'] = 0$. The optimal estimate can be derived as above by replacing y with $y - b_O$, and we see in this manner that the bias will be propagated through the equations to cause a corresponding bias in the solution. For a linear model $F(x) = kx$, the analytical solution given by (11.19) will be biased by gb_O.

So far we have limited ourselves to one single measurement. We can reduce the error on the optimal estimate by making m independent measurements y_i, each adding a term to the cost function $J(x)$ in (11.17). Assuming for illustrative purpose the same observational error variance and the same linear forward model parameter k for each measurement, and further assuming that the successive measurements are not only independent but uncorrelated, we have the following expression for $J(x)$:

$$J(x) = \frac{(x - x_A)^2}{\sigma_A^2} + \sum_{i=1}^{m} \frac{(y_i - kx)^2}{\sigma_O^2} = \frac{(x - x_A)^2}{\sigma_A^2} + \frac{\overline{(y_i - kx)^2}}{\sigma_O^2/m} \tag{11.30}$$

where the overbar denotes the average value and σ_O^2/m is the variance of the error on $\overline{(y_i - kx)^2}$. By taking m measurements, we have reduced the observational error variance on the average value by m; this is the *central limit theorem*. By increasing m, we could thus approach the true solution: $m \to \infty \Rightarrow \hat{x} \to \overline{y_i}/k$ and $\hat{\sigma} \to 0$. However, this works only if (1) the observational error has an expected value of zero (no mean bias), and (2) the m observations are *independent and identically distributed* (*IID*), meaning that they all sample the same PDF in an uncorrelated way.

With regard to (1), systematic error (mean bias) will not be reduced by increasing the number of measurements and will still propagate to affect the solution as discussed here. As the number of observations increases and the importance of the random error component decreases, the effect of bias on the solution increases in relative importance. With regard to (2), instrumental errors (as from photon counting) may be uncorrelated; however, forward model errors rarely are. In our example, two successive measurements at a site may sample the same air mass and thus be subject to the same model transport error in the CTM used as forward model. It is thus important to determine the *error correlation* between the different observations. This error correlation can best be described by assembling the measurements into a vector and constructing the *observational error covariance matrix* (Section 11.4.1). Dealing with error correlations, and more generally dealing with a multi-component state vector, requires that we switch to a vector-matrix formalism for the inverse problem. This vector-matrix formalism is central to any practical application of inverse modeling and we introduce the relevant mathematical tools in the next section.

One last word about bias before we move on. Bias in the observing system is the bane of inverse modeling. As we saw, it propagates through the inverse model to bias the solution. Random error in the observing system can be beaten down by making many measurements, but bias is irreducible. Inversions sometimes include a prior estimate for the *pattern* of the bias (for example, latitude-dependent bias in a satellite retrieval) and optimize it as part of the inversion. But for this we need to know that a bias is there and what form it has, and we generally are not that well informed. Minimizing bias in the observing system through independent calibration is a crucial prelude to inverse modeling. Bias in the instrument can be determined by analysis of known standards or by comparison with highly accurate independent measurements (for example, validation of satellite observations with vertical aircraft profiles during the satellite overpass). Bias in the forward model can be determined by applying the model to conditions where the state is known, though this is easier said than done. See Chapter 10 for discussion on quantifying errors in models. In the rest of this chapter, and unless otherwise noted, we will assume that errors are random.

11.4 Vector-Matrix Tools

Consider the general problem of a state vector $\mathbf{x}$ of dimension n with prior estimate $\mathbf{x_A}$ and associated error $\boldsymbol{\varepsilon_A}$, for which we seek an optimal estimate $\widehat{\mathbf{x}}$ on the basis of an ensemble of observations assembled into an observation vector $\mathbf{y}$ of dimension m. $\mathbf{y}$ is related to $\mathbf{x}$ by the forward model $\mathbf{F}$:

$$\mathbf{y} = \mathbf{F}(\mathbf{x}) + \boldsymbol{\varepsilon_O} \qquad (11.31)$$

where $\boldsymbol{\varepsilon_O}$ is the observational error vector as in (11.1). We have omitted the model parameters $\mathbf{p}$ in the expression for $\mathbf{F}$ to simplify notation. Inverse analysis requires definition of error statistics and PDFs for vectors. The error statistics are expressed as *error covariance matrices*, and the PDFs are constructed in a manner that accounts for covariance between vector elements. Solution of the inverse problem may involve construction of the *Jacobian matrix* and the *adjoint* of the forward model. We describe here these different objects. Their application to solving the inverse problem will be presented in the following sections.

11.4.1 Error Covariance Matrix

The error covariance matrix for a vector is the analogue of the error variance for a scalar. Consider an n-dimensional vector that we estimate as $\mathbf{x} + \boldsymbol{\varepsilon}$, where $\mathbf{x} = (x_1, \ldots x_n)^T$ is the true value and $\boldsymbol{\varepsilon} = (\varepsilon_1, \ldots \varepsilon_n)^T$ is the error vector representing the errors on the individual components of $\mathbf{x}$. The error covariance matrix $\mathbf{S}$ for $\mathbf{x}$ has as diagonal elements (s_{ii}) the error variances of the individual components of $\mathbf{x}$, and as off-diagonal elements (s_{ij}) the error covariances between components of $\mathbf{x}$:

$$s_{ii} = \text{var}(\varepsilon_i) = E\left[\varepsilon_i^2\right] \tag{11.32}$$

$$s_{ij} = \text{cov}(\varepsilon_i, \varepsilon_j) = E\left[\varepsilon_i\,\varepsilon_j\right] = r(\varepsilon_i, \varepsilon_j)\sqrt{\text{var}(\varepsilon_i)\text{var}(\varepsilon_j)} \tag{11.33}$$

where $r(\varepsilon_i, \varepsilon_j)$ is Pearson's correlation coefficient between ε_i and ε_j:

$$r(\varepsilon_i, \varepsilon_j) = \frac{\text{cov}(\varepsilon_i, \varepsilon_j)}{\sqrt{\text{var}(\varepsilon_i)\text{var}(\varepsilon_j)}} \tag{11.34}$$

The error covariance matrix is thus constructed as:

$$\mathbf{S} = \begin{pmatrix} \text{var}(\varepsilon_1) & \cdots & \text{cov}(\varepsilon_1, \varepsilon_n) \\ \vdots & \ddots & \vdots \\ \text{cov}(\varepsilon_1, \varepsilon_n) & \cdots & \text{var}(\varepsilon_n) \end{pmatrix} \tag{11.35}$$

and can be represented in compact form as $\mathbf{S} = E[\boldsymbol{\varepsilon}\boldsymbol{\varepsilon}^T]$. It is symmetric since the covariance operator is commutative: $\text{cov}(\varepsilon_i, \varepsilon_j) = \text{cov}(\varepsilon_j, \varepsilon_i)$. The covariance structure is often derived from error correlation coefficients, as expressed by the *error correlation matrix* $\mathbf{S}'$:

$$\mathbf{S}' = \begin{pmatrix} 1 & \cdots & r(\varepsilon_1, \varepsilon_n) \\ \vdots & \ddots & \vdots \\ r(\varepsilon_1, \varepsilon_n) & \cdots & 1 \end{pmatrix} \tag{11.36}$$

The error covariance matrix is then constructed from the error correlation matrix by multiplying the terms by the error variances of the corresponding elements (square roots for the off-diagonal terms).

Eigenanalysis of an error covariance matrix can be useful for identifying the dominant error patterns. The matrix has full rank n, since otherwise would imply that an element (or combination of elements) is perfectly known. It therefore has n orthonormal eigenvectors $\mathbf{e_i}$ with eigenvalues λ_I, and can be decomposed along its eigenvectors as follows:

$$\mathbf{S} = \sum_{i=1}^{n} \lambda_i\, \mathbf{e_i}\, \mathbf{e_i}^T = \mathbf{E}\,\boldsymbol{\Lambda}\,\mathbf{E}^T \tag{11.37}$$

where $\mathbf{E}$ is the matrix of eigenvectors arranged by columns and $\boldsymbol{\Lambda}$ is the diagonal matrix of eigenvalues:

$$\boldsymbol{\Lambda} = \begin{pmatrix} \lambda_1 & \cdots & 0 \\ \vdots & \ddots & \vdots \\ 0 & \cdots & \lambda_n \end{pmatrix} \tag{11.38}$$

Box 11.1   **Eigendecomposition of an Error Covariance Matrix**

To illustrate the eigendecomposition of an error covariance matrix, consider the matrix $\mathbf{S}$:

$$\mathbf{S} = \begin{bmatrix} 2 & -1 & 0.3 & 0.5 \\ -1 & 2 & -0.2 & -0.5 \\ 0.3 & -0.2 & 1 & -0.2 \\ 0.5 & -0.5 & -0.2 & 1 \end{bmatrix} \quad (11.39)$$

Its eigenvectors and eigenvalues are

$$\mathbf{E} = \begin{matrix} & \mathbf{e_1} & \mathbf{e_2} & \mathbf{e_3} & \mathbf{e_4} \\ & \begin{bmatrix} 0.67 & -0.24 & 0.64 & -0.28 \\ -0.67 & -0.15 & 0.71 & 0.15 \\ 0.12 & -0.78 & -0.17 & 0.59 \\ 0.29 & 0.56 & 0.23 & 0.74 \end{bmatrix} \\ eigenvalues & 3.3 & 1.2 & 1.0 & 0.5 \end{matrix} \quad (11.40)$$

The four eigenvectors define four orthogonal error patterns with error variances given by the eigenvalues. The total error variance is $3.3 + 1.2 + 1.0 + 0.5 = 6.0$. The first error pattern defined by $\mathbf{e_1}$ contributes an error variance of 3.3, more than half of the total error variance. This error pattern is dominated by the first two elements 1 and 2, as would be expected since they contribute most of the error variance in the diagonal of $\mathbf{S}$. The error pattern has opposite dependences for elements 1 and 2, as would be expected from the negative error correlation between the two $(r = s_{12}/\sqrt{s_{11}s_{22}} = -0.5)$. The second error pattern as defined by $\mathbf{e_2}$ amounts to 20% of the total error variance and accounts for the error patterns associated with elements 3 and 4, again with opposite dependences reflecting their negative error covariance. The third and fourth error patterns are less straightforward to interpret but account together for only 25% of the total error variance.

In the base for $\mathbf{x}$ defined by the eigenvectors, eigenvector $\mathbf{e_i}$ has a value of 1 for its ith element and a value of zero for all its other elements; we then see from (11.37) that the error covariance matrix in that base is $\mathbf{\Lambda}$. The eigenvalue λ_i thus represents the error variance associated with the error pattern $\mathbf{e_i}$. Eigendecomposition of $\mathbf{S}$ and ranking of eigenvalues identifies the dominant orthogonal error patterns and their contributions to the overall error. Box 11.1 gives an example. The eigenvalues of an error covariance matrix are all positive since they represent error variances. It follows that any error covariance matrix $\mathbf{S}$ is *positive definite*, a condition defined by the property that $\mathbf{x}^T \mathbf{S} \mathbf{x} \geq 0$ for any vector $\mathbf{x}$ of real numbers.

Bayesian solution to the inverse problem requires construction of the prior error covariance matrix $\mathbf{S_A} = \left[\boldsymbol{\varepsilon_A} \boldsymbol{\varepsilon_A^T} \right]$ and of the observational error covariance matrix $\mathbf{S_O} = \left[\boldsymbol{\varepsilon_O} \boldsymbol{\varepsilon_O^T} \right]$ as input to the problem. The observational error vector $\boldsymbol{\varepsilon_O}$ is the sum of the instrument error vector $\boldsymbol{\varepsilon_I}$, the representation error vector $\boldsymbol{\varepsilon_R}$, and the forward model error vector $\boldsymbol{\varepsilon_M}$, in the same way as for the scalar problem (11.12). These errors are generally uncorrelated so that $\mathbf{S_O}$ is the sum of the instrument error covariance matrix $\mathbf{S_I} = \begin{bmatrix} \boldsymbol{\varepsilon_I} \boldsymbol{\varepsilon_I}^T \end{bmatrix}$, the representation error covariance matrix $\mathbf{S_R} = \begin{bmatrix} \boldsymbol{\varepsilon_R} \boldsymbol{\varepsilon_R}^T \end{bmatrix}$, and the forward model error covariance matrix $\mathbf{S_M} = \begin{bmatrix} \boldsymbol{\varepsilon_M} \boldsymbol{\varepsilon_M}^T \end{bmatrix}$:

$$\mathbf{S_O} = \mathbf{S_I} + \mathbf{S_R} + \mathbf{S_M} \tag{11.41}$$

Note the similarity to the addition of variances in the scalar problem (11.13).

It is generally difficult to go beyond rough estimates in specifying the error covariance matrices $\mathbf{S_A}$ and $\mathbf{S_O}$. Box 11.2 gives some simple construction procedures. Particular uncertainty applies to constructing the off-diagonal terms (covariance structure). Simple assumptions are usually made, such as an error correlation length scale that relates adjacent vector elements and populates the off-diagonals nearest to the diagonal, producing a *band matrix* (Box 11.2). However, there is no guarantee that such an ad-hoc construction will yield a *bona fide* error covariance matrix, as the assumed error correlations between different elements may not be consistent across the whole vector. This problem is more likely to arise if the covariance structure is extensive. The validity of the construction can be checked by computing the eigenvalues and verifying that they are all positive. If they are not then the matrix needs to be corrected.

Box 11.2   **Construction of Prior and Observational Error Covariance Matrices**

Accurate knowledge of the prior and observational error covariance matrices $\mathbf{S_A}$ and $\mathbf{S_O}$ is in general not available and rough estimates are often used. It is good practice in those cases to repeat the inversion with a range of estimates of $\mathbf{S_A}$ and $\mathbf{S_O}$ – for example, changing their magnitudes by a factor of 2 – to assess the implied uncertainty on inversion results.

Estimating $\mathbf{S_A}$ often relies on expert judgment regarding the quality of the prior information. In the absence of better knowledge, simple estimates are often used. For example, one might assume a uniform 50% error on the individual components of $\mathbf{x_A}$ with no error correlation between the components. In that case, $\mathbf{S_A}$ is a diagonal matrix with elements $0.25x_{A,i}^2$. Error correlation between adjacent components of $\mathbf{x_A}$ is often approximated by an e-folding length scale, populating the off-diagonals of $\mathbf{S_A}$ adjacent to the diagonal and with a cut-off beyond which the off-diagonal terms are zero. This produces a *band matrix* where the presence of a large population of zero elements (*sparse matrix*) allows the use of fast algorithms for matrix inversion. For example, let us assume 50% error on the individual components of $\mathbf{x_A}$, an error correlation coefficient $r = 0.5$ for adjacent components, and zero error correlation for non-adjacent components. The resulting prior error covariance matrix is given by

$$S = \begin{pmatrix} 0.25x_{A,1}^2 & 0.125x_{A,1}x_{A,2} & 0 & \cdots & 0 \\ 0.125x_{A,1}x_{A,2} & 0.25x_{A,2}^2 & 0.125x_{A,2}x_{A,3} & 0 & \vdots \\ 0 & \ddots & & \ddots & \\ \vdots & & & & \\ 0 & & & & \end{pmatrix} \tag{11.42}$$

The observational error covariance matrix $\mathbf{S_O}$ can be constructed by adding the contributions from the instrument error ($\mathbf{S_I}$), representation error ($\mathbf{S_R}$), and forward model error ($\mathbf{S_M}$) estimated

Box 11.2 (cont.)

independently. S_I is typically a diagonal matrix that can be obtained from knowledge of the instrument precision relative to calibration standards. S_R can be constructed from knowledge of the subgrid variability of observations and can also in general be assumed diagonal. Construction of S_M is more difficult as calibration data for the forward model are generally not available. An estimate can be made by comparing different independent forward models.

An alternate approach for constructing S_O is the residual error method (Heald *et al.*, 2004). In this method, we conduct a forward model simulation using the prior estimate of the state vector, compare to observations, and subtract the mean bias to obtain the observational error:

$$\varepsilon_O = \mathbf{y} - \mathbf{F}(\mathbf{x_A}) - \overline{\mathbf{y} - \mathbf{F}(\mathbf{x_A})}$$ (11.43)

where the averaging can be done over the ensemble of observations or just a subset (for example, the observation time series at a given location). Here we assume that the systematic component of the error in $\mathbf{y} - \mathbf{F}(\mathbf{x_A})$ is due to error in the state vector $\mathbf{x}$ to be corrected through the inversion, while the random component is the observational error. The statistics of ε_O are then used to construct S_O. An example is shown in Box 11.2 Figure 1. The assumption that the systematic error is due solely to $\mathbf{x}$ may not be correct, as there may also be bias in the observing system; however, it is consistent with the premise of the inverse analysis that errors be random. From independent knowledge of S_I and S_R one can infer the forward model error covariance matrix as $S_M = S_O - S_I - S_R$, and from there diagnose the dominant terms contributing to the observational error.

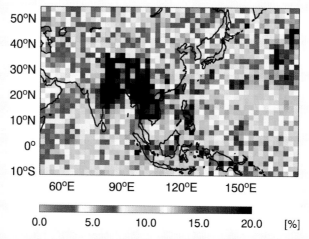

Diagonal terms of the observational error covariance matrix constructed for an inversion of carbon monoxide (CO) sources in East Asia in March–April 2001 using MOPITT satellite observations of CO columns and a CTM as forward model. The daily observations are averaged over $2° \times 2.5°$ CTM grid squares and compared to the CTM simulation using the prior estimate of sources, producing a time series of CTM-MOPITT differences in each grid square. The mean of that time series is subtracted and the residual difference defines the observational error for that grid square. The resulting error variance is normalized to the mean CO column for the grid square, thus defining a relative error expressed as percentage. The off-diagonal terms of the observational error covariance matrix are derived from an estimated 180-km error correlation length scale. From Heald *et al.* (2004).

11.4.2 Gaussian Probability Density Function for Vectors

Application of Bayes' theorem to the vector-matrix formalism requires formulation of PDFs for vectors. We derive here the general Gaussian PDF formulation for a vector $\mathbf{x}$ of dimension n with expected value $E[\mathbf{x}]$ and error covariance matrix $\mathbf{S}$. If the errors $x_i - E[x_i]$ for the individual elements of $\mathbf{x}$ are uncorrelated (i.e., if $\mathbf{S}$ is diagonal), then the PDF of the vector is simply the product of the PDFs for the individual elements. This simple solution can be obtained by projecting $\mathbf{x}$ on the basis of eigenvectors $\mathbf{e_i}$ of $\mathbf{S}$ with $i = [1, \ldots n]$. The error variances in that base are the eigenvalues λ_i of $\mathbf{S}$ (see derivation in Section 11.4.1). Let $\mathbf{z} = \mathbf{E}^T (\mathbf{x} - E[\mathbf{x}])$ be the value of $\mathbf{x} - E[\mathbf{x}]$ projected on the eigenvector basis, where $\mathbf{E}$ is the matrix of eigenvectors arranged by columns. The PDF of $\mathbf{z}$ is then

$$P(\mathbf{z}) = \prod_i \left[\frac{1}{(2\pi\lambda_i)^{1/2}} \exp\left[-\frac{z_i^2}{2\lambda_i} \right] \right] = \frac{1}{(2\pi)^{n/2} \prod_i \lambda_i^{1/2}} \exp\left[-\sum_i \frac{z_i^2}{2\lambda_i} \right] \quad (11.44)$$

which can be rewritten as

$$P(\mathbf{z}) = \frac{1}{(2\pi)^{n/2} |\mathbf{S}|^{1/2}} \exp\left[-\frac{1}{2} \mathbf{z}^T \mathbf{\Lambda}^{-1} \mathbf{z} \right] \quad (11.45)$$

Here $|\mathbf{S}|$ is the determinant of $\mathbf{S}$, equal to the product of its eigenvalues:

$$|\mathbf{S}| = \prod_i \lambda_i \quad (11.46)$$

and $\mathbf{\Lambda}$ is the diagonal matrix of eigenvalues (11.38). Replacing $\mathbf{z}$ in (11.45), we obtain:

$$P(\mathbf{x}) = \frac{1}{(2\pi)^{n/2} |\mathbf{S}|^{1/2}} \exp\left[-\frac{1}{2} (\mathbf{x} - E[\mathbf{x}])^T \mathbf{E} \mathbf{\Lambda}^{-1} \mathbf{E}^T (\mathbf{x} - E[\mathbf{x}]) \right] \quad (11.47)$$

Recall the matrix spectral decomposition $\mathbf{S} = \mathbf{E}\mathbf{\Lambda}\mathbf{E}^T$ (11.37). A matrix and its inverse have the same eigenvectors and inverse eigenvalues so that $\mathbf{S}^{-1} = \mathbf{E}\mathbf{\Lambda}^{-1}\mathbf{E}^T$. Replacing into (11.47) we obtain the general PDF expression for the vector $\mathbf{x}$:

$$P(\mathbf{x}) = \frac{1}{(2\pi)^{n/2} |\mathbf{S}|^{1/2}} \exp\left[-\frac{1}{2} (\mathbf{x} - E[\mathbf{x}])^T \mathbf{S}^{-1} (\mathbf{x} - E[\mathbf{x}]) \right] \quad (11.48)$$

11.4.3 Jacobian Matrix

The Jacobian matrix is the derivative of the forward model. We denote it $\mathbf{K}$ in this chapter to avoid confusion with the standard notation for the cost function (J). The Jacobian gives the local sensitivity of the observation variables $\mathbf{y}$ to the state variables $\mathbf{x}$ as described by the forward model:

$$\mathbf{K} = \nabla_{\mathbf{x}} \mathbf{F} = \frac{\partial \mathbf{y}}{\partial \mathbf{x}} \quad (11.49)$$

with individual elements $k_{ij} = \partial y_i / \partial x_j$. It is used in inverse modeling to compute the minimum of the Bayesian cost function (see (11.27) for application to the simple scalar problem). It is also used in the analytical solution to the inverse problem as a linearization of the forward model (see (11.28) for application to the simple scalar problem). If the forward model is linear, then $\mathbf{K}$ does not depend on $\mathbf{x}$ and fully describes the forward model for the purpose of the inversion. If the forward model is nonlinear, then $\mathbf{K}$ needs to be calculated initially for the prior estimate $\mathbf{x_A}$, representing the best initial guess for $\mathbf{x}$, and then re-calculated as needed for updated values of $\mathbf{x}$ during iterative convergence to the solution.

Construction of the Jacobian matrix may be done analytically if the forward model is simple, as for example in a 0-D chemical model where the evolution of concentrations for the n different species is determined by first-order kinetic rate expressions. If the forward model is complicated, such as a 3-D CTM, then the Jacobian must be constructed numerically. This can be done column by column if the dimension of the state vector is not so large as to make it computationally prohibitive. The task involves first conducting a base forward model calculation using the prior estimate $\mathbf{x_A}$ over the observation period, and then successively perturbing the individual elements x_i of the state vector by small increments Δx_i to calculate the resulting perturbation $\Delta \mathbf{y}$. This yields the sensitivity vector $\Delta \mathbf{y}/\Delta x_i \approx \partial \mathbf{y}/\partial x_i$, which is the ith column of the Jacobian. A total of $n + 1$ forward model calculations are required to fully construct the Jacobian matrix.

If the observations are sparse and the state vector is large, such as in a receptor-oriented problem where we wish to determine the sensitivity of concentrations at a few selected locations to a large array of surface fluxes, then a more effective way to construct the Jacobian is row by row using the adjoint of the forward model; this is described below. If both the state vector and the observation vector are large, then one can bypass the calculation of the Jacobian matrix in the minimization of the cost function by using the adjoint of the forward model; this will be described in Section 11.6.

11.4.4 Adjoint

The *adjoint* of a forward model is the transpose $\mathbf{K}^T$ of its Jacobian matrix (Section 11.4.3). It turns out to be very useful in inverse modeling applications for atmospheric chemistry where observed concentrations are used to constrain a state vector of emissions or concentrations at previous times. In that case, the *adjoint model* does not necessarily involve explicit construction of $\mathbf{K}^T$, but instead the application of $\mathbf{K}^T$ to vectors called *adjoint forcings*. We will discuss this in Section 11.6. The adjoint model can also be useful for numerical construction of the Jacobian matrix when $\dim(\mathbf{y}) \ll \dim(\mathbf{x})$. As we will see, by using the adjoint we can construct the Jacobian matrix row by row, instead of column by column, and the number of model simulations needed for that purpose is $\dim(\mathbf{y})$ rather than $\dim(\mathbf{x})$. A common application is in *receptor-oriented problems* where we seek, for example, to determine the sensitivity of the model concentration at a particular point to the ensemble of concentrations or emissions at previous times over the 3-D model domain. In that example, $\dim(\mathbf{y}) = 1$ but $\dim(\mathbf{x})$ can be very large, and a single pass of the adjoint model delivers the full vector of sensitivities. Box 11.3 illustrates the construction of the adjoint in a simple case.

Box 11.3	**Simple Adjoint Construction**

Consider a three-element state vector $(x_0, y_0 z_0)^T$ on which an operation $x = y^2 + z$ is applied. The resulting vector $(x_1, y_1, z_1)^T$ is

$$x_1 = y_0^2 + z_0$$
$$y_1 = y_0$$
$$z_1 = z_0$$

The Jacobian matrix for that operation is given by

$$\mathbf{K} = \begin{pmatrix} \partial x_1/\partial x_0 & \partial x_1/\partial y_0 & \partial x_1/\partial z_0 \\ \partial y_1/\partial x_0 & \partial y_1/\partial y_0 & \partial y_1/\partial z_0 \\ \partial z_1/\partial x_0 & \partial z_1/\partial y_0 & \partial z_1/\partial z_0 \end{pmatrix} = \begin{pmatrix} 0 & 2y_0 & 1 \\ 0 & 1 & 0 \\ 0 & 0 & 1 \end{pmatrix}$$

and the adjoint is then

$$\mathbf{K}^T = \begin{pmatrix} 0 & 0 & 0 \\ 2y_0 & 1 & 0 \\ 1 & 0 & 1 \end{pmatrix}$$

The null value of the first row of $\mathbf{K}^T$ means that $(x_1, y_1, z_1)^T$ has no sensitivity to x_0.

To understand how the adjoint works, consider a CTM discretized over time steps $[t_0, \ldots t_i, \ldots t_p]$. Let $\mathbf{y}_{(p)}$ represent the vector of gridded concentrations of dimension m at time t_p. We wish to determine its sensitivity to some state vector $\mathbf{x}_{(0)}$ of dimension n at time t_0. For example, $\mathbf{x}$ could be the gridded emissions. The corresponding Jacobian matrix is $\mathbf{K} = \partial \mathbf{y}_{(p)}/\partial \mathbf{x}_{(0)}$. By the chain rule,

$$\mathbf{K} = \frac{\partial \mathbf{y}_{(p)}}{\partial \mathbf{x}_{(0)}} = \frac{\partial \mathbf{y}_{(p)}}{\partial \mathbf{y}_{(p-1)}} \frac{\partial \mathbf{y}_{(p-1)}}{\partial \mathbf{y}_{(p-2)}} \cdots \frac{\partial \mathbf{y}_{(1)}}{\partial \mathbf{y}_{(0)}} \frac{\partial \mathbf{y}_{(0)}}{\partial \mathbf{x}_{(0)}} \tag{11.50}$$

where the right-hand side is a product of matrices. The adjoint model applies the transpose:

$$\mathbf{K}^T = \left(\frac{\partial \mathbf{y}_{(p)}}{\partial \mathbf{y}_{(p-1)}} \frac{\partial \mathbf{y}_{(p-1)}}{\partial \mathbf{y}_{(p-2)}} \cdots \frac{\partial \mathbf{y}_{(1)}}{\partial \mathbf{y}_{(0)}} \frac{\partial \mathbf{y}_{(0)}}{\partial \mathbf{x}_{(0)}} \right)^T = \left(\frac{\partial \mathbf{y}_{(0)}}{\partial \mathbf{x}_{(0)}} \right)^T \left(\frac{\partial \mathbf{y}_{(1)}}{\partial \mathbf{y}_{(0)}} \right)^T \cdots \left(\frac{\partial \mathbf{y}_{(p-1)}}{\partial \mathbf{y}_{(p-2)}} \right)^T \left(\frac{\partial \mathbf{y}_{(p)}}{\partial \mathbf{y}_{(p-1)}} \right)^T \tag{11.51}$$

where we have made use of the property that the transpose of a product of matrices is equal to the product of the transposed matrices in reverse order: $(\mathbf{AB})^T = \mathbf{B}^T \mathbf{A}^T$.

Consider now the application of $\mathbf{K}^T$ as expressed by (11.51) to a unit vector $\mathbf{v} = (1, 0, \ldots 0)^T$ taken as adjoint forcing. Following (11.51), we begin by applying matrix $\left(\partial \mathbf{y}_{(p)}/\partial \mathbf{y}_{(p-1)} \right)^T$ to $\mathbf{v}$:

$$\left(\frac{\partial \mathbf{y}_{(p)}}{\partial \mathbf{y}_{(p-1)}} \right)^T \begin{pmatrix} 1 \\ 0 \\ \vdots \\ 0 \end{pmatrix} = \begin{pmatrix} \partial y_{(p),1}/\partial y_{(p-1),1} \\ \partial y_{(p),1}/\partial y_{(p-1),2} \\ \vdots \\ \partial y_{(p),1}/\partial y_{(p-1),m} \end{pmatrix} = \frac{\partial y_{(p),1}}{\partial \mathbf{y}_{(p-1)}} \tag{11.52}$$

This yields a vector of *adjoint variables* $\partial y_{(p),1}/\partial \mathbf{y}_{(p-1)}$ that represents the sensitivity of $y_{(p),1}$ to $\mathbf{y}_{(p-1)}$. Let us now apply the next matrix $\left(\partial \mathbf{y}_{(p-1)}/\partial \mathbf{y}_{(p-2)}\right)^T$ in (11.51) to this vector of adjoint variables:

$$\left(\frac{\partial \mathbf{y}_{(p-1)}}{\partial \mathbf{y}_{(p-2)}}\right)^T \begin{pmatrix} \partial y_{(p),1}/\partial y_{(p-1),1} \\ \partial y_{(p),1}/\partial y_{(p-1),2} \\ \vdots \\ \partial y_{(p),1}/\partial y_{(p-1),m} \end{pmatrix} = \begin{pmatrix} \dfrac{\partial y_{(p-1),1}}{\partial y_{(p-2),1}}\dfrac{\partial y_{(p),1}}{\partial y_{(p-1),1}} + \dfrac{\partial y_{(p-1),2}}{\partial y_{(p-2),1}}\dfrac{\partial y_{(p),1}}{\partial y_{(p-1),2}} + \ldots \\ \dfrac{\partial y_{(p-1),1}}{\partial y_{(p-2),2}}\dfrac{\partial y_{(p),1}}{\partial y_{(p-1),1}} + \dfrac{\partial y_{(p-1),2}}{\partial y_{(p-2),2}}\dfrac{\partial y_{(p),1}}{\partial y_{(p-1),2}} + \ldots \\ \vdots \\ \dfrac{\partial y_{(p-1),1}}{\partial y_{(p-2),m}}\dfrac{\partial y_{(p),1}}{\partial y_{(p-1),1}} + \dfrac{\partial y_{(p-1),2}}{\partial y_{(p-2),m}}\dfrac{\partial y_{(p),1}}{\partial y_{(p-1),2}} + \ldots \end{pmatrix}$$

$$= \begin{pmatrix} \partial y_{(p),1}/\partial y_{(p-2),1} \\ \partial y_{(p),1}/\partial y_{(p-2),2} \\ \vdots \\ \partial y_{(p),1}/\partial y_{(p-2),m} \end{pmatrix} = \frac{\partial y_{(p),1}}{\partial \mathbf{y}_{(p-2)}} \tag{11.53}$$

where we have made use of

$$\frac{\partial y_{(p),1}}{\partial y_{(p-2),j}} = \sum_{k=1}^{m} \frac{\partial y_{(p),1}}{\partial y_{(p-1),k}} \frac{\partial y_{(p-1),k}}{\partial y_{(p-2),j}} \tag{11.54}$$

We thus obtain $\partial y_{(p),1}/\partial \mathbf{y}_{(p-2)}$. Application of the next matrix $\left(\partial \mathbf{y}_{(p-2)}/\partial \mathbf{y}_{(p-3)}\right)^T$ to this vector yields $\partial y_{(p),1}/\partial \mathbf{y}_{(p-3)}$ and so on. By sequential application of the suite of matrices in (11.51) we thus obtain $\partial y_{(p),1}/\partial \mathbf{x}_{(0)}$, which is a row of the Jacobian matrix. Repeating this exercise for the m unit vectors $\mathbf{v}$ representing the different elements of $\mathbf{y}$ yields the full matrix $\mathbf{K} = \partial \mathbf{y}_{(p)}/\partial \mathbf{x}_{(0)}$.

Notice from the above description that a single pass with the adjoint yields the sensitivity vectors $\partial y_{(p),1}/\partial \mathbf{y}_{(p-1)}$, $\partial y_{(p),1}/\partial \mathbf{y}_{(p-2)}$, $\ldots \partial y_{(p),1}/\partial \mathbf{y}_{(0)}$. This effectively integrates the CTM back in time, providing the sensitivity of the concentration at a given location and time (here $y_{(p),1}$) to the complete field of concentrations at prior times, i.e., the backward influence function. The same single pass with the adjoint can also provide the sensitivities of $y_{(p),1}$ to the state vector at any prior time; thus:

$$\left(\frac{\partial \mathbf{y}_{(p)}}{\partial \mathbf{x}_{(p)}}\right)^T \begin{pmatrix} 1 \\ 0 \\ \vdots \\ 0 \end{pmatrix} = \begin{pmatrix} \partial y_{(p),1}/\partial x_{(p),1} \\ \partial y_{(p),1}/\partial x_{(p),2} \\ \vdots \\ \partial y_{(p),1}/\partial x_{(p),n} \end{pmatrix} = \frac{\partial y_{(p),1}}{\partial \mathbf{x}_{(p)}} \tag{11.55}$$

$$\left(\frac{\partial \mathbf{y}_{(p-1)}}{\partial \mathbf{x}_{(p-1)}}\right)^T \left(\frac{\partial \mathbf{y}_{(p)}}{\partial \mathbf{y}_{(p-1)}}\right)^T \begin{pmatrix} 1 \\ 0 \\ \vdots \\ 0 \end{pmatrix} = \begin{pmatrix} \partial y_{(p),1}/\partial x_{(p-1),1} \\ \partial y_{(p),1}/\partial x_{(p-1),2} \\ \vdots \\ \partial y_{(p),1}/\partial x_{(p-1),n} \end{pmatrix} = \frac{\partial y_{(p),1}}{\partial \mathbf{x}_{(p-1)}} \tag{11.56}$$

$$
\left(\frac{\partial \mathbf{y}_{(p-2)}}{\partial \mathbf{x}_{(p-2)}}\right)^T \left(\frac{\partial \mathbf{y}_{(p-1)}}{\partial \mathbf{y}_{(p-2)}}\right)^T \left(\frac{\partial \mathbf{y}_{(p)}}{\partial \mathbf{y}_{(p-1)}}\right)^T \begin{pmatrix} 1 \\ 0 \\ \vdots \\ 0 \end{pmatrix} = \begin{pmatrix} \partial y_{(p),1}/\partial x_{(p-2),1} \\ \partial y_{(p),1}/\partial x_{(p-2),2} \\ \vdots \\ \partial y_{(p),1}/\partial x_{(p-2),n} \end{pmatrix} = \frac{\partial y_{(p),1}}{\partial \mathbf{x}_{(p-2)}}
$$

$$(11.57)$$

and so on. For example, if the state vector represents the emission field, we can obtain in this manner the sensitivity of the concentration $y_{(p),1}$ to the emissions at all prior time steps. Box 11.4 illustrates such an application.

The sensitivities computed by the adjoint method are true local derivatives. For a nonlinear problem they are sometimes called *adjoint sensitivities*. This is to contrast them with the sensitivities obtained by finite differencing calculations, i.e., by

---

Box 11.4   **Computing Adjoint Sensitivities**

We illustrate the computation of adjoint sensitivities with an example from Kim *et al.* (2015), shown in Box 11.4 Figure 1. Here the adjoint of a CTM is used to compute the sensitivity of mean smoke particle concentrations in Singapore in July–November 2006 to fires in different locations of equatorial Asia. The CTM has $0.5° \times 0.67°$ horizontal grid resolution and simulates smoke concentrations on the basis of a fire emission inventory that has the same grid resolution as the CTM and daily temporal resolution. The fires emit smoke particles that are transported by the model winds and are eventually removed by wet and dry deposition. Panel 1 shows the mean emissions and winds used in the CTM, and Panel 2 shows the resulting distribution of smoke concentrations in surface air.

We now want to use the CTM adjoint to determine the emissions contributing to the mean smoke concentrations in Singapore in July–November 2006. Fire emissions were limited to that period (dry season). We define $\mathbf{x}_{(i)}$ as the vector of 2-D gridded fire emissions at CTM time $t_i \in [t_1, t_p]$ where t_1 refers to 00:00 local time on July 1 and t_p refers to 00:00 on December 1. We define $\mathbf{y}_{(i)}$ as the vector of 3-D smoke concentrations simulated by the model at time step i, and choose the first element of that vector $y_{(i),1}$ to represent the smoke concentration in surface air at Singapore.

Following (11.52), we apply the model adjoint over one time step $[t_p, t_{p-1}]$ to a unit forcing $\mathbf{v} = (1, 0, \ldots 0)^T$ at time t_p. This yields the sensitivity vector $\partial y_{(p),1}/\partial \mathbf{y}_{(p-1)}$ that describes the sensitivity of concentrations at Singapore at time t_p to the 3-D field of concentrations at time t_{p-1}. It also yields the sensitivity vector $\partial y_{(p),1}/\partial \mathbf{x}_{(p)}$ that describes the sensitivity of concentrations at Singapore to the 2-D field of emissions at time t_p. We archive $\partial y_{(p),1}/\partial \mathbf{x}_{(p)}$, add a unit forcing $\mathbf{v}$ to the sensitivity vector $\partial y_{(p),1}/\partial \mathbf{y}_{(p-1)}$, and apply the adjoint over the next time step $[t_{p-1}, t_{p-2}]$. From there we get $\partial\left(y_{(p),1} + y_{(p-1),1}\right)/\partial \mathbf{y}_{(p-2)}$ and $\partial\left(y_{(p),1} + y_{(p-1),1}\right)/\partial \mathbf{x}_{(p-1)}$. We archive $\partial\left(y_{(p),1} + y_{(p-1),1}\right)/\partial \mathbf{x}_{(p-1)}$, which is the sensitivity of concentrations in Singapore to the emission field at time t_{p-1}, and proceed in that manner backward in time until time step 1.

A single pass of the adjoint simulation over the time interval $[t_p, t_1]$ thus yields sensitivities of the mean concentration in Singapore $\bar{y} = (1/p)\sum_1^p y_{(i),1}$ over the period $[t_1, t_p]$ to the emission field at every time step over that period:

Box 11.4 *(cont.)*

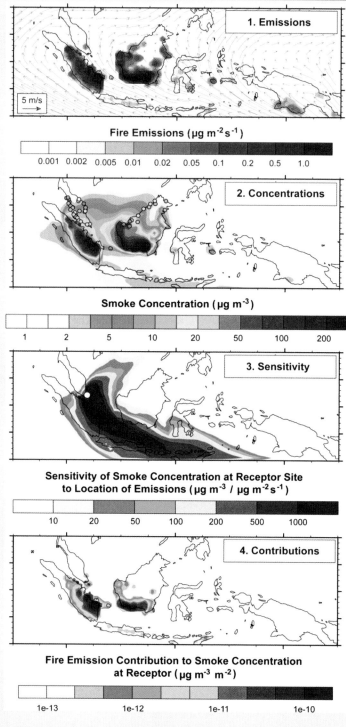

Fire Emissions (μg m⁻² s⁻¹) — Fire Emissions ($\mu g\ m^{-2}\ s^{-1}$)

0.001 0.002 0.005 0.01 0.02 0.05 0.1 0.2 0.5 1.0

Smoke Concentration ($\mu g\ m^{-3}$)

1 2 5 10 20 50 100 200

Sensitivity of Smoke Concentration at Receptor Site to Location of Emissions ($\mu g\ m^{-3} / \mu g\ m^{-2} s^{-1}$)

10 20 50 100 200 500 1000

Fire Emission Contribution to Smoke Concentration at Receptor ($\mu g\ m^{-3}\ m^{-2}$)

1e-13 1e-12 1e-11 1e-10

Box 11.4 Figure 1 Application of the adjoint method to determine the sensitivity of smoke concentrations in Singapore in July–November 2006 to fire emissions across equatorial Asia. Panel 1 shows mean July-November fire emissions and 0–1 km winds. Panel 2 shows the mean smoke concentrations simulated by the CTM (circles show observations). Panel 3 shows the sensitivity of smoke concentrations in Singapore to fire emissions in different regions. Panel 4 shows the contributions of different fire regions to the smoke concentrations in Singapore. Adapted from Kim *et al.* (2015).

$$\frac{\partial \bar{y}}{\partial \mathbf{x}_{(p)}} = \frac{1}{p} \frac{\partial y_{(p),1}}{\partial \mathbf{x}_{(p)}}$$

$$\frac{\partial \bar{y}}{\partial \mathbf{x}_{(p-1)}} = \frac{1}{p} \frac{\partial \left(y_{(p),1} + y_{(p-1),1} \right)}{\partial \mathbf{x}_{(p-1)}}$$

$$\vdots$$

$$\frac{\partial \bar{y}}{\partial \mathbf{x}_{(1)}} = \frac{1}{p} \frac{\partial \sum_{i=1}^{p} y_{(i),1}}{\partial \mathbf{x}_{(1)}}$$
(11.58)

where we recognize that emissions at a given time can only affect concentrations after that time. Panel 3 shows the mean adjoint sensitivities $(1/p) \sum_{i=1}^{p} \partial \bar{y}/\partial \mathbf{x}_{(i)}$. These indicate the potential of fires occurring in different locations to affect Singapore. We can express the smoke concentration at Singapore in July–November 2006 as the sum of these adjoint sensitivities weighted by the actual emissions:

$$\bar{y} = \sum_{j} \sum_{i=1}^{p} \frac{\partial \bar{y}}{\partial x_{(i),j}} x_{(i),j}$$
(11.59)

where the index j refers to the 2-D emission grid. Panel 4 shows the contributions $\sum_{i=1}^{p} \frac{\partial \bar{y}}{\partial x_{(i),j}} x_{(i),j}$ of emissions in individual model grid squares j to the mean smoke concentration at Singapore.

perturbing the state vector elements by Δx_i and diagnosing the difference in output $\Delta \mathbf{y}$. Finite differencing entails some effect of nonlinearity, which can be reduced by selecting a small Δx_i but at the cost of numerical noise.

Henze *et al.* (2007) describe in detail the steps involved in constructing the adjoint of a CTM. The main difficulty is linearization to express the CTM as a product of matrices. Linearization involves differentiation of the model (cf. (11.2)). One can differentiate either the model equations (*continuous adjoint*) or the model code (*discrete adjoint*). The discrete adjoint is more consistent with the actual CTM. The differentiated model is the Jacobian of the CTM and is called the *tangent linear model* (*TLM*). Construction of the TLM can be an arduous task and commercial software packages are available for this purpose.

We present here an elementary example of CTM adjoint construction to illustrate the basic tasks involved. The CTM calculates the evolution of concentrations over a time step $[t_i, t_{i+1}]$ by successive application of operators describing the different model processes. Consider a CTM including 3-D advection (operator A), chemistry (operator C), and emissions (operator E), with operator splitting described by

$$\mathbf{y}_{(i+1)} = A \cdot C \cdot E \left(\mathbf{y}_{(i)} \right)$$
(11.60)

where the • symbol means "applied to." The operators may or may not be linear in $\mathbf{y}$. If not, they need to be linearized by differentiation, as we did with the forward model in (11.49). Let $\mathbf{A}$, $\mathbf{C}$, $\mathbf{E}$ be the matrices of the linear operators. We have:

$$\mathbf{y}_{(i+1)} = \mathbf{A}\,\mathbf{C}\,\mathbf{E}\,\mathbf{y}_{(i)} \tag{11.61}$$

so that

$$\frac{\partial \mathbf{y}_{(i+1)}}{\partial \mathbf{y}_{(i)}} = \mathbf{A}\,\mathbf{C}\,\mathbf{E} \tag{11.62}$$

The transpose is given by

$$\left(\frac{\partial \mathbf{y}_{(i+1)}}{\partial \mathbf{y}_{(i)}}\right)^T = \mathbf{E}^T\mathbf{C}^T\mathbf{A}^T \tag{11.63}$$

The case of linear operators offers insight into the physical meaning of the adjoint. Let us begin with the advection operator. Three-dimensional advection is generally described by operator splitting with 1-D operators. Consider then a 1-D advection algorithm using a linear upstream scheme on an Eulerian grid (see Section 7.3.2):

$$y_{(i+1),j} = \alpha y_{(i),j-1} + (1-\alpha)y_{(i),j} \tag{11.64}$$

where α is the Courant number, (i) is the time index, and the flow is from gridbox $j-1$ to j. Let us take as an example a uniform cyclical flow over a domain $j = [1, 3]$. The advection operator is written in matrix form as

$$\mathbf{A} = \left(\frac{\partial \mathbf{y}_{(i+1)}}{\partial \mathbf{y}_{(i)}}\right)_{advection} = \begin{pmatrix} 1-\alpha & 0 & \alpha \\ \alpha & 1-\alpha & 0 \\ 0 & \alpha & 1-\alpha \end{pmatrix} \tag{11.65}$$

and its transpose is

$$\mathbf{A}^T = \left(\frac{\partial \mathbf{y}_{(i+1)}}{\partial \mathbf{y}_{(i)}}\right)^T_{advection} = \begin{pmatrix} 1-\alpha & \alpha & 0 \\ 0 & 1-\alpha & \alpha \\ \alpha & 0 & 1-\alpha \end{pmatrix} \tag{11.66}$$

We see that the transpose describes the *reverse* of the actual flow:

$$y_{(i+1),j} = \alpha y_{(i),j+1} + (1-\alpha)y_{(i),j} \tag{11.67}$$

This result is readily generalizable to any number of gridboxes and non-uniform flow. Thus the adjoint of a linear transport operator is simply the reverse flow, and this is also found in the continuous adjoint by differentiating the advection equation (Henze *et al.*, 2007). Advection operators may not be exactly linear because of safeguards for stability, positivity, or mass conservation. Nevertheless, the approximation of reverse flow is frequently used to construct the adjoint because of its simplicity.

Consider now a first-order loss chemistry operator $dy/dt = -ky$ where k is a loss rate constant. Application of this operator over a time step Δt is expressed in matrix form as follows:

$$\mathbf{C} = \left(\frac{\partial \mathbf{y}_{(i+1)}}{\partial \mathbf{y}_{(i)}}\right)_{chemistry} = \begin{pmatrix} \exp\left[-k\Delta t\right] & & 0 \\ & \ddots & \\ 0 & & \exp\left[-k\Delta t\right] \end{pmatrix} \tag{11.68}$$

which is a diagonal matrix. In this case the transpose operator is the same as the original operator; we refer to the operator as *self-adjoint*. It makes sense that the sensitivity going back in time should decay with the same time constant as the first-order chemical loss. From a coding standpoint, it means that the adjoint can use the same chemical operator as the forward model.

Finally, consider the emission operator $\mathbf{E}$. Its application over a time step Δt modifies the concentration field as $\mathbf{y}_{(i+1)} = \mathbf{y}_{(i)} + \mathbf{x}_{(i+1)}\Delta t$ where $\mathbf{x}$ is an emission flux vector that is non-zero only for gridboxes receiving emissions. In terms of sensitivity to concentrations at the previous time step, the emission operator is the identity matrix $\mathbf{I_m}$ and thus self-adjoint:

$$\mathbf{E} = \left(\frac{\partial \mathbf{y}_{(i+1)}}{\partial \mathbf{y}_{(i)}}\right)_{emissions} = \mathbf{I_m} \tag{11.69}$$

while the sensitivity to emissions is also self-adjoint:

$$\frac{\partial \mathbf{y}_{(i+1)}}{\partial \mathbf{x}_{(i+1)}} = \mathbf{I_m}\Delta t \tag{11.70}$$

We have thus shown how the matrices $\mathbf{A}^T$, $\mathbf{C}^T$, $\mathbf{E}^T$ can be computed in simple cases to define the adjoint model. In this manner the adjoint model marches back in time to describe the sensitivity of concentrations to concentrations and emissions at prior times. See Box 11.4 for an example application.

Another simple application of the adjoint is to linear multi-box models, often used in geochemical modeling to simulate the evolution of concentrations in m different coupled reservoirs (boxes). The model is described by

$$\frac{d\mathbf{y}}{dt} = \mathbf{K}\mathbf{y} + \mathbf{s} \tag{11.71}$$

where $\mathbf{y}$ is the vector of concentrations or masses in the different boxes, $\mathbf{K}$ is a Jacobian matrix of transfer coefficients k_{ij} describing the transfer between boxes, and $\mathbf{s}$ is a source vector. Starting from initial conditions at time t_0, the evolution of the system for one time step $\Delta t = t_1 - t_0$ is given in forward finite difference form by

$$\mathbf{y}_{(1)} = \mathbf{M}\mathbf{y}_{(0)} + \mathbf{s}_{(0)}\Delta t \tag{11.72}$$

where $\mathbf{M} = \mathbf{I_m} + \mathbf{K}\Delta t$. We see that $\partial \mathbf{y}_{(1)}/\partial \mathbf{y}_{(0)} = \mathbf{M}$ and $\partial \mathbf{y}_{(1)}/\partial \mathbf{s}_{(0)} = \mathbf{I_m}\Delta t$; the corresponding adjoint operators are $\mathbf{M}^T$ and $\mathbf{I_m}\Delta t$ (the source operator is self-adjoint). Consider a time period of interest $[t_0, t_p)$ (say from pre-industrial to present time). A single pass of the adjoint backward in time over $[t_p, t_0]$ yields the sensitivity of the concentrations in a given box at a given time to the concentrations and sources at previous times for all other boxes.

11.5 Analytical Inversion

The vector-matrix tools presented in Section 11.4 allow us to apply Bayes' theorem (Section 11.2) to obtain an optimal estimate of a state vector $\mathbf{x}$ (dim n) on the basis of the observation vector $\mathbf{y}$ (dim m), the prior information $\mathbf{x_A}$, the forward model $\mathbf{F}$, and the error covariance matrices $\mathbf{S_A}$ and $\mathbf{S_O}$. This is done by finding the minimum of a cost function describing the observational and prior constraints. Here we present the analytical solution to this minimization problem assuming Gaussian errors. A major advantage of the analytical approach, as we will see, is that it provides complete error characterization as part of the solution. It can also be fast and well suited for conducting an ensemble of inversions with varying assumptions. But it has three limitations:

1. It requires construction of the Jacobian $\mathbf{K} = \nabla_{\mathbf{x}}\mathbf{F}$, which may be computationally impractical for a very large state vector or for a nonlinear problem where the Jacobian would have to be re-constructed at each iteration toward the solution.
2. It requires the assumption of Gaussian errors, which may not always be appropriate and in particular does not guarantee positivity of the solution.
3. It does not accommodate prior constraints other than specified through Bayes' theorem.

Other approaches to solving the inverse problem that lift these limitations will be presented in subsequent sections. The reader is encouraged to consult the simple scalar example of Section 11.3 in order to develop intuition for the material presented here. Many of the equations derived here have scalar equivalents in Section 11.3 that are easier to parse and understand.

11.5.1 Optimal Estimate

Assuming Gaussian distribution of errors, the PDFs to be used for application of Bayes' theorem are given by (11.48):

$$-2 \ln P(\mathbf{x}) = (\mathbf{x} - \mathbf{x_A})^T \mathbf{S_A}^{-1}(\mathbf{x} - \mathbf{x_A}) + c_1 \qquad (11.73)$$

$$-2 \ln P(\mathbf{y}|\mathbf{x}) = (\mathbf{y} - \mathbf{F}(\mathbf{x}))^T \mathbf{S_O}^{-1}(\mathbf{y} - \mathbf{F}(\mathbf{x})) + c_2 \qquad (11.74)$$

from which we obtain by application of Bayes' theorem, $P(\mathbf{x}|\mathbf{y}) \propto P(\mathbf{x})P(\mathbf{y}|\mathbf{x})$:

$$-2 \ln P(\mathbf{x}|\mathbf{y}) = (\mathbf{x} - \mathbf{x_A})^T \mathbf{S_A}^{-1}(\mathbf{x} - \mathbf{x_A}) + (\mathbf{y} - \mathbf{F}(\mathbf{x}))^T \mathbf{S_O}^{-1}(\mathbf{y} - \mathbf{F}(\mathbf{x})) + c_3$$
$$(11.75)$$

Here c_1, c_2, c_3 are constants. The optimal estimate is defined by the maximum of $P(\mathbf{x}|\mathbf{y})$, or equivalently by the minimum of the scalar-valued χ^2 cost function $J(\mathbf{x})$:

$$J(\mathbf{x}) = (\mathbf{x} - \mathbf{x_A})^T \mathbf{S_A}^{-1}(\mathbf{x} - \mathbf{x_A}) + (\mathbf{y} - \mathbf{F}(\mathbf{x}))^T \mathbf{S_O}^{-1}(\mathbf{y} - \mathbf{F}(\mathbf{x})) \qquad (11.76)$$

We find this minimum by solving $\nabla_{\mathbf{x}} J(\mathbf{x}) = 0$:

$$\nabla_{\mathbf{x}} J(\mathbf{x}) = 2\mathbf{S_A}^{-1}(\mathbf{x} - \mathbf{x_A}) + 2\mathbf{K}^T \mathbf{S_O}^{-1}(\mathbf{F}(\mathbf{x}) - \mathbf{y}) = 0 \qquad (11.77)$$

where $\mathbf{K}^T = \nabla_{\mathbf{x}}\mathbf{F}^T$ is the transpose of the Jacobian matrix. Equations (11.26) and (11.27) in Section 11.3 are the scalar analogues.

Let us assume that $\mathbf{F}(\mathbf{x})$ is linear or can be linearized as given by (11.2), i.e., $\mathbf{F}(\mathbf{x}) = \mathbf{Kx} + \mathbf{c}$ where $\mathbf{c}$ is a constant, and for simplicity of notation let $\mathbf{c} = 0$ (this can always be enforced by replacing $\mathbf{y}$ by $\mathbf{y} - \mathbf{c}$). We then have

$$\nabla_{\mathbf{x}} J(\mathbf{x}) = 2\mathbf{S}_{\mathbf{A}}^{-1}(\mathbf{x} - \mathbf{x}_{\mathbf{A}}) + 2\mathbf{K}^T \mathbf{S}_{\mathbf{O}}^{-1}(\mathbf{K}\mathbf{x} - \mathbf{y}) = 0 \qquad (11.78)$$

The solution of (11.78) is straightforward and can be expressed in compact form as

$$\hat{\mathbf{x}} = \mathbf{x}_{\mathbf{A}} + \mathbf{G}(\mathbf{y} - \mathbf{K}\mathbf{x}_{\mathbf{A}}) \qquad (11.79)$$

where $\mathbf{G}$ is the *gain matrix* given by

$$\mathbf{G} = \mathbf{S}_{\mathbf{A}}\mathbf{K}^T \left(\mathbf{K}\mathbf{S}_{\mathbf{A}}\mathbf{K}^T + \mathbf{S}_{\mathbf{O}}\right)^{-1} \qquad (11.80)$$

$\mathbf{G}$ describes the sensitivity of the optimal estimate to the observations, i.e., $\mathbf{G} = \partial\hat{\mathbf{x}}/\partial\mathbf{y}$. It is a valuable diagnostic for the inversion as it tells us which observations contribute most to constrain specific components of the optimal estimate. Equations (11.78), (11.79), and (11.80) have scalar analogues (11.17), (11.19), and (11.20) in Section 11.3.

The *posterior error covariance matrix* $\hat{\mathbf{S}}$ of $\hat{\mathbf{x}}$ can be calculated in the same manner as in Section 11.3 by rearranging the right-hand side of (11.75) with $\mathbf{F}(\mathbf{x}) = \mathbf{K}\mathbf{x}$ to be of the form $(\mathbf{x}-\hat{\mathbf{x}})^T \hat{\mathbf{S}}^{-1}(\mathbf{x} - \hat{\mathbf{x}})$. This yields

$$\hat{\mathbf{S}} = \left(\mathbf{K}^T \mathbf{S}_{\mathbf{O}}^{-1}\mathbf{K} + \mathbf{S}_{\mathbf{A}}^{-1}\right)^{-1} \qquad (11.81)$$

Again, note the similarity of this equation to its simple scalar equivalent (11.24) in Section 11.3. An important feature of the analytical solution to the inverse problem is that it provides a full characterization of errors on the optimal estimate through $\hat{\mathbf{S}}$, as well as a diagnostic of the influence of different observations through $\mathbf{G}$.

11.5.2 Averaging Kernel Matrix

Error characterization in the analytical solution to the inverse problem allows us to measure the capability of the observing system to constrain the true value of the state vector. This is done with the *averaging kernel matrix* $\mathbf{A} = \partial\hat{\mathbf{x}}/\partial\mathbf{x}$, representing the sensitivity of the optimal estimate $\hat{\mathbf{x}}$ to the true state $\mathbf{x}$. $\mathbf{A}$ is the product of the gain matrix $\mathbf{G} = \partial\hat{\mathbf{x}}/\partial\mathbf{y}$ and the Jacobian matrix $\mathbf{K} = \partial\mathbf{y}/\partial\mathbf{x}$:

$$\mathbf{A} = \mathbf{G}\mathbf{K} \qquad (11.82)$$

Replacing (11.82) and $\mathbf{y} = \mathbf{K}\mathbf{x} + \varepsilon_{\mathbf{O}}$ into (11.79) we obtain an alternate form for $\hat{\mathbf{x}}$:

$$\hat{\mathbf{x}} = \mathbf{A}\mathbf{x} + (\mathbf{I}_{\mathbf{n}} - \mathbf{A})\mathbf{x}_{\mathbf{A}} + \mathbf{G}\varepsilon_{\mathbf{O}} \qquad (11.83)$$

or equivalently

$$\hat{\mathbf{x}} = \mathbf{x} + (\mathbf{I}_{\mathbf{n}} - \mathbf{A})(\mathbf{x}_{\mathbf{A}} - \mathbf{x}) + \mathbf{G}\varepsilon_{\mathbf{O}} \qquad (11.84)$$

where $\mathbf{I}_{\mathbf{n}}$ is the identity matrix of dimension n. Equations (11.21) and (11.22) in Section 11.3 are scalar analogues. $\mathbf{A}$ is a weighting factor for the relative contribution to the optimal estimate from the true state vs. the prior estimate. $\mathbf{A}\mathbf{x}$ represents the contribution of the true state to the solution, $(\mathbf{I}_{\mathbf{n}} - \mathbf{A})\mathbf{x}_{\mathbf{A}}$ represents the contribution from the prior estimate, and $\mathbf{G}\varepsilon_{\mathbf{O}}$ represents the contribution from the random observational error mapped onto state space by the gain matrix $\mathbf{G}$. A perfect observational system would have $\mathbf{A} = \mathbf{I}_{\mathbf{n}}$. $(\mathbf{I}_{\mathbf{n}} - \mathbf{A})(\mathbf{x}_{\mathbf{A}} - \mathbf{x})$ is called the *smoothing error*.

From (11.84) we can also derive an alternate expression for the error covariance matrix $\widehat{\mathbf{S}}$:

$$\widehat{\mathbf{S}} = E\left[(\mathbf{x} - \widehat{\mathbf{x}})(\mathbf{x} - \widehat{\mathbf{x}})^T\right] = E\left[(\mathbf{I_n} - \mathbf{A})(\mathbf{x} - \mathbf{x_A})(\mathbf{x} - \mathbf{x_A})^T(\mathbf{I_n} - \mathbf{A})^T\right] + E\left[\mathbf{G}\varepsilon_{\mathbf{O}}\varepsilon_{\mathbf{O}}^T\mathbf{G}^T\right]$$

$$= (\mathbf{I_n} - \mathbf{A})\mathbf{S_A}(\mathbf{I_n} - \mathbf{A})^T + \mathbf{GS_O}\mathbf{G}^T \tag{11.85}$$

from which we see that $\widehat{\mathbf{S}}$ can be decomposed into the sum of a *smoothing error covariance matrix* $(\mathbf{I_n} - \mathbf{A})\mathbf{S_A}(\mathbf{I_n} - \mathbf{A})^T$ and an *observational error covariance matrix in state space* $\mathbf{GS_O}\mathbf{G}^T$. The smoothing error covariance matrix describes the smoothing of the solution by the prior constraints. The observational error covariance matrix describes the noise in the observing system.

Algebraic manipulation yields an alternate form of the averaging kernel matrix as

$$\mathbf{A} = \mathbf{I_n} - \widehat{\mathbf{S}}\mathbf{S_A}^{-1} \tag{11.86}$$

which relates the improved knowledge of the state vector measured by $\mathbf{A}$ to the variance reduction previously discussed for the scalar problem (see (11.25) for the scalar analogue). This is a convenient way to derive $\mathbf{A}$ from knowledge of $\widehat{\mathbf{S}}$.

The averaging kernel matrix constructed from knowledge of $\mathbf{S_A}$, $\mathbf{S_O}$, and $\mathbf{K}$ is a very useful thing to know about an observing system. When designing the observing system it can be used to evaluate and compare the merits of different designs for quantifying $\mathbf{x}$. By relating the observed state to the true state, it enables comparison of data from different instruments (Rodgers and Connor, 2003; Zhang et al., 2010) Box 11.5 illustrates the utility of the averaging kernel matrix in interpreting satellite data.

11.5.3 Degrees of Freedom for Signal

The averaging kernel matrix quantifies the number of pieces of information in an observing system toward constraining an n-dimensional state vector. This is called the *degrees of freedom for signal* (*DOFS*) (Rodgers, 2000). Before making the observations we had n unknowns representing the state vector elements as constrained solely by the prior error covariance matrix. We express that number of unknowns as

$$E\left[(\mathbf{x} - \mathbf{x_A})^T\mathbf{S_A}^{-1}(\mathbf{x} - \mathbf{x_A})\right] = n \tag{11.87}$$

After making the observations the error on $\mathbf{x}$ is decreased, and we express this decrease as a reduction in the number of unknowns to $E\left[(\mathbf{x} - \widehat{\mathbf{x}})^T\mathbf{S_A}^{-1}(\mathbf{x} - \widehat{\mathbf{x}})\right]$. The number of pieces of information from the observations is the reduction in the number of unknowns:

$$\mathrm{DOFS} = E\left[(\mathbf{x} - \mathbf{x_A})^T\mathbf{S_A}^{-1}(\mathbf{x} - \mathbf{x_a})\right] - E\left[(\mathbf{x} - \widehat{\mathbf{x}})^T\mathbf{S_A}^{-1}(\mathbf{x} - \widehat{\mathbf{x}})\right]$$

$$= n - E\left[(\mathbf{x} - \widehat{\mathbf{x}})^T\mathbf{S_A}^{-1}(\mathbf{x} - \widehat{\mathbf{x}})\right] \tag{11.88}$$

The quantity $(\mathbf{x}-\widehat{\mathbf{x}})^T \mathbf{S}_{\mathbf{A}}^{-1} (\mathbf{x} - \widehat{\mathbf{x}})$ is a scalar and is thus equal to its trace in matrix notation:

$$(\mathbf{x}-\widehat{\mathbf{x}})^T \mathbf{S}_{\mathbf{A}}^{-1}(\mathbf{x} - \widehat{\mathbf{x}}) = \mathrm{tr}\left((\mathbf{x} - \widehat{\mathbf{x}})^T \mathbf{S}_{\mathbf{A}}^{-1}(\mathbf{x} - \widehat{\mathbf{x}})\right) = \mathrm{tr}\left((\mathbf{x} - \widehat{\mathbf{x}})(\mathbf{x}-\widehat{\mathbf{x}})^T \mathbf{S}_{\mathbf{A}}^{-1}\right) \quad (11.89)$$

where we have taken advantage of the general property $\mathrm{tr}(\mathbf{AB}) = \mathrm{tr}(\mathbf{BA})$. Thus

$$E\left[(\mathbf{x} - \widehat{\mathbf{x}})^T \mathbf{S}_{\mathbf{A}}^{-1}(\mathbf{x} - \widehat{\mathbf{x}})\right] = E\left[\mathrm{tr}\left((\mathbf{x} - \widehat{\mathbf{x}})(\mathbf{x}-\widehat{\mathbf{x}})^T \mathbf{S}_{\mathbf{A}}^{-1}\right)\right] = \mathrm{tr}\left(\widehat{\mathbf{S}}\mathbf{S}_{\mathbf{A}}^{-1}\right) \quad (11.90)$$

so that

$$\mathrm{DOFS} = n - \mathrm{tr}\left(\widehat{\mathbf{S}}\mathbf{S}_{\mathbf{A}}^{-1}\right) = \mathrm{tr}\left(\mathbf{I_n} - \widehat{\mathbf{S}}\mathbf{S}_{\mathbf{A}}^{-1}\right) = \mathrm{tr}(\mathbf{A}) \quad (11.91)$$

The number of pieces of information in an observing system is the trace of its averaging kernel matrix. This concept is analogous to the relative error variance reduction introduced in Section 11.3 with (11.25). If the matrices $\widehat{\mathbf{S}}$ and $\mathbf{S_A}$ are diagonal, then we see from (11.91) that the DOFS is simply the sum of the relative reductions in error variances σ^2 for the individual state vector elements:

$$\mathrm{DOFS} = n - \sum_{i=1}^{n} \frac{\widehat{\sigma}_i^2}{\sigma_{A,i}^2} = \sum_{i=1}^{n} \frac{\sigma_{A,i}^2 - \widehat{\sigma}_i^2}{\sigma_{A,i}^2} \quad (11.92)$$

Box 11.5   **Averaging Kernel Matrix for an Observing System**

We illustrate the analytical solution to the inverse problem with the retrieval of carbon monoxide (CO) vertical profiles from the MOPITT satellite instrument (Deeter *et al.*, 2003). The instrument makes nadir measurements of the temperature-dependent IR terrestrial emission at and around the 4.6 µm CO absorption band. Atmospheric CO is detected by its temperature contrast with the surface. The radiances measured at different wavelengths constitute the observation vector for the inverse problem. The state vector is chosen to include CO mixing ratios at seven different vertical levels from the surface to 150 hPa, plus surface temperature and surface emissivity. The forward model is a radiative transfer model (RTM) computing the radiances as a function of the state vector values. The observational error covariance matrix is constructed by summing the instrument error covariance matrix (obtained from knowledge of instrument noise) and the forward model error covariance matrix (obtained from comparison of the RTM to a highly accurate but computationally prohibitive line-by-line model). The prior CO vertical profile and its error covariance matrix are climatological values derived from a worldwide compilation of aircraft measurements. The prior surface temperature is specified from local assimilated meteorological data and the prior surface emissivity is taken from a geographical database. The forward model is nonlinear, in particular because the sensitivity to the CO vertical profile depends greatly on the surface temperature; thus a local Jacobian matrix needs to be computed for each scene.

Box 11.5 Figure 1 (a) shows the averaging kernel matrix **A** constructed in this manner for a typical ocean scene. Here, **A** is plotted row by row for the CO vertical profile elements only, with each line corresponding to a given vertical level indicated in the legend. The line for level i gives $\partial \widehat{x}_i / \partial \mathbf{x}$, the sensitivity of the retrieval at that level to the true CO mixing ratios at different

Box 11.5 (*cont.*)

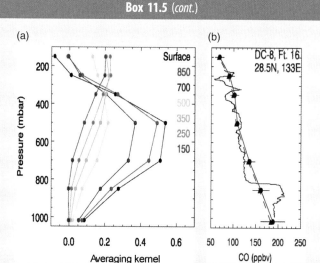

(a)   (b)

Box 11.5 Figure 1   Retrieval of CO mixing ratios by the MOPITT satellite instrument for a scene over the North Pacific. Lines with different colors in (a) show the rows of the averaging kernel matrix for seven vertical levels from the surface to 150 hPa. (b) Shows the MOPITT retrieval (solid line with symbols and posterior error standard deviations) together with a validation profile measured coincidently from aircraft. The dashed line represents the smoothing of the aircraft profile by the MOPITT averaging kernel matrix. From Jacob *et al.* (2003).

levels. A perfect observing system ($\mathbf{A} = \mathbf{I_n}$) would show unit sensitivity to that level ($\partial \hat{x}_i / \partial x_i = 1$) and zero sensitivity to other levels. However, we see from Box 11.5 Figure 1 that the averaging kernel elements are much less than 1 and that the information is smoothed across vertical levels.

Consider the retrieval of the CO mixing ratio at 700 hPa (blue line). We see that the retrieved value at 700 hPa is actually sensitive to CO at all altitudes, so that it is not possible from the retrieval to narrowly identify the CO mixing ratio at 700 hPa (or at any other specific altitude). The temperature contrast between vertical levels is not sufficient. We retrieve instead a broad CO column weighted toward the middle troposphere (700–500 hPa). In fact, the retrieval at 700 hPa is more sensitive to the CO mixing ratio at 500 hPa than at 700 hPa. Physically, this means that a given mixing ratio of CO at 500 hPa will give a spectral response similar to a larger mixing ratio at 700 hPa, because 500 hPa has greater temperature contrast with the surface.

Consider now the retrieval of CO in surface air (black line). There is some thermal contrast between surface air and the surface itself, but the signal is very faint. If we had a perfect observing system we could retrieve it; because of observational error, however, the sensitivity of the surface air retrieval to surface air concentrations is close to zero. In fact, the surface air retrieval is very similar to that at 700 hPa and exhibits the same maximum sensitivity at 700–500 hPa.

Inspection of the averaging kernel matrix in Box 11.5 Figure 1 suggests that the retrieval only provides two independent pieces of information on the vertical profile, one for 700–500 hPa (from the retrievals up to 500 hPa) and one for above 300 hPa (from the retrievals above 350 hPa). We can quantify the DOFS by the trace of the averaging kernel matrix, reading and adding up from the figure the $\partial \widehat{x}_i / \partial x_i$ values for the seven vertical levels. Starting from the lowest level, we find a DOFS of $0.09 + 0.23 + 0.33 + 0.21 + 0.20 + 0.22 + 0.20 = 1.4$. Thus the retrieval mostly provides a CO column weighted toward the middle troposphere, with a smaller additional piece of information in the upper troposphere.

The right panel of Box 11.5 Figure 1 shows the vertical profile of CO retrieved by MOPITT and compares it to a coincident vertical profile of CO measured from aircraft. The aircraft observations have high accuracy and can be regarded as defining the true vertical profile $\mathbf{x}$. They show a layer of elevated CO at 900–800 hPa that MOPITT does not detect, as would be expected because of the vertical smoothing. To determine if the MOPITT retrieval is consistent with the vertical profile measured from aircraft, we need to smooth the aircraft observations with the averaging kernel matrix in order to simulate what MOPITT should actually observe. Smoothing defines a vertical profile $\mathbf{x}'$, shown as the dashed line in the right panel of Box 11.5 Figure 1:

$$\mathbf{x}' = \mathbf{A}\mathbf{x} + (\mathbf{I_n} - \mathbf{A})\mathbf{x_A} \qquad (11.93)$$

This is the expected profile ($\boldsymbol{\varepsilon_0} = \mathbf{0}$) that MOPITT should see if its capability is as advertised by the error analysis that led to the averaging kernel matrix. We see that the smoothed vertical profile from the aircraft agrees closely with the MOPITT observation, supporting MOPITT's error characterization and implying that MOPITT provides an accurate measurement of the weighted tropospheric column (and not much more). Such aircraft validation of satellite instruments is critical for identifying retrieval biases and inadequate characterization of retrieval errors.

11.5.4 Evaluation of the Inverse Solution

Some basic checks need to be made to evaluate the quality of the optimal estimate obtained from an inverse model. A first check is that the inversion has actually decreased the cost function: $J(\widehat{\mathbf{x}}) < J(\mathbf{x_A})$. One should also check that $J(\widehat{\mathbf{x}}) \approx m + n$, verifying that the inverse solution is consistent with the specification of errors. If $J(\widehat{\mathbf{x}}) >> m + n$, the inversion was unable to achieve a solution consistent with the specification of errors. This can happen if errors were greatly underestimated. If $J(\widehat{\mathbf{x}}) << m + n$, by contrast, errors were likely overestimated. Another important test is to apply the forward model to the optimal estimate and compare the field of $\mathbf{F}(\widehat{\mathbf{x}}) - \mathbf{y}$ (*optimal estimate minus observations*) to that of $\mathbf{F}(\mathbf{x_A}) - \mathbf{y}$ (*prior estimate minus observations*). The differences with observations should be reduced when using the optimal estimate (this follows from the decrease in the cost function), and the field of $\mathbf{F}(\widehat{\mathbf{x}}) - \mathbf{y}$ should ideally be uniformly distributed as white noise around zero. Large coherent patterns with $\mathbf{F}(\widehat{\mathbf{x}}) - \mathbf{y}$ of consistent sign suggest model bias, poor characterization of errors, or a poor choice of state vector leading to *aggregation error*. Aggregation error is discussed in Section 11.5.5.

The relative weights of prior and observational errors play an important role in determining the optimal estimate. This weighting is determined not only by the specifications of $\mathbf{S_A}$ and $\mathbf{S_O}$, but also by the relative dimensions of n and m, as these affect the relative weighting of prior and observation terms in the computation of the cost function (see (11.30) and discussion for the scalar example in Section 11.3). If $m \gg n$, the solution may be insensitive to the prior estimate because the number of observational terms in the cost function overwhelms the number of prior terms. An implicit assumption in such a construction is that the observations are independent and identically distributed across the observational error PDF. However, we often have little confidence that this is the case. Autocorrelation between observations not captured by the covariance structure of $\mathbf{S_O}$ will result in excessive weight given to the observations. Beyond this concern, there is often large uncertainty in the specifications of $\mathbf{S_A}$ and $\mathbf{S_O}$. A way of testing the sensitivity to error specification is to introduce a *regularization factor* γ in the cost function:

$$J(\mathbf{x}) = (\mathbf{x} - \mathbf{x_a})^T \mathbf{S_A}^{-1}(\mathbf{x} - \mathbf{x_a}) + \gamma(\mathbf{y} - \mathbf{F}(\mathbf{x}))^T \mathbf{S_O}^{-1}(\mathbf{y} - \mathbf{F}(\mathbf{x})) \qquad (11.94)$$

which amounts to scaling the observational error covariance matrix $\mathbf{S_O}$ by $1/\gamma$. The solution $\widehat{\mathbf{x}}$ can be calculated for different values of γ spanning the range of confidence in error characterization. By plotting the cost function $J(\widehat{\mathbf{x}})$ versus γ, we may find that a value of γ other than 1 leads to an improved solution.

We pointed out in Section 11.3 the danger of over-interpreting the reduction in error variance that results from the accumulation of a large number of observations. Idealized assumption of random and representatively sampled observational error may cause $\widehat{\mathbf{S}}$ to greatly underestimate the actual error on $\widehat{\mathbf{x}}$. A more realistic way of assessing the error in $\widehat{\mathbf{x}}$ is to conduct an ensemble of inverse calculations with various perturbations to model parameters and error covariance statistics (such as through the regularization factor γ) within their expected uncertainties. Model parameters are often a recognized potential source of bias, so that producing an ensemble based on uncertainties in these parameters can be an effective way to address the effect of biases on the optimal estimate.

11.5.5 Limitations on State Vector Dimension: Aggregation Error

One would ideally like to use a state vector as large as possible in order to maximize the amount of information from the inversion. There are two limitations to doing so, one statistical and one computational. There are no such limitations on the size of the observation vector (see Box 11.6).

| Box 11.6 | Sequential Updating in Sampling the Observation Vector |

There is in general no computational limitation on the size $m = \dim(\mathbf{y})$ of the observation vector in an analytical inversion, even though one needs to invert a $m \times m$ matrix in the construction of $\mathbf{G}$. The reason is that it is usually possible to partition the observation vector into small uncorrelated "packets" of observations that are successively ingested into the inverse analysis. Rodgers (2000) calls this procedure *sequential updating*. The solution $(\widehat{\mathbf{x}}, \widehat{\mathbf{S}})$ obtained after ingesting one packet is used as the prior estimate for the next packet, and so on. The final solution is exactly the same as if the entire observation vector were ingested at once. The only limitation is that there must be no observational error correlation between packets; in other words, $\mathbf{S_O}$ for the ensemble of observations must be a block diagonal matrix where the blocks are the individual packets. It is indeed most computationally efficient to ingest uncorrelated data packets sequentially in the inversion. In the extreme case where individual observations have no error correlation, each single observation can be ingested successively and separately.

The statistical limitation in the size of the state vector is imposed by the amount of information actually provided by the observations. As the size of the state vector increases, the number of prior terms in the cost function (11.76) increases while the number of observation terms stays the same. As a result, the optimal estimate is more constrained by the prior estimate, and the smoothing error increases. This is not necessarily a problem if prior error correlations are properly quantified, so that information from the observations can propagate between state vector elements. However, this is generally not the case. In a data assimilation problem where the state vector dimension is by design much larger than the observational vector dimension, the effect of the prior constraints can be moderated by allowing observations to modify state variables only locally, or by using a regularization factor as in (11.94) to balance the contributions of the prior estimate and the observations in the cost function.

The computational limitation arises from the task of constructing the Jacobian matrix $\mathbf{K}$ and the gain matrix $\mathbf{G}$. Analytical solution to the inverse problem requires these matrices, but the computational cost of constructing them becomes prohibitive as the state vector dimension becomes very large. This constraint can be lifted by using a numerical (*variational*) rather than analytical method to solve the inverse problem, as described in Sections 11.6 and 11.8. However, numerical methods may not provide error characterization as part of the solution. A major advantage of the analytical solution is to provide a closed form of the posterior error covariance matrix and from there the averaging kernel matrix and the DOFS.

Say that we wish to reduce the state vector dimension in order to decrease the smoothing error or to enable an analytical solution. Starting from an initially large state vector $\mathbf{x}$, we can use various clustering schemes to reduce the state vector dimension (Turner and Jacob, 2015). Clustering introduces additional observational error by not allowing the relationship between the clustered state vector elements to change in the forward model. This is called the *aggregation error* and is part of the forward model error (the relationship between clustered elements is now a model parameter rather than resolved by the state vector). As the state vector dimension decreases, the smoothing error decreases while the aggregation error increases. We expect therefore an optimum state vector dimension where the total error is minimum. As long as aggregation error is not excessive, it may be advantageous to decrease the state vector dimension below that optimum in order to facilitate an analytical inversion with full error characterization.

The aggregation error for a given choice of reduced-dimension state vector can be characterized following Turner and Jacob (2015). Consider the clustering of an initial state vector $\mathbf{x}$ of dimension n to a reduced state vector $\mathbf{x}_\omega$ of dimension p. The clustering is described by $\mathbf{x}_\omega = \mathbf{\Gamma}_\omega \mathbf{x}$ where the $p \times n$ matrix $\mathbf{\Gamma}_\omega$ is called the *aggregation matrix*. The Jacobian matrix of the forward model is $\mathbf{K}$ in the original inversion and $\mathbf{K}_\omega$ in the reduced inversion. For the same ensemble of observations $\mathbf{y}$, the observational errors in the original and reduced inversions are

$$\varepsilon = \mathbf{y} - \mathbf{K}\mathbf{x} \tag{11.95}$$

$$\varepsilon_\omega = \mathbf{y} - \mathbf{K}_\omega \mathbf{x}_\omega \tag{11.96}$$

The only difference between the two errors is due to aggregation, causing ε_ω to be greater than ε. Thus the aggregation error ε_a is

$$\varepsilon_a = \varepsilon_\omega - \varepsilon = \mathbf{K}\mathbf{x} - \mathbf{K}_\omega\mathbf{x}_\omega = (\mathbf{K} - \mathbf{K}_\omega\mathbf{\Gamma}_\omega)\mathbf{x} \qquad (11.97)$$

We see from (11.97) that the aggregation error is a function of the true state $\mathbf{x}$, and the aggregation error statistics therefore depends on the PDF of the true state. Let $\overline{\mathbf{x}}$ be the mean value of the true state; the covariance matrix of the true states is $\mathbf{S}_e = E[(\mathbf{x} - \overline{\mathbf{x}})(\overline{\mathbf{x}} - \overline{\mathbf{x}})^T]$. The *aggregation bias* is the mean value $\overline{\varepsilon}_a$ of the aggregation error:

$$\overline{\varepsilon}_a = E[\varepsilon_a] = (\mathbf{K} - \mathbf{K}_\omega\mathbf{\Gamma}_\omega)\overline{\mathbf{x}} \qquad (11.98)$$

and the aggregation error covariance matrix $\mathbf{S}_a$ is

$$\mathbf{S}_a = E\left[(\varepsilon_a - \overline{\varepsilon}_a)(\varepsilon_a - \overline{\varepsilon}_a)^T\right] = (\mathbf{K} - \mathbf{K}_\omega\mathbf{\Gamma}_\omega)\mathbf{S}_e(\mathbf{K} - \mathbf{K}_\omega\mathbf{\Gamma}_\omega)^T \qquad (11.99)$$

In general we have no good knowledge of $\overline{\mathbf{x}}$ and $\mathbf{S}_e$. However, we can still estimate the aggregation error covariance matrix when designing the inversion system to select an optimum dimension for the state vector. For this purpose we use our prior knowledge $\mathbf{x}_A$ and $\mathbf{S}_A$ as the best estimates for $\overline{\mathbf{x}}$ and $\mathbf{S}_e$. When aggregating state vector elements, the relationship between state vector elements in the forward model is not allowed to depart from the prior estimate. It follows that if $\overline{\mathbf{x}} = \mathbf{x}_A$ then there is no aggregation bias since the prior relationship between state vector elements is true and hence $\mathbf{K} = \mathbf{K}_\omega\mathbf{\Gamma}_\omega$; the forward model with the reduced state vector is identical to that with the original state vector. The aggregation error covariance matrix is given by

$$\mathbf{S}_a = (\mathbf{K} - \mathbf{K}_\omega\mathbf{\Gamma}_\omega)\mathbf{S}_A(\mathbf{K} - \mathbf{K}_\omega\mathbf{\Gamma}_\omega)^T \qquad (11.100)$$

The corresponding aggregation error covariance matrix in state space (error on $\mathbf{x}$) is $\mathbf{G}\mathbf{S}_a\mathbf{G}^T$.

We previously derived in (11.85) the smoothing and observational error covariance matrices for $\mathbf{x}$. We can now write a complete error budget for $\mathbf{x}$ including the aggregation error for a reduced-dimension state vector:

$$\widehat{\mathbf{S}}_\omega = \underbrace{(\mathbf{I_n} - \mathbf{A}_\omega)\mathbf{S}_{A,\omega}(\mathbf{I_n} - \mathbf{A}_\omega)^T}_{\text{smoothing error}} + \underbrace{\mathbf{G}_\omega(\mathbf{K} - \mathbf{K}_\omega\mathbf{\Gamma}_\omega)\mathbf{S}_A(\mathbf{K} - \mathbf{K}_\omega\mathbf{\Gamma}_\omega)^T\mathbf{G}_\omega^T}_{\text{aggregation error}} + \underbrace{\mathbf{G}_\omega\mathbf{S_O}\mathbf{G}_\omega^T}_{\text{observational error}}$$

$$(11.101)$$

where $\widehat{\mathbf{S}}_\omega$, $\mathbf{S}_{A,\omega}$, $\mathbf{A}_\omega$, and $\mathbf{G}_\omega$ apply to the reduced-dimension state vector. Equation (11.101) separates the aggregation error from the observational error by having $\mathbf{S_O}$ include forward model error only for the original-dimension state vector (not including the effect of aggregation). We can also express the posterior error covariance matrix $\widehat{\mathbf{S}}_\omega^*$ in observation space as describing the error on $\mathbf{K}_\omega\mathbf{x}_\omega$:

$$\widehat{\mathbf{S}}_\omega^* = \underbrace{\mathbf{K}_\omega(\mathbf{I_n} - \mathbf{A}_\omega)\mathbf{S}_{A,\omega}(\mathbf{I_n} - \mathbf{A}_\omega)^T\mathbf{K}_\omega^T}_{\text{smoothing error}} + \underbrace{\mathbf{K}_\omega\mathbf{G}_\omega(\mathbf{K} - \mathbf{K}_\omega\mathbf{\Gamma}_\omega)\mathbf{S}_A(\mathbf{K} - \mathbf{K}_\omega\mathbf{\Gamma}_\omega)^T\mathbf{G}_\omega^T\mathbf{K}_\omega^T}_{\text{aggregation error}} + \underbrace{\mathbf{K}_\omega\mathbf{G}_\omega\mathbf{S_O}\mathbf{G}_\omega^T\mathbf{K}_\omega^T}_{\substack{\text{observational} \\ \text{error}}}$$

$$(11.102)$$

Figure 11.3 illustrates how the different error components of (11.102) contribute to the overall posterior error covariance matrix. The smoothing error decreases with decreasing state vector size while the aggregation error increases. There is an

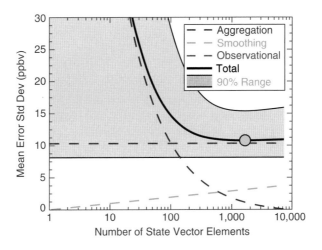

Figure 11.3   Total error budget from the aggregation of state vector elements in an inverse model. The application here is to an inversion of methane emissions over North America using satellite observations of methane and with $n = 7.366$ native-resolution state vector elements representing methane emissions on a 50×50 km^2 grid. Results are shown as the square roots of the means of the diagonal terms (mean error standard deviation) for the aggregation, smoothing, observational, and total (posterior) error covariance matrices following (11.102) There is an optimum state vector size for which the total error is minimum and this is shown as the circle. However, the aggregation error remains small compared to the observational error down to $n \approx 300$ and this could be a suitable choice for a reduced state vector dimension. Gray shading indicates 90% confidence intervals for the total error as diagnosed from the 5th and 95th quantiles of diagonal elements in the posterior error covariance matrix. From Turner and Jacob (2015).

optimum dimension of **x** where the posterior error is minimum. Reducing the state vector dimension beyond this minimum may still be desirable for computational reasons and to achieve an analytical solution, and incurs little penalty as long as the aggregation error remains small relative to the observational error. Beyond a certain reduction the aggregation error grows rapidly to exceed the observational error.

Figure 11.4 shows the impact of smoothing errors in an inversion of satellite observations of methane columns to constrain methane fluxes over North America at high spatial resolution. The top-left panel gives results from an attempt to constrain emissions at the 50×50 km^2 native grid resolution of the forward model (state vector with $n = 7906$). Correction to the prior emissions is less than 50% anywhere because the information from the satellite data is insufficient to constrain emissions at such a high resolution. There results a large smoothing error – the solution is strongly anchored by the prior estimate. The top-right panel gives results from an inversion where the state vector has been reduced to $n = 1000$ elements by hierarchical clustering of the native grid. Corrections to the prior emissions are much larger. The bottom-right panel compares the quality of inversions with different levels of hierarchical clustering (n ranging from 3 to 7906) in terms of their ability to fit the satellite data. This fit is measured by the observational term

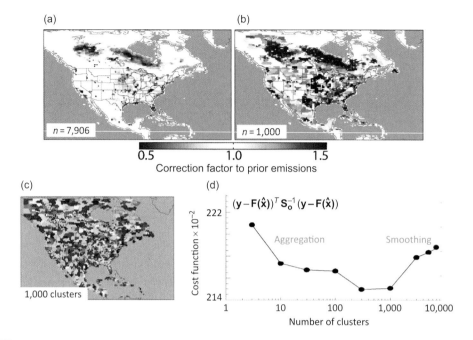

Figure 11.4 Effect of smoothing and aggregation errors in a high-resolution inversion of methane emissions using satellite observations of methane columns. (a) The correction factors to prior emissions when attempting to constrain emissions at the native 50×50 km^2 grid resolution of the forward model ($n = 7906$). (b) The same inversion but with a reduced state vector ($n = 1000$) constructed by hierarchical clustering of the native-resolution grid cells. The clustering is shown in (c) with arbitrary colors for individual clusters. Panel (d) shows the ability of the inversion to fit the satellite observations as the state vector dimension is decreased from $n = 7906$ to $n = 3$ by hierarchical clustering. The quality of the fit is measured by the observational terms of the cost function for the inversion. Optimal results are achieved for n in the range 300–1000. Finer resolution incurs large smoothing errors, while coarser resolution incurs large aggregation errors. From Wecht *et al.* (2014).

$(\mathbf{y} - \mathbf{F}(\widehat{\mathbf{x}}))^T \mathbf{S}_\mathbf{O}^{-1} (\mathbf{y} - \mathbf{F}(\widehat{\mathbf{x}}))$ in the cost function for the inversion. We find that $n = 300$–1000 provides the best fit. Coarser clustering (smaller n) incurs large aggregation error.

11.6 Adjoint-Based Inversion

Analytical solution to the cost function minimization problem $\nabla_\mathbf{x} J(\mathbf{x}) = 0$ as stated by equation (11.77) requires that the forward model be linear with respect to the state vector and places a practical computational limit on the size of the state vector for the inversion. These limitations can be lifted by minimizing J numerically rather than analytically. Such numerical methods, called *variational methods*, compute $\nabla J(\mathbf{x})$ for successive guesses and converge to the solution by a steepest-descent algorithm. Figure 11.5 illustrates the general strategy. In the

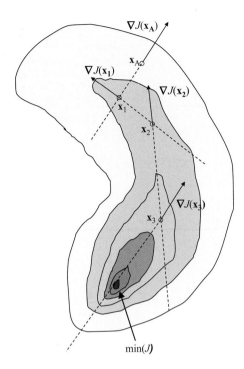

Figure 11.5 Steepest-descent algorithm to find the minimum of a cost function. The cost function gradient ∇ is first computed for the prior estimate $\mathbf{x_A}$ and this is used to obtain an improved estimate $\mathbf{x_1}$. The cost function gradient is then re-calculated for $\mathbf{x_1}$ and this is used to obtain an improved estimate $\mathbf{x_2}$, and so on until convergence. Adapted from an original figure by David Baker (Colorado State University).

adjoint-based inversion, the adjoint of the forward model is used to compute $\nabla J(\mathbf{x})$ efficiently for successive iterations.

Figure 11.6 gives a graphical representation of the procedure for computing $\nabla J(\mathbf{x})$ with the adjoint. The observations $\mathbf{y}$ collected over a period $[t_o, t_p]$ constrain a state vector $\mathbf{x}_{(0)}$ evaluated at t_o. Using the notation introduced in Section 11.4.4 and the expression for $\mathbf{K}^T$ given in (11.51), we have

$$\mathbf{K}^T = \left(\frac{\partial \mathbf{y}_{(0)}}{\partial \mathbf{x}_{(0)}} \right)^T \prod_{i=1}^{p} \left(\frac{\partial \mathbf{y}_{(i)}}{\partial \mathbf{y}_{(i-1)}} \right)^T \tag{11.103}$$

where $\mathbf{y}_{(i)}$ denotes the ensemble of observations at time step i. Starting from the prior estimate $\mathbf{x_A}$ and following (11.77), we write the cost function gradient $\nabla_{\mathbf{x}} J(\mathbf{x_A})$ as

$$\nabla_{\mathbf{x}} J(\mathbf{x_A}) = 2\mathbf{K}^T \mathbf{S_O}^{-1}(\mathbf{F}(\mathbf{x_A}) - \mathbf{y}) \tag{11.104}$$

where the adjoint is applied to the adjoint forcing $\mathbf{S_O}^{-1}(\mathbf{F}(\mathbf{x_A}) - \mathbf{y})$ representing the error-weighted differences between the forward model and the observations. We make one pass of the forward model $\mathbf{F}(\mathbf{x_A})$ through the observational period $[t_o, t_p]$ and collect the corresponding adjoint forcing terms $\mathbf{S_O}^{-1}(\mathbf{F}(\mathbf{x_A}) - \mathbf{y})$, which may be scattered over the period. Starting from the observations $\mathbf{y}_{(p)}$ at t_p, we apply the first adjoint operator $\left(\partial \mathbf{y}_{(p)} / \partial \mathbf{y}_{(p-1)} \right)^T$ from (11.103) to the adjoint forcing terms

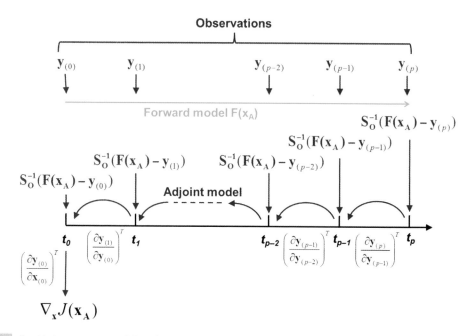

Graphical representation of the adjoint method for computing the cost function gradient $\nabla_{\mathbf{x}}J(\mathbf{x_A}) = 2\mathbf{K}^T\mathbf{S_O}^{-1}(\mathbf{F}(\mathbf{x_A}) - \mathbf{y}))$. We consider here an ensemble of observations over the period $[t_o, t_p]$ to constrain a state vector $\mathbf{x}$ evaluated at time t_o. We start with a pass of the forward model $\mathbf{F}(\mathbf{x_A})$ over the observation period $[t_o, t_p]$. From there we collect adjoint forcings $\mathbf{S_O}^{-1}(\mathbf{F}(\mathbf{x_A}) - \mathbf{y})$ (in red) for the ensemble of observations. We then force the adjoint model with the adjoint forcings at time t_p and propagate these forcings back in time with the adjoint model (in blue), adding new forcings as we march backward in time from t_p to t_o and pick up new observations along the way. At time t_o we apply the final operation $\left(\partial\mathbf{y}_{(0)}/\partial\mathbf{x}_{(0)}\right)^T$ to the adjoint variables to obtain the cost function gradient $\nabla_{\mathbf{x}}J(\mathbf{x_A})$.

$\mathbf{S_O}^{-1}\left(\mathbf{F}(\mathbf{x_A}) - \mathbf{y}_{(p)}\right)$ and obtain the adjoint variables $\left(\partial\mathbf{y}_{(p)}/\partial\mathbf{y}_{(p-1)}\right)^T \mathbf{S_O}^{-1}\left(\mathbf{F}(\mathbf{x_A}) - \mathbf{y}_{(p)}\right)$ as a 3-D field on the model grid. We then add to these adjoint variables the adjoint forcings $\mathbf{S_O}^{-1}\left(\mathbf{F}(\mathbf{x_A}) - \mathbf{y}_{(p-1)}\right)$ from the observations at time t_{p-1}, apply the next adjoint operator $\left(\partial\mathbf{y}_{(p-1)}/\partial\mathbf{y}_{(p-2)}\right)^T$ to the resulting quantities, and so on until time t_o when the final application of the adjoint operator $\left(\partial\mathbf{y}_{(0)}/\partial\mathbf{x}_{(0)}\right)^T$ returns the quantity $\mathbf{K}^T\mathbf{S_O}^{-1}(\mathbf{F}(\mathbf{x_A}) - \mathbf{y}) = \nabla_{\mathbf{x}}J(\mathbf{x_A})$ from (11.104). The procedure can be readily adapted to obtain the cost function gradient for a time-invariant state vector; see related discussion in Section 11.4.4.

The value of $\nabla_{\mathbf{x}}J(\mathbf{x_A})$ obtained in this manner is passed to the steepest-descent algorithm to make an updated guess $\mathbf{x_1}$ for the state vector. We then recalculate $\nabla_{\mathbf{x}}J(\mathbf{x_1})$ for that updated guess,

$$\nabla_{\mathbf{x}}J(\mathbf{x_1}) = 2\mathbf{S_A}^{-1}(\mathbf{x_1} - \mathbf{x_A}) + 2\mathbf{K}^T\mathbf{S_O}^{-1}(\mathbf{F}(\mathbf{x_1}) - \mathbf{y}) \qquad (11.105)$$

using the adjoint as before and adding the terms $\mathbf{S_A}^{-1}(\mathbf{x_1} - \mathbf{x_A})$ which are now non-zero. We pass the result to the steepest-descent algorithm, which makes an updated guess $\mathbf{x_2}$, and so on until convergence to the optimal estimate $\hat{\mathbf{x}}$. Each iteration involves one pass of the forward model over $[t_o, t_p]$ followed by one pass of the adjoint model over $[t_p, t_0]$.

An important feature of the adjoint-based inversion method is that the Jacobian matrix $\mathbf{K}$ is never actually constructed. We do not in fact need the sensitivity of individual observations to $\mathbf{x}$, which is what $\mathbf{K}$ would give us (and would be very expensive to compute for a large number of observations). All we need is the summed sensitivity $\mathbf{K}^T \mathbf{S}_O^{-1}(\mathbf{F}(\mathbf{x_A}) - \mathbf{y})$, which is what the adjoint method provides. Increasing the number of observations does not induce additional computing costs, as it just amounts to updating the adjoint variables on the forward model grid to add new adjoint forcings (Figure 11.5). Increasing the number of state variables (that is, the size of $\mathbf{x}$) also does not incur additional computing costs other than perhaps requiring more iterations to achieve convergence.

Constructing the adjoint requires differentiation of the forward model (Section 11.4.4). Approximations are often made in that differentiation, such as assuming reverse flow for a nonlinear advection operator. The accuracy of the cost function gradients $\nabla_{\mathbf{x}}J(\mathbf{x})$ produced by the adjoint method can be checked with finite difference testing (Henze et al., 2007). The test involves applying the forward model to $\mathbf{x_A}$, calculating the cost function $J(\mathbf{x_A})$, and repeating for a small perturbation to one of the elements $\mathbf{x_A} + \Delta \mathbf{x_A}$ where $\Delta \mathbf{x_A}$ has value Δx_i for element i and zero for all other elements. The resulting finite difference approximation

$$\nabla_{\mathbf{x}}J(\mathbf{x_A}) \approx \frac{J(\mathbf{x_A} + \Delta \mathbf{x_A}) - J(\mathbf{x_A})}{\Delta x_i} \tag{11.106}$$

is then compared to the value obtained with the adjoint model.

Box 11.7 illustrates the adjoint-based inversion with an example using satellite observations of atmospheric concentrations to optimize emissions. The size of the emissions state vector is limited solely by the grid resolution of the forward model. Comparison to a coarse-resolution analytical inversion shows large advantages in the amount of information retrieved.

A drawback of the adjoint method is that it does not provide the posterior error covariance matrix as part of the solution. The matrix can be estimated by constructing the *Hessian* (second derivative) of the cost function. By differentiating equation (11.77) we obtain:

$$\nabla_{\mathbf{x}}^2 J(\mathbf{x}) = 2\mathbf{S}_A^{-1} + 2\mathbf{K}^T \mathbf{S}_O^{-1}\mathbf{K} \tag{11.107}$$

from which we see that the posterior error covariance matrix $\widehat{\mathbf{S}}$ (11.81) is the inverse of the Hessian:

$$\widehat{\mathbf{S}} = 2\left(\nabla_x^2 J\right)^{-1} \tag{11.108}$$

The adjoint method allows an estimate of the Hessian by finite-difference sampling and calculation of $\nabla_{\mathbf{x}}J(\mathbf{x})$ around the optimal estimate solution. Full construction of the Hessian would require $n + 1$ calculations (where n is the state vector dimension) and this is not practical for large-dimension state vectors. Targeted sampling can provide the leading eigenvalues and eigenvectors to approximate the Hessian. See Bousserez et al. (2015) for a discussion of methods.

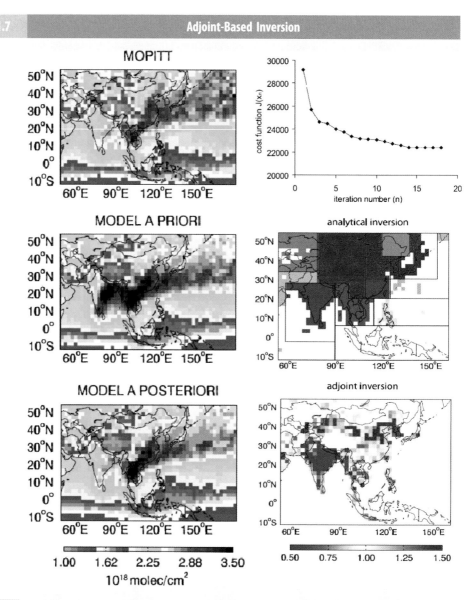

Box 11.7

Adjoint-Based Inversion

Box 11.7 Figure 1 Adjoint inversion of CO emissions in East Asia using CO column observations from the MOPITT satellite instrument in March–April 2001. The top left panel shows the mean MOPITT observations for the period, the middle left panel shows the forward model with prior emissions, and the bottom panel shows the forward model with posterior emissions optimized through the inversion. The top right panel shows the evolution of the cost function with the number of iterations by the adjoint method. The bottom right panel shows the mean multiplicative factors to the prior emissions from the adjoint inversion, and the middle right panel shows the same factors for an analytical inversion with only 11 emission regions as state vector elements. From Kopacz *et al.* (2009).

Box 11.7 Figure 1 from Kopacz *et al.* (2009) illustrates the adjoint-based inversion with an optimization of CO emissions over East Asia using satellite observations of CO columns from the

MOPITT instrument in March–April 2001 (Box 11.5). Here, 21 569 observations are used to constrain mean scale factors to prior CO emissions on the $2° \times 2.5°$ grid of the CTM taken as forward model. The state vector of emission fluxes has 3013 elements. The CTM with prior emissions shows large differences with observations, and the inversion cost function is consequently large. Successive iterations bring the cost function down to its expected value for a successful inversion. The corrections to the prior emissions from the adjoint-based inversion show fine structure associated with geographical boundaries and the type of source (fuel combustion or open fires). An analytical inversion of the same observations with coarse spatial averaging of the emission state vector not only misses the fine structure but also incurs a large aggregation error (Section 11.5.5), as apparent for example in the wrong-direction correction for Korea.

11.7 Markov Chain Monte Carlo (MCMC) Methods

Markov chain Monte Carlo (*MCMC*) methods construct the posterior PDF $P(\mathbf{x}|\mathbf{y}) \sim P(\mathbf{x}) P(\mathbf{y}|\mathbf{x})$ from Bayes' theorem by directly computing $P(\mathbf{x})$ and $P(\mathbf{y}|\mathbf{x})$, and their product, for a very large ensemble of values of $\mathbf{x}$ sampling strategically the n-dimensional space defined by the dimension of $\mathbf{x}$. MCMC methods can use any form of the prior and observational PDFs. They allow for non-Gaussian errors, a nonlinear forward model, and any prior constraints. With sufficient sampling, they can return the full structure of $P(\mathbf{x}|\mathbf{y})$ with no prior assumption as to its form. A drawback of MCMC methods is their computational cost. In addition, because $P(\mathbf{x})$ and $P(\mathbf{x}|\mathbf{y})$ are not Gaussian, one cannot calculate an averaging kernel matrix to quantify the information content of the observations.

The basis for MCMC methods is a *Markov chain* where successive values of $\mathbf{x}$ are selected in a way that the next value depends on the current value but not on previous values. This is done by randomly sampling a transition PDF $T(\mathbf{x}'|\mathbf{x})$ where $\mathbf{x}$ is the current value and $\mathbf{x}'$ is the next value. T is often taken to be a Gaussian form so that values close to $\mathbf{x}$ are more likely to be selected as the next value. Through this Markov chain and by including additional criteria to adopt or reject candidate next values, we achieve a targeted random sampling of $\mathbf{x}$ to map the function $P(\mathbf{x}|\mathbf{y})$.

The general strategy for MCMC methods is as follows. We start from a first choice for $\mathbf{x}$, such as the prior estimate $\mathbf{x_A}$, and calculate $P(\mathbf{x_A}|\mathbf{y})$. We then choose randomly a candidate for the next value $\mathbf{x_1}$ by sampling $T(\mathbf{x_1}|\mathbf{x_A})$ and calculate the corresponding $P(\mathbf{x_1}|\mathbf{y})$. Comparison of $P(\mathbf{x_1}|\mathbf{y})$ to $P(\mathbf{x_A}|\mathbf{y})$ tells us whether $\mathbf{x_1}$ is more likely than $\mathbf{x_A}$ or not, and on that basis we may choose to adopt $\mathbf{x_1}$ as our next value or reject it. If we adopt it, then we use $\mathbf{x_1}$ as a starting point to choose a candidate $\mathbf{x_2}$ for the next value by applying $T(\mathbf{x_2}|\mathbf{x_1})$. If we reject it, then we come back to $\mathbf{x_A}$ and make another tentative choice for $\mathbf{x_1}$ by random sampling of $T(\mathbf{x_1}|\mathbf{x_A})$. In this manner we sample the PDF $P(\mathbf{x}|\mathbf{y})$ in a representative way. Figure 11.7 gives an illustration.

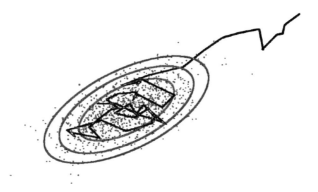

Figure 11.7   Sampling of a PDF by an MCMC method. The blue points represent the population and the contours are the isolines of the PDF. The black line is the MCMC sampling trajectory. For a sufficiently large sample, the PDF of the sample tends toward the PDF of the population. From Iain Murray (University of Edinburgh).

A frequently used MCMC method is the *Metropolis–Hastings* algorithm. In this algorithm, following on the above sampling strategy, we calculate the ratio $P(\mathbf{x_1}|\mathbf{y})/P(\mathbf{x_A}|\mathbf{y})$. If the ratio is greater than 1, we are moving in the right direction toward the most likely value; $\mathbf{x_1}$ is then adopted as the next value of the Markov chain and we proceed to choose a candidate for $\mathbf{x_2}$ on the basis of $\mathbf{x_1}$. If the ratio is less than 1, this means that $\mathbf{x_1}$ is less likely than $\mathbf{x_A}$. In that case, the ratio $P(\mathbf{x_1}|\mathbf{y})/P(\mathbf{x_A}|\mathbf{y})$ defines the probability that $\mathbf{x_1}$ should be selected as the next iteration and a random decision is made based on that probability. If the decision is made to adopt $\mathbf{x_1}$, then we proceed as above. If the decision is made to reject $\mathbf{x_1}$, then we go back to $\mathbf{x_A}$ and make another tentative choice for $\mathbf{x_1}$. Given a sufficiently large sampling size, one can show that this sampling strategy will eventually generate the true structure of $P(\mathbf{x}|\mathbf{y})$.

11.8 Other Optimization Methods

Standard Bayesian optimization regularizes the fitting of the state vector to observations by applying the prior PDF of the state vector as an additional constraint. We may wish to use a different type of prior constraint, or ignore prior information altogether. We briefly discuss these approaches here.

Ignoring prior information is the effective outcome of a Bayesian inversion when the prior terms $\mathbf{S_A^{-1}(x - x_A)}$ are small relative to the observational terms $\mathbf{K}^T\mathbf{S_O^{-1}(F(x) - y)}$ in the computation of the cost function gradient by equation (11.77). This happens when the prior errors are large relative to the observational errors, and/or when the number of state vector elements is very small relative to the number of observations. If we know this from the outset, then there is little point in going through the trouble of including prior information in the cost function. The cost function including only the observational terms amounts to an error-weighted fit to the observations and this is called the *maximum likelihood estimator*:

$$J(\mathbf{x}) = (\mathbf{y} - \mathbf{F}(\mathbf{x}))^T \mathbf{S_O^{-1}} (\mathbf{y} - \mathbf{F}(\mathbf{x})) \tag{11.109}$$

The minimum in J can be found by computing the gradient as in (11.77) without the prior terms. Sequential updating can be applied in computing the gradient by successive ingestion of small data packets (Box 11.6). The optimal estimate after ingesting one data packet is used as the prior estimate for ingesting the next data packet.

If all elements in equation (11.109) have the same observational error variances and zero covariances, so that $\mathbf{S_O}$ is a multiple of the identity matrix, then the cost function becomes a simple least-squares fit: $J(\mathbf{x}) = \|\mathbf{y} - \mathbf{F}(\mathbf{x})\|^2$ where $\|.\|$ denotes the Euclidean norm. If in addition, the forward model is linear and thus fully described by its Jacobian matrix $\mathbf{K}$, then we have the familiar linear least-squares optimization

$$J(\mathbf{x}) = \|\mathbf{y} - \mathbf{Kx}\|^2 \tag{11.110}$$

One can show that the corresponding minimum of $J(\mathbf{x})$ is for

$$\mathbf{x} = \mathbf{K}^+\mathbf{y} \tag{11.111}$$

where $\mathbf{K}^+ = (\mathbf{K}^T\mathbf{K})^{-1}\mathbf{K}^T$ is the *Moore–Penrose pseudoinverse* of $\mathbf{K}$.

A danger of not including prior information in the solution to the inverse problem is that the resulting solution may exhibit non-physical attributes such as negative values or unrealistically large swings between adjacent elements (*checkerboard noise*). An important role of the prior information is to prevent such non-physical behavior by regularizing the solution. Bayesian regularization as described in Section 11.2 is not the only method for imposing prior constraints. In the *Tikhonov regularization*, a term is added to the linear least-squares minimization to enforce desired attributes of the solution:

$$J(\mathbf{x}) = \|\mathbf{y} - \mathbf{Kx}\|^2 + \|\mathbf{\Gamma x}\|^2 \tag{11.112}$$

Here $\mathbf{\Gamma}$ is the *Tikhonov matrix* and carries the prior information. For example, choosing for $\mathbf{\Gamma}$ a multiple of the identity matrix ($\mathbf{\Gamma} = \gamma\mathbf{I_n}$) enforces smallness of the solution. Off-diagonal terms relating adjacent state vector elements enforce smoothness of the solution. Bayesian inference can be seen as a particular form of Tikhonov regularization when we write it as

$$J(\mathbf{x}) = \|\mathbf{y} - \mathbf{Kx}\|^2_{\mathbf{S_O}} + \|\mathbf{x} - \mathbf{x_A}\|^2_{\mathbf{S_A}} \tag{11.113}$$

where $\|\mathbf{a}\|_{\mathbf{S}} = \mathbf{a}^T\mathbf{S}^{-1}\mathbf{a}$ is the error-weighted norm for vector $\mathbf{a}$ with error covariance matrix $\mathbf{S}$.

In some inverse problems, we have better prior knowledge of the patterns between state vector elements than of the actual magnitudes of the elements. For example, when using observed atmospheric concentrations to optimize a 2-D spatial field of emissions, we may know that emissions relate to population density for which we have good prior information, even if we don't have good information on the emissions themselves. This type of knowledge can be exploited through a *geostatistical inversion*. Here we express the cost function as

$$J(\mathbf{x}, \mathbf{\beta}) = (\mathbf{x} - \mathbf{P}\mathbf{\beta})^T\mathbf{S}^{-1}(\mathbf{x} - \mathbf{P}\mathbf{\beta}) + (\mathbf{y} - \mathbf{F}(\mathbf{x}))^T\mathbf{S_O}^{-1}(\mathbf{y} - \mathbf{F}(\mathbf{x})) \tag{11.114}$$

where the $n \times q$ matrix $\mathbf{P}$ describes the q different state vector patterns, with each column of $\mathbf{P}$ describing a normalized pattern. The unknown vector $\mathbf{\beta}$ of dimension

q gives the mean scaling factor for each pattern over the ensemble of state vector elements. Thus $\mathbf{P}\boldsymbol{\beta}$ represents a prior model for the mean, with $\boldsymbol{\beta}$ to be optimized as part of the inversion. The covariance matrix $\mathbf{S}$ gives the prior covariances of $\mathbf{x}$, rather than the error covariances.

11.9 Positivity of the Solution

Inverse problems in atmospheric chemistry frequently require positivity of the solution. This is the case, in particular, when the state vector consists of concentrations or emission fluxes (Table 11.1). The standard assumption of Gaussian errors is at odds with the positivity requirement since it allows for the possibility of negative values, but this is not a serious issue as long as the probability of negative values remains small. When small negative values are incurred in the solution for a state vector element, it may be acceptable to simply aggregate them with adjacent elements to restore positivity. In some cases, maintaining positivity in the solution requires stronger measures. Miller *et al.* (2014) review different approaches for enforcing positivity in inverse modeling of emission fluxes using observations of atmospheric concentrations.

A straightforward way to enforce positivity of the solution is to transform the state variables into their logarithms and optimize the logarithms using Gaussian error statistics. This assumes that the errors on the state variables are lognormally distributed, which is often a realistic assumption for a quantity constrained to be positive. For example, when we say that a state variable is uncertain by a 1-σ factor of 2, we effectively make the statement that the natural logarithm of that variable has a Gaussian error standard deviation of ln 2. Any of the inversion methods described above can be applied to the logarithms of the state variables with Gaussian error statistics, in the same way as for the original state variables. However, if the forward model was linear with respect to the original state variables, then transformation to logarithms loses the linearity. Analytical solution to the inverse problem then requires an iterative approach with reconstruction of the Jacobian matrix at each iteration. This is not an issue with an adjoint-based inverse method since the Jacobian is never explicitly constructed in that case. An adjoint-based inverse method incurs no computational penalty when transforming the state variables into their logarithms (or any other transformation). The methane flux inversion in Figure 11.4, for example, used an adjoint-based method with logarithms of emission scaling factors as the state variables.

MCMC methods (Section 11.7) can easily handle the requirement of positivity by restricting the Markov chain to positive values of the state vector elements. However, these methods are computationally expensive. An additional drawback is that they may lead to an unrealistic structure of the posterior error PDF with exaggerated probability density close to zero.

Another approach to enforce positivity of the solution is by applying *Karush–Kuhn–Tucker* (*KKT*) *conditions* to the cost function. KKT conditions are a general method for enforcing inequality constraints in the cost function and are an extension of the *Lagrange multipliers* method that enforces equality constraints. Here we describe the specific application of minimizing the cost function $J(\mathbf{x})$ subject to the positivity constraints $x_i \geq 0$ for all n elements of the state vector $\mathbf{x}$. This is done by minimizing the *Lagrange function* $L(\mathbf{x}, \boldsymbol{\mu})$ with respect to $\mathbf{x}$:

$$L(\mathbf{x}, \boldsymbol{\mu}) = J(\mathbf{x}) + \sum_{i=1}^{n} \mu_i x_i \tag{11.115}$$

where $\boldsymbol{\mu} = (\mu_1, \ldots \mu_n)$ is the vector of unknown *KKT multipliers*. The KKT multipliers are constrained to be non-negative ($\mu_i \geq 0$) and satisfy the *complementary slackness* conditions:

$$\mu_i x_i = 0 \quad \forall i = 1, \ldots n \tag{11.116}$$

The minimum of L with respect to $\mathbf{x}$ is given by

$$\nabla_{\mathbf{x}} L = \nabla_{\mathbf{x}} J + \boldsymbol{\mu} = 0 \tag{11.117}$$

Combination of (11.116) and (11.117) gives us $2n$ equations to solve for the elements of $\mathbf{x}$ and $\boldsymbol{\mu}$ in a way that forces all elements of $\mathbf{x}$ to be positive or zero. To see this, consider that the solution for x_i must either be positive, in which case (11.115) imposes $\mu_i = 0$, or it must be zero to satisfy (11.116). Application of the KKT conditions will thus produce a solution for the state vector where some elements will be zero and others will be positive. Two drawbacks of this method are that (1) it does not provide characterization of errors on the solution, (2) the solution of zero (or any positive threshold value) for a subset of elements may not be realistic and bias the solution for other elements.

11.10 Data Assimilation

Data assimilation is sometimes used in the literature to refer to any inverse problem. Standard practice in atmospheric chemistry is to refer to *chemical data assimilation* as a particular kind of inverse problem where we seek to optimize a gridded time-dependent 3-D model field of concentrations based on observations of these concentrations or related variables. This is similar to *meteorological data assimilation* where the 3-D state of the meteorological variables is optimized for the purpose of initializing weather forecasts or creating a consistent meteorological data archive. The optimized state resulting from the assimilation of observations is called the *analysis*. Chemical data assimilation can be used to improve weather forecasts by accounting for chemical effects on weather (for example, aerosols affecting clouds), or by providing indirect information on meteorological variables (for example, ozone as a tracer of stratospheric motions). Other applications of chemical data assimilation include air quality forecasting, construction of chemical data archives (*reanalyses*), and assessment of consistency between instruments viewing different species for different scenes and with different schedules. Figure 11.8 gives as an example an assimilated field for stratospheric ozone. Data assimilation generally involves a

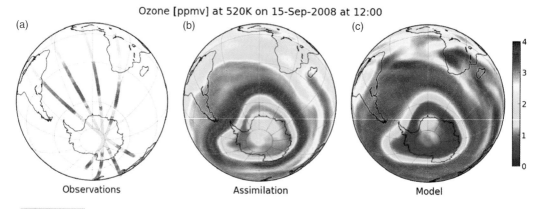

Figure 11.8 Illustration of data assimilation principles. An analysis of stratospheric ozone on September 15, 2008 (b) is obtained by optimizing a stratospheric transport model (c) to fit observations (a). Observations are from the Aura MLS satellite instrument.on September 15, 2008 between 9 and 15 UT, and the analysis is for 12 UT. The right panel shows the free-running stratospheric model initialized on April 1, 2008 with no subsequent data assimilation. Data from the Belgian Assimilation System for Chemical ObsErvations (BASCOE, Errera *et al.*, 2008; Errera and Ménard 2012). Provided by Quentin Errera.

chemical forecast to serve as the prior estimate and this will be referred to here as the *forecast model*. This forecast model is usually a numerical weather prediction model initialized with assimilated meteorological data and including simulation of the chemical variables to be assimilated.

The state vector $\mathbf{x}$ in chemical data assimilation is the gridded 3-D field of concentrations at a given time, and it evolves with time as determined by the forecast model and by the assimilated observations. It is typically very large. Solution of the optimization problem usually requires numerical methods for minimizing the Bayesian cost function J. Such methods, involving iterative computation of $\nabla_{\mathbf{x}} J$ as part of a minimization algorithm, are called *variational methods*. The adjoint-based inversion in Section 11.6 is an example of a variational method.

In the standard chemical data assimilation problem, the forecast model initialized with $\mathbf{x}_{(0)}$ at time t_0 produces a forecast $\mathbf{x}_{(1)}$ at time $t_1 = t_0 + h$ where h is the *assimilation time step*, typically of the order of a few hours to a day. Two different strategies can be used to assimilate the observations (Figure 11.9). In *3-D variational data assimilation (3DVAR)*, observed concentrations are collected and assimilated at fixed time intervals h. The observations $\mathbf{y}_{(0)}$ at time t_0 are used to optimize $\mathbf{x}_{(0)}$, and the forecast model is then integrated over the time interval $[t_0, t_1 = t_0 + h]$ to obtain a prior estimate for $\mathbf{x}_{(1)}$. The observations at time t_1 are used to optimize the estimate of $\mathbf{x}_{(1)}$, and so on. In some cases and with simplifying assumptions, the minimization of J can be done analytically as described in Section 11.5 instead of with a variational method; the assimilation is then called a *Kalman filter*. In *4-D variational data assimilation (4DVAR)*, the ensemble of observations spread over the time interval $[t_0, t_1]$ are used to optimize $\mathbf{x}_{(0)}$ by applying the adjoint of the forecast model backward in time over $[t_1, t_0]$.

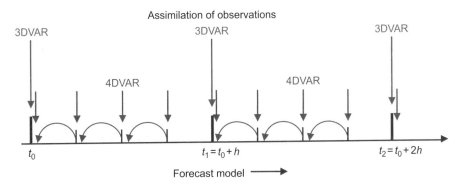

Assimilation of observations

Figure 11.9
Schematic of 3DVAR and 4DVAR data assimilation. The long tick marks indicate the assimilation time steps h, and the short tick marks indicate the internal time steps of the forecast model. In 3DVAR, observations assembled at discrete assimilation time steps are used to optimize the state vector at the corresponding times. In 4DVAR, observations spread over the forecast interval are used to optimize the state vector at the beginning of the forecast by propagating the information backward in time with the adjoint of the forecast model (see Figure 11.6).

The forecast model is then used to obtain a prior estimate for $\mathbf{x}_{(1)}$, observations over $[t_1, t_1 + h]$ are used to optimize $\mathbf{x}_{(1)}$, and so on.

11.10.1 3DVAR Data Assimilation and the Kalman Filter

The 3DVAR approach uses observations at discrete assimilation time steps to optimize the state vector at the corresponding times. It is called "3D" because the optimization operates on the 3-D state vector without consideration of time. Consider an ensemble of observations collected at assimilation time steps t_0, t_1, t_2, etc. Let $\mathbf{y}_{(0)}$ be the ensemble of observations collected at time t_0, and $\mathbf{x}_{(0)}$ the value of the state vector at that time. Starting from some prior knowledge $(\mathbf{x}_{A,(0)}, \mathbf{S}_{A,(0)})$, we use the observations $\mathbf{y}_{(0)}$ to minimize the cost function as in (11.76):

$$J\left(\mathbf{x}_{(0)}\right) = \left(\mathbf{x}_{(0)} - \mathbf{x}_{A,(0)}\right)^T \mathbf{S}_{A,(0)}^{-1} \left(\mathbf{x}_{(0)} - \mathbf{x}_{A,(0)}\right) + \left(\mathbf{F}\left(\mathbf{x}_{(0)}\right) - \mathbf{y}_{(0)}\right)^T \mathbf{S}_O^{-1} \left(\mathbf{F}\left(\mathbf{x}_{(0)}\right) - \mathbf{y}_{(0)}\right)$$

$$(11.118)$$

Here the forward model $\mathbf{F}$ is not the forecast model, but instead a mapping of the state vector to the observations. If the observations are of the same quantity as the state vector and on the same grid, then $\mathbf{F}$ is the identity matrix. If the observations are of the same quantity as the state vector but offset from the grid in space or time, then $\mathbf{F}$ is an interpolation operator. If the observations are of different variables than the state vector, then $\mathbf{F}$ is a separate model needed to relate the two; it could be for example a 0-D chemical model relating the concentrations of different species through a chemical mechanism.

The minimum for $J(\mathbf{x}_{(0)})$ is obtained by

$$\nabla_{\mathbf{x}(0)}J\left(\mathbf{x}_{(0)}\right) = 2\mathbf{S}_{\mathbf{A},(0)}^{-1}\left(\mathbf{x}_{(0)} - \mathbf{x}_{\mathbf{A},(0)}\right) + 2\nabla\mathbf{F}^T\mathbf{S}_{\mathbf{O}}^{-1}\left(\mathbf{F}\left(\mathbf{x}_{(0)}\right) - \mathbf{y}_{(0)}\right) = \mathbf{0} \quad (11.119)$$

In 3DVAR, (11.119) is solved numerically by starting from initial guess $\mathbf{x}_{\mathbf{A},(0)}$, computing $\nabla_{\mathbf{x}(0)}J\left(\mathbf{x}_{\mathbf{A},(0)}\right)$, and using a steepest-descent algorithm to iterate until convergence (Figure 11.4). The computations of J and ∇J require simplifications to keep the matrices to a manageable size and this is typically done by limiting the spatial extent of error correlation. Simplifications may also be needed in the form of the forward model.

In the case of a linear forward model (such as the identity matrix or a linear interpolator) analytical solution to (11.119) is possible. This analytical approach is called the *Kalman filter* and it has the advantage of characterizing the error on the solution through computation of the posterior error covariance matrix. Starting from the prior estimate $(\mathbf{x}_{(0)}, \mathbf{S}_{\mathbf{A},(0)})$, assimilation of observations $\mathbf{y}_{(0)}$ yields the optimal estimate and its error covariance matrix $(\widehat{\mathbf{x}}_{(0)}, \widehat{\mathbf{S}}_{(0)})$ through application of (11.79)–(11.81):

$$\widehat{\mathbf{x}}_{(0)} = \mathbf{x}_{\mathbf{A},(0)} + \mathbf{G}\left(\mathbf{y}_{(0)} - \mathbf{K}\mathbf{x}_{\mathbf{A},(0)}\right) \quad (11.120)$$

$$\mathbf{G} = \mathbf{S}_{\mathbf{A},(0)}\mathbf{K}^T\left(\mathbf{K}\mathbf{S}_{\mathbf{A},(0)}\mathbf{K}^T + \mathbf{S}_{\mathbf{O}}\right)^{-1} \quad (11.121)$$

$$\widehat{\mathbf{S}}_{(0)} = \left(\mathbf{K}^T\mathbf{S}_{\mathbf{O}}^{-1}\mathbf{K} + \mathbf{S}_{\mathbf{A},(0)}^{-1}\right)^{-1} \quad (11.122)$$

where the Jacobian matrix $\mathbf{K} = \nabla_{\mathbf{x}}\mathbf{F}$ defines the forward model. We then apply the forecast model to compute the evolution of $(\widehat{\mathbf{x}}_{(0)}, \widehat{\mathbf{S}}_{(0)})$ over the forecasting time step $[t_0, t_1]$, leading to a prior estimate $\left(\mathbf{x}_{\mathbf{A},(1)}, \mathbf{S}_{\mathbf{A},(1)}\right)$ at time t_1 (calculation of $\mathbf{S}_{\mathbf{A},(1)}$ is described below). Assimilation of observations $\mathbf{y}_{(1)}$ at time t_1 is then done following the above equations to yield an optimal estimate $(\widehat{\mathbf{x}}_{(1)}, \widehat{\mathbf{S}}_{(1)})$. From there, we apply the forecast model over $[t_1, t_2]$, and so on.

The prior error covariance matrix $\mathbf{S}_{\mathbf{A},(1)}$ is the sum of the forecast model error covariance matrix $\mathbf{S}_{\mathbf{M}} = E[\boldsymbol{\varepsilon}_{\mathbf{M}}\boldsymbol{\varepsilon}_{\mathbf{M}}^T]$ and the error covariance matrix on the initial state $\widehat{\mathbf{S}}_{(0)} = E\left[\widehat{\boldsymbol{\varepsilon}}_{(0)}\widehat{\boldsymbol{\varepsilon}}_{(0)}^T\right]$ modified by the forecast model over $[t_0, t_1]$. It can be fully computed if the forecast model is linear, i.e., represented by a matrix $\mathbf{M}$. In that case we have

$$\mathbf{x}_{\mathbf{A},(1)} = \mathbf{x}_{(1)} + \mathbf{M}\widehat{\boldsymbol{\varepsilon}}_{(0)} + \boldsymbol{\varepsilon}_{\mathbf{M}} \quad (11.123)$$

where $\mathbf{x}_{(1)}$ is the true value at time t_1. Thus

$$\mathbf{S}_{\mathbf{A},(1)} = E\left[\mathbf{M}\widehat{\boldsymbol{\varepsilon}}_{(0)}\widehat{\boldsymbol{\varepsilon}}_{(0)}^T\mathbf{M}^T\right] + E\left[\boldsymbol{\varepsilon}_{\mathbf{M}}\boldsymbol{\varepsilon}_{\mathbf{M}}^T\right] = \mathbf{M}\widehat{\mathbf{S}}_{(0)}\mathbf{M}^T + \mathbf{S}_{\mathbf{M}} \quad (11.124)$$

The term $\mathbf{M}\widehat{\mathbf{S}}_{(0)}\mathbf{M}^T$ in (11.124) transports the posterior error covariance matrix from one assimilation time step to the next, thus propagating information on errors. In the trivial case of a *persistence model* assumed as forecast, $\mathbf{x}_{\mathbf{A},(1)} = \widehat{\mathbf{x}}_{(0)}$ and $\mathbf{M}$ is the identity matrix. This may be an adequate assumption if the assimilation time step is short relative to the timescale over which $\mathbf{x}$ evolves.

Transport of the full error covariance matrices through $\mathbf{M}\widehat{\mathbf{S}}_{(0)}\mathbf{M}^T$ is often not computationally practical and various approximations are used. One can produce an ensemble of estimates for $\mathbf{x}_{(1)}$ by randomly sampling the probability density function $\left(\widehat{\mathbf{x}}_{(0)}, \widehat{\mathbf{S}}_{(0)} \right)$, transporting this ensemble over the assimilation time step $[t_0, t_1]$, and estimating $\mathbf{M}\widehat{\mathbf{S}}_{(0)}\mathbf{M}^T$ from the ensemble of values at time t_1. This is called an *ensemble Kalman filter*. Other alternatives are to transport only the error variances (diagonal terms of $\widehat{\mathbf{S}}_{(0)}$) or ignore the transport of prior errors altogether and just assign $\mathbf{S}_{\mathbf{A},(1)} = \mathbf{S}_{\mathbf{M}}$. The Kalman filter is then called *suboptimal*.

11.10.2 4DVAR Data Assimilation

The 3DVAR approach assimilates observations collected at discrete assimilation time steps, say once per day, and applies a forecast model to update the state vector over that interval. However, observations are often taken at all times of day and therefore scattered in time over the forecast time interval. In *4DVAR data assimilation*, observations scattered at times t_i over the forecast time interval $[t_0, t_1]$ are used to optimize the initial state $\mathbf{x}_{(0)}$. This involves minimizing the cost function

$$J\left(\mathbf{x}_{(0)}\right) = \left(\mathbf{x}_{(0)} - \mathbf{x}_{\mathbf{A},(0)}\right)^T \mathbf{S}_{\mathbf{A},(0)}^{-1} \left(\mathbf{x}_{(0)} - \mathbf{x}_{\mathbf{A},(0)}\right) + \sum_{t_i=t_0}^{t_1} \left(\mathbf{F}\left(\mathbf{x}_{(i)}\right) - \mathbf{y}_{(i)}\right)^T \mathbf{S}_{\mathbf{O},(i)}^{-1} \left(\mathbf{F}\left(\mathbf{x}_{(i)}\right) - \mathbf{y}_{(i)}\right)$$

$$(11.125)$$

and is done by applying the adjoint of the forecast model over $[t_0, t_1]$. The optimization approach is exactly as described in Section 11.6. It returns an optimized estimate $\widehat{\mathbf{x}}_{(0)}$ but no associated error covariance matrix $\widehat{\mathbf{S}}_{(0)}$ (although one can estimate $\widehat{\mathbf{S}}_{(0)}$ from numerical construction of the Hessian $\nabla^2_{\mathbf{x}_{(0)}} J$; see Section 11.6). The optimized estimate $\widehat{\mathbf{x}}_{(0)}$ is used to produce a forecast $\mathbf{x}_{\mathbf{A},(1)}$ at time t_1 with error covariance matrix $\mathbf{S}_{\mathbf{A},(1)}$ defined by the error in the forecast model over $[t_0, t_1]$: $\mathbf{S}_{\mathbf{A},(1)} = \mathbf{S}_{\mathbf{M}}$. Observations over the next forecasting time interval are then applied with the model adjoint to derive an optimized estimate $\widehat{\mathbf{x}}_{(1)}$, and so on. The temporal discretization in the assimilation of the observations is limited solely by the internal time step of the forecast model.

11.11 Observing System Simulation Experiments

Observation system simulation experiments (OSSEs) are standard tests conducted during the design phase of a major observing program such as a satellite mission. Their purpose is to determine the ability of the proposed measurements to deliver on the scientific objectives of the program. These objectives can often be stated in terms of improving knowledge of a state vector $\mathbf{x}$ by using the proposed observations $\mathbf{y}$ in combination with a state-of-science forward model to relate $\mathbf{x}$ to $\mathbf{y}$. For example, we might want to design a geostationary satellite mission to measure CO_2 concentrations in order to constrain CO_2 surface fluxes, and ask what measurement precision is needed to improve the constraints on the fluxes beyond those achievable from the existing CO_2 observation network.

If the state vector is sufficiently small that an analytical solution to the inverse problem is computationally practical (Section 11.5), then a first assessment of the usefulness of the proposed observing system can be made by characterizing the error covariance matrices $\mathbf{S_A}$ and $\mathbf{S_O}$, and constructing the Jacobian matrix $\mathbf{K} = \partial \mathbf{y}/\partial \mathbf{x}$ of the forward model relating the state vector $\mathbf{x}$ to the observations $\mathbf{y}$. Here $\mathbf{S_A}$ is the prior error covariance matrix describing the current state of knowledge of $\mathbf{x}$ in the absence of the proposed observations, and $\mathbf{S_O}$ is the observational error covariance matrix for the proposed observations. $\mathbf{S_O}$ describes the properties of the proposed observing system including instrument errors, forward model errors, and observation density. From knowledge of $\mathbf{S_A}$, $\mathbf{S_O}$, and $\mathbf{K}$ we can calculate the averaging kernel matrix $\mathbf{A}$ and posterior error covariance matrix $\widehat{\mathbf{S}}$ as described in Section 11.5. The trace of $\mathbf{A}$ defines the DOFS of the observing system (Section 11.5.3) and provides a simple metric for characterizing the improved knowledge of $\mathbf{x}$ to result from the proposed observations. Analysis of the rows of $\mathbf{A}$ determines the ability of the observing system to retrieve the individual components of $\mathbf{x}$ (Box 11.5). Results of such an OSSE will tend to be over-optimistic because of the idealized treatment of observational errors (Section 11.5.1). In particular, the defining assumptions of unbiased Gaussian error statistics and independent and identically distributed (IID) sampling of the PDFs means that posterior error variances will decrease roughly as the square root of the number of observations. This will typically not hold in the actual observing system because of observation bias, non-random sampling of errors, and unrecognized error correlations. Still, this simple approach for error characterization of the proposed observed system is very valuable for defining the *potential* of the proposed observations to better quantify the state variables, and to compare the merits of different configurations for the proposed observing system (such as instrument precision and observing strategy)

A more stringent and general OSSE is to use two independent forward models simulating the same period (Figure 11.10). For an observing system targeting atmospheric composition these would be two independent CTMs: CTM 1 and CTM 2. The CTMs should both be state-of-science but otherwise be as different as possible, so that difference between the two provides realistic statistics of the forward model error ε_M. In particular, they should use different assimilated meteorological data sets for the same period. We take CTM 1 as representing the "true" atmosphere, with "true" values of the state vector $\mathbf{x}$ generating a "true" 3-D field of atmospheric concentrations to be sampled by the observing system. This simulation is often called the *Nature Run*. We sample the "true" atmosphere with the current and proposed components of the observing system to produce synthetic data sets for the observed species, locations, and schedules, including random errors based on instrument precision. This generates a synthetic ensemble of current observations $\mathbf{y}$ and proposed observations $\mathbf{y}'$ over the time period. We then use CTM 2 as the forward model over the same time period, starting from a prior estimate $(\mathbf{x_A}, \mathbf{S_A})$ for the state vector and assimilating the synthetic observations $\mathbf{y}$ from the current observing system to achieve an optimal estimate $\widehat{\mathbf{x}}_1$. This is called the *Control Run*. The posterior error covariance matrix $\widehat{\mathbf{S}}_1$ associated with $\widehat{\mathbf{x}}_1$ may be generated from the Control Run or have to be estimated, as discussed in Section 11.10. We then

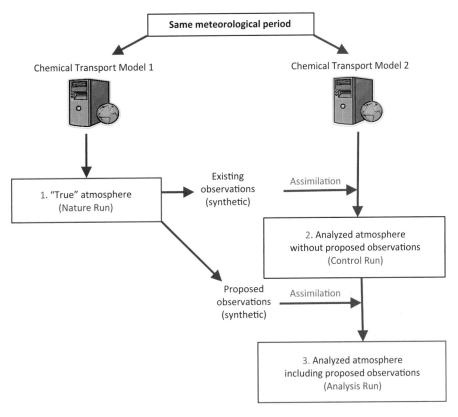

Figure 11.10 General structure of an observing system simulation experiment for atmospheric composition. Two independent chemical transport models (1 and 2) are used to simulate the same meteorological period. The first model is used to simulate a synthetic "true" atmosphere (Nature Run) to be sampled by the observing system. The second model assimilates current observations in a first step (Control Run), and the additional proposed observations in a second step (Analysis Run). Comparison of the paired differences (Analyzed–Nature) and (Control–Nature) measures the improvement in knowledge to be contributed by the proposed observations.

take the prior state defined by $(\widehat{\mathbf{x}}_1, \widehat{\mathbf{S}}_1)$ and assimilate the proposed synthetic observations $\mathbf{y}'$ to obtain a new optimal estimate $(\widehat{\mathbf{x}}_2, \widehat{\mathbf{S}}_2)$ reflecting the benefit of the proposed observations. This is called the *Analysis Run*. The difference $\widehat{\mathbf{x}}_2 - \mathbf{x}$ measures the departure of our knowledge from the truth *after* the proposed observations have been made, while $\widehat{\mathbf{x}}_1 - \mathbf{x}$ measures the departure from the truth *before* the proposed observations have been made. Comparison of $\widehat{\mathbf{x}}_2 - \mathbf{x}$ to $\widehat{\mathbf{x}}_1 - \mathbf{x}$ allows us to quantify the value of the proposed observations for constraining $\mathbf{x}$. Compared to the simple error characterization approach presented in the previous paragraph, this more advanced OSSE system has two critical advantages: (1) a more realistic description of forward model errors and their sampling by the observing system; and (2) the use of a variational data assimilation system to more realistically mimic the use of the observations. Figure 11.11 gives an example.

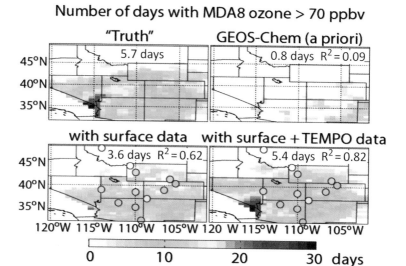

Figure 11.11 Observing system simulation experiment evaluating the utility of a proposed geostationary satellite instrument for ozone measurements (TEMPO) to detect high-ozone events in surface air over the remote western USA (Intermountain West). The OSSE structure is as shown in Figure 11.10, and the objective is to map the number of days exceeding the ozone air quality standard defined as a 70 ppb maximum daily 8-h average (MDA8). The MOZART and GEOS-Chem chemical transport models are used to simulate the same three-month meteorological period of April–June 2010. These two CTMs use different assimilated meteorological data sets and are also different in their chemical mechanisms, emissions, etc. The MOZART simulation is taken as the "truth" (Nature Run) and the GEOS-Chem model is used as the forecast model for the analysis. The top-left panel shows the "true" atmosphere simulated by MOZART, and the top-right panel shows the atmosphere simulated by GEOS-Chem without data assimilation. The two are very different. The bottom-left panel shows the GEOS-Chem atmosphere after assimilation of data from the current network of surface sites, where the data (circles) are synthetic observations of the "true" atmosphere. This defines the Control Run. The bottom-right panel shows the GEOS-Chem atmosphere after assimilation of additional TEMPO synthetic observations. This defines the Analysis Run. We see that the Analysis Run does much better than the Control Run in simulating the number of days with MDA8 ozone exceeding 70 ppb. Values inset are the average number of days with MDA8 ozone exceeding 70 ppb, and the coefficient of determination R^2 relative to the "true" atmosphere. From Zoogman *et al.* (2014).

References

Bousserez N., Henze D. K., Perkins K. W., *et al.* (2015) Improved analysis-error covariance matrix for high-dimensional variational inversions: application to source estimation using a 3D atmospheric transport model, *QJRMS*, doi: 10.1002/qj.2495.

Deeter M. N., Emmons L. K., Francis G. L., *et al.* (2003) Operational carbon monoxide retrieval algorithm and selected results for the MOPITT instrument, *J. Geophys. Res.*, **108**, 4399.

Errera Q. and Ménard R. (2012) Technical note: Spectral representation of spatial correlations in variational assimilation with grid point models and application to the Belgian Assimilation System for Chemical Observations (BASCOE), *Atmos. Chem. Phys.*, **12**, 10015–10031.

Errera Q., Daerden F., Chabrillat S., *et al.* (2008) 4D-Var assimilation of MIPAS chemical observations: Ozone and nitrogen dioxide analyses, *Atmos. Chem. Phys.*, **8**, 6169–6187.

Heald C. L., Jacob D. J., Jones D., *et al.* (2004) Comparative inverse analysis of satellite (MOPITT) and aircraft (TRACE-P) observations to estimate Asian sources of carbon monoxide, *J. Geophys. Res.*, **109**, D23306.

Henze D. K., Hakami A., and Seinfeld J. H. (2007) Development of the adjoint of GEOS-Chem, *Atmos. Chem. Phys.*, **7**, 2413–2433.

Jacob D. J., Crawford J., Kleb M., *et al.* (2003) The Transport and Chemical Evolution over the Pacific (TRACE-P) aircraft mission: Design, execution, and first results, *J. Geophys. Res.*, **108**, 9000.

Kim P. S., Jacob D. J., Mickley L., *et al.* (2015) Sensitivity of population smoke exposure to fire locations in Equatorial Asia, *Atmos. Environ.*, **102**, 11–17.

Kopacz M., Jacob D. J., Henze D., *et al.* (2009) Comparison of adjoint and analytical Bayesian inversion methods for constraining Asian sources of carbon monoxide using satellite (MOPITT) measurements of CO columns, *J. Geophys. Res.*, **114**, D04305.

Miller S. M., Michalak A. M., and Levi P. J. (2014) Atmospheric inverse modeling with known physical bounds: An example for trace gas emissions, *Geosci. Mod. Dev.*, **7**, 303–315.

Rodgers C. D. (2000) *Inverse Methods for Atmospheric Sounding*, World Sci., Tokyo.

Rodgers C. D., and Connor B. J. (2003) Intercomparison of remote sounding instruments, *J. Geophys. Res.*, **108**, 4116.

Turner A. J. and Jacob D. J. (2015) Balancing aggregation and smoothing errors in inverse models, *Atmos. Chem. Phys.*, **15**, 7039–7048.

Wecht K. J., Jacob D. J., Frankenberg C., *et al.* (2014) Mapping of North America methane emissions with high spatial resolution by inversion of SCIAMACHY satellite data, *J. Geophys. Res.*, **119**, 7741–7756.

Zhang L., Jacob D. J., Liu X., *et al.* (2010) Intercomparison methods for satellite measurements of atmospheric composition: Application to tropospheric ozone from TES and OMI, *Atmos. Chem. Phys.*, **10**, 4725–4739.

Zoogman P. W., Jacob D. J., Chance K., *et al.* (2014) Monitoring high-ozone events in the US Intermountain West using TEMPO geostationary satellite observations, *Atmos. Chem. Phys.*, **14**, 6261–6271.

Physical Constants and Other Data

A.1 General and Universal Constants

Base of natural logarithms	2.71828
π (Pi)	3.14159
Boltzmann's constant	1.38066×10^{-23} J K^{-1}
Molar gas constant	8.3144 J K^{-1} mol^{-1}
Stefan–Boltzmann's constant	5.67032×10^{-8} W m^{-2} K^{-4}
Planck's constant	6.62618×10^{-34} J s
Speed of light in vacuum	2.99792×10^{8} m s^{-1}
Gravitational constant	6.67259×10^{-11} m^3 s^{-2} kg^{-1}
Electron mass	9.1096×10^{-31} kg
Electron charge	1.6022×10^{-19} C
Atomic mass unit (amu)	1.66054×10^{-27} kg
Avogadro number	6.0221×10^{23} mol^{-1}

A.2 Earth

Average radius	6.371×10^{6} m
Surface area	5.10×10^{14} m^2
Surface area (continents)	1.49×10^{14} m^2
Surface area (oceans)	3.61×10^{14} m^2
Average height of land	840 m
Average depth of oceans	3730 m
Acceleration of gravity (surface)	9.80665 m s^{-2}
Mass of Earth	5.983×10^{24} kg
Mass of atmosphere	5.3×10^{18} kg
Eccentricity of Earth's orbit	0.016750
Inclination of rotation axis	$23.45°$ or 0.409 rad
Mean angular rotation rate	7.292×10^{-5} rad s^{-1}
Earth orbital period	365.25463 days
Solar constant	1367 ± 2 W m^{-2}

A.3 Dry Air

Average molar mass	28.97 g mol^{-1}
Specific gas constant	287.05 J K^{-1} kg^{-1}
Standard surface pressure	1.01325 $\times$ 10^5 Pa
Mass density at 0 °C and 101325 Pa	1.293 kg m^{-3}
Number density at 0 °C and 101325 Pa	2.69 $\times$ 10^{25} m^{-3}
Molar volume 0 °C and 101325 Pa	22.414 $\times$ 10^{-3} m^3 mol^{-1}
Specific heat at constant pressure (c_p)	1004.64 J K^{-1} kg^{-1}
Specific heat at constant volume (c_v)	717.6 J K^{-1} kg^{-1}
Poisson constant (c_p/c_v)	1.4
Index of refraction for air	1.000277
Dry diabatic lapse rate	9.75 K km^{-1}
Speed of sound for standard conditions	343.15 m s^{-1}

A.4 Water

Molecular weight	18.016 g mol^{-1}
Gas constant for water vapor	461.6 J K^{-1} kg^{-1}
Density of pure liquid water at 0 °C	1000 kg m^{-3}
Density of ice at 0 °C	917 kg m^{-3}
Density of water vapor at STP (0 °C, 1 atm)	0.803 kg m^{-3}
Specific heat of water vapor at constant pressure	1952 J K^{-1} kg^{-1}
Specific heat of water vapor at constant volume	1463 J K^{-1} kg^{-1}
Specific heat of liquid water at 0 °C	4218 J K^{-1} kg^{-1}
Specific heat of ice at 0 °C	2106 J K^{-1} kg^{-1}
Latent heat of vaporization at 0 °C	2.501 $\times$ 10^6 J kg^{-1}
Latent heat of vaporization at 100 °C	2.25 $\times$ 10^6 J kg^{-1}
Latent heat of fusion at 0 °C	3.34 $\times$ 10^5 J kg^{-1}
Latent heat of sublimation at 0 °C	2.83 $\times$ 10^6 J kg^{-1}
Index or refraction for liquid water	1.336
Index of refraction for ice	1.312
Triple-point temperature of water	273.16 K

B   Units, Multiplying Prefixes, and Conversion Factors

B.1 International System of Units

Quantity	Name of Unit	Symbol	Definition
Length	Meter	m	
Mass	Kilogram	kg	
Time	Second	s	
Electrical current	Ampere	A	
Temperature	Kelvin	K	
Force	Newton	N	$kg\ m\ s^{-2}$
Pressure	Pascal	Pa	$N\ m^{-2}$
Energy	Joule	J	$kg\ m^2\ s^{-2}$
Power	Watt	W	$J\ s^{-1}$
Electrical potential	Volt	V	$W\ A^{-1}$
Electrical charge	Coulomb	C	$A\ s$
Electrical resistance	Ohm	Ω	$V\ A^{-1}$
Electrical capacitance	Farad	F	$A\ s\ V^{-1}$
Frequency	Hertz	Hz	s^{-1}
Moles	Mole	mol	

B.2 Multiplying Prefixes

Multiple	Prefix	Symbol	Multiple	Prefix	Symbol
10^{-1}	Deci	d	10^1	Deca	da
10^{-2}	Centi	c	10^2	Hecto	h
10^{-3}	Milli	m	10^3	Kilo	k
10^{-6}	Micro	μ	10^6	Mega	M
10^{-9}	Nano	n	10^9	Giga	G
10^{-12}	Pico	p	10^{12}	Tera	T
10^{-15}	Femto	f	10^{15}	Peta	P
10^{-18}	Atto	a	10^{18}	Exa	E

Wavelengths are typically expressed in micrometers (μm) or nanometers (nm). Wavenumbers are expressed in inverse centimeters (cm^{-1}). Atmospheric pressure is often expressed in hectopascals (hPa), number densities in molecules per cubic centimeter (cm^{-3}). Molar mixing ratios are given in percent, parts per million (ppm), parts per billion (ppbv), parts per trillion (pptv), or parts per quadrillion (ppqv). Mass mixing ratios are expressed in kilograms per kilogram or grams per kilogram.

B.3 Conversion Factors

Area	$1 \text{ ha} = 10^4 \text{ m}^2$
Volume	$1 \text{ liter} = 10^{-3} \text{ m}^3$
Velocity	$1 \text{ m s}^{-1} = 3.6 \text{ km h}^{-1} = 2.237 \text{ mi h}^{-1}$
Force	$1 \text{N} = 10^5 \text{ dyn}$
Pressure	$1 \text{ bar} = 10^5 \text{ Pa} = 10^3 \text{ mb} = 750.06 \text{ mm Hg}$
	$1 \text{ atm} = 1.01325 \times 10^5 \text{ Pa} = 760 \text{ Torr}$
Energy	$1 \text{ cal} = 4.1855 \text{ J}$
	$1 \text{ eV} = 1.6021 \times 10^{-19} \text{ J}$
	$1 \text{ J} = 1 \text{ N m} = 10^7 \text{ erg} = 0.239 \text{ cal}$
Power	$1 \text{ W} = 14.3353 \text{ cal min}^{-1}$
Temperature	$T(^\circ\text{C}) = T(\text{K}) - 273.15$
	$T(^\circ\text{F}) = 1.8 \ T(^\circ\text{C}) + 32$
Mixing ratios	$1 \text{ ppb} = 10^{-3} \text{ ppm}$
	$1 \text{ ppt} = 10^{-3} \text{ ppb} = 10^{-6} \text{ ppm}$
Logarithms	$\ln x = 2.3026 \log_{10} x$

B.4 Commonly Used Units for Atmospheric Concentrations

Number density	molecules cm^{-3}
Mass density	kg m^{-3}
Mixing ratio (molar)[a]	$\text{ppm} \equiv \text{ppmv} \equiv \mu\text{mol mol}^{-1} \equiv 10^{-6} \text{ mol mol}^{-1}$
	$\text{ppb} \equiv \text{ppbv} \equiv \text{nmol mol}^{-1} \equiv 10^{-9} \text{ mol mol}^{-1}$
	$\text{ppt} \equiv \text{pptv} \equiv \text{pmol mol}^{-1} \equiv 10^{-12} \text{ mol mol}^{-1}$
	$\text{ppq} \equiv \text{ppqv} \equiv \text{fmol mol}^{-1} \equiv 10^{-15} \text{ mol mol}^{-1}$
Mixing ratio (mass)	$\text{g g}^{-1} \equiv \text{g per g of air}$
	$\text{g kg}^{-1} \equiv \text{g per kg of air}$
Partial pressure	Pa, Torr (1 Torr = 133 Pa)
Column	molecules cm^{-2},
	Dobson Unit (1 DU $= 2.69 \times 10^{16}$ molecules cm^{-2})[b]

[a] Mixing ratios in the atmospheric chemistry literature denote molar fractions unless otherwise specified. Mol mol^{-1} is the SI unit but ppm, ppb, etc. are conventionally used. To avoid confusion with mass mixing ratios the conventional units are often written as ppmv, ppbv, etc. where v refers to volume (in an ideal gas such as the atmosphere, the number of moles is proportional to volume)

[b] The Dobson Unit was originally introduced and is still mainly used as a measure of the thickness of the ozone layer, with 1 DU corresponding to a 0.01 mm thick layer of pure ozone under standard conditions of temperature and pressure (0 °C, 1 atm).

International Reference Atmosphere

Elevation	Temperature	Pressure	Relative density	Kinematic viscosity	Thermal conductivity	Speed of sound
z [m]	T [K]	p [Pa] $\times 10^5$	ρ/ρ_0	ν [m^2 s^{-1}] $\times 10^{-5}$	κ [W m^{-1} K^{-1}] $\times 10^{-2}$	c [m s^{-1}]
−1500	297.9	1.2070	1.1522	1.301	2.611	346.0
−1000	294.7	1.1393	1.0996	1.352	2.585	344.1
−500	291.4	1.0748	1.0489	1.405	2.560	342.2
0	288.15	1.01325	1.0000	1.461	2.534	340.3
500	284.9	0.9546	0.9529	1.520	2.509	338.4
1000	281.7	0.8988	0.9075	1.581	2.483	336.4
1500	278.4	0.8456	0.8638	1.646	2.457	334.5
2000	275.2	0.7950	0.8217	1.715	2.431	332.5
2500	271.9	0.7469	0.7812	1.787	2.405	330.6
3000	268.7	0.7012	0.7423	1.863	2.379	328.6
3500	265.4	0.6578	0.7048	1.943	2.353	326.6
4000	262.2	0.6166	0.6689	2.028	2.327	324.6
4500	258.9	0.5775	0.6343	2.117	2.301	322.6
5000	255.7	0.5405	0.6012	2.211	2.275	320.5
5500	252.4	0.5054	0.5694	2.311	2.248	318.5
6000	249.2	0.4722	0.5389	2.416	2.222	316.5
6500	245.9	0.4408	0.5096	2.528	2.195	314.4
7000	242.7	0.4111	0.4817	2.646	2.169	312.3
7500	239.5	0.3830	0.4549	2.771	2.142	310.2
8000	236.2	0.3565	0.4292	2.904	2.115	308.1
8500	233.0	0.3315	0.4047	3.046	2.088	306.0
9000	229.7	0.3080	0.3813	3.196	2.061	303.8
9500	226.5	0.2858	0.3589	3.355	2.034	301.7
10 000	223.3	0.2650	0.3376	3.525	2.007	299.8
10 500	220.0	0.2454	0.3172	3.706	1.980	297.4
11 000	216.8	0.2270	0.2978	3.899	1.953	295.2
11 500	216.7	0.2098	0.2755	4.213	1.952	295.1
12 000	216.7	0.1940	0.2546	4.557	1.952	295.1
12 500	216.7	0.1793	0.2354	4.930	1.952	295.1
13 000	216.7	0.1658	0.2176	5.333	1.952	295.1
13 500	216.7	0.1533	0.2012	5.768	1.952	295.1
14 000	216.7	0.1417	0.1860	6.239	1.952	295.1
14 500	216.7	0.1310	0.1720	6.749	1.952	295.1
15 000	216.7	0.1211	0.1590	7.300	1.952	295.1
15 500	216.7	0.1120	0.1470	7.895	1.952	295.1

			(cont.)			
Elevation	Temperature	Pressure	Relative density	Kinematic viscosity	Thermal conductivity	Speed of sound
16 000	216.7	0.1035	0.1359	8.540	1.952	295.1
16 500	216.7	0.09572	0.1256	9.237	1.952	295.1
17 000	216.7	0.08850	0.1162	9.990	1.952	295.1
17 500	216.7	0.08182	0.1074	10.805	1.952	295.1
18 000	216.7	0.07565	0.09930	11.686	1.952	295.1
18 500	216.7	0.06995	0.09182	12.639	1.952	295.1
19 000	216.7	0.06467	0.08489	13.670	1.952	295.1
19 500	216.7	0.05980	0.07850	14.784	1.952	295.1
20 000	216.7	0.05529	0.07258	15.989	1.952	295.1
22 000	218.6	0.04047	0.05266	22.201	1.968	296.4
24 000	220.6	0.02972	0.03832	30.743	1.985	297.7
26 000	222.5	0.02188	0.02797	42.439	2.001	299.1
28 000	224.5	0.01616	0.02047	58.405	2.018	300.4
30 000	226.5	0.01197	0.01503	80.134	2.034	301.7

Chemical Mechanism

This appendix lists important chemical and photolysis reactions occurring in the troposphere and stratosphere, including rate constants and typical photolysis frequencies. It is based on a mechanism described by Emmons *et al.* (2010) and Lamarque *et al.* (2012). Many rate constants in the mechanism are simplified and uncertain. More comprehensive and detailed information with references can be found in various compilations, including the regularly updated NASA Jet Propulsion Laboratory (JPL) *Chemical Kinetics and Photochemical Data for Use in Atmospheric Studies* and the International Union of Pure and Applied Chemistry (IUPAC) *Evaluated Kinetic Data for Atmospheric Chemistry*. See also *The Atmospheric Chemist's Companion* by P. Warneck and J. Williams (Springer, 2012). Chemical mechanisms used in models often vary in their lumping of larger organic species and their oxidation products. The present mechanism lumps alkanes and alkenes with four or more carbon atoms, lumps aromatic hydrocarbons as toluene, and also includes isoprene and a lumped terpene (α-pinene). The chemical reactivity of the lumped species is chosen to approximately represent the average reactivity of the different individual hydrocarbons that are accounted for by the lumped species. The mechanism symbols identify individual species in the computer code for the mechanism and are sometimes used in atmospheric chemistry jargon (PAN, for example) but have otherwise no meaning. The "common names" may depart from IUPAC nomenclature but represent standard usage in the atmospheric chemistry literature.

D.1 Chemical Species and Definitions of Symbols

D.1.1 Inorganic Gas-Phase Species

Chemical formula	Common name
$O(^3P)$	Ground state "triplet-P" atomic oxygen
$O(^1D)$	Excited state "singlet-D" atomic oxygen
O_3	Ozone
N_2O	Nitrous oxide
NO	Nitric oxide
NO_2	Nitrogen dioxide
NO_3	Nitrate radical
$HONO$	Nitrous acid

(cont.)	
Chemical formula	*Common name*
HNO_3	Nitric acid
HNO_4	Pernitric acid
N_2O_5	Dinitrogen pentoxide
H	Atomic hydrogen
H_2	Molecular hydrogen
OH	Hydroxyl radical
HO_2	Hydroperoxy radical
H_2O_2	Hydrogen peroxide
CO	Carbon monoxide
SO_2	Sulfur dioxide
NH_3	Ammonia
Cl	Chlorine atom
ClO	Chlorine monoxide
OClO	Chlorine dioxide
Cl_2O_2	Chlorine monoxide dimer
Cl_2	Molecular chlorine
HCl	Hydrogen chloride
HOCl	Hypochlorous acid
$ClONO_2$	Chlorine nitrate
$ClNO_2$	Nitryl chloride
Br	Bromine atom
BrO	Bromine monoxide
Br_2	Molecular bromine
BrCl	Bromine monochloride
HBr	Hydrogen bromide
HOBr	Hypobromous acid
$BrONO_2$	Bromine nitrate

D.1.2 Organic Gas-Phase Species

Mechanism symbol	Chemical formula	Common name
C_1 *species*		
CH_4	CH_4	Methane
CH_3O_2	CH_3O_2	Methylperoxy radical
CH_3OOH	CH_3OOH	Methylhydroperoxide
CH_2O	CH_2O	Formaldehyde
CH_3OH	CH_3OH	Methanol
HCOOH	HCOOH	Formic acid
C_2 *species*		
C_2H_4	C_2H_4	Ethene
C_2H_6	C_2H_6	Ethane

	(cont.)	
Mechanism symbol	Chemical formula	Common name
C_2H_2	C_2H_2	Acetylene
CH_3CHO	CH_3CHO	Acetaldehyde
C_2H_5OH	C_2H_5OH	Ethanol
EO	$HOCH_2CH_2O$	Hydroxy ethene oxy radical
EO_2	$HOCH_2CH_2O_2$	Hydroxy ethene peroxy radical
CH_3COOH	CH_3COOH	Acetic acid
GLYOXAL	HCOCHO	Glyoxal
GLYALD	$HOCH_2CHO$	Glycolaldehyde
$C_2H_5O_2$	$C_2H_5O_2$	Ethylperoxy radical
C_2H_5OOH	C_2H_5OOH	Ethylhydroperoxide
CH_3CO_3	CH_3CO_3	Peroxyacetyl radical
CH_3COOOH	$CH_3C(O)OOH$	Peracetic acid
PAN	$CH_3C(O)OONO_2$	Peroxyacetyl nitrate
DMS	$(CH_3)_2S$	Dimethylsulfide
C_3 species		
C_3H_6	C_3H_6	Propene
C_3H_8	C_3H_8	Propane
$C_3H_7O_2$	$C_3H_7O_2$	Propylperoxy radical
C_3H_7OOH	C_3H_7OOH	Propylhydroperoxide
PO_2	e.g., $CH_3CH(OO)CH_2OH$	Hydroxyl propene peroxy radicals
POOH	e.g., $CH_3CH(OOH)CH_2OH$	Hydroxyl propene peroxide
CH_3COCH_3	CH_3COCH_3	Acetone
HYAC	CH_3COCH_2OH	Hydroxyacetone
CH_3COCHO	CH_3COCHO	Methylglyoxal
AO_2	$CH_3COCH_2O_2$	Acetone peroxy radical
AOOH	CH_3COCH_2OOH	Acetone hydroperoxide
ONIT	$CH_3COCH_2ONO_2$	Organic nitrate
C_4 species		
BIGENE	C_4H_8	Lumped $>C_3$ alkene
$ENEO_2$	e.g., $CH_3CH(OH)CH(OO)CH_3$	Lumped alkene peroxy radical
MEK	$CH_3C(O)CH_2CH_3$	Methyl ethyl ketone
$MEKO_2$	$CH_3COCH(OO)CH_3$	MEK peroxy radical
MEKOOH	$CH_3COCH(OOH)CH_3$	MEK hydroperoxide
MVK	$CH_2CHCOCH_3$	Methyl vinyl ketone
MACR	CH_2CCH_3CHO	Methacrolein
MPAN	$CH_2CCH_3CO_3NO_2$	Methacryloyl peroxynitrate
$MACRO_2$	e.g., $CH_3COCH(OO)CH_2OH$	MVK + MACR peroxy radical
MACROOH	e.g., $CH_3COCH(OOH)CH_2OH$	MVK + MACR hydroperoxide
MCO_3	$CH_2CCH_3CO_3$	MACR peroxyacyl radical
C_5 species		
BIGALK	C_5H_{12}	Lumped $>C_3$ alkane
$ALKO_2$	$C_5H_{11}O_2$	Lumped alkyl peroxy radical
ALKOOH	$C_5H_{11}OOH$	Lumped alkyl hydroperoxide
ISOP	C_5H_8	Isoprene

Mechanism symbol	Chemical formula	Common name
ISOPO$_2$	e.g., HOCH$_2$C(OO)CH$_3$CHCH$_2$	Isoprene peroxy radical
ISOPOOH	e.g., HOCH$_2$C(OOH)CH$_3$CHCH$_2$	Isoprene hydroperoxide
HYDRALD	e.g., HOCH$_2$CCH$_3$CHCHO	Lumped unsaturated Hydroxycarbonyl
XO$_2$	e.g., HOCH$_2$C(OO)CH$_3$CH(OH)CHO	HYDRALD peroxy radical
XOOH	e.g., HOCH$_2$C(OOH)CH$_3$CH(OH)CHO	HYDRALD hydroperoxide
BIGALD	C$_5$H$_6$O$_2$	Unsaturated dicarbonyl
ISOPNO$_3$	e.g., CH$_2$CHCCH$_3$OOCH$_2$ONO$_2$	Peroxy radical from NO$_3$ + ISOP
ONITR	e.g., CH$_2$CCH$_3$CHONO$_2$CH$_2$OH	Lumped isoprene nitrate
C$_7$ species		
TOLUENE	C$_6$H$_5$(CH$_3$)	Lumped aromatic hydrocarbon
CRESOL	e.g., C$_6$H$_4$(CH$_3$)(OH)	Phenols and cresols
TOLO$_2$	C$_6$H$_5$(CH$_3$OO)	Aromatic peroxy radical
TOLOOH	C$_6$H$_5$(CH$_3$OOH)	Aromatic hydroperoxide
XO$_2$	C$_7$H$_7$O$_2$	CRESOL peroxy radical
C$_{10}$ species		
TERPENE	C$_{10}$H$_{16}$	Lumped monoterpenes, as α-pinene
TERPO$_2$	C$_{10}$H$_{16}$(OH)(OO)	Terpene peroxy radical
TERPOOH	C$_{10}$H$_{16}$(OH)(OOH)	Terpene hydroperoxide

D.1.3 Bulk Aerosols

Mechanism symbol	Chemical formula	Name
SO$_4$	S(VI) $\equiv$ SO$_4^{2-}$ + HSO$_4^-$ + H$_2$SO$_4$(aq)	Sulfate
NH$_4$	NH$_4^+$	Ammonium
NO$_3$A	NO$_3^-$	Ammonium nitrate
SOA		Secondary organic aerosol
OC		Organic carbon
EC		Elemental carbon

D.2 Photolysis

The following table lists photolysis reactions of importance for the troposphere and the stratosphere. The photolysis frequency (often called *J-value*) for a given molecule A is calculated as a function of altitude z and solar zenith angle χ by spectral integration over all wavelengths λ of the product of (1) the solar actinic flux density $q_\lambda(\lambda;z,\chi)$; (2) the absorption cross-section $\sigma_A(\lambda)$ of the molecule; and (3) the quantum efficiency $\varepsilon_A(\lambda)$:

$$J_A(z,\chi) = \int_0^\infty \varepsilon_A(\lambda)\ \sigma_A(\lambda)q_\lambda(\lambda;z,\chi)d\lambda$$

The actinic flux at a given altitude and for a given solar zenith angle is calculated with a radiative transfer model; see Chapter 5. The photolysis products reported in the table are the ones used in the chemical mechanism above and assume in some cases fast reactions of the immediate photolysis products; for example, $CCl_4 + hv \rightarrow CCl_3 + Cl$ is given as $CCl_4 + hv \rightarrow 4Cl$. Values of the photolysis frequency J for different molecules, calculated by the TUV-5.1 model (Madronich, personal communication), are provided at sea level and at 25 km altitude for the following conditions: ozone column 300 DU, solar zenith angle 30°, surface albedo 5%, no clouds, no aerosol effects. They can be viewed as typical clear-sky daytime values. Some of the photolysis processes listed here are of importance for the upper stratosphere but negligible at lower altitudes, in which case the photolysis frequency is given as "0.0." The symbol X e-Y stands for $X \times 10^{-Y}$.

D.2.1 Inorganic Species

Reaction	J at sea level [s^{-1}]	J at 25 km [s^{-1}]
Oxygen species		
$O_2 + hv \rightarrow 2O(^3P)$	0.0	1.5e-11
$O_3 + hv \rightarrow O(^1D) + O_2$	3.2e-05	1.3e-04
$O_3 + hv \rightarrow O(^3P) + O_2$	4.1e-04	4.9e-04
Hydrogen species		
$H_2O + hv \rightarrow OH + H$	0.0	0.0
$H_2O + hv \rightarrow H_2 + O(^1D)$	0.0	0.0
$H_2O_2 + hv \rightarrow 2OH$	7.4e-06	1.3e-05
Nitrogen species		
$N_2O + hv \rightarrow O(^1D) + N_2$	0.0	2.8e-08
$NO + hv \rightarrow N + O$	0.0	0.0
$NO_2 + hv \rightarrow NO + O$	9.3e-03	1.2e-02
$N_2O_5 + hv \rightarrow NO_2 + NO_3$	4.3e-05	7.4e-05
$HONO + hv \rightarrow NO + OH$	1.5e-03	2.2e-03
$HNO_3 + hv \rightarrow NO_2 + OH$	6.0e-07	6.3e-06
$NO_3 + hv \rightarrow NO_2 + O$	1.7e-01	1.8e-01
$NO_3 + hv \rightarrow NO + O_2$	2.2e-02	2.4e-02
$HO_2NO_2 + hv \rightarrow OH + NO_3$ (20%) or $NO_2 + HO_2$ (80%)	6.6e-06	2.3e-05
Halogen species		
$Cl_2 + hv \rightarrow 2Cl$	2.3e-03	3.6e-03
$OClO + hv \rightarrow O + ClO$	8.2e-02	1.2e-01

Reaction	J at sea level $[s^{-1}]$	J at 25 km $[s^{-1}]$
(cont.)		
$ClOOCl + hv \rightarrow 2\ Cl$	1.7e-03	2.9e-03
$HOCl + hv \rightarrow OH + Cl$	2.7e-04	4.5e-04
$HCl + hv \rightarrow H + Cl$	0.0	2.4e-08
$ClONO_2 + hv \rightarrow Cl + NO_3$	3.9e-05	5.7e-05
$ClONO_2 + hv \rightarrow ClO + NO_2$	7.7e-06	1.6e-05
$BrCl + hv \rightarrow Br + Cl$	1.1e-02	1.4e-02
$BrO + hv \rightarrow Br + O$	3.6e-02	6.0e-02
$HOBr + hv \rightarrow Br + OH$	2.2e-03	3.1e-03
$BrONO_2 + hv \rightarrow Br + NO_3$	4.0e-04	5.9e-04
$BrONO_2 + hv \rightarrow BrO + NO_2$	9.8e-04	1.5e-03
$CCl_4 + hv \rightarrow 4Cl$	0.0	1.1e-06
$CFCl_3 + hv \rightarrow 3Cl$	0.0	5.9e-07
$CF_2Cl_2 + hv \rightarrow 2Cl$	0.0	6.9e-08
$CCl_2FCClF_2 + hv \rightarrow 3Cl$	0.0	9.6e-08
$CF_3Br + hv \rightarrow Br$	0.0	2.6e-07
$CF_2ClBr + hv \rightarrow Br + Cl$	0.0	2.5e-06
$CH_3Cl + hv \rightarrow Cl + CH_3O_2$	0.0	1.5e-08
$CH_3CCl_3 + hv \rightarrow 3Cl$	0.0	8.8e-07
$CHF_2Cl + hv \rightarrow Cl$	0.0	1.7e-10
$CH_3Br + hv \rightarrow Br + CH_3O_2$	0.0	1.5e-06

D.2.2 Organic Species (Chemical Mechanism)

$CH_3OOH + hv \rightarrow CH_2O + H + OH$

$CH_2O + hv \rightarrow CO + 2H$

$CH_2O + hv \rightarrow CO + H_2$

$CH_4 + hv \rightarrow H + CH_3O_2$

$CH_4 + hv \rightarrow 1.44\ H_2 + 0.18CH_2O + 0.18O + 0.66\ O\ H + 0.44\ CO_2 + 0.38\ CO$
      $+0.05H_2O$

$CH_3\ CHO + hv \rightarrow CH_3O_2 + CO + HO_2$

$POOH + hv \rightarrow CH_3CHO + CH_2O + HO_2 + OH$

$CH_3COOOH + hv \rightarrow CH_3O_2 + OH + CO_2$

$PAN + hv \rightarrow 0.6\ CH_3CO_3 + 0.6\ NO_2 + 0.4\ CH_3O_2 + 0.4\ NO_3 + 0.4\ CO_2$

$MPAN + hv \rightarrow MCO_3 + NO_2$

$MACR + hv \rightarrow 0.67\ HO_2 + 0.33\ MCO_3 + 0.67\ CH_2O + 0.67\ CH_3CO_3 + 0.33\ OH$
      $+0.67\ CO$

$MVK + hv \rightarrow 0.7\ C_3H_6 + 0.7\ CO + 0.3\ CH_3O_2 + 0.3\ CH_3CO_3$

$C_2H_5OOH + hv \rightarrow CH_3CHO + HO_2 + OH$

$C_3H_7OOH + hv \rightarrow 0.82\ CH_3COCH_3 + OH + HO_2$

$ROOH + hv \rightarrow CH_3CO_3 + CH_2O + OH$

$CH_3COCH_3 + hv \rightarrow CH_3CO_3 + CH_3O_2$

$CH_3COCHO + hv \rightarrow CH_3CO_3 + CO + HO_2$

$XOOH + hv \rightarrow OH$

$ONITR + hv \rightarrow HO_2 + CO + NO_2 + CH_2O$

$ISOPOOH + hv \rightarrow 0.402\ MVK + 0.288\ MACR + 0.69\ CH_2O + HO_2$

$HYAC + hv \rightarrow CH_3CO_3 + HO_2 + CH_2O$

$GLYALD + hv \rightarrow 2\ HO_2 + CO + CH_2O$

$MEK + hv \rightarrow CH_3CO_3 + C_2H_5O_2$

$BIGALD + hv \rightarrow 0.45\ CO + 0.13\ GLYOXAL + 0.56\ HO_2 + 0.13\ CH_3CO_3$
$\qquad + 0.18\ CH_3COCHO$

$GLYOXAL + hv \rightarrow 2\ CO + 2\ HO_2$

$C_5H_{11}OOH + hv \rightarrow 0.4\ CH_3CHO + 0.1\ CH_2O + 0.25\ CH_3COCH_3 + 0.9\ HO_2$
$\qquad + 0.8\ MEK + OH$

$MEKOOH + hv \rightarrow OH + CH_3CO_3 + CH_3CHO$

$TOLOOH + hv \rightarrow OH + 0.45\ GLYOXAL + 0.45\ CH_3COCHO + 0.9\ BIGALD$

$TERPOOH + hv \rightarrow OH + 0.1\ CH_3COCH_3 + HO_2 + MVK + MACR$

D.2.3 Organic Species (Photolysis Frequencies)

Reaction	J at sea level [s^{-1}]	J at 25 km [s^{-1}]
$CH_3OOH + hv \rightarrow CH_3O + OH$	5.4e-06	1.1e-05
$CH_2O + hv \rightarrow HCO + H$	3.3e-05	6.5e-05
$CH_2O + hv \rightarrow CO + H_2$	3.8e-05	1.1e-04
$CH_4 + hv \rightarrow products$	0.0	0.0
$CH_3CHO + hv \rightarrow CH_3 + HCO$	4.7e-06	5.4e-05
$CH_3COOOH + hv \rightarrow CH_3O_2 + OH + CO_2$	7.4e-07	1.9e-06
$PAN + hv \rightarrow CH_3CO_3 + NO_2$	4.8e-07	4.1e-06
$PAN + hv \rightarrow CH_3 + CO_2 + NO_3$	2.1e-07	1.7e-06
$MACR + hv \rightarrow products$	5.0e-06	8.1e-06
$MVK + hv \rightarrow products$	4.1e-06	3.5e-05
$C_2H_5OOH + hv \rightarrow CH_3CH_2O + OH$	5.4e-06	1.1e-05
$HOCH_2OOH + hv \rightarrow HOCH_2O + OH$	4.5e-06	9.4e-06
$C_3H_7OOH + hv \rightarrow CH_3CH(O)CH_3 + OH$	5.4e-06	1.1e-05
$CH_3COCH_3 + hv \rightarrow CH_3CO_+ CH_3$	8.5e-07	1.0e-05
$CH_3COCHO + hv \rightarrow CH_3CO_+ HCO$	1.4e-04	5.6e-04
$CH_3ONO_2 + hv \rightarrow NO_2 + CH_3O$	8.5e-07	1.5e-05
$HYAC + hv \rightarrow CH_3CO + CH_2(OH)$	9.1e-07	2.5e-06
$HYAC + hv \rightarrow CH_2(OH)CO + CH_3$	9.1e-07	2.5e-06
$GLYALD + hv \rightarrow CH_2OH + HCO$	9.1e-06	2.5e-05
$GLYALD + hv \rightarrow CH_3OH + CO$	1.1e-06	3.1e-06
$GLYALD + hv \rightarrow CH_2CHO + OH$	7.7e-07	2.1e-06
$MEK + hv \rightarrow CH_3CO + C_2H_5$	6.1e-06	4.0e-05
$C_2H_5CHO + hv \rightarrow C_2H_5 + HCO$	1.7e-05	8.8e-05
$GLYOXAL + hv \rightarrow HCO + HCO$	7.4e-05	1.1e-04
$GLYOXAL + hv \rightarrow H_2 + 2\ CO$	1.6e-05	3.3e-05
$GLYOXAL + hv \rightarrow CH_2O + CO$	2.9e-05	5.6e-05

D.3 Gas-Phase Reactions

The following table lists the rate constants k for gas-phase reactions. In the case of two-body (bimolecular) reactions, written as X + Y $\rightarrow$ products, the temperature-dependent rate constant [cm^3 s^{-1}] is generally expressed as

$$k(T) = A \exp \left[\frac{-B}{T} \right]$$

where A [cm^3 s^{-1}] is the Arrhenius factor, B [K] the activation temperature equal to the activation energy E_a [J mol^{-1}] divided by the gas constant R=8.3144 J K^{-1} mol^{-1}, and T is the temperature [K]. The table also includes single-body (unimolecular) thermolysis reactions, written as X $\rightarrow$ products, with rate coefficients expressed in [s^{-1}].

In the case of three-body (termolecular) reactions, written as X + Y + M $\rightarrow$ XY + M where M is an inert third body (typically N$_2$ or O$_2$), the pressure- and temperature-dependent coefficients k [cm^3 s^{-1}] are derived by the Troe formula

$$k = \frac{k_\infty k_0 [\boldsymbol{M}]}{k_\infty + k_0 [\boldsymbol{M}]} f^{\left\{ 1 + \left(\log_{10} \frac{k_0 [\boldsymbol{M}]}{k_\infty} \right)^2 \right\}^{-1}}$$

Here, [$\boldsymbol{M}$] denotes the air number density [cm^{-3}] and f = 0.6 if it is not otherwise specified in the tables below. The temperature dependence of coefficients k_0 (low-pressure limit) and k_∞ (high-pressure limit) is often expressed as $C(T/300)^{-n}$ where C and n are constants. For these types of reactions, the table provides the values of k_0 [cm^6 s^{-2}] and k_∞ [cm^3 s^{-1}]; only k_0 is given when the low-pressure limit dominates throughout the atmosphere. The rate of the reverse reaction, XY + M $\rightarrow$ X + Y + M, is given as the rate of the forward reaction times an equilibrium constant.

D.3.1 Oxygen–Hydrogen–Nitrogen Chemistry

Oxygen reactions	Rate constant
Two-body reactions	
O + O$_3$ $\rightarrow$ 2O$_2$	8.0e-12 $\times$ exp($-$2060/T)
O(^{1}D) + N$_2$ $\rightarrow$ O + N$_2$	2.1e-11 $\times$ exp(115/T)
O(^{1}D) + O$_2$ $\rightarrow$ O + O$_2$	3.2e-11 $\times$ exp(70/T)
O(^{1}D) + H$_2$O $\rightarrow$ 2OH	2.2e-10
O(^{1}D) + H$_2$ $\rightarrow$ HO$_2$ + OH	1.1e-10
O(^{1}D) + N$_2$O $\rightarrow$ N$_2$ + O$_2$	4.9e-11
O(^{1}D) + N$_2$O $\rightarrow$ 2NO	6.7e-11
O(^{1}D) + CH$_4$ $\rightarrow$ CH$_3$O$_2$ + OH	1.1e-10
O(^{1}D) + CH$_4$ $\rightarrow$ CH$_2$O + H + HO$_2$	3.0e-11
O(^{1}D) + CH$_4$ $\rightarrow$ CH$_2$O + H$_2$	7.5e-12
O(^{1}D) + HCN $\rightarrow$ OH	7.7e-11 $\times$ exp(100/T)
Three-body reactions	
O + O + M $\rightarrow$ O$_2$ + M	k_0 = 2.8e-34 $\times$ exp(720/T)
O + O$_2$ + M $\rightarrow$ O$_3$ + M	k_0 = 6.0e-34 $\times$ (T/300)$^{-2.4}$

Hydrogen oxide reactions	Rate constant
Two-body reactions	
$H + O_3 \rightarrow OH + O_2$	$1.4e\text{-}10 \times \exp(-470/T)$
$H + HO_2 \rightarrow 2OH$	$7.2e\text{-}11$
$H + HO_2 \rightarrow H_2 + O_2$	$6.9e\text{-}12$
$H + HO_2 \rightarrow H_2O + O$	$1.6e\text{-}12$
$OH + O \rightarrow H + O_2$	$2.2e\text{-}11 \times \exp(120/T)$
$OH + O_3 \rightarrow HO_2 + O_2$	$1.7e\text{-}12 \times \exp(-940/T)$
$OH + HO_2 \rightarrow H_2O + O_2$	$4.8e\text{-}11 \times \exp(250/T)$
$OH + OH \rightarrow H_2O + O$	$1.8e\text{-}12$
$OH + H_2 \rightarrow H_2O + H$	$2.8e\text{-}12 \times \exp(-1800/T)$
$OH + H_2O_2 \rightarrow H_2O + HO_2$	$1.8e\text{-}12$
$HO_2 + O \rightarrow OH + O_2$	$3.0e\text{-}11 \times \exp(200/T)$
$HO_2 + O_3 \rightarrow OH + 2O_2$	$1.0e\text{-}14 \times \exp(-490/T)$
$HO_2 + HO_2 \rightarrow H_2O_2 + O_2$	$(k_A+k_B) + 1.4e\text{-}21 \times [H_2O] \times \exp(2200/T)$
	$k_A = 3.0e\text{-}13 \times \exp(460/T)$
	$k_B = 2.1e\text{-}33 \times [M] \times \exp(920/T)$
$H_2O_2 + O \rightarrow OH + HO_2$	$1.4e\text{-}12 \times \exp(-2000/T)$
Three-body reactions	
$H + O_2 + M \rightarrow HO_2 + M$	$k_0 = 4.4e\text{-}32 \times (T/300)^{-1.3}$
	$k_\infty = 4.7e\text{-}11 \times (T/300)^{-0.2}$
$OH + OH + M \rightarrow H_2O_2 + M$	$k_0 = 6.9e\text{-}31 \times (T/300)^{-1.0}$
	$k_\infty = 2.6e\text{-}11$

Nitrogen oxide reactions	Rate constant
Two-body reactions	
$N + O_2 \rightarrow NO + O$	$1.5e\text{-}11 \times \exp(-3600/T)$
$N + NO \rightarrow N_2 + O$	$2.1e\text{-}11 \times \exp(100/T)$
$N + NO_2 \rightarrow N_2O + O$	$5.8e\text{-}12 \times \exp(220/T)$
$NO + HO_2 \rightarrow NO_2 + OH$	$3.5e\text{-}12 \times \exp(250/T)$
$NO + O_3 \rightarrow NO_2 + O_2$	$3.0e\text{-}12 \times \exp(-1500/T)$
$NO_2 + O \rightarrow NO + O_2$	$5.1e\text{-}12 \times \exp(210/T)$
$NO_2 + O_3 \rightarrow NO_3 + O_2$	$1.2e\text{-}13 \times \exp(-2450/T)$
$HNO_3 + OH \rightarrow NO_3 + H_2O$	$k = k_0 + k_3[M]/(1 + k_3[M]/k_2)$ with
	$k_0 = 2.4e\text{-}14 \times \exp(460/T)$
	$k_2 = 2.7e\text{-}17 \times \exp(2199/T)$
	$k_3 = 6.5e\text{-}34 \times \exp(1335/T)$
$NO_3 + NO \rightarrow 2NO_2$	$1.5e\text{-}11 \times \exp(170/T)$
$NO_3 + O \rightarrow NO_2 + O_2$	$1.0e\text{-}11$
$NO_3 + OH \rightarrow HO_2 + NO_2$	$2.2e\text{-}11$
$NO_3 + HO_2 \rightarrow OH + NO_2 + O_2$	$3.5e\text{-}12$
$HO_2NO_2 + OH \rightarrow H_2O + NO_2 + O_2$	$1.3e\text{-}12 \times \exp(380/T)$
Three-body and reverse reactions	
$NO + O + M \rightarrow NO_2 + M$	$k_0 = 9.0e\text{-}32 \times (T/300)^{-1.5}$
	$k_\infty = 3.0e\text{-}11$

(cont.)	

Nitrogen oxide reactions	Rate constant
$NO_2 + O + M \rightarrow NO_3 + M$	$k_0 = 2.5e\text{-}31 \times (T/300)^{-1.8}$
	$k_\infty = 2.2e\text{-}11 \times (T/300)^{-0.7}$
$NO_2 + NO_3 + M \rightarrow N_2O_5 + M$	$k_0 = 2.0e\text{-}30 \times (T/300)^{-4.4}$
	$k_\infty = 1.4e\text{-}12 \times (T/300)^{-0.7}$
$N_2O_5 + M \rightarrow NO_2 + NO_3 + M$	$k_{NO2+NO3} \times 3.7e + 26 \times \exp(-11000/T)$
$NO_2 + HO_2 + M \rightarrow HO_2NO_2 + M$	$k_0 = 2.0e\text{-}31 \times (T/300)^{-3.4}$
	$k_\infty = 2.9e\text{-}12 \times (T/300)^{-1.1}$
$HO_2NO_2 + M \rightarrow HO_2 + NO_2 + M$	$k_{HO2+NO2} \times 4.8e + 26 \times \exp(-10900/T)$
$NO_2 + OH + M \rightarrow HNO_3 + M$	$k_0 = 1.8e\text{-}30 \times (T/300)^{-3.0}$
	$k_\infty = 2.8e\text{-}11$

D.3.2 Organic Chemistry

C-1 degradation (methane CH_4)	Rate constant
Two-body reactions	
$CH_4 + OH \rightarrow CH_3O_2 + H_2O$	$2.5e\text{-}12 \times \exp(-1775/T)$
$CH_3O_2 + NO \rightarrow CH_2O + NO_2 + HO_2$	$2.8e\text{-}12 \times \exp(300/T)$
$CH_3O_2 + HO_2 \rightarrow CH_3OOH + O_2$	$4.1e\text{-}13 \times \exp(750/T)$
$CH_3OOH + OH \rightarrow CH_3O_2 + H_2O$	$3.8e\text{-}12 \times \exp(200/T)$
$CH_2O + NO_3 \rightarrow CO + HO_2 + HNO_3$	$6.0e\text{-}13 \times \exp(-2058/T)$
$CH_2O + OH \rightarrow CO + H_2O + H$	$5.5e\text{-}12 \times \exp(125/T)$
$CH_2O + O \rightarrow HO_2 + OH + CO$	$3.4e\text{-}11 \times \exp(-1600/T)$
$CH_3O_2 + CH_3O_2 \rightarrow 2CH_2O + 2HO_2$	$5.0e\text{-}13 \times \exp(-424/T)$
$CH_3O_2 + CH_3O_2 \rightarrow CH_2O + CH_3OH$	$1.9e\text{-}14 \times \exp(706/T)$
$CH_3OH + OH \rightarrow HO_2 + CH_2O$	$2.9e\text{-}12 \times \exp(-345/T)$
$CH_3OOH + OH \rightarrow 0.7\ CH_3O_2 + 0.3\ OH + 0.3\ CH_2O + H_2O$	$3.8e\text{-}12 \times \exp(200/T)$
$CH_2O + HO_2 \rightarrow HOCH_2OO$	$9.7e\text{-}15 \times \exp(625/T)$
$HOCH_2OO \rightarrow CH_2O + HO_2$	$2.4e\text{+}12 \times \exp(-7000/T)$
$HOCH_2OO + NO \rightarrow HCOOH + NO_2 + HO_2$	$2.6e\text{-}12 \times \exp(265/T)$
$HOCH_2OO + HO_2 \rightarrow HCOOH$	$7.5e\text{-}13 \times \exp(700/T)$
$HCOOH + OH \rightarrow HO_2 + CO_2 + H_2O$	$4.5e\text{-}13$
$CO + OH \rightarrow CO_2 + H$	$1.5e\text{-}13 \times (1.0 + 6.e\text{-}7\ p)$
	(p = air pressure in Pa)

C-2 degradation (acetylene C_2H_2, ethylene C_2H_4 and ethane C_2H_6)	Rate constant
$C_2H_2 + OH + M \rightarrow 0.65\ GLYOXAL + 0.65\ OH + 0.35\ HCOOH + 0.35\ HO_2 + 0.35\ CO + M$	$k_0 = 5.5e\text{-}30$
	$k_\infty = 8.3e\text{-}13 \times (T/300)^{2.0}$
$GLYOXAL + OH \rightarrow HO_2 + CO + CO_2$	$1.1e\text{-}11$
$C_2H_4 + O_3 \rightarrow CH_2O + 0.12\ HO_2 + 0.5\ CO + 0.12\ OH + 0.5\ HCOOH$	$1.2e\text{-}14 \times \exp(-2630/T)$

(cont.)	

C-2 degradation (acetylene C_2H_2, ethylene C_2H_4 and ethane C_2H_6)	Rate constant
$C_2H_4 + OH + M \rightarrow 0.75\ EO_2 + 0.5\ CH_2O + 0.25\ HO_2 + M$	$k_0 = 1.0e\text{-}28 \times (T/300)^{-0.8}$ $k_\infty = 8.8e\text{-}12$
$EO_2 + NO \rightarrow EO + NO_2$	$4.2e\text{-}12 \times \exp(180/T)$
$EO + O_2 \rightarrow GLYALD + HO_2$	$1.0e\text{-}14$
$EO \rightarrow 2\ CH_2O + HO_2$	$1.6e\text{+}11 \times \exp(-4150/T)$
$GLYALD + OH \rightarrow HO_2 + 0.2\ GLYOXAL$ $+ 0.8\ CH_2O + 0.8\ CO_2$	$1.0e\text{-}11$
$C_2H_6 + OH \rightarrow C_2H_5O_2 + H_2O$	$8.7e\text{-}12 \times \exp(-1070/T)$
$C_2H_5O_2 + NO \rightarrow CH_3CHO + HO_2 + NO_2$	$2.6e\text{-}12 \times \exp(365/T)$
$C_2H_5O_2 + HO_2 \rightarrow C_2H_5OOH + O_2$	$7.5e\text{-}13 \times \exp(700/T)$
$C_2H_5O_2 + CH_3O_2 \rightarrow 0.7\ CH_2O + 0.8\ CH_3CHO + HO_2$ $+ 0.3\ CH_3OH + 0.2\ C_2H_5OH$	$2.0e\text{-}13$
$C_2H_5O_2 + C_2H_5O_2 \rightarrow 1.6\ CH_3CHO + 1.2\ HO_2$ $+ 0.4\ C_2H_5OH$	$6.8e\text{-}14$
$C_2H_5OOH + OH \rightarrow 0.5\ C_2H_5O_2 + 0.5\ CH_3CHO + 0.5\ OH$	$3.8e\text{-}12 \times \exp(200/T)$
$CH_3CHO + OH \rightarrow CH_3CO_3 + H_2O$	$5.6e\text{-}12 \times \exp(270/T)$
$CH_3CHO + NO_3 \rightarrow CH_3CO_3 + HNO_3$	$1.4e\text{-}12 \times \exp(-1900/T)$
$CH_3CO_3 + NO \rightarrow CH_3O_2 + CO_2 + NO_2$	$8.1e\text{-}12 \times \exp(270/T)$
$CH_3CO_3 + HO_2 \rightarrow 0.75\ CH_3COOOH$ $+ 0.25\ CH_3COOH + 0.25\ O_3$	$4.3e\text{-}13 \times \exp(1040/T)$
$CH_3CO_3 + CH_3O_2 \rightarrow 0.9\ CH_3O_2 + CH_2O + 0.9\ HO_2$ $+ 0.9\ CO_2 + 0.1\ CH_3COOH$	$2.0e\text{-}12 \times \exp(500/T)$
$CH_3CO_3 + CH_3CO_3 \rightarrow 2\ CH_3O_2 + 2\ CO_2$	$2.5e\text{-}12 \times \exp(500/T)$
$CH_3COOH + OH \rightarrow CH_3O_2 + CO_2 + H_2O$	$7.0e\text{-}13$
$CH_3COOOH + OH \rightarrow 0.5\ CH_3CO_3 + 0.5\ CH_2O$ $+ 0.5\ CO_2 + H_2O$	$1.0e\text{-}12$
$C_2H_5OH + OH \rightarrow HO_2 + CH_3CHO$	$6.9e\text{-}12 \times \exp(-230/T)$
$CH_3CO_3 + NO_2 + M \rightarrow PAN + M$	$k_0 = 8.5e\text{-}29 \times (T/300)^{-6.5}$ $k_\infty = 1.1e\text{-}11 \times (T/300)^{-1}$
$PAN + M \rightarrow CH_3CO_3 + NO_2 + M$	$k_{CH3CO3+NO2} \times 1.1e\text{+}28 \times$ $\exp(-14000/T)$
$PAN + OH \rightarrow CH_2O + NO_3$	$4.0e\text{-}14$

C-3 degradation (propene C_3H_6 and propane C_3H_8)	Rate constant
$C_3H_6 + OH + M \rightarrow PO_2 + M$	$k_0 = 8.0e\text{-}27 \times (T/300)^{-3.5}$ $k_\infty = 3.0e\text{-}11$ $f = 0.5$
$C_3H_6 + O_3 \rightarrow 0.54\ CH_2O + 0.19\ HO_2 + 0.33\ OH$ $+ 0.08\ CH_4 + 0.56\ CO + 0.5\ CH_3CHO + 0.31\ CH_3O_2$ $+ 0.25\ CH_3COOH$	$6.5e\text{-}15 \times \exp(-1900/T)$
$C_3H_6 + NO_3 \rightarrow ONIT$	$4.6e\text{-}13 \times \exp(-1156/T)$
$PO_2 + NO \rightarrow CH_3CHO + CH_2O + HO_2 + NO_2$	$4.2e\text{-}12 \times \exp(180/T)$
$PO_2 + HO_2 \rightarrow POOH + O_2$	$7.5e\text{-}13 \times \exp(700/T)$
$POOH + OH \rightarrow 0.5\ PO_2 + 0.5\ OH + 0.5\ HYAC + H_2O$	$3.8e\text{-}12 \times \exp(200/T)$

(cont.)	
C-3 degradation (propene C_3H_6 and propane C_3H_8)	*Rate constant*
$ROOH + OH \rightarrow RO_2 + H_2O$	$3.8e\text{-}12 \times \exp(200/T)$
$HYAC + OH \rightarrow CH_3COCHO + HO_2$	$3.0e\text{-}12$
$CH_3COCHO + OH \rightarrow CH_3CO_3 + CO + H_2O$	$8.4e\text{-}13 \times \exp(830/T)$
$CH_3COCHO + NO_3 \rightarrow HNO_3 + CO + CH_3CO_3$	$1.4e\text{-}12 \times \exp(-1860/T)$
$ONIT + OH \rightarrow NO_2 + CH_3COCHO$	$6.8e\text{-}13$
$C_3H_8 + OH \rightarrow C_3H_7O_2 + H_2O$	$1.0e\text{-}11 \times \exp(-665/T)$
$C_3H_7O_2 + NO \rightarrow 0.82\ CH_3COCH_3 + NO_2 + HO_2$ $+ 0.27\ CH_3CHO$	$4.2e\text{-}12 \times \exp(180/T)$
$C_3H_7O_2 + HO_2 \rightarrow C_3H_7OOH + O_2$	$7.5e\text{-}13 \times \exp(700/T)$
$C3H_7O_2 + CH_3O_2 \rightarrow CH_2O + HO_2 + 0.82\ CH_3COCH_3$	$3.8e\text{-}13 \times \exp(-40/T)$
$C_3H_7OOH + OH \rightarrow H_2O + C_3H_7O_2$	$3.8e\text{-}12 \times \exp(200/T)$
$CH_3COCH_3 + OH \rightarrow RO_2 + H_2O$	$3.8e\text{-}11 \times \exp(-2000/T)$ $+1.3e\text{-}13$
$RO_2 + NO \rightarrow CH_3CO_3 + CH_2O + NO_2$	$2.9e\text{-}12 \times \exp(300/T)$
$RO_2 + HO_2 \rightarrow ROOH + O_2$	$8.6e\text{-}13 \times \exp(700/T)$
$RO_2 + CH_3O_2 \rightarrow 0.3\ CH_3CO_3 + 0.8\ CH_2O + 0.3\ HO_2$ $+ 0.2\ HYAC + 0.5\ CH_3COCHO + 0.5\ CH_3OH$	$7.1e\text{-}13 \times \exp(500/T)$

C-4 degradation (lumped species BIGENE represented by butene C_4H_8)	*Rate constant*
$BIGENE + OH \rightarrow ENEO_2$	$5.4e\text{-}11$
$ENEO_2 + NO \rightarrow CH_3CHO + 0.5\ CH_2O + 0.5\ CH_3COCH_3$ $+ HO_2 + NO_2$	$4.2e\text{-}12 \times \exp(180/T)$

C-5 degradation (isoprene C_5H_8 and lumped species BIGALK represented by pentane C_5H_{12})	*Rate constant*
$BIGALK + OH \rightarrow ALKO_2$	$3.5e\text{-}12$
$ALKO_2 + NO \rightarrow 0.4\ CH_3CHO + 0.1\ CH_2O$ $+ 0.25\ CH_3COCH_3 + 0.9\ HO_2 + 0.8\ MEK + 0.9\ NO_2$ $+ 0.1\ ONIT$	$4.2e\text{-}12 \times \exp(180/T)$
$ALKO_2 + HO_2 \rightarrow ALKOOH$	$7.5e\text{-}13 \times \exp(700/T)$
$ALKOOH + OH \rightarrow ALKO_2$	$3.8e\text{-}12 \times \exp(200/T)$
$C_5H_8 + OH \rightarrow ISOPO_2$	$2.5e\text{-}11 \times \exp(410/T)$
$C_5H_8 + O_3 \rightarrow 0.4\ MACR + 0.2\ MVK + 0.07\ C_3H_6$ $+ 0.27\ OH + 0.06\ HO_2 + 0.6\ CH_2O + 0.3\ CO + 0.1\ O_3$ $+ 0.2\ MCO_3 + 0.2\ CH_3COOH$	$1.1e\text{-}14 \times \exp(-2000/T)$
$C_5H_8 + NO_3 \rightarrow ISOPNO_3$	$3.0e\text{-}12 \times \exp(-446/T)$
$ISOPO_2 + NO \rightarrow 0.08\ ONITR + 0.92\ NO_2 + HO_2$ $+ 0.51\ CH_2O + 0.23\ MACR + 0.32\ MVK$ $+ 0.37\ HYDRALD$	$4.4e\text{-}12 \times \exp(180/T)$
$ISOPO_2 + NO_3 \rightarrow HO_2 + NO_2 + 0.6\ CH_2O$ $+ 0.25\ MACR + 0.35\ MVK + 0.4\ HYDRALD$	$2.4e\text{-}12$
$ISOPO_2 + HO_2 \rightarrow ISOPOOH$	$8.0e\text{-}13 \times \exp(700/T)$

	(cont.)

C-5 degradation (isoprene C_5H_8 and lumped species BIGALK represented by pentane C_5H_{12})	Rate constant
ISOPOOH + OH → 0.8 XO$_2$ + 0.2 ISOPO$_2$	1.5e-11 × exp(200/T)
ISOPO$_2$ + CH$_3$O$_2$ → 0.25 CH$_3$OH + HO$_2$ + 1.2 CH$_2$O + 0.19 MACR + 0.26 MVK + 0.3 HYDRALD	5.0e-13 × exp(400/T)
ISOPO$_2$ + CH$_3$CO$_3$ → CH$_3$O$_2$ + HO$_2$ + 0.6 CH$_2$O + 0.25 MACR + 0.35 MVK + 0.4 HYDRALD	1.4e-11
ISOPNO$_3$ + NO → 1.206 NO$_2$ + 0.794 HO$_2$ + 0.072 CH$_2$O + 0.167 MACR + 0.039 MVK + 0.794 ONITR	2.7e-12 × exp(360/T)
ISOPNO$_3$ + NO$_3$ → 1.206 NO$_2$ + 0.072 CH$_2$O + 0.167 MACR + 0.039 MVK + 0.794 ONITR + 0.794 HO$_2$	2.4e-12
ISOPNO$_3$ + HO$_2$ → XOOH + 0.206 NO$_2$ + 0.794 HO$_2$ + 0.008 CH$_2$O + 0.167 MACR + 0.039 MVK + 0.794 ONITR	8.0e-13 × exp(700/T)
ONITR + OH → HYDRALD + 0.4 NO$_2$ + HO$_2$	4.5e-11
ONITR + NO$_3$ → HO$_2$ + NO$_2$ + HYDRALD	1.4e-12 × exp(−1860/T)
HYDRALD + OH → XO$_2$	1.9e-11 × exp(175/T)
XO$_2$ + NO → NO$_2$ + HO$_2$ + 0.5 CO + 0.25 GLYOXAL + 0.25 HYAC + 0.25 CH$_3$COCHO + 0.25 GLYALD	2.7e-12 × exp(360/T)
XO$_2$ + NO$_3$ → NO$_2$ + HO$_2$ + 0.5 CO + 0.25 HYAC + 0.25 GLYOXAL + 0.25 CH$_3$COCHO + 0.25 GLYALD	2.4e-12
XO$_2$ + HO$_2$ → XOOH	8.0e-13 × exp(700/T)
XO$_2$ + CH$_3$O$_2$ → 0.3 CH$_3$OH + 0.8 HO$_2$ + 0.7 CH$_2$O + 0.2 CO + 0.1 HYAC + 0.1 GLYOXAL + 0.1 CH3COCHO + 0.1 GLYALD	5.0e-13 × exp(400/T)
XO$_2$ + CH$_3$CO$_3$ → 0.5 CO + CH$_3$O$_2$ + HO$_2$ + CO$_2$ + 0.25 GLYOXAL + 0.25 HYAC + 0.25 CH$_3$COCHO + 0.25 GLYALD	1.3e-12 × exp(640/T)
XOOH + OH → H$_2$O + XO$_2$	1.9e-12 × exp(190/T)
XOOH + OH → H$_2$O + OH	7.7e-17 × T^2 × exp(253/T)
MVK + OH → MACRO$_2$	4.1e-12 × exp(452/T)
MVK + O$_3$ → 0.8 CH$_2$O + 0.95 CH$_3$COCHO + 0.08 OH + 0.2 O$_3$ + 0.06 HO$_2$ + 0.05 CO + 0.04 CH$_3$CHO	7.5e-16 × exp(−1521/T)
MEK + OH → MEKO$_2$	2.3e-12 × exp(−170/T)
MEKO$_2$ + NO → CH$_3$CO$_3$ + CH$_3$CHO + NO$_2$	4.2e-12 × exp(180/T)
MEKO$_2$ + HO$_2$ → MEKOOH	7.5e-13 × exp(700/T)
MEKOOH + OH → MEKO$_2$	3.8e-12 × exp(200/T)
MACR + OH → 0.5 MACRO$_2$ + 0.5 H$_2$O + 0.5 MCO$_3$	1.9e-11 × exp(175/T)
MACR + O$_3$ → 0.8 CH$_3$COCHO + 0.275 HO$_2$ + 0.2 CO + 0.2 O$_3$ + 0.7 CH$_2$O + 0.215 OH	4.4e-15 × exp(−2500/T)
MACRO$_2$ + NO → NO$_2$ + 0.47 HO$_2$ + 0.25 CH$_2$O + 0.53 GLYALD + 0.25 CH$_3$COCHO + 0.53 CH$_3$CO$_3$ + 0.22 HYAC + 0.22 CO	2.7e-12 × exp(360/T)
MACRO$_2$ + NO → 0.8 ONITR	1.3e-13 × exp(360/T)
MACRO$_2$ + NO$_3$ → NO$_2$ + 0.47 HO$_2$ + 0.25 CH$_2$O + 0.25 CH$_3$COCHO + 0.22 CO + 0.53 GLYALD + 0.22 HYAC + 0.53 CH$_3$CO$_3$	2.4e-12

(cont.)

C-5 degradation (isoprene C_5H_8 and lumped species BIGALK represented by pentane C_5H_{12})	Rate constant
$MACRO_2 + HO_2 \rightarrow MACROOH$	8.0e-13 $\times$ exp(700/T)
$MACRO_2 + CH_3O_2 \rightarrow 0.73\ HO_2 + 0.88\ CH_2O + 0.11\ CO$ $+ 0.24\ CH_3COCHO + 0.26\ GLYALD + 0.26\ CH_3CO_3$ $+ 0.25\ CH_3OH + 0.23\ HYAC$	5.0e-13 $\times$ exp(400/T)
$MACRO_2 + CH_3CO_3 \rightarrow 0.25\ CH_3COCHO + CH_3O_2$ $+ 0.22\ CO + 0.47\ HO_2 + 0.53\ GLYALD + 0.22\ HYAC$ $+ 0.25\ CH_2O + 0.53\ CH_3CO_3$	1.4e-11
$MACROOH + OH \rightarrow 0.5\ MCO_3 + 0.2\ MACRO_2$ $+ 0.1\ OH + 0.2\ HO_2$	2.3e-11 $\times$ exp(200/T)
$MCO_3 + NO \rightarrow NO_2 + CH_2O + CH_3CO_3$	5.3e-12 $\times$ exp(360/T)
$MCO_3 + NO_3 \rightarrow NO_2 + CH_2O + CH_3CO_3$	5.0e-12
$MCO_3 + HO_2 \rightarrow 0.25\ O_3 + 0.25\ CH_3COOH$ $+ 0.75\ CH_3COOOH + 0.75\ O_2$	4.3e-13 $\times$ exp(1040/T)
$MCO_3 + CH_3O_2 \rightarrow 2\ CH_2O + HO_2 + CO_2 + CH_3CO_3$	2.0e-12 $\times$ exp(500/T)
$MCO_3 + CH_3CO_3 \rightarrow 2\ CO_2 + CH_3O_2 + CH_2O + CH_3CO_3$	4.6e-12 $\times$ exp(530/T)
$MCO_3 + MCO_3 \rightarrow 2\ CO_2 + 2\ CH_2O + 2\ CH_3CO_3$	2.3e-12 $\times$ exp(530/T)
$MCO_3 + NO_2 + M \rightarrow MPAN + M$	1.1e-11 $\times$ (300/T)/[M]
$MPAN + M \rightarrow MCO_3 + NO_2 + M$	$k_{MCO3+NO2} \times$ 1.1e+28 $\times$ exp(−14000/T)
$MPAN + OH + M \rightarrow 0.5\ HYAC + 0.5\ NO_3$ $+ 0.5\ CH_2O + 0.5\ HO_2 + 0.5\ CO_2 + M$	k_0 = 8.0e-27 $\times$ $(T/300)^{-3.5}$ k_∞ = 3.0e-11 f = 0.5

C-7 degradation (lumped aromatics represented by toluene C_7H_8)	Rate constant
$TOLUENE + OH \rightarrow 0.25\ CRESOL + 0.25\ HO_2 + 0.7\ TOLO_2$	1.7e-12 $\times$ exp (352/T)
$TOLO_2 + NO \rightarrow 0.45\ GLYOXAL + 0.45\ CH_3COCHO$ $+ 0.9\ BIGALD + 0.9\ NO_2 + 0.9\ HO_2$	4.2e-12 $\times$ exp (180/T)
$TOLO_2 + HO_2 \rightarrow TOLOOH$	7.5e-13 $\times$ exp (700/T)
$TOLOOH + OH \rightarrow TOLO_2$	3.8e-12 $\times$ exp (200/T)
$CRESOL + OH \rightarrow XOH$	3.0e-12
$XOH + NO_2 \rightarrow 0.7\ NO_2 + 0.7\ BIGALD + 0.7\ HO_2$	1.0e-11

C-10 degradation (terpenes lumped as α-pinene $C_{10}H_{16}$)	Rate constant
$TERPENE + OH \rightarrow TERPO2$	1.2e-11 $\times$ exp(444/T)
$TERPENE + O_3 \rightarrow 0.7\ OH + MVK + MACR + HO_2$	1.0e-15 $\times$ exp(−732/T)
$TERPENE + NO_3 \rightarrow TERPO2 + NO_2$	1.2e-12 $\times$ exp(490/T)
$TERPO2 + NO \rightarrow 0.1\ CH_3COCH_3 + HO_2 + MVK + MACR$ $+ NO_2$	4.2e-12 $\times$ exp(180/T)
$TERPO2 + HO_2 \rightarrow TERPOOH$	7.5e-13 $\times$ exp(700/T)
$TERPOOH + OH \rightarrow TERPO2$	3.8e-12 $\times$ exp(200/T)

D.3.3 Halogen Chemistry

$O(^1D)$ reactions with halogens	Rate constant
$O^1D + CFCl_3 \rightarrow 3\ Cl$	1.7e-10
$O^1D + CF_2Cl_2 \rightarrow 2\ Cl$	1.2e-10
$O^1D + CCl_2FCClF_2 \rightarrow 3\ Cl$	1.5e-10
$O^1D + CHF_2Cl \rightarrow Cl$	7.2e-11
$O^1D + CCl_4 \rightarrow 4\ Cl$	2.8e-10
$O^1D + CH_3Br \rightarrow Br$	1.8e-10
$O^1D + CF_2ClBr \rightarrow Cl + Br$	9.6e-11
$O^1D + CF_3Br \rightarrow Br$	4.1e-11

Inorganic chlorine reactions	Rate constant
$Cl + O_3 \rightarrow ClO + O_2$	2.3e-11 $\times$ exp($-200/T$)
$Cl + H_2 \rightarrow HCl + H$	3.1e-11 $\times$ exp($-2270/T$)
$Cl + H_2O_2 \rightarrow HCl + HO_2$	1.1e-11 $\times$ exp($-980/T$)
$Cl + HO_2 \rightarrow HCl + O_2$	1.8e-11 $\times$ exp($170/T$)
$Cl + HO_2 \rightarrow OH + ClO$	4.1e-11 $\times$ exp($-450/T$)
$Cl + CH_2O \rightarrow HCl + HO_2 + CO$	8.1e-11 $\times$ exp($-30/T$)
$Cl + CH_4 \rightarrow CH_3O_2 + HCl$	7.3e-12 $\times$ exp($-1280/T$)
$ClO + O \rightarrow Cl + O_2$	2.8e-11 $\times$ exp($85/T$)
$ClO + OH \rightarrow Cl + HO_2$	7.4e-12 $\times$ exp($270/T$)
$ClO + OH \rightarrow HCl + O_2$	6.0e-13 $\times$ exp($230/T$)
$ClO + HO_2 \rightarrow O_2 + HOCl$	2.7e-12 $\times$ exp($220/T$)
$ClO + NO \rightarrow NO_2 + Cl$	6.4e-12 $\times$ exp($290/T$)
$ClO + ClO \rightarrow 2Cl + O_2$	3.0e-11 $\times$ exp($-2450/T$)
$ClO + ClO \rightarrow Cl_2 + O_2$	1.0e-12 $\times$ exp($-1590/T$)
$ClO + ClO \rightarrow Cl + OClO$	3.5e-13 $\times$ exp($-1370/T$)
$HCl + OH \rightarrow H_2O + Cl$	2.6e-12 $\times$ exp($-350/T$)
$HCl + O \rightarrow Cl + OH$	1.0e-11 $\times$ exp($-3300/T$)
$HOCl + O \rightarrow ClO + OH$	1.7e-13
$HOCl + Cl \rightarrow HCl + ClO$	2.5e-12 $\times$ exp($-130/T$)
$HOCl + OH \rightarrow H_2O + ClO$	3.0e-12 $\times$ exp($-500/T$)
$ClONO_2 + O \rightarrow ClO + NO_3$	2.9e-12 $\times$ exp($-800/T$)
$ClONO_2 + OH \rightarrow HOCl + NO_3$	1.2e-12 $\times$ exp($-330/T$)
$ClONO_2 + Cl \rightarrow Cl_2 + NO_3$	6.5e-12 $\times$ exp($135/T$)

Three-body and reverse reactions

$ClO + ClO + M \rightarrow Cl_2O_2 + M$

$$k_0 = 1.6e\text{-}32 \times (T/300)^{-4.5}$$
$$k_\infty = 2.0e\text{-}12 \times (T/300)^{-2.4}$$

$Cl_2O_2 + M \rightarrow ClO + ClO + M$

$$k_{ClO+ClO} \times 5.8e26 \times \exp(-8649/T)$$

$ClO + NO_2 + M \rightarrow ClONO_2 + M$

$$k_0 = 1.8e\text{-}31 \times (T/300)^{-3.4}$$
$$k_\infty = 1.5e\text{-}11 \times (T/300)^{-1.9}$$

Inorganic bromine reactions	Rate constant
Two-Body Reactions	
$Br + O_3 \rightarrow BrO + O_2$	$1.7e\text{-}11 \times \exp(-800/T)$
$Br + HO_2 \rightarrow HBr + O_2$	$4.8e\text{-}12 \times \exp(-310/T)$
$Br + CH_2O \rightarrow HBr + HO_2 + CO$	$1.7e\text{-}11 \times \exp(-800/T)$
$BrO + O \rightarrow Br + O_2$	$1.9e\text{-}11 \times \exp(230/T)$
$BrO + OH \rightarrow Br + HO_2$	$1.7e\text{-}11 \times \exp(250/T)$
$BrO + HO_2 \rightarrow HOBr + O_2$	$4.5e\text{-}12 \times \exp(460/T)$
$BrO + NO \rightarrow Br + NO_2$	$8.8e\text{-}12 \times \exp(260/T)$
$BrO + ClO \rightarrow Br + OClO$	$9.5e\text{-}13 \times \exp(550/T)$
$BrO + ClO \rightarrow Br + Cl + O_2$	$2.3e\text{-}12 \times \exp(260/T)$
$BrO + ClO \rightarrow BrCl + O_2$	$4.1e\text{-}13 \times \exp(290/T)$
$BrO + BrO \rightarrow 2Br + O_2$	$1.5e\text{-}12 \times \exp(230/T)$
$HBr + OH \rightarrow Br + H_2O$	$5.5e\text{-}12 \times \exp(200/T)$
$HBr + O \rightarrow Br + OH$	$5.8e\text{-}12 \times \exp(-1500/T)$
$HOBr + O \rightarrow BrO + OH$	$1.2e\text{-}10 \times \exp(-430/T)$
$BrONO_2 + O \rightarrow BrO + NO_3$	$1.9e\text{-}11 \times \exp(215/T)$
Three-body reactions	
$BrO + NO_2 + M \rightarrow BrONO_2 + M$	$k_0 = 5.2e\text{-}31 \times (T/300)^{-3.2}$
	$k_\infty = 6.9e\text{-}12 \times (T/300)^{-2.9}$

Organic halogen reactions with Cl, OH	Rate constant
$CH_3Cl + Cl \rightarrow HO_2 + CO + 2\,HCl$	$2.2e\text{-}11 \times \exp(-1130/T)$
$CH_3Cl + OH \rightarrow Cl + H_2O + HO_2$	$2.4e\text{-}12 \times \exp(-1250/T)$
$CH_3CCl_3 + OH \rightarrow H_2O + 3\,Cl$	$1.6e\text{-}12 \times \exp(-1520/T)$
$CHF_2Cl + OH \rightarrow Cl + H_2O + CF_2O$	$1.1e\text{-}12 \times \exp(-1600/T)$
$CH_3Br + OH \rightarrow Br + H_2O + HO_2$	$2.4e\text{-}12 \times \exp(-1300/T)$

D.4 Heterogeneous Reactions

The following table lists the dimensionless reactive uptake probabilities γ of different heterogeneous reactions. The symbol OC stands for organic carbon, SO_4 for sulfate, NH_4NO_3 for ammonium nitrate and SOA for secondary organic aerosols. r denotes the particle radius.

Heterogeneous reactions on tropospheric aerosols	Reactive uptake probability γ
$N_2O_5 \rightarrow 2\,HNO_3$	0.1 on OC, SO_4, NH_4NO_3, SOA
$NO_3 \rightarrow HNO_3$	0.001 on OC, SO_4, NH_4NO_3, SOA
$NO_2 \rightarrow 0.5\,OH + 0.5\,NO + 0.5\,HNO_3$	0.0001 on OC, SO_4, NH_4NO_3, SOA
$HO_2 \rightarrow 0.5\,H_2O_2$	0.2 on OC, SO_4, NH_4NO_3, SOA

Stratospheric sulfate aerosol reactions	Reactive uptake probability γ
$N_2O_5 \rightarrow 2\ HNO_3$	0.04
$ClONO_2 \rightarrow HOCl + HNO_3$	f(sulfuric acid wt%)
$BrONO_2 \rightarrow HOBr + HNO_3$	$f(T, p, [HCl], [H_2O], r)$
$ClONO_2 + HCl \rightarrow Cl_2 + HNO_3$	$f(T, p, [H_2O], r)$
$HOCl + HCl \rightarrow Cl_2 + H_2O$	$f(T, p, [HCl], [H_2O], r)$
$HOBr + HCl \rightarrow BrCl + H_2O$	$f(T, p, [HCl], [HOBr], [H_2O], r)$

Nitric acid trihydrate reactions	Reactive uptake probability γ
$N_2O_5 \rightarrow 2\ HNO_3$	0.0004
$ClONO_2 \rightarrow HOCl + HNO_3$	0.004
$ClONO_2 + HCl \rightarrow Cl_2 + HNO_3$	0.2
$HOCl + HCl \rightarrow Cl_2 + H_2O$	0.1
$BrONO_2 \rightarrow HOBr + HNO_3$	0.3

Ice aerosol reactions	Reactive uptake probability γ
$N_2O_5 \rightarrow 2\ HNO_3$	0.02
$ClONO_2 \rightarrow HOCl + HNO_3$	0.3
$BrONO_2 \rightarrow HOBr + HNO_3$	0.3
$ClONO_2 + HCl \rightarrow Cl_2 + HNO_3$	0.3
$HOCl + HCl \rightarrow Cl_2 + H_2O$	0.2
$HOBr + HCl \rightarrow BrCl + H_2O$	0.3

References

Emmons L. K., Walters S., Hess P. G., *et al.* (2001) Description and evaluation of the Model for Ozone and Related Chemical Tracers, version 4 (MOZART-4), *Geosci. Model Dev.*, **3**, 43–67.

Lamarque J.-F., Emmons L. K., Hess P. G., *et al.* (2012) CAM-chem: Description and evaluation of interactive atmospheric chemistry in the Community Earth System Model, *Geosci. Model Dev.*, **5**, 369–411.

E | Brief Mathematical Review

E.1 Mathematical Functions

A *function f* from set S to set T is a rule that associates with each element x of set S a unique element y of set T. One writes

$$y = f(x) \qquad (1)$$

where y is said to be a function of x, or to be the image of x under f. Function f maps therefore x on y. Element x is called the *independent variable* and element y the *dependent variable*. The concept can be extended to multiple independent variables:

$$y = f(x_1, x_2, .., x_n)$$

A function written as (1) is said to be explicit. If expressed as

$$f(x, y) = 0$$

it is implicit.

Partial Derivatives

We define the partial derivative $\partial f / \partial x$ of a function f versus independent variable x as the variation of f relative to infinitesimal variation in x, while keeping all other independent variables constant.

Total Differential

The total differential of function f is defined as the variation in f for an infinitesimal perturbation of all independent variables. If f depends on variables x_i (with $i = 1, n$), the total differential is thus

$$df = \sum_{i=1}^{n} \frac{\partial f}{\partial x_i} \, dx_i \qquad (2)$$

Total Derivative

By direct application of relation (2), the total derivative of function f versus any independent variable x_k is

$$\frac{df}{dx_k} = \sum_{i=1}^{n} \left(\frac{\partial f}{\partial x_i} \frac{dx_i}{dx_k} \right)$$

Time Derivative: Eulerian Versus Lagrangian

For a function f that depends, for example, on three spatial variables (x, y, z) and on time (t), the total derivative is

$$\frac{df}{dt} = \frac{\partial f}{\partial x}\frac{dx}{dt} + \frac{\partial f}{\partial y}\frac{dy}{dt} + \frac{\partial f}{\partial z}\frac{dz}{dt} + \frac{\partial f}{\partial t}$$

or

$$\frac{df}{dt} = \frac{\partial f}{\partial x}u + \frac{\partial f}{\partial y}v + \frac{\partial f}{\partial z}w + \frac{\partial f}{\partial t}$$

where $u = dx/dt$, $v = dy/dt$, and $w = dz/dt$ are the velocity components. The total derivative on the left-hand side is called the *Lagrangian* time derivative because it expresses the change of function f following a moving parcel. The partial derivative $\partial f/\partial t$ on the right-hand side defines the change in the function at a given point of the domain in response to local sources and sinks. It is called the *Eulerian* time derivative. The remaining terms on the right-hand side represent the change in f versus time due to the advection of air parcels from other locations in the domain where the value of f is different.

Notations for Differentiation

If equation $y = f(x)$ represents a mathematical relationship between a dependent variable y and an independent variable x, the first and second derivatives of y versus x can be expressed in different equivalent forms:

- Leibniz's notation: $\quad \dfrac{dy}{dx} \qquad \dfrac{d^2y}{dx^2}$
- Lagrange's notation: $\quad f'(x) \qquad f''(x)$
- Euler's notation: $\quad D_x y \qquad D_x^2 y$
- Newton's notation: $\quad \dot{y} \qquad \ddot{y}$

The notation can be generalized for higher-order derivatives. In this book we generally use Leibniz's notation but adopt the Lagrangian notation in some cases. The "dot" notation of Newton is used in the fluid dynamics literature to express time derivatives.

For a function $y = f(x, y)$ of two independent variables x and y, the first and second partial derivatives are often expressed using the following notations:

$$\frac{\partial f}{\partial x} = f_x = \partial_x f$$

$$\frac{\partial^2 f}{\partial x^2} = f_{xx} = \partial_{xx} f$$

$$\frac{\partial^2 f}{\partial x \partial y} = f_{xy} = \partial_{xy} f$$

Differential operators for a scalar field φ and a vector field $\mathbf{a}$ are expressed as follows:

- Gradient: *grad* φ or $\nabla \varphi$
- Divergence: *div* $\mathbf{a}$ or $\nabla \cdot \mathbf{a}$
- Laplacian: *div grad* φ or $\nabla^2 \varphi$ or $\Delta \varphi$
- Rotation: *curl* $\mathbf{a}$ or   *rot* $\mathbf{a}$ or $\nabla \times \mathbf{a}$

where ∇ is a differential operator expressed as a vector whose Cartesian components are

$$\left(\frac{\partial}{\partial x}, \frac{\partial}{\partial y}, \frac{\partial}{\partial z} \right)^T .$$

Higher-Order Derivatives

The second-order derivatives of function $f(x, y)$ versus independent variables x and y are expressed by

$$\frac{\partial^2 f}{\partial x \partial y} = \frac{\partial}{\partial x} \left(\frac{\partial f}{\partial y} \right) = \frac{\partial}{\partial y} \left(\frac{\partial f}{\partial x} \right)$$

This can be generalized to all higher-order derivatives, for example

$$\frac{\partial^n f}{\partial x^n} = \frac{\partial}{\partial x} \left(\frac{\partial^{n-1} f}{\partial x^{n-1}} \right)$$

Taylor Expansion of $f(x)$

If $f(x)$ is a function whose successive derivatives exist, it can be expressed by an infinite polynomial series

$$f(x) = f(x_0) + \sum_{i=1}^{\infty} \frac{1}{i!} \left[\frac{\partial^i f}{\partial x^i} \right]_{x_0} (x - x_0)^i$$

where $f(x_0)$ is the value of f estimated at a point x_0, and $\left[\partial^i f / \partial x^i \right]_{x_o}$ represents the ith order derivative of function f evaluated at x_o. In many applications, function $f(x)$ is approximated in the vicinity of point x_0 by a finite Taylor expansion limited to order n. Terms with higher order derivatives are neglected. The accuracy of the approximation increases with the value of n. Nonlinear functions can be *linearized* around x_0 by limiting the Taylor expansion to the first order

$$f(x) = f(x_0) + \left[\frac{\partial f}{\partial x} \right]_{x_0} (x - x_0)$$

Taylor expansions are the basis for many numerical methods.

E.2 Scalars and Vectors

Scalars

A *scalar* is a physical quantity that is completely defined by a single number. Examples are temperature or pressure at a given location and time. A *scalar field* associates a scalar value to every point in space. In many atmospheric applications, a scalar field such as the temperature or the concentration of a chemical species is provided by a real function $f(x, y, z, t)$ expressed as a function of three independent spatial variables (x, y, z) and time (t).

Vectors

A *vector* $\mathbf{x}$ (noted by a bold symbol) of dimension n is an ordered collection of n elements called components:

$$\mathbf{x} = \begin{pmatrix} x_1 \\ \vdots \\ x_i \\ \vdots \\ x_n \end{pmatrix}$$

The elements of a vector are usually numbers but can also be functions. A *vector field* associates a vector to every point of a Euclidean space. Vectors are used in physics to represent physical quantities that have both a magnitude and a direction such as force, velocity, acceleration, flux. Vectors are also basic tools of matrix algebra, used in atmospheric chemistry for statistical applications. In such applications, one may for example define a concentration vector $\mathbf{x}$ where the components x_i represent the concentrations of the different species i.

Vector elements are usually arranged as a column, as shown in the example above; one sometimes refers to *column vectors*, but "column" is generally assumed by default. When vector elements are arranged in a row one refers to a *row vector*. A column vector is an $(n \times 1)$ matrix (n rows $\times$ 1 column). When column vectors are written out horizontally in text, we express the vector as its transpose $(1 \times n)$ to avoid ambiguity with row vectors. Thus we write $\mathbf{x} = (x_1, x_2, \ldots x_n)^T$ for the column vector in the example above.

In a Cartesian coordinate frame, a vector $\mathbf{a}$ can be described by its orthogonal projections on the axes of the reference frame. In a 3-D space (x, y, z), for example, we write

$$\mathbf{a} = a_x \mathbf{i} + a_y \mathbf{j} + a_z \mathbf{k}$$

where a_x, a_y, a_z are the three components of vector $\mathbf{a}$ and $(\mathbf{i}, \mathbf{j}, \mathbf{k})$ are the unit vectors along the axes (x, y, z) of the coordinate frame. The norm of the vector is

$$\|\boldsymbol{a}\| = \left[a_x^2 + a_y^2 + a_z^2 \right]^{\frac{1}{2}}$$

and measures the length (or magnitude, or amplitude) of the vector. The sum of two vectors **a** and **b** is a *resultant vector* **c** represented by the diagonal of a parallelogram whose adjacent sides are the two vectors. The components of the resultant vector are

$$c_x = a_x + b_x \qquad\qquad c_y = a_y + b_y \qquad\qquad c_z = a_z + b_z$$

Scalar Product

The *scalar product* (also known as the inner or dot product) of vectors **a** and **b** is a scalar whose value is

$$\mathbf{a} \cdot \mathbf{b} = \|\mathbf{a}\|\|\mathbf{b}\| \cos\theta = a_x b_x + a_y b_y + a_z b_z$$

Here θ denotes the angle between the two vectors. The following rules are satisfied by scalar products (m is a scalar):

$$\mathbf{a} \cdot \mathbf{b} = \mathbf{b} \cdot \mathbf{a}$$
$$\mathbf{a} \cdot (\mathbf{b} + \mathbf{c}) = \mathbf{a} \cdot \mathbf{b} + \mathbf{a} \cdot \mathbf{c}$$
$$m(\mathbf{a} \cdot \mathbf{b}) = (m\mathbf{a}) \cdot \mathbf{b} = \mathbf{a} \cdot (m\mathbf{b})$$

Vector Product

The *vector product* (also known as the outer or cross product) of vectors **a** and **b** is a vector **c** = **a** × **b** directed perpendicularly to the plane defined by **a** and **b**. Stretch out the three fingers of your right hand so that your middle finger is perpendicular to the plane defined by your thumb and your index finger (the "right hand rule"). If your thumb is vector **a**, and your index finger is vector **b**, then **c** is oriented in the direction of your middle finger. The amplitude of **c** is

$$\|\mathbf{c}\| = \|\mathbf{a} \times \mathbf{b}\| = \|\mathbf{a}\|\|\mathbf{b}\|\sin\theta$$

The components of the vector product are

$$c_x = a_y b_z - a_z b_y \qquad\quad c_y = a_z b_x - a_x b_z \qquad\quad c_z = a_x b_y - a_y b_x$$

The following rules are satisfied by the vector product (m is a scalar):

$$\mathbf{a} \times \mathbf{b} = -\mathbf{b} \times \mathbf{a}$$
$$\mathbf{a} \times (\mathbf{b} + \mathbf{c}) = \mathbf{a} \times \mathbf{b} + \mathbf{a} \times \mathbf{c}$$
$$m(\mathbf{a} \times \mathbf{b}) = (m\mathbf{a}) \times \mathbf{b} = \mathbf{a} \times (m\mathbf{b})$$
$$\mathbf{a} \times (\mathbf{b} \times \mathbf{c}) = (\mathbf{a} \cdot \mathbf{c})\mathbf{b} - (\mathbf{a} \cdot \mathbf{b})\mathbf{c}$$

Triple Products

$$\mathbf{a} \cdot (\mathbf{b} \times \mathbf{c}) = \mathbf{c} \cdot (\mathbf{a} \times \mathbf{b}) = \mathbf{b} \cdot (\mathbf{c} \times \mathbf{a}) = (\mathbf{a} \times \mathbf{b}) \cdot \mathbf{c}$$
$$(\mathbf{a} \times \mathbf{b}) \times (\mathbf{c} \times \mathbf{d}) = (\mathbf{a} \cdot (\mathbf{c} \times \mathbf{d}))\mathbf{b} - (\mathbf{b} \cdot (\mathbf{c} \times \mathbf{d}))\mathbf{a}$$
$$= (\mathbf{a} \cdot (\mathbf{b} \times \mathbf{d}))\mathbf{c} - (\mathbf{a} \cdot (\mathbf{b} \times \mathbf{c}))\mathbf{d}$$
$$(\mathbf{a} \times \mathbf{b}) \cdot (\mathbf{c} \times \mathbf{d}) = (\mathbf{a} \cdot \mathbf{c})(\mathbf{b} \cdot \mathbf{d}) - (\mathbf{a} \cdot \mathbf{d})(\mathbf{b} \cdot \mathbf{c})$$

Derivative of a Vector

The derivative $d\mathbf{a}/d\xi$ of vector $\mathbf{a}$ (a_x, a_y, a_z) with respect to independent variable ξ is obtained by computing the derivative of each component. For example, in a Cartesian frame,

$$\frac{d\mathbf{a}}{d\xi} = \frac{da_x}{d\xi}\mathbf{i} + \frac{da_y}{d\xi}\mathbf{j} + \frac{da_z}{d\xi}\mathbf{k}$$

We have

$$\frac{d(\mathbf{a}\cdot\mathbf{b})}{d\xi} = \frac{d\mathbf{a}}{d\xi}\cdot\mathbf{b} + \mathbf{a}\cdot\frac{d\mathbf{b}}{d\xi}$$

$$\frac{d(\mathbf{a}\times\mathbf{b})}{d\xi} = \frac{d\mathbf{a}}{d\xi}\times\mathbf{b} + \mathbf{a}\times\frac{d\mathbf{b}}{d\xi}$$

E.3 Matrices

A *matrix* of size $(m \times n)$ is a rectangular table of elements arranged as m rows and n columns. A $(m \times n)$ matrix with elements $a_{i,j}$ $(i = 1, n; j = 1, m)$, is written as

$$\mathbf{A} = \begin{pmatrix} a_{1,1} & \cdots & a_{1,n} \\ \vdots & \ddots & \vdots \\ a_{m,1} & \cdots & a_{m,n} \end{pmatrix} \tag{3}$$

A matrix with the same number of rows and columns $(m = n)$ is called a *square* matrix. A matrix with zero elements below (above) the diagonal $a_{i,i}$ is an upper (lower) *triangular* matrix. A matrix with non-zero elements on the main diagonal and zero off-diagonal elements is called a *diagonal* matrix. A matrix with non-zero elements only on the main diagonal and on the first diagonals below and above the main diagonal is called a *tridiagonal* matrix. Matrices that include a large number of zero elements (often encountered in chemical modeling) are referred to as *sparse* matrices.

Matrices of the same size $(m \times n)$ can be added or subtracted element by element. If $a_{i,j}$ are the elements of matrix $\mathbf{A}$ and $b_{i,j}$ the elements of matrix $\mathbf{B}$, the elements of matrix $\mathbf{C} = \mathbf{A} + \mathbf{B}$ are

$$c_{i,j} = a_{i,j} + b_{i,j}$$

The multiplication of a matrix $\mathbf{A}$ by a scalar γ is a matrix $\mathbf{B}$ whose elements are $b_{i,j} = \gamma\, a_{i,j}$.

Matrices can be multiplied when the number of columns (equal to n) of the first matrix $(\mathbf{A})$ is equal to the number of rows of the second matrix $(\mathbf{B})$. If $\mathbf{C} = \mathbf{AB}$

$$c_{i,j} = \sum_{k=1}^{n} a_{i,k} b_{k,j}$$

Matrix multiplication satisfies several rules:

$$(\mathbf{AB})\mathbf{C} = \mathbf{A}(\mathbf{BC})$$

$$(\mathbf{A} + \mathbf{B})\mathbf{C} = \mathbf{AC} + \mathbf{BC}$$

$$\mathbf{A}(\mathbf{B} + \mathbf{C}) = \mathbf{AB} + \mathbf{AC}$$

Matrix multiplication is not commutative

$$\mathbf{AB} \neq \mathbf{BA}$$

The *trace* of a square matrix $\mathbf{A}$ is the sum of its diagonal elements.

The transpose $\mathbf{A}^T$ of a $(m \times n)$ matrix $\mathbf{A}$ is a $(n \times m)$ matrix in which the rows have been turned into the columns and the columns into the rows. For example, the transpose of the $(m \times n)$ matrix $\mathbf{A}$ in (3) is

$$\mathbf{A}^T = \begin{pmatrix} a_{1,1} & \cdots & a_{m,1} \\ \vdots & \ddots & \vdots \\ a_{1,n} & \cdots & a_{m,n} \end{pmatrix}$$

An $(n \times n)$ square matrix $\mathbf{A}$ is said to be symmetric if $\mathbf{A} = \mathbf{A}^T$ such that $a_{i,j} = a_{j,i}$. It is said to be antisymmetric if $\mathbf{A} = -\mathbf{A}^T$ with $a_{i,j} = -a_{j,i}$.

If the matrix includes complex elements, the complex conjugate matrix $\mathbf{A}^*$ of matrix $\mathbf{A}$ is a matrix whose elements are the complex conjugates of the elements of matrix $\mathbf{A}$.

For a complex matrix $\mathbf{A}$ with elements $a_{i,j}$, the *adjoint* $\mathbf{A}^\dagger$ with elements $a^\dagger_{i,j}$ is the complex conjugate of the transpose $\mathbf{A}^T$ (with elements $a_{j,i}$)

$$a^\dagger_{i,j} = a^*_{j,i}$$

The adjoint of a real matrix is thus its transpose.

The identity matrix $\mathbf{I}$ of size n (denoted $\mathbf{I}_n$) is a $(n \times n)$ square matrix in which all diagonal elements are equal to 1 and all off-diagonal elements are 0. For example $(n = 2)$

$$\mathbf{I} = \begin{pmatrix} 1 & 0 \\ 0 & 1 \end{pmatrix}$$

The multiplication of a matrix $\mathbf{A}$ by the identity matrix $\mathbf{I}$ is

$$\mathbf{IA} = \mathbf{A}$$

The inverse matrix of a square matrix $\mathbf{A}$ is a matrix of the same size denoted $\mathbf{A}^{-1}$ that satisfies the relation

$$\mathbf{A}^{-1}\mathbf{A} = \mathbf{A}\,\mathbf{A}^{-1} = \mathbf{I}$$

The inverse of a diagonal matrix is a diagonal matrix of reciprocal elements.

Determinants

The *determinant* of a square matrix $\mathbf{A}$ of size n with elements $a_{i,j}$ is denoted $\det(\mathbf{A})$ or $|\mathbf{A}|$ and is calculated as

$$\det(\mathbf{A}) = \sum_{\substack{k_i \neq k_j \\ i,j = 1,n}} \varepsilon(k_1 \ldots k_n)\, a_{1,k_1} a_{2,k_2} \ldots a_{n,k_n} \tag{4}$$

The Levi-Civita symbol $\varepsilon(k_1 \ldots k_n)$ denotes the permutation sign function defined by

$$\varepsilon(k_1 \ldots k_n) = (-1)^{\pi(K)}$$

where $\pi(K)$ is the number of pairwise exchanges among the indices of sequence $K = \{k_1, k_2, \ldots k_n\}$ needed to reorder the sequence of distinct indices ranging from 1 to n into an ascending order given by $[1, 2, \ldots, n]$. Thus, for example, $\varepsilon(1, 2, 3) = +1$ and $\varepsilon(2, 1, 3, 4) = -1$. The summation in (4) is performed over all non-repeated combinations of indices $1, 2, \ldots n$.

For example, the determinant of a matrix of size $n = 2$ is

$$\det(\mathbf{A}) = a_{1,1}\, a_{2,2} - a_{1,2}\, a_{2,1}$$

and for a matrix of size $n = 3$:

$$\det(\mathbf{A}) = a_{1,1}\,(a_{2,2}\, a_{3,3} - a_{2,3}\, a_{3,2}) + a_{1,2}(a_{2,3}\, a_{3,1} - a_{2,1}\, a_{3,3})$$
$$+ a_{1,3}(a_{2,1}\, a_{3,2} - a_{2,2}\, a_{3,1})$$

A matrix whose determinant is equal to zero is said to be *singular* or *degenerate*. A matrix whose determinant is non-zero is said to be *non-singular* or of *full rank*.

Important properties of determinants are

$$\det(\mathbf{AB}) = \det(\mathbf{BA})$$
$$\det(\mathbf{A}^T) = \det(\mathbf{A})$$
$$\det(\mathbf{A}^{-1}) = 1/\det(\mathbf{A})$$

The determinant of a triangular or diagonal matrix $\mathbf{T}$ with elements $t_{i,j}$ is the product of its diagonal elements

$$\det(\mathbf{T}) = \prod_{1}^{n} t_{i,i}$$

Linear Systems

A system of m linear equations with n independent variables

$$a_{1,1}x_1 + a_{1,2}x_2 + \cdots + a_{1,n}x_n = b_1$$
$$a_{2,1}x_1 + a_{2,2}x_2 + \cdots + a_{2,n}x_n = b_2$$
$$\cdots$$
$$a_{m,1}x_1 + a_{m,2}x_2 + \cdots + a_{m,n}x_n = b_m \tag{5}$$

can be represented in a matrix form by

$$\begin{pmatrix} a_{1,1} & \cdots & a_{1,n} \\ \vdots & \ddots & \vdots \\ a_{m,1} & \cdots & a_{m,n} \end{pmatrix} \begin{pmatrix} x_1 \\ \vdots \\ x_n \end{pmatrix} = \begin{pmatrix} b_1 \\ \vdots \\ b_m \end{pmatrix}$$

or

$$\mathbf{Ax} = \mathbf{b}$$

Here, $\mathbf{A}$ is an $m \times n$ matrix, $\mathbf{x}$ is a vector of dimension n, and $\mathbf{b}$ is a vector of dimension m. If $m = n$, the number of unknowns in system (5) is equal to the number of equations, and matrix $\mathbf{A}$ is a square matrix. If $\mathbf{A}$ is also non-singular ($\det(\mathbf{A}) \neq 0$) then the system can be solved as

$$\mathbf{x} = \mathbf{A}^{-1}\mathbf{b}$$

A generic expression for the inverse of a non-singular matrix of size $n \times n$ is given by

$$\mathbf{A}^{-1} = \frac{1}{\det(\mathbf{A})}\mathbf{B}$$

In this expression, $\mathbf{B}$, called the *adjugate* of $\mathbf{A}$, is a matrix of size $n \times n$ whose coefficients $b_{i,j}$, referred to as cofactors of elements $a_{i,j}$, are given by

$$b_{i,j} = (-1)^{i+j}\det(\mathbf{A}(i,j))$$

where matrix $\mathbf{A}(i, j)$ of size $(n - 1 \times n - 1)$, called (i,j)-th redact, is the same as matrix $\mathbf{A}$ in which row i and column j have been removed. In practical applications, more efficient algorithms are adopted to seek the solution of linear systems. Among these are direct methods (e.g., Gauss–Jordan elimination, LU decomposition) or iterative methods (e.g., Jacobi, Gauss–Seidel, least-square, conjugate gradients methods). See Press *et al.* (2007) and Co (2013) for more details. The Thomas algorithm (see Box 4.4) is used to solve tridiagonal systems.

Jacobian Matrix

Let $\mathbf{x} = (x_1, x_2, \ldots x_n)^T$ be a vector of dimension n and $\mathbf{f} = (f_1, f_2, \ldots f_m)^T$ be a *vector-valued* function of dimension m whose elements f_i are scalar-valued functions of $\mathbf{x}$. The Jacobian matrix $\mathbf{J}(\mathbf{f})$ of $\mathbf{f}$ defined by

$$\mathbf{J} = \frac{\partial \mathbf{f}}{\partial \mathbf{x}}$$

is the $m \times n$ matrix whose elements $j_{i,j}$ are partial derivatives of f_i with respect to vector elements x_j.

$$j_{i,j} = \frac{\partial f_i}{\partial x_j}$$

$j_{i,j}$ represents the sensitivity of function f_i to variable x_j. The Jacobian generalizes the notion of gradient to describe the sensitivity to a vector. The m components of $\mathbf{f}$ are assumed to be continuous over the entire domain under consideration. The Jacobian matrix is often used as the linearized expression of a mathematical model. In Chapter 11, the Jacobian is denoted $\mathbf{K}$ to avoid confusion with the usual notation for the cost function (J).

The *Hessian* $\mathbf{H}(f)$ of a scalar-valued function f is a square matrix whose elements $h_{i,j}$ are the second-order partial derivatives of function f

$$h_{i,j} = \frac{\partial^2 f}{\partial x_i \partial x_j}$$

It describes the local curvature of a function of many variables. The Hessian matrix is related to the Jacobian matrix by

$$\mathbf{H}(f) = \mathbf{J}(\nabla f)$$

Eigenvalues and Eigenvectors

An *eigenvector* $\mathbf{e}$ of a square matrix $\mathbf{A}$ $(n \times n)$ is a vector that satisfies

$$\mathbf{A}\mathbf{e} = \lambda \mathbf{e} \qquad (6)$$

where the scalar λ (real or complex) is the *eigenvalue* corresponding to the eigenvector. It is often useful to know the eigenvectors of a matrix because they represent the vectors for which operation by the matrix modifies the amplitude without changing the direction. If $\mathbf{e}$ is an eigenvector of $\mathbf{A}$, then any scalar multiplier of $\mathbf{e}$ must also be an eigenvector as seen from (6). It follows that (6) must yield an infinite number of solutions for $\mathbf{e}$ and hence that

$$\det(\mathbf{A} - \lambda \mathbf{I}_n) = 0 \qquad (7)$$

The eigenvalues of matrix $\mathbf{A}$ are thus the solutions to (7) and replacement into (6) yields the corresponding eigenvectors $\mathbf{e}_1, \mathbf{e}_2, \ldots$

One can show that the trace of a matrix is equal to the sum of its eigenvalues, and that its determinant is equal to the product of its eigenvalues.

E.4 Vector Operators

The Gradient Operator

If $f(x, y, z)$ is a scalar field at position (x, y, z) in three dimensions, the gradient of f denoted grad(f) or∇f is defined in Cartesian coordinates by

$$\nabla f(x, y, z) = \text{grad}(f) = \frac{\partial f}{\partial x}\mathbf{i} + \frac{\partial f}{\partial y}\mathbf{j} + \frac{\partial f}{\partial z}\mathbf{k}$$

where, as above, $\mathbf{i}$, $\mathbf{j}$, and $\mathbf{k}$ represent the basis of unit vectors in the coordinate system, and ∇ (*nabla* symbol) the vector differential operator. At any point P, the gradient of a scalar-valued function f is a vector that points in the direction of the greatest change of f at point P. In spherical coordinates (Figure E.1), the gradient is expressed by

$$\nabla f(r, \theta, \varphi) = \text{grad}(f) = \frac{\partial f}{\partial r}\mathbf{i}_r + \frac{1}{r}\frac{\partial f}{\partial \theta}\mathbf{i}_\theta + \frac{1}{r \sin \theta}\frac{\partial f}{\partial \varphi}\mathbf{i}_\varphi$$

where $\mathbf{i}_r$, $\mathbf{i}_\theta$, and $\mathbf{i}_\varphi$ are unit vectors pointing along coordinate directions, and where φ is the azimuth angle and θ the zenith angle.

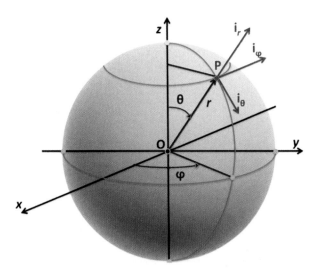

Figure E.1 Definition of spherical coordinates. Point P is defined by the radial distance r, the azimuthal angle φ, and the zenith angle θ. If $\mathbf{i}$, $\mathbf{j}$, and $\mathbf{k}$ are the unit vectors along rectangular coordinates x, y, and z, the unit vectors in spherical coordinates $\mathbf{i}_r$, $\mathbf{i}_\varphi$, and $\mathbf{i}_\theta$ are defined by $\mathbf{i}_r = \mathbf{r}/\|\mathbf{r}\| = (x\,\mathbf{i} + y\,\mathbf{j} + z\,\mathbf{k})/(x^2 + y^2 + z^2)^{1/2}$, $\mathbf{i}_\varphi = \mathbf{k} \times \mathbf{i}_r$, and $\mathbf{i}_\theta = \mathbf{i}_\varphi \times \mathbf{i}_r$.

The gradient of a vector-valued function $\mathbf{f}(\mathbf{x})$ is its Jacobian matrix $\mathbf{J} = \partial\mathbf{f}/\partial\mathbf{x}$. This is often used in a steepest-descent algorithm to find the minimum of $\mathbf{f}$.

Divergence of a Vector Field

If $\mathbf{f} = (f_x, f_y, f_z)^T$ is a vector function in a 3-D Euclidean space, its divergence is a scalar field defined in Cartesian coordinates by

$$\nabla \cdot \mathbf{f} = \mathrm{div}(\mathbf{f}) = \frac{\partial f_x}{\partial x} + \frac{\partial f_y}{\partial y} + \frac{\partial f_z}{\partial z}$$

The divergence of a vector field represents the flux generation per unit volume at each point of the field. In spherical coordinates, the divergence of a vector $\mathbf{f}(f_r, f_\theta, f_\varphi)$ is written

$$\nabla \cdot \mathbf{f}(r, \theta, \varphi) = \mathrm{div}(\mathbf{f}) = \frac{1}{r^2}\frac{\partial}{\partial r}\left(r^2 f_r\right) + \frac{1}{r\sin\theta}\frac{\partial}{\partial \theta}\left(\sin\theta f_\theta\right) + \frac{1}{r\sin\theta}\frac{\partial f_\varphi}{\partial \varphi}$$

Curl of a Vector Field

The curl of a vector field $\mathbf{f} = (f_x, f_y, f_z)^T$ is a vector defined as the cross product of the operator ∇ with vector $\mathbf{f}$. In Cartesian coordinates,

$$\nabla \times f = \text{curl}(f) = \begin{vmatrix} \mathbf{i} & \mathbf{j} & \mathbf{k} \\ \dfrac{\partial}{\partial x} & \dfrac{\partial}{\partial y} & \dfrac{\partial}{\partial z} \\ f_x & f_y & f_z \end{vmatrix}$$

$$= \left(\frac{\partial f_z}{\partial y} - \frac{\partial f_y}{\partial z} \right) \mathbf{i} + \left(\frac{\partial f_x}{\partial z} - \frac{\partial f_z}{\partial x} \right) \mathbf{j} + \left(\frac{\partial f_y}{\partial x} - \frac{\partial f_x}{\partial y} \right) \mathbf{k}$$

The curl of a vector field **f** represents the vorticity or circulation per unit area of the field. In spherical coordinates, it is expressed by

$$\nabla \times \mathbf{f}(r, \theta, \varphi,) = \text{curl}(\mathbf{f}) = \frac{1}{r \sin \theta} \left[\frac{\partial}{\partial \theta} \left(f_\varphi \sin \theta \right) - \frac{\partial f_\theta}{\partial \varphi} \right] \mathbf{i}_r$$

$$+ \frac{1}{r} \left[\frac{1}{\sin \theta} \frac{\partial f_r}{\partial \varphi} - \frac{\partial}{\partial r} \left(r f_\varphi \right) \right] \mathbf{i}_\theta$$

$$+ \frac{1}{r} \left[\frac{\partial}{\partial r} \left(r f_\theta \right) - \frac{\partial f_r}{\partial \theta} \right] \mathbf{i}_\varphi$$

Laplacian of a Scalar Field

The Laplacian Δf of a scalar field f is the divergence of its gradient.

$$\Delta f = \nabla \cdot (\nabla f)$$

It is a scalar field. In Cartesian coordinates, it is expressed as

$$\Delta f = \frac{\partial^2 f}{\partial x^2} + \frac{\partial^2 f}{\partial y^2} + \frac{\partial^2 f}{\partial z^2}$$

In spherical coordinates, it is written

$$\Delta f = \frac{1}{r^2} \frac{\partial}{\partial r} \left(r^2 \frac{\partial f}{\partial r} \right) + \frac{1}{r^2 \sin \theta} \frac{\partial}{\partial \theta} \left(\sin \theta \frac{\partial f}{\partial \theta} \right) + \frac{1}{r^2 \sin^2 \theta} \frac{\partial^2 f}{\partial \varphi^2}$$

Important Relations

If f is a scalar field and **f** a vector field, we have

$$\nabla \times (\nabla f) = 0$$

$$\nabla \cdot (\nabla \times \mathbf{f}) = 0$$

$$\nabla \times (\nabla \times \mathbf{f}) = \nabla(\nabla \cdot \mathbf{f}) - \nabla^2 \mathbf{f}$$

$$\nabla \cdot (f \, \mathbf{f}) = f(\nabla \cdot \mathbf{f}) + (\nabla f) \cdot \mathbf{f}$$

$$\nabla \times (f \, \mathbf{f}) = f(\nabla \times \mathbf{f}) + (\nabla f) \times \mathbf{f}$$

Scalar and Vector Potentials

Any vector field $\mathbf{f}$ whose curl is equal to zero can be expressed as the gradient of a scalar potential V

$$\mathbf{f} = -\nabla V$$

Any vector field $\mathbf{f}$ whose divergence is equal to zero can be expressed as the curl of a vector field $\mathbf{a}$

$$\mathbf{f} = \nabla \times \mathbf{a}$$

Integration Theorems

If $\mathbf{f} = (f_x, f_y, f_z)^T$ is a vector field defined in a given region, Γ a curve in this region, and $d\mathbf{r} = (dx, dy, dz)^T$ an elementary displacement along this curve, the circulation C between points P_1 and P_2 is given by the line integral

$$C = \int_{\Gamma, P_1}^{P_2} \mathbf{f}\, d\mathbf{r} = \int_{\Gamma, P_1}^{P_2} \left[f_x dx + f_y dy + f_z dz \right]$$

For a vector $\mathbf{f}$ whose curl equals zero, the circulation can be expressed as a function of potential V, and

$$C = -\int_{\Gamma, P_1}^{P_2} \frac{\partial V}{\partial x}\, dx + \frac{\partial V}{\partial y}\, dy + \frac{\partial V}{\partial z}\, dz = -\int_{\Gamma, P_1}^{P_2} dV = V_{P_1} - V_{P_2}$$

In this case, the circulation is thus independent of the path between points P_1 and P_2. For a closed curve, the circulation of a vector whose curl is equal to zero is equal to zero.

The *Kelvin–Stokes' theorem*, also called the *curl theorem*, relates the surface integral of the curl of vector $\mathbf{f}$ over a surface A in the 3-D Euclidean space to the line integral of the vector field over its boundary Γ

$$\oint_{\Gamma} \mathbf{f} \cdot d\mathbf{r} = \iint_A \mathbf{n} \cdot (\nabla \times \mathbf{f}) dS$$

where $\mathbf{n}$ is the outward-pointing unit normal vector on the surface boundary. The Gauss–Ostrogradsky's theorem, also called the *divergence theorem*, states that the outward flux of a vector field $\mathbf{f}$ through a closed surface A is equal to the volume integral of the divergence of $\mathbf{f}$ over the volume V inside the surface

$$\oiint_A \mathbf{f} \cdot \mathbf{n}\, dS = \iiint_V (\nabla \cdot \mathbf{f}) dV$$

The 2-D version (in a plane) of the divergence theorem, called *Green's theorem*, is expressed as

$$\oint_{\Gamma} \mathbf{f} \cdot \mathbf{n}\, ds = \iint_A (\nabla \cdot \mathbf{f}) dS$$

E.5 Differential Equations

A differential equation is a mathematical expression that relates some function to its derivatives. The order of a differential equation is the highest order of the derivative(s) found in the equation. We consider here two types of differential equations: ordinary differential equations (ODEs) and partial differential equations (PDEs).

Ordinary Differential Equations

An ODE prescribes a function and its derivative(s) relative to a single independent variable. A simple example of an initial value ODE for a vector-valued function $\mathbf{y}(t)$ is the first-order equation

$$\frac{d\mathbf{y}(t)}{dt} = \mathbf{f}(\mathbf{y}, t)$$

which describes the rate at which the vector-valued function $\mathbf{y}(t)$ varies with time t, for an applied forcing $\mathbf{f}(\mathbf{y}, t)$. The system is said to be *autonomous* if function $\mathbf{f}$ is not explicitly dependent on time $[\mathbf{f} = \mathbf{f}(\mathbf{y})]$ and *non-autonomous* otherwise $[\mathbf{f} = \mathbf{f}(\mathbf{y}, t)]$. The solution $\mathbf{y}(t)$ is expressed for a specified initial value $\mathbf{y}(t_0)$ by

$$\mathbf{y}(t) = \mathbf{y}(t_0) + \int_{t_0}^{t} \mathbf{f}(\mathbf{y}, t')\, dt'$$

A linear ODE of order N, written here for a scalar function $y(x)$, is expressed as

$$a_0 \frac{d^N y}{dx^N} + a_1 \frac{d^{N-1} y}{dx^{N-1}} + a_2 \frac{d^{N-2} y}{dx^{N-2}} + \cdots + a_{N-1} \frac{dy}{dx} + a_N y = f(x)$$

It is said to be *homogeneous* if the forcing term $f(x) = 0$ and *inhomogeneous* otherwise. The solution of this equation requires that N independent conditions be specified. These can be, for example, values of function y and its $N - 1$ first derivatives at one end of the interval under consideration (e.g., at point $x = 0$). These conditions are referred to as initial conditions when the independent variable is time. Another option is to provide conditions at each boundary of the interval.

Partial Differential Equations

A PDE prescribes a function and its partial derivatives relative to several independent variables. If y is a dependent variable and $\mathbf{x}(x_1, x_2, \ldots, x_n)$ are n independent variables, the general form of a first-order partial differential equation is

$$F\left(\mathbf{x}, y, \frac{\partial y}{\partial x_1}, \frac{\partial y}{\partial x_2}, \ldots, \frac{\partial y}{\partial x_n}\right) = 0$$

One distinguishes between different forms of first-order PDEs:

Quasi-linear equation: $\displaystyle\sum_{i=1}^{n} a_i(\mathbf{x}, y)\frac{\partial y}{\partial x_i} = f(\mathbf{x}, y)$

Linear equation: $\displaystyle\sum_{i=1}^{n} a_i(\mathbf{x})\frac{\partial y}{\partial x_i} = f(\mathbf{x}) + b(\mathbf{x})y$

Strictly linear equation: $\displaystyle\sum_{i=1}^{n} a_i(\mathbf{x})\frac{\partial y}{\partial x_i} = f(\mathbf{x})$

The second-order linear PDE for a field $\psi(x, y)$

$$A\frac{\partial^2 \psi}{\partial x^2} + B\frac{\partial^2 \psi}{\partial x \partial y} + C\frac{\partial^2 \psi}{\partial y^2} + D\frac{\partial \psi}{\partial x} + E\frac{\partial \psi}{\partial y} + F\,\psi = G$$

is said to be

- *elliptic* if $B^2 - 4AC < 0$
- *parabolic* if $B^2 - 4AC = 0$
- *hyperbolic* if $B^2 - 4AC > 0$

by formal analogy with the name of the conics (*ellipse, parabola,* and *hyperbola*) represented for these same conditions when applied to the quadratic equation

$$Ax^2 + Bxy + Cy^2 + Dx + Ey + F = 0$$

An example of an *elliptic* PDE is the Laplace equation

$$\nabla^2 \psi = 0$$

describing a time-independent "boundary value" problem. The determination of the solution requires that a condition be prescribed at each point of the boundary of the domain in which this equation is to be solved. This condition can be a specified value of the function y (Dirichlet or first-type condition) or a specified value of the normal derivative of y (Neumann or second-type condition).

An example of a *parabolic* PDE is the diffusion equation

$$\frac{\partial \psi}{\partial t} = D\nabla^2 \psi$$

The solution of this linear parabolic PDE requires that an initial condition and a condition at each point of the boundary of the spatial domain be specified.

Finally, the wave equation

$$\frac{\partial^2 \psi}{\partial t^2} = c^2 \nabla^2 \psi$$

is an example of a *hyperbolic* PDE. The advection equation

$$\frac{\partial \psi}{\partial t} + c\frac{\partial \psi}{\partial x} = 0$$

can also be classified as a hyperbolic *PDE* since its solution verifies the wave equation. The solution of the linear advection equation requires that initial conditions be specified along with boundary conditions at the upstream boundary of the spatial domain.

Another characterization of PDEs that is more useful from a computational point of view is to distinguish between *initial value problems* (such as the advection and the diffusion equations) in which the equations are integrated forward in time from a specified initial condition, and *boundary value problems* (such as the Laplace equation) in which the correct solution must be found everywhere at once (Press *et al.*, 2007).

Different analytical methods are available to solve PDEs (see, e.g., Durran, 2010; Co, 2013) including the method of characteristics, discussed in different textbooks, the use of the Laplace or Fourier transforms (see Section E.6), etc. Numerical approaches to solve the advection and diffusion equations are presented in Chapters 7 and 8.

E.6 Transforms

Orthogonal Transforms

We consider a linear transformation of n-element vector $\mathbf{x}$ to a new n-element vector $\mathbf{y}$:

$$\mathbf{y} = \mathbf{M}\,\mathbf{x}$$

where $\mathbf{M}$ is an $n \times n$ matrix. The transformation is said to be *orthogonal* if matrix $\mathbf{M}$ is orthogonal, i.e., its inverse $\mathbf{M}^{-1}$ is equal to its transpose $\mathbf{M}^{T}$:

$$\mathbf{M}^{T}\mathbf{M} = \mathbf{I}$$

where $\mathbf{I}$ is the identity matrix. The transformation is said to be *unitary* if the adjoint matrix $\mathbf{M}^{\ddagger}$ (see Section E.3) is equal to the inverse matrix $\mathbf{M}^{-1}$:

$$\mathbf{M}^{\ddagger}\mathbf{M} = \mathbf{I}$$

Under these conditions, the norm of vector $\mathbf{y}$ is equal to the norm of vector $\mathbf{x}$. A unitary transformation (such as a Fourier transform, see next section) preserves the norm of a vector.

Laplace and Fourier Transforms

If function $f(x)$ of a real variable x is equal to zero for $x < 0$, the *Laplace transform* of $f(x)$ is defined by function $F(p)$ of a complex variable p

$$F(p) = \int_{0}^{\infty} f(x)\,e^{-px}dx$$

where $F(p)$ is said to be the image of $f(x)$. One shows easily that the Laplace transform of the derivative of function f is given by $p\,F(p) - f(0)$ and, more generally, the Laplace transform of the nth derivative of $f(x)$ denoted $f^{(n)}$ is $p^{n}F(p) - \sum_{m=0}^{n-1} p^{n-m-1}f^{(m)}(0)$. Here $f^{(m)}$ is the mth derivative of function f at $x = 0$.

The Laplace transform can thus be applied to transform an ODE into an algebraic equation, or a PDE into an ODE. The solution of these transformed equations is generally easier to obtain than for the original equations, and the solution $f(x)$ is then obtained by applying an inverse transform.

The *Fourier transform* of $f(x)$ is defined by

$$F(k) = \frac{1}{\sqrt{2\pi}} \int_{-\infty}^{+\infty} f(x)\, e^{-i\,k\,x} dx$$

and its inverse by

$$f(x) = \frac{1}{\sqrt{2\pi}} \int_{-\infty}^{+\infty} F(k)\, e^{i\,k\,x} dk$$

where k is a real variable and $i^2 = -1$. In many global modeling applications, information is repeatedly transferred between spectral and grid point representations (see Chapter 4) by applying Fourier transforms. If independent variable x represents space [m], variable k represents wavenumber [m^{-1}]. If the independent variable x represents time [s], variable k represents frequency [s^{-1}]. The Fourier transform of the derivative of $f(x)$ is equal to $i\,k\,F(k)$ if $f(0) = 0$.

In grid point models, the values of function $f(x)$ are known only at N discrete points x_j, ($j = 0, 1, \ldots, N-1$) of the spatial domain, for example, on regularly spaced grid points along a longitude circle on a sphere. In this case, the transfer of data between the grid point and spectral spaces is achieved by applying a *discrete Fourier transform* (*DFT*), which takes the form

$$F_k \doteq \sum_{j=0}^{N-1} f_j\, e^{-2\pi\,i\,j\,k/N} \tag{8}$$

$$f_j \doteq \sum_{k=0}^{N-1} F_k\, e^{2\pi\,i\,j\,k/N} \tag{9}$$

Here, f_j denotes the value of function $f(x)$ at the geometric point x_j, F_k is proportional to the value of function $F(k)$ for a discrete value of the wavenumber k.

The application of expressions (8) and (9) requires N multiplications and additions. Since these operations have to be applied N times, the total number of operations required is of the order of N^2. The method can become computationally impractical when the value of N is large. The *fast Fourier transform* (*FFT*) algorithm, introduced in its modern version by Cooley and Tukey (1965), but already considered by Gauss in 1805, circumvents this problem by breaking the N-point transform into two $N/2$ point transforms, one for the even ($2\,j$) points and one for the odd ($2\,j+1$) points. The splitting is further repeated until the problem is broken into N single-point transforms. The number of steps required to calculate the discrete Fourier Transform is then of the order of $N \log_2 N$ operations, and the computational cost is considerably smaller.

Principal Component Analysis and Empirical Orthogonal Functions

Principal component analysis (PCA) is a standard method to identify the characteristic patterns of variability in a data set. Consider a time series of observations for an ensemble of K variables assembled into a vector $\mathbf{x} = (x_1, x_2, \ldots, x_K)^T$. The time series consists of successive values of $\mathbf{x}$ at discrete times. The variability of $\mathbf{x}$ with time is characterized by the $(K \times K)$ covariance matrix

$$\mathbf{Cov(x)} = \begin{pmatrix} \mathrm{var}(x_1) & \cdots & \mathrm{cov}(x_1, x_K) \\ \vdots & \ddots & \vdots \\ \mathrm{cov}(x_K, x_1) & \cdots & \mathrm{var}(x_K) \end{pmatrix}$$

This is a symmetric matrix of full rank and thus has K eigenvalues λ_k and K corresponding eigenvectors $\mathbf{e}_k$ forming an orthonormal basis. The eigenvectors represent uncorrelated patterns in the data and are called the *empirical orthogonal functions* (*EOFs*) for the data set.

For a given realization of $\mathbf{x}$ at an individual time, one defines the *principal components* (PCs) as the projections of $\mathbf{x}$ onto the orthonormal basis defined by the EOFs. Thus the k-th principal component y_k is the projection of $\mathbf{x}$ onto the k-th eigenvector $\mathbf{e}_k$:

$$y_k = \mathbf{e}_k^T \mathbf{x} = \sum_{i=1}^{K} e_{k,i} x_i$$

Each realization of the vector $\mathbf{x}$ can thus be projected onto the basis of EOFs using the PCs:

$$\mathbf{x} = \sum_{k=1}^{K} y_k \mathbf{e_k}$$

The PCs form a time series corresponding to the time series of $\mathbf{x}$, and the variance of a given PC is given by the corresponding eigenvalue: $\mathrm{var}(y_k) = \lambda_k$. We thus see that λ_k measures the variance associated with the EOF pattern defined by eigenvector $\mathbf{e}_k$. In this manner, we can define the dominant independent patterns accounting for the variability in a data set.

E.7 Probability and Statistics

A *random* (or *stochastic*) *variable* X is a variable whose values cannot be predicted deterministically, but can be described probabilistically. These values can be either discrete or continuous. In the first case, the variable may take only a countable number of distinct values x_i, while in the second case it may take an infinite number of possible values x. Here, x_i and x refer to specific values taken by X.

A random variable is described by its *probability density*. The probability density of a discrete random variable is given by a list of N probabilities p_i associated with

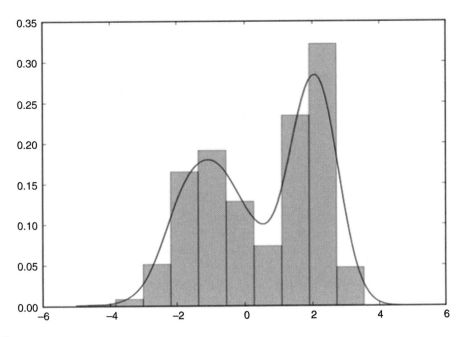

Figure E.2   Probability distribution (blue histogram) and probability density function (red curve) for a random variable. Reproduced from http://glowingpython.blogspot.com.

each of its possible values. All values of p_i must be non-negative and normalized so that their sum is equal to 1:

$$\sum_{1}^{N} p_i = 1$$

Continuous random variables are described by a *probability density function* (*PDF*), denoted $p(x)$, such that $p(x)dx$ is the probability for X being in the range $[x, x + dx]$, normalized to add up to 1 for all possible realizations of X:

$$\int_{-\infty}^{+\infty} p(x)\, dx = 1$$

The PDF for a discrete random variable is often represented by a histogram (in blue in Figure E.2) that displays the probability p_i ($i = 1, N$) that X lies in certain ranges over a given domain. The red curve shows the corresponding continuous PDF $p(x)$.

Joint Probability Distribution

If $p(x)\,dx$ is the probability that a random variable X takes a value between x and $x + dx$, and $p(y)\,dy$ is the probability that variable Y takes a value between y and $y + dy$, the *joint probability density function $p(x, y)$* is defined as the probability that the values x and y be in the range x and $x + dx$, and $y + dy$, respectively. This joint PDF is normalized so that

$$\int_{-\infty}^{+\infty} \int_{-\infty}^{+\infty} p(x,y)\, dx\, dy = 1$$

Further, it is also normalized with respect to each variable

$$p(x) = \int_{-\infty}^{+\infty} p(x,y)\, dy \qquad \text{and} \qquad p(y) = \int_{-\infty}^{+\infty} p(x,y)\, dx$$

Conditional Probability Distribution

The probability that X takes a value between x and $x + dx$, given a *known* value y of Y, is expressed by the conditional probability $p(x|y)$

$$p(x|y) = \frac{p(x,y)}{p(y)}$$

If random variables X and Y are *independent*,

$$p(x,y) = p(x)\, p(y)$$

and consequently

$$p(x|y) = p(x)$$

Mean of a Distribution

We consider a random variable X that is distributed according to a PDF $p(x)$. The mean value m of a *population*, also called the *expected value* of x and denoted by $E[x]$, is given by the first-order population moment:

$$m = \int_{-\infty}^{+\infty} x\, p(x)\, dx$$

The mean value m_{xy} of the product of two *independent* random variables x and y is given by

$$m_{xy} = \int_{-\infty}^{+\infty} \int_{-\infty}^{+\infty} x\, y\, p(x,y)\, dx\, dy = \int_{-\infty}^{+\infty} x\, p(x)\, dx \int_{-\infty}^{+\infty} y\, p(y)\, dy = m_x\, m_y$$

Variance and Standard Deviation

The *variance* of a variable X whose distribution is expressed by the PDF $p(x)$ relative to its mean value m is defined as the second population moment

$$\sigma^2 = \int_a^b (x - m)^2 p(x)\, dx$$

The *standard deviation* σ is the square root of the variance.

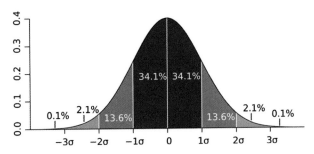

Gaussian probability density function. Symbol σ denotes the standard deviation of the distribution. The fraction [percent] of the population included in different intervals is also shown. Reproduced from Wikimedia Commons.

Normal Density Function

A common PDF form is the *Gaussian function* (also called *normal function*)

$$p(x) = \frac{1}{\sigma\sqrt{2\pi}} \exp\left[-\frac{(x-m)^2}{2\sigma^2}\right]$$

where m and σ are the mean and standard deviation of the distribution, respectively. One standard deviation from the mean accounts for 68.2% of the population, two standard deviations for 95.4%, and three standard deviations for 99.7% (Figure E.3).

Sample Data

We now consider a *sample* of random data containing N points $x_i(i = 1, N)$ (e.g., N observations of variable x). The sample *mean value* (denoted by $\bar{x}$) is given by

$$\bar{x} = \sum_{i=1}^{N} x_i p_i$$

where p_i is the weight given to point i with

$$\sum_{i=1}^{N} p_i = 1$$

For variables of equal weights, the sample mean is

$$\bar{x} = \frac{1}{N}\sum_{i=1}^{N} x_i$$

The sum of the deviation of variables x_i from their arithmetic mean $\bar{x}$ is by definition equal to zero

$$\sum_{i=1}^{N} (x_i - \bar{x}) = 0$$

For a scalar random variable x that takes N discrete values $x_1, x_2, \ldots, x_N$ with the corresponding probabilities $p_1, p_2, \ldots, p_N$, the sample variance of x relative to its sample mean is given by

$$s^2 = \sum_{i=1}^{N} p_i (x_i - \bar{x})^2$$

For equally probable values, we have

$$s^2 = \frac{1}{N} \sum_{i=1}^{N} (x_i - \bar{x})^2 \tag{10}$$

In many practical applications, one often examines a limited number of individual data points, and number N corresponds to a sample of the entire data population. One can show that the variance of the entire population is equal to the average of the variances derived for all possible samples if, in expression (10), the squared distance $(x_i - \bar{x})^2$ is divided by $N-1$ rather than by N. In this case, the sample variance represents an unbiased estimate of the population variance. Therefore, it is often recommended to divide the squared distance by $N - 1$ when calculating the sample variance. The correction, however, is very small when the number of data in the sample becomes large.

Covariance

The *covariance* between two scalar-valued random variables x and y, each characterized by N sampled data points $(x_1, x_2, \ldots, x_N)$ and $(y_1, y_2, \ldots y_N)$, is defined by

$$\mathrm{cov}(x, y) = \overline{(x - \bar{x})(y - \bar{y})}$$

or

$$\mathrm{cov}(x, y) = \frac{1}{N} \sum_{i=1}^{N} (x_i - \bar{x})(y_i - \bar{y}) \quad .$$

where the overbar is again a representation of the sample mean. The variance is the covariance of two identical variables. One shows easily that

$$\mathrm{cov}(x, y) = \overline{xy} - \bar{x}\,\bar{y}$$

If $\mathbf{z}$ is a random vector (vector whose components are random numbers), the *covariance matrix*

$$\mathbf{Cov(z)} = \overline{(\mathbf{z} - \bar{\mathbf{z}})(\mathbf{z} - \bar{\mathbf{z}})^T}$$

represents a multivariate generalization of the variance and covariance defined above in the case of a scalar. If $z_1, z_2, z_3, \ldots z_k$ denote the K elements of vector $\mathbf{z}$ (assumed to be random variables), the $K \times K$ covariance matrix (also called variance–covariance matrix) has the following structure:

$$\mathbf{Cov(z)} = \begin{pmatrix} \mathrm{var}(z_1) & \cdots & \mathrm{cov}(z_1, z_K) \\ \vdots & \ddots & \vdots \\ \mathrm{cov}(z_K, z_1) & \cdots & \mathrm{var}(z_K) \end{pmatrix} \tag{11}$$

The inverse of this matrix is called the precision matrix or the concentration matrix. The diagonal elements represent variances of the individual elements of vector $\mathbf{z}$, and the non-diagonal elements a measure of the correlation between the different elements of the vector. The covariance matrix is symmetric since the covariance operator is commutative ($\mathrm{cov}(z_i,z_j) = \mathrm{cov}(z_j,z_i)$).

The cross-covariance between the two random vectors $\mathbf{x}$ and $\mathbf{y}$ of dimension m and n, respectively, is a matrix of dimension $m \times n$

$$\mathbf{Cov}(\mathbf{x}, \mathbf{y}) = \overline{(\mathbf{x} - \bar{\mathbf{x}})(\mathbf{y} - \bar{\mathbf{y}})^T}$$

Ordinary Regression Analysis

Relationships among different variables, specifically between a dependent variable y and an independent variable x, can be estimated by using a statistical process called *regression analysis*. We consider here the simple case in which we fit N given data points (x_1, y_1), (x_2, y_2), ..., (x_N, y_N), of two correlated random variables x and y by a parameter-dependent regression function

$$\tilde{y}(x) = f(x, a, b, ...)$$

We use the method of least squares to derive the parameters $a, b, \ldots$ that minimize the distance between the values of the data points and the corresponding values of the dependent variable y provided by function f. Thus, we minimize over $a, b, \ldots$ the function

$$S(a, b, ...) = \sum_{i=1}^{n} [y_i - f(x_i, a, b...)]^2$$

and write

$$\frac{\partial S}{\partial a} = 0 \qquad \frac{\partial S}{\partial b} = 0 \qquad .....$$

The first step is to discover the form of the relationship that exists between variables x and y. A scatterplot diagram for y versus x displaying the data points (x_i, y_i) may be useful. If the diagram suggests that a linear relationship is a suitable approximation, one expresses the regression curve as a linear statistical model (see Figure E.4)

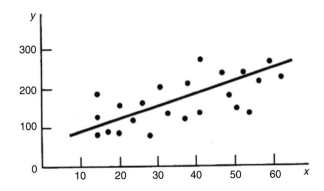

Figure E.4   Scatterplot with a number of data points (x_i, y_i) and the corresponding regression line (statistical model).

$$\tilde{y}(x) = a + bx$$

More complex polynomial regressions may have to be considered if the scatterplot points to a nonlinear relationship.

In the linear ordinary regression method (ORM), coefficients a and b are derived from the minimization of the sum of the residuals between the data points y_i and the corresponding predicted values $\tilde{y}_i$ for $x = x_i$

$$S = \sum_{i=1}^{N} [y_i - \tilde{y}_i]^2$$

This is expressed by

$$\frac{\partial S}{\partial a} = -2 \sum_{i=1}^{N} (y_i - a - bx_i) = 0 \qquad \text{and} \qquad \frac{\partial S}{\partial b} = -2 \sum_{i=1}^{N} (y_i - a - bx_i) x_i = 0$$

and results in linear equations in the parameters a and b called the normal equations of least-squares

$$Na + b \sum_{i=1}^{N} x_i = \sum_{i=1}^{N} y_i$$

$$a \sum_{i=1}^{N} x_i + b \sum_{i=1}^{N} x_i^2 = \sum_{i=1}^{N} y_i x_i$$

The solution of these equations is for the intercept

$$a = \bar{y} - b\bar{x}$$

and the slope

$$b = \frac{\sum_{i=1}^{N} (x_i - \bar{x})(y_i - \bar{y})}{\sum_{i=1}^{N} (x_i - \bar{x})^2} = \frac{N \sum_{i=1}^{N} x_i y_i - \left(\sum_{i=1}^{N} x_i\right)\left(\sum_{i=1}^{N} y_i\right)}{N \sum_{i=1}^{N} x_i^2 - \left(\sum_{i=1}^{N} x_i\right)^2} = \frac{\text{cov}(x, y)}{s^2(x)} = r(x, y) \frac{s(y)}{s(x)}$$

where

$$\bar{x} = \frac{1}{N} \sum_{i=1}^{N} x_i \qquad \text{and} \qquad \bar{y} = \frac{1}{N} \sum_{i=1}^{N} y_i$$

denote the averages of the sampled x_i and y_i values, and $s(x)$ and $s(y)$ denote the corresponding standard deviations. Figure E.4 shows an example of a regression line.

When quantity y depends on several independent variables x_1, x_2, x_3, ..., the simple regression model described here must be generalized and replaced by a *multiple regression approach*.

The strength and sign of the linear relationship between the two random variables x and y are expressed by the Pearson *correlation coefficient r*

$$r = \frac{\sum_{i=1}^{N} (x_i - \bar{x})(y_i - \bar{y})}{\left[\sum_{i=1}^{N} (x_i - \bar{x})^2\right]^{\frac{1}{2}} \left[\sum_{i=1}^{N} (y_i - \bar{y})^2\right]^{\frac{1}{2}}}$$

or by the alternative formula

$$r = \frac{N \sum_{i=1}^{N} x_i y_i - \left(\sum_{i=1}^{N} x_i \right) \left(\sum_{i=1}^{N} y_i \right)}{\left[N \sum_{i=1}^{N} x_i^2 - \left(\sum_{i=1}^{N} x_i \right)^2 \right]^{\frac{1}{2}} \left[N \sum_{i=1}^{N} y_i^2 - \left(\sum_{i=1}^{N} y_i \right)^2 \right]^{\frac{1}{2}}}$$

This coefficient is equal to the covariance of x and y divided by the standard deviation of the sampled values of variables x and y:

$$r(x,y) = \frac{\mathrm{cov}(x,y)}{s(x)\, s(y)}$$

Its value ranges from -1 to $+1$. The sign of the correlation coefficient indicates whether the random variables x and y are positively or negatively correlated.

The *coefficient of determination* R^2 provides information on the goodness of fit of a statistical model. It represents the degree by which a regression line or curve represents the data. It provides the proportion of the variance of variable y that is predictable from the value of the other variable (x). The coefficient is expressed by

$$R^2 = 1 - \frac{\sum_{i=1}^{N} (y_i - \tilde{y}_i)^2}{\sum_{i=1}^{N} (y_i - \bar{y})^2}$$

where $\tilde{y}_i$ represents again the value of function y approximated by the regression model for $x = x_i$, and $\bar{y}$ is the mean value of all data y_i. If $R^2 = 1$, all data points perfectly fit the regression model. For linear least squares regression with an estimated intercept term, R^2 equals the square r^2 of the Pearson correlation coefficient.

In multilinear regression analyses, the value of R^2 increases automatically when extra explanatory variables are added to the model. The adjusted R^2 corrects for this spurious behavior by taking into account the number of explanatory variables p relative to the number of data points n. It is defined as

$$R^2(\mathrm{adj}) = R^2 - \left(1 - R^2 \right) \frac{p}{n - p - 1}$$

Reduced Major Axis Regression Analysis

The ordinary regression method described above minimizes the distance between the measured and predicted values of variable y (vertical distance, see Figure E.5a), but ignores errors in the measured values of variable x (horizontal distance). The uncertainties in both variables can be accounted for by applying the reduced major axis (RMA) regression method. In this case, the regression line is defined by minimizing the error in both Cartesian directions x and y, and specifically the area of the triangle shown in Figure E.5b.

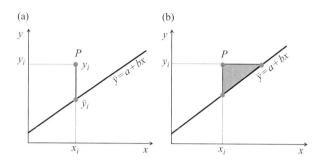

Figure 8.5 Geometric representation of the ordinary regression (a) and of the reduced major axis regression (b). Point P represents a data point (x_i, y_i). In the ordinary regression method, the sum (for all data points) of the squared distance (vertical red line) between the data points and the corresponding values of y on the regression line is minimized. In the reduced major axis regression method, the sum (for all data points) of the area of the red triangle is minimized.

The resulting value for the slope of the regression line is

$$b = \pm \frac{\sum_{i=1}^{N} (y_i - \bar{y})^2}{\sum_{i=1}^{N} (x_i - \bar{x})^2} = \pm \frac{s(y)}{s(x)}$$

where the sign is chosen to be the same as the sign of the correlation coefficient. Note that, in this approach, the slope is provided by the ratio between the standard deviations of the y and x variables, respectively. The slope b_{RMA} from the RMA regression is related to the slope b_{ORM} from the ordinary regression method by $b_{RMA} = b_{ORM}/r(x, y)$, thus the RMA regression will produce a steeper slope.

Student's t-Test

Introduced by the Irish chemist William Sealy Gosset under the pseudonym of Student, the t-test is used to determine whether two sets of data are significantly different from each other. We distinguish between the *one-sample test*, used to test whether the mean of the population from which a sample is drawn randomly differs significantly from the mean of a reference population, and the *two-sample test*, used to test whether two population means are significantly different from each other on the basis of randomly drawn samples. The test allows for some chance of error, and this is measured by the *level of significance* α, indicating a probability of error (*p-value*) not to be exceeded. Standard practice is to choose $\alpha = 0.05$, meaning that we require the t-test to have less than a 5% chance of being in error. We then say that the *p-value* must be less than 0.05. Significance depends on the *degrees of freedom* (*df*) in the data set, defined as the number of variables that are free to vary. Computation of the *t-value* for the test and comparison to a *critical value* $t_c(\alpha, df)$ gives us the result. Critical values $t_c(\alpha, df)$ are tabulated here.

	Critical values of Student's *t*-distribution with different degrees of freedom (*df*)										
One-tail	0.50	0.25	0.20	0.15	0.10	0.05	0.025	0.01	0.005	0.001	0.0005
Two-tails	1.00	0.50	0.50	0.30	0.20	0.10	0.05	0.02	0.01	0.002	0.001
df											
1	0.000	1.000	1.376	1.963	3.078	6.314	12.71	31.82	63.66	318.31	636.62
2	0.000	0.816	1.061	1.386	1.886	2.920	4.303	6.965	9.925	22.327	31.599
3	0.000	0.765	0.978	1.250	1.638	2.353	3.182	4.541	5.841	10.215	12.924
4	0.000	0.741	0.941	1.190	1.533	2.132	1.776	3.747	4.604	7.173	8.610
5	0.000	0.727	0.920	1.156	1.476	2.015	2.571	3.365	4.032	5.893	6.869
6	0.000	0.718	0.906	1.134	1.440	1.943	2.447	3.143	3.707	5.208	5.959
7	0.000	0.711	0.896	1.119	1.415	1.895	2.365	2.998	3.499	4.785	5.408
8	0.000	0.706	0.889	1.108	1.397	1.860	2.306	2.896	3.355	4.501	5.041
9	0.000	0.703	0.883	1.100	1.383	1.833	2.262	2.821	3.250	4.297	4.781
10	0.000	0.700	0.879	1.093	1.372	1.812	2.228	2.764	3.169	4.144	4.587
11	0.000	0.697	0.876	1.088	1.363	1.796	2.201	2.718	3.106	4.025	4.437
12	0.000	0.695	0.873	1.083	1.356	1.782	2.179	2.681	3.055	3.930	4.318
13	0.000	0.694	0.870	1.079	1.350	1.771	2.160	2.650	3.012	3.852	4.221
14	0.000	0.692	0.868	1.076	1.345	1.761	2.145	2.624	2.977	3.787	4.140
15	0.000	0.691	0.866	1.074	1.341	1.753	2.131	2.602	2.947	3.733	4.073
16	0.000	0.690	0.865	1.071	1.337	1.746	2.120	2.583	2.921	3.686	4.015
17	0.000	0.689	0.863	1.069	1.33	1.740	2.110	2.567	2.898	3.646	3.965
18	0.000	0.688	0.862	1.067	1.330	1.734	2.101	2.552	2.878	3.610	3.922
19	0.000	0.688	0.861	1.066	1.328	1.729	2.093	2.539	2.861	3.579	3.883
20	0.000	0.687	0.860	1.064	1.325	1.725	2.086	2.528	2.845	3.552	3.850
21	0.000	0.686	0.859	1.063	1.323	1.721	2.080	2.518	2.831	3.527	3.819
22	0.000	0.686	0.858	1.061	1.321	1.717	2.074	2.508	2.819	3.505	3.792
23	0.000	0.685	1.060	1.319	1.714	2.069	2.500	2.807	3.485	3.485	3.768
24	0.000	0.685	0.857	1.059	1.318	1.711	2.064	2.492	2.797	3.467	3.745
25	0.000	0.684	0.856	1.056	1.316	1.708	2.060	2.485	2.787	3.450	3.725
26	0.000	0.684	0.856	1.056	1.315	1.706	2.056	2.479	2.779	3.435	3.707
27	0.000	0.684	0.855	1.057	1.314	1.703	2.052	2.473	2.771	3.421	3.690
28	0.000	0.683	0.855	1.056	1.313	1.701	2.048	2.467	2.763	3.408	3.674
29	0.000	0.683	0.854	1.055	1.311	1.699	2.045	2.462	2.756	3.396	3.659
30	0.000	0.683	0.854	1.055	1.310	1.697	2.042	2.457	2.750	3.385	3.646
40	0.000	0.681	0.851	1.050	1.303	1.684	2.021	2.423	2.704	3.307	3.551
60	0.000	0.679	0.848	1.045	1.296	1.671	2.000	2.390	2.660	3.232	3.460
80	0.000	0.678	0.846	1.043	1.292	1.664	1.990	2.374	2.639	3.195	3.416
100	0.000	0.677	0.845	1.042	1.290	1.660	1.984	2.364	2.626	3.174	3.390
1000	0.000	0.675	0.842	1.037	1.282	1.646	1.962	2.330	2.581	3.098	3.300
∞	0.000	0.674	0.842	1.036	1.282	1.645	1.960	2.326	2.576	3.090	3.290
Confidence level	0%	50%	60%	70%	80%	90%	95%	98%	99%	99.8%	99.9%

One-sample t-test. Here, we compare the sample to a reference population with mean *m*. We denote by $\bar{x}$ and s^2 the mean and variance of the N independent data points in the sample, and compute *t* as

$$t = \frac{(\bar{x} - m)}{s/\sqrt{N}}$$

with degrees of freedom $df = N - 1$. We formulate the hypothesis that the difference between the mean m for the reference population and the mean of the population from which the sample was drawn is not significant (null hypothesis). If the value of $|t|$ calculated from the above expression is larger than the critical value $t_c(\alpha, df)$, then we reject that null hypothesis. The test can be *one-tailed*, in which we check for significant difference in only one direction (higher or lower), or *two-tailed*, in which case we allow for the possibility of significant difference at either end. Using $\alpha = 0.05$, a one-tailed test checks whether the mean of the sample is within the lower 95th quantile of the PDF for it to be consistent with m, while a two-tailed test checks whether the mean of the sample is between the 2.5th and 97.5th quantiles of the PDF.

For example, if the mean value m of the reference population is 100, and if a sample provides 17 data points ($df = 16$), with a mean value $\bar{x}$ of 90 and a standard deviation s of 10, we calculate for $|t|$ a value of 4. If we adopt for α a value of 0.05, the table provides for t_c a value of 2.12 (two-tailed test). We reject therefore the null hypothesis since $|t| > t_\alpha$, and conclude that the mean of the population from which the sample was drawn is significantly different ($p < 0.05$) from the reference mean m.

Two-sample t-test. The t-distribution may also be used to test whether the means of two populations from which samples are drawn are the same. The means of the populations are m_x and m_y, respectively. The number of data points in each sample is N_x and N_y, and the corresponding sample means are $\bar{x}$ and $\bar{y}$. The sample variances are s_x^2 and s_y^2. The t-value is computed as

$$t = \frac{(\bar{x} - \bar{y}) - (m_x - m_y)}{\sqrt{N_x s_x^2 + N_y s_y^2}} \sqrt{\frac{N_x N_y (N_x + N_y - 2)}{N_x + N_y}}$$

with degrees of freedom $df = N_x + N_y - 2$. We test again the null hypothesis (i.e., the hypothesis that $m_x = m_y$) by comparing the t-value with the critical value t_c. If $|t| > t_\alpha$ the null hypothesis is discarded, and the difference in the means of the two populations is significant.

References

Co T. B. (2013) *Methods of Applied Mathematics for Engineers and Scientists*, Cambridge University Press, Cambridge.

Durran D. R. (2010) *Numerical Methods for Fluid Dynamics*, Springer-Verlag, Berlin.

Press W. H., Teukolsky S. A., Vetterling W. T., and Flannery B. P. (2007) *Numerical Recipes: The Art of Scientific computing*, Cambridge University Press, Cambridge.

Further Reading

Earth System and Climate Science

Baird C. and Cann M. (2008) *Environmental Chemistry*, Bookman, Taipei.

Houghton J. (2009) *Global Warming: The Complete Briefing*, Cambridge University Press, Cambridge.

Kump L. R., Kasting J. F., and Crane R. G. (2010) *The Earth System*, Prentice Hall, Upper Saddle River, NJ.

Marshall J. and Plumb R. A. (2008) *Atmosphere, Ocean and Climate Dynamics: An Introductory Text*, Academic Press, New York.

Atmospheric Science

Frederick J. E. (2008) *Principles of Atmospheric Science*, Jones & Bartlett Learning, Sudbury, MA.

Goody R. (1995) *Principles of Atmospheric Physics and Chemistry*, Oxford University Press, Oxford.

Salby M. L. (2012) *Physics of the Atmosphere and Climate*, Cambridge University Press, Cambridge.

Wallace J. M. and Hobbs P. V. (2006) *Atmospheric Science: An Introductory Survey*, Academic Press, New York.

Atmospheric Chemistry

Barker, J. R. (ed.) (1995) *Progress and Problems in Atmospheric Chemistry*, World Scientific, River Edge, NJ.

Brasseur, G. P., Orlando J. J., and Tyndall G. S. (eds.) (1999) *Atmospheric Chemistry and Global Change*, Oxford University Press, Oxford.

Brasseur G. P., Prinn R. G., and Pszenny A. A. P. (eds.) (2003) *Atmospheric Chemistry in a Changing World: An Integration and Synthesis of a Decade of Tropospheric Chemistry Research*, Springer, New York.

Burrows, J. P., Platt U., and Borrell P. (eds.) (2011) *The Remote Sensing of Tropospheric Composition from Space*, Springer, New York.

Finlayson-Pitts B. J, and Pitts Jr. J. N. (2000) *Chemistry of the Upper and Lower Atmosphere: Theory, Experiments, and Applications*, Academic Press, New York.

Hobbs P. V. (1995) *Basic Physical Chemistry for the Atmospheric Sciences*, Cambridge University Press, Cambridge.

Hobbs P. V. (2000) *Introduction to Atmospheric Chemistry*, Cambridge University Press, Cambridge.

Jacob D. J. (1999) *Introduction to Atmospheric Chemistry*, Princeton University Press, Princeton, NJ.

Jaeschke, W. (ed.) (1986) *Chemistry of Multiphase Atmospheric Systems*, Springer-Verlag, Berlin.

Seinfeld J. H. and Pandis S. N. (2006) *Atmospheric Chemistry and Physics: From Air Pollution to Climate Change*, Wiley, New York.

Singh, H. B. (ed.) (1995) *Composition, Chemistry, and Climate of the Atmosphere*, Van Nostrand Reinhold, New York.

Sportisse B. (2008) *Fundamentals in Air Pollution. From Processes to Modelling*, Springer, New York.

Warneck P. (1999) *Chemistry of the Natural Atmosphere*, Academic Press, New York.

Warneck P. and Williams J. (2012) *The Atmospheric Chemist's Companion: Numerical Data for Use in the Atmospheric Sciences*, Springer, New York.

Wayne R. P. (1985) *Chemistry of Atmospheres: An Introduction to the Chemistry of the Atmospheres of Earth, the Planets, and their Satellites*, Oxford University Press, Oxford.

Yung Y. L. and DeMore W. B. (1999) *Photochemistry of Planetary Atmospheres*, Oxford University Press, Oxford.

Atmospheric Physics and Dynamics

Andrews D. G. (2010) *An Introduction to Atmospheric Physics*, Cambridge University Press, Cambridge.

Holton J. R. (2004) *An Introduction to Dynamic Meteorology*, Academic Press, New York.

Houghton J. (2002) *The Physics of Atmospheres*, Cambridge University Press, Cambridge.

Mak M. (2011) *Atmospheric Dynamics*, Cambridge University Press, Cambridge.

McWilliams J. C. (2006) *Fundamentals of Geophysical Fluid Dynamics*, Cambridge University Press, Cambridge.

Neufeld Z. and Hernández-Garcia E. (2010) *Chemical and Biological Processes in Fluid Flows: A Dynamical Systems Approach*, Imperial College Press, London.

Pruppacher H. R., and Klett J. D. (1997) *Microphysics of Clouds and Precipitation*, Kluwer, Dordrecht.

Riegel C. A. and Bridger A. F. C. (1992) *Fundamentals of Atmospheric Dynamics and Thermodynamics*, World Scientific, River Edge, NJ.

Salby M. L. (1996) *Fundamentals of Atmospheric Physics*, Academic Press, New York.

Vallis G. K. (2006) *Atmospheric and Oceanic Fluid Dynamics, Fundamentals and Large-Scale Circulation*, Cambridge University Press, Cambridge.

Radiative Transfer

Goody R. M. and Yung Y. L. (1989) *Atmospheric Radiation, Theoretical Basis*, Oxford University Press, Oxford.

Liou K. N. (2002) *An Introduction to Atmospheric Radiation*, Academic Press, New York.

Petty G. W. (2006) *A First Course in Atmospheric Radiation*, Sundog Publishing, Madison, WI.

Thomas G. E., and Stamnes K. (1999) *Radiative Transfer in the Atmosphere and Ocean*, Cambridge University Press, Cambridge.

Turbulence, Boundary Layer Meteorology, and Surface Exchanges

Bonan G. (2008) *Ecological Climatology, Concepts and Applications*, Cambridge University Press, Cambridge.

Garratt J. R. (1992) *The Atmospheric Boundary Layer*, Cambridge University Press, Cambridge.

Granier, C., Artaxo P., and Reeves C. E. (eds.) (2004) *Emissions of Atmospheric Trace Compounds*, Kluwer, Dordrecht.

Monson R. and Baldocchi D. (2009) *Terrestrial Biosphere–Atmosphere Fluxes*, Cambridge University Press, Cambridge.

Stull R. B. (1988) *An Introduction to Boundary Layer Meteorology*, Kluwer, Dordrecht.

Wyngaard J. C. (2010) *Turbulence in the Atmosphere*, Cambridge University Press, Cambridge.

Vilà-Guerau de Arellano J., van Heerwaarden C. C., van Stratum B. J. H., and van den Dries K. (2015) *Atmospheric Boundary Layer: Integrating Air Chemistry and Land Interactions*, Cambridge University Press, Cambridge.

Middle Atmosphere

Andrews D. G., Holton J. R., and Leovy C. B. (1987) *Middle Atmosphere Dynamics*, Academic Press, New York.

Brasseur G. P. and Solomon S. (2005) *Aeronomy of the Middle Atmosphere: Chemistry and Physics of the Stratosphere and Mesosphere*, Springer, New York.

Dessler A. (2000) *Chemistry and Physics of Stratospheric Ozone*, Academic Press, New York.

Müller, R. (ed.) (2012) *Stratospheric Ozone Depletion and Climate Change*, RSC Publishing, Cambridge.

Numerical Methods, Modeling, and Data Assimilation

Co T. B. (2013) *Methods of Applied Mathematics for Engineers and Scientists*, Cambridge University Press, Cambridge.

Courant R. and Hilbert D. (1962), *Methods of mathematical Physics*, vols. **1** and **2**, Wiley, New York.

Daley R. (1991) *Atmospheric Data Analysis*, Cambridge University Press, Cambridge.

DeCaria A. J. and Van Knowe G. E. (2014) *A First Course in Atmospheric Numerical Modeling*, Sundog Publishing, Madison, WI.

Durran D. R. (2010) *Numerical Methods for Fluid Dynamics, with Applications to Geophysics*, Springer, New York.

Gear W. (1971) *Numerical Initial Value Problems in Ordinary Differential Equations*, Prentice-Hall, Englewood Cliffs, NJ.

Fox R. O. (2003) *Computational Models for Turbulent Reacting Flows*, Cambridge University Press, Cambridge.

Gershenfeld N. (1999) *The Nature of Mathematical Modeling*, Cambridge University Press, Cambridge.

Hairer E. and Wanner G. (1996) *Solving Ordinary Differential Equations II: Stiff and Differential-Algebraic Problems*, Springer-Verlag, Berlin.

Jacobson M. Z. (1999) *Fundamentals of Atmospheric Modeling*, Cambridge University Press, Cambridge.

Kalnay E. (2003) *Atmospheric Modeling, Data Assimilation and Predictability*, Cambridge University Press, Cambridge.

Kiehl, J. T. and Ramanathan V. (eds.) (2006) *Frontiers of Climate Modeling*, Cambridge University Press, Cambridge.

Lahoz, W., Khattatov B., and Ménard R. (eds.) (2010) *Data Assimilation: Making Sense of Observations*, Springer, New York.

Lauritzen, P. H., Jablonowski C., Taylor M. A., and Nair R. D. (eds.) (2011) *Numerical Techniques for Global Atmospheric Models: Tutorials*, Springer, New York.

Müller P. and von Storch H. (2004) *Computer Modelling in Atmospheric and Oceanic Sciences: Building Knowledge*, Springer, New York.

Potter D. (1977) *Computational Physics*, Wiley, New York.

Press W. H., Teukolsky S. A., Vetterling W. T., and Flannery B. P. (2007) *Numerical Recipes: The Art of Scientific Computing*, Cambridge University Press, Cambridge.

Rodgers C. D. (2000) *Inverse Methods for Atmospheric Sounding: Theory and Practice*, World Scientific, River Edge, NJ.

Slingerland R. and Kump L. (2011) *Mathematical Modeling of Earth's Dynamical Systems: A Primer*, Princeton University Press, Princeton, NJ.

Stensrud D. J. (2007) *Parameterization Schemes: Keys to Understanding Numerical Weather Prediction Models*, Cambridge University Press, Cambridge.

Stoer J. and Bulirsch R. (1993) *Introduction to Numerical Analysis*, 2nd edition, Springer-Verlag, Berlin.

Trenberth K. E. (ed.) (1992) *Climate System Modeling*, Cambridge University Press, Cambridge.

Warner T. T. (2011) *Numerical Weather and Climate Prediction*, Cambridge University Press, Cambridge.

Washington W. M. and Parkinson C. L. (2005) *An Introduction to Three-Dimensional Climate Modeling*, University Science Books, Sausalito, CA.

Wilks D. S. (2011) *Statistical Methods in the Atmospheric Sciences*, Academic Press, New York.

Index